T0387227

Flow Around Circular Cylinders

The source

> '*The arrangement of the vortices in the wake is connected with my name: it is usually called a* Kármán vortex street *or a* Kármán vortex trail. *But I do not claim to have discovered these vortices: they were known long before I was born. The earliest picture in which I have seen them is one in a church in Bologna, Italy, where St Christopher is shown carrying the child Jesus across a flowing stream. Behind the saint's naked foot the painter indicated alternating vortices.*'

Th. von Kármán, *Aerodynamics*, (1954B), Cornell University Press, also McGraw–Hill, (1963B) paperback, p. 68.

The search

> '*I bought Kármán's* Aerodynamics *in 1966 and was captivated by the story about St Christopher. During the 70s, I went twice to Bologna and was unable to trace St Christopher in any of the churches. In the 80s, Professor Buresti of the University of Pisa suggested the help of his relative, a cardinal in the Vatican. The answer came that, although a long list of St Christopher's paintings exists, none is in Bologna. My wife spotted St Christopher in a 1420–30 manuscript held at the Fitzwilliam museum in Cambridge. It was a book illumination on vellum and behind St Christopher's feet only two alternating eddies are depicted.*
> *In the late 90s, Professor Mizota of Fukuoka Institute of Technology came on sabbatical to England. He was fascinated and decided to renew the search. He found a life-size St Christopher in the badly damaged fifteenth century mural in the Basilica di San Domenico in Bologna. Miraculously, the best preserved part is the wake in the mural, see above. The procession of eddies along the wake is now called the Kármán–Bénard eddy street. (Bénard's sketches and photo from (1908J) and (1913J) were reproduced in Vol. 1, Figs 1.7 and 18.4, p. 11 and 511, respectively.)*'

Mizota *et al.* (2000J), *Nature*, **404**, No. 6775, p. 226.

Madonna col Bambino tra i Santi Demenico, Pietro Martire e Cristoforo, remnants of fifteenth century mural in the Basilica di San Domenico, Bologna. (By courtesy of the Italian Ministry of Culture.)

FLOW AROUND CIRCULAR CYLINDERS

A Comprehensive Guide Through Flow Phenomena, Experiments, Applications, Mathematical Models, and Computer Simulations

---

M. M. ZDRAVKOVICH

former: Visiting Post-Doctoral Fellow
University of Cambridge, UK
Visiting Reader, Imperial College, London
Visiting Professor, Kanazawa University, Japan

VOLUME 2

This book has been printed digitally and produced in a standard specification in order to ensure its continuing availability

OXFORD
UNIVERSITY PRESS

Great Clarendon Street, Oxford OX2 6DP

Oxford University Press is a department of the University of Oxford.
It furthers the University's objective of excellence in research, scholarship,
and education by publishing worldwide in

Oxford New York

Auckland Cape Town Dar es Salaam Hong Kong Karachi
Kuala Lumpur Madrid Melbourne Mexico City Nairobi
New Delhi Shanghai Taipei Toronto
With offices in
Argentina Austria Brazil Chile Czech Republic France Greece
Guatemala Hungary Italy Japan South Korea Poland Portugal
Singapore Switzerland Thailand Turkey Ukraine Vietnam

Published in the United States
by Oxford University Press Inc., New York

Reprinted 2009

ISBN 978-0-19-856561-1

PREFACE

This book is intended to be a useful guide for researchers in institutes, practising engineers in industry, academics and students in the fields of aerospace, civil, hydraulic, electric transmission lines, mechanical, nuclear, offshore, and wind engineering. The book may be useful reading for applied mathematicians and physicists engaged in fluid mechanics.

The unique feature of the book is that it deals exclusively with a single class of bluff bodies with circular cross-section. This class has attracted a great deal of research due to the highly complex flow structures generated. Since most bluff bodies used in engineering applications are circular cylinders, I felt they warranted this in-depth treatment.

The first volume of *Flow Around Circular Cylinders* was published in 1997. It dealt with Fundamentals in three parts, A–C:

(A) disturbance-free flow around a nominally two-dimensional cylinder across the whole Reynolds number range;

(B) various theoretical models such as free streamline, vortex, vortex-sheet, and other miscellaneous less well-known models;

(C) real free stream effects such as turbulence, shear, compressibility, sound, cavitation, as well as non-Newtonian fluids.

The present second volume covers:

(D) effects of geometry, such as surface roughness, aspect ratio, taper, blockage, finite cylinder, yaw, rotation, wall proximity, pairs, clusters, and arrays of cylinders.

The future third volume is not yet written and will cover:

(E) unsteady free stream, such as acceleration, deceleration, reversal, oscillatory flow, waves and combinations of waves and current, as well as streaming;

(F) flow-induced oscillations, free and forced oscillations, synchronization, influencing and governing parameters, means for suppressing oscillations, interfering cylinders, arrays, theoretical models, and ovalling mode of oscillation.

The two volumes may be compared to a strange and incomplete jig-saw puzzle of a peculiar kind. It is composed like the jig-saw puzzle of hundreds and hundreds 'pieces', which vary in size and do not fit nicely together. This reflects the complexity of Fluid Mechanics, where more research is always needed to fit together the remaining pieces. If the reader bears this in mind, it will help him or her to appreciate fully this book.

It has been my life's work to study and research into all aspects of flow around cylinders. This book has grown out of this long-term effort. It was not my intention to write two books but I felt it necessary to make extensive use of descriptions since knowledge of flow past bluff bodies is still largely acquired from observations. The main emphasis is on the physical explanation of phenomena. Each description of what is happening is accompanied by a possible explanation as to why it is happening. Every effort has been made to deal with all possible aspects of flow and the books can therefore lay claim to being comprehensive.

Many papers written before World War II, which were milestones in their time, are included in order to revive their fading significance in reaching the current overall understanding. The reader is reminded that papers and information are produced nowadays at a rate greater than the capability of and time available to scientists and engineers to absorb them. Hence, a guide like this book is an attempt to offer an overview and insight in a time-saving manner.

I have introduced in each reference important descriptions to allow further insight into the contents. In order to facilitate the handling of the large number of references, they are classified into four categories:

(i) books, including collected works (only parts of which are relevant);

(ii) reviews—highly relevant and usable as initial reading;

(iii) papers published in learned journals (easy access);

(iv) papers published in conference proceedings, reports, theses, etc. (harder access).

Last, but not least, I only hope that part of my enthusiasm, and devotion to the subject, will pass to you.

Prestwich, Lancashire
25 May
Anno Domini MMI

M. M. Z.

ACKNOWLEDGEMENTS

A large number of my colleagues and friends have shared their ideas with me and have given me invaluable suggestions on various topics. I owe an immense debt of gratitude to all of them for the comments, frank criticisms, and for generously giving their precious time. I cannot, unfortunately, include here all the names of those who have helped or advised me in whatever ways over the years because the list would be too large. However, I am sincerely grateful to:

Peter BEARMAN, Eberhard BERGER, Jack GERRARD, Hiroyuki HONJI, Tamotsu IGARASHI, Mark MORKOVIN, Eduard NAUDASCHER, Atsushi OKAJIMA, Michael PAIDOUSSIS, Geoff PARKINSON, Turgut SARPKAYA, and Samir ZIADA.

The permission to use figures was granted by the following journals and copyright holders:

Aeronautical Journal, Royal Aeronautical Society; *AIAA Journal and Journal of Aircraft*, American Institute of Aeronautics and Astronautics; *Applied Scientific Research*, Kluwer; *ASCE Journals: Engineering Mechanics Division, Structural Division, and Water Division*, American Society of Civil Engineers; *ASME Journals: Basic Engineers, Fluids Engineering, Heat Transfer, Offshore Mechanics and Arctic Engineering, Power, Pressure Vessel Technology*, and *Mechanical Engineer*, American Society of Mechanical Engineers; *Atmospheric Environment*, Pergamon Press; *BMT Reports*, British Maritime Technology Ltd; *Bulletin JSME* and *Transactions JSME*, Japanese Society of Mechanical Engineers; *Bulletin Research Institute of Applied Mechanics*, Kyushu University; *Experiments in Fluids*, Springer Verlag; *Fluid Mechanics Research*, Begell House Inc.; *Flow Visualization Conferences*, Hemisphere Publications Inc.; *IEEE Conferences*, Institute of Electrical and Electronic Engineers; *International Journal of Heat and Mass Transfer*, Pergamon Press; *Heat Transfer Japanese Research*, Society of Chemical Engineers of Japan; *Journal of Fluid Mechanics*, Cambridge University Press; *Journal of Fluids and Structures*, Academic Press; *Journal of Mechanical Engineering Science*, Institution of Mechanical Engineers; *Journal of Physical Society of Japan*, Physical Society of Japan; *Journal of Science*, Hiroshima University; *Journal of Sound and Vibration*, Academic Press; *Journal of Wind Engineering and Industrial Aerodynamics*, Elsevier; *L'Aeronautique et L'Astronautique*, Elsevier; *Max Planck Institute Reports*, Göttingen; *OTC Papers*, Offshore Technology Conference, Houston, Texas; *Physics of Fluids*, American Institute of Physics; *Proceedings Ferrybridge Cooling Towers and After*, Institution of Civil Engineers; *Transactions ICE*, Institution of Chemical Engineers; *Transactions RINA*, Royal Institute of Naval Architects; *Wärme und Stoffübertragung*, Springer Verlag; *Zeitschrift für Flugwissenschaften und Luftfahrtforschung*, Springer Verlag.

The final LaTeX version was produced by Dr S. G. Sajjadi and the typing of the References and the List of Captions was done by Mrs D. Millward, both of Salford University. My younger daughter Sandra read the manuscript and corrected it many times. I would like to express my gratitude for invaluable help and significant improvements during the careful editing to Dr Julie Harris and Mr Richard Lawrence, both of OUP.

M. M. Z.

CONTENTS OF VOLUME 1: FUNDAMENTALS

A NOMINALLY TWO-DIMENSIONAL CYLINDER IN AN ALMOST DISTURBANCE-FREE FLOW

1 Conceptual overview 3
2 Steady laminar wake 19
3 Periodic laminar regime 33
4 Transition-in-wake state 79
5 Transition-in-shear-layers state 94
6 Transition-in-boundary-layers state 163
7 Fully turbulent state 201

B THEORETICAL MODELS

8 Solutions of the N–S equations 245
9 Boundary layer approximation 276
10 Free streamline models 291
11 Vortex models and stability 317
12 Vortex sheet models 351
13 Miscellaneous models 373

C REAL FLOW EFFECTS

14 Free stream turbulence 431
15 Non-uniform free stream 458
16 Compressible flow 497
17 Heat transfer 523
18 Aerodynamic sound 566
19 Cavitation 599
20 Non-Newtonian fluids 614

CONTENTS OF VOLUME II: APPLICATIONS

D GEOMETRY EFFECTS

NOMENCLATURE 675

21 ASPECT RATIO 679

21.1 Introduction 679

21.2 Horseshoe-swirl system 680
21.2.1 Laminar boundary layer 680
21.2.2 Oscillating horseshoe-swirl system 682
21.2.3 Turbulent boundary layer 683

21.3 Closed test section 687
21.3.1 Steady laminar wake, L2 regime 687
21.3.2 Periodic laminar wake, L3 regime 688

21.4 Cylinder spanning the free jet 691
21.4.1 Transition-in-shear layer, TrSL state 691

21.5 End plates 694
21.5.1 Laminar periodic wake, L3 regime 694
21.5.2 Effect of a single end plate 695
21.5.3 Effect of two end plates 697
21.5.4 Transition-in-wake, TrW, state 699
21.5.5 Transition-in-shear-layer, TrSL, state 701
21.5.6 Small aspect ratio 702
21.5.7 Fluctuating force 705
21.5.8 Transition-in-boundary layer, TrBL, state 706

21.6 Free water surface 708
21.6.1 Towed cylinder in water at low *Re* 708

21.7 Theoretical modelling 710
21.7.1 The Landau model 710
21.7.2 Extension of Landau's model 712
21.7.3 Other theoretical models 712

21.8 Free end 714
21.8.1 Periodic laminar wake, L3 regime 714
21.8.2 Secondary flow at the free end 716
21.8.3 Spanwise variation in mean pressure 720
21.8.4 Spanwise variation in the local drag coefficient 722

21.8.5 Spanwise fluctuating pressure and lift 724
21.8.6 Strouhal number variation along the span 726
21.8.7 Symmetric eddy street 728
21.8.8 Short cylinder in a boundary layer 730
21.8.9 Finite cylinder in the TrBL4 regime 730
21.8.10 Eddy shedding near the free end 733
21.8.11 Local fluctuating lift and drag 734
21.8.12 Finite cylinder in natural wind 734
21.8.13 Fuel storage tank 738

21.9 Two free ends 740
21.9.1 Drag variation in terms of the aspect ratio 740
21.9.2 Asymmetric pressure distribution 744
21.9.3 Small aspect ratio, $L/D < 1$ 744

22 SURFACE ROUGHNESS AND CHANGE IN DIAMETER 748

22.1 Introduction 748
22.1.1 Nature of surface roughness 749
22.1.2 Fage and Warsap's glass paper tests 751
22.1.3 Skin friction distribution 753
22.1.4 Strouhal number variation 755
22.1.5 Correlation length and vorticity dispersion 756

22.2 Surface roughness textures 758
22.2.1 Pyramidal roughness 758
22.2.2 Brick-wall roughness 760
22.2.3 Wire-gauze roughness 763
22.2.4 Marine roughness 763
22.2.5 Partially roughened surface 765
22.2.6 Roughness Reynolds number 767

22.3 Tripping wires 769
22.3.1 Historical introduction 769
22.3.2 Fage and Warsap's tripping wire tests 769
22.3.3 Effect of tripping wire location 769
22.3.4 Classification of flow regimes 772
22.3.5 Staggered separation wires 773
22.3.6 Tripping and separation wires 777
22.3.7 Helical wires and strakes 778
22.3.8 Stranded cables and conductors 779

22.4 Tripping spheres 781
22.4.1 Pairs of spheres 781
22.4.2 Spanwise row of spheres 784

22.5 Other surface disturbances 784
22.5.1 Streamwise eddy generators 785
22.5.2 Serrated saw-blade 788
22.5.3 Dimpled surface 789
22.5.4 Spanwise slit 790
22.5.5 Fins 793
22.5.6 Circumferential grooves 794
22.5.7 Skin friction and boundary layer 795
22.5.8 Partly grooved surface 796
22.5.9 Spanwise grooves 798

22.6 Change in diameter 798
22.6.1 Introduction 798
22.6.2 Laminar periodic wake, the L3 regime 799
22.6.3 Transition-in-shear-layer, the TrSL state 801
22.6.4 Step interference in the TrSL3 regime 801

22.7 Tapered cylinder 802
22.7.1 Laminar periodic wake, the L3 regime 803
22.7.2 Shedding cells along the span 806
22.7.3 Theoretical model 807
22.7.4 Turbulent wake, the TrSL state 808
22.7.5 Tapered cylinder with free end 809

22.8 Non-linear change in diameter 809
22.8.1 Introduction 809
22.8.2 Cooling towers 811
22.8.3 Model tests 813
22.8.4 Validity of *Re* extrapolation 813
22.8.5 Surface roughness 815
22.8.6 Meridional ribs 816
22.8.7 Cooling tower model in a gust 819
22.8.8 Full-scale tests in natural wind 820
22.8.9 Possible causes of the Ferrybridge failure 822

23 BLOCKAGE AND WALL PROXIMITY 827

23.1 Introduction 827

23.2 Laminar, L, state of flow 828
23.2.1 Creeping flow, the L1 regime 828
23.2.2 Closed near-wake, the L2 regime 829
23.2.3 Instability of the near-wake 831
23.2.4 Laminar periodic wake, the L3 regime 833

23.3 Transition in shear layers, the TrSL state of flow 838
23.3.1 Mechanics of blockage 838

23.3.2 Mean pressure distribution and drag 838
23.3.3 Strouhal number and fluctuating pressure 841
23.3.4 Suppression of eddy shedding 842
23.3.5 Strength and correlation of eddies 846
23.3.6 Effect of free stream turbulence 848

23.4 Transition in the boundary layer, the TrBL state 850
23.4.1 Introduction 850
23.4.2 Drag variation with blockage 850
23.4.3 Strouhal number and fluctuating force 852

23.5 Theoretical correction models 855
23.5.1 Introduction 855
23.5.2 Fage's blockage correction 856
23.5.3 Lock's method of images 857
23.5.4 Glauert's semi-empirical formula 860
23.5.5 Allen and Vincenti's source model 860
23.5.6 Maskell's correction model 864
23.5.7 Modi and El-Sherbiny's streamline model 867

23.6 Asymmetric blockage 869
23.6.1 Laminar wake 869
23.6.2 Turbulent wake 869

23.7 Proximity to a boundary 871
23.7.1 Introduction 871
23.7.2 Classification of flow regimes 872
23.7.3 Contact regime 874
23.7.4 Potential flow for a circle on a boundary 877
23.7.5 Narrow-gap regime 878
23.7.6 Wide-gap regime 880
23.7.7 Effect of wall boundary layer 884

23.8 Erodible boundary, scour 888
23.8.1 Scouring mechanism 888
23.8.2 Forces and Strouhal number 889

24 BOUNDARY LAYER CONTROL 893

24.1 Introduction 893

24.2 Rotating cylinder 893
24.2.1 Magnus effect 893
24.2.2 Classification of flow patterns 894
24.2.3 Prandtl's concept of circulation 896
24.2.4 Potential flow theory 896
24.2.5 Bickley's potential model 899

24.3 Effect of Reynolds number 902
24.3.1 Laminar, L3, and transitional, TrW, wakes 902
24.3.2 Pressure distribution in the TrSL state 905
24.3.3 Inversion of the Magnus effect 909
24.3.4 Boundary layer 913
24.3.5 Strouhal number 916
24.3.6 Effect of end plates 917
24.3.7 Effect of surface roughness and fins 919
24.3.8 Far-wake development 920

24.4 Applications 923
24.4.1 Flettner's rotor ship 923
24.4.2 Rotor windmill 925
24.4.3 Madaras Power Plant Project 926
24.4.4 Wallis's 'dam-buster' 927

24.5 Rotary angular oscillation of a surface 929
24.5.1 Physical background 929
24.5.2 Laminar L2 and L3 regimes 930
24.5.3 A solution of Navier–Stokes equations 932
24.5.4 Forced rotary oscillation eddy shedding 935

24.6 Concentric rotating cylinders 937
24.6.1 Introduction 937
24.6.2 Taylor's theory and experiment 938
24.6.3 Coles' further transitions 940
24.6.4 Classification of flow regimes 942

24.7 Boundary layer control by suction and blowing 944
24.7.1 Suction 944
24.7.2 Porous surface suction 945
24.7.3 Thwaites' flap 946
24.7.4 Jet-blowing 949
24.7.5 Lift and drag forces 950
24.7.6 Dunham's theoretical model 950

25 YAWED CYLINDERS 954

25.1 Introduction 954
25.1.1 Independence principle 954

25.2 Laminar wakes in the L2 and L3 regimes 956
25.2.1 Strouhal number 956
25.2.2 Effect of end plate 959
25.2.3 Free-ended and yawed cylinders 960

25.3 Transition-in-shear layers, TrSL, state 963
25.3.1 Eddy formation region and base pressure 963

25.3.2 Elliptic cross-section 965
25.3.3 Effect of end plates 966
25.3.4 Effect of the aspect ratio 969
25.3.5 Skin friction 969
25.3.6 Strouhal number 973
25.3.7 Drag coefficient 975

25.4 Turbulent wakes in the TrBL state 976
25.4.1 Marine surface roughness 978
25.4.2 Stranded cables and conductors 980

25.5 High angle of incidence 981
25.5.1 Classification of flow regimes 981
25.5.2 Impulsive cross-flow analogy 984
25.5.3 Strength of detached vortices 985
25.5.4 Normal and side force components 986
25.5.5 Effect of Reynolds number 989
25.5.6 Effect of Mach number 990
25.5.7 Detachment instability 992
25.5.8 Suppression of eddy detachment 992

26 TWO CYLINDERS 994

26.1 Introduction 994
26.1.1 Basic interference flow regimes 995

26.2 Tandem arrangements 998
26.2.1 Creeping flow regime, L1 998
26.2.2 Kármán–Bénard street, L3 regime 1000
26.2.3 Early research in the TrSL state of flow 1001
26.2.4 Modification of pressure distribution 1003
26.2.5 Strouhal number 1007
26.2.6 Drag coefficients 1009
26.2.7 Transition-in-boundary-layer, TrBL, state 1011
26.2.8 Effect of free stream turbulence 1012
26.2.9 Effect of surface roughness 1015
26.2.10 Effect of finite height 1015
26.2.11 Effect of heat transfer 1016

26.3 Side-by-side arrangements 1018
26.3.1 Classification of interference regimes 1018
26.3.2 Laminar wakes 1018
26.3.3 Strouhal number 1022
26.3.4 Drag and lift forces 1023
26.3.5 Origin of biased gap flow 1026
26.3.6 Effect of partition plate and sound 1027
26.3.7 Landweber's theoretical model 1029

26.3.8 Other theoretical models 1031
26.3.9 Transition-in-boundary-layer, TrBL, state 1032
26.3.10 Effect of free stream turbulence 1035

26.4 Staggered arrangements 1035
26.4.1 Classification of interference flows 1035
26.4.2 Laminar, L, state of flow 1038
26.4.3 Mean pressure distribution in the TrSL state 1038
26.4.4 Lift and drag in the TrSL state 1039
26.4.5 Gap flow interference regime 1042
26.4.6 Wake displacement interference regime 1044
26.4.7 Strouhal number 1044
26.4.8 Transition-in-boundary-layer, TrBL, state 1047
26.4.9 Effect of free stream turbulence 1047
26.4.10 Stranded conductors 1049
26.4.11 Effect of the finite aspect ratio 1050
26.4.12 Twin cooling towers 1051

26.5 Two cylinders of unequal diameter 1052
26.5.1 Categorization of arrangements 1052
26.5.2 Tandem cylinders, $D_1/D_2 < 1$ 1052
26.5.3 Tandem cylinders, $D_1/D_2 > 1$ 1054
26.5.4 Strouhal number for tandem cylinders 1056
26.5.5 Synchronization of eddy shedding 1056
26.5.6 Unequal side-by-side cylinders 1059
26.5.7 Control cylinder upstream 1060
26.5.8 Control cylinder outside laminar wake 1061
26.5.9 Boundary layer control 1064
26.5.10 Free shear layer control 1065

26.6 Two cylinders crossing at right angles 1067
26.6.1 Introduction 1067
26.6.2 Local pressure and forces along cylinders 1068
26.6.3 Tentative topology 1070
26.6.4 Two intersecting cylinders 1073
26.6.5 Effect of gap between cylinders 1074

27 CYLINDER CLUSTERS 1077

27.1 Introduction 1077

27.2 Three cylinders 1077
27.2.1 In-line clusters 1077
27.2.2 Effect of tripping wires 1081
27.2.3 Three cylinders of different diameters 1082
27.2.4 Side-by-side and staggered clusters 1083
27.2.5 Triangle clusters at low *Re* 1085

27.2.6 Triangle clusters, forces 1089
27.2.7 Triangle clusters, Strouhal number 1090
27.2.8 Irregular triangle clusters 1092

27.3 Four cylinders 1095
27.3.1 In-line clusters 1095
27.3.2 Heat transfer 1098
27.3.3 Side-by-side clusters 1098
27.3.4 Square clusters, forces 1099
27.3.5 Square clusters, Strouhal number 1101

27.4 Cluster of $n > 4$ cylinders 1104
27.4.1 Five cylinders, side-by-side clusters 1104
27.4.2 Seven cylinders, side-by-side clusters 1106
27.4.3 Six and eight cylinders, polygonal clusters 1106
27.4.4 $3 \times 3,\ 4 \times 4, \ldots,\ n \times n$ clusters 1108

27.5 Satellite clusters 1109
27.5.1 Introduction 1109
27.5.2 Interference parameters 1110
27.5.3 Force on satellite clusters 1111
27.5.4 Effect of satellite tube spacing 1113

28 MULTI-TUBE ARRAYS 1118

28.1 Introduction 1118
28.1.1 Categorization of tube arrays 1118

28.2 Single row of tubes 1120
28.2.1 Gap flow jets 1120
28.2.2 Structure of non-uniform flow 1122
28.2.3 Mean pressure distribution and forces 1125
28.2.4 Transition to turbulence in the TrBL state 1126
28.2.5 Metastable states of flow 1127
28.2.6 Suppression of metastable states 1128
28.2.7 Strouhal number 1129
28.2.8 Effect of tube proximity 1130

28.3 In-line tube arrays 1133
28.3.1 Mean pressure distribution 1133
28.3.2 Fluctuating forces 1137
28.3.3 Effect of surface roughness 1138
28.3.4 Acoustic resonance; historical background 1139
28.3.5 Speed of sound in tube arrays 1140
28.3.6 Acoustic excitation and suppression 1141
28.3.7 Owen's buffeting model 1143
28.3.8 Interstitial flow, transition eddies 1146

28.3.9 Instability of jet shear layers 1147
28.3.10 Acoustic synchronization mechanism 1149
28.3.11 Interstitial flow and turbulence 1153
28.3.12 Classification of in-line tube arrays 1155

28.4 Staggered tube arrays 1156
28.4.1 Proximity effects 1156
28.4.2 Mean and fluctuating pressure 1156
28.4.3 Structure of interstitial flow 1159
28.4.4 Effect of tube displacement 1160
28.4.5 Effect of surface roughness 1163
28.4.6 Strouhal number; historical 1163
28.4.7 Parallel triangle tube arrays 1165
28.4.8 Rotated square arrays 1168
28.4.9 Normal triangle arrays 1169
28.4.10 New universal St; a proposal 1171
28.4.11 Maximum sound level and its prediction 1173

28.5 Non-uniform flow in and behind arrays 1173
28.5.1 Historical introduction 1173
28.5.2 Non-uniform interstitial flow 1175

APPENDIX 1177

A1 Glossary of terms 1178

A2 Non-dimensional similarity parameters 1179

A3 Epitome of disturbance-free flow regimes 1180

A4 Abbreviations 1181

D. REFERENCES 1182

AUTHOR INDEX 1249

SUBJECT INDEX 1259

> *'...Academic training of our young engineers often mesmerising with idealized beauties of science and mathematics, which seduces them away from the realities of engineering practice. Playing games with words, numbers, diagrams, graphs, and computer programs has become an end in itself. The fact that they are models, approximations only of the real physical world has often been lost the sight of...'*

T. Furman (1981B)

PART D

GEOMETRY EFFECTS

	Nomenclature	675
21	Aspect ratio, end plates, and free end(s)	679
22	Surface roughness, step-change in diameter, taper, and cooling towers	748
23	Blockage and wall proximity	827
24	Boundary layer control: rotating cylinder, concentric rotating cylinders, suction, and blowing	893
25	Yawed cylinder and missiles at high incidence	954
26	Two cylinders: tandem, side-by-side, and staggered arrangements, different diameters and crossing at right angles	994
27	Cylinder clusters: three, four, and $n > 4$ cylinders, polygonal, and satellite clusters	1077
28	Tube arrays: in-line, staggered, non-uniform flow inside and outside tube arrays	1118
	Appendix	1177
	D. References	1182

NOMENCLATURE

a	distance between consecutive eddies in the same rows, speed of sound, long axis of ellipse
A	constant, cross-sectional area
A_m, A_w	area of model, wind tunnel
$A_T, A/D$	amplitude, relative amplitude of oscillation
b	distance between two rows of eddies, short axis of ellipse
b/a	spacing ratio of eddies, Kármán number
B	breadth of test section, distance between end plates
B/D	aspect ratio between end plates
c	empirical factor
C	constant, coefficient
C_d, C'_d	mean, fluctuating local drag coefficient
C_D, C'_D	mean, fluctuating overall drag coefficient
C_{Dp}, C_{Df}	pressure, friction drag coefficient
C_f, C'_f	mean, fluctuating friction coefficient
C_l, C'_l	mean, fluctuating local lift coefficient
C_L, C'_L	mean, fluctuating overall lift coefficient
$C_{\Delta p}$	pressure drop across array, adverse pressure difference coefficient
C_p, C'_p	mean, fluctuating pressure coefficient
C_{pb}, C_{po}	base, stagnation pressure coefficients
C^*_p	pressure coefficient based on stagnation pressure
C_{pm}, C_{ps}	minimum, separation pressure coefficients
C_R	resistance coefficient (low Re)
$D, D/B$	diameter, blockage ratio
D_b, D_c	base, control cylinder diameter
D_e, D_r	end plate diameter, root diameter
D_m	mean diameter of cooling tower
e	eccentricity of cylinder in test section
E_k, E_m	kinetic turbulent, mean energy
f, f_{Tr}	frequency of eddy shedding, transition eddies
$G, G/D$	gap, relative gap between cylinder and plane boundary
h	height of groove tip
$H, H/D$	height, aspect ratio for free end cylinder
$k = (1 - C_{pb})^{1/2}$	constant separation velocity factor
k_d, k_0	local, overall drag factors (pre-1928)
k_l, k_L	local, overall lift factors (pre-1928)
$K, K/D$	roughness, relative surface roughness
K, K_0	strength of eddy, total generated vorticity
K_r	rib height

l	eddy scale
L	laminar state of flow
$L, L/D$	length of cylinder span, aspect ratio
L_c	correlation length
L_F, L_W	eddy formation, near-wake length
L_t	transition to turbulence length
m	blockage velocity factor
m_a	added mass coefficient
n, n_c	number of cycles, blockage drag factor
N	number of rows
p, p_s	static, free stream pressure
$P, P/D$	pitch, pitch ratio
q	fluctuating kinetic energy
Q	source strength
R/D	distance between rows ratio
R_{12}	correlation coefficient between points 1 and 2
Re	Reynolds number
Re_s, Re_{osc}	separation, oscillation
s	distance between source and sink in oval
$S, S/D$	streamwise spacing, spacing ratio
$\underline{S}/D, \alpha$	radial pitch, polar angle
St	Strouhal number
t	thickness of oval
T	turbulent state of flow
T/D	transverse spacing ratio
Ti, Ts	intensity, scale of turbulence
TrW, TrSL, TrBL	transition in wake, shear layer, boundary layer
u, v, w	velocity components in the x, y, and z directions
u', v', w'	fluctuating velocity components in the x, y, and z directions
u_c	convection velocity
V, V_0	free stream velocity
V', V''	velocity at stations $'$ and $''$
V_g	gap velocity
$V_r, V_r/V$	rotational velocity, rotational velocity ratio
V_s	separation velocity
X, Y, Z	streamwise, transverse, spanwise coordinates

Greek symbols

α	angle of incidence
β	eddy filament slant angle
γ	intermittency factor, strand angle, groove angle
δ, δ_B	cylinder boundary layer thickness, flat boundary
ε	blockage factor
Λ	yaw angle

μ, ν	absolute, kinematic viscosity
θ	circumferential angle
θ_0, θ_s	stagnation angle, separation angle
θ_{ls}, θ_{ts}	laminar, turbulent separation angle
θ_r, θ_{tr}	reattachment angle, transition angle
$\theta_d, \theta_e, \theta_w$	tripping sphere, eddy generator, tripping wire angle
σ	solidity of tube array

Subscripts

c	critical
n	normal velocity component

21

ASPECT RATIO

21.1 Introduction

We should recognize that both experimentalists and theoreticians have shared until recently the comfortable dream-world of two-dimensionality and that the vast majority of the reported measurements on flow around cylinders contained an unknown dose of bias and distortion via two-dimensional interpretations. Marc V. Morkovin (1964R).

It was firmly believed that flow around a long and nominally two-dimensional cylinder submerged in a uniform free stream had to be two-dimensional. The high value of length to diameter ratio, L/D, which was termed the *aspect ratio*, was taken a priori as a guarantee for the two-dimensional flow along the span. Also, it had been assumed that the cylinder ends were the sole cause of three-dimensional flow. These rather arbitrary presumptions considerably delayed research into aspect ratio effects.

The cylinders may be attached to the side walls of a closed test section, may span the free jet, or have either one or both ends protruding free into the flow. The convenient categorization of aspect ratio effects will be based on the kind of end-effect involved:

(i) Most common-type wind tunnels have a closed test section, and the cylinder stretches from one wall to the other. The wall boundary layers separate, roll-up upstream of the cylinder ends, and form *horseshoe*-swirl[100] systems. The swirl represents the helicoidal flow that becomes aligned in the streamwise direction downstream from the cylinder.

(ii) Less common-type wind tunnels have an open test section, and the cylinder protrudes from the free jet. Both cylinder ends are subjected to the jet shear layers.

(iii) The well-established method to minimize end effects is the use of *end plates*. They are designed to prevent transverse flow along the span, and preserve the same size and shape of the near-wake. However, for small spacings between the end plates, $L/D < 2$, an entirely different near-wake in size and shape may be formed in some flow regimes.

(iv) The common engineering applications of *finite* cylinders with one *free end* are chimney stacks, TV towers, lighthouses, fuel-storage tanks, etc. The free

[100]The term *swirl* has been adopted to emphasize the three-dimensional nature of the horseshoe eddy.

end may be open (with or without an efflux) or closed (flat, rounded, hemispherical, conical, etc.). The free end induces a strong three-dimensional flow along the span.

(v) Yet another possibility is a short cylinder with two free ends, such as wheels, coins, AWACS,[101] frisbees, etc. The three-dimensional flows produced at two free ends strongly interact when $L/D < 5$.

This chapter describes the effect of the aspect ratio on the flow along and around cylinders in the above categories.

21.2 Horseshoe-swirl system

The occurrence of the horseshoe-swirl systems in engineering applications is widespread and can take different forms. For example, the scour under bridge piers in river beds is caused by high shear stress beneath the horseshoe-swirl. Other examples are the junction of wings and fuselage on an aircraft, the root of turbomachinery blades, struts in pipelines, high-rise buildings in atmospheric boundary layers, etc. The suppression of the horseshoe-swirl around high-rise buildings is the main concern of architects with a view to improving pedestrian comfort on windy plazas and squares.

21.2.1 *Laminar boundary layer*

Boundary layers along the walls of a wind tunnel test section are subjected to a zero pressure gradient everywhere except some distance upstream of the cylinder spanning the test section. The cylinder ends are submerged in the retarded wall boundary layers, and that means an adverse pressure gradient. This causes a three-dimensional separation followed by a crescent-like roll-up of the separated boundary layers into a system of swirls. The swirl system is swept around the base of the cylinder and assumes a characteristic shape, which has led to its name, the *horseshoe-swirl.*

The earliest sketch of the horseshoe-swirl was made by Leonardo da Vinci; see the frontispiece in Vol. 1. An early photograph taken by Sutton in Cambridge[102] is reproduced in Fig. 21.1(a). The wall boundary layer was laminar, the finite cylinder had $H/D = 0.5$, and the swirl system was made visible by smoke filaments injected upstream into the free stream.

Several swirls are visible upstream of the cylinder in Fig. 21.1(a), and a reflection of the flow pattern in the lower wall can also be seen. There are two large swirls rotating in the clockwise sense, and two triangular ones in between rotating in the anticlockwise sense. The flow topology of the swirl system is sketched in Fig. 21.1(b). The primary separation of the wall boundary layer is denoted by S, and the secondary separation induced by the swirls is denoted by S_1 and S_2. The confluence and reattachment points are marked by SP_1, SP_2, and A_0, A_1, A_2, respectively.

[101] AWACS stands for Airborne Warning And Command System.

[102] The flow visualization photograph was printed as the frontispiece in Thwaites (1960B).

(a)

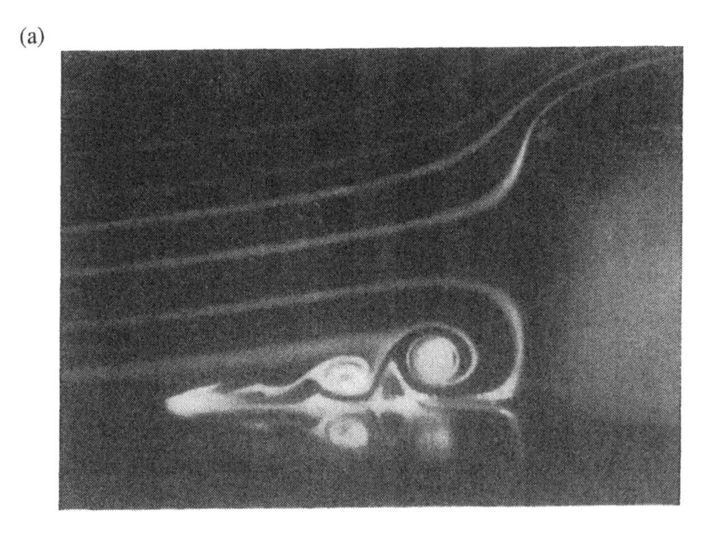

(b)

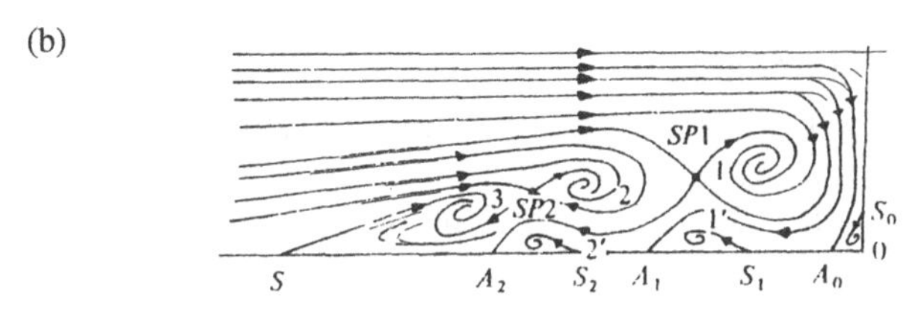

FIG. 21.1. Sutton's flow visualization upstream of finite cylinder $H/D = 0.5$ attached to a wall and submerged in a laminar boundary layer (a) photograph, (b) topology, Baker (1979J)

The physical parameters governing this phenomenon are the thickness δ_B of the wall boundary layer and the cylinder diameter D. The wall boundary layer may be laminar or turbulent, depending on $Re_B = V\delta_B/\nu$. The cylinder is specified by the aspect ratio, H/D, and $Re = VD/\nu$. There are two common non-dimensional parameters, namely D/δ_B and H/δ_B (the latter may be ignored when $H \gg \delta_B$). Most researchers use the cylinder Re, which is relevant to the cylinder boundary layers, while Re_B is relevant to the wall boundary layer.

The number of swirls formed in the horseshoe system depends on the flow velocity and thickness of the wall boundary layer. Figure 21.2 shows Baker's (1979J) observation of 2,4,6-eddy systems followed by the unsteady region. More swirls appear as the cylinder Re increases and the boundary layer thickness decreases. Note that the variation in D/δ_B is caused by the decrease in δ_B with increasing velocity. Beyond the six-swirl region, the entire swirl system begins to oscillate in a regular manner at first. At higher velocities, the flow becomes very unsteady and turbulent with no traces of periodicity.

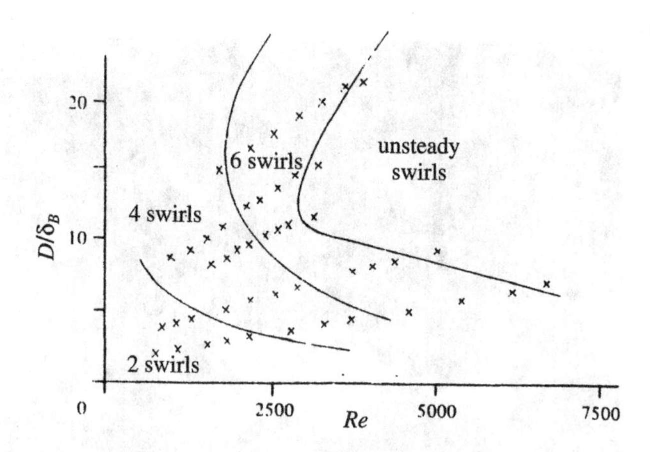

FIG. 21.2. Regions of multiple swirls in the $D/\delta_B, Re$ plane, Baker (1979J)

21.2.2 *Oscillating horseshoe-swirl system*

It was shown in Fig. 21.2 that at sufficiently high *Re*, and when $D/\delta_B > 10$, the laminar horseshoe-swirl system becomes unstable and starts to oscillate. Baker (1979J) made a cine-film of the oscillating horseshoe system; some sequences are reproduced in Fig. 21.3. The primary swirl initially moves towards the cylinder (frames 2 and 4 in Fig. 21.3). The feeding smoke filament is cut (frames 6 and 8) and the swirl appears isolated and shrinking (frames 10 and 12). Twelve frames represent half of the oscillation cycle, after which the swirl starts to move upstream, where it is again fed by the smoke filament from the shear layer. Baker (1979J) found that the oscillating motion starts when $Re_B = 125$.

A similar oscillating horseshoe-swirl was observed by Schwind (1962P), who attempted to classify flow regimes. The difficulty was an enormous overlapping of flow regimes. At any velocity, up to four different waveforms could be detected by the hot wire at different times:

(i) steady flow with no trace of oscillation;

(ii) low-frequency horseshoe-swirl system oscillation, $St \simeq 0.26$ (based on D);

(iii) high-frequency horseshoe-swirl system oscillation, $St \approx 0.4 - 0.6$;

(iv) irregular turbulent trace.

Each of the above flow regimes might exist for any period of time between 1 s and 5 min, after which the flow switches from one regime to another in a completely random manner. Baker (1979J) demonstrated that the addition of a splitter plate behind the cylinder or the replacement of the circular cylinder by a streamlined strut had no effect on the oscillation.

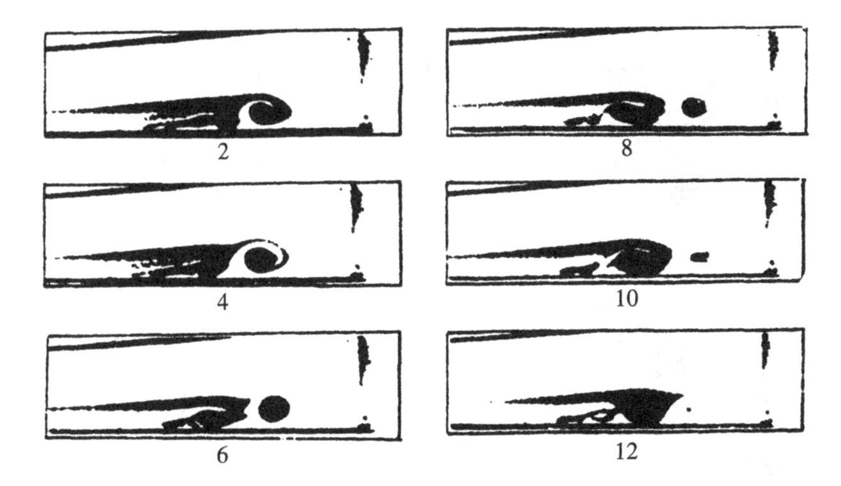

FIG. 21.3. Sequences of oscillation of the swirl system, Baker (1979J)

Two hypotheses concerning the nature of horseshoe-swirl oscillations were put forward by Baker (1991J):

(i) instability of the swirl core;

(ii) instability of the wall boundary layer.

21.2.3 *Turbulent boundary layer*

The imprint of the horseshoe-swirl is always left on the wall when surface flow visualization is performed. Figure 21.4(a,b,c) shows the wall pattern and C_p distribution for $Re = 34\text{k}$, $D/\delta_B = 3.9$, and $H/D = 1, 3, 5$, respectively. The distinct horseshoe-shaped pattern is similar for all three values of H/D. However, the kidney-like pattern in the near-wake of the cylinder changes from widely spaced, $H/D = 1$, to narrowly spaced, $H/D = 3$, and into a merged lump, $H/D = 5$.

Taniguchi *et al.* (1981Ja) also measured the static pressure, and plotted the isobar C_p curves in Fig. 21.4(a′–c′). The 'islets' of $C_{p\min}$ correspond to the observed kidney-like patterns. For all three aspect ratios tested, the adverse pressure gradient exists on the upstream side. The downstream C_p curves do not correspond to the surface visualization. This may be due to the displacement of the horseshoe-swirl away from the wall as it is carried downstream. Sakamoto and Arie (1983Ja) argued that the kidney-like pattern at $H/D = 1$ represents the attachment of an arch-like eddy, which is symmetrically shed. This hypothesis will be discussed later.

The well-defined imprint of the time-averaged turbulent horseshoe-swirl in Fig. 21.4 gives a false impression of the actual flow, which is unsteady and turbulent. Figure 21.5(a,b) shows smoke visualization across a meridional plane up-

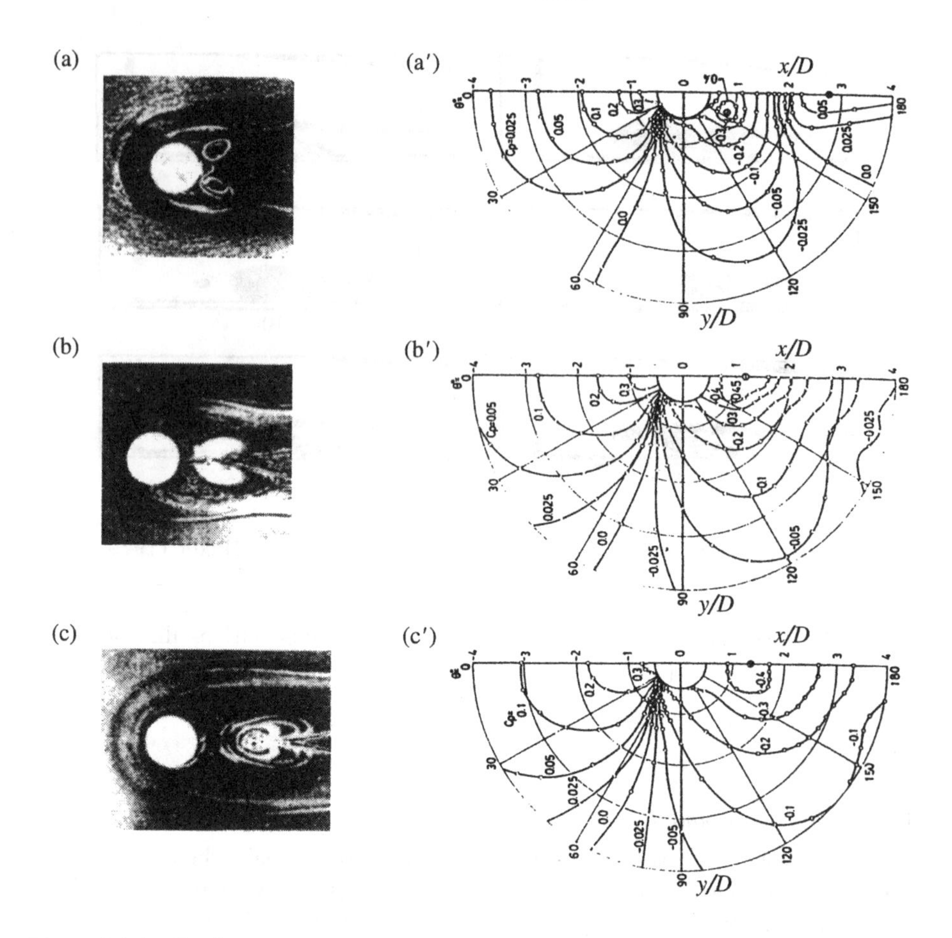

FIG. 21.4. Surface visualization and measured isobars on the wall: (a,a′) $H/D = 1$, (b,b′) $H/D = 3$, (c,c′) $H/D = 5$, Taniguchi *et al.* (1981Ja)

stream of the cylinder. A short-exposure (1/500 s) for $H/D = 1.75$ at $Re = 7.6$k and $D/\delta_B = 2.5$ in Fig. 21.5(a) reveals highly irregular small-scale eddy structures typical for turbulent flows. However, a long-exposure (1/15 s) of the same flow shows the blurred image of the horseshoe-swirl in Fig. 21.5(b). This demonstrates that the turbulent horseshoe-swirl exists only intermittently, and can be quantified as a time-averaged phenomenon.

Belik (1973J) measured C_p upstream of the cylinder in the meridional plane for $H/D = 3.3$, $D/\delta_B = 2$, and $36\text{k} < Re < 220\text{k}$. Figure 21.6 shows the C_p distribution (points) and the theoretical C_p (curve) for potential two-dimensional flow given by

$$C_p = 2\cos(2\theta)(2r/D)^{-2} - (2r/D)^{-4} \tag{21.1}$$

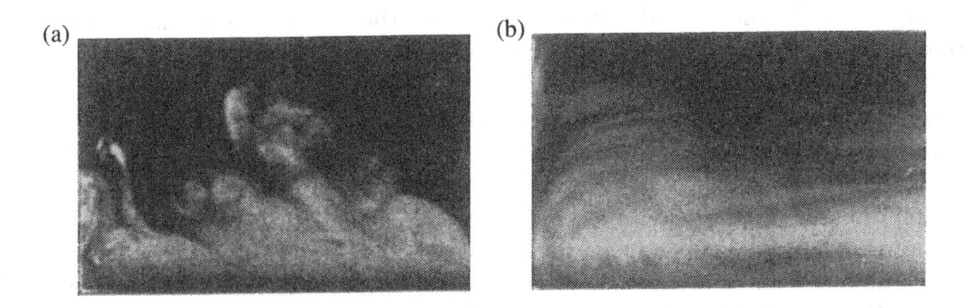

FIG. 21.5. Smoke visualization of the horseshoe-swirl for $H/D = 1.75$ at $Re = 7.6$k (flow from right to left): (a) instant 1/500 s, (b) 1/15 s, Baker (1980J)

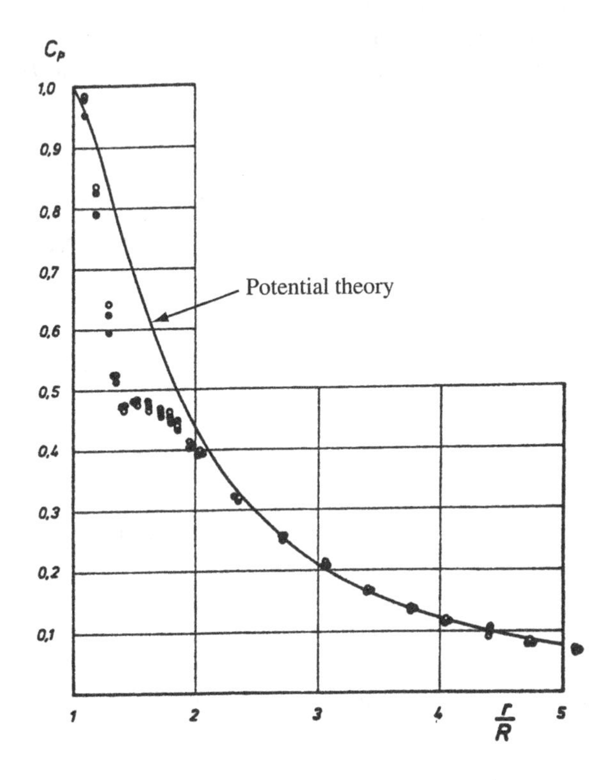

FIG. 21.6. Pressure distribution in the meridional plane upstream from the cylinder (full line, theory), Belik (1973J)

where r and θ are the polar coordinates. The departure of the experimental points from the theoretical curve is caused by the wall boundary layer separation. The dip in the experimental points corresponds to the location of the axis of the primary horseshoe-swirl. This was proved by Eckerle and Awad (1991J), who carried out simultaneously flow visualization, hot-wire traverses, and pressure measurements in the meridional plane. Belik (1973J) also estimated the distance of the boundary layer separation from the cylinder. The relative separation distance was in the range $0.45 < x_s/D < 0.65$.

Further insight into the nature of the horseshoe-swirl system can be gained from the mean shear stress and turbulence intensity variation in the meridional plane, Dargahi (1989J). Figure 21.7 shows the start of the separation region at $c_f = 0$, which corresponds to $Ti_{\max}$ in Fig. 21.8. The horseshoe-swirl induces a reverse velocity at the wall and at $\tau_{\min}/\tau_{\max} = -1.5$. This high shear is responsible for the erosion of the river bed upstream of the bridge pier.

The distance from the boundary layer separation to the cylinder depends on the height and diameter of the cylinder. When the height H is greater than the wall boundary layer thickness δ_B, the separation length x_s is independent of H. Baker (1991J) measured the separation length x_s and location of swirl axis x_{s1} at $Re = 14.3$k and $D/\delta_B = 5.1$. Figure 21.9 confirms that for $H/D > 1$ both x_s/D and x_{s1}/D become independent of H/D.

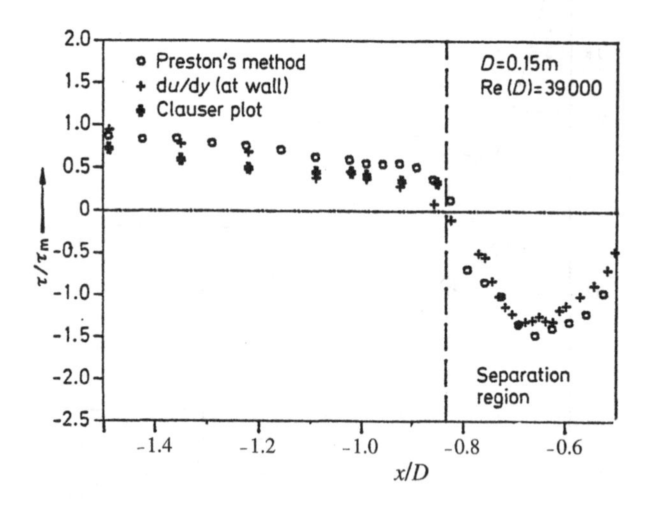

FIG. 21.7. Shear stress distribution at $Re = 39$k in the meridional plane, Dargahi (1989J)

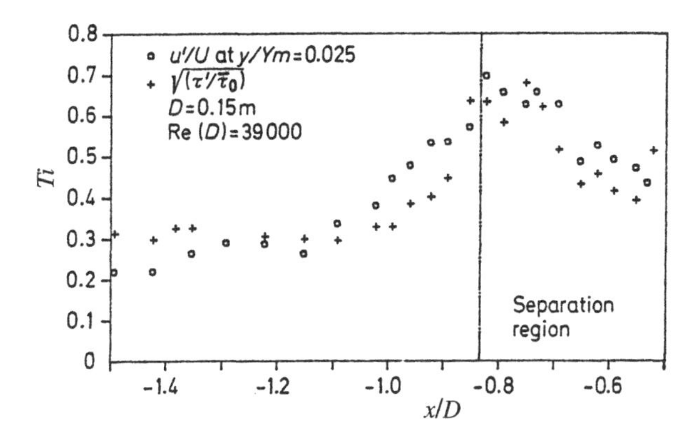

FIG. 21.8. Comparison of turbulence intensity, ∘, and shear ratio, +, Dargahi (1989J)

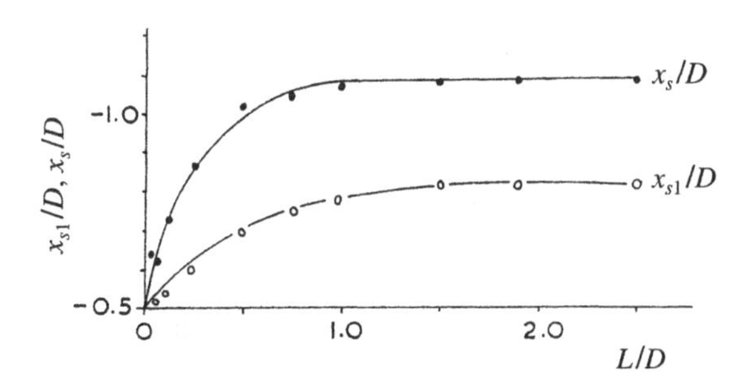

FIG. 21.9. Distance of boundary layer separation from cylinder: x_s-primary, x_{s1}-secondary, Baker (1991J)

21.3 Closed test section

21.3.1 *Steady laminar wake, L2 regime*

The steady closed laminar near-wake was designated as the L2 regime in Vol. 1, Chapter 1.[103] The effect of low aspect ratio on flow around cylinders spanning closed test sections is considered next. A prevalent three-dimensional creeping flow around the short cylinder was found by Nisi[104] (1925J) in the L2 regime. He used a 'miniature' wind tunnel having a test section of 2.4 cm × 2.4 cm and a 'cylinder' of $D = 2.75$ mm and $L/D = 8.7$. Figure 21.10(a,b) shows the near-wake flow visualization at $Re \simeq 15$ viewed from the side and along the span,

[103]The adopted abbreviation of flow regimes is given in the Appendix.

[104]According to the present spelling of Japanese names he should be Nishi.

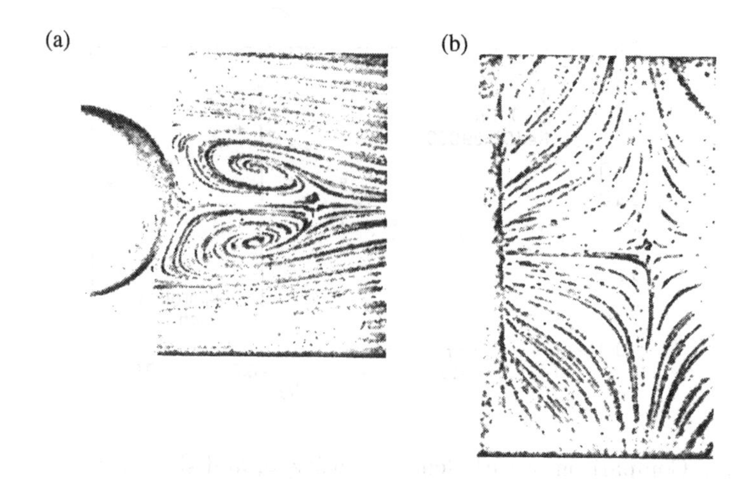

FIG. 21.10. Flow visualization at $Re \simeq 15$ for $L/D = 8.7$, (a) across near-wake, (b) spanwise, Nisi (1925J)

respectively. The peculiar feature of the flow topology is that the free shear layers separated from the cylinder do not meet at the confluence point, i.e. the near-wake is not closed. Nisi (1925J) suggested that 'the air flowing out from the eddy centres must be supplied from elsewhere. This implies that the motion cannot be purely two-dimensional.'

Figure 21.10(b) shows the same flow but the cylinder is illuminated horizontally along its span. Two streams, which flow from the walls along the span of the cylinder, meet at the confluence point along the midplane, as seen in Fig. 21.10(a). The motion in the adjacent planes is more complicated, and the photograph covers only $4D$ of the span. Nisi (1925J) concluded: 'The form of the eddy behind the cylinder, especially in the midplane is considerably influenced by the boundaries.'

Another example where the small aspect ratio became a governing parameter was provided by Nishioka and Sato (1974J). They used large movable end plates ($25D \times 20D$), and examined the initiation of eddy shedding. Figure 21.11 shows that the Re_{osc} was unaffected for $L/D > 25$. For $L/D < 25$, however, an exponential increase in Re_{osc} occurred. For example, the formation of an eddy street was suppressed up to $Re_{osc} = 195$ for $L/D = 4$.

21.3.2 *Periodic laminar wake, L3 regime*

The laminar periodic flow with the Kármán–Bénard eddy street has been designated as the L3 regime. It was discussed in Vol. 1, Chapter 3.2, how the laminar flow instability initiates a wavy trail along a near-wake at $Re_{osc} = 35$–40. At this early stage, the wave crests and troughs are straight and parallel to the cylinder axis. As soon as the eddy filaments are formed at $Re \approx 50$–55, the side walls

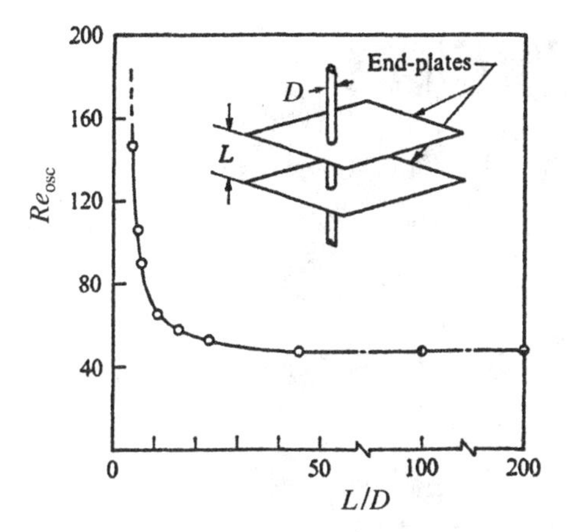

FIG. 21.11. Commencement of oscillation Re_{osc} in terms of aspect ratio L/D, Nishioka and Sato (1974J)

induce an end effect. The eddy filaments adjacent to the walls are subjected to a lower velocity and become bowed. The end effect slowly spreads across the entire span and the eddy filaments, though parallel to each other, are slanted relative to the cylinder axis.

A typical eddy pattern near the wall at $Re = 75$ is shown in Fig. 21.12(a,b). There are three 'cells' separated by two 'knots'. The slantwise filament angle β is smallest in the central cell β_c, increases in the side cell β_s, and is largest in the end cell β_e adjacent to the wall. The frequency of eddy shedding differs from cell to cell in such a way that it is highest in the central cell and lowest in the end cell. The difference in eddy shedding means that the number of eddy filaments cannot be the same from cell to cell. Figure 21.12(b) shows the topology of the 'mismatch' of eddy filaments causing the appearance of the knots, where two eddy filaments from the high-frequency side are 'tied' or 'linked' to one eddy filament on the low-frequency side. The same phenomenon appears as eddy 'splitting' when viewed from the low-frequency side. This caused different terminologies such as knot, eddy 'splitting' or 'linking', 'dislocation',[105] etc.

König *et al.* (1990J) identified four cells of different frequencies separated by three knots in the range $55 < Re < 65$. As Re increases, the extent of the central cell decreases until it disappears at $Re \simeq 65$. The second range, $65 < Re < 85$, which has three cells separated by two knots, was shown in Fig. 21.12(a). Again, the central cell decreases and disappears at around $Re = 85$. These events can be observed in Fig. 21.13(a,b), which shows St measured by the hot wire located at the midspan. The three cells having shedding frequencies f_c, f_s, and f_e are

[105]The term *dislocation* has been used by physicists for crystal deformation in solids and so is less appropriate for highly deformable eddy filaments.

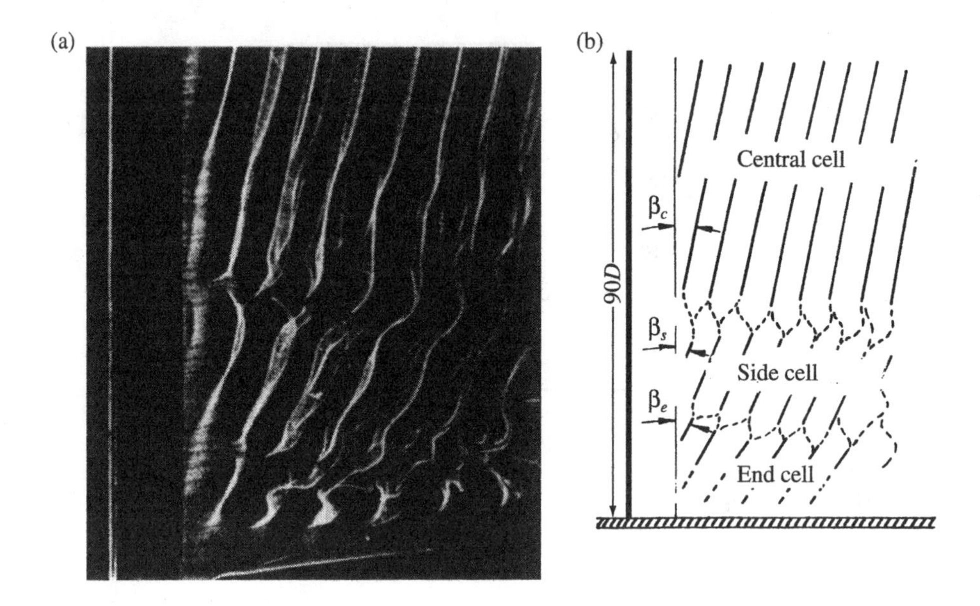

FIG. 21.12. Slanted eddy filaments at $Re = 75$, (a) flow visualization, (b) flow topology, König *et al.* (1990J)

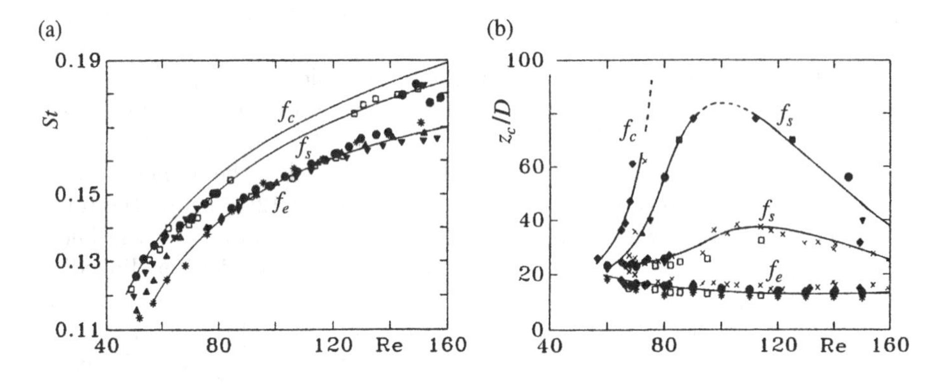

FIG. 21.13. (a) Strouhal number versus *Re*, (b) relative cell length, z_c/D, König *et al.* (1990J)

depicted by three *St* curves. Some experimental points for $L/D = 56, 70, 80, 112,$ and 140 do fall from high to low *St* curves at certain *Re*, but others do not. For example, for $L/D = 70$ and 80 the experimental points start on the second *St* curve, while all points for $L/D = 56$ remain on the lowest *St* curve throughout. The reason is the variation in the cell size with increasing *Re*. Figure 21.13(b) shows that the end cell f_e has the smallest spanwise length, and does not change with Re. The end cell f_e corresponds to the lowest St curve in Fig. 21.13(a), and varies slightly with Re. For the next cell f_s, the middle St curve does vary considerably with Re. The jumps from the bottom to the middle St curve are related to the decrease in cell size. Finally, the initially central cell f_c shows an exponential increase in cell length z_c, which explains the discontinuous fall from the top to the middle St curve in Fig. 21.13(a) for $L/D = 80, 112$, and 140. Hence, the puzzling discontinuous falls and jumps in St across a wide range of Re have a simple origin: disappearance and appearance of shedding cells, respectively.

21.4 Cylinder spanning the free jet

Wind tunnels with open test sections utilize a free jet discharged from a nozzle. The free jet is at ambient pressure, and free shear layers are formed along the jet boundaries. A typical distribution of the stagnation pressure C_{po} across the free jet without the cylinder (full line), and with the cylinder spanning the jet (dashed line) is shown in Fig. 21.14. The free jet boundary is estimated where $C_{po} = 0.5$, and the rate of the free jet expansion by a subtended angle of 6°. When a cylinder of aspect ratio[106] $B/D = 11$ is placed at a right angle across the jet at $B/2$ distance downstream from the nozzle exit, a considerable change in C_{po} distribution takes place (dashed line). The free shear layers widen, and the jet expansion angle increases to 8°. The interference between the free shear layers and flow around the cylinder will be discussed in detail in subsequent sections.

21.4.1 *Transition-in-shear layer, TrSL state*

Kraemer (1965P) carried out extensive tests in a free jet for aspect ratios

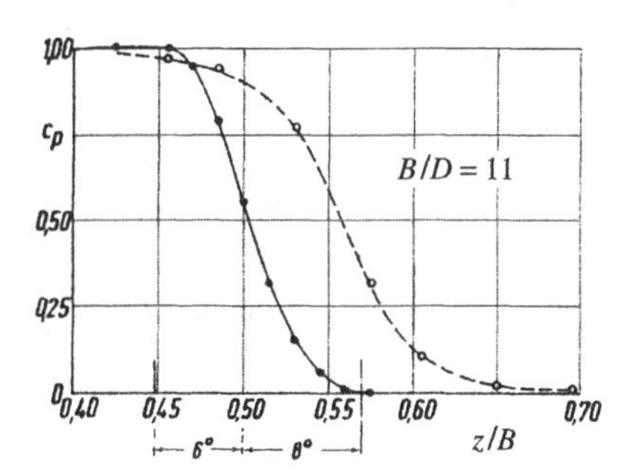

FIG. 21.14. Stagnation pressure distribution across a free jet (full circles), and for cylinder, $B/D = 11$ (open circles), Kraemer (1965P)

[106]The breadth of the open jet is denoted by B and the cylinder length is denoted by L (where $L > B$). Hence, the aspect ratio is related to the jet breadth and is defined by B/D.

$0.5 < B/D < 220$. A typical C_{pb} distribution along the half-span is shown in Fig. 21.15(a,b) for the TrSL2 and TrSL3 flow regimes, respectively. There are three common features:

(i) The lower the aspect ratio B/D, the higher the $-C_{pb}$ values. For $Re = 2.5\text{k}$, at least $B/D = 65$ is required for a uniform C_{pb} along the span, while for $Re = 50\text{k}$, even $B/D = 86$ is insufficient.

(ii) The lowest $-C_{pb}$ does not appear at the midspan but inside the free shear layers at the jet boundary. This unexpected feature may be due to an intensive spanwise flow.

(iii) The uniform C_{pb} region does not extend more than $z/B = \pm 0.3$. This indicates that end plates must be used irrespective of B/D value.

The difference between C_{pb} distributions at $Re = 2.5\text{k}$ and 50k, which are the long and short eddy formation regions, respectively, appears to be small in Fig. 21.15(a,b). This false impression is dispelled by plotting C_{pb} in terms of z/D for $B/D > 66$ in the range $1.2\text{k} < Re < 100\text{k}$. Figure 21.16 shows a considerable variation in C_{pb} in the TrSL2 regime ($1.2\text{k} < Re < 20\text{k}$), and a small one in the TrSL3 regime ($20\text{k} < Re < 100\text{k}$). The large variation in C_{pb} in the TrSL2 regime is caused by shortening of the eddy formation region with increasing Re. The short eddy formation region in the TrSL3 regime does not vary with Re.

An amazing qualitative similarity is seen in Fig. 21.17 for C_{pb} distributions measured in the open test section (open circles) and closed test section (full circles) at $Re = 23\text{k}$ (the start of TrSL3). There are two slightly displaced $C_{pb\max}$ followed by less displaced $C_{pb\min}$. The larger width of the free shear layer in comparison with the boundary layer thickness may explain the different displacement

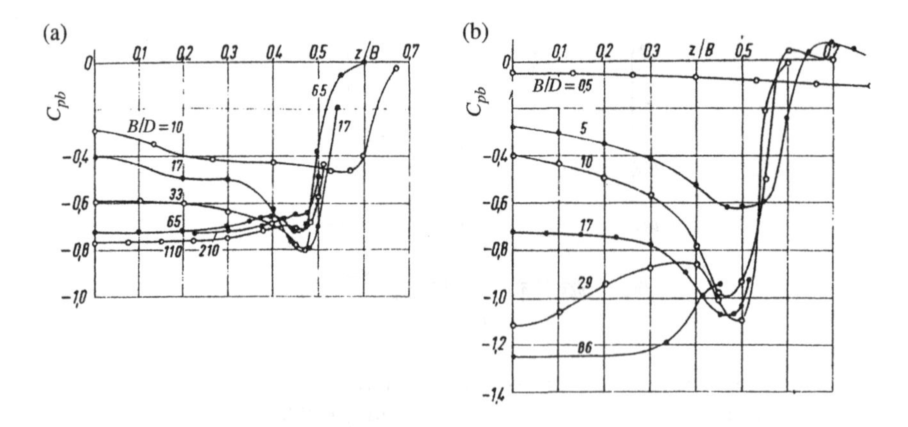

FIG. 21.15. Base pressure distribution along a half-span, (a) $Re = 2.5\text{k}$, (b) $Re = 50\text{k}$, Kraemer (1965P)

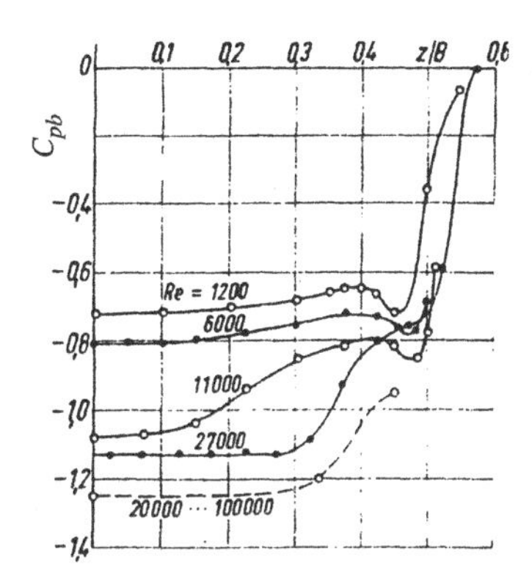

FIG. 21.16. Base pressure in terms of half-span for various *Re*, Kraemer (1965P)

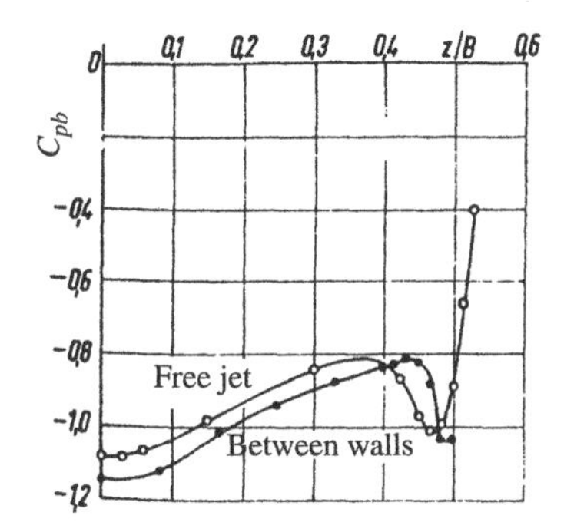

FIG. 21.17. Base pressure distribution at $Re = 23\text{k}$ along the cylinder spanning free jet (open circles), closed test section (closed circles), Kraemer (1965P)

of the peaks. The non-uniformity in C_{pb} along the entire span indicates that $B/D = 30$ is not sufficient to maintain a nominally two-dimensional flow.

Kraemer (1965P) also summarized his measured C_{pb} at the midspan in the range $600 < Re < 600\text{k}$ and $0.5 < B/D < 220$. Figure 21.18 shows the variation in C_{pb} with *Re*, and may be classified as follows:

(i) for small aspect ratio, $0.5 < B/D < 10$, C_{pb} is independent of *Re*;

(ii) for medium aspect ratios, $11 < B/D < 100$, there is a combined effect of both *Re* and B/D on C_{pb};

(iii) for large aspect ratios, $B/D > 100$, there is a dominant effect of *Re* and a negligible effect of B/D.

Special features of C_{pb} variation are also included in Fig. 21.18, such as the

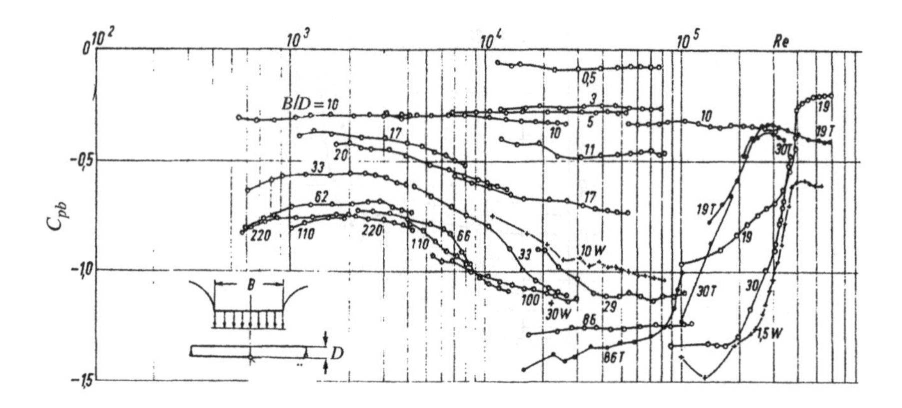

FIG. 21.18. Mid-span base pressure in terms of *Re* and aspect ratio, Kraemer (1965P)

closed test section (denoted by W) and the free-stream turbulence (denoted by T). Curves 19T and 30T are displaced towards lower *Re* in comparison with curves 19 and 30, due to the effect of the free-stream turbulence.

21.5 End plates

It has been a long-established practice to fit short aspect ratio cylinders with end plates for tests in free jets. The aim of the end plates was to eliminate the effect of the free-jet shear layers in open test sections. The end plates were also used to isolate the horseshoe-swirls in closed test sections. It was presumed that the end effects would be eliminated or isolated, and a *nominally* two-dimensional flow along the span should be established. Validation of these presumptions has been carried out only recently.

The improvement in two-dimensional flow by using end plates depends on several influencing parameters, such as the end-plate shape, size, alignment, etc. Two shapes of the end plates are widely used: circular and rectangular. The size of the end plates may vary in a wide range $2 < D_e/D < 150$, where D_e is the end-plate diameter. The excessively large end plates are used for laminar flows. Another important aspect is the location of the cylinder relative to the end-plate leading and trailing edges. The optimal size and shape of the end plates are distinct for different flow regimes. Graham (1969J) found that the end plates should not be placed parallel in order to allow for the boundary layer growth.

21.5.1 *Laminar periodic wake, L3 regime*

Low *Re* flows have been simulated by testing thin wires in air. Their small diameter ensured that even in a small test section an aspect ratio of several hundred diameters could be obtained. Most researchers carried out measurements at the cylinder midspan, and assumed that a large L/D guaranteed a nominally two-dimensional flow.

Teissié-Solier *et al.* (1937J) were an exception because their tests were carried out in a water duct, and the cylinder had $L/D = 50$; see Vol. 1, Chapter 3, Fig. 3.39. They discovered two different frequencies of eddy shedding at the midspan and near the side walls in the range $73 < Re < 214$, as shown in Fig. 3.39. The existence of the low shedding frequency near the side walls might have been caused by a thick boundary layer (unspecified) and related to the non-uniform free stream.

These possible causes were eliminated by Gerrard (1966Ja), who spanned the cylinder between two end plates in the form of RAF 30 aerofoils. He also found a 16.7% lower frequency when the hot wire was located 7D from the end plate at $Re = 85$ and $L/D = 87$. Evidently, neither the non-uniform flow nor the thick boundary layer were present in his experiment.

21.5.2 *Effect of a single end plate*

Gerich and Eckelmann (1982J) placed a hot wire at $x/D = 10$ and $y/D = 2$, and by using a movable end plate $D_e/D = 50$ varied L/D. As long as the end plate was sufficiently far from the hot wire, the velocity fluctuations detected by the hot wire were steady both in frequency f_c and amplitude. When the end plate was moved nearer to the hot wire the velocity fluctuations became modulated, and a lower frequency f_e appeared. Typical hot-wire signals measured between two shedding cells at z_c are reproduced in Fig. 21.19 for two Re values. The signals showed a superposition of two different frequencies f_c and f_e.

The frequency spectrum for $Re = 100$ is shown in Fig. 21.20 with two nearly equal peaks at $f_c = 230\,\text{Hz}$ and $f_e = 201\,\text{Hz}$. Their difference, $\Delta f = 29\,\text{Hz}$, as well as a host of various combinations thereof, is also seen. The occurrence of Δf indicates two eddy streets with different shedding frequencies. The end plate presence induces the end cell at z_c/D. Note that the size of the end cell depends on D_e/D. Figure 21.21 shows an increase in cell width to $z_c/D = 12$ for $D_e/D = 150$.

Gerich (1986P) carried out a detailed flow visualization of both low- and

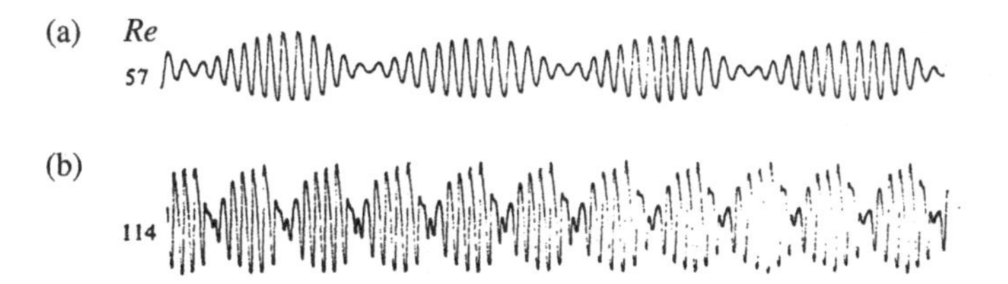

FIG. 21.19. Hot-wire signals measured between cells, (a) $Re = 57$, (b) 114, Gerich and Eckelmann (1982J)

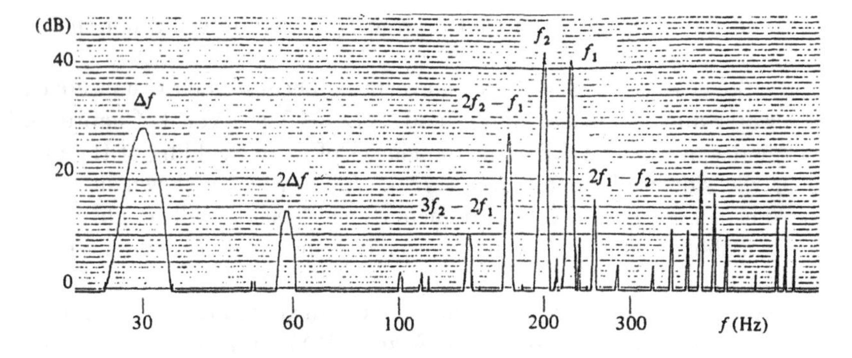

FIG. 21.20. Frequency power spectrum for Re = 100, Gerich and Eckelmann (1982J)

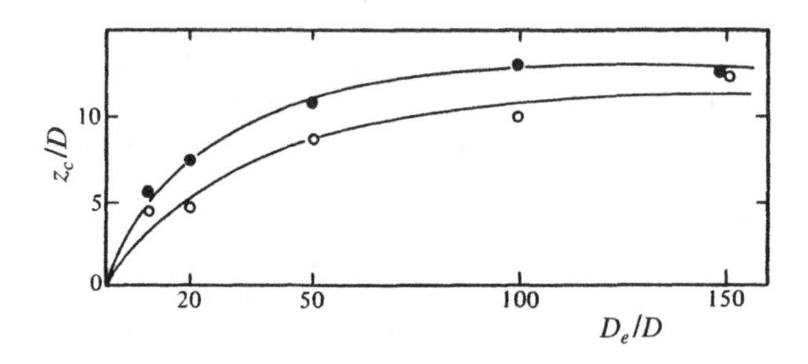

FIG. 21.21. End cell size in terms of end plate size, two end plates (full circles), one end plate (open circles), Gerich and Eckelmann (1982J)

high-frequency cells at $Re = 125$. Figure 21.22(a–c) shows a selection of three side views: (a) low-frequency cell at $z/D = 3$, (b) cell boundary at the knot, $z_c/D = 9$, and (c) high-frequency cell at $z/D = 16$.

The low- and high-frequency eddy streets are regular, while flow at the knot is irregular. There is another feature worth commenting on. The width of the near-wake adjacent to the end plate is disproportionately large in Fig. 21.22(a) compared with that at the knot and in the other cell, Fig. 21.22(b,c), respectively. The irregular appearance of the near-wake at the knot is caused by the splitting and linking of eddy filaments from the two cells.

A quantitative insight into the flow structures in two cells can be gained from the measured mean and fluctuating velocity profiles. Gerich (1986P) placed the hot wire at $x/D = 10$ and $Re = 136$, and the movable end plate provided the range $4.2 < L/D < 33$. Only a selection of velocity profiles is reproduced in Fig. 21.23(a,b,c) at $z/D = 4.6$, 13, and 33, respectively. There is a remarkable dissimilarity in mean velocity profiles measured at the chosen stations. The mean

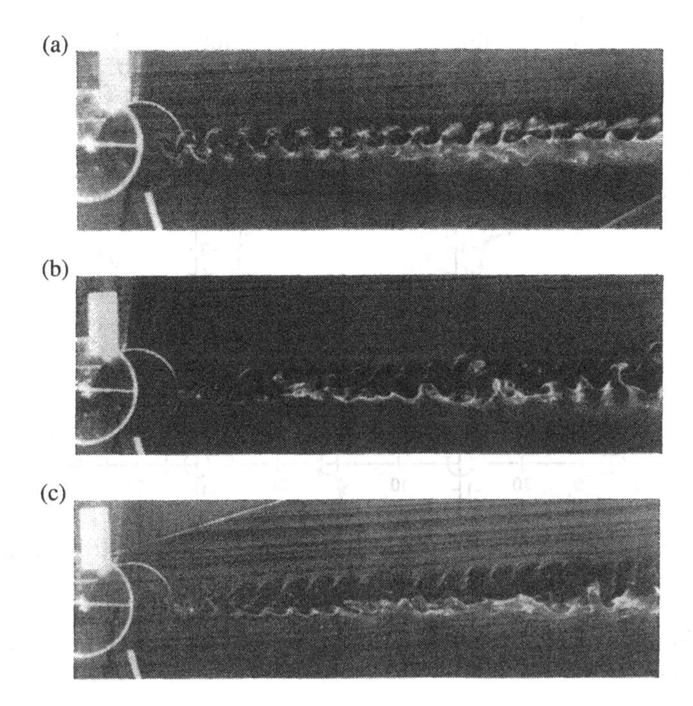

FIG. 21.22. Flow visualization behind cylinder at Re = 125, B/D = 44, $D_c/D = 20$, (a) $z/D = 3$, (b) $z_c/D = 9$, (c) $z/D = 16$, Gerich (1986P)

velocity profiles at z/D = 4.6 show a small velocity defect at y/D = 0. At $z/D = 13$ and 33, the mean velocity defects are 65% and 70%, respectively. The dissimilarity is also seen in the turbulence intensity profiles. Figure 21.23(a′) shows two peaks at $y/D = \pm 1.0$ for $z/D = 4.6$, and a minimum at $y/D = 0$. At the knot, $z/D = 13$, Fig. 21.23(b′), $Ti_{\max}$ appears at $y/D = 0$ and two kinks at $y/D = \pm 1.4$. Finally, at $z/D = 33$, Fig. 21.23(c′), two maxima are formed again at $y/D = \pm 1$ and a minimum at $y/D = 0$.

An overall view of the mean velocity and turbulence variation measured at $y/D = 0$, $x/D = 10$, and $Re = 136$ along the span is shown in Fig. 21.24(a,b). The mean velocity profile is dominated by an unexpected peak at $z/D = 4$ where the velocity defect disappears. Beyond the knot at $z_c/D = 9$, there is a decrease in the mean velocity up to a minimum at $z/D = 13$. Figure 21.24(b) shows the variation in the turbulence intensity with $Ti_{\max}$ at the knot.

21.5.3 *Effect of two end plates*

When a second end plate is added at the other end of the cylinder and the distance between them is $L/D > 2z_c/D$, there is still a side cell of frequency f_s and two end cells of frequency f_e. However, when $L/D \leqslant 2z_c/D$, the central cell

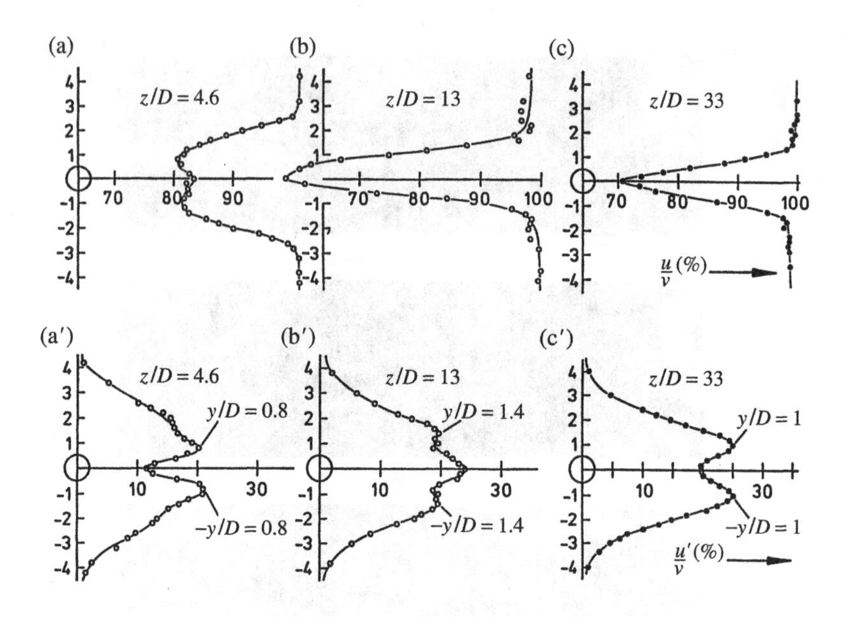

FIG. 21.23. Mean and fluctuating velocity profiles, $Re = 136$, (a,a′) $z/D = 4.6$, (b,b′) $z/D = 13$, (c,c′) $z/D = 33$, Gerich (1986P)

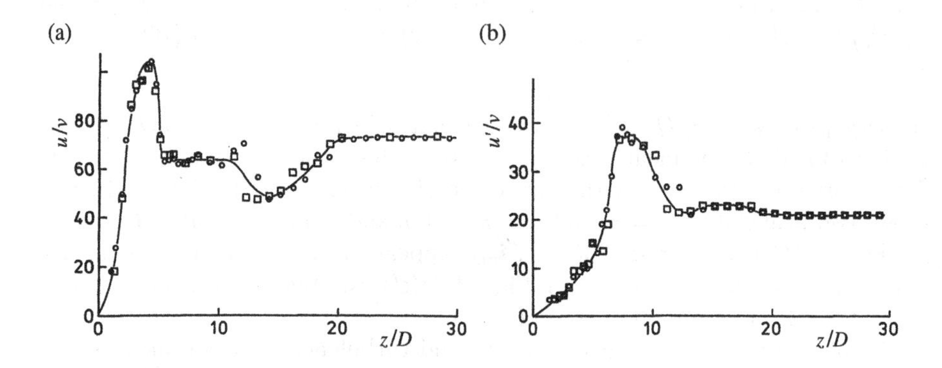

FIG. 21.24. Spanwise variation at $Re = 136$ at $y/D = 0$, $z_c/D = 9$, (a) mean, (b) fluctuating velocity, Gerich (1986P)

disappears and a single cell between the end plates is formed having frequency f_e'. All three frequencies expressed through Ro versus Re have been shown in Vol. 1, Fig. 3.40. The single-cell eddy-street visualization is shown in Fig. 21.25 for $B/D = 16$ and $Re = 125$. Gerich (1986P) visualized eight cross-sections in the range $0 < z/D \leqslant 8$ (midspan). Only four of his photographs are reproduced

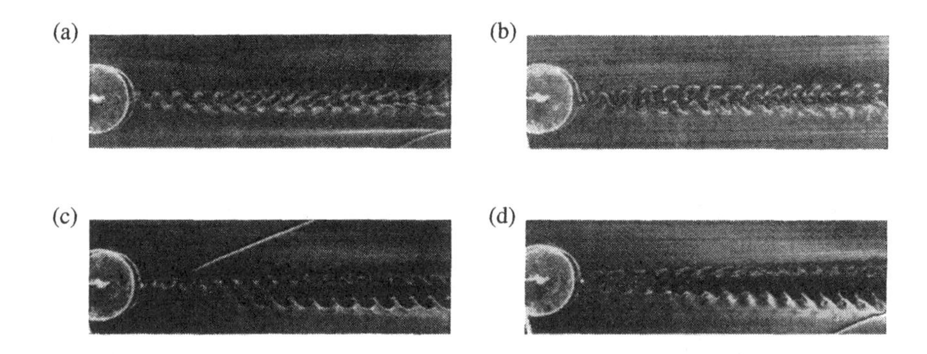

FIG. 21.25. Flow visualization behind a cylinder at $Re = 125$, $B/D = 16$, (a) $z/D = 2$, (b) $z/D = 4$, (c) $z/D = 6$, (d) $z/D = 8$, Gerich (1986P)

in Fig. 21.25(a,b,c,d) for $z/D = 2$, 4, 6, and 8, respectively.

When B/D is reduced to 25 there is a discontinuous fall in St caused by the disappearance of the central shedding cell. The two end cells merge and the shedding frequency f_e' becomes slightly lower than for the former end cell f_e for $L/D > 25$. A further decrease in B/D leads to a continuous decrease in St until eddy shedding eventually ceases at $B/D = 5$. Gerich and Eckelmann (1982J) also examined the effect of the end-plate size on St. Figure 21.26 shows the variation in St in terms of Re for $D_e/D = 1$, 20, 50, and 100. The top curve (full circles) corresponds to $D_e/D = 1$, no end plates.

21.5.4 *Transition-in-wake, TrW, state*

The transition-in-wake, TrW, state of flow is characterized by the random appearance and persistence of fingers along the span. It was argued in Vol. 1, Chapter 4, that they might have been caused by a combination of the change in shedding mode and the appearance of the spanwise shear layer instability. The fingers seem to be forerunners of a fine multicell structure along the span, as seen in Vol. 1, Fig. 4.1, p. 80. Similar flow structures were found in high-image PIV

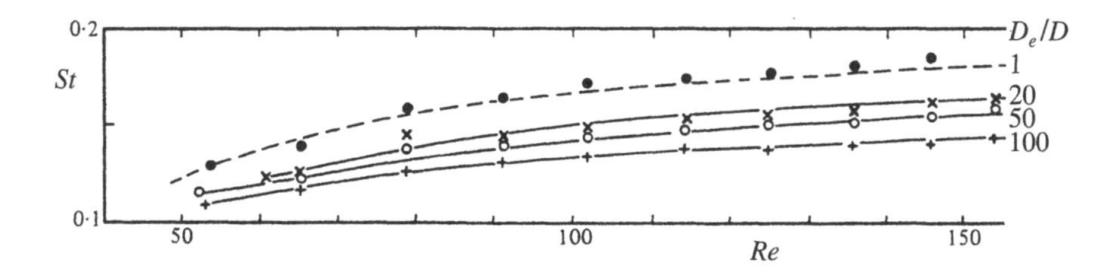

FIG. 21.26. Strouhal number variation with Re and D_e/D, Gerich and Eckelmann (1982J)

by Wu *et al.* (1994J, 1996J) and Brede *et al.* (1996J). The topological sketches are given in Fig. 21.27(a,b). There are two important features to be noted:

(i) The distortion of the free shear layers leads to the formation of fingers and counter-rotated streamwise eddies, as depicted by the arrows in Fig. 21.27(a). The spanwise wavelength of the eddy pairs is irregular and spreads from $3 < \lambda_z/D < 7$ in the TrW1 regime (Re < 200–220).

(ii) The formation of streamwise counter-rotated eddies becomes regular with $\lambda_z/D \approx 1$ in the TrW2 regime (Re > 200–220), as depicted in Fig. 21.27(b). The idealized depiction of the 'immutable' streamwise eddies does not take into account the diffusion of vorticity. The latter will also alter the topology along the wake downstream. Both topological sketches are tentative.

Yokoi and Kamemoto (1993J) measured the wavelength of an undulated separation line along the span, $L/D = 20$, in the TrW and TrSL1 regimes. They found a continuous λ_z/D decrease from 7 to 2, which is in striking contrast to data taken in the near-wake in Fig. 21.28(a). This indicates that the spanwise instability starts in free shear layers within the eddy formation region and not in the boundary layers near the separation line. There is a considerable reduction in eddy strength in the TrW2 regime and beyond, as seen in Fig. 21.28(b). It is notable that a constant λ_z/D in the TrW2 regime in Fig. 21.28(a) is not matched by a constant $K/\pi DV$ in Fig. 21.28(b). The reason for this discrepancy is not known.

Brede *et al.* (1996J) also noted a significant effect of the aspect ratio on the spanwise wavelength and circulation of streamwise eddies in the TrW1 regime. Figure 21.28(a,b) shows the measured λ_z/D and $K/\pi DV$ in terms of Re for three aspect ratios. The wide scatter in λ_z/D in the TrW1 regime is found only for $L/D \geqslant 72$. Figure 21.28(a) shows that $\lambda_z/D \simeq 1$ throughout the TrW2 regime and does not depend on the aspect ratio.

Another important insight into the effect of the aspect ratio on St and C_{pb} in the L3 and TrW regimes was provided by Norberg (1994J). He covered a wide range of aspect ratios, $20 < L/D < 2000$, and size of end plates, $15 < D_e/D <$

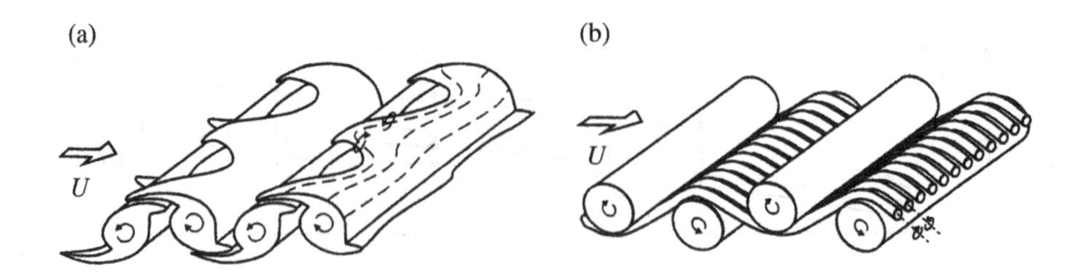

FIG. 21.27. Sketch of tentative spanwise flow structures between eddy filaments, (a) fingers, (b) fine streamwise eddies, Brede *et al.* (1996J)

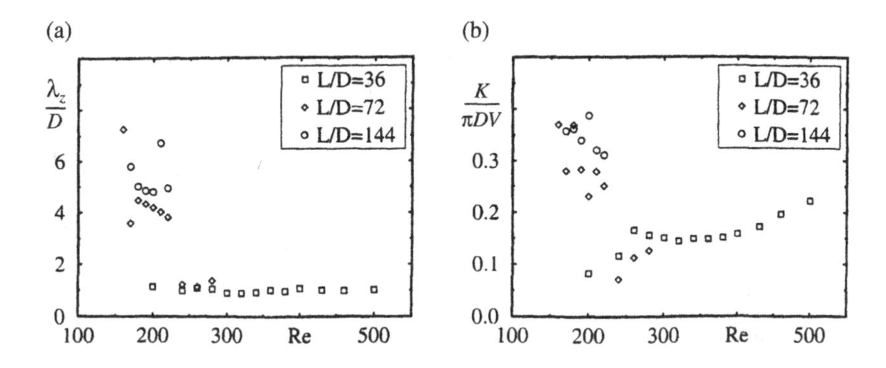

FIG. 21.28. (a) Wavelength of spanwise structures in terms of Re, (b) strength of streamwise eddies, Brede *et al.* (1996J)

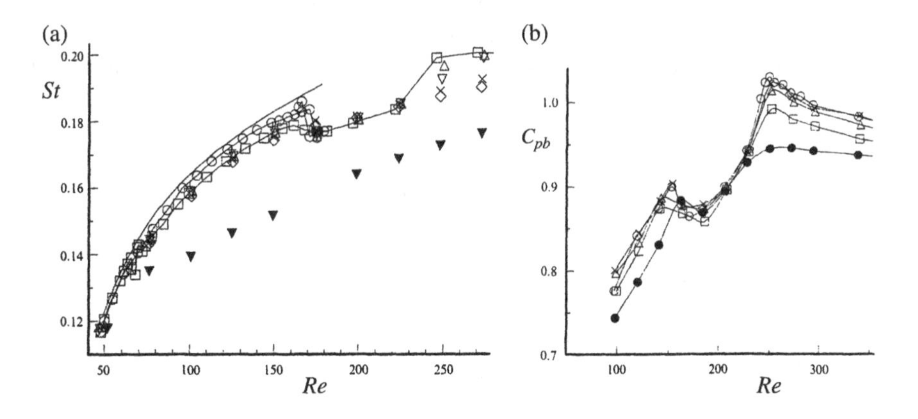

FIG. 21.29. (a) Strouhal number versus Re for various L/D, (b) base pressure versus Re, Norberg (1994J)

30. Figure 21.29(a) shows that in the TrW1 regime St is independent of L/D. The smallest $L/D = 20$ produces an extreme departure in the St curve from the other curves. The C_{pb} curves in Fig. 21.29(b) for $50 < L/D < 480$ are similar to the St curves. The peak in $-C_{pb}$ at the end of the TrW2 regime indicates the shortest eddy formation region.

21.5.5 *Transition-in-shear-layer, TrSL, state*

The cellular eddy shedding found in laminar wakes is likely to be significantly modified in turbulent wakes. Gerich (1986P) measured frequency spectra in the range $400 < Re < 4\text{k}$ and found the same peaks at $z/D = 1$ and 20 for $D_e/D = 10$. However, there were two peaks at $z/D = 5$ when larger end plates $D_e/D = 60$

were fitted. The existence and non-existence of the shedding cells was reflected in St variation for different sizes of end plates D_e/D. Stäger and Eckelmann (1991J) also found two cells at $Re = 1.6$k for $D_e/D = 21$. A further increase in Re to 4.8k produced a single frequency peak.

The effect of aspect ratio variation on St was also investigated by Gerich (1986P). For $L/D > 30$, the St value was constant. The reduction in L/D caused a slight increase in St with a maximum at $L/D = 14$. A further reduction in L/D led to the full suppression of eddy shedding for $L/D < 5$.

21.5.6 *Small aspect ratio*

Norberg (1994J) investigated the effect of aspect ratios[107] $5 < B/D < 50$ on St and C_{pb} in the L3, TrW, and TrSL states of flow. Figure 21.30(a,b) shows $B/D = 10$ and 5, fitted with end plates $D_e/D = 15$ (open symbols) and 10 (closed symbols). Figure 21.30(a) shows a considerable departure of data obtained for $B/D = 10$ from $B/D = 50$ (dashed line). For $B/D = 5$, a continuous curve for $D_e/D = 15$ is seen in Fig. 21.30(b) and a discontinuous one for $D_e/D = 10$. The jump in St from 0.15 to 0.19 takes place at $Re = 2.5$k, and indicates a sudden change from wide to narrow wake. The bistable nature of this jump will be discussed later.

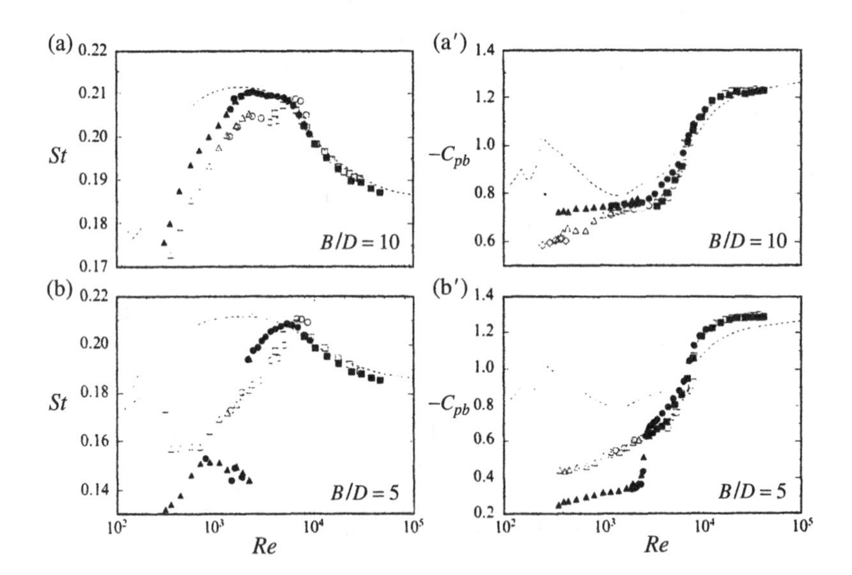

FIG. 21.30. Strouhal number and base pressure variation in terms of Re, (a,a′) $B/D = 10$, (b,b′) $B/D = 5$, Norberg (1994J)

[107]The aspect ratio was varied by changing the end plates distance B, while the actual length of the cylinder was L.

Systematic measurements of C_{pb} for $B/D = 1$ and $D_e/D = 5$ in a wide range $4\text{k} < Re < 400\text{k}$ were carried out by Kraemer (1965P). Figure 21.31 exhibits a discontinuous jump and fall in C_{pb} as follows:

(a) regime (I); high $C_{pb} = -0.4$ in the ranges $5\text{k} < Re < 15\text{k}$ and $Re > 80\text{k}$;

(b) regime (II); low $-1.3 < C_{pb} < -1.6$ in the range $15\text{k} < Re < 80\text{k}$.

The bistable flow regimes (I) and (II) switch over intermittently in the overlapping Re ranges. The bistable phenomenon strongly depends on the size of the end plates, as shown in Fig. 21.32. The aspect ratio varies in the range $0.3 < B/D < 20$ and $1 < D_e/D \leqslant 8$ ($D_e/D = 1$ designates no end plates). The measured C_{pb} at the midspan at $Re = 27\text{k}$ increases with rising D_e/D up to 5. Only for $D_e/D = 5$ do two flow regimes exist with $C_{pb} = -0.4$ and -1.3, as seen in Fig. 21.32 for $B/D < 2$.

Norberg (1994J) carried out flow visualization of bistable flows at $Re = 2\text{k}$ and $D_e/D = 10$ for $B/D = 5$. Figure 21.33(a) shows a long eddy formation region typical of the TrSL2 regime. An entirely different wake is seen in Fig. 21.33(b) with suppressed eddy shedding. The size of the end plates is just sufficient to cover the eddy formation region in Fig. 21.33(a). However, its occasional excursion downstream may introduce an outside flow into the wake. The eddy formation is postponed in an analogous manner when the base bleed is introduced behind the cylinder, Bearman (1967J). The occasional roll-up of free shear layers may re-start eddy shedding and switch the flow regime. This bistable flow is restricted within $5 < B/D < 6$, $1.8\text{k} < Re < 2.2\text{k}$ for $D_e/D = 10$, Norberg (1994J).

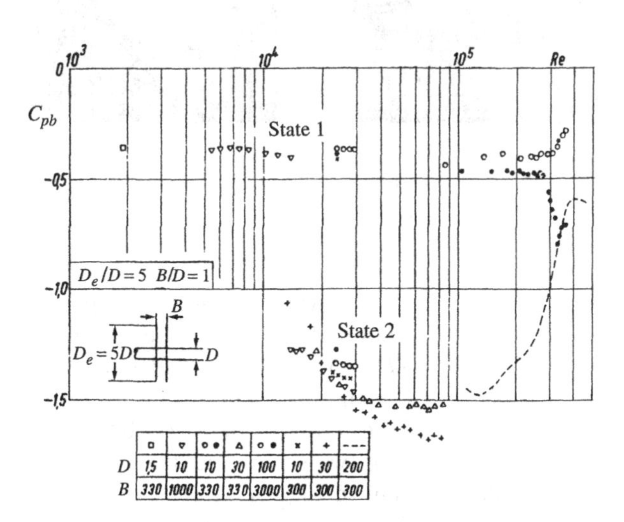

FIG. 21.31. Mid-span base pressure in terms of Re for $D_e/D = 5$ and $B/D = 1$, Kraemer (1965P)

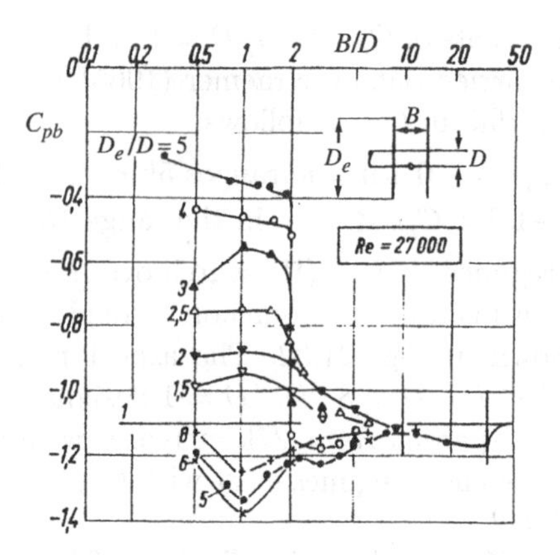

FIG. 21.32. Mid-span base pressure in terms of B/D for different D_e/D at $Re = 27\text{k}$, Kraemer (1965P)

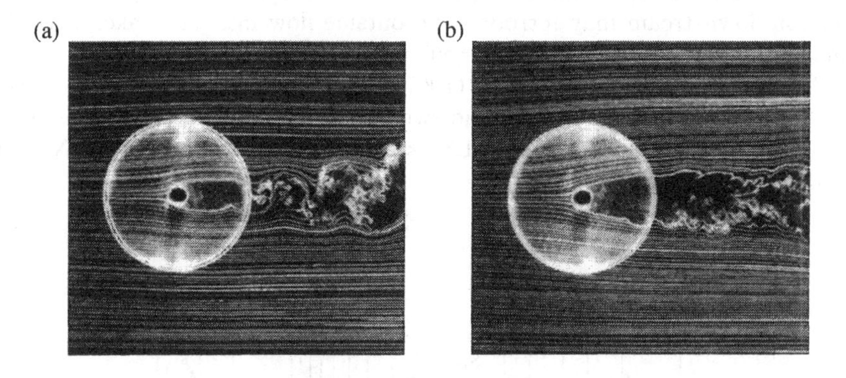

FIG. 21.33. Flow visualization at $Re = 2\text{k}$, $D_e/D = 10$, $B/D = 5$, (a) narrow wake, (b) wide wake, Norberg (1994J)

A considerable difference in C_{pb} on two sides of an end plate was found by Kraemer (1965P). Figure 21.34 shows the C_{pb} distribution along the span as affected by a single end plate $D_e/D = 5$, $Re = 72\text{k}$, and $B/D = 11$ in a free jet. C_{pb} is the same on both sides of the end plate only when the latter is located at the midspan. The displacement of the end plate to $z/D = 0.23$ and 0.46 (near the jet edge) produces a different C_{pb} on two sides. The distribution of C_{pb} without end plates is also given (dashed line) in Fig. 21.34 for comparison.

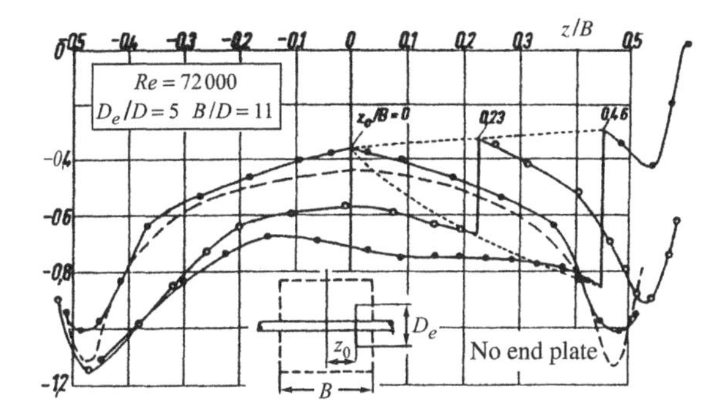

FIG. 21.34. Base pressure distribution at $Re = 72\text{k}$, $D_e/D = 5$, $B/D = 11$ on two sides of a single end plate, Kraemer (1965P)

21.5.7 *Fluctuating force*

It has been discussed in Vol. 1, Chapter 5, p.137, that a significant increase in C'_L and C'_D was found when a small aspect ratio cylinder was fitted with end plates, Keefe (1962J). Recent measurements of C'_L and C'_D by Szepessy and Bearman (1992J) were carried out in the range $8\text{k} < Re < 140\text{k}$. They moved the end plates ($D_e/D = 7$) in the range $0.25 < B/D < 12$. The fluctuating C'_L and C'_D were obtained by using special pneumatic pressure averaging at the midspan. Figure 21.35 shows C'_L in terms of B/D and Re as a parameter. The Re effect on C'_L is seen only for small B/D; $Re = 71\text{k}$ represents the TrSL3 regime, and a dissimilar curve for $Re = 130\text{k}$ belongs to the TrBL0 regime. The

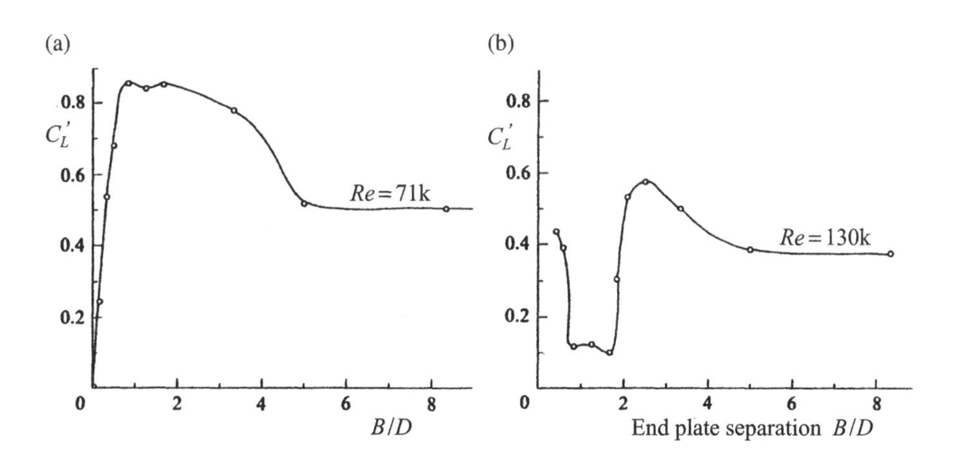

FIG. 21.35. Fluctuating lift in terms of B/D for different Re, (a) $Re = 71\text{k}$, (b) $Re = 130\text{k}$, Szepessy and Bearman (1992J)

pronounced drop in C'_L in the range $1 < B/D < 3$ for $Re = 130$k might have been caused by the same mechanism as that observed by Kraemer (1965P) and Norberg (1994J), Fig. 21.33. Szepessy and Bearman (1992J) also measured the spanwise correlation for $B/D = 1$, 6, and 7 in terms of Re.

A compilation of C'_L data measured by various researchers is shown in Fig. 21.36. There is an orderly decrease in C'_L from $B/D = 1$, Szepessy and Bearman (1992J), $B/D = 3$, Keefe (1962J), $B/D = 6.7$, Szepessy and Bearman (1992J), $B/D = 18$, Keefe (1962J), and $L/D = 10$ (without end plates), Schewe (1983J).

Extensive measurements of C'_p correlation were carried out by West and Apelt (1997J) in the range $2 < B/D < 50$, $22\text{k} < Re < 170\text{k}$, $D_e/D = 7$ (square end plates), and $D/B = 0.042$, $0.2\% < Ti < 7.5\%$. Figure 21.37(a,b) shows that the effect of B/D is medium on C'_L and small on C'_D down to $B/D = 10$. The local pressure fluctuations (top curve) are converted to C'_L and C'_D by using measured correlation curves for each B/D. The other five curves in Fig. 21.37 are obtained simply by multiplying the local coefficients (top curves) by the appropriate correlation factor.

21.5.8 *Transition-in-boundary layer, TrBL, state*

The TrBL state of flow consists of several distinct flow regimes: TrBL0, pre-critical; TrBL1, single-bubble; TrBL2, two-bubble; TrBL3, supercritical; and TrBL4, postcritical. The formation of separation bubble(s) in the TrBL1 and 2 regimes, respectively, causes a considerable narrowing of the near-wake by a delayed turbulent separation. The irregular fragmentation of separation bubbles in the TrBL3 suppresses eddy shedding, which reappears in the TrBL4 regime when fragmented bubbles are obliterated.

An early study of the effect of the aspect ratio on flows in the TrBL state was carried out by Kraemer (1965P). Figure 21.38 (a,b) show the C_p distribution at the midspan of the cylinder fitted with end plates $D_e/D = 6.3$ at $Re = 307$k

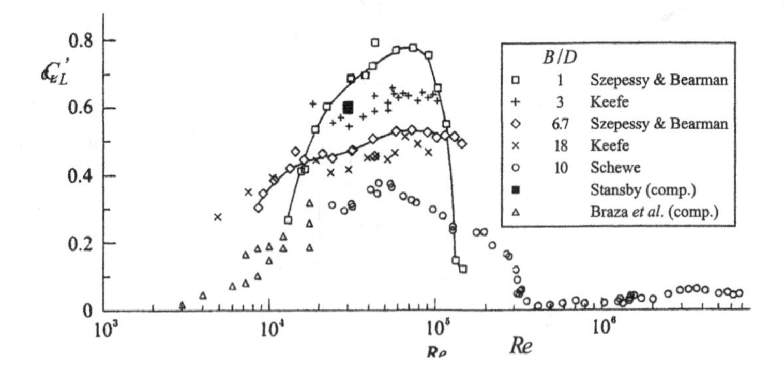

FIG. 21.36. Compilation of fluctuating lift in the range of $20\text{k} < Re < 8\text{M}$, Szepessy and Bearman (1992J)

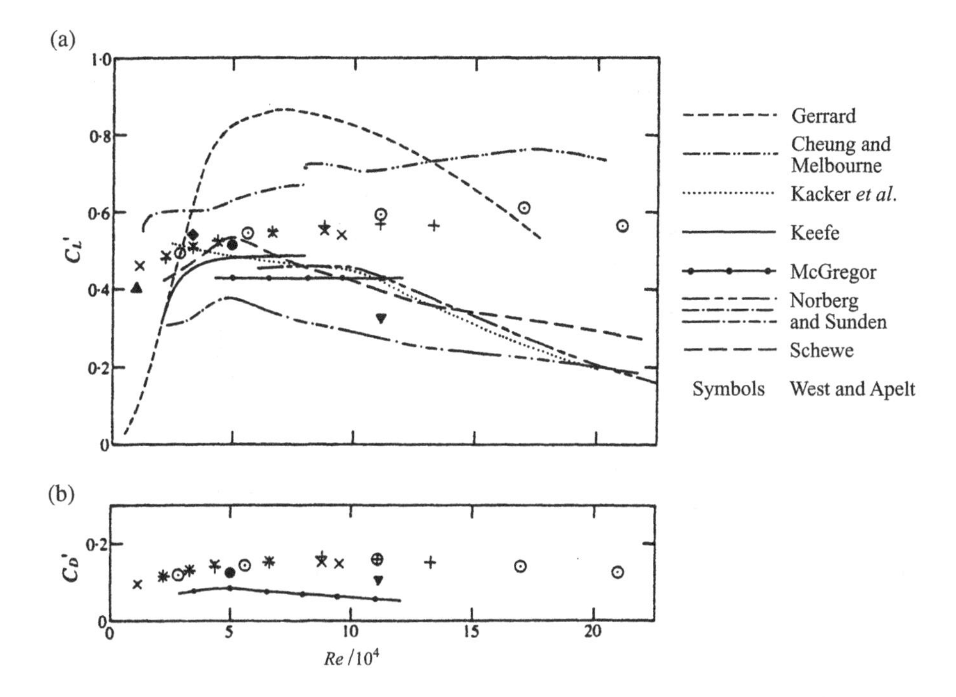

FIG. 21.37. Local fluctuating, (a) lift, (b) drag, in terms of Re for $D/B = 0.042$, $0.2\% < Ti < 7.5\%$, West and Apelt (1993J)

and 435k, respectively. There are two distinct features seen in Fig. 21.38 (a):

(i) the C_p distribution is asymmetric for all tested aspect ratios;

(ii) for low B/D values, the curves are similar to those in the TrBL3 regime, while for high $B/D = 16$ and 19 the curves are similar to the TrBL0 regime.

An opposite trend is seen in Fig. 21.38(b) for $Re = 435$k:

(i) there is no asymmetric C_p distribution for any B/D tested;

(ii) for low $B/D = 0.5$ and 1.0, the C_p curves are similar to the TrBL0 regime, while for medium and high $B/D = 1.9$, 3.2, and 19 the curves are as in the TrBL2 regime.

Kraemer (1965P) also traversed the near-wake along the x/D axis at the midspan ($z = 0$) using a thin hollow circular tube provided with pressure tappings at $\theta = 0°$ and 180°. The boundary of the reversed wake flow was estimated to be where $C_{p0} = C_{pb}$. The end plates were $D_e/D = 6.3$, and the range of aspect ratios was $1 < B/D < 19$. The reversed flow region extended from $1.7D$ to $2.0D$ for $Re = 307$k and from $0.9D$ to $1.3D$ for $Re = 435$k.

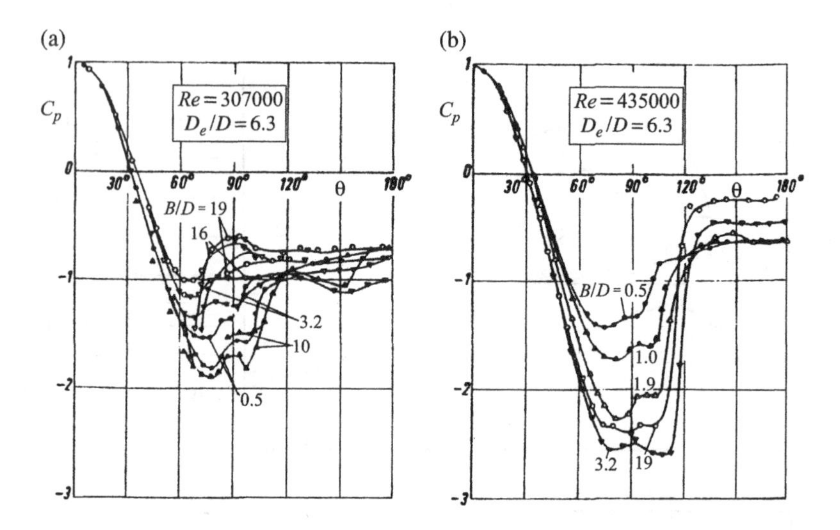

FIG. 21.38. Mean pressure distribution for $D_e/D = 6.3$, $0.05 < B/D < 19$, (a) $Re = 307$k, (b) $Re = 435$k, Kraemer (1965P)

21.6 Free water surface

A free surface is always present in open water channels and rivers. When a stationary vertical cylinder is partly submerged in a water current or towed along a flume the free surface will interfere with the flow. The Helmholtz law states that eddy filaments must form closed loops. This implies that each eddy filament in water should penetrate through the free surface and continue unabated in air above the water. The no-slip condition is not applicable at the free surface because the rotation of water under the free surface can induce the rotation of air above it. Each eddy in the cylinder wake induces rotational motion that deforms the free surface and forms dimples. This feature was used by Bénard (1908J, 1913J) to obtain the first photographs of Kármán–Bénard eddy streets.

21.6.1 *Towed cylinder in water at low Re*

Slaouti and Gerrard (1981J) carried out extensive flow visualization to examine the free surface effect on the cylinder wake. They found that the free surface behaved differently depending on the degree of water contamination. This made the water–air interface act either as a fluid interface capable of rotation or as a solid surface.

The water surface was clean when fresh tap water was used, and it remained so for a few hours before surface contamination developed (film of algae). Typical examples of eddy streets near the clean and contaminated free surface are shown in Fig. 21.39(a,b), respectively, for $Re = 127$ and $L/D = 30$. When the water surface was clean the eddy filaments remained perpendicular to the surface. The

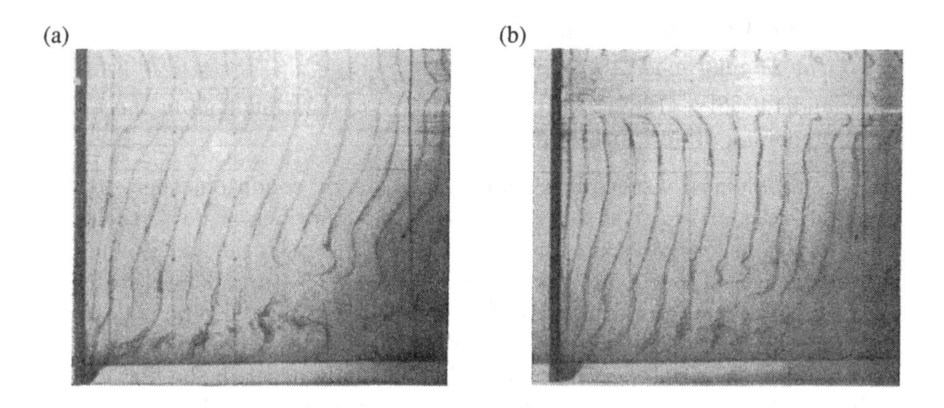

FIG. 21.39. Spanwise eddy filaments at $Re = 127$ for $L/D = 30$, water, (a) clear, (b) contaminated, Slaouti and Gerrard (1981J)

inclination of eddies towards the bottom end of the cylinder was due to the wall cell forming a knot. When the water was contaminated, Fig. 21.39(b), the eddy filaments did not reach the surface.

Slaouti and Gerrard (1981J) also examined the effect of a gap between the free surface and the cylinder end. Figure 21.40(a,b) shows eddy filaments at $Re = 109$ for gaps $0.2D$ and $0.45D$, respectively. For small gaps, the eddy filaments were unaffected and remained perpendicular to the free surface. For $0.45D$ gaps, the knot formation was evident, and the water surface behaved as a solid wall.

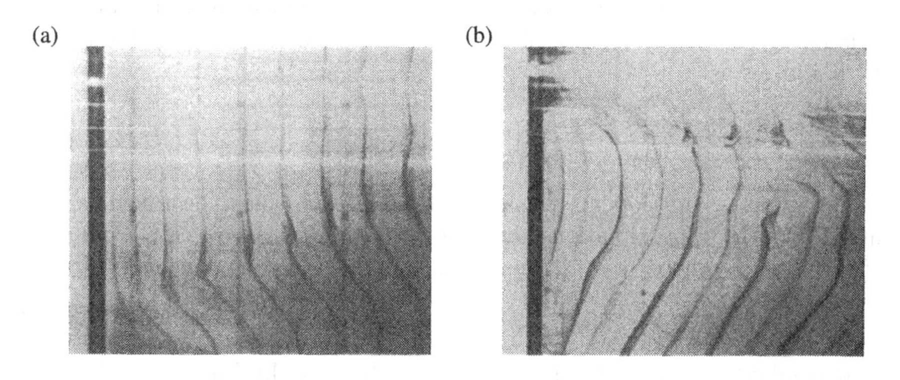

FIG. 21.40. Spanwise eddy filaments at $Re = 109$ with top gaps, (a) $0.2D$, (b) $0.45D$, Slaouti and Gerrard (1981J)

21.7 Theoretical modelling

21.7.1 *The Landau model*

Landau and Lifschitz (1987B) stated the basic concepts involved in the mathematical modelling of flow instabilities:

> For any problem of viscous flow under given steady conditions there must in principle exist an exact steady solution of the equations of fluid mechanics. These solutions formally exist for all *Re*. Yet not every solution of the equations of motion, even if it is exact, can actually occur in Nature. Those, which do, must not only obey the equations of fluid mechanics, but also be stable. Any small perturbations which arise must decrease in the course of time. If, on the contrary, the small perturbations which initially occur in the flow tend to increase in time, the flow is absolutely unstable at a given location, and cannot actually exist.

Landau considered periodic wake flows that are established as a result of the instability of a steady flow. For $Re < Re_{osc}$, the imaginary part of the complex frequencies, $\omega = \omega_1 + i\gamma_1$, for all possible small perturbations, is negative ($\gamma_1 < 0$). For $Re = Re_{osc}$, there is one frequency at which the imaginary part of this frequency becomes $\gamma_1 > 0$ and the flow is periodic. The velocity varies with time as

$$v = A(t) f(x, y, z) \tag{21.2}$$

where $A(t) = C e^{\gamma_1 t} e^{-\omega_1 t}$ is the amplitude of oscillation. This expression for $A(t)$ is only valid during a short interval of time after the disturbance of the steady flow; the factor $e^{\gamma t}$ increases rapidly with time. In reality, the modulus $|A|$ of $A(t)$ does not increase in the non-steady flow indefinitely, but tends to a finite value. Landau suggested an expansion of the time derivative of the squared amplitude $|A|^2$ in a series as

$$\frac{d|A|^2}{dt} = 2\gamma_1 |A|^2 - \alpha |A|^4 - \beta |A|^6 - \ldots \tag{21.3}$$

As $|A|$ increases, subsequent terms in this expression must be taken into account. When the approximation is limited to the first two terms eqn (21.3) yields $|A|^2_{max} = 2\gamma_1/\alpha$, where $\gamma_1 = C(Re - Re_{osc})$. By substituting this into $|A|^2_{max}$

$$|A|_{max} \propto (Re - Re_{osc})^{1/2} \tag{21.4}$$

This dependence is depicted in Fig. 21.41(a). When the third term, $-\beta|A|^6$, is retained, the curve becomes displaced, as shown in Fig. 21.41(b). When $Re > Re_{osc}$ there can be no steady flow; when $Re = Re_{osc}$ the perturbation discontinuously reaches a finite amplitude. In the range $Re'_{osc} < Re < Re_{osc}$, the unperturbed flow is *metastable*, that is being stable with respect to infinitesimal perturbations but unstable with respect to those with finite amplitudes (full line). The hysteretic loop is established.

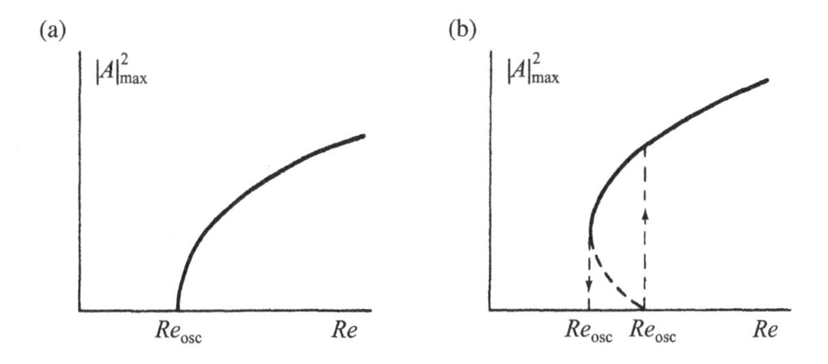

FIG. 21.41. Maximum amplitude square in terms of Re, (a) first term retained, (b) second, Landau and Lifschitz (1987B)

Provansal *et al.* (1987J) carried out experiments around Re_{osc} and determined the constant C in the expression for γ_1 as

$$Ro = (Re - Re_{osc})/5 \tag{21.5}$$

where $Ro = \gamma_1 D^2/\nu$.

Provansal *et al.* (1987J) also examined transient impulsive regimes, the effect of B/D on Re_{osc}, and beyond. Figure 21.42 shows the linear relationship and gradual displacement as B/D decreases.

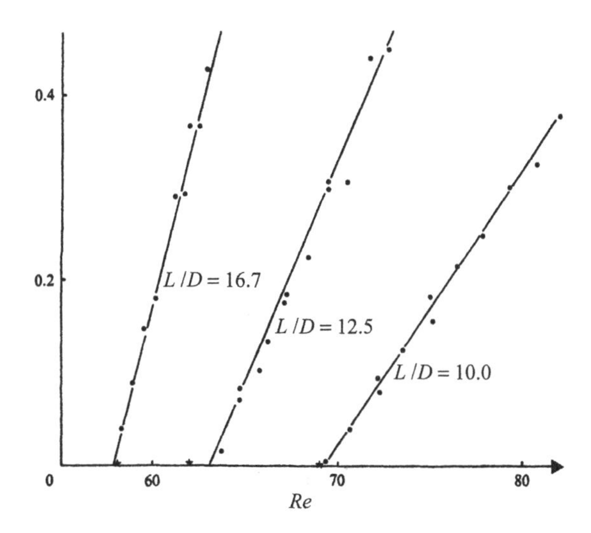

FIG. 21.42. Transient impulsive regimes in terms of Re_{osc} for three L/D, Provansal *et al.* (1987J)

21.7.2 *Extension of Landau's model*

Landau and Lifschitz (1987B) considered further instabilities for laminar periodic wakes when Re is increased beyond Re_{osc}. As Re increases, a point is eventually reached where the flow becomes unstable again. The unperturbed flow is now periodic $v_0(\boldsymbol{r}, t)$ with frequency ω_1, when $v = v_0 + v_2$ (where v_2 is a small secondary perturbation) is substituted into the equations of motion. During the period T_1, the perturbation changes by a factor $\mu = e^{-\omega_1 T_1}$. This factor is called the *multiplier* of the periodic flow, and represents the amplification or damping of perturbations in that flow. The flow ceases to be stable at the value Re'_{osc} for which one or more factors become positive. The loss of stability of the periodic flow is accompanied by a particular qualitative change in the flow regime. This local *bifurcation* leads to the secondary flow regime with a new independent frequency ω_2 leading to a quasi-periodic flow with two unrelated frequencies.

Landau and Lifschitz (1987B) also formulated a mathematical concept of a *space of states* each corresponding to a particular flow regime. An adjacent range of states then corresponds to adjacent flow regimes. A steady flow is represented by a point, and a periodic flow by a limited range in the space of states. These are called a *limit point or critical point* and a *limit cycle*, respectively. If the flows are stable, then adjacent curves representing the established flow tend to a limit point or cycle as $t \to \infty$.

A limit cycle (or point) has in the space of states a certain *domain of attraction*, and paths, which begin in that region, will eventually reach the limit cycle. In this context, the limit cycle is an *attractor*. It should be emphasized that for flow in a given volume with given boundary conditions (and given value of Re) there may be more than one attractor. That is, when $Re > Re_{cr}$ there may be more than one stable flow regime. The different flow regimes occur in accordance with the way in which the Re value is reached (hysteresis).

21.7.3 *Other theoretical models*

Karniadakis and Triantafillou (1992J) simulated numerically three-dimensional Navier–Stokes equations in the range $175 < Re < 500$. They found that the wake first becomes three-dimensional as a result of a secondary instability of the two-dimensional eddy street. This secondary instability appears at $Re_{osc2} = 200$. For slightly higher Re, a harmonic state develops in which the flow oscillates at its fundamental frequency (St) around a spanwise modulated time-average flow. Figure 21.43 shows a projection on the yz plane at $x = 2D$ of the velocity vector field at $Re = 225$. The pitch of the spanwise periodicity is between $0.7D$ and $0.8D$, which is less than the $4D$ or $1D$ found at $Re = 225$. This might be due to the narrow strip of $B/D = 1.6$ used for the computation. Further computations at $Re = 300$ lead to a periodic state with a period of oscillation equal to twice the fundamental one, which has not been observed experimentally.

A global three-dimensional stability analysis of steady and periodic laminar wakes was carried out by Noack and Eckelmann (1994Ja,b) by employing Galerkin's method. The steady flow was found to be asymptotically stable

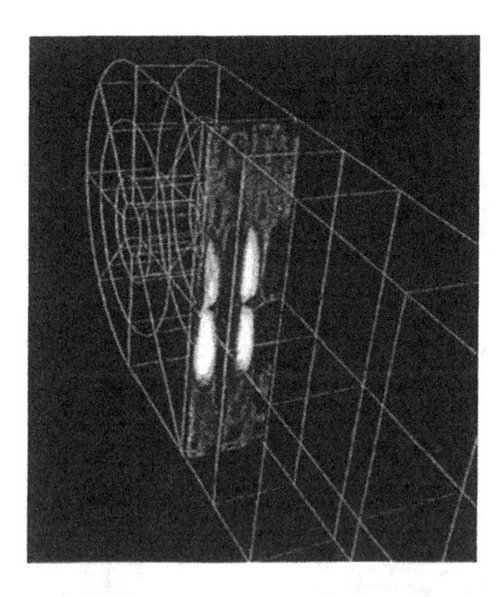

FIG. 21.43. Computation of secondary instability of streamwise eddies at $Re = 225$, Karniadakis and Triantafillou (1992J)

to all kinds of perturbations for $Re < 54$, which is higher than the observed $Re < 40$. The periodic solution was neutrally stable in the range $54 < Re < 170$. While two-dimensional perturbations of the eddy street rapidly decay, the three-dimensional perturbations with long spanwise wavelengths neither grow nor decay. Finally, the periodic solution becomes unstable at $Re = 170$ by a perturbation with the spanwise wavelength of $1.8D$. This is lower than the observed $4D$ at that Re value, and $St = 0.24$ is higher than the value 0.17 observed at $Re = 200$.

Further refinement of the theoretical model by Zhang *et al.* (1995J) predicted two different spanwise near-wake instabilities, fingers, and streamwise counter-rotated eddy pairs. Figure 21.44(a,a′) shows the computational and experimental top views, respectively, of the near-wake at $Re = 160$. The near-wake area covered $-0.5 < x/D < 16$ and $0 < z/D < 24$, and the aspect ratio of the cylinder was $B/D = 93$. The main feature was the irregular and aperiodic spanwise appearance of fingers. The location of fingers was artificially triggered by imposing a disturbance on the computed periodic flow.

The regular and periodic formation of streamwise eddies is seen in Fig. 21.44(b,b′) for $Re = 250$. Note that the flow visualization was obtained with the smoke-wire at $x/D = 2$ and $y/D = 1$, and hence only eddies on one side of the near-wake were visible. The computed wavelengths $\lambda_z/D > 1$ and 4 are greater than $\lambda_z/D = 1$, as shown in Fig. 21.28(a).

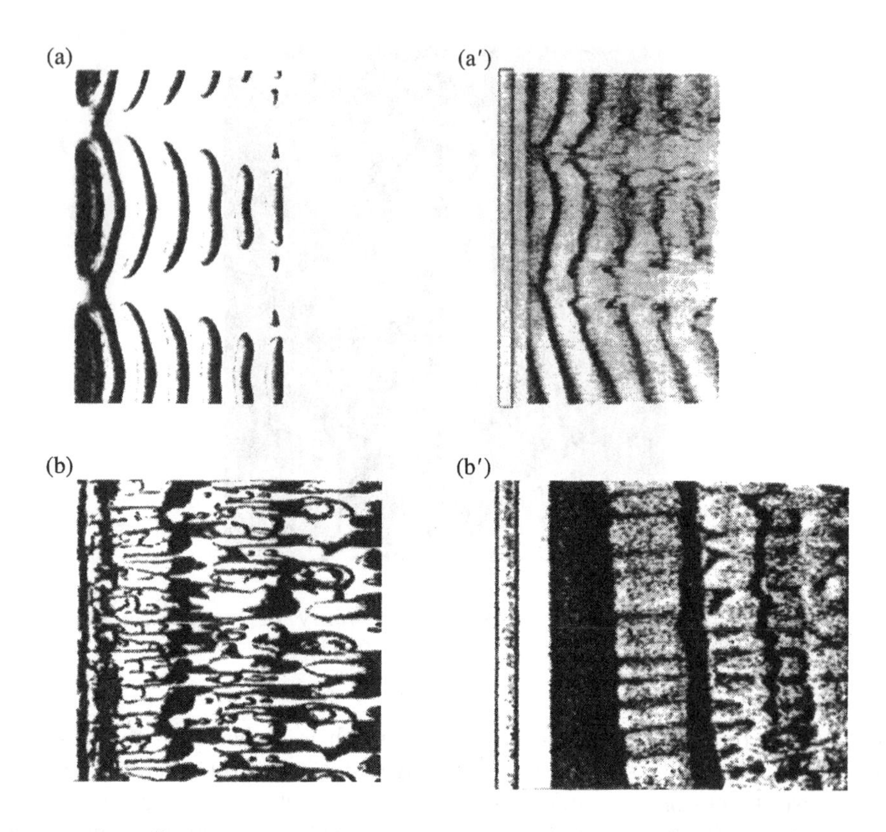

FIG. 21.44. Computational three-dimensional stability analysis compared with flow visualization, (a,a′) $Re = 160$, (b,b′) $Re = 250$, Zhang *et al.* (1995J)

21.8 Free end

Finite cylinders with a *free end*, such as chimney stacks, impose a fundamental and perplexing question, namely how do the eddy filaments end? For *ideal* fluids, it has been postulated by Helmholtz that vortex lines cannot begin or end in the fluid. If they do not form a closed curve, each end of the vortex line is either attached to the boundary or ends at infinity. Note that when the vortex filament is bent by 180° the sense of rotation is changed, and the linking of two counter-rotated vortex filaments is possible at their ends.

21.8.1 *Periodic laminar wake, L3 regime*

Taneda (1952P) carried out pioneering tests by towing a finite cylinder ($H/D = 40$) in a small water tank (40 cm × 45 cm deep), and used aluminium particles to make the flow visible. A light was passed through a narrow slit, which illuminated a thin layer of flow at any cross-section perpendicular to the cylinder axis. It was also possible to turn the slit by 90° and illuminate planes either passing through the cylinder axis or parallel to it.

Figure 21.45(a) shows a sketch of the illuminated wake slice. The light reflected from the aluminium particles produces bright and dark strips. The particle orientation depends on whether they move towards or away from the point of observation, denoted by a and b, respectively. Figure 21.45(b) shows the periodic waviness of the wake, and the unexpected tilt of the bright strip near the free end (bottom). Note that eddy filaments are invisible and located between the bright and dark strips. Hence, the strips tilted away from the cylinder free end preclude the Helmholtz hypothesis that eddy filaments might be attached to the cylinder.

The complementary flow visualization in the planes perpendicular to the cylinder axis is shown in Fig. 21.46(a,b,c) for the illuminated strip at $z/D = 7$, 2, and 1 from the free end, respectively. The distorted eddy street at $z/D = 7$ is replaced by a zig-zag pattern at $z/D = 2$, which disappears far downstream. Finally, at $z/D = 1$ from the free end, only faint lumps are seen downstream. The disappearing flow patterns at the last two stations indicate a spanwise con-

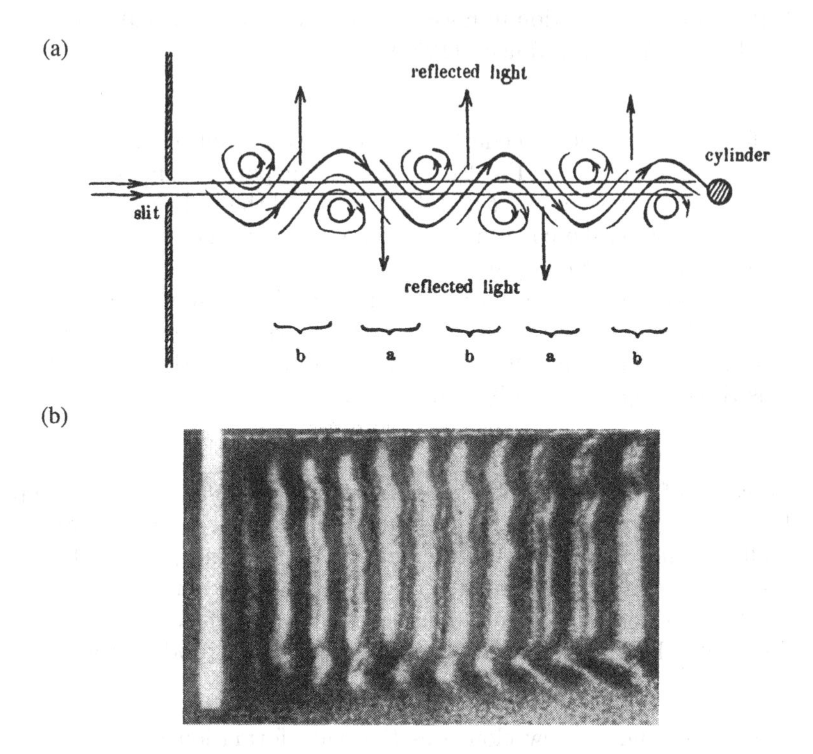

FIG. 21.45. Flow past a cylinder with a free end, (a) sketch of the Kármán–Bénard street, (b) spanwise eddy 'filaments', Taneda (1952P)

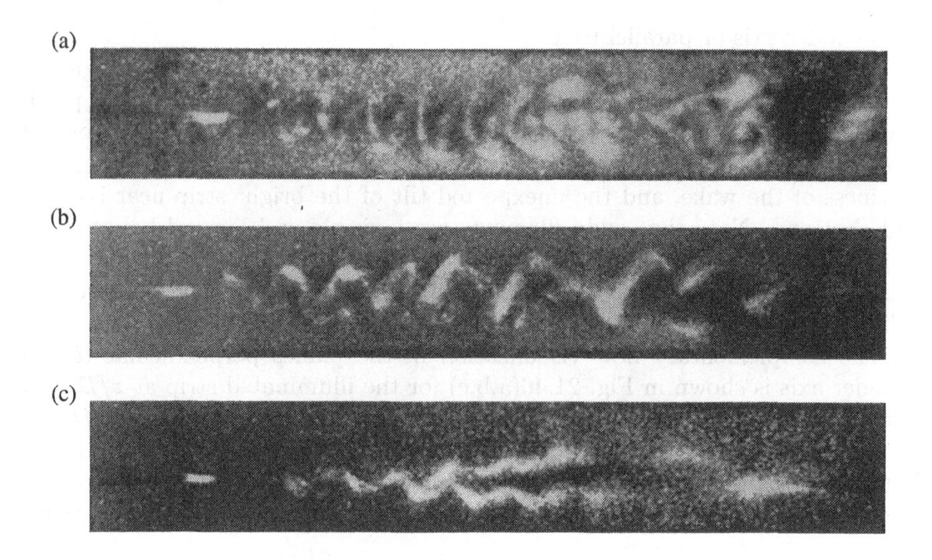

FIG. 21.46. Flow visualization in normal planes, (a) $z/D = 7$, (b) $z/D = 2$, (c) $z/D = 1$, from free end, Taneda (1952P)

traction of the eddy street. Taneda (1952P) offered a topological interpretation of the observed flow pattern in Fig. 21.47(i to v). The parallel eddy filaments *a*, *c*, and *e* link[108] behind the free end *b* and *d*. Once linked the eddy structure induces its own deformation as sketched in sequences (ii) to (v). The final sequence (v) corresponds to the deformed cluster *a*–*e*.

Slaouti and Gerrard (1981J) carried out dye visualization behind a finite cylinder with a free end (bottom). Figure 21.48 shows the dye pattern originally injected behind the free end. It is first drawn upwards in the near-wake, and later forms tilted loops induced by eddy linking.

21.8.2 *Secondary flow at the free end*

The main cause of a high drag of a two-dimensional cylinder is a low base pressure. This region of low pressure is physically separated from the free stream by tunnel walls or end plates. For finite cylinders, there is no physical boundary between the free stream pressure ($C_p = 0$) and the base pressure C_{pb} at the free end. The considerable pressure difference will induce a secondary flow over the free end into the near-wake. The secondary flow may produce the following effects:

(i) the weak secondary flow elongates the eddy formation region L_F downstream and widens the near-wake. Thus, St decreases;

[108] This was the first example of the eddy-link topology.

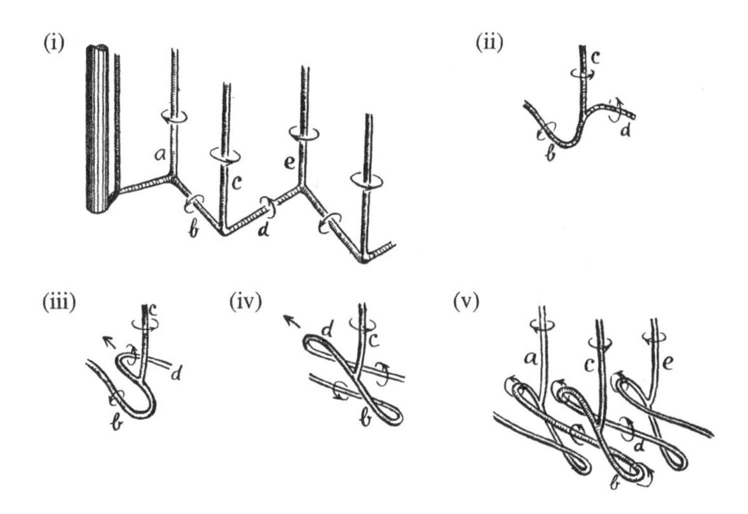

FIG. 21.47. Topology of flow patterns (i) initial, (ii) to (v) temporal developments, Taneda (1952P)

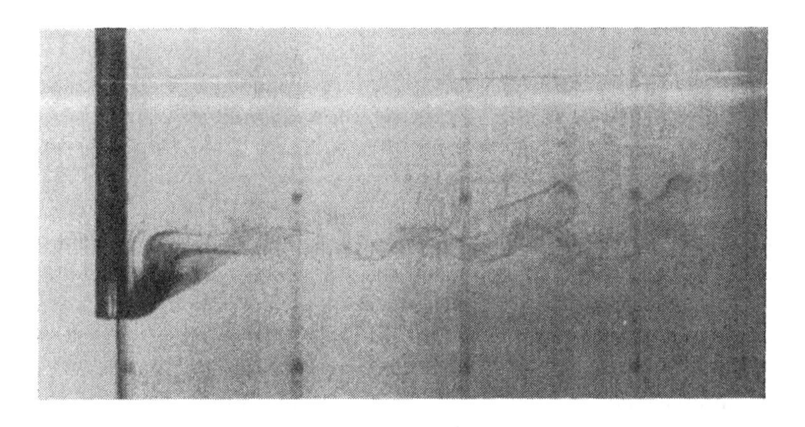

FIG. 21.48. Flow visualization near the free end for $Re = 108$, $G/D = 2$, Slaouti and Gerrard (1981J)

(ii) the intensive secondary flow may displace the start of eddy shedding downstream, analogously to the effect of a base bleed, Bearman (1967J);

(iii) for small aspect ratios, eddy shedding does not take place, and the secondary flow becomes the primary flow.

Etzold and Fiedler (1976J) carried out flow visualization at Re = 30k and $H/D = 4$. Figure 21.49(a) shows long exposure particle paths produced by the

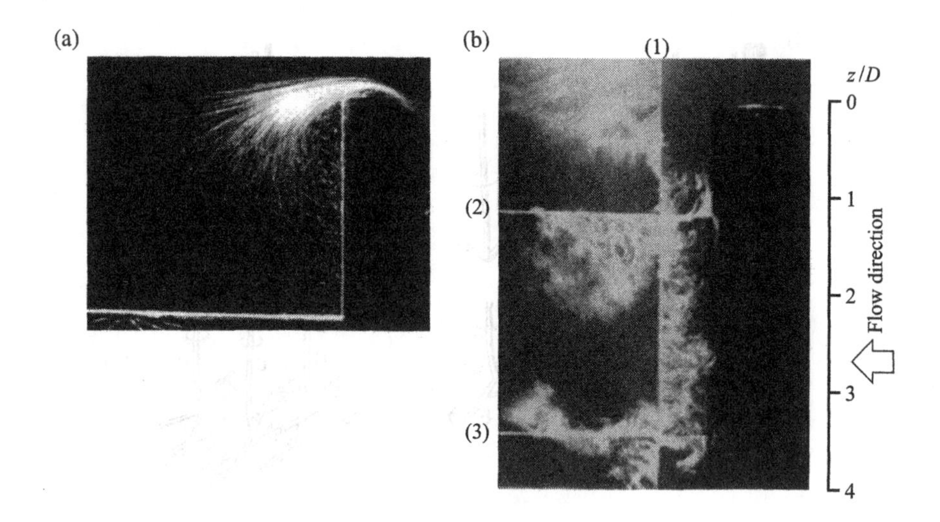

FIG. 21.49. Flow (from right to left) visualization past free end, $Re = 30\text{k}$, $H/D = 4$, (a) downwash in meridional plane, Etzold and Fiedler (1976J), (b) downwash at top $z/D = 1.3$ and 3.5, Kawamura *et al.* (1984J)

secondary flow in the symmetry plane. The downwash gradually becomes invisible farther downstream. A complementary flow visualization by Kawamura *et al.* (1984Ja) at $Re = 32\text{k}$ and $H/D = 4$ used three smoke-wires (1), (2), and (3). Figure 21.49(b) clearly shows not only the secondary flow at the top (1) and middle (2), but also the upward flow near the bottom (3).

It has been found that for a nominal two-dimensional cylinder the velocity defect has a maximum value along the wake axis. The secondary flow behind the free end reduces the velocity defect. Okamoto and Sunabashiri (1992J) traversed the wake at the midspan behind the finite cylinders having $H/D = 1$, 2, 4, and 7. Figure 21.50(a–c) shows the profiles of the velocity defect behind finite cylinders and a two-dimensional cylinder, $H/D = 24$ (dashed line). At $x/D = 7$ and 10, Fig. 21.50(a,b), all four velocity defects are significantly reduced in comparison with $H/D = 24$. Figure 21.50(c) shows that the secondary flow at $x/D = 5$ reduces the velocity defect for $H/D = 1$ and 2, does not change it for $H/D = 4$, and exceeds it for $H/D = 7$. Okamoto and Sunabashiri (1992J) also measured the velocity correlation and the fluctuating velocities u', v', and w'.

Etzold and Fiedler (1976J) traversed the smoke-wire along the nearwake axis ($y/D = 0$) up to the point where the reversed flow vanished. This position is plotted in Fig. 21.51(a) for six finite cylinders $H/D = 2$, 4, 5, 6.5, 7.2, and 10.1. The nominal two-dimensional cylinder $H/D \to \infty$ is shown as a vertical line for comparison. The length of the reversed flow region increases as H/D decreases down to $H/D = 6.5$. For $H/D < 6$, the trend reverses, as seen in Fig. 21.51(a).

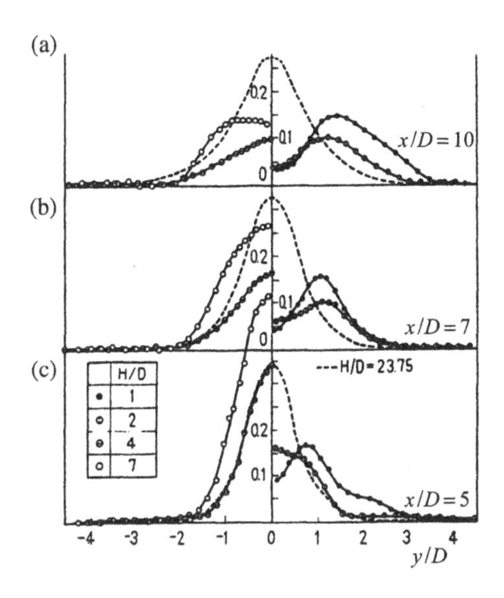

FIG. 21.50. Velocity profiles across half wakes for $1 < H/D < 7$, (a) $x/D = 10$, (b) $x/D = 7$, (c) $x/D = 5$, Okamoto and Sunabashiri (1992J)

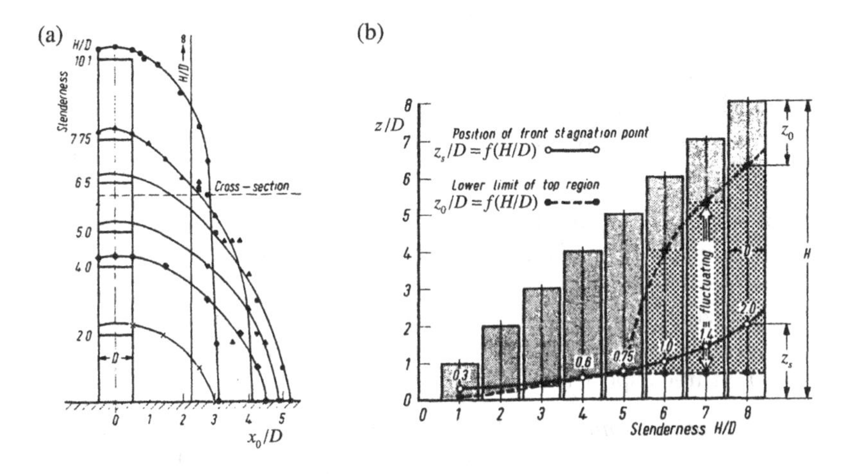

FIG. 21.51. (a) Reversed flow region behind finite cylinders, (b) flow direction near separation, grey – upwards, white – downwards, dotted – fluctuating, Etzold and Fiedler (1976J)

Another insight into the secondary flow is gained by visualizing the flow direction near the separation along the span. Figure 21.51(b) shows the extent of three regions:

(i) the grey region is where the smoke is directed upwards towards the free end;

(ii) the white region is where the smoke is directed downwards towards the root;

(iii) the mixed region (dotted) is where the smoke fluctuates in both directions.

There are only two flow regions up to $H/D = 5$, namely the free end and horseshoe-swirl. For $H/D > 5$, there are three flow regions as will be discussed in the next subsection.

21.8.3 *Spanwise variation in mean pressure*

The flow at different levels along finite cylinders with a free end can be described as follows:

(i) The *free-end region* extends in the spanwise direction $0 < z/D < 2$ to 3. It is characterized by a three-dimensional flow, interlinking of eddy filaments, and spanwise contraction of the near-wake.

(ii) The *middle region* represents a quasi two-dimensional flow with regular eddy shedding. It appears for $H/D > 5$, and is located between the free end and wall regions. For high values of H/D, the middle region dominates the flow.

(iii) The *wall region* is related to the horseshoe-swirl, see section 21.2. The extent of this region depends on the cylinder diameter and the thickness of the wall boundary layer D/δ_B. This region is restricted to about $\frac{1}{3}D$ from the wall.

Okamoto and Yagita (1973J) displayed their extensive measurements of mean pressure distributions around and along the finite cylinders in the form of interpolated C_p isobars. Figure 21.52(a,b,c) shows developed surfaces of a half-cylinder perimeter (horizontal axis) and relative height z/D (vertical axis) for $H/D = 1$, 5, and 7, respectively. There are similarities and dissimilarities in the C_p distribution for the three aspect ratios chosen:

(i) The *stagnation region*, $0 < \theta < 30°$. For a nominal two-dimensional cylinder ($H/D \to \infty$), $C_{p0} = 1$ all along the stagnation line. There is a considerable departure from that for $H/D = 1$. The maximum departure occurs at the free end and near the horseshoe-swirl.

(ii) The *side region*, $30° < \theta < 100°$. This appears to be similar for all finite cylinders. The spanwise C_{pb} variation along $\theta = 70°$ is small for $H/D = 1$. The pressure difference maintains the secondary flow up to the minimum side pressure seen as an islet in Fig. 21.52 at around $\theta = 70°$ and $z/D = 2/3$ near the free end.

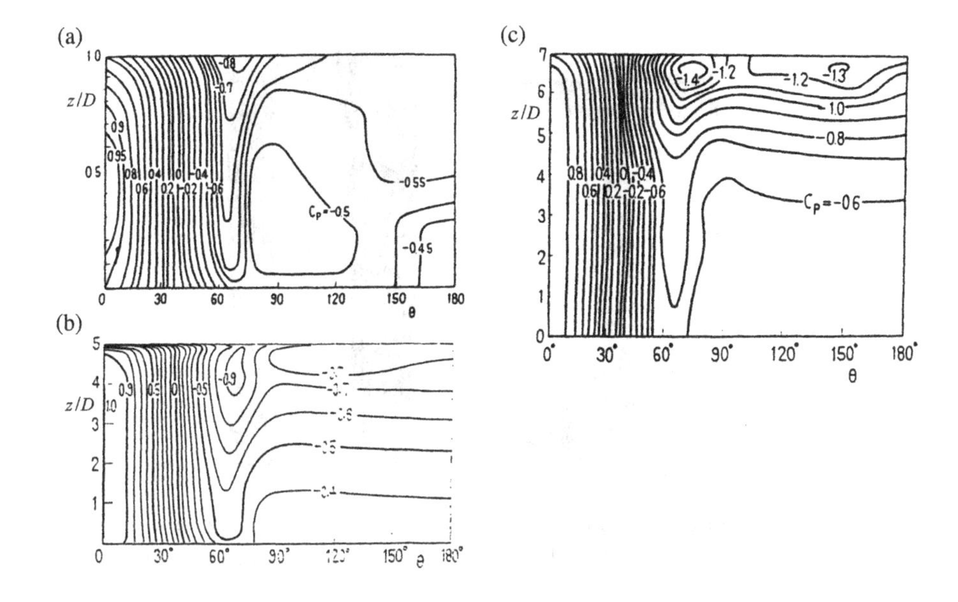

FIG. 21.52. Circumferential and spanwise mean pressure distribution, (a) $H/D = 1$, (b) $H/D = 5$, (c) $H/D = 7$, Okamoto and Yagita (1973J)

(iii) The *base region*, $110° < \theta \leqslant 180°$. This is marked by dissimilar isobar patterns near the free end for $H/D = 1, 5$, and 7. A second islet of minimum pressure appears only for $H/D = 7$ at $\theta = 135°$ and $z/D = 0.7$ from the free end. This unexpected second $C_{p\min}$ raises considerably the local drag near the free end, as will be shown shortly.

Another important feature is the separation line along the finite cylinder span. Figure 21.53(a,b) shows the surface flow pattern for a finite cylinder $H/D = 12$ in water at $Re = 13.3$k and $H/D = 6$ in air at $Re = 190$k, respectively. The separation line is straight up to about $z/D = 2$, where it starts to bend towards higher θ angles. Fine details of the surface pattern near the free end are seen in Fig. 21.53(b). The secondary upwards flow behind separation deposits a white strip culminating in an 'eye' near the free end. The secondary downwards flow in the $y = 0$ plane is invisible.

Figure 21.54(a–c) shows surface flow, smoke-wire visualization behind $H/D = 8$, $\delta_B = 1D$, and windmill rotation behind a short $H/D = 1$ cylinder at $Re = 32$k. Figure 21.54(a) shows a view from the downstream direction, where $\theta = 180°$ is a symmetry line. Two swirl patterns are seen which correspond to the second $C_{p\min}$ in Fig. 21.52. Figure 21.54(b) shows a smoke-wire visualization placed at $z/D = 0.3$ elevation and $x/D = 0.75$ downstream. The secondary flow is directed downwards in the near-wake and upwards behind the separation. Figure 21.54(c) confirms the existence of the swirl. The short cylin-

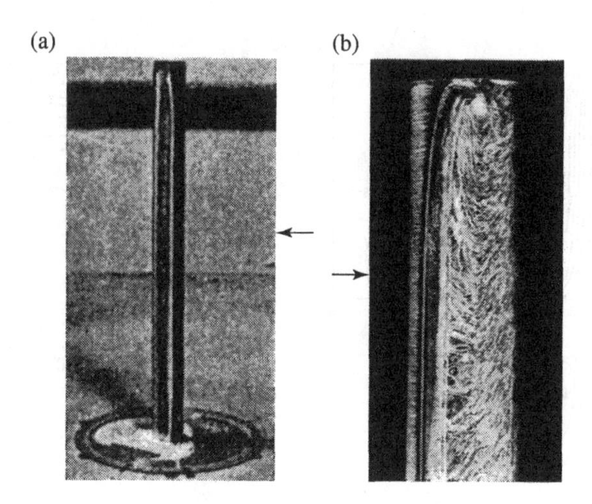

FIG. 21.53. Separation line along the cylinder, (a) $H/D = 12$, $Re = 13.3$k, Okamoto and Yagita (1973J), (b) $H/D = 6$, $Re = 190$k, Gould *et al.* (1968P)

der $H/D = 1$ (dash-dot line) submerged in a thick ($\delta_B/D = 1$) boundary layer also generates the horseshoe-swirl seen in Fig. 21.54(c). The windmill speed of rotation n indicates that the horseshoe-swirl is at least three times weaker than the trailing swirl near the free end.

On the basis of various flow visualizations, it is possible to trace the origin of the two measured $C_{p\min}$:

(i) The first $C_{p\min}$ is around $\theta = 70°$ and $z/D = 2/3$ below the free end. It is due to the displacement of the separation to higher θ. A similar increase in $C_{p\min}$ occurred in the TrBL2 regime where separation bubbles displace the separation downstream, see Vol. 1, Chapter 6.

(ii) The origin of the second $C_{p\min}$ at $\theta = 135°$ and $z/D = 1/3$ below the free end is probably due to the trailing swirl attachment. The swirl is formed by the combined action of the upward flow beyond the separation and secondary downwards flow near the symmetry plane.

21.8.4 *Spanwise variation in the local drag coefficient*

The considerable variation in C_{pb} along the span of finite cylinders is reflected in a similar variation in the local drag coefficient C_d. The first type of $C_{p\min}$ at $\theta = \pm 75°$ affects C_d slightly, while the second type at $C_{p\min} = \pm 135°$ has a significant effect. Figure 21.55 shows a series of C_d curves in the range $1 \leqslant H/D \leqslant 12.5$. There are two distinct groupings of C_d curves:

(i) For short cylinders in the range $1 \leqslant H/D < 6$, there is a rounded C_d peak at around $z/D = 1$ followed by a small variation towards the wall

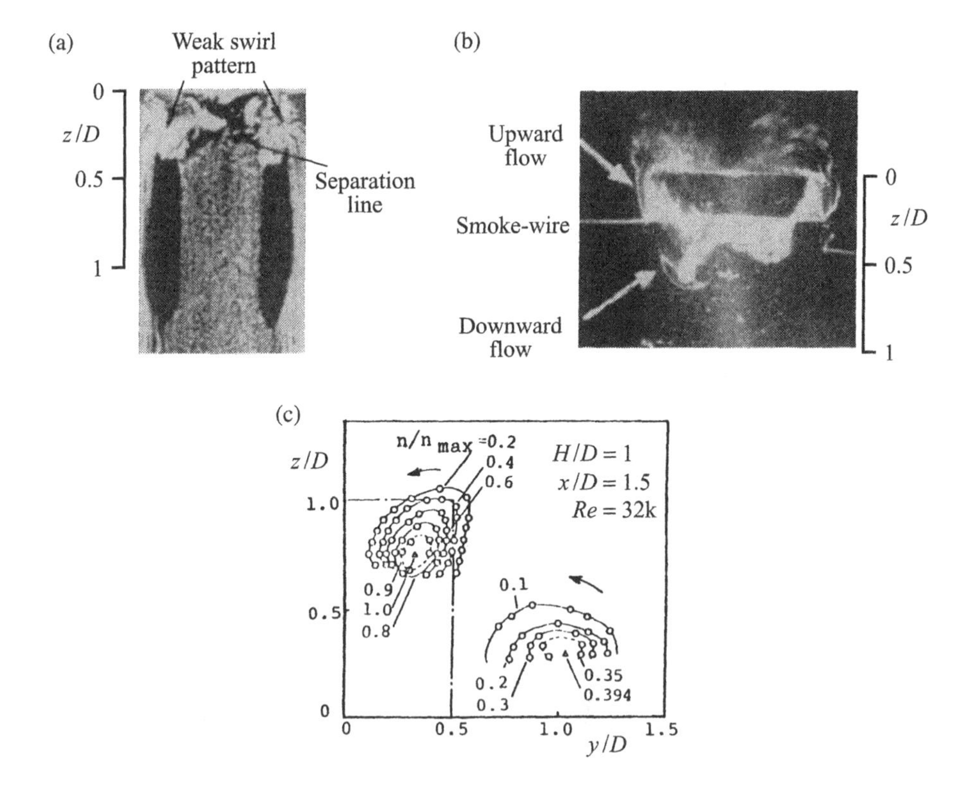

FIG. 21.54. Finite cylinder $H/D = 8$, $\delta_B/D = 1$, $Re = 32k$, (a) surface visualization, (b) smoke-wire, (c) vorticity contours $H/D = 1$, Kawamura *et al.* (1984J)

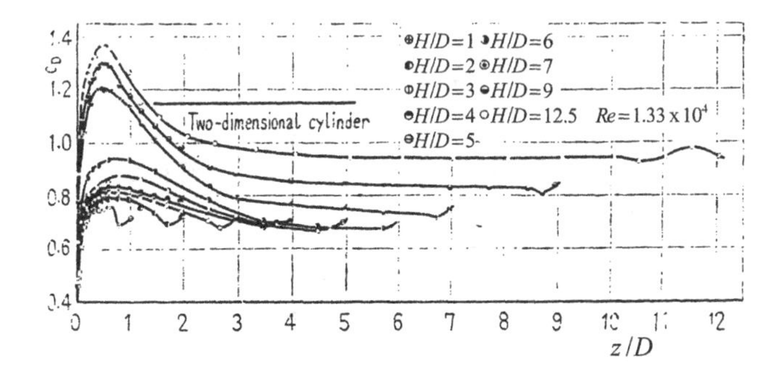

FIG. 21.55. Spanwise variation in local drag, Okamoto and Yagita (1973J)

where the horseshoe-swirl is formed. The latter becomes more pronounced as H/D decreases.

(ii) For long cylinders (in the range $7 < H/D < 12.5$), there are more pronounced $C_{d\max}$ followed by a rapid decrease in C_d. The rest of the span is characterized by an almost flat C_d curve up to the horseshoe-swirl at the wall. Note that $C_{d\max}$ near the free end exceeds C_d for the nominal two-dimensional cylinder (horizontal line).

21.8.5 *Spanwise fluctuating pressure and lift*

Farivar (1981J) measured both the mean and fluctuating pressures around and along several finite cylinders in the range $2.78 < H/D < 12.5$. Figure 21.56(a,b) shows distributions of C_p and C_p' for $H/D = 5$ and 10, respectively. For the short cylinder, C_p' is negligible, while for the long cylinder, C_p' is considerable. This is particularly the case near the free end. It is remarkable that C_p' near the free end exceeds that along the rest of the span where eddy shedding exists.

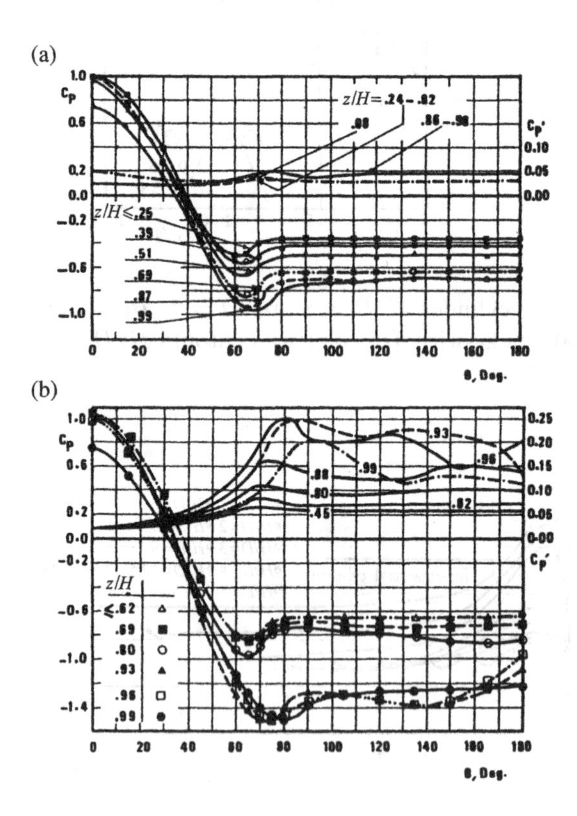

FIG. 21.56. Mean and fluctuating pressure distributions, (a) $H/D = 5$, (b) $H/D = 10$, Farivar, © (1981J) AIAA – reprinted with permission

There is a coincidence between the low $C_{p\min}$ islet ($0.93 < z/D < 0.99$) and corresponding high C_p' curves. It is noteworthy that *all* $C_{p\max}'$ are in the vicinity of the separation lines as for the nominal two-dimensional cylinders.

Another way of displaying C_p' variations is shown in Fig. 21.57(a). The C_p' distribution near the free end ($z/D = 1/3$) is re-plotted for all finite cylinders tested, $2.78 \leqslant H/D \leqslant 12.5$. There is a distinct grouping of C_p' curves for short and long cylinders. The origin of the second smaller peak is different for finite and two-dimensional cylinders at about 130° and 150°, respectively. The second peak for the nominal two-dimensional cylinder is caused by the proximity of the eddy formation region to the base of the cylinder. For finite cylinders, it is caused by the fluctuating swirl attachment. Figure 21.57(b) shows the variation of $C_{p\max}'$ along the cylinder span. Note that $C_{p\max}'$ away from the free end is more than four times smaller than behind the nominal two-dimensional cylinder.

Fox and West (1993Jc) measured local fluctuating lift C_L' along the span of finite cylinders $4 < H/D < 30$. Figure 21.58 shows the following features:

(a) C_L' is negligible for short finite cylinders, $H/D < 6$;

(b) high values of C_L' occur at $z/D = 1$ for $H/D > 6$;

(c) for long cylinders, C_L' increases up to $H/D = 20$;

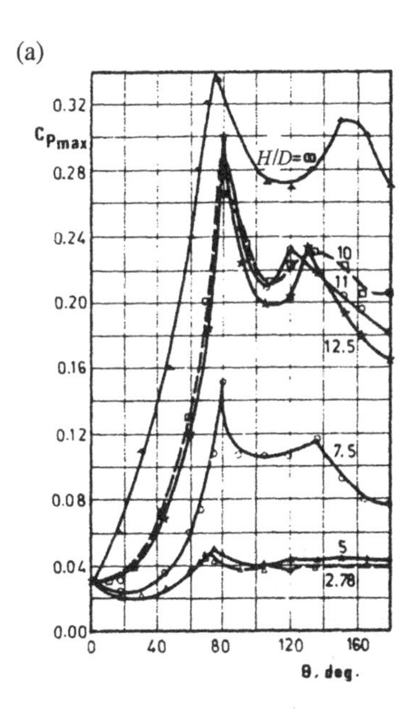

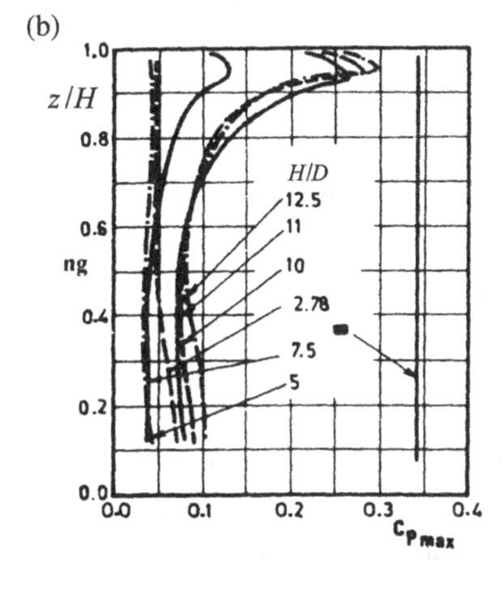

FIG. 21.57. Fluctuating pressure, (a) distribution, (b) maximum along span, Farivar, © (1981J) AIAA – reprinted with permission

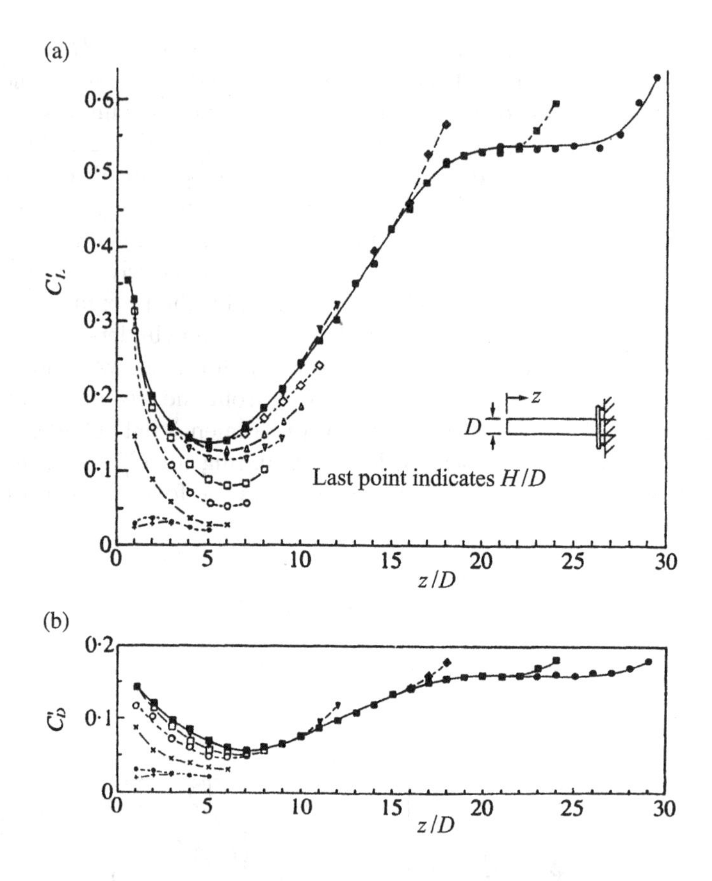

FIG. 21.58. Fluctuating (a) lift spanwise distribution, (b) drag, Fox and West (1993Jc)

(d) for $H/D > 20$, C'_L is constant.[109]

21.8.6 *Strouhal number variation along the span*

Early measurements of eddy shedding frequency by Okamoto and Yagita (1973J) were evaluated from hot-wire signals at $x/D = 2.5$ and $y/D = 1.5$. They did not find a frequency peak behind the free end of the finite cylinder in the range $0 < z/D < 2$. When $H/D < 6$, there was no eddy shedding all along the span. For long finite cylinders ($H/D > 6$), the eddy shedding frequency varied along the top third of the span. Okamoto and Yagita (1973J) drew continuous St curves for $H/D = 7, 9$, and 12.5 at $Re = 13$k.

[109]Fox and West (1993Jc) did not measure C'_L for the nominal two-dimensional cylinder.

Subsequently, Farivar (1981J) measured the fluctuating pressure on the cylinder surface. Figure 21.59(a) reveals that eddy shedding near the free end is at a distinctly different frequency than that around the rest of the cylinder. The discontinuous jumps in frequency occur at about $z/D = 1$ and 4 from the free end. The noteworthy feature of the end cell is that its frequency and St are subharmonics of those in the adjacent cell.

Fox and West (1993Jb) repeated measurements using both hot-wire and pressure signals for $H/D = 30$. They found a wideband frequency peak for hot-wire signals, which reflected disturbed eddy shedding near the free end. In the region $z/D > 6$, a narrow band peak leads to a constant St along the rest of the span. Figure 21.59(b) shows St evaluated from the pressure fluctuations. Fox and West (1993Jb) found three frequency cells along the span in addition to the end cell. In contrast to Farivar's finding in Fig. 21.59(a), the long cell had a lower St than the adjacent cell. However, in Fig. 21.59(a,b) the end-cell frequency is the subharmonic of the main frequency.

Fox and West (1993Jb) commented that apart from the discontinuity at $z/D = 15$, there is no evidence of cellular eddy shedding in St variation in hot-wire signals. This explains why Okamoto and Yagita (1973J) presented a continuous St curve near the free end. The velocity fluctuations tend to mask the discrete steps in frequency peaks measured at the surface near the free end.

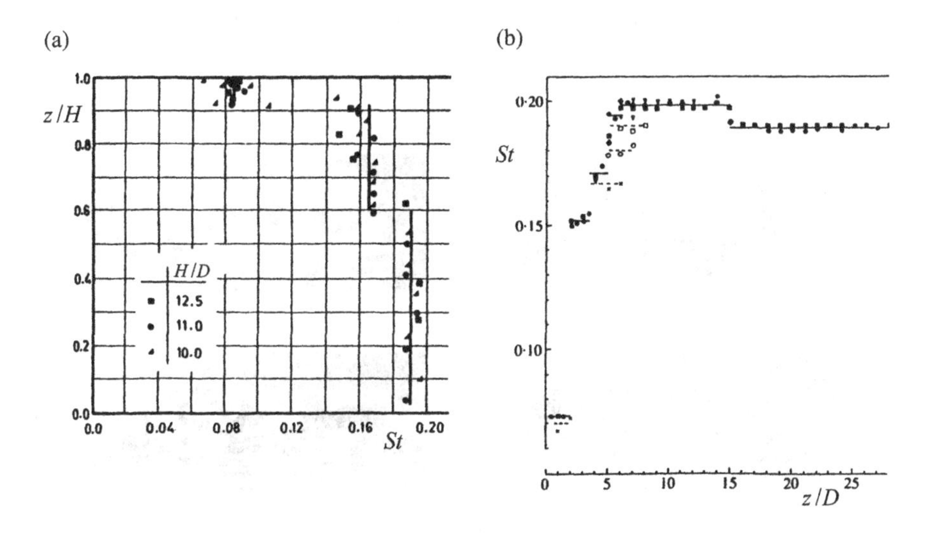

FIG. 21.59. Strouhal number, spanwise variation, (a) Farivar, © (1981J) AIAA – reprinted with permission, (b) Fox and West (1993Jb)

21.8.7 *Symmetric eddy street*

It might be expected that the combined effect of the free end and shortening of the aspect ratio of finite cylinders should lead to a gradual weakening and eventual suppression of eddy shedding. This was reported by Okamoto and Yagita (1973J), and confirmed by subsequent researchers until Sakamoto and Arie (1983J) discovered that the alternate eddy shedding mode changed to a symmetric one for $H/D < 2.5$. They used a smoke-wire flow visualization in the range $270 < Re < 920$ in the TrW2 and TrSL1 regimes. Short finite cylinders were submerged in a thick laminar wall boundary layer, $1 < H/\delta_B < 1.5$.[110] However, the eddy shedding frequency measurements were carried out at $Re = 50$k in the TrSL3 regime.

Figure 21.60 shows a selection of the observed eddy streets at the midspan behind short finite cylinders. For the shortest $H/D = 1$, the separated shear layers are wavy and mirror each other, Fig. 21.60(a). For $H/D = 1.5$, the shear layers roll up and symmetric eddy pairs are formed in Fig. 21.60(b). For $H/D = 2.5$, both symmetric and alternate modes are possible, Fig. 21.60(c,d). Hence, the alternate eddy shedding does not cease for $H/D < 5$ but changes to the symmetric mode.

Sakamoto and Arie (1983J) suggested that two adjacent vertical eddy filaments are linked at the top and form an arch. This has been confirmed by Okamoto and Sunabashiri (1992J). Figure 21.61(a,b) shows dye visualization in water at $Re = 1.1$k for $H/D = 1$ and 4 of the symmetric arch mode and alternate mode, respectively.

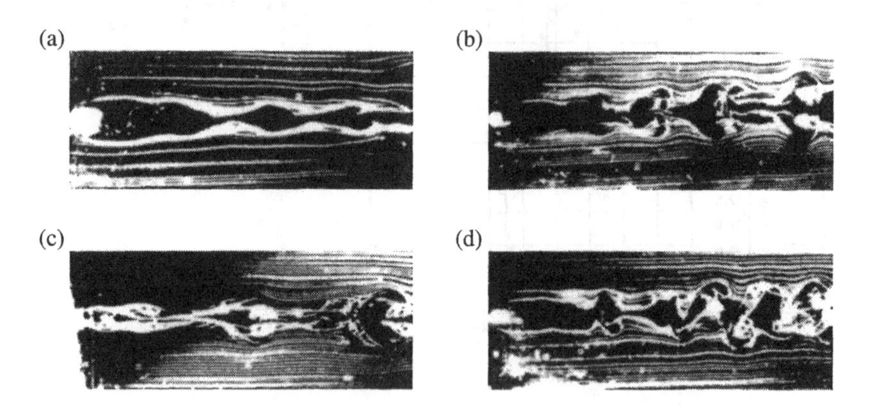

FIG. 21.60. Flow visualization of wakes for $Re = 600$, (a) $H/D = 1$, (b) $H/D = 1.5$, (c,d) $H/D = 2.5$, Sakamoto and Arie (1983J)

[110]The effect of shear velocity on the free end flow will be discussed in the next section.

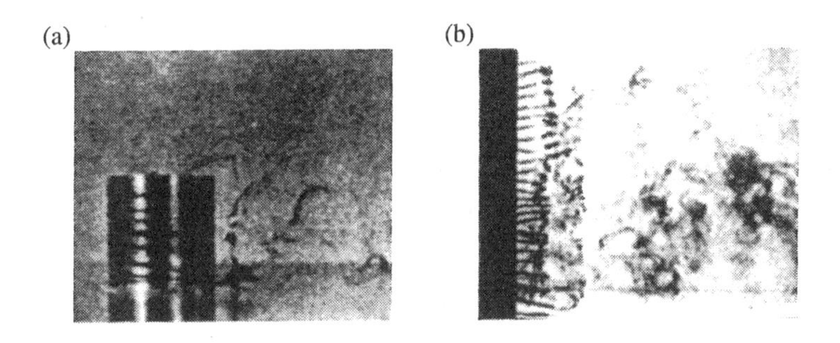

FIG. 21.61. Dye streaklines for $Re = 1.1$k, (a) $H/D = 1$, (b) $H/D = 4$, Okamoto and Sunabashiri (1992J)

The frequency of eddy shedding was measured by using a hot wire located at $x/D = 2$, $y/D = 0.5$, and $z/D = 0.6$ in the range 2.6k $< Re <$ 57k in the TrSL2 and 3 regimes. The different St curves in terms of H/D in Fig. 21.62 correspond to different wall boundary layer thicknesses, $0.35 < H/\delta_B < 1.48$. All St curves show a change in the slope around $H/D = 2.5$, where the change of mode occurred in the TrW2 regime.

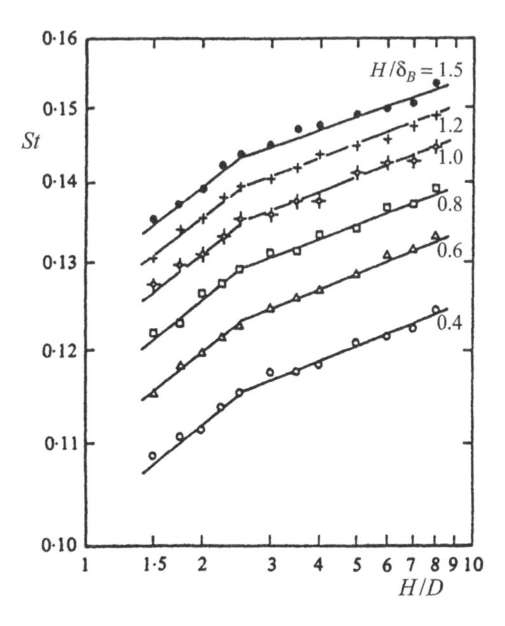

FIG. 21.62. Strouhal number variation in terms of H/D, Sakamoto and Arie (1983J)

21.8.8 *Short cylinder in a boundary layer*

The height of a short finite cylinder may be comparable to the thickness of a wall boundary layer. The velocity profile across the boundary layer is non-uniform. This implies that the local dynamic pressure $\frac{1}{2}\rho V^2$ varies along the span.

Figure 21.63 shows the variation of local dynamic pressure q (full lines) and stagnation pressure coefficient measured along $\theta = 0°$ (open circles). The departures of C_{p0} from $\frac{1}{2}\rho V^2$ occur at the wall produced by the horseshoe-swirl, and near the free end where the velocity is deflected upwards.

Figure 21.64 is a re-plot of C_{p0} for the $H/D = 1$ cylinder submerged in $0.11 < H/\delta_B < 0.85$ boundary layers, where $C_{p0}/C_{p0\max}$ is related to $C_{p0\max} = 0.8$. There is a remarkable similarity of all C_{p0} curves, which indicates that the flow does not depend on H/δ_B. The similarity extends to C_{pb} reduced by $C_{pb\max}$. Finally, Fig. 21.65 shows the variation of the overall C_D. The point of application of the mean drag force on the $H/D = 1$ cylinder is found from C_M to be at $z/H = 1/2$ irrespective of H/δ_B.

21.8.9 *Finite cylinder in the TrBL4 regime*

The transition to turbulence in boundary layers before separation leads to a narrow near-wake and low C_D. This decreases the wind loading on stacks at high *Re*. However, there is an additional loading on the finite cylinder near the free end related to a local $C_{d\max}$. In the TrSL3 regime $C_{d\max}$ is 20% above the C_d value at midspan. Gould *et al.* (1968P) found that $C_{d\max}$ was 60% above

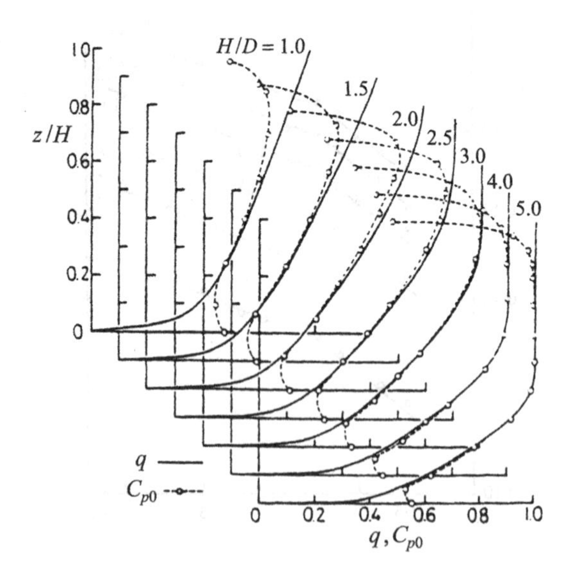

FIG. 21.63. Variation of dynamic pressure across boundary layer (full line), and measured on an $H/D = 1$ cylinder, Taniguchi *et al.* (1981Ja)

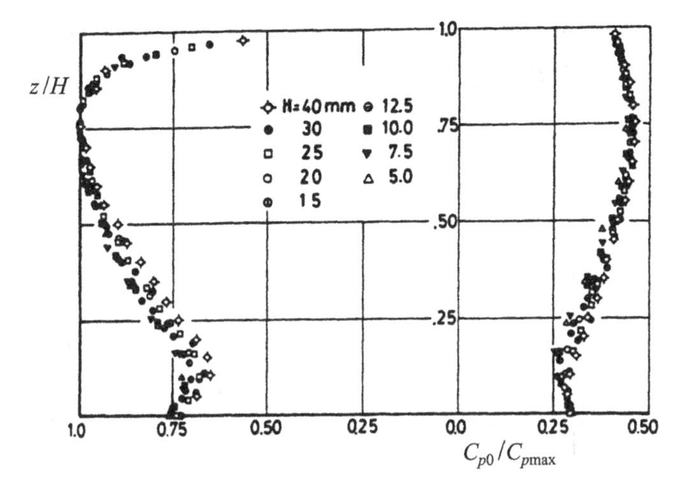

FIG. 21.64. Relative stagnation pressure and base pressure coefficients for different boundary layers, Taniguchi *et al.* (1981Jb)

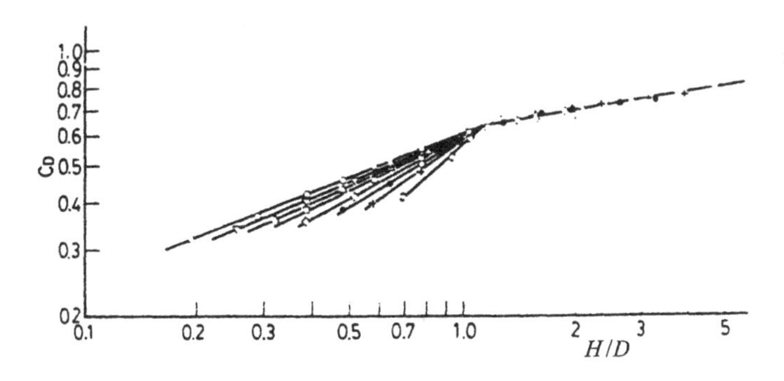

FIG. 21.65. Overall drag coefficient in terms of H/D, Taniguchi *et al.* (1981Ja)

the value at the midspan of the stack at $Re = 2.7$M in the postcritical regime. They carried out tests in a high-pressure wind tunnel (25 bar) on finite cylinders $H/D = 6$, 9, and 12 in the range $2.7\text{M} < Re < 5.4\text{M}$. Figure 21.66 shows the variation in C_d evaluated from C_p measured at 12 elevations. The values of $C_{d\max}$ are different for the open and closed free ends. The open end increases $C_{d\max}$ while the efflux lowers it, and moves away from the free end.

Gould *et al.* (1968P) also examined the effect of the free stream turbulence and velocity profiles on the C_d distribution along the span. They found an overall decrease in C_d for $Ti = 6\%$ and a further decrease for velocity profile $V \propto (H/D)^{0.2}$.

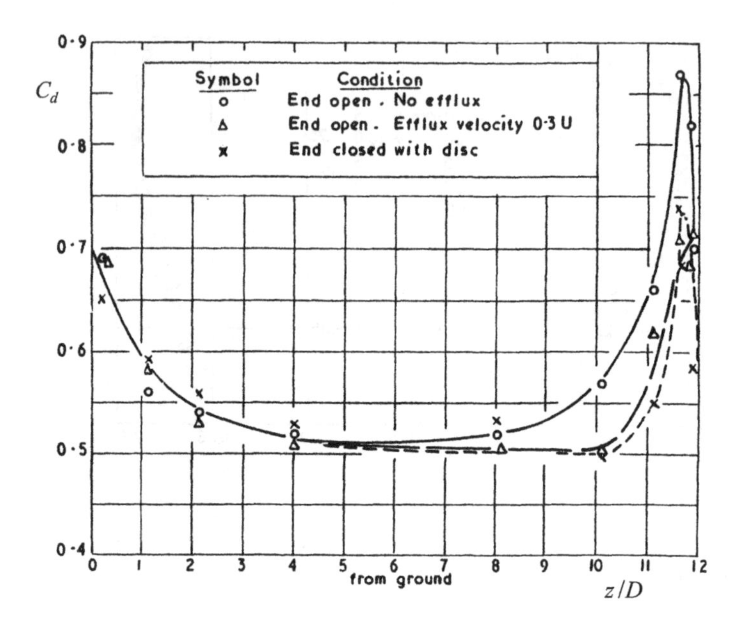

FIG. 21.66. Local drag along finite cylinder for Re = 2.7M and H/D = 12, Gould *et al.* (1968P)

Figure 21.67 shows measured C_p' in terms of θ at various elevations. For z/D = 2.96 half way from the free end, the C_p' distribution is similar to that

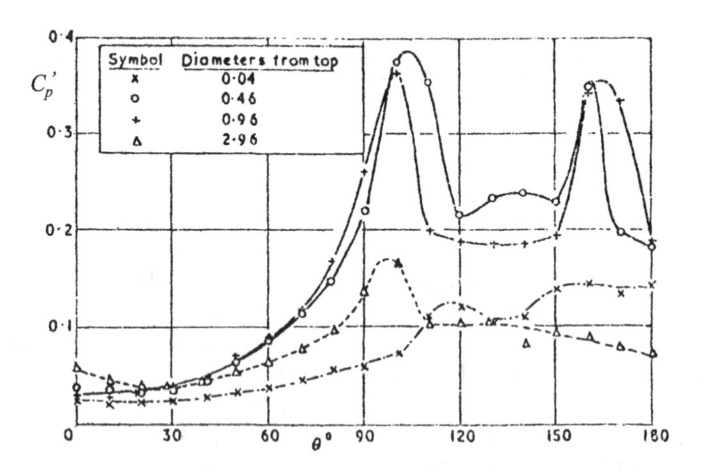

FIG. 21.67. Fluctuating pressure distribution near free end, Gould *et al.* (1968P)

of a nominal two-dimensional cylinder. There is a large increase in C'_p, and the appearance of a second $C'_{p\max}$ at $\theta = 160°$ for $z/D = 0.46$ and 0.96. Near the free end at $z/D = 0.09$, the C'_p curve is low, and has a single $C'_{p\max}$ at $\theta = 120°$.

Surface flow visualization at Re = 5.4M for $H/D = 6$ is shown in Fig. 21.68(a,b) without and with efflux, respectively. The separation line is straight up to $z/D = 2$. Near to the free end it becomes curved towards higher θ_s angles. The secondary flow along the downstream side of the cylinder is highly three-dimensional. The secondary separation of the reversed upwards flow is marked by a distinct white strip. By comparing the surface flow visualization and C'_p distribution it seems that $C'_{p\max}$ at $\theta = 105°$ corresponds to the flow separation, while $C'_{p\max}$ at $\theta = 160°$ corresponds to the attachment of the swirl flow. Gould *et al.* (1968P) found a similar surface pattern for all three finite cylinders tested, $H/D = 6$, 8, and 12. However, when efflux was added a lowering of the swirl location occurred, as seen in Fig. 21.68(b).

The most puzzling observation by Gould *et al.* (1968P) was the following:

> The correlation coefficients between p' at points 180° apart show, at Re = 5.4M, the pressure fluctuations around the sides near the free end are phased to cause some lateral oscillation, whereas those below the direct influence of the chimney end are phased to cause some ovaling oscillations.

21.8.10 *Eddy shedding near the free end*

Ayoub and Karamcheti (1982J) investigated the free-end region for $H/D = 12$ and Re = 180k and 770k, i.e. in the TrBL0 and TrBL3 regimes, respectively. Simultaneous measurements of surface pressures and wake velocity fluctuations

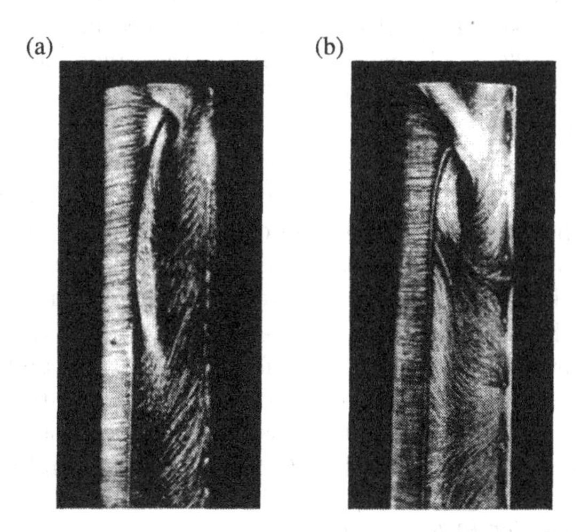

FIG. 21.68. Surface flow visualization for $H/D = 6$ and Re = 5.4M, (a) without efflux, (b) with efflux $0.3V$, Gould *et al.* (1968P)

were analysed by using frequency spectra, auto- and cross-correlations.

For Re = 180k, the frequency spectra near the free end show prominent spectral peaks associated with different frequencies. Each periodic component occurs only intermittently, and is modulated in amplitude. When the frequency peak occurs, it does so simultaneously at different locations along the span. The peaks are about 180° out-of-phase between the points on the upper and lower parts of the cylinder. When eddy shedding takes place, it does so at one frequency in the free end region but the frequency is not the same at different times. Ayoub and Karamcheti (1982J) wrote

> The flow in the end region is constantly attempting to lock itself into a shedding regime only to find, when it does manage to do so, that the regime reached is unstable, and consequently goes into a chaotic state for a while before repeating the process.

For Re = 770k, no eddy shedding is observed behind the main portion of the cylinder, as expected in the TrBL3 regime. However, the free end region shows a frequency peak corresponding to $St = 0.2$ by both pressure and velocity fluctuations. The cross-correlation of p' from two sides of the cylinder confirms alternate eddy shedding near the free end. The notable feature is that the spectral peaks remained centred at the same frequency near the free end.

21.8.11 *Local fluctuating lift and drag*

Schmidt (1965J, 1966P) tested an $H/D = 8.1$ cylinder in the range 380k < Re < 5M straddling through the TrBL2, 3, and 4 regimes. He measured C_p and C_p' at a fixed elevation $z/D = 3.2$, and the velocity fluctuation in the range $2.1 < z/D < 3$ from the free end. Schmidt (1965J, 1966P) evaluated local C_ℓ' and C_d' from the measured C_p'. His data were compiled in Vol. 1, Figs 6.32 and 33, Chapter 6, pp. 193–4. The variation was in the range $0.04 < c_\ell' < 0.1$ and $0.03 < c_d' < 0.04$. The local C_d evaluated from C_p is shown in Vol. 1, Fig. 6.22, Chapter 6, p. 185.

A spanwise correlation measurement of C_p' was carried out by using a relocatable pressure tapping ring. The evaluated correlations R_z in terms of C_l' and C_d' from C_p' are shown in Fig. 21.69. The correlation curves are different for C_l' and C_d' at Re = 750k. The odd point (open square) corresponds to a disturbance produced by a weak blowing through the pressure tapping. The correlation length L_c is defined by the area under the cross-correlation curve and normalized by the cylinder diameter D. Schmidt (1965J) found $L_c/D = 0.5$ and 0.15 for C_l' and C_d', respectively. These values presumably depend on the location of the pressure ring from the free end. Schmidt (1966P) repeated correlation measurements at Re = 1.1M and 5M, and found $L_c/D = 0.2$ and 0.1 for C_l' and C_d', respectively.

21.8.12 *Finite cylinder in natural wind*

Dryden and Hill's (1930J) report (41 pages) was a milestone in finite cylinder research. The first part of the report described model experiments $H/D = 7.5$

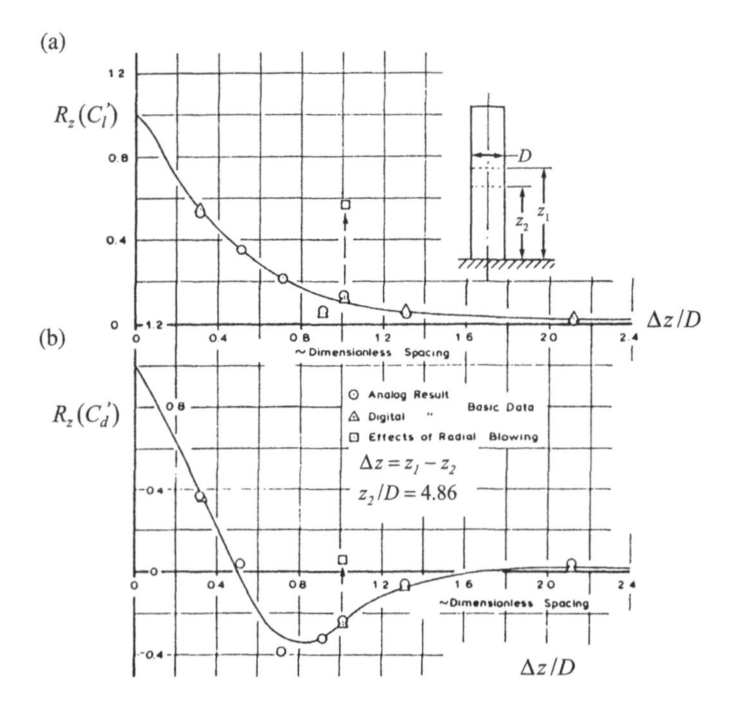

FIG. 21.69. Correlation of fluctuating (a) lift, (b) drag, for $H/D = 8.1$, $Re = 750\text{k}$, Schmidt, © (1966J) AIAA – reprinted with permission

and 4.8 in a wind tunnel in the range $170\text{k} < Re < 640\text{k}$. The mean circumferential C_p was measured at 12 stations along the span, four of which were near the free end in the range $0 < z/D < 1$. Local drag factors k_d[111] were evaluated from C_p distributions. Figure 21.70 shows that all four curves have peak $C_{d\max}$ near the free end. Dryden and Hill (1930J) concluded:

> The Reynolds number corresponding to a chimney 20 ft (6 m) in diameter in a wind of 100 mph (160 km/h) is 9.25M. Extrapolation to such a value cannot safely be made from model experiments at $Re = 642\text{k}$.

This prompted a second phase of experiments in natural wind:

> Such experiments have somewhat greater appeal to the engineer than model experiments although the difficulties of obtaining even reasonably accurate results are very great, and the interpretation of observations is a very troublesome matter.

[111] Up to 1930, the drag factor k_d was used, being drag$/\rho V^2$, i.e. equals $\frac{1}{2}C_d$.

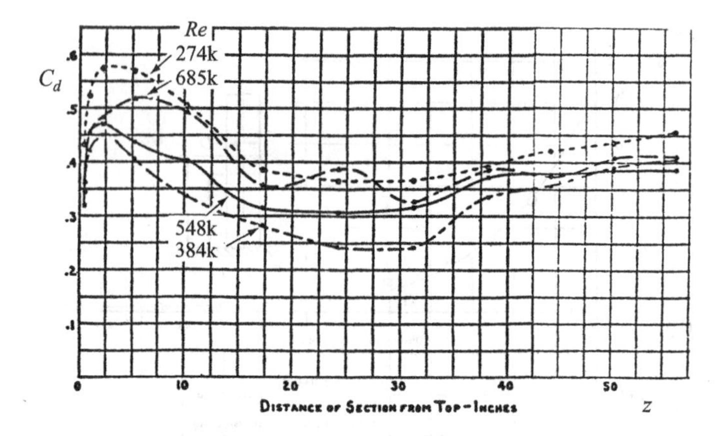

FIG. 21.70. Local drag factor along $H/D = 4.8$, Dryden and Hill (1930P)

A cylindrical stack, $D = 3\,\text{m}$, $H = 9\,\text{m}$, and $H/D = 3$ was erected on the roof of the West building at the Bureau of Standards in Washington, DC, as seen in Figure 21.71(a). At an elevation 1D below the free end, 24 pressure tappings were equally spaced around the circumference, and drag force was also measured on four attachments.

> The windy days were unpredictable; for example the first measurements started on April 28, 1928, and continued on Nov 8, 1928. The mean value of the overall drag was $C_D = 0.39 \pm 0.04$. As a result of continual racking by the wind and the decreasing resistance to overturning caused by the four-point support, the stack was blown from the roof on March 7, 1929, fortunately without serious damage to the building. The wind gust at that time exceeded 60 mph (96 km/h).

This accident led to a third phase of research into the instrumented power plant chimney $H/D = 11.4$ ($H = 60\,\text{m}$, $D = 3.6\,\text{m}$, tapered), as seen in Fig. 21.71(b). The comparison of the C_p distribution obtained in three phases of research is reproduced in Fig. 21.72. The marked difference in the three curves led Dryden and Hill to conclude that *both Re* and H/D are equally important governing parameters.

Subsequent tests in natural wind were reported by Pechstein (1942J). He tested a smooth and roughened finite cylinder $H/D = 4$ ($D = 2\,\text{m}$), and measured C_p at three stations. The gust periods change from 24 s to 79 s and fluctuations from 1 s to 3 s. The highest mean velocity corresponded to $Re = 4.5\text{M}$.

A large number of full-scale measurements of cylindrical structures and smoke stacks was subsequently performed in natural wind. For example, Tunstall (1974P), Ruscheweyh (1976J), in Vol. 1, Fig. 14.26, p. 455, Christensen and Askergaard (1978J), Melbourne *et al.* (1983J), Waldeck (1992J), Sanada and

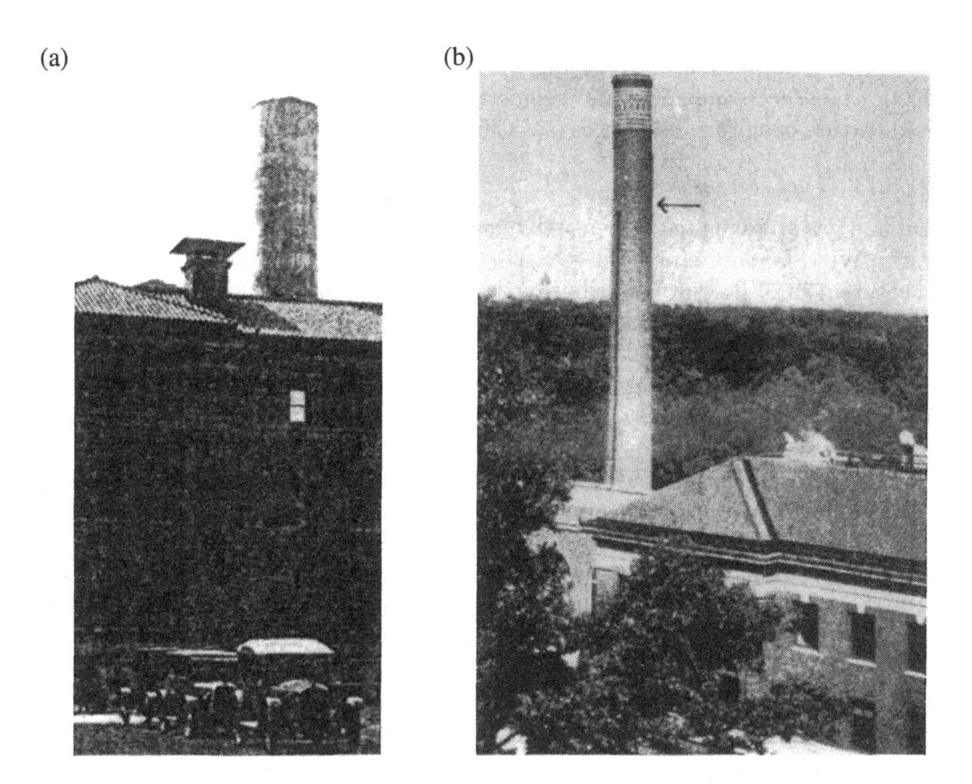

FIG. 21.71. Experimental stacks, (a) $H/D = 3$ on the roof of West Building, (b) $H/D = 11.4$, the Bureau of Standards, Washington, DC, Dryden and Hill (1930P)

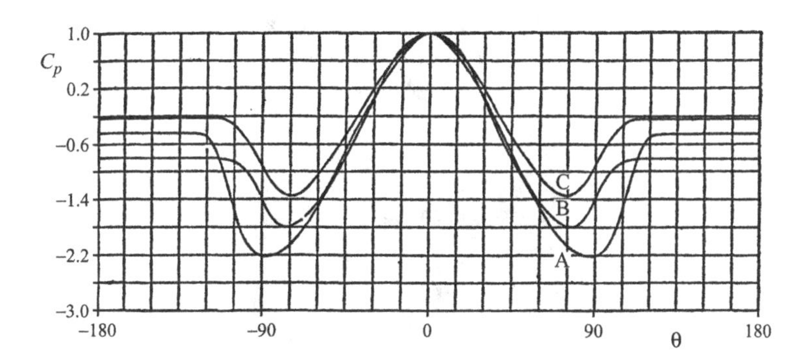

FIG. 21.72. Mean pressure distribution comparison of A, wind tunnel model, $z/D = 1.4$, B, roof model, $z/D = 1$, and C, real stack, $z/D = 3.6$, Dryden and Hill (1930P)

Matsumoto (1992J), Vol 1., Figs 7.2 and 7.3, pp. 203–4, etc. Each full-scale measurement is applicable only to the particular wind characteristics on the site: $V(t)$, $\alpha(t)$, $Ti(t)$, and $Ts(t)$ and geometry of the structure defined by H/D, size, and texture of surface roughness (see Chapter 22), taper, and attachments if any.

21.8.13 *Fuel storage tank*

Oil and gas storage tanks are usually large cylindrical structures of small aspect ratio. Wind loading may be dangerous when the tanks are empty. A pioneering measurement of pressure distribution around a gas storage tank was reported by Ackeret (1934J). A 1:100 model having $H/D = 1.09$ with slightly rounded top, $\Delta H/D = 0.07$, and elaborate scaffolding all around the model is shown in Fig. 21.73. The model was mounted on the board in an open circular jet wind tunnel. Pressure tappings were arranged around the circumference at 11 elevations and along five circles on the top.

All 16 C_p distributions are given in the paper and additionally an ingenious C_p display on the model replicas is reproduced in Fig. 21.74(a,b) without and with scaffolding, respectively. Positive C_p around the stagnation regions are shown light, and negative C_p regions by dark (blue and red on the original model). There is little variation in C_p along the height except near the top at $Re = 1.2$M. The display shows excessive negative C_p near the upstream part of the rim. However, the effect of scaffolding on the wind loading in Fig. 21.74(a,b) shows a significant decrease in C_p all around the model circumference and top.

The comparison of C_p distribution at the midspan without and with scaffolding is given in Fig. 21.75 for $Re = 1.2$M. There is a large decrease in $C_{p\min}$ when

FIG. 21.73. 1:100 model of gas storage, $H/D = 1.09$ with scaffolding around, Ackeret (1934J)

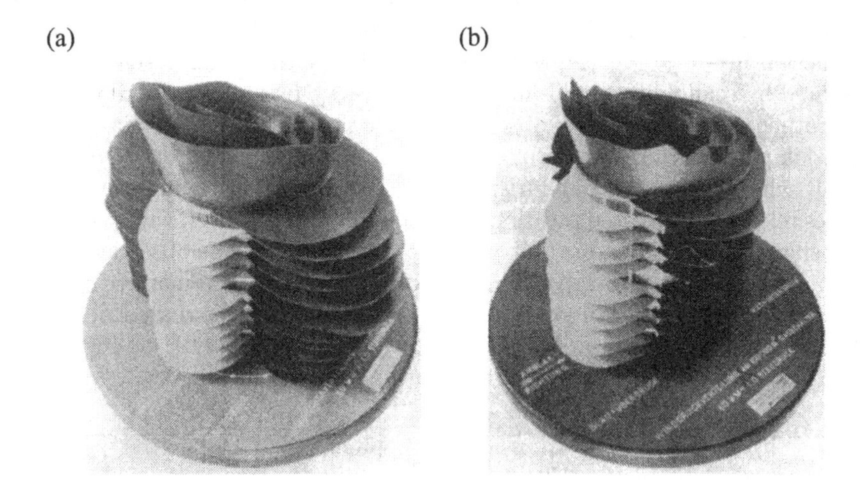

FIG. 21.74. Three-dimensional presentation of mean pressure distribution at $Re = 1.2$M around gas storage model, (a) without scaffolding, (b) with scaffolding, Ackeret (1934J)

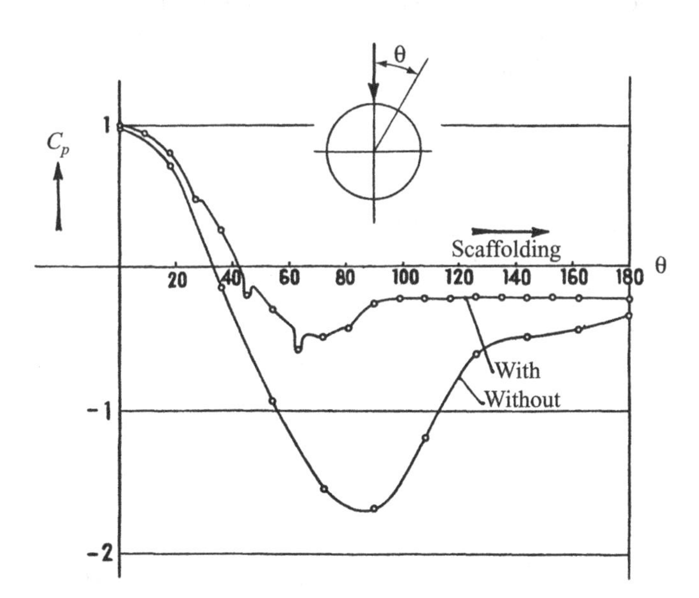

FIG. 21.75. Mean pressure distribution at mid-height level with, 1, and without scaffolding, 2, Ackeret (1934J)

scaffolding is fitted to the model, and a distinct change in the shape of the two C_p distributions. The most notable is the shift from $\theta_s = 85°$ to $115°$, which causes a reduction in $-C_{pb}$ to 0.22. It is puzzling why at $Re = 1.2$M the turbulence generated by scaffolding could have such a strong effect on C_p distribution.

Another *in situ* qualitative observation of high Re wind flow around a storage tank took place during a short blizzard storm. It was reported by Probert *et al.* (1973J) who also carried out further wind observations by using soap-solution bubbles released at selected positions near the tank, $H/D = 1.4$, with a closed top in the shape of a shallow cone (15° pointing upwards). In these observations, only the wind magnitude was estimated; the other parameters remained unknown. Figure 21.76 shows the estimated θ_s in terms of Re. As Re increased from 5M to 22M the separation angle at the midheight increased from 110° to 160° (±5°). This indicates wake narrowing and decreasing C_D (not measured).

21.9 Two free ends

21.9.1 *Drag variation in terms of the aspect ratio*

The earliest experiments on finite circular cylinders with two free ends were reported by Wieselsberger (1922J). He covered an enormous range of Re from 400 to 800k for $L/D = 5$. Figure 21.77 shows C_D curves for the nominally two-dimensional cylinder $L/D = \infty$ and finite $L/D = 5$ in terms of Re. Both C_D curves are similar in shape but the $L/D = 5$ curve is displaced towards the lower C_D values. For example, the plateau in the TrSL3 regime occurs at $C_D = 0.78$ instead at $C_D = 1.25$ for $L/D = \infty$. The rapid fall in C_D values in the TrBL1 and 2 regimes ($Re > 400$k) takes place at the same Re towards the same value of $C_D = 0.3$.

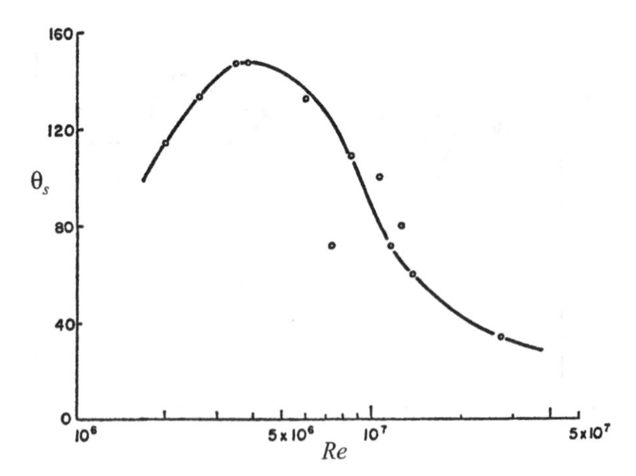

FIG. 21.76. Observation of separation in natural wind on gas storage $H/D = 1.4$ in terms of Re, Probert *et al.* (1973J)

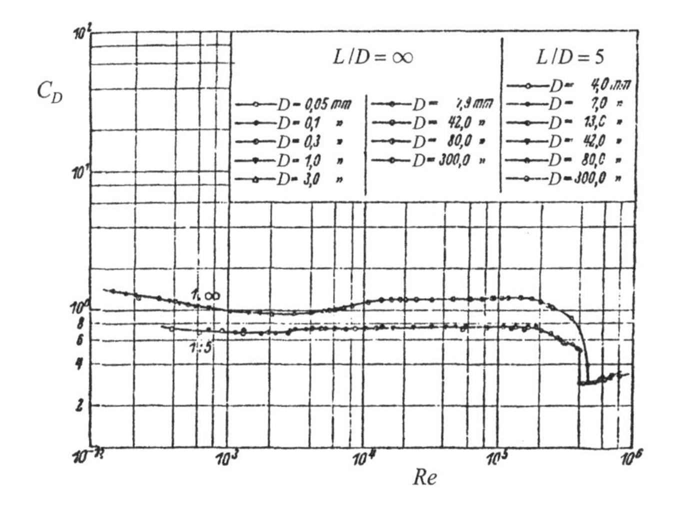

FIG. 21.77. Drag coefficient on two-dimensional (top curve) and $L/D = 5$ cylinder (bottom curve) with two free ends in terms of Re, Wieselsberger (1922J)

Subsequently, Wieselsberger (as reported by Muttray, 1932B) examined the effect of the aspect ratio in the range $1 < L/D < \infty$ on C_D at a single $Re =$ 88k. Figure 21.78 shows the variation in C_D in terms of both L/D and less commonly used D/L. Subsequent measurements by Okamoto and Yagita (1973J) on a cylinder with one free end and Zdravkovich *et al.* (1989J), are also added, together with data obtained with a cylinder fitted with two hemispherical ends.

The sphere may be thought of as a special case of the circular cylinder with hemispherical ends having $L/D = 1$. The dashed line in Fig. 21.78 shows C_D measured on three short circular cylinders with hemispherical ends. The values of C_D are significantly lower than those for short cylinders with flat ends. This shows convincingly that for short cylinders ($L/D < 5$) the shape of the free ends becomes a governing parameter.

Muttray (1932B) cited Wieselsberger's conclusion:

> The explanation for the fall in drag coefficient with decreasing aspect ratio should be sought in the venting of a 'dead-water' (near-wake) behind a bluff body, i.e. an inflow of the fluid into the near-wake space on its way around the cylinder ends. Hence, the pressure over the back side of the body would rise, and so the drag would be smaller.

This was further updated by Zdravkovich *et al.* (1989J) in Fig. 21.79.

> The inflow into the near-wake does not merely increase the base pressure but also displaces the eddy formation region further downstream, as found by Etzold and Fiedler (1976J), and widens the separated shear layers before roll

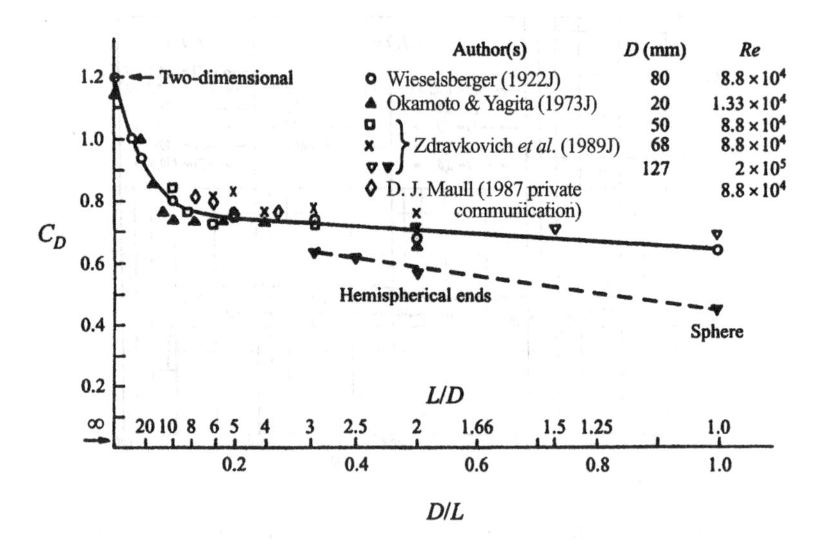

FIG. 21.78. Drag coefficient in terms of L/D and D/L, Zdravkovich *et al.* (1989J)

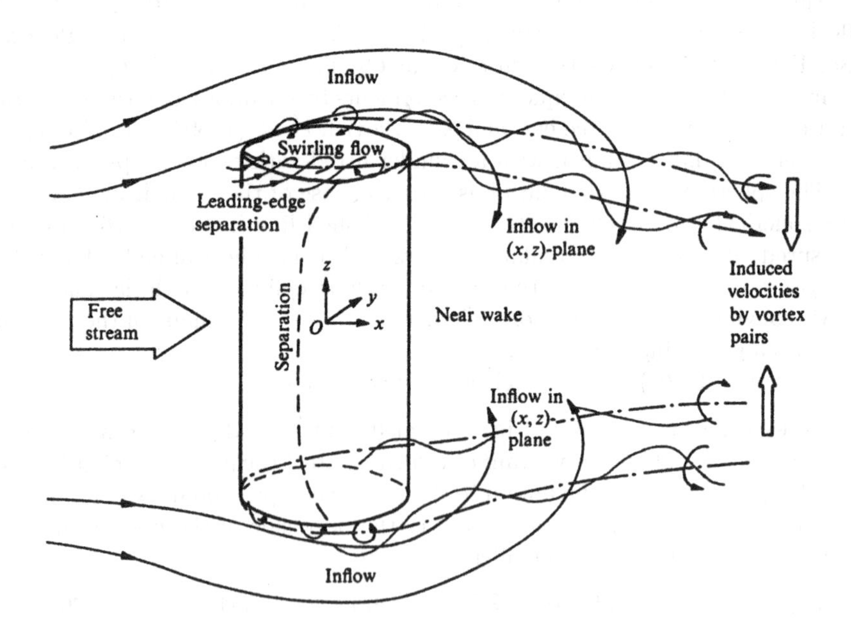

FIG. 21.79. Topological sketch of flow around cylinder with two free ends, Zdravkovich *et al.* (1989J)

up into eddies. The former additionally reduces C_D, and the latter causes a decrease in St, as found by Farivar (1981J).

An 'anomalous' increase in C_D is found as L/D is reduced below 1, see Fig. 21.80(a). As projected area $LD \to 0$, $C_D \to \infty$ because the drag force is reduced to a finite friction force on the flat sides. This may be circumvented by replacing the projected area in the equation for C_D by the side area $D^2\pi/4$

$$C_{DS} = C_D \frac{L}{D}\frac{4}{\pi} \tag{21.6}$$

Figure 21.80(b) shows $C_{DS} \to C_{Df}$, as $L/D \to 0$. Hence, the 'anomaly' is only apparent, and due to the inadequate formulation of the reference area.

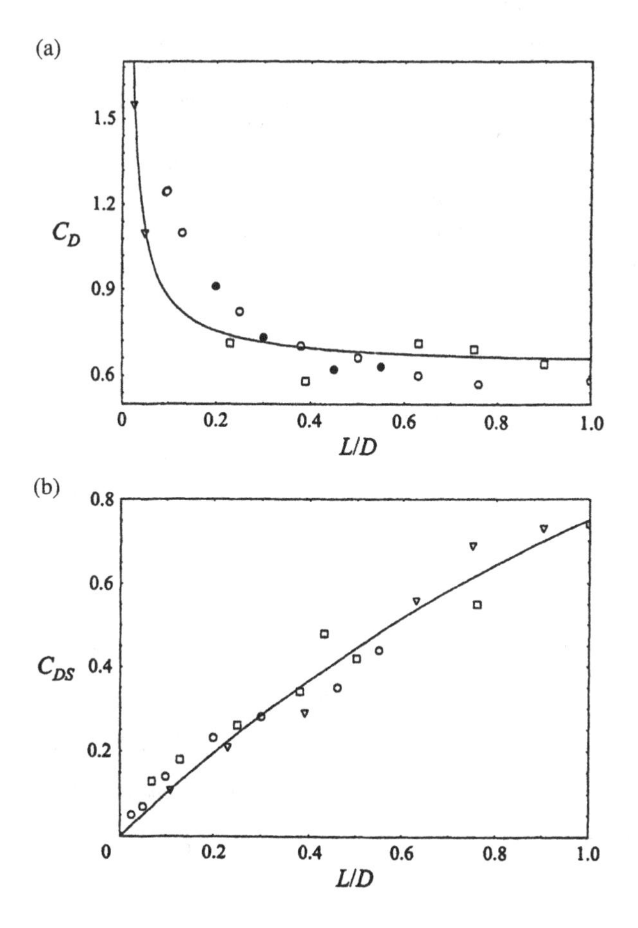

FIG. 21.80. Drag coefficient based on (a) L/D, (b) $D^2\pi/4$, Zdravkovich *et al.* (1998J)

21.9.2 *Asymmetric pressure distribution*

It has been shown that separation lines bend downstream near the free end. When $L/D \leqslant 4$, the separation lines are also bowed towards two free ends. However, the minimum θ_s is displaced towards one end. The spanwise flow appears asymmetric relative to the midspan axis. This unexpected feature prompted simultaneous pressure measurements over both halves of the short cylinder. Figure 21.81 shows an asymmetric isobar pattern for $L/D = 1$ and $Re = 205\text{k}$. The minimum C_p appears on one side in one run, and in subsequent runs it appears on the opposite side. It appears that the asymmetry is bistable, and can be biased to either side away from the midspan.

21.9.3 *Small aspect ratio, $L/D < 1$*

Flow around coin-like cylinders is dominated by the salient separation from the sharp edges at both free ends. The separated shear layers may or may not reattach onto two flat sides. In the first case, the region between the primary separation from the sharp edge and subsequent reattachment forms the separation bubble, as reviewed by Kiya and Sasaki (1982J, 1983J). The reversed flow along the flat side underneath the separation bubble eventually detaches from the surface as the secondary separation.

Surface flow visualization reveals an almost steady secondary separation line and the unsteady reattachment region is shown in Fig. 21.82(a) on the flat side for the $L/D = 0.11$ model at $Re = 214\text{k}$. The secondary separation forms a distinct

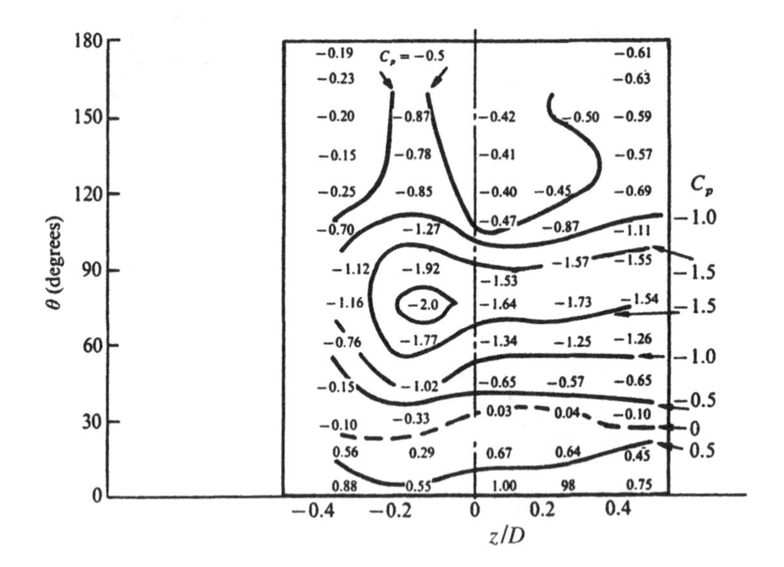

FIG. 21.81. Isobars on the developed cylinder surface, $L/D = 1$, $Re = 205\text{k}$, Zdravkovich *et al.* (1998J)

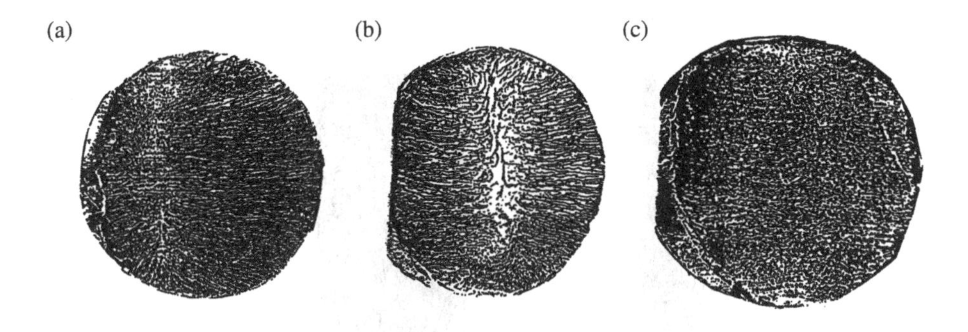

FIG. 21.82. Surface flow visualization on the flat sides at Re = 214k, (a) $L/D = 0.11$, (b) $L/D = 0.39$, (c) $L/D = 0.80$, Zdravkovich *et al.* (1998J)

crescent-like area. The reattachment may be inferred from the spanwise surface pattern covering a straight vertical strip. An estimate of the position of the secondary separation and reattachment along the equatorial axis of the flat side gives x/D = 0.11 and 0.31, respectively. Figure 21.82(b) shows a considerable downstream displacement of the reattachment for $L/D = 0.39$ in comparison with $L/D = 0.11$. The surface pattern for $L/D = 0.89$ reveals that the separated shear layers do not reattach to the flat sides in Fig. 21.82(c).

Figure 21.83(a,b) shows the visualization of a cylindrical surface for $L/D = 0.39$ up to $\theta = \pm 140°$. The separation 'islets' are seen at $\theta = \pm 96°$ as they replace the separation lines at the $\pm 86°$ for $L/D = 0.75$ in Fig. 21.83(b).

The qualitative features of the surface flow patterns are confirmed by the measured local static pressure along the flat side. The area under the separation bubble is associated with low pressure, the reattached region causes adverse pressure recovery, and the reattached flow is reflected by the closeness to the free stream pressure, see Roshko (1993P). Figure 21.84 shows C_p distribution measured along the flat side of coin-like cylinders in the region $0.11 \leqslant L/D \leqslant 0.89$ at Re = 214k. The lowest $L/D = 0.11$ shows the earliest pressure recovery due to the shortest separation bubble. There is a considerable displacement of the pressure recovery for $L/D = 0.23$, and it proceeds at a slower rate as the aspect ratio increases. The free stream pressure ($C_p = 0$) is not reached for $L/D \geqslant 0.75$, which indicates that reattachment does not occur.

Two horseshoe-shaped eddies are formed along the upstream edges of the coin-like cylinder, as sketched in Fig. 21.85. They are carried downstream as two streamwise pairs of counter-rotated eddies. The location of separation islets on the cylindrical surface coincides with the detachment of the horseshoe eddies. Similar horseshoe eddies were visualized behind a stationary frisbee in a wind tunnel by Nakamura and Fukamachi (1991J).

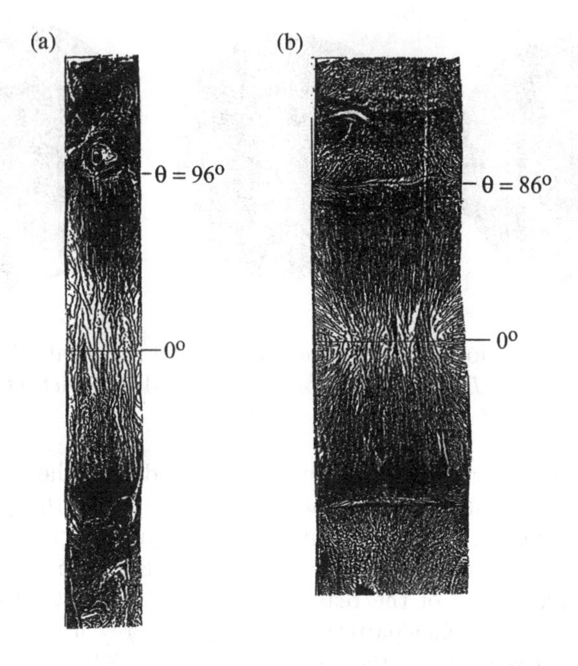

FIG. 21.83. Surface flow visualization along the developed cylindrical surface, (a) $L/D = 0.39$, (b) $L/D = 0.75$, Zdravkovich *et al.* (1998J)

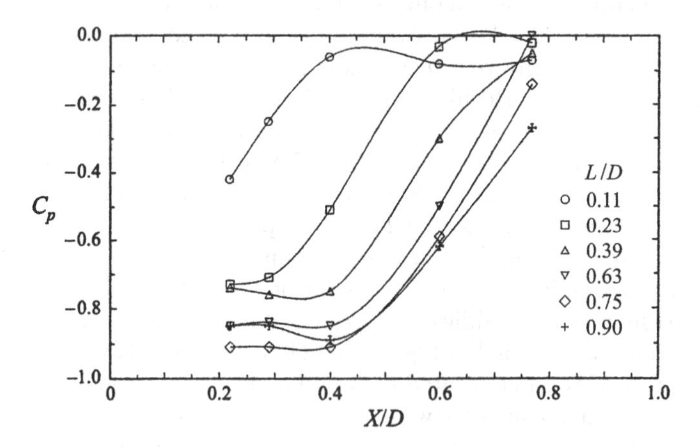

FIG. 21.84. Mean pressure distribution along the flat side at Re = 214k for $0.11 < L/D < 0.80$, Zdravkovich *et al.* (1998J)

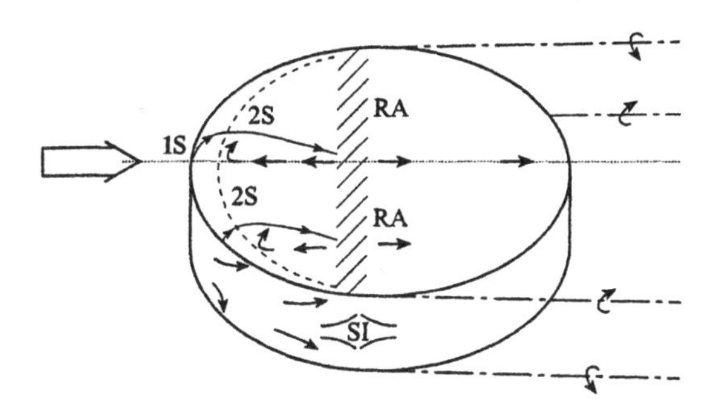

FIG. 21.85. Topological sketch of flow structures around coin-like cylinder; 1S – primary separation, 2S – secondary separation, SI – separation islet, and RA – reattachment, Zdravkovich *et al.* (1998J)

22

SURFACE ROUGHNESS AND CHANGE IN DIAMETER

22.1 Introduction

It is customary to regard a surface as smooth when it is so even that no roughness or points are perceptible to the touch. It is known, however, that an examination of such a surface under a microscope would reveal the presence of small excrescences and undulations, that is the surface would appear 'rough'. The smoothness or roughness of a surface is a relative quantity expressed as the ratio of the mean size of excrescences K to the diameter of a cylinder, D. In Fluid Mechanics, the surface of a cylinder can be regarded as smooth when the excrescences and undulations are small, and of such a character that they do not affect to any measurable extent the flow around the cylinder. Fage and Warsap (1929P).

Prandtl (1961B) remarked:

Slightly rough surfaces may be regarded as effectively smooth when the irregularities are completely embedded in the laminar boundary layer. At high *Re*, when the laminar boundary layer becomes thinner, such roughness may become effective causing an increase in drag.

Kármán (1956B) elaborated further:

All surfaces may be considered rough when *Re* is sufficiently high because the frictional resistance of the rough surface is made up essentially of the resistances of all excrescences, and those become independent of the skin friction at high values of *Re*.

Fage and Warsap (1929P) noted that surface roughness generates turbulence and should affect the flow around the cylinder in an analogous manner as the free stream turbulence. The difference is that the boundary layer is disturbed by the free stream turbulence from 'outside' while the surface roughness turbulence acts from 'inside'. The intensity Ti and scale Ts of the roughness turbulence should be related to:

(i) relative roughness K/D, where K is the average height of excrescences;

(ii) the shape of excrescences, i.e. the most frequently occurring shape;

(iii) the distribution of excrescences, i.e. irregular, regular, partial, etc.

The last two parameters might be combined and termed the *texture*, Te, of surface roughness.

The mechanism of roughness-generated turbulence has not yet been investigated in detail. Research was directed towards the effect of K/D, and to a less extent to Te on flow around cylinders. The roughness turbulence *per se* was ignored. It might be hypothesized that K/D is dominant in generating the turbulence intensity Ti, while Te determines the turbulence scale Ts and its dissipation. Hence, both K/D and Te are required to define the surface roughness.

It has been shown in Vol. 1, Chapter 14, that the flow around a cylinder in the transition-in-boundary-layer, TrBL, state is strongly influenced by a very low intensity of turbulence. Analogously, the TrBL state of flow should also be sensitive to roughness turbulence. The effect of roughness turbulence strongly depends on the thickness and state of the boundary layer. For example, when the laminar boundary layer is stable and $K < \delta$ the roughness turbulence is ineffective in triggering transition, while the same K/D becomes highly effective in the TrBL state of flow. The high efficiency of roughness turbulence might be attributed to small Ts being less than δ. Free stream turbulence usually has high Ts of the order 10δ–100δ, and the boundary layer is disturbed primarily by Ti.

22.1.1 *Nature of surface roughness*

An infinite variety of roughness textures has been found in engineering applications and nature. The rough surfaces in engineering are due to corrosion (metals), erosion (cavitation, sand), manufacture (concrete), machining (knurling), etc. The irregularity of the corroded area and unevenness in size of excrescences makes it hard for simulation. The widely adopted method is to test artificially roughened surfaces instead such as those simulated by various abrasive papers wrapped around the cylinder.

The most common abrasive paper used to simulate surface roughness in wind tunnel testing is sand paper. Figure 22.1(a,b) shows coarse and fine sand paper magnified 40×, respectively. The crushed sand particles differ in size and orientation, and appear fairly uniformly distributed. The similarity in texture for coarse and fine sand is remarkable.

Another example is emery paper made of extremely hard crystallized alumina. Figure 22.1(c,d) shows coarse and fine emery paper magnified 40×, respectively. The particles are markedly smaller and different in shape in comparison with the sand paper. Hence, it appears that the various textures are self-similar only for the same kind of abrasive papers.

The common feature of all abrasive papers is irregularity of excrescences in size and the relative uniformity in distribution over the surface. A different kind is a pyramidal roughness machined by knurling the surface. Figure 22.2 shows a sketch of pyramidal excrescences regular in shape and evenly distributed. The pyramidal roughness is not a simulation but a real surface, developed to enhance heat transfer, Achenbach (1977J).

Natural kinds of surface roughness occur on offshore structures by marine fouling. Figure 22.3(a,b) shows hard and soft roughness, respectively, highly irregular in size and shape. The hard roughness is made of barnacles, mussels, etc.

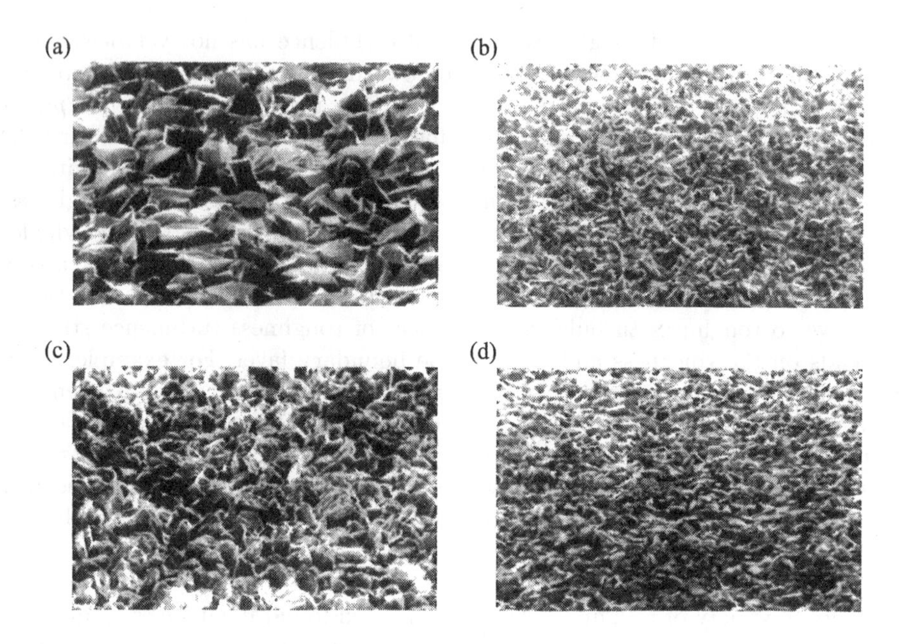

FIG. 22.1. Texture of surface roughness amplified 40×, (a) coarse sand paper, (b) fine sand paper, (c) coarse emery paper, (d) fine emery paper

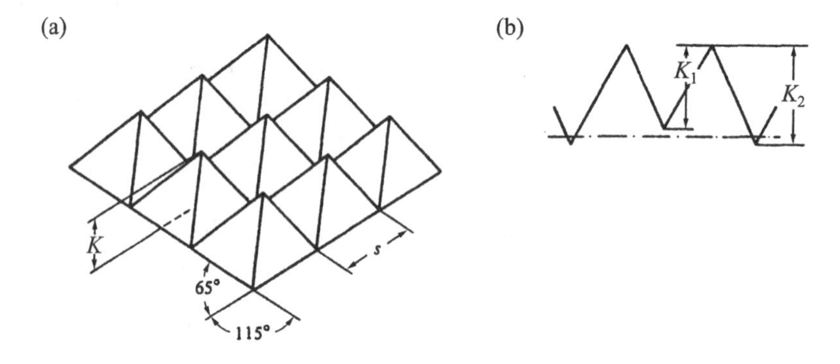

FIG. 22.2. Texture of pyramidal surface roughness (knurling), Achenbach (1977J)

while the soft roughness is made of seaweed, kelps, anemone, etc. The two intrinsically different roughness textures generate dissimilar roughness turbulence and related effects.

All surface roughnesses described so far consist of innumerable small excrescences of various shapes and sizes distributed over the whole surface or over some

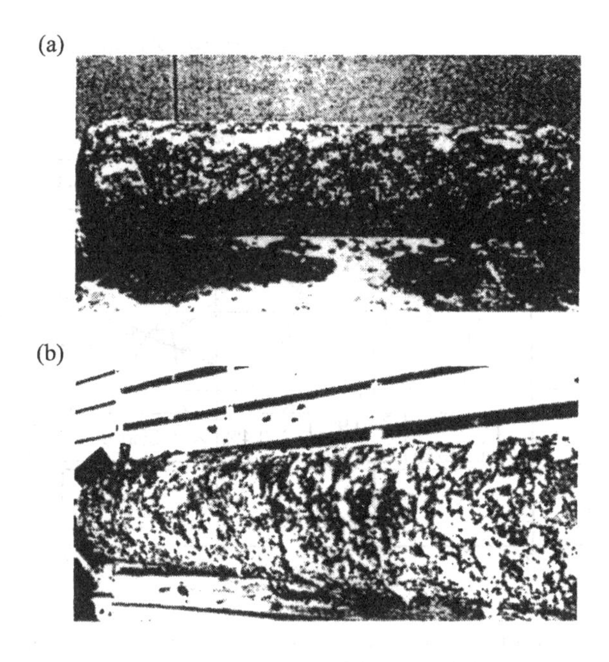

FIG. 22.3. Texture of marine fouling, (a) hard, (b) soft, Miller (1976J)

part of it. Another category of surface roughness consists of a few excrescences attached to a smooth surface at some strategically chosen locations. For example, tripping wires, spheres, eddy generators, grooves, dimples, etc.

22.1.2 *Fage and Warsap's glass paper tests*

Fage and Warsap (1929P) carried out pioneering experiments to determine how the drag over a cylinder was influenced by surface roughness. The roughness was simulated by wrapping the cylinder in John Oakey's glass paper of five grades (the relative roughness K/D was subsequently estimated by others). The drag was measured on two cylinders having $L/D = 8$ and 20.2 and $D/B = 0.125$ and 0.05, respectively.[112] The central cylinder section was suspended from the balance, and two dummy sections were attached to the side walls.

Figure 22.4 shows the original Fage and Warsap's (1929P) graph of k_D ($\frac{1}{2}C_D$) versus Re for five grades of glass paper.

> The orderly change in shape of k_D, Re curves is clearly illustrated. As K/D increases, the fall in k_D occurs at a lower value of Re. There are indications that if the roughening is continued, eventually the fall in k_D would disappear. It appears then, that as the surface is roughened the boundary layer is retarded,

[112]Blockage corrections were not made and free stream turbulence was estimated later by Baines and Peterson (1951J) to be about 0.9% and 1%.

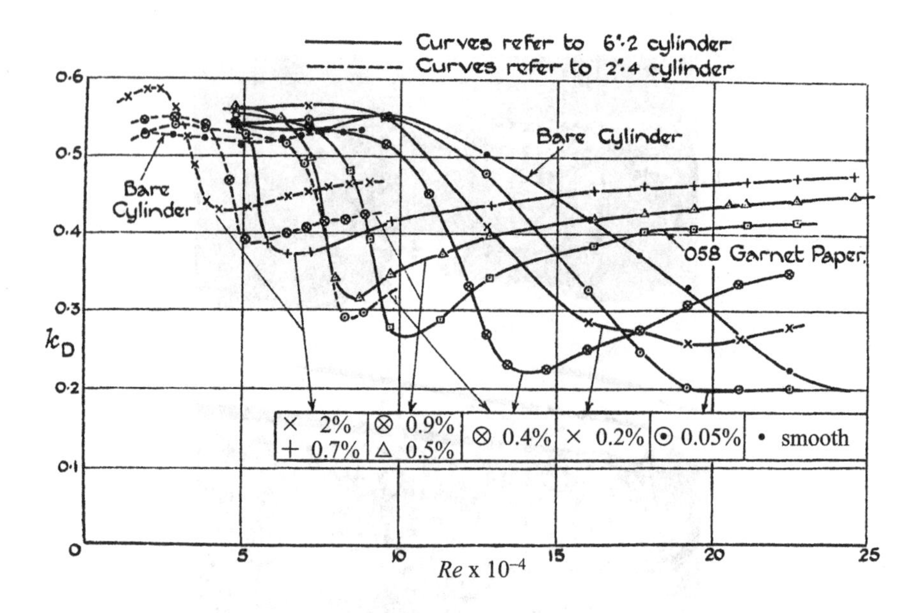

FIG. 22.4. Effect of glass paper roughness on k_D in the TrBL state, Fage and Warsap (1929P)

so that the separation moves forward, and as a consequence the drag of the cylinder increases. Fage and Warsap (1929P).

It is necessary to update the above description by linking the k_D, Re curve variation to the flow regimes. The TrBL1, 2, and 3 regimes found for smooth cylinders are disturbed by the surface roughness for small K/D. For higher K/D, all three flow regimes are obliterated, and the TrBL0 regime is followed by the TrBL4. It will be shown that eddy shedding does not cease, hence the TrBL3 regime is also obliterated.

The slope of the k_D curves increases with rising K/D. This indicates that the roughness-generated turbulence becomes more effective in promoting transition in the boundary layer.

A compilation of C_D data by subsequent measurements up to 1980 is reproduced in Fig. 22.5. Fage and Warsap's (1929P) curves for three K/D are re-plotted (dash-dot lines), together with Achenbach's (1971Jc) data corrected for blockage (full line), Szechenyi's (1975J) (broken lines and open circles), and Guven's *et al.* (1980J) (all other points). The large scatter of experimental points may be attributed to influencing parameters such as the aspect ratio, blockage, free stream turbulence, and the different texture of abrasive papers used by researchers.

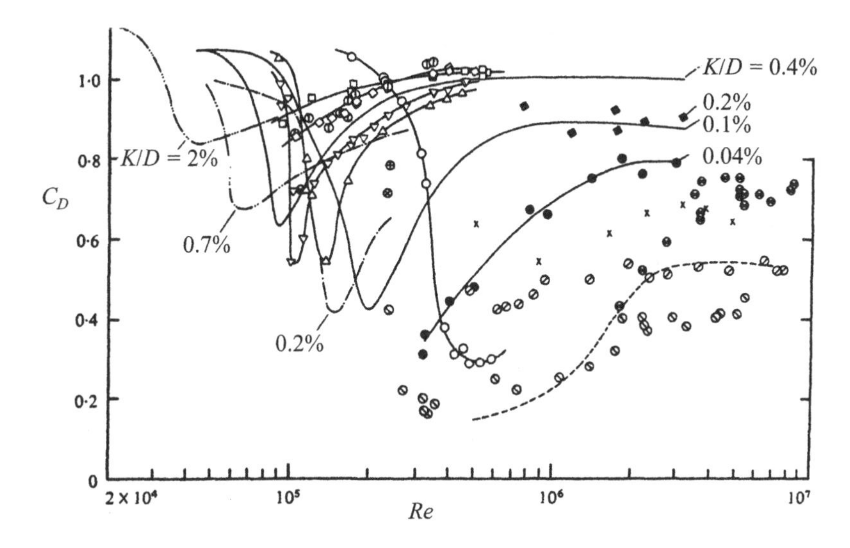

FIG. 22.5. Compilation of drag coefficients as affected by surface roughness, Guven *et al.* (1980J)

22.1.3 *Skin friction distribution*

The roughening of the cylinder surface by glueing abrasive paper onto it presents difficulties in installing and calibrating the skin friction probe. Achenbach (1971Jc) used a probe in the form of a miniature fence ($h = 0.01\,\mathrm{mm}$). This required a smooth area around the probe during calibration and measurement. The calibration was done by transplanting the probe into a square duct roughened with the same emery paper. The measured pressure drop along the duct was compared with Nikuradze's (1933P) measurements for sand roughness. This gave an *equivalent* sand roughness for the emery paper, K_s/D.

Achenbach (1971Jc) measured skin friction and static pressure distributions around a cylinder ($L/D = 3.3$, $D/B = 0.16$) wrapped in three emery papers of the equivalent sand roughness $K_s/D = 0.11\%$, 0.45%, and 0.9%. Figure 22.6(a,b) shows C_f and C_p distributions, respectively, for five Re. For $K_s/D = 0.11\%$, the $C_{D\min}$ value occurs at $Re = 200\text{k}$. Thus, the lowest tested $Re = 130\text{k}$ corresponds to the TrBL0 regime with laminar separation at $C_f = 0$ ($\theta_s = 82°$). The next $Re = 240\text{k}$, shows $C_f = 0$ at $\theta_s = 135°$. The absence of laminar separation and subsequent reattachment indicates the TrBL3 regime. The letter T denotes the start of transition in the boundary layer at around $\theta_{Tr} = 100°$. These drastic changes in C_f are reflected in the C_p adverse pressure recovery being displaced toward higher θ. The other three curves for $Re = 430\text{k}$, 650k, and 3M belong to the TrBL4 regime with the turbulent separation at $\theta_s = 100°$, and gradual upstream displacement of transition in the boundary layer (marked by T).

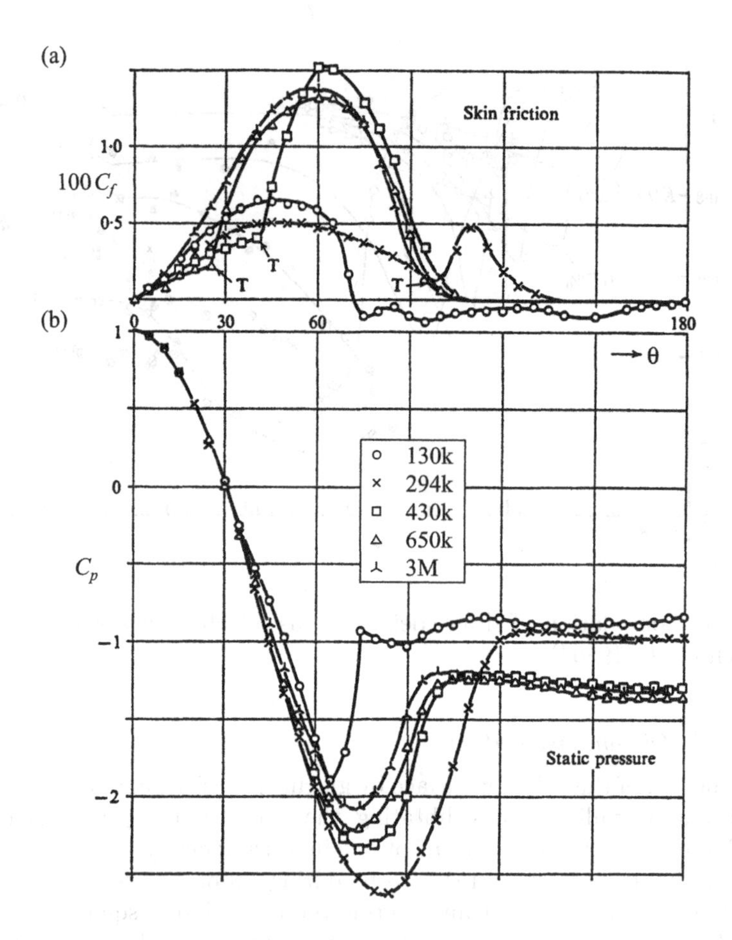

FIG. 22.6. Local (a) skin friction distribution for $K/D = 0.11\%$, (b) pressure, Achenbach (1971Jc)

For $K_s/D = 0.45\%$, $C_{D\min}$ is displaced to $Re = 90$k, and all tested Re exceed that value. All C_f curves are similar having $\theta_s = 100°$ in the TrBL4 regime. The value of $C_{f\max}$ at $\theta = 55°$ produced by the turbulent boundary layer along the rough surface increases threefold in comparison with the laminar boundary layer. Figure 22.7 shows the advancement of transition in the boundary layer, θ_{Tr}, towards the stagnation point for the three K_s/D tested. The turbulent separation is displaced upstream at $Re = 4$M, reaching $\theta_s = 100°, 93°$, and $91°$ for $K_s/D = 0.11\%, 0.45\%$, and 0.90%, respectively. The measured θ_s values might have been affected by the probe protruding into and disturbing the sensitive transitional boundary layer.

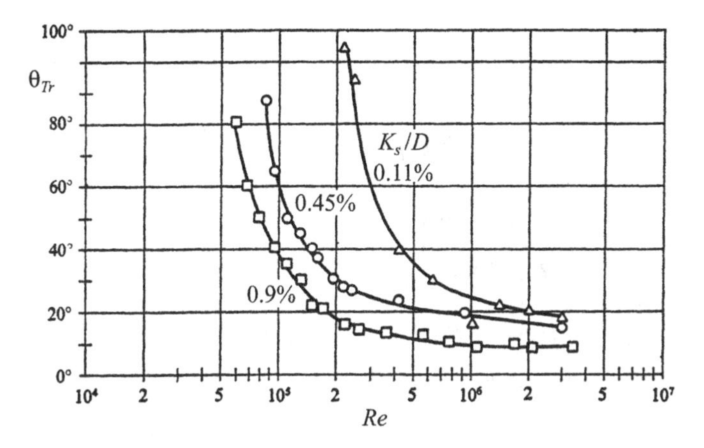

FIG. 22.7. Angular position of transition in the boundary layer in terms of *Re* for three K/D values, Achenbach (1971Jc)

22.1.4 *Strouhal number variation*

It is expected that surface roughness should have a disturbing effect on the mechanism of eddy shedding. However, it was puzzling why rough chimney stacks were excited to oscillate in high wind in the range 400k $< Re <$ 4M. This was a supercritical TrBL3 regime for smooth cylinders where eddy shedding ceased.

Buresti (1981J) carried out tests over the range $0.09\% < K_s/D < 1.23\%$ in an open test section using cylinders with $L/D = 22$ and 12. Local C_d was evaluated from the measured C_p distribution at the midspan. Figure 22.8(a,b) shows St and C_d versus Re, respectively, as affected by $0.3\% < K/D < 1.2\%$. The orderly displacement in the C_d curves towards lower Re is matched by the displacement in the St jump. The gradual rise in C_d beyond $C_{d\min}$ and the decrease in St reflect widening of the near-wake. The low St values are followed by the disappearance of the peak in the frequency spectra (dashed line), typical of the TrBL3 regime. However, for $K/D = 1.2\%$, the jump from low to high St is not separated by the cessation of eddy shedding.

The gradual shrinkage and disappearance of the TrBL3 regime is depicted in Fig. 22.9. The two converging curves designate the loss and reappearance of the frequency peak, respectively; that is the extent of the TrBL3 regime. The third line designates a plateau in C_d and St; that is the fully turbulent flow regime T1. The constant values of C_d and St in the T1 regime indicate that the separation has reached an extreme upstream position. Note that all three C_d, St, and θ_s values are invariant in the T1 regime, but the actual values are different for each K/D.

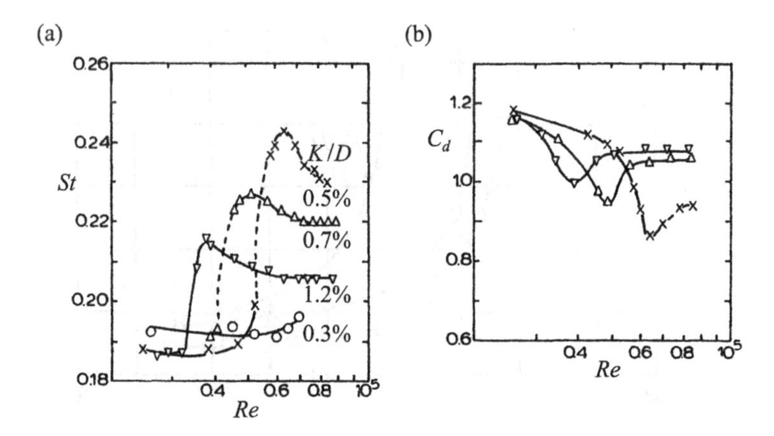

FIG. 22.8. Variation in (a) Strouhal number, (b) drag coefficient, in terms of *Re* for $L/D = 22$, Buresti (1981J)

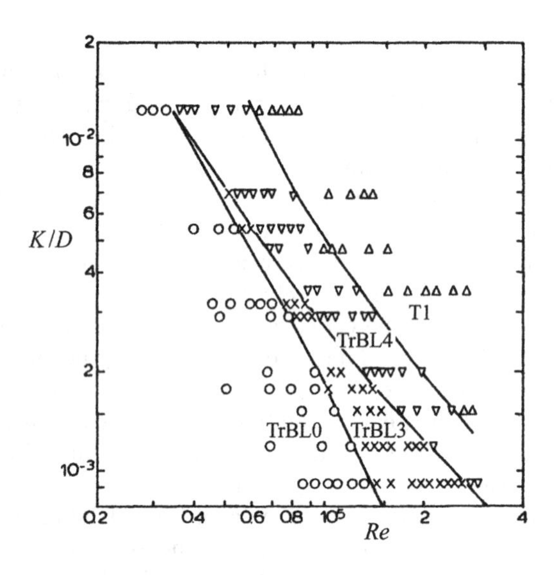

FIG. 22.9. Boundaries of flow regimes in the K/D, *Re* plane, Buresti (1981J)

22.1.5 *Correlation length and vorticity dispersion*

Batham (1973J) carried out comparative tests using smooth and rough ($K/D = 0.22\%$) cylinders, $L/D = 6.65$, $D/B = 0.05$, and $Ti = 0.5\%$ and 13%, at only two $Re = 111\text{k}$ and 235k. His measurements are summarized in Table 22.1.

Some expected and unexpected trends are evident in Table 22.1:

Table 22.1 *Summary of Batham's (1973J) experiment*

Surface	Re	Ti%	C_d	C_l'	C_d'	St	L_c/D
Smooth	111k	0.5	1.17	0.33	0.10	0.20	2.1
Smooth	234k	0.5	0.78	0.08	0.04	-	1.5
Smooth	111k	13	0.41	0.13	0.05	-	0.3
Smooth	235k	13	0.38	0.09	0.02	-	0.3
$K/D = 0.22\%$	111k	0.5	0.22	0.09	0.03	-	1.7
$K/D = 0.22\%$	235k	0.5	0.21	0.15	0.03	0.18	3.1
$K/D = 0.22\%$	111k	13	0.85	0.15	0.07	0.21	1.5
$K/D = 0.22\%$	228k	13	0.88	0.27	0.05	0.23	2.3

(i) The smooth cylinder at $Ti = 0.5\%$ is in the TrSL3 regime for $Re = 111$k (high values of C_d, C_L', and strong eddy shedding). For $Re = 235$k, it is in the TrBL0 regime (medium C_d, low C_L', and weak eddy shedding).

(ii) For the smooth cylinder at $Ti = 13\%$, the flow is in the TrBL3 regime for both Re (low C_d, C_L', and no eddy shedding).

(iii) The rough cylinder at $Ti = 0.5\%$ is in the TrBL3 regime for low Re (very low C_d, C_L', and no eddy shedding). For $Re = 235$k, the flow is in the TrBL3 regime (high C_L' and no eddy shedding).

(iv) The combination of roughness and high $Ti = 13\%$ leads to the TrBL4 regime for both Re (high C_d, C_L', and St).

The spanwise correlation length is high for the smooth cylinder at low Ti and very low for high Ti. The rough cylinder shows high L_c/D for both low and high Ti. The roughness turbulence triggers transition around separation more evenly and in doing so the separation line is straightened. This is illustrated by surface visualization by Szechenyi (1974P) in Fig. 22.10.

Pearcey *et al.* (1985P) proposed a new interpretation of the effect of adverse pressure recovery, as discussed for smooth cylinders in Vol. 1, Chapter 6.14, p. 198. They considered the balance between the vorticity generated in the boundary layer, and the manner in which it is subsequently diffused and shed in the near-wake.

The vorticity dispersion ratio VDR was defined as

$$VDR = \frac{C_{pb} - C_{p\min}}{1 - C_{p\min}} = \frac{V_{\max}^2 - V_s^2}{V_{\max}^2} \tag{22.1}$$

It was assumed that $C_{pb} = C_{ps}$, $V_{\max}$ and V_s were the velocities at $C_{p\min}$ and separation, respectively. Figure 22.11 shows VDR variation in terms of Re for three relative roughnesses. The rise in VDR is gradually displaced towards the lower Re, and the magnitude decreases accordingly. This was caused by the obliteration of separation bubbles, that is the TrBL1 and TrBL2 regimes. The

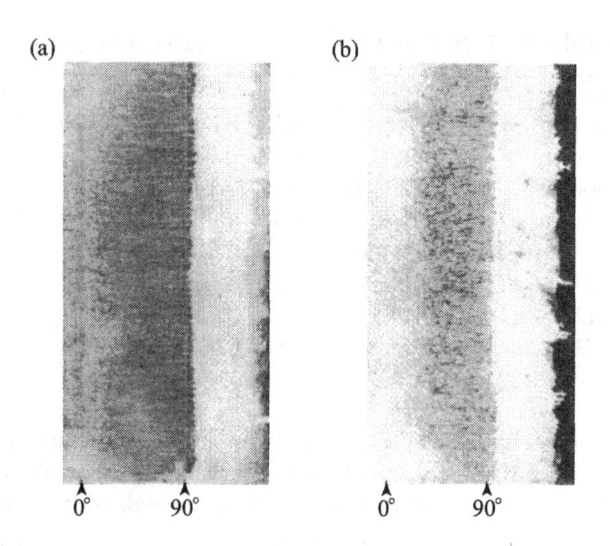

FIG. 22.10. Surface flow separation line, (a) $Re = 1\text{M}$, $K/D = 0.04\%$, (b) $Re = 1.33\text{M}$, $K/D = 0.1\%$, Szechenyi (1974P)

TrBL3 regime designated by the decrease in VDR also gradually disappears, as found by Buresti (1981J). The variation in $C_{pb} - C_{p\text{min}}$ in terms of Re for the relative roughnesses has already been given in Vol. 1, Fig. 6.37, p. 199.

22.2 Surface roughness textures

22.2.1 *Pyramidal roughness*

For a long time, researchers have sought to increase the heat transfer from circular cylinders without increasing drag. Surface roughness always increases heat transfer but it also increases drag. Achenbach (1977J) attempted to enhance heat transfer by using pyramidal roughness produced by knurling the surface, see Fig. 22.2.

Achenbach (1977J) suggested a simple way of evaluating an equivalent sand roughness for the pyramidal roughness. Fage and Warsap's (1929P) and Achenbach's (1977J) Re values associated with $C_{D\text{min}}$ are plotted in terms of the sand paper relative roughness K_s/D in Fig. 22.12 (open circles). The three Re values for $C_{D\text{min}}$ for pyramidal roughness are depicted with arrows, and the equivalent sand roughness is obtained by referring to the vertical axis. Note that both sets of data were obtained for high blockage and low aspect ratio. Hence, the empirical equation suggested by Achenbach and Heinecke (1981J)

$$Re_{(C_{D\text{min}})} = \frac{6000}{(K_s/D)^{1/2}} \tag{22.2}$$

requires correction.

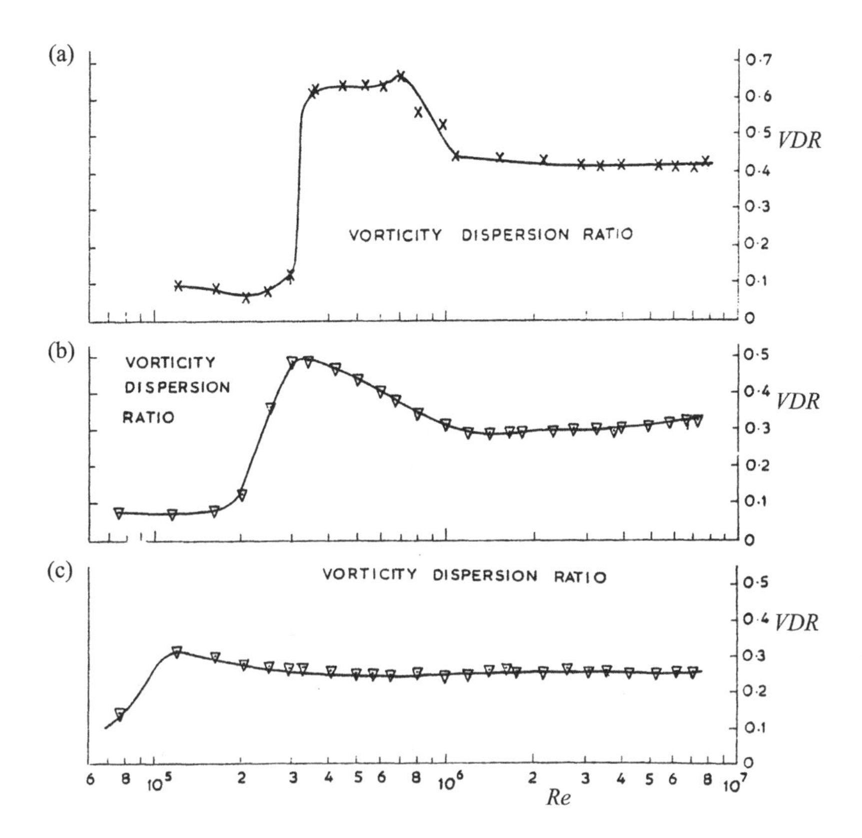

FIG. 22.11. Vorticity dispersion ratio VDR in terms of Re for (a) $K_s/D = 0.04\%$, (b) $K_s/D = 0.14\%$, (c) $K_s/D = 0.28\%$, Pearcey *et al.* (1985P)

Achenbach and Heinecke (1981J) measured St and evaluated C_d from C_p at the midspan[113] of two cylinders, $L/D = 3.4$ and 6.7 and $D/B = 0.16$ in the range $10\text{k} < Re < 4\text{M}$ for $K_s/D = 0.075\%$, 0.3%, 0.9%, and 3%. Figure 22.13(a,b) shows two types of C_d and St distributions, respectively:

(i) for smooth and small $K_s/D \leqslant 0.075\%$, the extent of $C_{d\min}$ indicates a TrBL2 regime characterized by the high value of St;

(ii) for $K_s/D \geqslant 0.3\%$, both the TrBL2 and TrBL3 regimes are obliterated. The TrBL0 is followed by the TrBL4 regime, and the St curve is uninterrupted.

[113]Note that $C_{D\min} = 0.5$ and $St = 0.51$ for the smooth cylinder were due to the high blockage and small aspect ratio. Pearcey *et al.* (1985P) carried out tests for $L/D = 7.8$ with end plates and found $C_{D\min} = 0.2$ for the smooth cylinder.

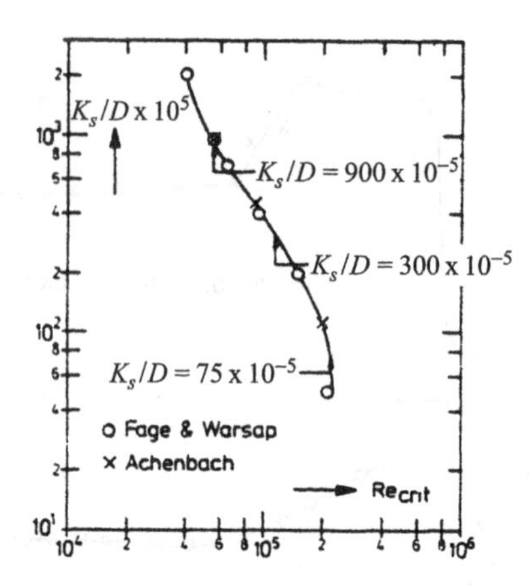

FIG. 22.12. Calibration of pyramidal roughness in terms of sand roughness, after Fage and Warsap (1929P), and Achenbach (1977J)

The effect of small pyramidal roughness on heat transfer is shown in Fig. 22.14. The local $Nu/Re^{1/2}$ distribution was measured for $K_s/D = 0.075\%$. The set of curves corresponds to: subcritical regime TrSL3 for Re = 92k, precritical TrBL0 for Re = 190k, two-bubble TrBL2 for Re = 410k, supercritical TrBL3 for Re = 590k, and postcritical TrBL4 for Re = 1.9M and 4M. Separation occurs near $(Nu/Re^{1/2})_{\min}$ at $\theta_s = 80°$ for Re = 92k and 190k. However, two minima at $\theta_s = 105°$ and 145° occur when the separation bubbles form for Re = 410k. Finally, there is a small displacement of θ_s from 120° to 105° for Re = 590k, 1.9M and 4M. The large $(Nu/Re^{1/2})_{\max}$ values are associated with the complete transition and fully turbulent boundary layers. For $K_s/D = 0.9\%$, the TrSL3 regime at Re = 48k is followed by the TrBL4 regime for Re > 190k.

Figure 22.15(a,b) shows the contribution of laminar, turbulent boundary layers, and separated flow to the overall heat transfer in terms of Re for K_s/D = 0.075% and 0.9%, respectively. The region associated with the turbulent boundary layer expands for high Re, and contributes 70% to the total heat transfer. The higher $k_s/D = 0.9\%$ shows the same trend at low Re as seen in Fig. 22.15(b). This is caused by the transition triggered by surface roughness.

22.2.2 *Brick-wall roughness*

The most common type of surface roughness in civil engineering is made by brick-laying. Standard brick-laying is half staggered. Ackeret (1936J) carried out wind tunnel tests using a brick chimney model, $L/D = 5$, fitted with two elliptical end plates, Fig. 22.16(a). Three sets of measurements were carried out:

(a) the joints between bricks were 4 mm wide and 2 mm deep, and the brick surface was smooth;

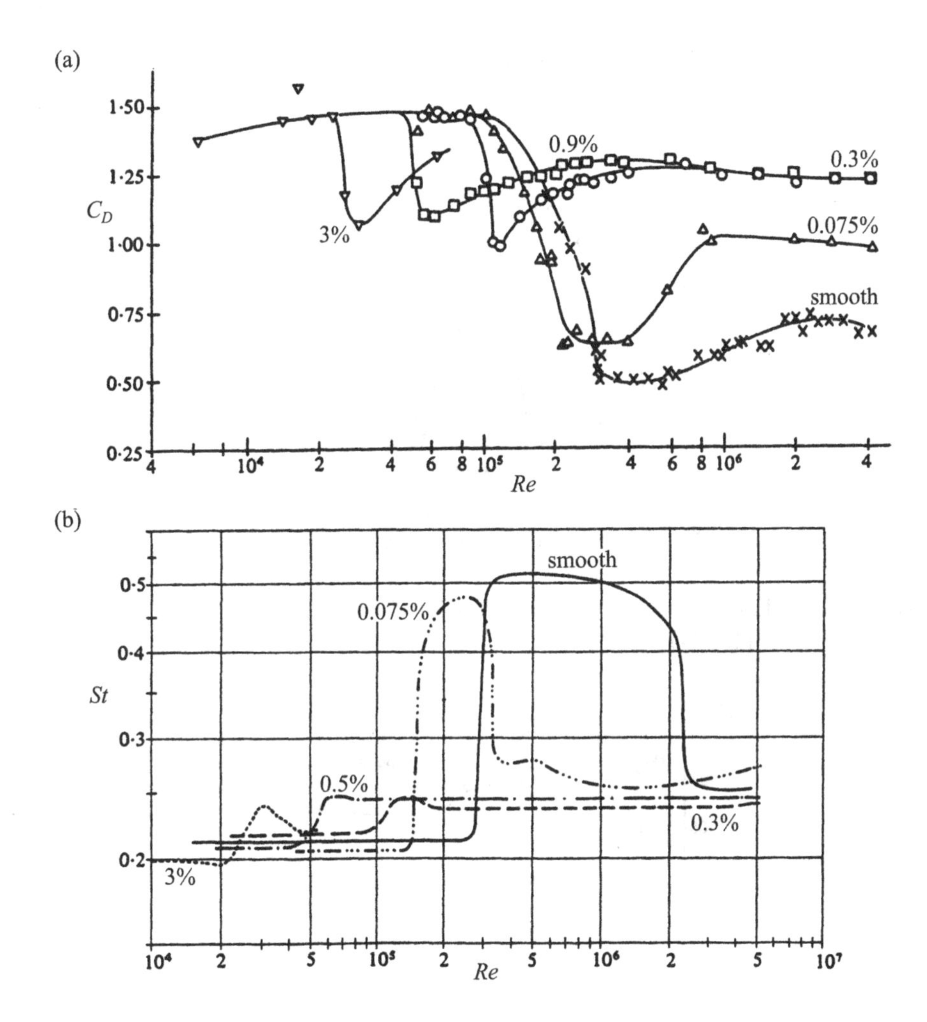

FIG. 22.13. Comparative measurements of (a) drag, (b) Strouhal number in terms of *Re*, Achenbach and Heinecke (1981J)

(b) the brick surface was roughened by 1–2 mm sand grains;

(c) all joints were filled and the model surface varnished.

Figure 22.16(b) shows that $C_{D\min}$ occurs at Re = 350k, 150k, and 100k for smooth, grooved joints, and rough bricks, respectively. The shape and sequence of the curves are similar to the other type of roughness.

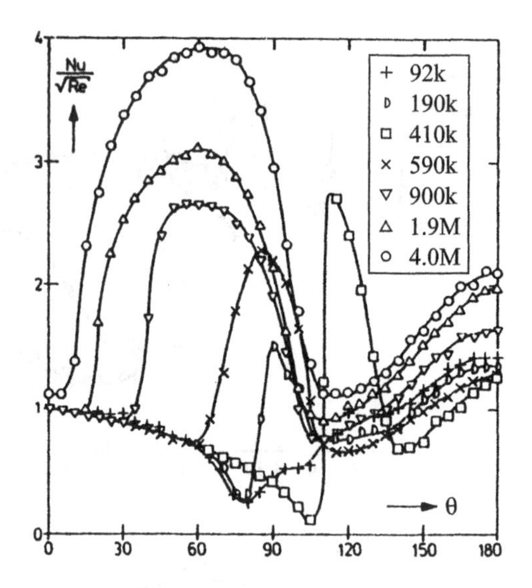

FIG. 22.14. Local heat transfer distribution for $K_s/D = 0.075\%$, $Pr = 0.72$, Achenbach (1977J)

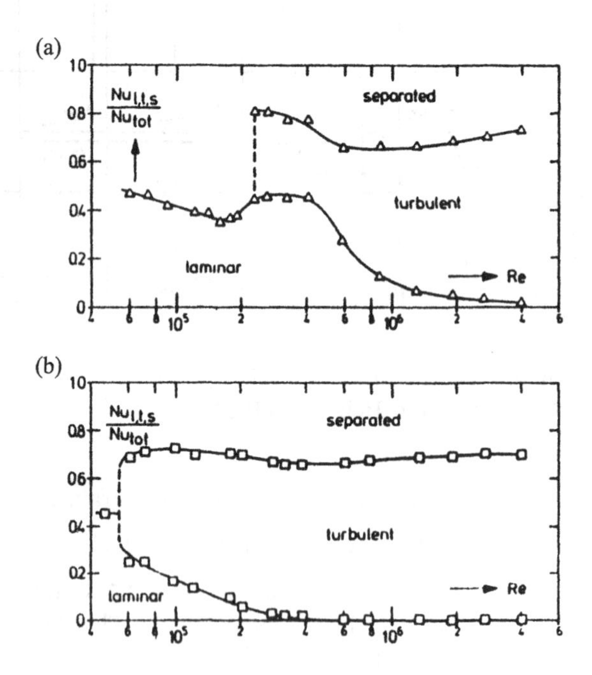

FIG. 22.15. Contribution of three flow states to overall heat transfer, (a) $K_s/D = 0.075\%$, (b) $K_s/D = 0.9\%$, Achenbach (1977J)

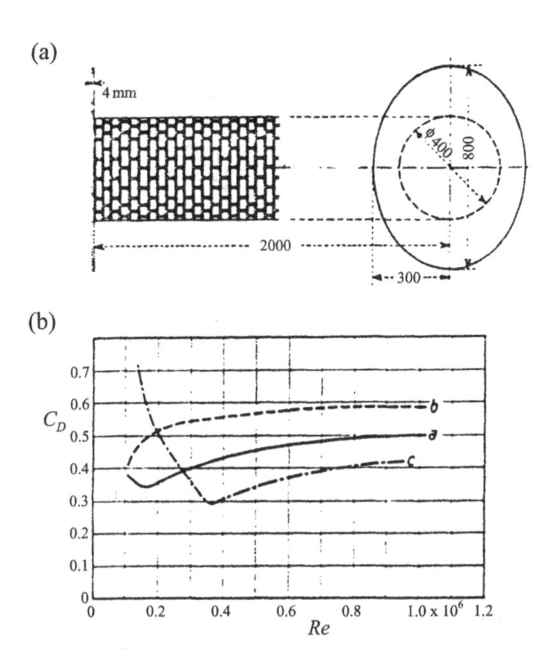

FIG. 22.16. Model of brick chimney stack, (a) brick pattern and end plates, (b) drag in terms of *Re*, Ackeret (1936J)

22.2.3 *Wire-gauze roughness*

Another kind of regular and uniform artificial roughness is wire-gauze cloth. A thin wire of constant diameter d is interwoven to form a square mesh. When a smooth cylinder of diameter D is wrapped in a wire-gauze, the relative roughness is described by d/D. An equivalent sand roughness may be found by using the same method as for the pyramidal roughness, see Fig. 22.12.

Farell and Fedeniuk (1988J), and Farell and Arroyave (1989P) used wire-gauze cloth, $d/D = 0.65\%$ and 1%, wrapped around a cylinder, $L/D = 6$ and $D/B = 0.117$. The effect of rectangular end plates $7D \times 5D$ was found to be significant. Figure 22.17 shows the variation of C_D with Re for two roughnesses tested. When the end plates are fitted there is a considerable rise in C_D across the entire Re range. A strong effect of end plates is also found on St values. Without the end plates, three shedding cells are found at $z/D = 0.5, 1$, and 1.5 with three St values. When the end plates are fitted a single St corresponds to a single cell. The decrease in St curves with increasing Re is similar to that found by Buresti (1981J), see Fig. 22.7.

22.2.4 *Marine roughness*

Offshore structures are partly or wholly immersed in sea, and most structural members are circular cylinders. After deployment in the sea, these structures

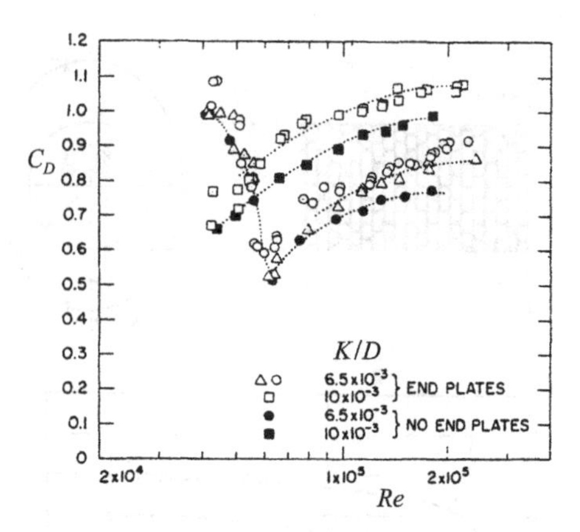

FIG. 22.17. Variation in drag with and without end plates, $K/D = 0.6\%$ and 1%, Farell and Fedeniuk (1988J)

become roughened by corrosion and marine fouling. Forces exerted by waves and currents may be substantially altered by the increase in diameter and added surface roughness.

It has already been mentioned in Section 22.1 that there are two kinds of marine fouling:

(i) the rigid roughness produced by rust, scale, barnacles, mussels, etc.;

(ii) the soft or flexible roughness made of seaweeds, kelps, anemone, etc.

The accumulated marine fouling sketched in Fig. 22.18 acts in two ways:

(i) an average roughness height K is defined as the representative height of all excrescences from an imaginary smooth surface;

(ii) the thickness of the accumulated growth ΔD increases the cylinder diameter from D to $D + 2K/D$, Fig. 22.18.

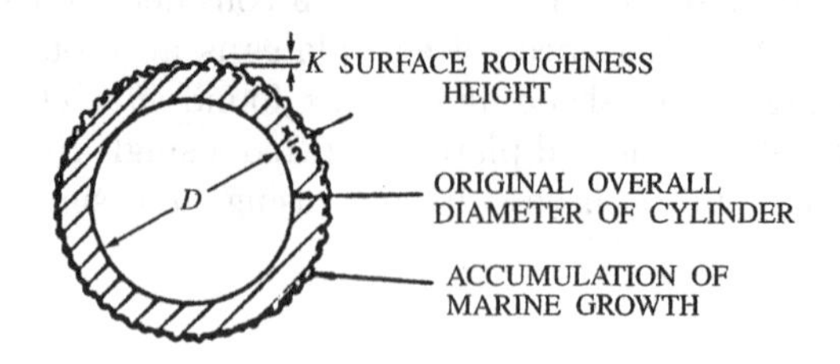

FIG. 22.18. Definition sketch of marine fouling, Miller (1976J)

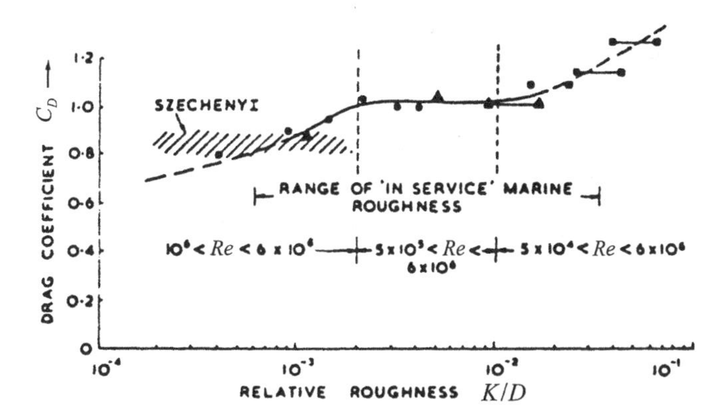

FIG. 22.19. Drag coefficient in terms of surface roughness K/D, Miller (1976J)

Miller (1976J) carried out tests with large artificial and natural marine roughness in the range $1.5\% < K/D < 6.3\%$ and $200\text{k} < Re < 4\text{M}$. The artificial marine roughness was made of pearl barley ($K = 2.4\,\text{mm}$) and dried peas ($K = 6.6\,\text{mm}$) glued onto the cylinder surface. The natural marine roughness was obtained by immersing three cylinders in the sea for almost a year. After drying, the soft marine fouling was removed leaving barnacles encrusted on the corroded cylinder surface.

The test showed that for high Re in the T1 regime, C_D was constant. However, the values of C_D for different K/D were not the same. The C_D value increased with K/D, and the highest value was for the roughest cylinder. This trend is seen in Fig. 22.19, where C_D is plotted[114] in terms of K/D. The rapid rise in C_D up to $K/D = 0.5$ is followed by a very small increase. This may indicate a significant change in the mechanism of roughness-generated turbulence. This feature has not attracted the attention of subsequent researchers.

22.2.5 *Partially roughened surface*

Fage and Warsap (1929P) wrote: 'It is of interest to inquire on what part of the surface areas a local roughness has the largest effect.' They partly covered a smooth cylinder with glass paper strips as depicted in Fig. 22.20. The least effective are the stagnation region $-37.5° < \theta < 37.5°$ and the rear region 120° to 240°. Fage and Warsap (1929P) concluded:

> It would appear that the flow is particularly sensitive to surface roughness in the regions $37.5° < \theta < 120°$, a result to be expected since they include those parts where the flow separates from the surface. It was not possible to obtain by direct measurements the effect of roughening by glass paper these regions

[114] The square root has been used to compress K/D range.

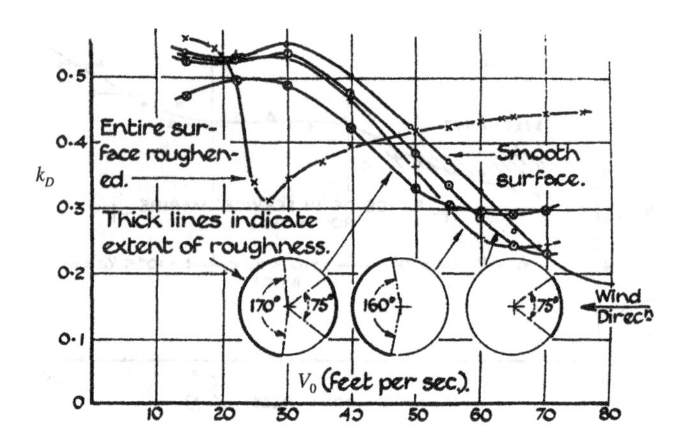

FIG. 22.20. Effect on k_D of limited rough surface, Fage and Warsap (1929P)

of separation because the front edge of the paper, even if carefully chamfered, would greatly influence the flow and so mask the effect of roughness.

Direct comparative measurements and confirmation of the above statement were provided by Okajima and Nakamura (1973P). They tested a short cylinder $L/D = 3.3$, $D/B = 0.105$, $Ti = 0.1\%$, and $Ts = 1D$, without end plates in the range 400k $< Re <$ 2M. Figure 22.21 shows the variation in C_D and St in terms of Re for a fully and partly roughened cylinder, $K/D = 0.25\%$. It is evident that experimental points for the fully and partly covered cylinders collapse onto single C_D and St curves. The roughness turbulence is ineffective under the stable

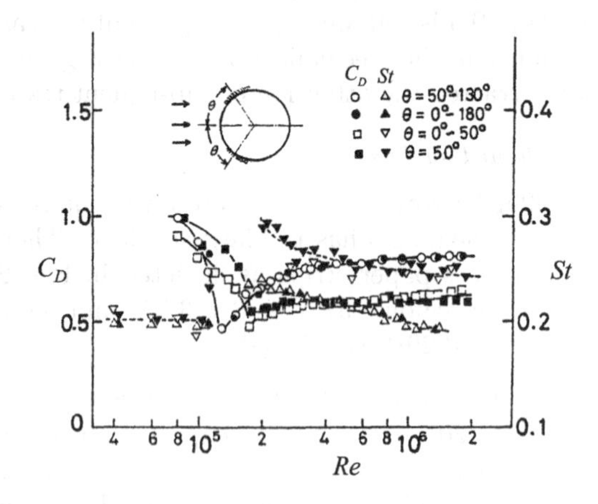

FIG. 22.21. Effect of partially rough surface on drag and Strouhal number in terms of Re, Okajima and Nakamura (1973P)

laminar boundary layer and in a turbulent near-wake. However, the roughness turbulence becomes more and more effective as the laminar boundary layer becomes transitional. It will be shown in Section 22.4 that a single tripping wire may be sufficient to trigger the early transition to turbulence in a boundary layer.

22.2.6 *Roughness Reynolds number*

Szechenyi (1974P, 1975J) carried out tests in two wind tunnels using four cylinders roughened with glued glass beads. The wide range of influencing parameters contributed to the scatter of measured data. The discussion that follows will be restricted to data obtained in one wind tunnel using one cylinder $L/D = 4.4$, $D/B = 0.23$,[115] $0.05\% < K_s/D < 0.17\%$. Szechenyi (1975J) argued that for sufficiently large K/D and high Re the mean C_D should be independent of D, but depends on K instead. He suggested replacing Re based on D by Re_K based on K and termed it the *roughness Reynolds number*. It might be expected that the new similarity number should also lead to a single value of Re_K for $C_{D\mathrm{min}}$. Table 22.2 shows that this was not the case, and that even for the same set of data, the Re_K value for $C_{D\mathrm{min}}$ varies widely. The replacement of D by K does not produce a new similarity parameter.

Szechenyi (1975J) measured St across a wide range $96\mathrm{k} < Re < 6.5\mathrm{M}$ and in a narrow range $0.015\% < K/D < 0.175\%$ of a relatively small roughness. Figure 22.22 shows a compilation of all his St data[116] in terms of Re_K (the dashed line designates the St jumps for all K/D tested). He suggested that $Re_K = 200$ is a criterion for the start of the TrBL4 regime. This criterion is valid only for his limited range of K/D values.

Szechenyi (1975J) also evaluated C'_L from the fluctuating pressure measured at 20 tappings distributed around the circumference at the cylinder midspan. The rms values of C'_L were obtained through a real time summation of C'_p at each tapping. The plot of C'_L versus Re showed wide scatter (different K/D values are shown using the same symbols). The jump in C'_L occurred in a wide range of $400\mathrm{k} < Re_K < 700\mathrm{k}$, and not at $Re_K = 200$ as found for the St jump.

Table 22.2 *Relationship between Re_K for $C_{D\mathrm{min}}$*

Author	Re	K_s/D (%)	Re_K
	51k	0.90	459
Achenbach (1971Jc)	90k	0.45	405
	200k	0.11	220
	38k	0.12	469
Buresti (1981J)	54k	0.23	397
	65k	0.53	346

[115] The effect of excessive blockage has not been corrected.

[116] The value $St = 0.23$ at low Re_K instead of 0.21 was due to high blockage, $D/B = 0.23$.

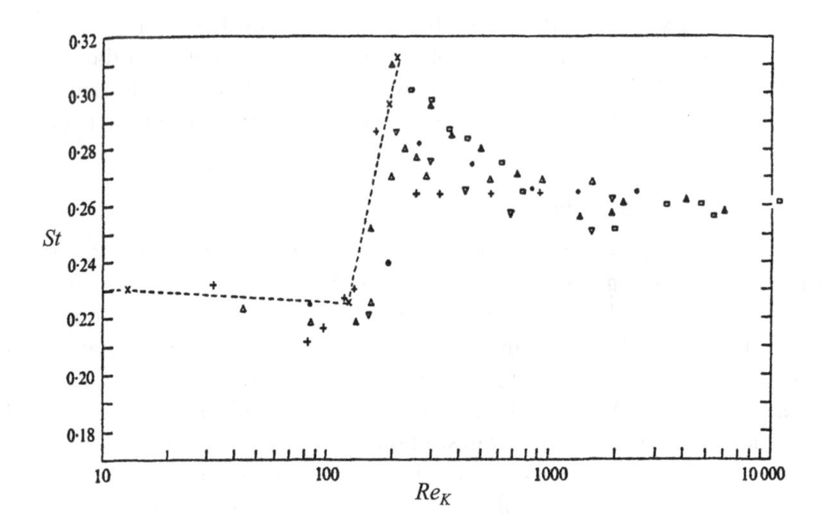

FIG. 22.22. Compilation of Strouhal number in terms of Re_K, Szechenyi (1975J)

Niemann and Hölscher (1990R) compiled the available data for $C_{D\min}$ and $C_{D\text{ult}}$ (the start of the T1 regime) over a wide range of $0.001\% < K/D < 1.2\%$. Figure 22.23 shows that neither $C_{D\min}$ nor $C_{D\text{ult}}$ reach final values.[117] The range of K/D covered by Szechenyi (1975J) is designated by Sz, and corresponds to the start of the $C_{D\min}$ rise. The relationship between $C_{D\min}$ and K/D is non-linear.

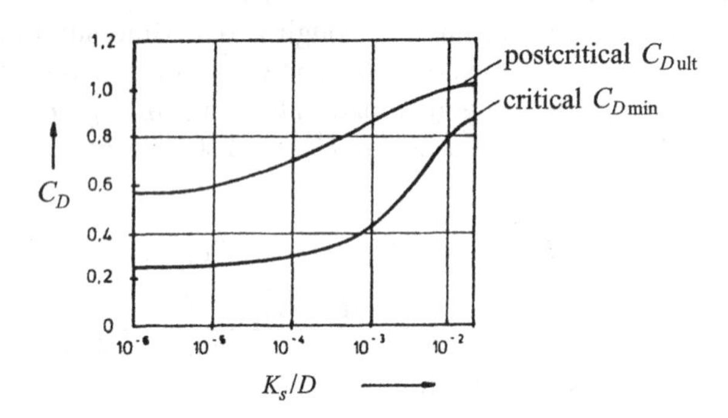

FIG. 22.23. Minimum drag and turbulent drag in terms of K_s/D, Niemann and Hölscher (1990R)

[117]Fage and Warsap (1929P) suggested that for high K/D the dip in the C_D curve should disappear, hence $C_{D\min} = C_{D\text{ult}}$.

22.3 Tripping wires

22.3.1 *Historical introduction*

Eiffel (1912J) measured the drag of spheres and other bodies by using a free-fall drop from the Eiffel tower in Paris. He calculated $C_D = 0.176$ for spheres while Prandtl measured $C_D = 0.44$ in a wind tunnel in Göttingen. Prandtl (1961B) attributed such a large discrepancy to turbulent and laminar separation, respectively. In order to prove his hypothesis, Prandtl (1961B) attached a tripping wire in the form of a ring onto a sphere. The tripping wire was supposed to trigger the laminar boundary layer to become turbulent before separation. The measured C_D on the sphere at low *Re* with the tripping wire became the same as found by Eiffel at high *Re* without the tripping wire.

Rough surfaces consist of innumerable excrescences of variable size, shape, and orientation distributed more or less uniformly over the cylinder. The cumulative effect due to the continuous generation and dissipation of roughness turbulence has been described in previous sections. A single pair of tripping wires generates some sort of 'two-dimensional' turbulence locally. Depending on d/D, the tripping wires may be fully submerged in or protruding from the boundary layer. This may trigger transition or cause separation, respectively. The optimal location θ_w depends on d/D and *Re*.

22.3.2 *Fage and Warsap's tripping wire tests*

Fage and Warsap (1929P) attached two straight tripping wires of diameter d at $\theta_w = \pm 65°$ onto a smooth cylinder.[118] The size of the tripping wire d was less than the local thickness of the cylinder boundary layer δ. A set of five tripping wires, $0.02\% < d/D < 0.3\%$, was tested in the range $90\text{k} < Re < 262\text{k}$. Figure 22.24 shows an orderly change in shape of k_D ($k_D = \frac{1}{2}C_D$) curves as d/D increases. It is notable that even the finest wire, $d/D = 0.02\%$ ($d/\delta = 3\%$), has a visible effect.

Large changes in k_D curves caused by the tripping wires are associated with a marked change in pressure distribution. Figure 22.25 shows the C_p distribution for smooth and rough cylinders (full and dashed lines, respectively). The three *Re* values for smooth cylinders corresponded to the TrSL3, TrBL0, and TrBL2 regimes, while all rough cylinders were in the TrBL4 regime collapsing onto the single C_p curve (dashed line). Note that $d/D = 0.54\%$ and 0.87% corresponded to $d/\delta = 0.98$ and 1.61. Fage and Warsap (1929P) concluded: 'It would seem that wires of $d \geqslant \delta$ when placed at $\theta_w = \pm 65°$ caused the boundary layer to leave the surface at the same angle.'

22.3.3 *Effect of tripping wire location*

James and Truong (1972J) attached a single tripping wire $0.6\% < d/D < 6.3\%$ in the range $15° < \theta_w < 90°$ along the cylinder span, $L/D = 14.3$ and 33, $D/B = 0.07$ and 0.03, respectively. Figure 22.26 shows the C_D variation in terms

[118] There was no mention why $\theta_w = \pm 65°$ was chosen; actually, it was an optimal location.

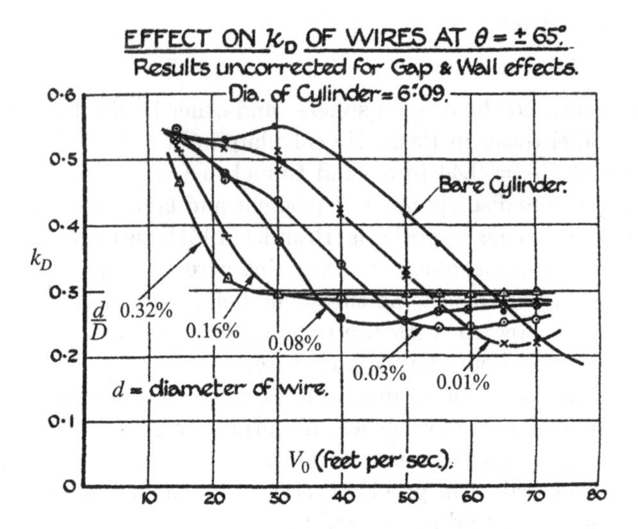

FIG. 22.24. Effect on k_D of tripping wires at $\theta_w = \pm 65°$ in terms of V, Fage and Warsap (1929P)

of *Re* and different θ_w for a fine $d/D = 0.6\%$ tripping wire. The effect of the tripping wire on C_D variation is negligible for $\theta_w = 15°$ and $30°$. The increase in effectiveness is seen for $\theta_w = 45°$, $55°$, and $60°$ where the C_D curves are displaced to lower *Re*. The maximum effect is at $\theta_w = 65°$, starts to decrease at $\theta_w = 70°$, and becomes ineffective at $\theta_w = 80°$ and $90°$ in the separated region. It is noteworthy that both Fage and Warsap (1929P) and Prandtl (1961B) adopted the optimal $\theta_w = \pm 65°$.

There was a radical departure from that trend when a stouter tripping wire $d/D = 3.2\%$ was attached. Figure 22.27 shows a small change in the C_D curves for $\theta_w = 15°, 120°$, and $180°$ in comparison with the smooth cylinder. However, for all other θ_w the C_D curves became dissimilar to those for $d/D = 0.6\%$. The most effective location appears to be $\theta_w = 45°$, while a high C_D is reached for $\theta_w = 65°$, which is the same as for $\theta_w = 90°$.

Fujita *et al.* (1985J) attached two tripping wires onto a smooth cylinder $L/D = 8$ and $D/B = 0.1$. The tripping wires were in the range $0.2\% < d/D < 4\%$ attached at $15° < \theta_w < 90°$, and tested at a single $Re = 50$k. Figure 22.28 shows the C_p distribution for $\theta_w = 30°$, $45°$, $60°$, and $90°$. The location of the tripping wires is designated by the dash-dot lines, and the full line curves are for the smooth cylinders at the same *Re*.

Figure 22.28(a), $\theta_w = 30°$, shows that the effect of the tripping wires is local, and restricted to the separation region behind the tripping wire itself. The length of the separation region (flat C_p) elongates as the wire size d/D increases. Figure 22.28(b), $\theta_w = 45°$, shows an increase in pressure recovery and C_{pb} for

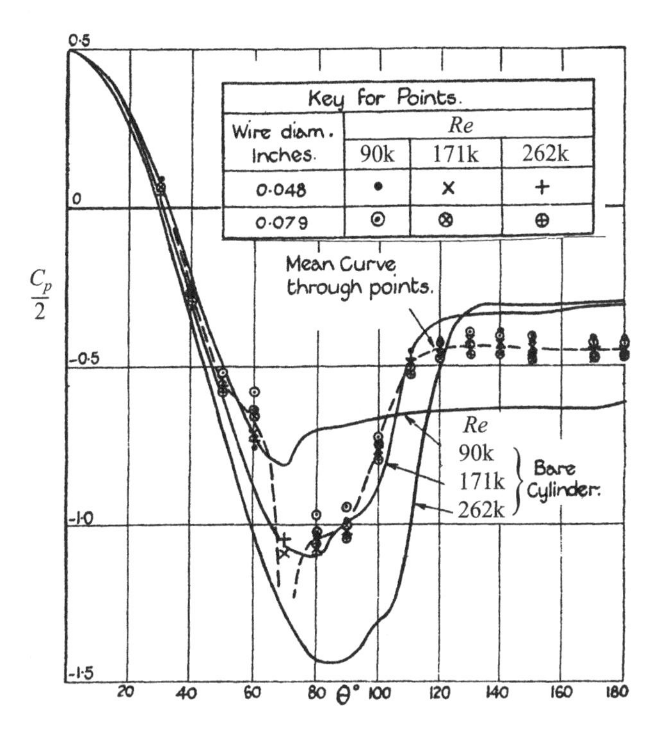

FIG. 22.25. Mean pressure distribution for smooth and roughened cylinder, Fage and Warsap (1929P)

stouter wires. There is a marked increase in the size of the separated region and a decrease in C_p behind the tripping wires. Figure 22.28(c), $\theta_w = 60°$, shows slightly disturbed C_p distribution for $d/D = 0.2\%$ in comparison with C_p for a smooth cylinder. For $0.4\% < d/D < 0.8\%$, there is a large pressure recovery at the end of the separated region which indicates a turbulent reattachment. Finally, there is a new C_p distribution for $2\% < d/D < 4\%$ marked by a low C_{pb}. Figure 22.28(d), $\theta_w = 75°$, shows that the new C_p distribution is present for all d/D but the finest $d/D = 0.2\%$. The new C_{pb} is caused by the full and permanent flow separation from the tripping wire. The latter acts as the separation wire, as will be discussed shortly.

Fujita *et al.* (1985J) estimated the angular location of separation θ_s from the C_p distribution. Figure 22.29 shows the variation in θ_s in terms of θ_w and d/D. For small θ_w values, θ_s is unaffected. For high θ_w, θ_s coincides with θ_w (full line). Figure 22.29(b) shows the measured St in the same range of θ_w and d/D. The high St corresponds to a narrow wake and large θ_s, while the low St is related to a wide wake and small θ_s. There is a remarkable similarity between the θ_s and St curves.

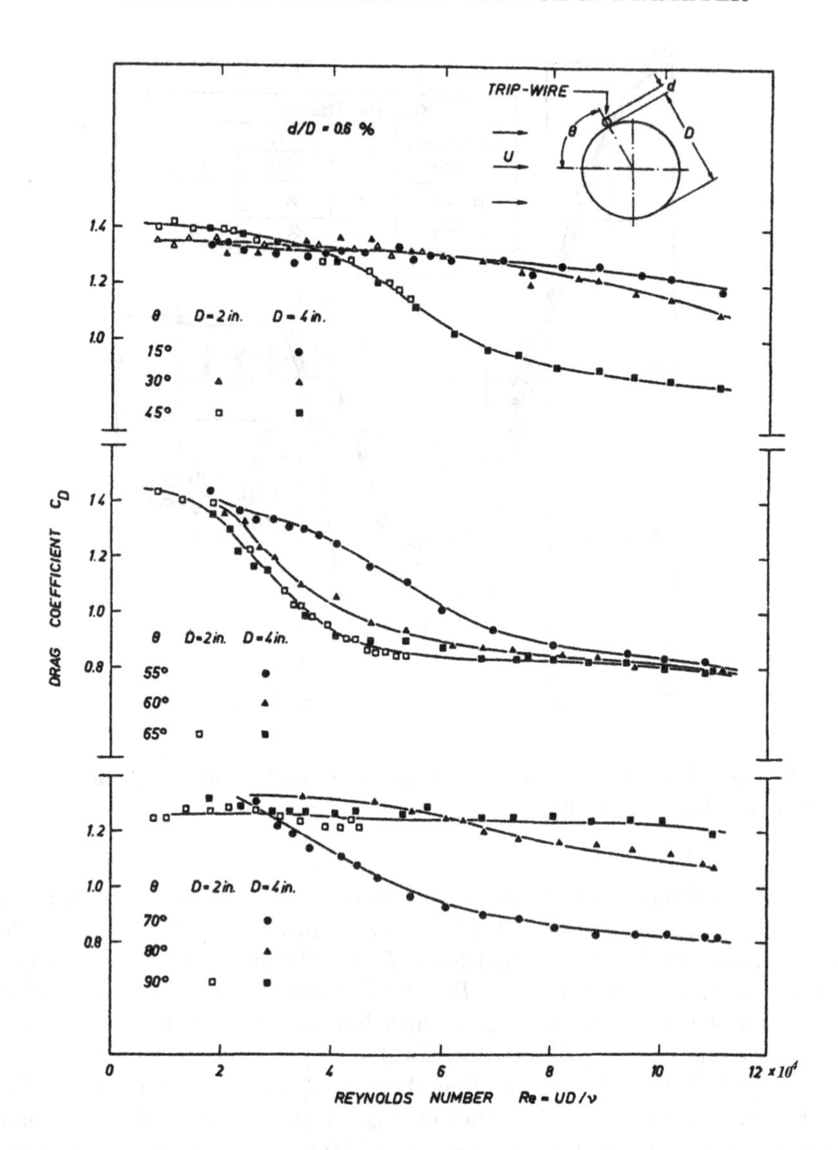

FIG. 22.26. Drag variation in terms of *Re* and angular location of tripping wire for $d/D = 0.6\%$, James and Truong (1972J)

22.3.4 *Classification of flow regimes*

It has been shown that flow around a circular cylinder fitted with tripping wires strongly depends on *Re*, d/D, and θ_w. The boundary layer upstream of the tripping wire first decelerates, then separates, and forms an upstream separated region. Another separation takes place from the tripping wire itself. Both the

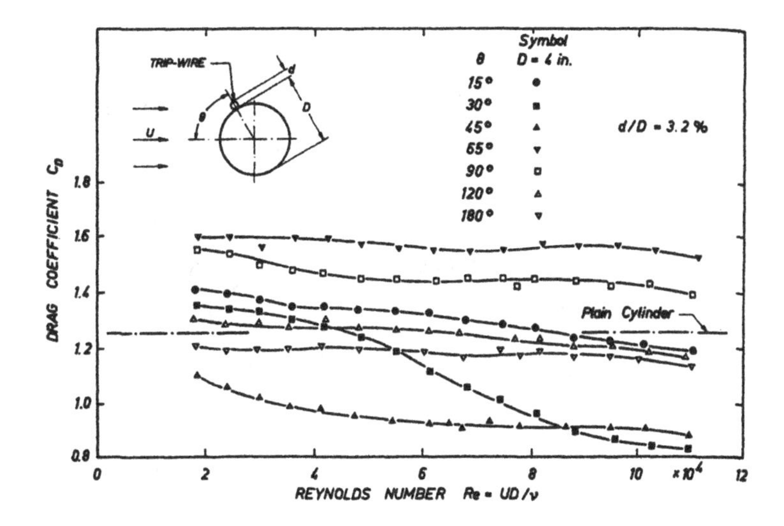

FIG. 22.27. Drag variation in terms of Re and angular location of tripping wire for $d/D = 3.2\%$, James and Truong (1972J)

upstream and downstream separation regions terminate at the reattachment onto the tripping wire and cylinder surface, respectively.

Figure 22.30(a,b) shows sketches of flow topology deduced from surface visualization by Igarashi (1986Ja). The downstream separation region ends with reattachments of the following types: laminar A, transitional B, turbulent C, or none D (when the tripping wire acts as the separation wire). Igarashi (1986Ja) also measured C_D and St in the range 30k $< Re <$ 65k for $\theta_w = 60°$ and several d/D. Figure 22.31(a,b) shows C_D and St variation in terms of Re for the C and D flow regimes.

Another attempt to delineate flow regimes induced by tripping wires is shown in Fig. 22.32(a,b). The plot of d/D versus Re shows the extent of the flow regimes for $\theta_w = 50°$ and $60°$, respectively. Flow regime D appeared as a dominant one for $\theta_w = 60°$. The dissimilarity of the two plots demonstrates the importance of the tripping wire angular location θ_w.[119]

22.3.5 *Staggered separation wires*

Straight tripping wires exert a strong two-dimensional effect on the flow around a cylinder. When tripping wires of sufficient size are attached near the separation line they induce straight separation. Hence, they become the separation wires.

Naumann *et al.* (1966P) demonstrated that straight wires attached at $\theta_w = \pm 115°$ triggered the post-critical TrBL4 regime with regular eddy shedding. They

[119]The effect of tripping wire on the local heat transfer has been discussed in Vol. 1, Chapter 17, p. 560.

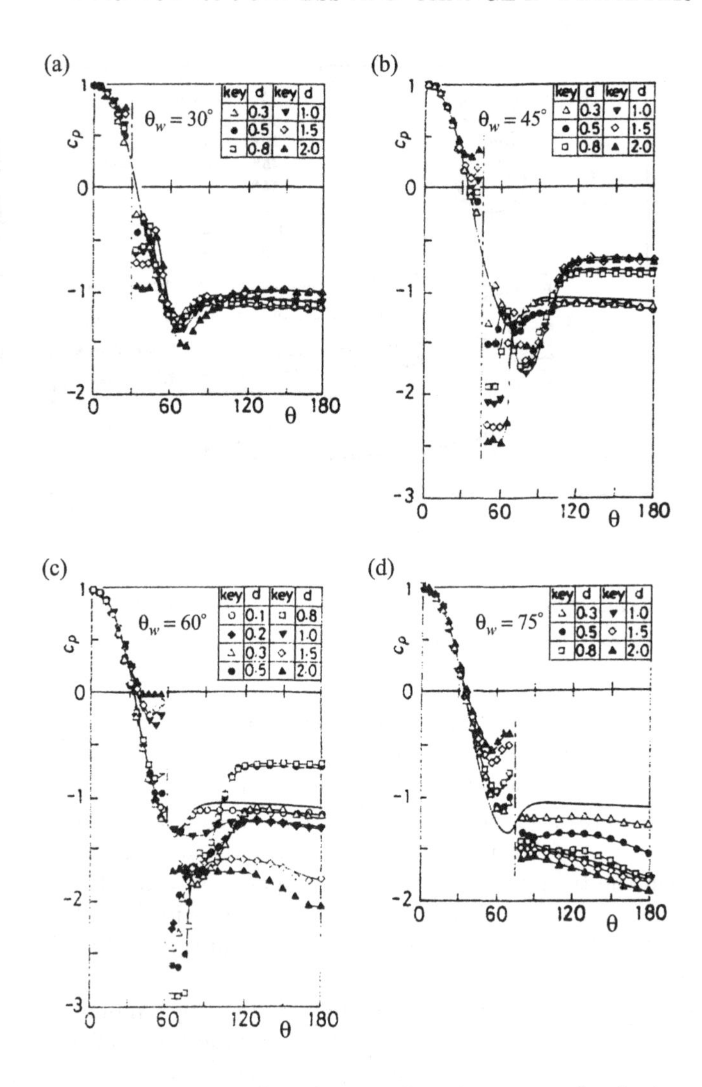

FIG. 22.28. Mean pressure distribution for six sizes of tripping wires at Re = 50k, (a) $\theta_w = 30°$, (b) $\theta_w = 45°$, (c) $\theta_w = 60°$, (d) $\theta_w = 75°$, Fujita *et al.* (1985J)

correctly concluded that the prerequisite for periodic eddy shedding was the straight separation line.

Naumann *et al.* (1966P) also examined the opposite effect of staggered separation wires on suppressing eddy shedding. The wire was cut and attached at two angular locations θ_{w1} and θ_{w2} in a staggered arrangement as shown in Fig. 22.33(a). The staggered separation wires induced a three-dimensional

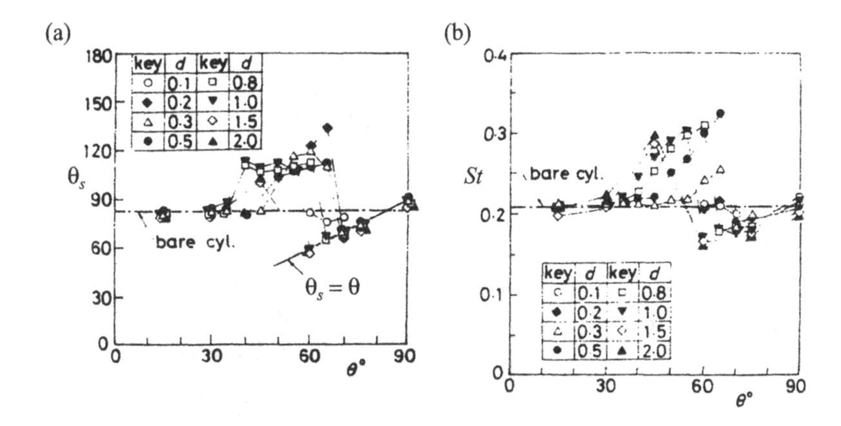

FIG. 22.29. (a) Angular separation θ_s, (b) Strouhal number at Re = 50k, for various tripping wire size and location, Fujita *et al.* (1985J)

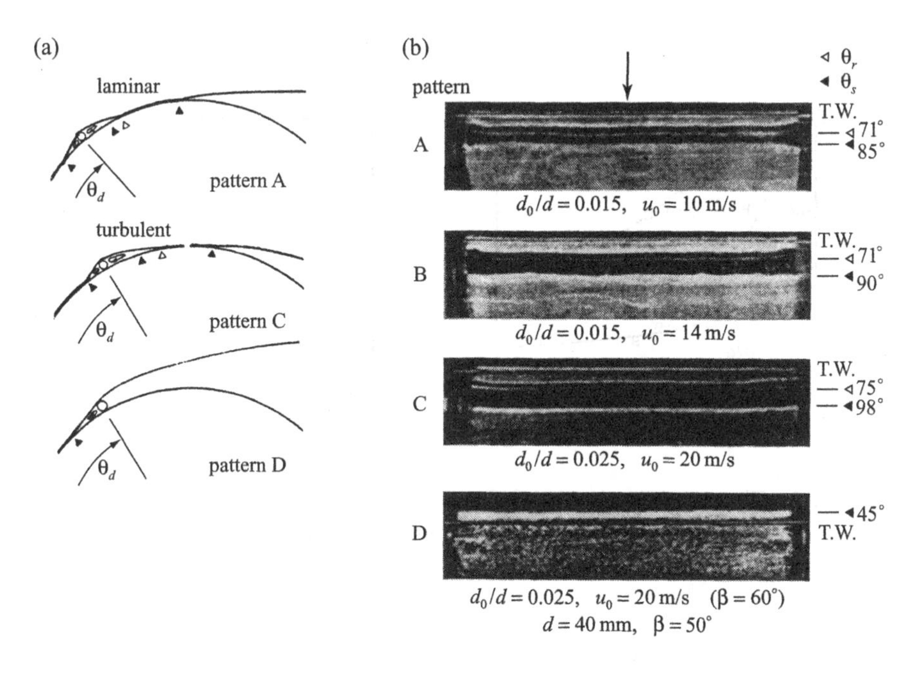

FIG. 22.30. (a) Flow topology sketches, (b) surface flow patterns, A laminar reattachment, B transitional, C turbulent reattachment, D no reattachment, Igarashi (1986Ja)

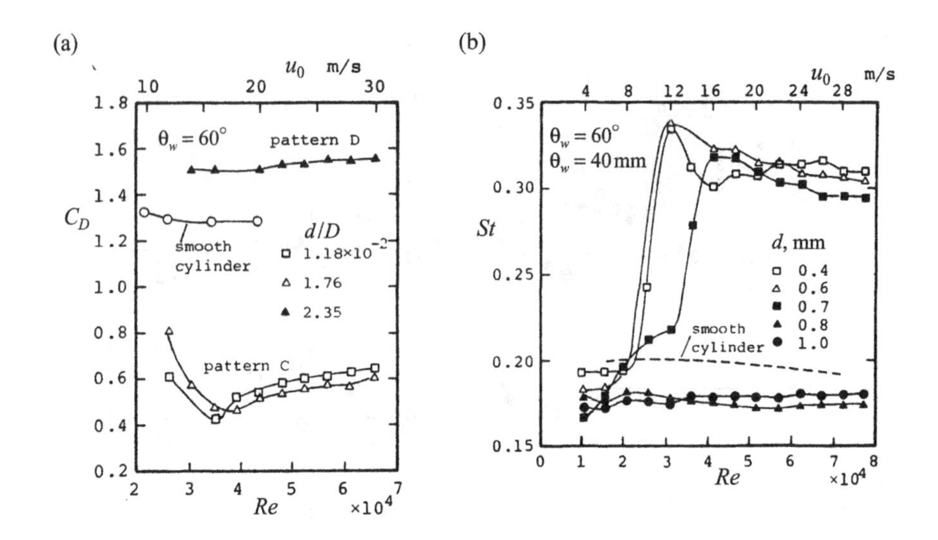

FIG. 22.31. (a) Drag, (b) Strouhal number, in terms of *Re* for $\theta_w = 60°$, Igarashi (1986Ja)

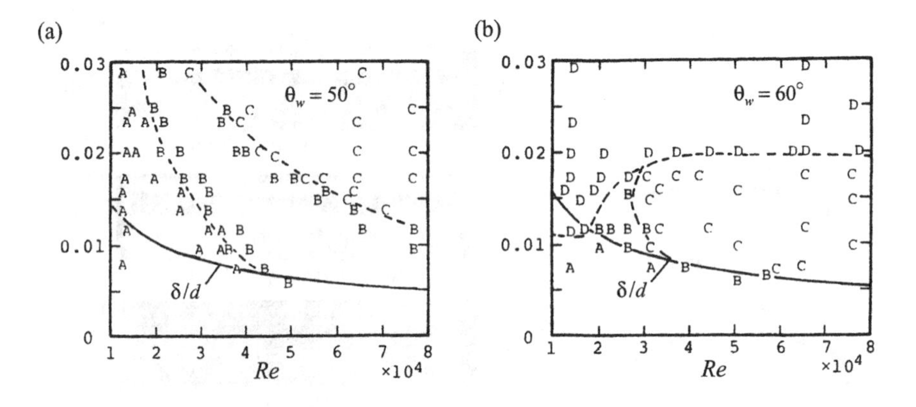

FIG. 22.32. Flow regimes in the d/D, *Re* plane, (a) $\theta_w = 50°$, (b) $\theta_w = 60°$, reattachment: A laminar, B transitional, C turbulent, D no reattachment, Igarashi (1986Ja)

disturbance. For a critical stagger $\Delta\theta_c = \theta_{w1} - \theta_{w2}$, eddy shedding was suppressed. For $d/D = 1.3\%$ the location of the wires was $105° < \theta_{w1} < 120°$ and $3° < \Delta\theta_c < 19°$. Figure 22.33(b) shows the $\Delta\theta_c$ curves in terms of the $b/2a$ ratio (a and b are defined in Fig. 22.33(a)). For the short span $b/2a = 1$, Fig. 22.33(c), $\Delta\theta_c > 3°$ is sufficient to suppress eddy shedding. However, when $\Delta\theta_w < \Delta\theta_c$,

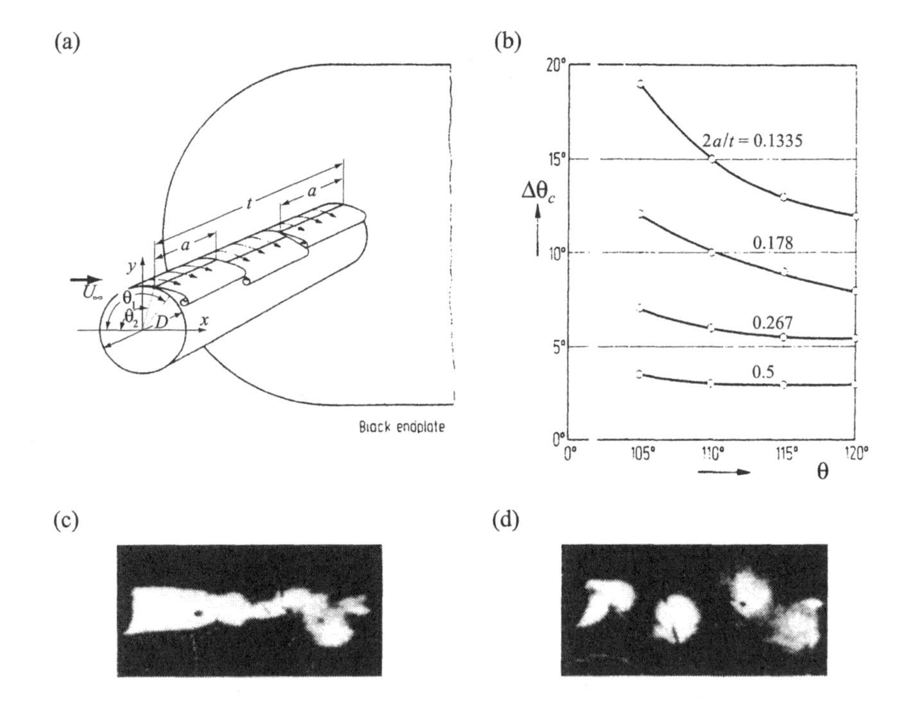

FIG. 22.33. Staggered separation wires, (a) definition sketch, (b) critical angular separation, $\Delta\theta_c$, in terms of θ_{w1}, (c) delayed eddy shedding, (d) immediate eddy shedding, Naumann *et al.* (1966P)

eddy shedding is unaffected, as seen in Fig. 22.33(d).

22.3.6 *Tripping and separation wires*

Most applications of circular cylinders in wind and offshore engineering have been associated with very high Re. The limited size of wind and water tunnels prompted various attempts to simulate high Re flow by using surface roughness. A unique combination of tripping wires to trigger transition and separation wires to fix the separation location was developed by Naumann and Quadflieg (1974P). They attached tripping wires, $d/D = 0.36\%$, at $\theta_w = \pm 70°$ and separation wires, $d/D = 1.5\%$, at $\theta_{ws} = 112°$. Figure 22.34 shows the C_p distribution at $Re = 296$k and 420k together with smooth cylinders at $Re = 3.6$M and 8.4M, Roshko (1961J).

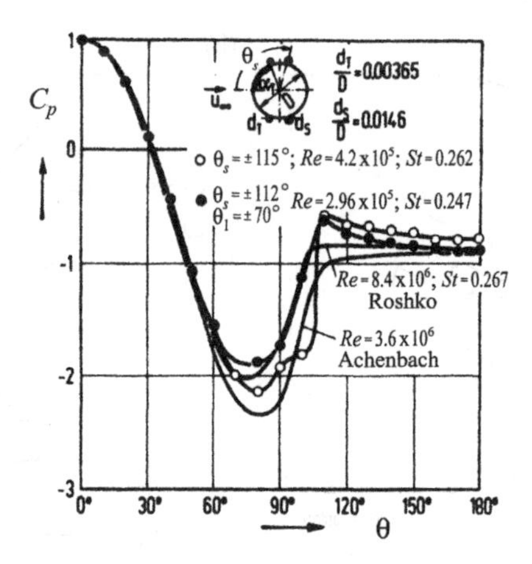

FIG. 22.34. Mean pressure distribution for tripping and separation wires as well as smooth cylinders, Naumann and Quadflieg (1974P)

22.3.7 *Helical wires and strakes*

One of the most successful and widely used means of suppressing eddy shedding, and related flow-induced oscillations of an elastic cylinder, are helical strakes.[120] The strakes are wound helically around the cylinder in order to disrupt straight separation, and in doing so to postpone and weaken eddy formation. The separation line becomes three-dimensional, and alternates from the strakes to the cylinder surface. A detailed description and discussion of helical strakes is given by Zdravkovich (1981R).

Nakagawa *et al.* (1963P) examined the effect of a fine helical wire, $d/D = 0.4\%$, attached to a smooth cylinder, $L/D = 5.7$, in an open test section, $Ti = 1.5\%$. The one-start helix had a $0.5D$ pitch, the four-start had $8D$ and $16D$ pitches, and the eight-start had $32D$ pitch. The first and last helices were ineffective in suppressing eddy shedding oscillation in the TrBL state. The other helices reduce C_L' from 0.16 to 0.03 for a smooth cylinder. Figure 22.35 shows the variation in C_D in terms of Re measured on a $4D$-long segment of the stationary cylinder. The effective four-start and eight-start helices produced a higher C_D despite the fact that eddy shedding was suppressed.

Weaver (1961J) attempted to establish an optimum number of helices N, a relative size of wires d/D, and pitch p/D. The tests were carried out on a stationary cylinder $15 < L/D < 40$, $0.02 < D/B < 0.05$, and $Ti = 0.5\%$. Figure 22.36(a) shows the ratio of C_L with and without helical wire in terms of the number of helix starts. The four-start helix produced $(C_{Lw}/C_{Lo})_{\min}$. Figure 22.36(b) shows that the optimum d/D for the $(C_{Lw}/C_{Lo})_{\min}$ occurred for $d/D =$

[120]The term 'strake' is used for a line plank along the boat hull attached to protect it from damage in harbour.

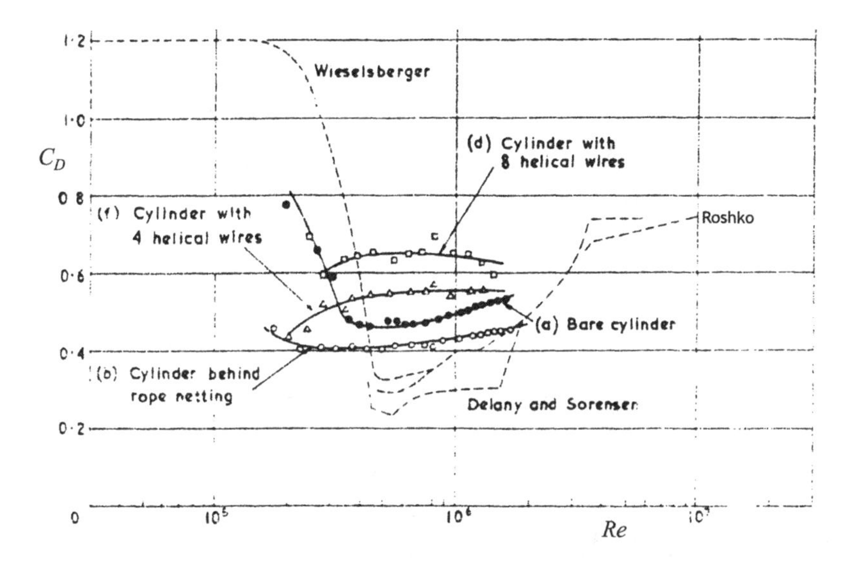

FIG. 22.35. Drag for cylinder with helical wires, $d/D = 0.4\%$ (full line) and smooth (dashed line) in terms of Re, Nakagawa *et al.* (1963P)

9.4%. The effect of the pitch variation was small $(C_{Lw}/C_{Lo})_{\min}$, and appeared in the range $12 < p/D < 14$. This was in agreement with Nakagawa *et al.* (1963P), $8 < p/D < 16$.

22.3.8 *Stranded cables and conductors*

Stranded cables have been widely used in oceanology, underwater acoustics, mooring, tethered offshore platforms, and conductors in overhead transmission lines. The surface undulation on a marine cable is produced by strands, as shown in Fig. 22.37(a). The relative roughness is $d/2D$ and the twist produces a three-dimensional surface. The stranded conductors shown in Fig. 22.37(b) appear similar to cables but have a larger number of strands N (up to 42).

All stranded cables and conductors possess a unique feature, namely the strands are orientated differently on opposite sides, $\pm\gamma$. When the cable or conductor is yawed to a current or wind at an angle Λ the relative helix pitch angle γ becomes $\Lambda \pm \gamma$ on the top and bottom side, respectively. The extreme is reached when $\Lambda = \gamma$; the wind flows along the strands on one side and across the strands on the other. The different surface texture on the two sides produces an asymmetric flow around the conductor, resulting in the mean lift force.

Votaw and Griffin (1971J) carried out comparative measurements on smooth and three-, four-, and six-start stranded cables having pitch angles $27° < \gamma < 64°$ for $Re = 200$ and 500. They did not find any difference in St values. The boundary layer thickness was estimated at $\delta/D = 0.33$ and 0.22, while $K/D = 0.16$.

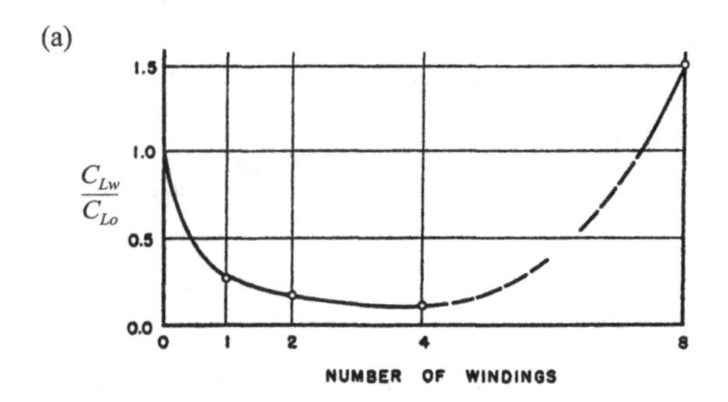

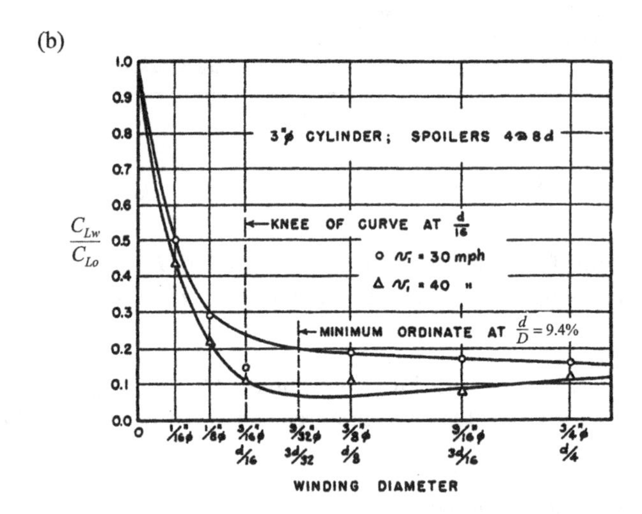

FIG. 22.36. Optimization of helical wires, (a) lift ratio in terms of number of windings, (b) in terms of wire diameter, Weaver (1961J)

Thus all strands were buried beneath a thick laminar boundary layer. The surface appeared hydrodynamically smooth according to Prandtl's definition cited in Section 22.1.

At high Re, the boundary layer thickness decreases, and strands start to protrude. Early measurements by Relf and Powell (1917P) at the National Physical Laboratory (England), AVA (Germany)[121] (1925P) at $Re = 50$k, and CAGI in Moscow (USSR)[122] (1928P) at $Re = 10$k, produced $C_D = 1$ for stranded cables and $C_D = 1.19$ for smooth cylinders. The decrease in C_D in the TrSL3 regime might be attributed to the elongation of the eddy formation region induced by the three-dimensional twist of strands.

[121] AVA, Aerodynamic Research Establishment, Göttingen

[122] CAGI, Central Aero- and Hydrodynamic Institute

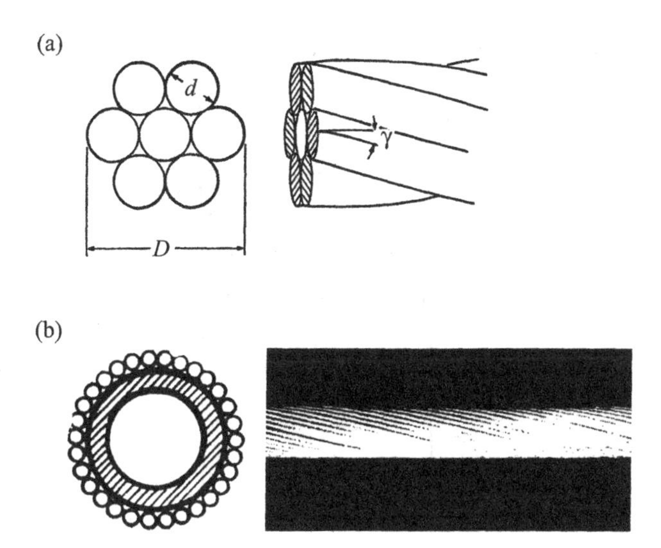

FIG. 22.37. Two views of (a) marine cable, Votaw and Griffin (1971J), (b) transmission conductor, Cooper (1974P)

Extensive tests on stranded conductors in the TrBL state were carried out by Counihan (1963P). He tested three stranded conductors, $N = 18$, 24, and 42, $L/D = 9.7$, 29 and 49, $D/B = 0.02$, and fitted with end plates. The flow was unyawed and yawed past a single conductor or groups of two and four conductors. The yawed and groups of conductors will be discussed in Chapters 24, 26, and 27. Figure 22.38 shows the measured C_D for three stranded conductors in terms of $\log_{10} Re$. Fage and Warsap's (1929P) curves for roughened cylinders were added for comparison. The stranded conductors appeared more effective in displacing $C_{D\min}$ towards lower Re than rough cylinders. Unexpectedly, there was no orderly displacement of C_D curves in terms of N. Subsequent tests by Wardlaw and Cooper (1973P) for the 'chucar' conductor, $N = 30$, are in good agreement with Counihan's (1963P) curves.

22.4 Tripping spheres

22.4.1 *Pairs of spheres*

Tripping wires are essentially two-dimensional fences that disturb the flow along the entire cylinder span. Tripping spheres are local three-dimensional disturbances attached to the smooth cylinder surface. They have attracted considerably less research than tripping wires.

Mizuno (1970P) investigated the effect of tripping spheres, $0.36\% < d/D < 0.9\%$, attached to the cylinder surface at $\pm 30° < \theta_d < \pm 80°$ in the range $60\text{k} < Re < 290\text{k}$. The cylinder, $L/D = 3$, $D/B = 0.23$, and $Ti = 0.1\%$, was without end plates. The first set of tests was restricted to two pairs of spheres fitted on

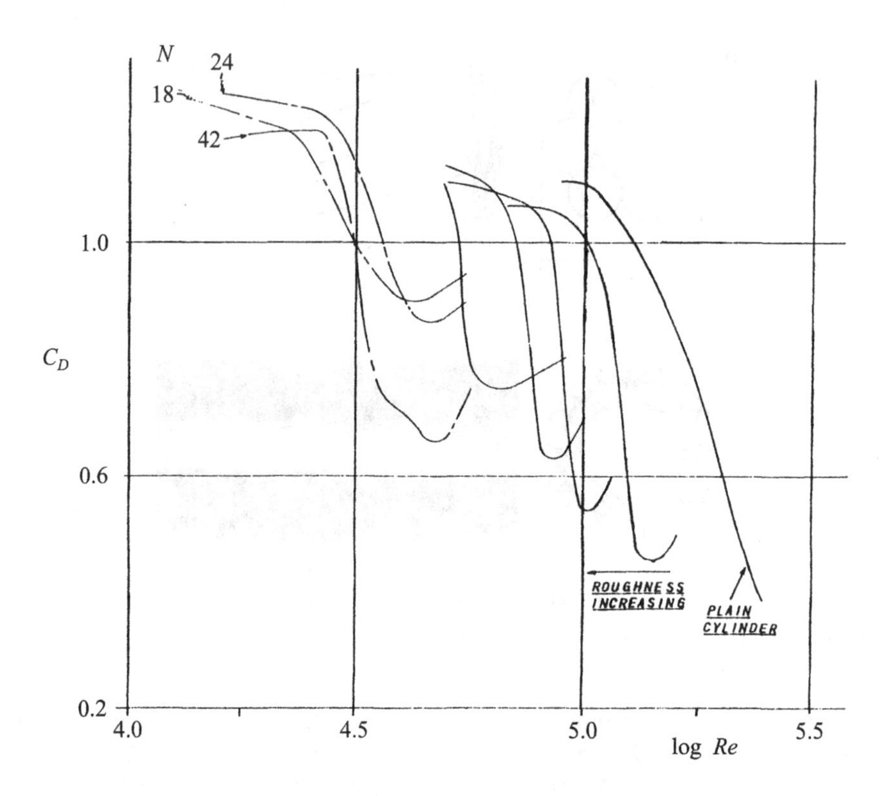

FIG. 22.38. Drag for stranded conductors in terms of Re, Counihan (1963P)

either side of the cylinder. The spacing between the two spheres on the same side was varied in the range $0.125 < L_s/D < 1.50$. The surface flow visualization revealed two types of interaction between the spheres and the cylinder laminar boundary layer:

(i) when the spheres were separated $L_s/D > 1.4$ and $Re > 80$k, a narrow turbulent wedge was formed behind each sphere, and the separation line remained straight along the cylinder span;

(ii) when the spheres were separated $L_s/D < 1.2$ the transition in the cylinder boundary layer spread between the spheres as well. The separation line became parabolic in shape.

The surface flow visualization used the method described by Smith and Murphy (1955J). The air–talcum mixture was blown towards the cylinder near $\theta = 180°$ by a gun-dispenser. The reverse flow in the separated region carried the talcum powder towards the separation line where a distinct white patch was formed. The surface flow upstream from the separation line could not be visualized.

Figure 22.39(a,b) shows the surface visualization for two spheres, $d/D = 0.6\%$, $L_s/D = 1$ at $Re = 150\text{k}$, and the assumed flow topology, respectively. The parabolic white patch was attributed to turbulent separation and the grey upstream patch to the turbulent wedge. The existence of the finite separation bubble could not be deduced from the visualization method used.

Mizuno (1970P) also measured the C_p distribution in the spherical plane and at the midspan. The C_p distribution corresponded to the TrBL0 and TrBL3 regimes, respectively. The hot-wire traverse at $x/D = 1$ showed that the near-wake was considerably narrower at the midspan, $y/D = \pm 0.3$, than at the spherical plane, $y/D = \pm 0.8$.

The occurrence of the parabolic separation line between the two spheres strongly depends on Re, d/D, L_s/D, and the angular location θ_d. Figure 22.40 shows the critical θ_{dc} for the occurrence of the parabolic separation line in terms of Re for $L_s/D = 1.2$ and 0.6, and $d/D = 0.6\%$. The minimum Re for the parabolic separation strongly depends on d/D. The parabolic separation occurs above the lower curve, and disappears above the upper curve in Fig. 22.40 when the separation line is straight again.

Mizuno (1970P) attempted to predict the lower bound of the region by using the roughness Re_K. The dotted curves for $Re_K = 700$ and 850 do not agree with the lower experimental curve in Fig. 22.40. However, a better agreement is seen in Fig. 22.40 when Re_δ is based on the boundary layer thickness

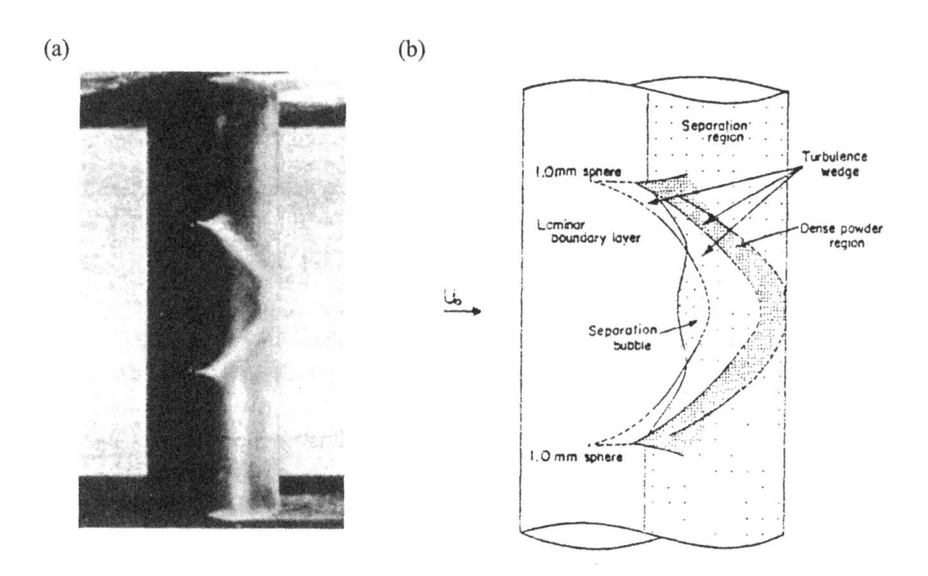

FIG. 22.39. Separation lines between two tripping spheres, $Re = 150\text{k}$, $L_s/D = 1$, (a) surface pattern, (b) sketched topology, Mizuno (1970P)

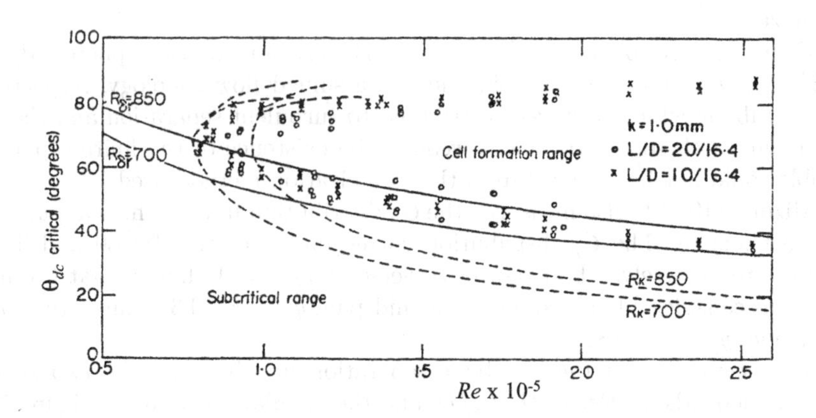

FIG. 22.40. Critical angular location of tripping spheres θ_d versus Re for $d/D = 0.6\%$, Mizuno (1970P)

$$Re_\delta = Re(\delta/D)(V/V_0) \tag{22.3}$$

where δ is calculated for the undisturbed flow at θ_d and velocity V is at θ_d. The full lines in Fig. 22.40 are in better agreement with experiment. The parabolic separation showed both bistable and hysteretic features. In the range $1.2 < L_s/D < 1.4$ both parabolic and straight separation are observed intermittently. The parabolic separation occurs at higher Re for increased velocity and at lower Re for decreased velocity. The change-over spreads across $125\text{k} < Re < 200\text{k}$ for $L_s/D = 1.4$.

22.4.2 *Spanwise row of spheres*

When a row of four spheres, $d/D = 0.6\%$, was attached at $\theta_d = \pm 60°$ and $L_s/D = 0.61$, only two parabolic separations were formed and the straight separation in between, as seen in Fig. 22.41(a). Adding the fifth sphere to a row having $L_s/D = 0.5$ produced a similar surface flow pattern consisting of two parabolic cells and the straight separation cell spanning three spheres, Fig. 22.41(b). When the number of spheres, $d/D = 0.9\%$ in the row is increased to $N = 11$, and $L_s/D = 0.25$, the surface pattern changes with Re. For $Re = 87\text{k}$, two parabolic cells appear behind four spheres and the straight cell behind five spheres, as shown in Fig. 22.41(c). For $Re = 92\text{k}$, the parabolic cells span five spheres each, and a single straight cell forms over three spheres. Not a single cell is found when $N = 23$ and $L_s/D = 0.125$ in the range $70\text{k} < Re < 250\text{k}$, and the separation line appears sinuous.

22.5 Other surface disturbances

A wide variety of other surface disturbances has been tested such as: streamwise eddy generators in the form of triangular spoilers, serrated saw-blades, dimples

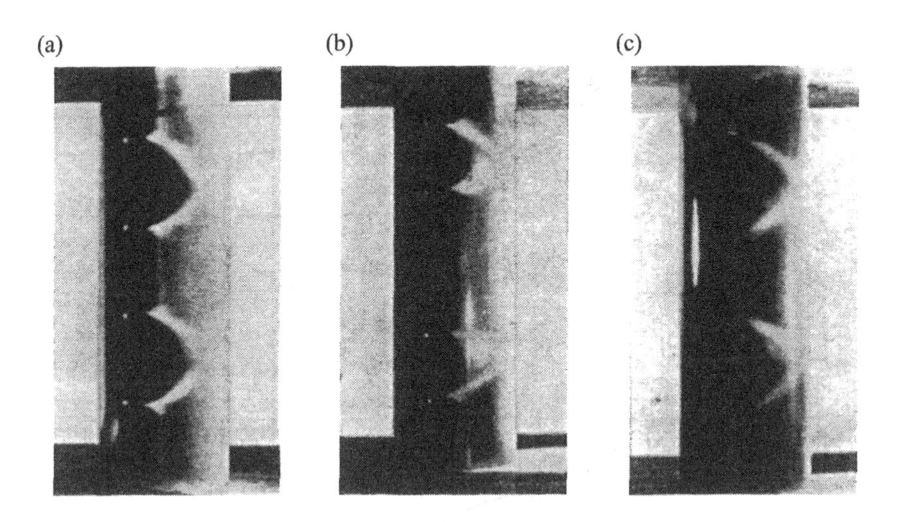

FIG. 22.41. Surface flow pattern, (a) $N = 4$, $L_s/D = 0.61$, $Re = 116$k, (b) $N = 5$, $L_s/D = 0.5$, $Re = 92$k, (c) $N = 11$, $L_s/D = 0.25$, $Re = 87$k, Mizuno (1970P)

impressed onto the surface, slit, fins, and circumferential and spanwise grooves.

22.5.1 *Streamwise eddy generators*

It has already been discussed that surface roughness generates turbulence in the form of small eddies of random orientation. The tripping wires attached along the span generate vorticity mostly parallel to the cylinder axis. The flow in the vicinity of separation is also sensitive to streamwise eddies and related momentum transfer to the boundary layer.

Joubert and Hoffman (1962J) and Johnson and Joubert (1969J) examined the influence of streamwise eddy generators on the drag of, and heat transfer from, a circular cylinder. The cylinder was fitted with rows of eddy generators shaped as triangles. Figure 22.42(a) shows the size, inclination, and pitch of the triangles. The triangular spoilers were placed symmetrically on both sides of the cylinder and parallel to the stagnation line at an angle $\pm\theta_e$. The relative 'roughness' produced by the triangles was $h/D = 3.3\%$, and the cylinder $L/D = 8.5$, $D/B = 0.1$ was tested in the range $40\text{k} < Re < 400\text{k}$. Figure 22.42(b) shows local C_d evaluated from the C_p distribution in terms of Re and θ_e. The shape and orderly displacement of the C_d curves are similar to those obtained with tripping wires. For $\theta_e > 40°$, $C_{d\min}$ becomes flat, which reflects the independence of Re.

A marked effect of streamwise eddies on separation can be seen in Fig. 22.43(a,b). The wavelength of the separation line coincided with the pitch of the triangles. The streamwise eddies caused a periodic spanwise variation in momentum of the boundary layer. The higher momentum sustained the greater

(a)

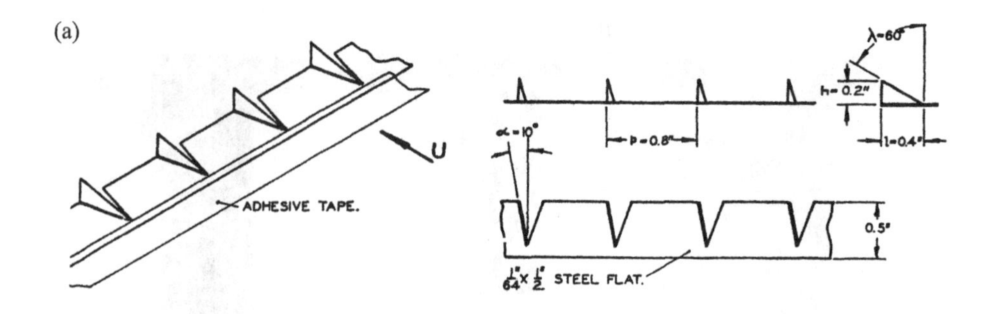

(b)

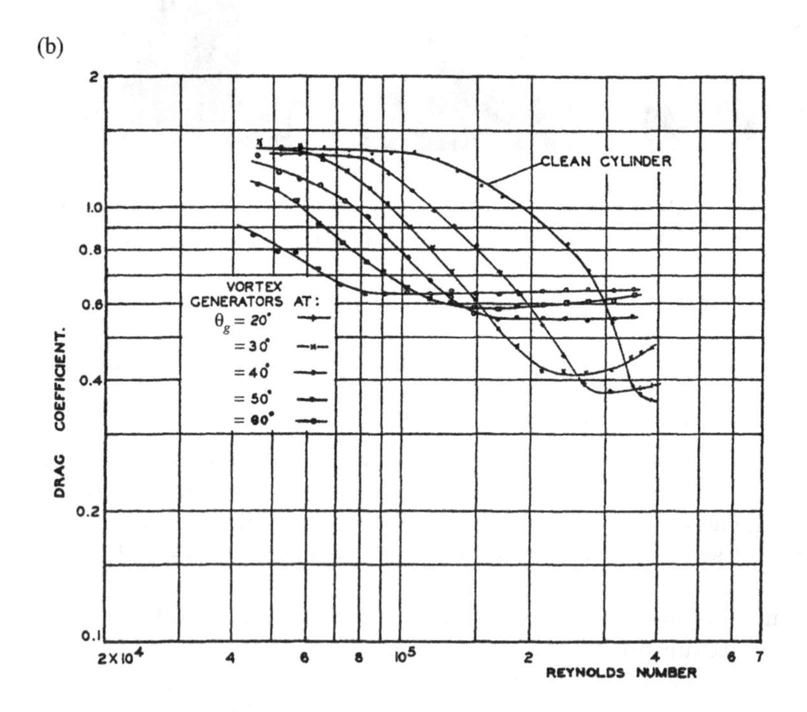

FIG. 22.42. (a) Three views of eddy generator, (b) effect on drag of angular location, Joubert and Hoffman (1962J)

adverse pressure recovery before separation. The maximum θ_s along the span coincided with the streamwise eddies.

Figure 22.44 shows measured Nu in terms of θ for various Re for a plane cylinder (dashed lines) fitted with eddy generators at $\theta_e = \pm 50°$ (full lines). The latter produced $Nu_{\min}$ at $\theta = 50°$, followed by a peak at $\theta = 60°$. The rise in Nu in the range $50° < \theta < 60°$ was induced by the streamwise eddy. The streamwise eddy momentum was transferred to the laminar boundary layer at $\theta = 60°$. The

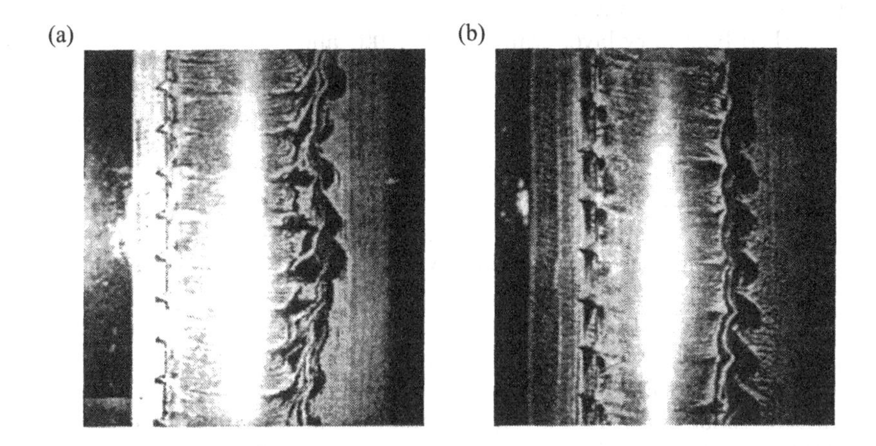

FIG. 22.43. Surface flow visualization behind generators at Re = 143k, (a) $\theta_e = 50°$, $C_d = 0.59$, (b) $\theta_e = 60°$, $C_d = 0.64$, Johnson and Joubert (1969J)

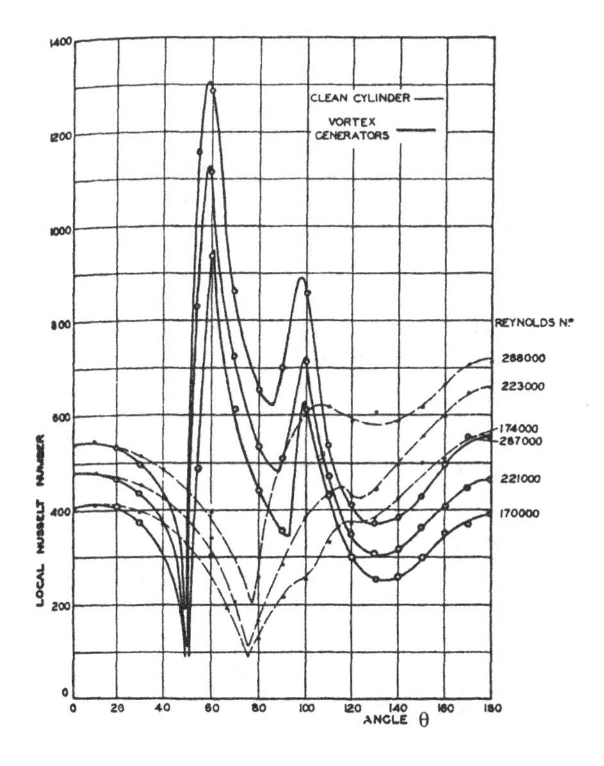

FIG. 22.44. Local Nusselt number distribution with (full line) and without (dashed line) eddy generators, Joubert and Hoffman (1962J)

rapid decay of the streamwise eddy led to the second Nu_{min}. The transition in the perturbed boundary layer started at $\theta = 90°$ and culminated at $\theta = 100°$ in the second Nu_{max}.

22.5.2 *Serrated saw-blade*

Igarashi (1985J) and Igarashi and Iida (1987P) examined the effect of a special 'three-dimensional fence' in the form of a serrated saw-blade. Figure 22.45(a) shows three views of the saw-blade. The serrations appear as triangles 0.8 mm high at 1.4 mm pitch twisted alternately in the opposite direction. Thus, the saw-blade represents a combination of an eddy generator and tripping fence. Figure 22.45(b) shows that the saw-blade could be partly submerged in the slot and partly protruded above the cylinder surface. The protruding part of the triangles resembles the eddy generator.

Igarashi (1985J) measured u', C_p, St, and evaluated C_d. The cylinder $L/D = 4.4$, $D/B = 0.085$, and $Ti = 0.5\%$ was fitted with a pair of saw-blades $\pm 40° < \theta_b < \pm 70°$ protruding in the range $0.6\% < h/D < 4.4\%$ at 8.7k < $Re <$ 69.7k. Figure 22.46(a,b) shows St and C_d in terms of Re for various flow regimes identified by Igarashi (1985J) as:

(A) laminar reattachment behind the saw-blade was associated with $St \simeq 0.21$ and $C_d = 1.25$;

(B) transition to turbulent reattachment;

(C) turbulent reattachment behind the saw-blade was associated with $St \simeq 0.26$ and $C_d = 0.7$;

(D) full separation behind the saw-blade was associated with $St = 0.16$ and $C_d = 1.7$.

The tests with saw-blades at $\theta_b = \pm 40°$ showed that the small h/D was ineffective, and for the large h/D the transition extended over a wide Re range.

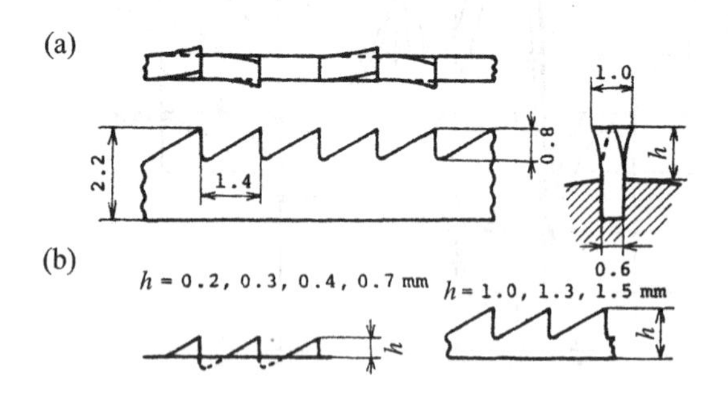

FIG. 22.45. Serrated blade, (a) three views, (b) partly submerged and fully protruded, Igarashi (1985J)

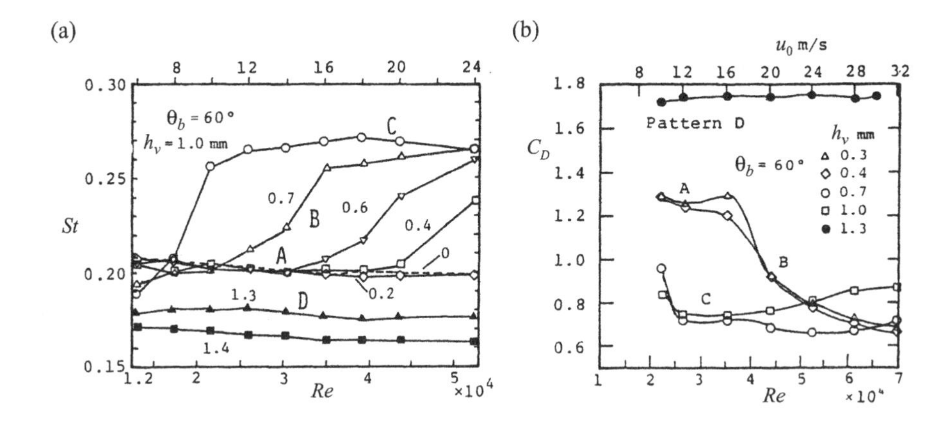

FIG. 22.46. Variation in (a) Strouhal number, (b) drag, in terms of *Re* for various h/D, Igarashi (1985J)

For $\theta_b = \pm 50°$ and $\pm 60°$, the St and C_d curves were affected. The most effective angular location was $\theta_b = \pm 60°$.

22.5.3 *Dimpled surface*

It is well known that a golf ball with dimples flies farther than a smooth ball, see Bearman and Harvey (1976J). The golf ball dimples may be moulded and/or additionally hammered by the club during play. The dimples may be thought of as a 'negative roughness' because the smooth surface is indented. This basic distinction from the other kind of surface roughness is also likely to lower the drag at high *Re*.

Bearman and Harvey (1993J) applied the golf ball dimples to a circular cylindrical surface. The spherically shaped dimples were made on the cylinder surface using a spherical cutter. They appear as ellipses, $b/a = 0.9$. The staggered arrangement of dimples had a spanwise pitch $T/b = 0.8$, a circumferential pitch of $S/a = 0.9$, and a negative relative roughness of $h/D = -1\%$. Twelve equally spaced dimples were machined around the circumference, and the density was such that 180 dimples appeared per one diameter span. The plan area of each dimple was 50% greater than the corresponding scaled area of a dimple on a conventional golf ball, and the density of dimples was half that of a standard golf ball. The cylinder has $L/D = 12.26$, $D/B = 0.082$, $Ti = 0.2\%$, and end plates $D_e/D = 2.1$.

Figure 22.47(a,b) shows the measured C_D and St, respectively. Three C_D curves are added for a smooth cylinder, sand roughness $K_s/D = 0.9\%$ and 0.45%. The C_D curve for the dimpled cylinder, $h/D = -1\%$, is between the two curves for the sand roughness only for the descending part of the curve. The subsequent rise in the C_D curve is well below the other two. Hence, the dimpled

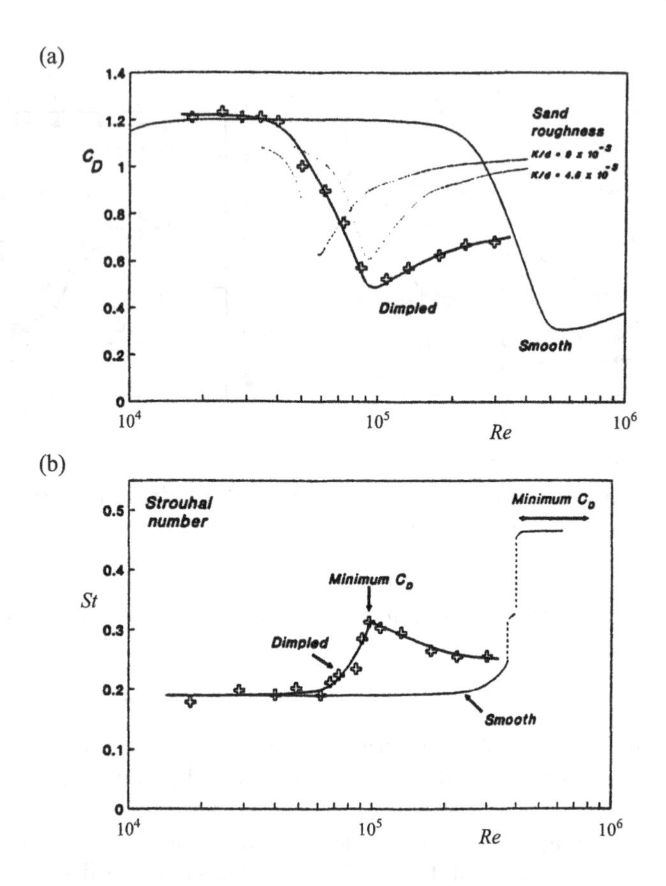

FIG. 22.47. Cylinder with indented dimples, $h/D = -1\%$, (a) drag, (b) Strouhal number, in terms of Re, Bearman and Harvey, © (1993J) AIAA – reprinted with permission

cylinder triggers early $C_{D\min}$ without causing high C_D at high Re. The minimum C_D coincides with $St_{\max}$, and both reflect the narrowest near-wake. The rotation of cylinder for half a dimple spacing has a small effect on C_D.

22.5.4 *Spanwise slit*

Prandtl (1961B) demonstrated that a fully separated flow in a wide-angled diffuser could be forced to attach to the walls by applying suction to boundary layers. Conversely, an attached flow along a flat plate could be forced to separate by applying blowing into boundary layers, as discussed in Chapter 24.

Igarashi (1978J) achieved alternate suction and blowing through a spanwise slit cut across a circular cylinder, as shown in Fig. 22.48(a). He examined the flow around a cylinder, $L/D = 4.4$, $D/B = 0.085$, provided with a spanwise slit

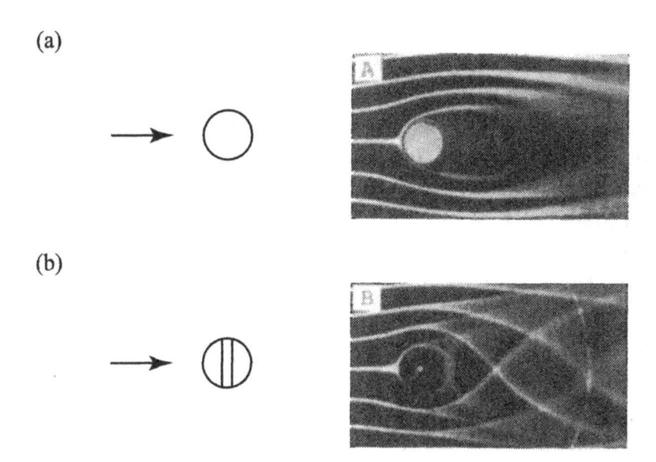

FIG. 22.48. (a) Smoke visualization with slit closed, (b) with slit open, Igarashi (1978J)

perpendicular to the free stream velocity. Two slit widths were tested, $s/D = 0.08$ and 0.18, in the range $13.8\text{k} < Re < 52\text{k}$. A remarkably strong effect of the slit on the mean length of the eddy formation region can be seen in Fig. 22.48(b,c) (long exposure photographs).

The short exposure sequences are shown in Fig. 22.49 for one half of the shedding cycle. Sequences 1 and 2 correspond to the final stage of the top eddy growth. The cut-off of the top free shear layer is associated with the eddy being shed, sequence 3. The lower free shear layer remains attached to the rear of the cylinder due to the slit suction on that side. Sequence 3 also shows the roll-up

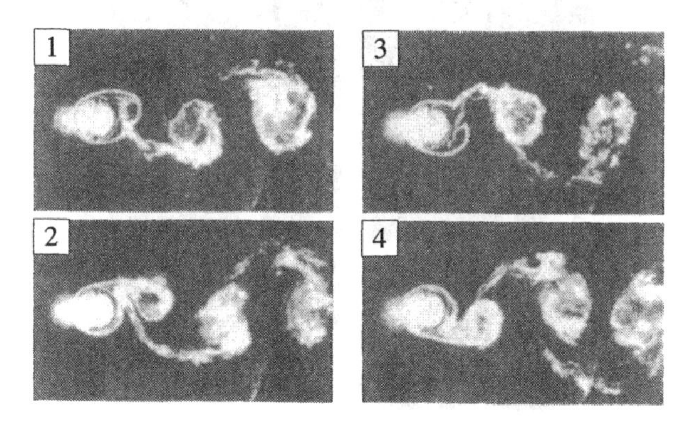

FIG. 22.49. Short-time exposure of smoke visualization sequences during half of the eddy shedding cycle at $Re = 4.4\text{k}$, Igarashi (1978J)

of the lower shear layer and formation of the lower eddy. The final sequence, 4, appears as a mirror-image of sequence 1 and marks the midpoint of the shedding cycle.

At Re = 4.8k, the length of the eddy formation region is about $3D$ for the plain cylinder without the slit. The extremely short eddy formation region, seen in Fig. 22.49(c), is induced by alternate boundary layer suction and blowing at the slit ends. Flow visualization sequences within the slit are shown in Fig. 22.50. The first three are suction sequences and show a gradual attachment of the free shear layer onto the downstream side of the cylinder. The separation induced by blowing is seen in the last two sequences, 4 and 5. The suction on one side of the slit is always associated with blowing on the opposite side, and vice versa.

The out-of-phase pressure fluctuations on the opposite cylinder's sides are induced by alternate eddy shedding. The resulting pressure difference on the two sides of the slit synchronizes alternate suction and blowing in the rhythm of eddy shedding. The synchronization of eddy shedding and alternate flow in the slit enhances spanwise correlation considerably, shortens the eddy formation region, increases the eddy strength, lowers $C_{p\mathrm{min}}$, magnifies $C'_{p\mathrm{max}}$, and increases C_D. The phenomenon appears analogous to an oscillating cylinder in the synchronization range. The transverse displacement of the oscillating cylinder enhances the favourable pressure gradient on one side and the adverse pressure gradient on the other.

Igarashi (1978J) also measured maximum C'_p for several bluff bodies: (A) a circular cylinder (plain); (B) a circular cylinder with slit; (C) a semi-circular cylinder; (D) a triangle; (E) a trapezoid; and (F) a square. Figure 22.51 shows

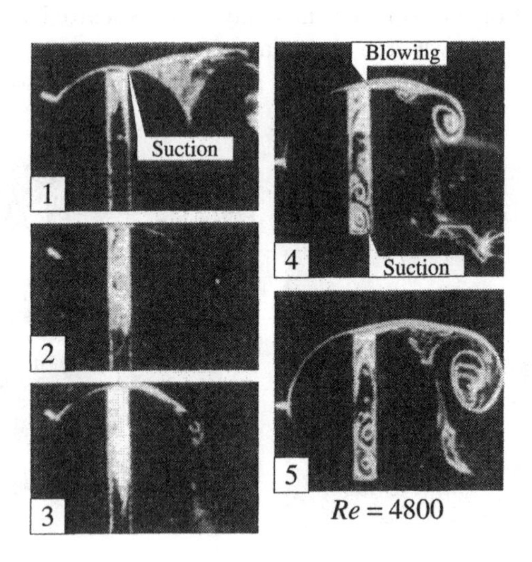

FIG. 22.50. Sequences of alternate suction 1, 2, 3, and blowing, 4, 5, through slit at Re = 4.8k, Igarashi (1978J)

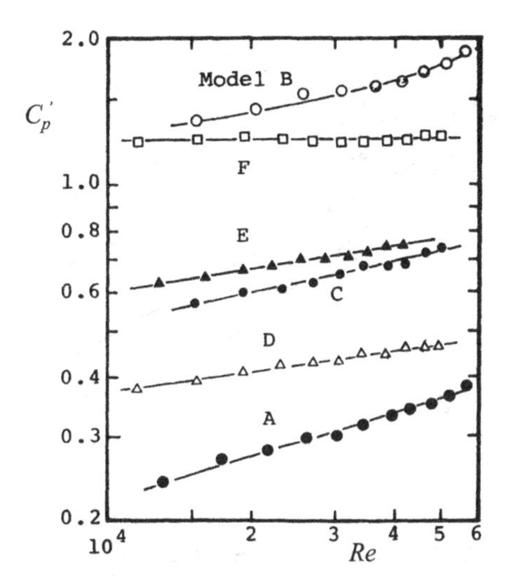

FIG. 22.51. Maximum fluctuating pressure in terms of *Re* for various bluff bodies, A plain cylinder, B with slit, C semi-cylinder, D triangle, E trapezoid, F square cylinder, Igarashi (1978J)

the lowest $C'_{p\max}$ for the plain circular cylinder and the highest $C'_{p\max}$ for the cylinder with the slit. The magnification of $C'_{p\max}$ is almost five-fold, and the magnitude exceeds all other bluff bodies with sharp edges. This demonstrates how powerful the alternate synchronization of suction and blowing induced by the slit is.

22.5.5 *Fins*

It is a well-established practice to add fins to tubes in heat exchangers in order to increase heat transfer. For the finned tubes most commonly used in heat exchangers, the ratio of the fin diameter, D_f, to the tube diameter, D, is $1.1 < D_f/D < 1.5$. The fin thickness, t, and the pitch, p, between the fins are in the ranges $0.02 < t/D < 0.07$ and $0.06 < p/D < 0.15$, respectively.

Mair *et al.* (1975J) measured the frequency of eddy shedding behind the finned tube. The tube had $L/D = 25$, and the finned segment was $L_f/D = 6.3$ bounded by end plates $D_e/D = 10.6$. Two finned tubes were used, $D_f/D = 1.2$ and 1.4, and three relative fin thicknesses $t/D = 0.022$, 0.038, and 0.064. The relative pitch ratio was varied $0.06 < p/D < 0.8$ in the range $16\text{k} < Re < 46\text{k}$.

Figure 22.52(a) shows the variation in St in terms of D/p for two D_f/D and three t/D values. The reason why D/p is used rather than the reciprocal p/D is because the former allows the tube with no fins to be included ($D/p = 0$). All St curves intersect at $D/p = 0$ where the plain tube appears as a limit. The monotonous decrease in St reflects the near-wake widening caused by the fin

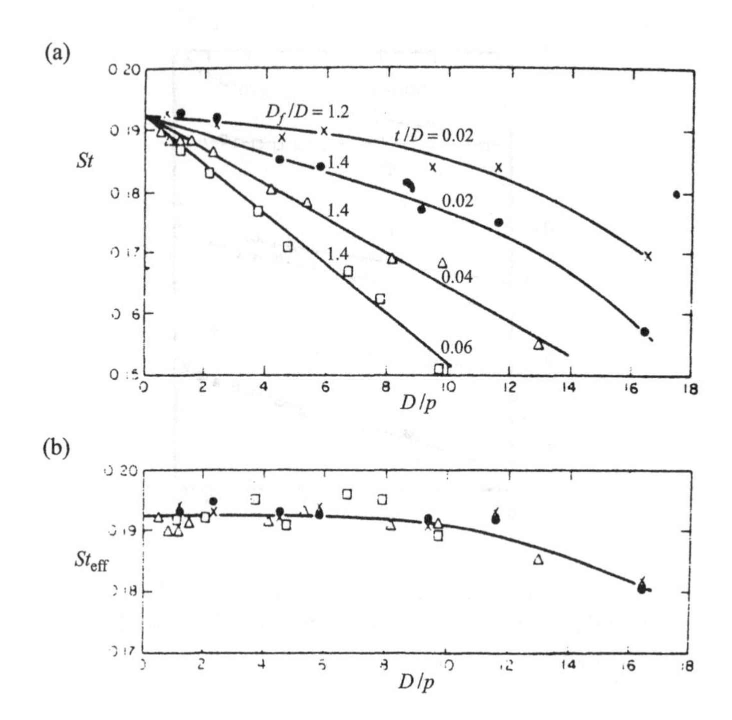

FIG. 22.52. Finned cylinder, (a) Strouhal number, (b) effective Strouhal number in terms of D/P, Mair *et al.* (1975J)

size and pitch. There seems to be an analogy with grooved tubes. However, the collapse of the St curve occurs in Fig. 22.52(b) when a new *equivalent* St_{eff} is introduced by using the effective D_{eff} as

$$D_{\text{eff}} = [(p - t)D + tD_f]/p \tag{22.4}$$

which is the frontal area of the finned tube per unit width.

The oscilloscope record of the hot-wire signal measured at $x/D = 6$ for the tube without fins showed some random variation in shedding frequency with time. When the fins were added, there was a marked reduction in the variation in frequency, and the peak in the power spectrum became sharper. This effect of fins is analogous to surface roughness.

22.5.6 *Circumferential grooves*

Flow past circular cylinders with a smooth circumferentially grooved surface unexpectedly appeared analogous to cylinders with rough surfaces. There was an orderly displacement of $C_{D\text{min}}$ towards low Re with increasing groove depth. The puzzling feature was that smooth grooves cut in the streamwise direction affected the flow around the cylinder in a similar manner as surface roughness and tripping wires arranged in the spanwise direction.

Ko *et al.* (1987J) carried out tests using smooth and circumferentially grooved cylinders. The V-groove was described by its maximum depth h and angle formed by the groove sides γ. These parameters were varied in the range $0.42\% < h/D < 1.23\%$ and $23° < \gamma < 44°$. The cylinder, $L/D = 9$ and $D/B = 0.087$, was fitted with small end plates, $De/D = 2$, and tested in the range $30\text{k} < Re < 72\text{k}$.

Ko *et al.* (1987J) evaluated C_d from the measured C_p at the midspan. Figure 22.53 shows the variation in C_d in terms of Re for various grooves h/D and surface roughnesses. They added curves by Fage and Warsap (1929P), Achenbach (1971Jc), and Guven *et al.* (1980J). There is a remarkable similarity in C_d curves obtained with grooves and surface roughness. The increase in either K_s/D or h/D leads to the displacement of $C_{d\min}$ towards lower Re.

22.5.7 *Skin friction and boundary layer*

Leung and Ko (1991J) measured skin friction with a miniature fence similar to Achenbach's (1971Jc). The grooved cylinder, $L/D = 4.5$, $D/B = 0.18$, $h/D = 1.3\%$, and $\gamma = 19°$, was tested at $Re = 39\text{k}$, 48k, and 150k. For these Re, $C_f = 0$ at $\theta_s = 85°$, 125°, and 118°, respectively. For $Re = 150\text{k}$, the second $\theta_s = 90°$ appeared, possibly triggered by the probe itself.

Leung and Ko (1991J) also examined the effect of grooves on boundary layer development. They traversed the boundary layer at several circumferential positions in the range $30° < \theta < 105°$ and $21\text{k} < Re < 80\text{k}$. Figure 22.54 shows the boundary layer growth δ/D in terms of θ for various Re and types of surfaces. For smooth cylinders, δ/D decreases with a rise in Re, and increases beyond

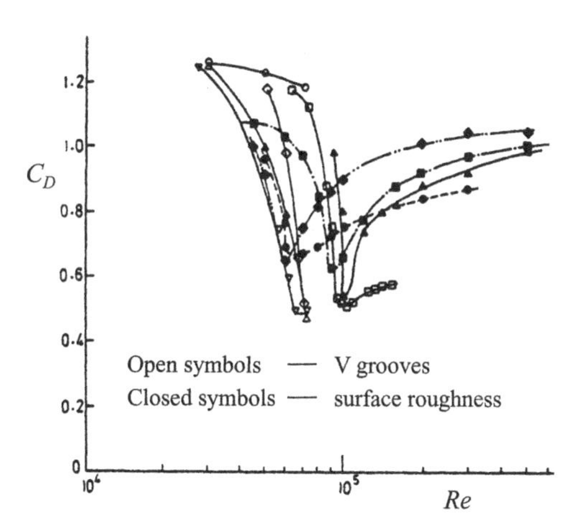

FIG. 22.53. Local drag in terms of Re for a circumferentially grooved cylinder, Ko *et al.*, © (1987J) AIAA – reprinted with permission

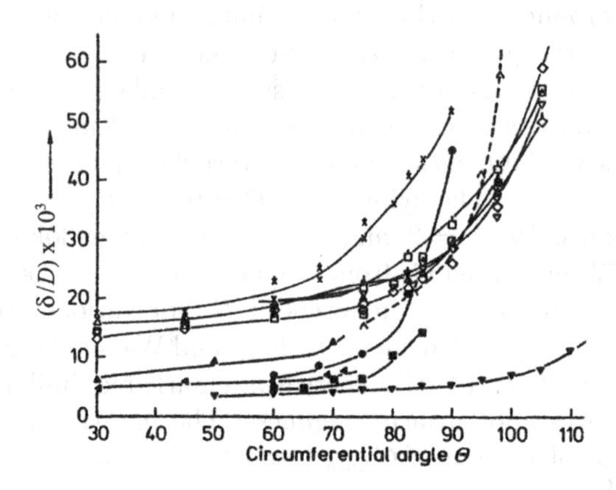

FIG. 22.54. Thickening of the boundary layer with grooves (open symbols) and for smooth cylinders (closed symbols), Leung and Ko (1991J)

separation. All δ/D curves for the grooved cylinders are well above the curves for smooth cylinders, and agree with δ/D for rough cylinders. The three-fold increase in δ/D for grooved cylinders in comparison with smooth cylinders corresponds to a similar increase in surface area by grooves and related skin friction. Hence, the analogy between the grooved and rough surfaces seems to stem from a similar increase in skin friction. Figure 22.54 also shows the size of the groove tip (dash-dot line). The boundary layer is submerged into the groove up to $\theta = 30°$, and emerged for $\delta/D > h/D$. The gradual retardation of flow inside the grooves with rising θ may lead to flow overspill at the tips. There might be some analogy between the action of the grooves and streamwise eddy generators.

22.5.8 *Partly grooved surface*

Additional insight into the mechanism of groove action was not gained by tests on partly grooved surfaces. The reason was that the effect of the end of grooving disturbed the flow more than the grooves themselves.

Leung *et al.* (1992J) attempted to examine which segment of the grooved circumference had a dominant effect on C_d reduction. A cylinder, $L/D = 4.3$ and $D/B = 0.16$, was grooved along segments from $\pm 30°$ to $\pm 165°$, and tested in the range $22\text{k} < Re < 130\text{k}$. The measured C_p distribution showed that the end of grooving caused a pressure recovery due to a sudden increase in the local flow area. When the grooves extended to $\theta_g = \pm 60°$ the value of $C_p = -2.1$, and for $\theta_g = \pm 75°$, $C_{p\min} = -2.6$. It appeared that the end of grooving has a greater effect on flow than the grooves themselves.

Figure 22.55(a,b) shows the variation in C_d (evaluated from C_p and St) in terms of Re for various grooves ($\theta_g = 0$ corresponds to the smooth cylinder). The most effective grooves are up to $\theta_g = \pm 75°$, for which $C_d = 0.6$; a further

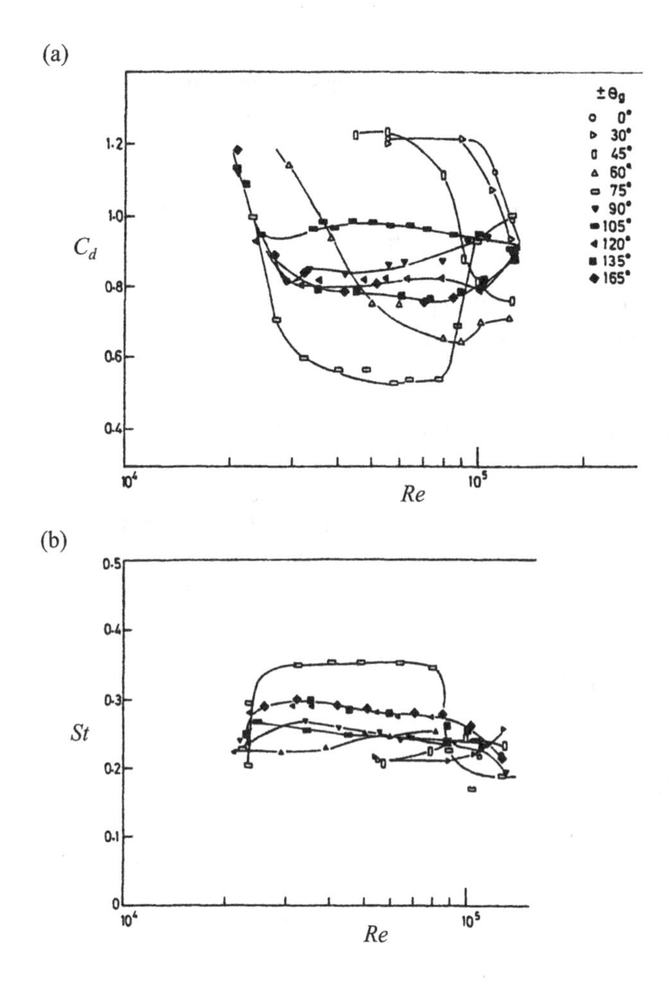

FIG. 22.55. Variation in (a) drag, (b) Strouhal number, in terms of *Re*, Leung *et al.* (1992J)

increase in the grooved surface area has less effect, $C_d = 0.8$. Figure 22.55(b) shows the variation in St, which appears as a mirror-image of the C_d curves.

Leung *et al.* (1992J) also carried out tests on a cylinder grooved on one side, $0° < \theta_g < 180°$, and smooth on the other, $180° < \theta < 360°$. The boundary layer development was independent on two sides, and the near-wake was biased towards the smooth side. The flow asymmetry resulted in a mean lift in the range $0.1 < C_\ell < 0.6$ for $105\text{k} < Re < 250\text{k}$.

22.5.9 *Spanwise grooves*

A circular-arc groove cut along a cylinder span has been shown to delay laminar separation in the TrSL1 regime. The laminar boundary layer separates from the leading edge of the groove, and reattaches at the trailing edge. The shear stress is reduced in the separated region above the groove.

Kimura and Tsutahara (1991J) carried out flow visualization in a water tank at $Re = 22\text{k}$ around an immersed finite cylinder, $H/D = 1.75$ and $D/B = 0.2$, fitted with an end plate, $D_e/D = 3$, at the free end. The width of the spanwise groove was $S/D = 0.08$ and depth $0.7\% < -h/D < 1.7\%$. The circumferential location of the groove was varied in the range $50° < \theta_g < 90°$. Figure 22.56(a,b) shows a typical flow visualization pattern around the groove, and at the separation for $\theta_g = 80°$ (optimal) and $90°$, respectively. The angular location of separation θ_s for a smooth cylinder at $Re = 2.2\text{k}$ was $92°$, and with the groove at $\theta_g = 80°$ it moved to $\theta_s = 94.5°$.

Kimura and Tsutahara (1991J) carried out a numerical simulation of the boundary layer and groove flows. The length of the eddy formation region L_F was short for the smooth cylinder and even shorter for the grooved cylinder. This was at variance with their experimental observation at $Re = 2.2\text{k}$.

22.6 Change in diameter

22.6.1 *Introduction*

Some cylindrical structures in engineering applications are of varying diameter along the span. According to how these changes in diameter occur, the following categorization is adopted:

(i) A discontinuous or 'step' change from large D to small d, like antenna members, multi-stage rockets, TV towers, etc. The influencing parameter is the ratio d/D. All step changes are confined in the range in $0 < d/D < 1$, where $d/D = 0$ and 1 correspond to the free end and constant diameter cylinder, respectively.

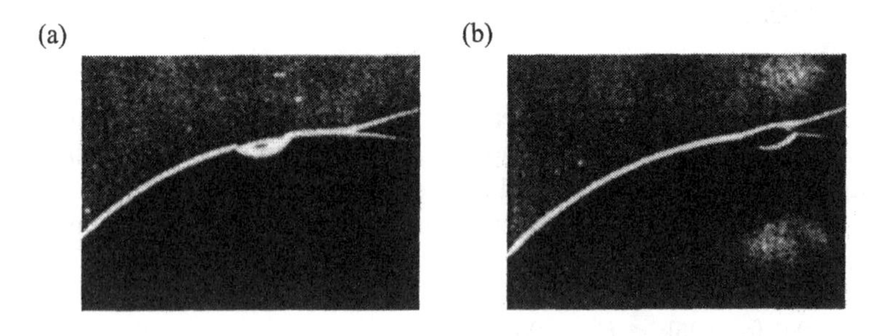

FIG. 22.56. Dye visualization around the spanwise groove at $Re = 2.2\text{k}$, (a) $\theta_g = 80°$, (b) $\theta_g = 90°$, Kimura and Tsutahara, © (1991J) AIAA – reprinted with permission

(ii) A continuous linear taper from D_r, at the root, to D_e at the narrow end, like most chimney stacks designed as truncated cones. The influencing parameter is the taper ratio, $TR = (D_r - D_e)/H$.

(iii) A non-linear change in diameter, like cooling towers designed as hyperboloids of revolution with both ends open. The reference diameter is usually the mean diameter $D_m = (D_r + D_e)/2$, and the influencing parameter is H/D_m.

The step change in cylinder diameter is of fundamental interest because two different eddy streets may be formed side by side. Two 'nominal' two-dimensional eddy streets are separated by a three-dimensional interference region behind the step.

22.6.2 *Laminar periodic wake, the L3 regime*

An exploration of a wake disturbed by a step change in diameter was carried out by Lewis and Gharib (1992J) in the range $35 < Re < 200$ for $0.57d/D < 0.88$. Aspect ratios were in the range $100 < L/D < 140$ and $126 < L/d < 194$. At low *Re* it was possible to generate a laminar periodic wake behind the large cylinder D_1, and either a stationary closed near-wake or a sinuous trail behind the small cylinder *d*. Figure 22.57(a,b) shows both possibilities:

(a) two wakes behind the small and large cylinder $d/D = 0.57$, $Re = 37/68$, are seen in Fig. 22.57(a), top and bottom, respectively. The eddy filaments are formed only behind the large cylinder. They link in a remarkably similar way as behind the free end of the finite cylinder, as found by Taneda (1952P) at $Re = 120$, Chapter 21;

(b) for $d/D = 0.75$ at $Re = 57/76$ parallel and slanted eddy filaments are seen in Fig. 22.57(b). There is only one knot seen behind the small cylinder. The roll-up of free shear layers at $Re = 57/76$ is very slow, as noted by

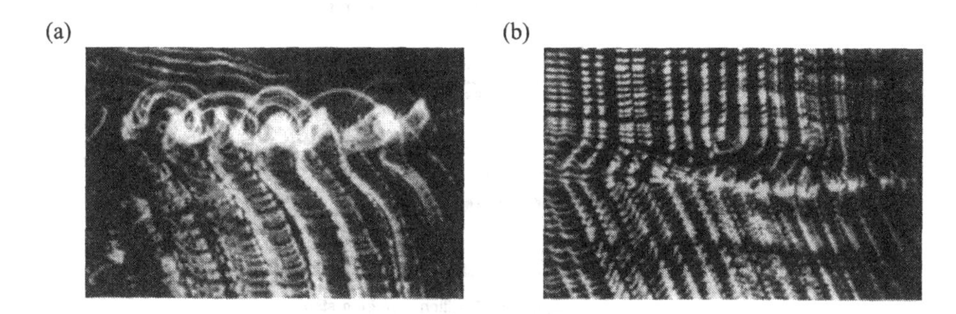

FIG. 22.57. Smoke visualization behind a step change in diameter, (a) $d/D = 0.57$, $Re = 37/58$ (top/bottom), (b) $d/D = 0.75$, $Re = 57/76$, Lewis and Gharib (1992J)

Kovasznay (1949J), Vol. 1, p. 51. This may explain the late start of the interference.

At higher $Re = 74/99$, two eddy streets are formed behind the step $d/D = 0.75$. The mean velocity defect and rms fluctuations versus z/D are shown in Fig. 22.58(a,b) at $x/D = 2$ and 15, respectively. There are three features to be noted:

(i) an almost two-dimensional flow persists behind both cylinders up to $z/D = \pm 5$ from the step;

(ii) the interference region behind the step is marked by a considerable variation in the defect velocity;

(iii) the variation in the fluctuating velocity is higher behind the large cylinder than behind the small cylinder.

Lewis and Gharib (1992J) also measured the angle of the slanted eddy filaments. Figure 22.59 shows the slanted angle β versus Re for $d/D = 0.65$ and 0.75. There is a scatter for the small cylinder data. However, there are two distinct upper curves for the large cylinder (open symbols). The difference in the slanted angle β for the two cylinders becomes small for $Re > 130$.

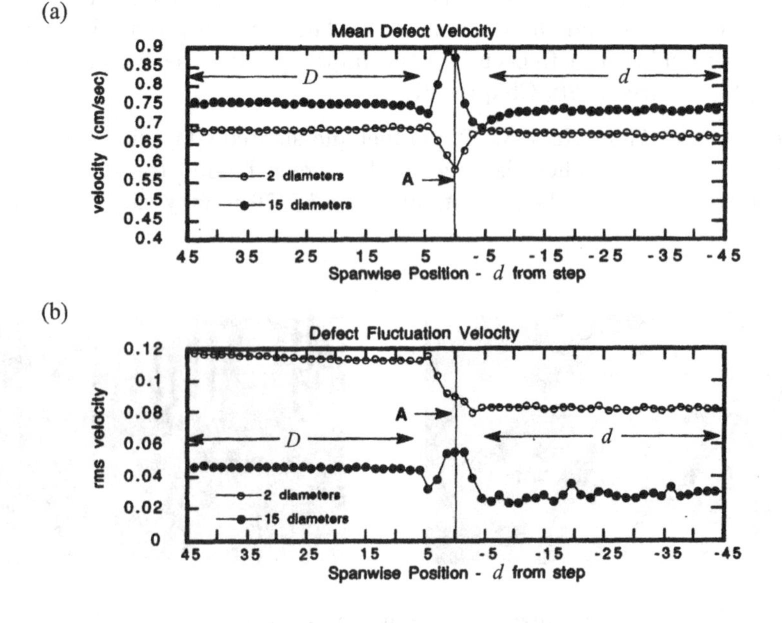

FIG. 22.58. (a) Mean velocity defect, (b) fluctuating velocity, for $d/D = 0.75$, $Re = 74/99$ at $x/D = 2$ (open), 15 (closed circles), Lewis and Gharib (1992J)

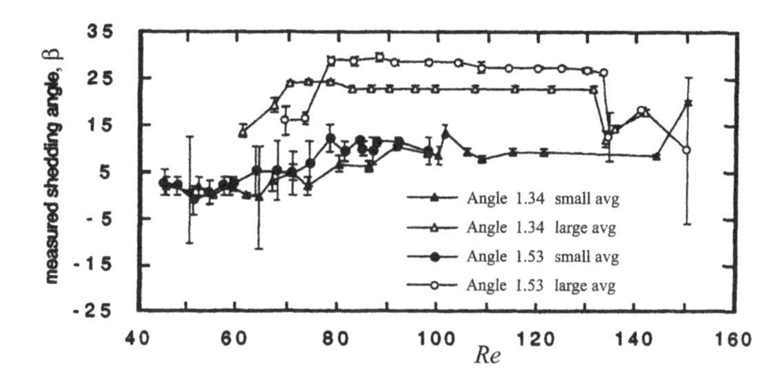

FIG. 22.59. Eddy filaments slanted angle β in terms of Re, small d (closed symbols), big D (open symbols) for $d/D = 0.75$ and 0.65, Lewis and Gharib (1992J)

22.6.3 *Transition-in-shear-layer, the TrSL state*

Yagita *et al.* (1984J) carried out St measurements for $0 \leqslant d/D < 1$ and $800 < Re < 10\text{k}$. They argued that the following parameters may affect eddy shedding:

$$St = f(Re, Re_d, d/D, L/D, L/d) \tag{22.5}$$

The eddy shedding frequency was measured behind the large cylinder (full symbols) $10D$ away from the step and small cylinder (open symbols). Figure 22.60(a,b) shows the variation in St in terms of Re and d/D for $L/D = 10$ and 27, respectively. For small $L/D = 10$, the increase in St is caused by both increasing d/D and Re. However, for $L/D = 27$, St appears almost independent of Re and d/D, and close to the cylinder without step, $d/D = 1$.

Norberg (1992P) carried out complementary St measurements in the range $0.5 < d/D < 1$, $L/D = 56$, and $3\text{k} < Re < 13\text{k}$. The St behind the small cylinder was not affected by the step, and followed closely the St curve for $d/D = 1$. The St curves for the large cylinder $L/D = 56$ were similar to Yagita's curves obtained for $L/D = 27$. Norberg (1992P) argued that the large cylinder acted as some sort of end plate to the small cylinder.

22.6.4 *Step interference in the TrSL3 regime*

Naumann and Quadflieg (1974P) carried out flow visualization behind a stepped cylinder at $Re = 140\text{k}$. Figure 22.61 shows three frames from a high-speed cine film (5750 frames/s) for $d/D = 1$, 0.87, 0.71, and 0.49. The development of the eddy filament distortion is evident with decreasing d/D. For $d/D = 0.71$ and 0.49, there were two eddy streets separated by the step.

Ko *et al.* (1982J) carried out tests on $d/D = 0.5$, $L/D = 5.5$, and $L/d = 11.1$ at $Re = 40\text{k}/80\text{k}$, in the TrSL3 regime. The fluctuating pressure C_p' was measured along the span $-2.5 < z/D < 1$ behind the small and large cylinders. The C_p'

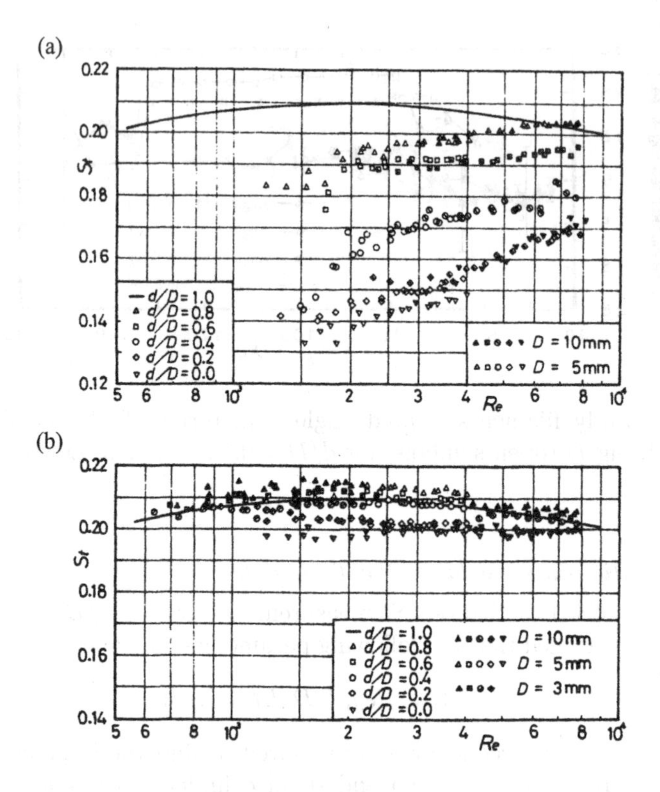

FIG. 22.60. Strouhal number variation in terms of *Re* for (a) $L/D = 10$, (b) $L/D = 27$, Yagita *et al.* (1984J)

spectra yielded two frequencies behind the small and large cylinders giving the same St value. Behind the step at $z/D = 0$, only the small cylinder frequency is found. The two frequencies overlapped in the range $-1.2 < z/D < 0.25$. This confirmed that the eddy street behind the small cylinder is less affected by the step than that behind the large cylinder.

22.7 Tapered cylinder

The effect of free stream shear on the flow around a constant diameter cylinder has been discussed in Vol. 1, Chapter 15. A similar variation in local *Re* along the span takes place along a tapered cylinder submerged in a uniform stream. The varying *Re* along the span precludes eddy shedding at a constant St. In the sheared free stream, cells of constant shedding frequency are found with St varying within each cell. It may be expected that similar shedding cells will also be found behind tapered cylinders because of the similarity between the two flows.

The taper ratio TR is defined as

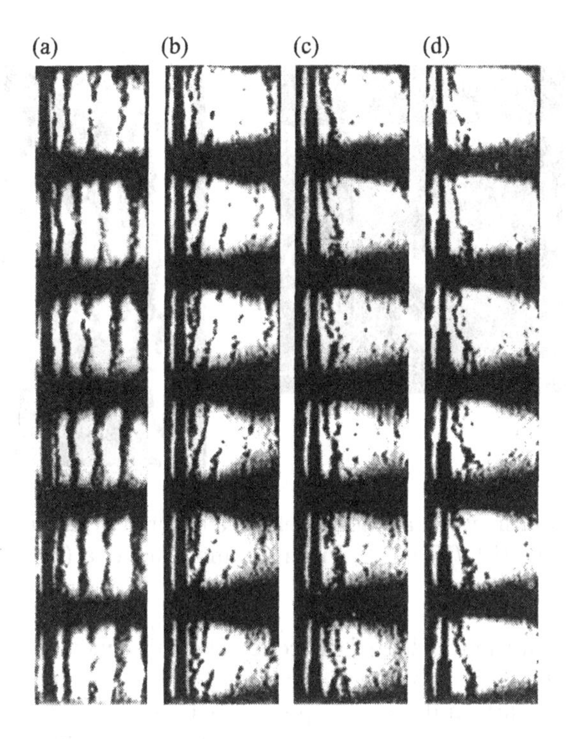

FIG. 22.61. Schlieren cine film visualization behind stepped cylinders, (a) $d/D = 1$, (b) $d/D = 0.87$, (c) $d/D = 0.71$, (d) $d/D = 0.49$, Naumann and Quadflieg (1974P)

$$TR = (D_r - D_e)/H \tag{22.6}$$

where D_r and D_e are the root and tip diameters, respectively, and H is the cylinder height.

22.7.1 *Laminar periodic wake, the L3 regime*

Eddy shedding from slender cones at low Re was examined by Gaster (1969J). Two sharp pointed cones were fitted with an end plate $D_e/D = 4.5$ at the root. The taper ratio TR was 0.023 and 0.055 for the two cones tested, respectively. For the first cone, $Re_{\max} = 150$ was at the root and zero at the pointed end. This meant that the periodic eddy shedding occurred only along two-thirds of the cone height. For the second cone, $Re_{\max} = 600$; hence, the TrSL1, TrW, L3, 2 and 1 regimes occurred side by side along the height.

Figure 22.62(a,b) shows the $TR = 0.055$ cone in water, visualized with a single fluorescent dye filament introduced from the upstream top. Passing the cone, the filament splits, revealing two regions:

(i) The horizontal downstream pattern, top in Fig. 22.62(a), shows no eddy

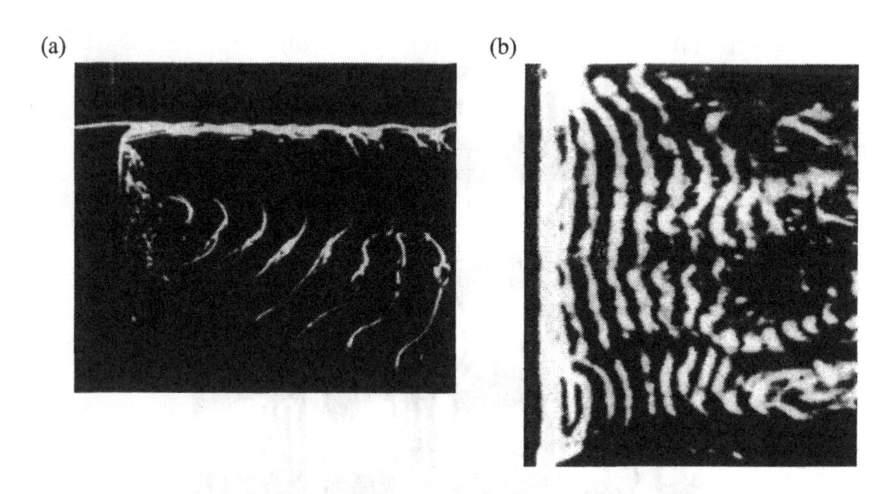

FIG. 22.62. Water flow visualization behind a tapered cylinder, (a) $TR = 0.055$, Gaster (1969J), (b) $TR = 0.2$, Piccirillo and Van Atta (1993J).

street behind the top $\frac{1}{4}H$ of the cone. The subsequent development of the entangled loops results from linking the neighbouring eddy filaments. The 'frozen' dye pattern does not represent actual flow farther downstream. It is caused by a large difference in the diffusivity of the vorticity and dye, see Vol. 1, Chapter 3.

(ii) The strong secondary flow in the form of a downwash is caused by base pressure lowering with increasing D. The phenomenon is analogous to the downwash produced by the free stream shear behind the constant diameter cylinder. The slanted eddy filaments farther downstream are also analogous to those observed in the free stream shear, see Vol. 1, Chapter 15.

A complementary flow visualization using a fluorescent oil painted cylinder washed slowly by water was carried out by Piccirillo and Van Atta (1993J). Figure 22.62(b) shows not only slanted eddy filaments but also three shedding cells. At the start of the motion, parallel eddy filaments were observed. Owing to the increased rate of shedding at the root of the tapered cylinder, the eddy filaments began to bend over and cells were formed. The knots then occurred in succession, moving from the small diameter to the cylinder root. The flow behind $TR = 0.2$ was not so well ordered.

Gaster (1969J) positioned the hot wire at $x/D = 7.5$, $y/D = 1$, and $z/H = 0.26$ behind the $TR = 0.055$ cone. Figure 22.63(a,b,c) shows a selection of oscillograms at $Re = 74.9$, 101, and 117, respectively, based on D at $z/H = 0.26$. At low Re, when the wake becomes just unstable, the hot-wire signals are regular and periodic, Fig. 22.63(a). An increase in Re brings a low-frequency modulation, Fig. 22.63(b), and further increases makes modulation deeper, Fig. 22.63(c).

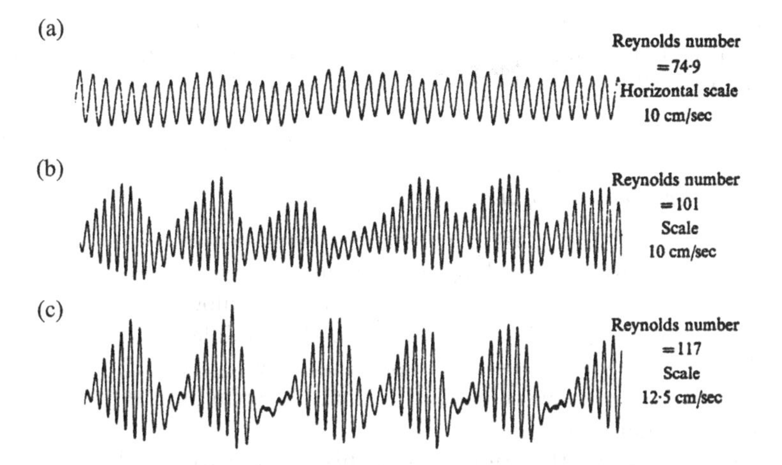

FIG. 22.63. Velocity fluctuation at $x/D = 7.5$, $y/D = 1$, $z/H = 0.26$, (a) $Re = 74.9$, (b) $Re = 101$, (c) $Re = 117$, Gaster (1969J)

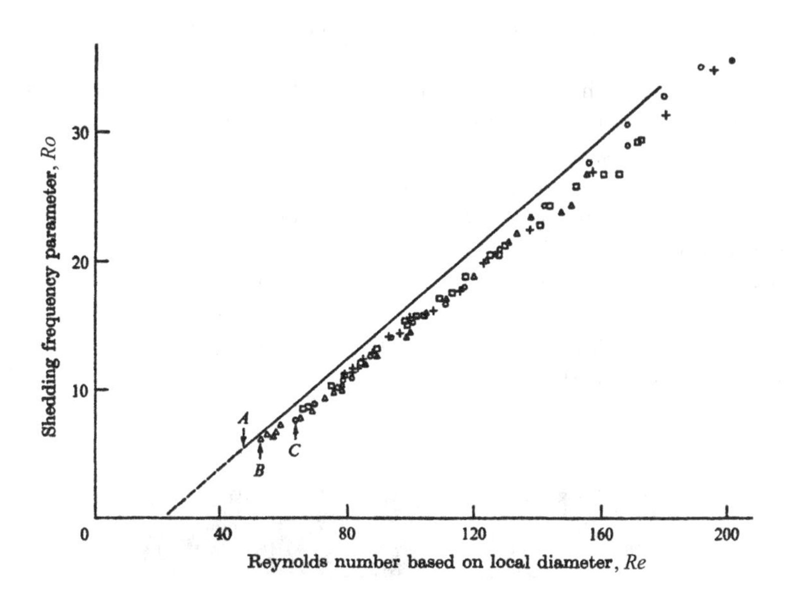

FIG. 22.64. Roshko number in terms of local Re for $TR = 0.023$ and 0.055, Gaster (1969J)

Figure 22.64 shows the comparison of measured $Ro = nD^2/\nu$ in terms of local Re. The full line corresponds to Roshko's measurements for a non-tapered cylinder. Subsequent measurements by Piccirillo and Van Atta (1993J) on truncated cones confirmed the lowered Ro value for the tapered cylinders. An empirical relationship for St was in the form

$$St_c = 0.195 - 5/Re \tag{22.7}$$

22.7.2 *Shedding cells along the span*

A typical feature of flow past a non-tapered cylinder in a free stream shear has been the formation of shedding cells along the span. The tapered cylinder submerged in the uniform free stream produces an analogous phenomenon. The existence of shedding cells is associated with the modulation of the signal and two peaks in the frequency spectra. A practical way to determine the boundary between the cells is to move the hot wire along the span until the two frequency peaks become equal in magnitude.

Piccirillo and Van Atta (1993J) used this method to establish the number of cells N along the TR = 0.01, 0.13, 0.17, and 0.20 cylinders. They found that N depends on TR and less on Re. In fact, N appears independent of Re for $Re_m > 90$. The cell size was also expressed through a non-dimensional parameter

$$(D_{c1} - D_{c2})/D_m = (z_{c1} - z_{c2})(D_r - D_e)/HD_m \tag{22.8}$$

where D_{c1} and D_{c2} are the diameters where the cell starts and ends, respectively, D_m is the local diameter at the midspan, and z_{c1} and z_{c2} are the elevations where the cell starts and ends, respectively.

The size parameter was comprised of the relative cell size $\Delta z_c/H$ and the relative change in diameter, $(D_r - D_e)/D_m$. The cell size became 0.15 and independent of TR and Re for $Re > 100$. The position and size of the shedding cells were not affected by the end conditions. This was established by Piccirillo and Van Atta (1993J) by adding small and large end plates to the initially free ends.

Piccirillo and Van Atta (1993J) also applied wavelet analysis to some data sets. As an illustration, Fig. 22.65(a) shows wavelets for TR = 0.02, at Re =

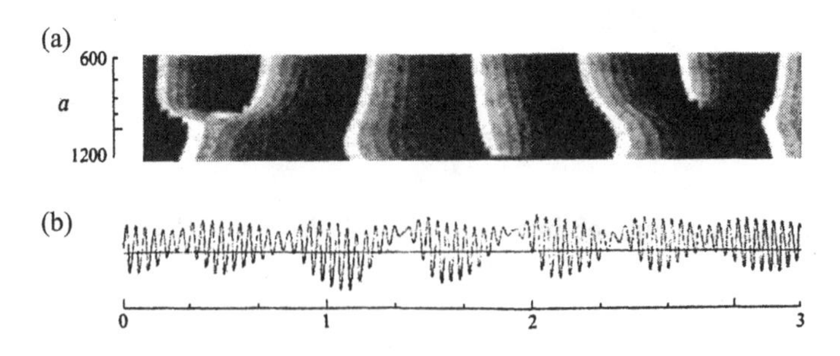

FIG. 22.65. (a) Wavelet profiles along span, (b) velocity modulation for $TR = 0.2$ and $Re = 99/146$, Piccirillo and Van Atta (1993J)

49/146, for the frequency scales corresponding to the shedding and modulation frequencies, respectively, Fig. 22.65(b). The phase is plotted at the top and the transformed signal at the bottom. The parameter a designates the scale of the wavelet, being small and large for the shedding and modulation frequencies, respectively.

22.7.3 *Theoretical model*

Gaster (1969J) attempted to model cellular eddy shedding from slender cones. The common feature of the eddy shedding mechanism is that the periodicity is self-excited and self-limited. Gaster (1969J) noted that the van der Pol oscillator with its weak linear excitation term has just such behaviour. Thus the observed modulated oscillation from a cone can be represented by a line of such oscillators having suitable coefficients to give different oscillation frequencies along the span. It is also necessary to incorporate some coupling between the cells of different frequencies by including a spanwise 'stiffness' term suitably scaled

$$D^2\left(\frac{\partial^2 v'}{\partial t^2} - A\frac{\partial^2 v'}{\partial z^2}\right) + n^2 v' = \varepsilon D \frac{\partial v'}{\partial t}\left(v'^2 - 1\right) \tag{22.9}$$

where D is the local length scale, n is the characteristic frequency, v' is the fluctuating velocity, A is the arbitrary constant, and ε is a small non-linear parameter determining the rate of growth of v'.

For very slender cones, eqn (22.9) becomes the simple van der Pol equation and the corresponding wake structure is shown in Fig. 22.66. This simulation of wake oscillators is applicable to a low Re flow where the shed eddies are coherent over a finite length along the span. The theoretical model does not predict the knot formation between cells arising from the difference in eddy shedding frequency in neighbouring cells.

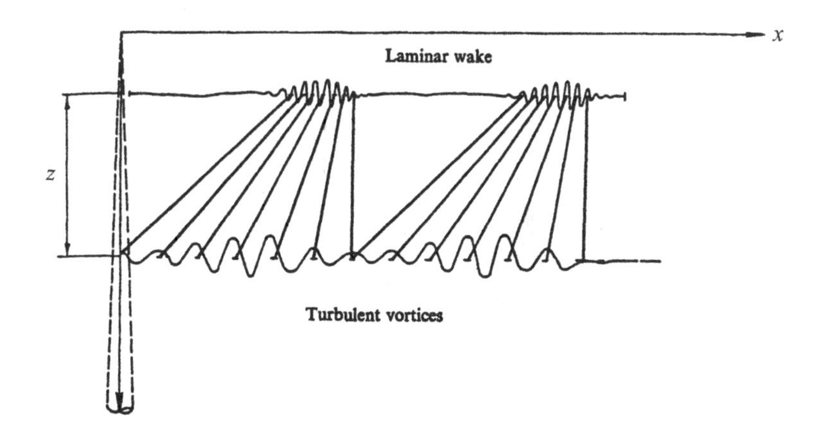

FIG. 22.66. Mathematical model of wake structure, Gaster (1969J)

22.7.4 *Turbulent wake, the TrSL state*

Naumann and Quadflieg (1968P, 1974P) used the schlieren optical flow visualization, which shows differences in air density. A high speed, $Ma = 0.3$, blow-down wind tunnel was used to test a non-tapered cylinder, $L/D = 7.5$, $D/B = 0.13$, and Re = 147k, and three tapered cylinders having $TR = 0.017$, 0.044, and 0.087. Figure 22.67(a–d) shows three frames from two cine films shot simultaneously in the x, z and x, y planes at Re = 147k. The straight and distinct eddy filaments for the non-tapered cylinder become slanted and distorted as the taper increases.

Hsiao *et al.* (1992P) measured the frequency of eddy shedding behind tapered cylinders at $0.025 < TR < 0.12$, $5.33 < L/D_m < 7.27$, and $10\text{k} < Re_m < 14\text{k}$. Figure 22.68(a,b,c) shows the effect of taper, TR = 0.025, 0.075, and 0.125, on shedding cells, respectively. For the smallest taper, two overlapping cells are found, and three for $TR = 0.075$. However, for $TR = 0.125$, this was reversed to two end cells. The spanwise extent of cells along the narrow side appears twice as long as along the wide side.

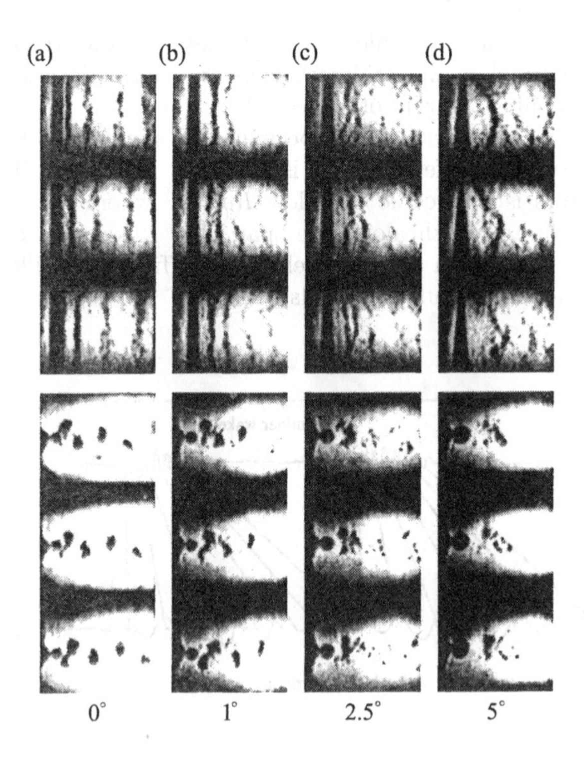

FIG. 22.67. Schlieren visualization along (top), and across tapered cylinder (bottom), (a) $TR = 0$, (b) $TR = 0.017$, (c) $TR = 0.044$, (d) $TR = 0.087$, Naumann and Quadflieg (1968P)

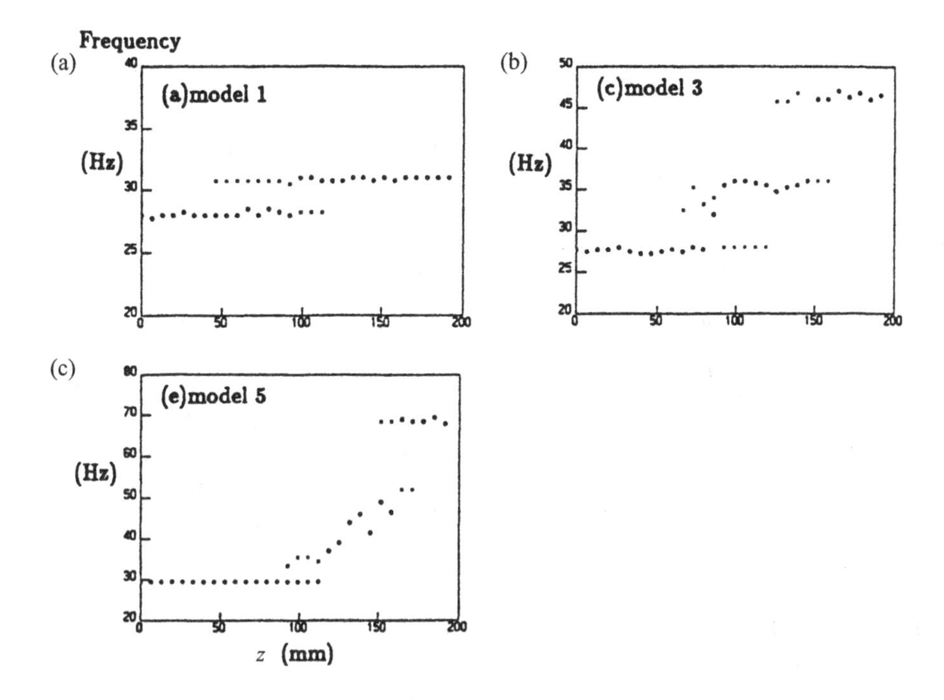

FIG. 22.68. Shedding frequency versus span, (a) $TR = 0.025$, (b) $TR = 0.075$, (c) $TR = 0.125$, Hsiao *et al.* (1992P)

22.7.5 *Tapered cylinder with free end*

Most chimney stacks are tapered and exposed to a shear turbulent wind. Vickery and Clark (1972J) carried out tests on a 1:500 stack model, $H/D_m = 20$, $TR = 0.04$, $20\text{k} < Re < 40\text{k}$, $Ti = 1\%$–2%, 8%–12%, smooth and turbulent flow, respectively. Figure 22.69(a) shows end cells at both ends of the tapered cylinder in the smooth flow. The chain-dot line and full line correspond to a constant $St = 0.2$. The free stream turbulence obliterated the root cell, and the top cell was split in two. The frequency peak band-width widened in comparison with the corresponding peak for the smooth flow. Figure 22.69(b) shows the local base pressure, drag, and fluctuating (rms) lift along the height of the model.

22.8 Non-linear change in diameter

22.8.1 *Introduction*

Thermodynamic processes utilized in power plants require cold water as a heat sink. This water is heated during the process, and has to be discharged straight into rivers, lakes, or the sea, provided the temperature increase in the latter does not endanger fish and other aquatic life. Where this may happen, cooling towers have to be built. The hot water from the plant is sprinkled from the top of

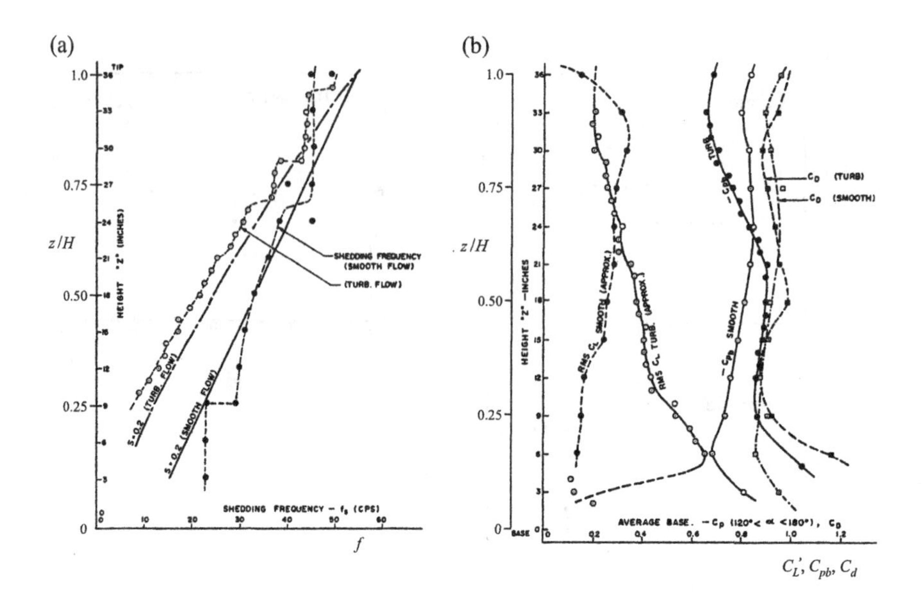

FIG. 22.69. Model of tapered stack $H/D = 20$, $TR = 0.04$, $Re = 40$k in smooth (closed circles) and turbulent (open circles) flow, (a) shedding frequency, (b) local fluctuating lift, base pressure, and drag, Vickery and Clark (1972J)

the cooling tower, and cascades downwards against the atmospheric air flowing upwards either due to a natural draught or forced by a fan. For efficient heat transfer, a vertical convergent–divergent nozzle is required placed some distance above the ground and open at both ends to the atmosphere. Hence, the shape and elevation of cooling towers are governed by the internal flow, and not by the external flow produced by the wind.

The importance of wind loading was highlighted after the collapse of three cooling towers at the Ferrybridge power station in Yorkshire, England. Eight cooling towers 117 m high were arranged in two staggered rows, four towers in each row, as seen in Fig. 22.70. Two spacing ratios between the towers divided by the mean diameter, D_m, were defined as the transverse ratio $T/D_m = 2.2$ and the longitudinal ratio between two rows, $L/D_m = 1.5$.

On Monday, 1 November 1966, three towers (marked with crosses) collapsed in gale-force winds at $Re \approx 130$M. Towers 1B and 1A collapsed within a period of 10 min, tower 2A failed 40 min later, and the wind direction was estimated to be as sketched in Fig. 22.70(a). All three collapsed towers were in the second 'shielded' row, and not in the wind-exposed first row. Another puzzle was that the subsequent inquiry established that the cause of the failure was the excessive tension in the shell on the windward side of the collapsed towers, and not on the sides where $C_{p\min}$ occurred. This disaster triggered, and maintained, worldwide

(a)

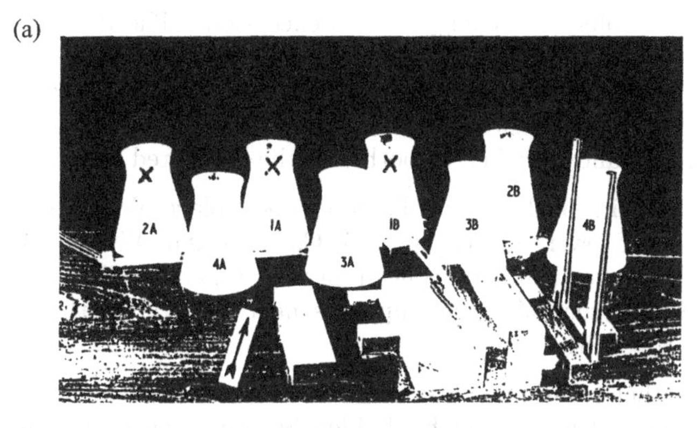

(b)

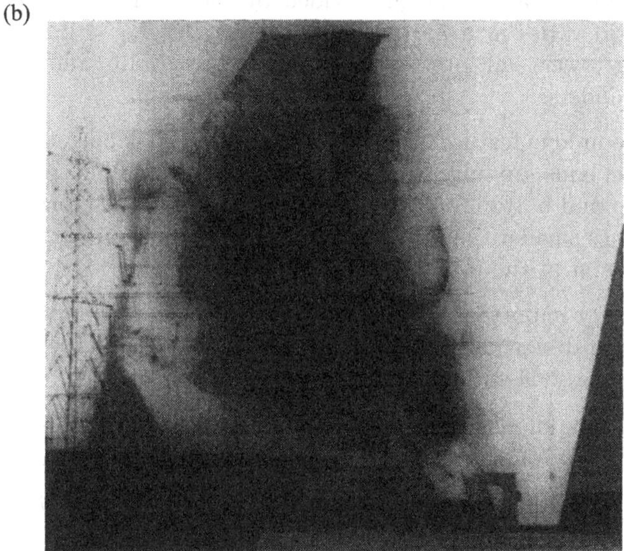

FIG. 22.70. (a) Ferrybridge (Yorkshire, UK) power station wind tunnel model, Armitt *et al.* (1967P), (b) collapse of cooling tower 1A

research.

22.8.2 *Cooling towers*

The formidable combination of a series of governing and influencing parameters is always present in wind flow around cooling towers. There are at least ten important parameters, which may be grouped into two main categories:

(1) Parameters related to the geometry of the cooling towers:

(a) the shape adopted for most cooling towers is an axially symmetric hyperboloid.[123] There are two exceptions. The first model for wind tunnel tests made by Cowdrey and O'Neill (1956P) consisted of a cylindrical 'throat' section attached to two truncated cones. Some later cooling towers were designed as a toroidal surface forming a throat and attached smoothly to the truncated cones on two sides;

(b) a non-linear change in diameter takes place along the height of the cooling tower. The exception is Cowdrey and O'Neill's (1956P) model where two truncated cones represent tapered cylinders. The distinct 'throat' section always appears near the top in the range $0.75 < z/H < 0.81$;

(c) another self-similar feature of cooling towers is the aspect ratio. It is expressed as the height divided by the mean diameter. The aspect ratio varies in a narrow range $1.80 < H/D_m < 2.28$. Hence, cooling towers fall into the category of short, finite, and double-tapered cylinders;

(d) the unique feature of all cooling towers is that both the top and bottom ends are fully opened for air intake and efflux, respectively. The top and bottom openings are unequal in size, the former being three times smaller than the latter. The sheared wind effect is considerably greater at the top due to the high elevation;

(e) a very important parameter, which plays a dominant role in cooling tower design, is surface roughness. There are two types of roughnesses on the shell surface:

  (i) meridional ribs protruding from the shell surface are a design feature to reduce the $C_{p\min}$ value on the tower in high winds;

  (ii) irregular surface roughness for plain concrete shells.

  There is a transition related to the thinning boundary layer with rising *Re* until the surface roughness starts to protrude above the boundary layer thickness. This causes a considerable increase in the local drag force;

(f) the size of cooling towers is enormous, $60\,\mathrm{m} < H < 207\,\mathrm{m}$ at present. This contributes to extreme values of *Re* at high winds. The typical *Re* range is $20\mathrm{M} < Re < 200\mathrm{M}$, where *Re* is based on the mean tower diameter D_m. This range is more than one order of magnitude beyond the reach of existing wind tunnels;

(g) the final feature, almost always present, is the grouping of cooling towers from two up to eight. The restricted land on sites is conducive

[123]The surface of a hyperboloid is generated when a slanted straight line rotates around the vertical axis.

to small spacing between cooling towers. This leads to considerable interference effects with possible catastrophic consequences;

(2) Parameters related to wind characteristics:

(a) the wind is an atmospheric boundary layer, and is always sheared and turbulent. This makes the wind velocity profile non-uniform and randomly fluctuating. Hence, there are three uncontrollable parameters: wind velocity profile, intensity of turbulence Ti, and scale of turbulence Ts;

(b) the full-scale measurements in a gusty natural wind are subject to large uncertainties caused by limited locations of anemometers for measuring wind magnitude, direction, distribution, and fluctuation. The ever-present buildings and structures on the site introduce additional interference effects.

22.8.3 *Model tests*

Early wind tunnel tests of cooling towers were initiated by Cowdrey and O'Neill (1956P), followed by Salter and Raymer (1962P). They approximated the hyperboloid by a short cylindrical throat joined onto two truncated cones, as seen in Fig. 22.71(a). The tests were carried out in a compressed wind tunnel with an open test section (no blockage effects) at 6.1M $< Re <$ 14.1M. The surface was smooth (estimated $K/D = 0.016\%$). The local C_{po} and C_{pb} are shown in Fig. 22.71(a) for uniform flow (full line), shear flow (crosses), and when the velocity at the top was taken as a reference one (dash-dot line).

The C_p distributions at A-A ($z/H = 0.79$) and at B-B ($z/H = 0.43$) are given in Fig. 22.71(b,c), respectively. There is some departure in C_{pb} between the uniform (full line) and shear flow (crosses) for the A–A elevation. The B–B elevation shows a difference in $C_{p\min}$ instead. The first elevation is associated with the throat and the second with the middle of the lower taper of the model. The C_p distribution for Re = 7.1M and 14.1M was found to be virtually the same.

Cowdrey and O'Neill (1956P) also examined the effect of the internal flow and its efflux on the internal pressure distribution. They found that C_p inside the model was independent of θ; the values are shown with arrows in Fig. 22.71(b,c). The difference between the pressure on the outer and inner surfaces is maximum along the stagnation line $\theta = 0°$, and zero along $\theta = 180°$. This fact will be recalled from the previous subsection.

22.8.4 *Validity of Re extrapolation*

The upper limit in Re for wind tunnels is around 14M at present. This is one order of magnitude below $Re \approx$ 140M for large cooling towers in high wind. The unavoidable extrapolation represents a big leap into the unknown. The extrapolation is 'supported' by a series of hypotheses.

One sweeping hypothesis is that although cooling towers have a low aspect

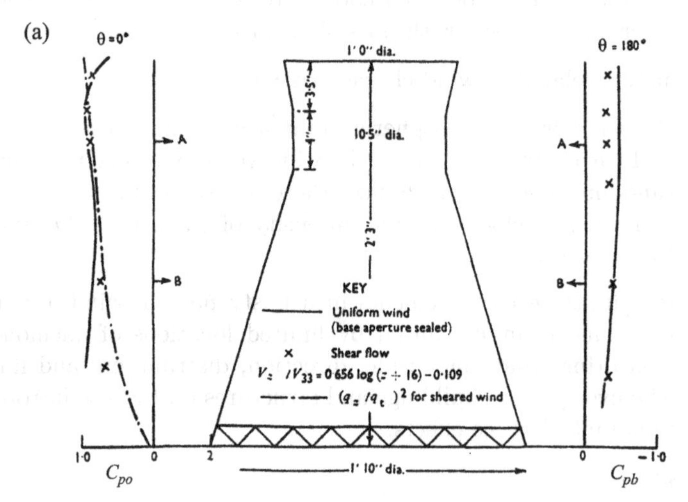

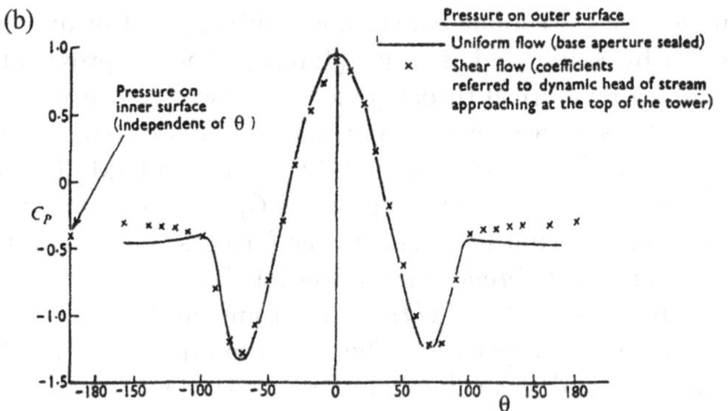

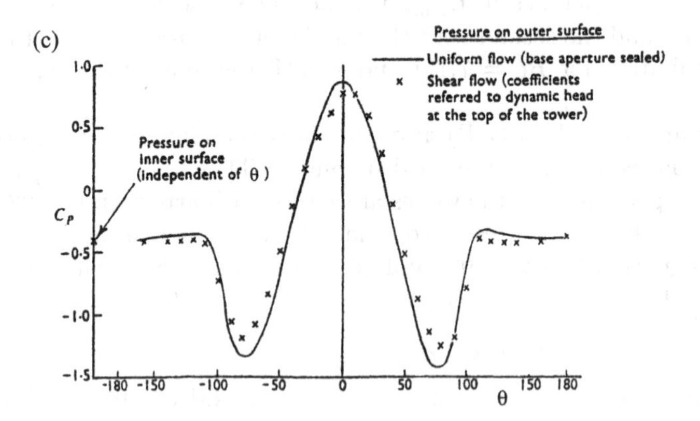

FIG. 22.71. Simplified cooling tower model, (a) local C_{po} and C_{pb} distributions, (b) C_p at $z/H = 0.79$, A–A, (c) C_p at $z/H = 0.49$, B–B, Scruton (1967P)

ratio, double taper, rough surface, and two open free ends they are expected to be 'similar' to finite, non-tapered cylinders. It has been discussed in detail in Vol. 1, Chapter 7, that nominal two-dimensional cylinders reach the fully turbulent T1 regime for $6M < Re < 10M$, where the final transition in boundary layers ends. The T1 regime may result in C_p, C_p', C_D, C_L, θ_s, etc. being independent of Re. However, this has not yet been proven for finite cylinders. The secondary flow near the free end depends on the aspect ratio and efflux even in the T1 regime.

An attempt to validate experimentally the Re-independence hypothesis for cooling towers was carried out for the Ferrybridge Committee of Inquiry. New measurements of the C_p distribution were made in the NPL[124] compressed air wind tunnel at rather low Re = 1.5M, 1.8M, and 2.9M. Figure 22.72(a,b) shows the C_p distribution at nine elevations $0.19 < z/H < 0.92$ in terms of θ. The C_p curves differ considerably for the three Re at all elevations. Particularly large differences are for $C_{p\min}$ and the start of the adverse pressure recovery. Lowe and Richards (1967P) concluded that Re independence was not reached at Re = 2.9M for the cooling towers tested.

22.8.5 *Surface roughness*

Another questionable hypothesis is that in the T1 regime $C_p, C_{p\min}, C_{pb}$, and C_D do not depend on the size and texture of the surface roughness. The turbulent boundary layer thickness δ becomes thinner with increasing Re, while the relative roughness thickness K/D remains the same. This means that for $K = \delta$ the roughness will start to protrude from the boundary layer. This mechanism of transition from submerged to protruding roughness will affect the boundary layer development, separation, and overall flow around cooling towers in the T1 regime.

Armitt (1968P) carried out tests in the range $85k < Re < 390k$ using Cowdrey and O'Neill's (1956P) model. The aim was to simulate the C_p distribution obtained at Re = 7.1M by using surface roughness. Sieved sand was glued onto the model producing K/D = 0.09%, 0.19%, and 0.38%, and two turbulence grids were used Ti = 3.5% and 10.5%. After some trial and error, Armitt (1968P) found that the best fit to Cowdrey and O'Neill's (1956P) data was obtained by using K/D = 0.19% and Ti = 3.5% at Re = 185k. Figure 22.73 shows good agreement between C_p curves at Re = 7.1M (full line) and 185k (open circles). Similarly, good agreement was obtained by using K/D = 0.09% at Re = 390k. This proved that the T1 regime can be simulated by using surface roughness. However, it remained unknown what the C_p distribution would be on a rough cooling tower at high Re. However, the generalization that the introduction of free stream turbulence and surface roughness always simulate a higher Re regime should be avoided.

[124]NPL stands for the National Physical Laboratory, Teddington, UK.

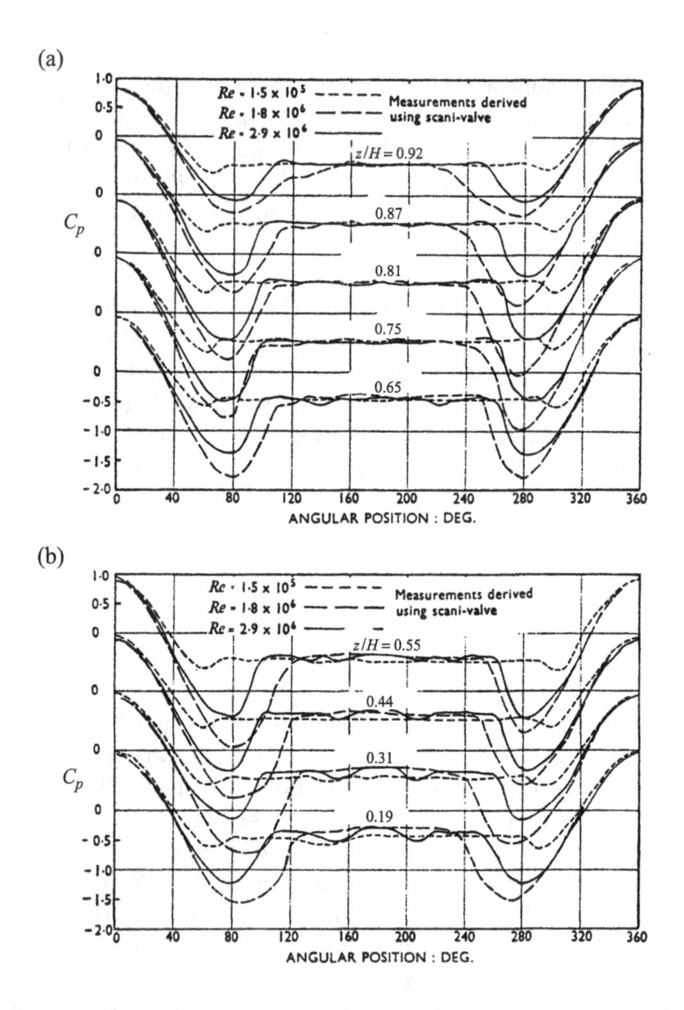

FIG. 22.72. Mean pressure distribution at Re = 150k, 1.8M, 2.9M, (a) $0.19 \leqslant z/H \leqslant 0.55$, (b) $0.65 \leqslant z/H \leqslant 0.92$, Lowe and Richards (1967P)

22.8.6 *Meridional ribs*

Cooling towers built in Germany and some in the USA and France were fitted with meridional ribs spaced at a constant pitch p around the shell. Figure 22.74(a) shows the model of the ribbed tower. Figure 22.74(b,c) shows the C_{po}, C_{pb}, and $C_{p\min}$ distributions along the height of a full-scale cooling tower at $Re = 1.1$M (ribbed) and $Re = 6.5$M (smooth), respectively. There is a significant reduction in $C_{p\min}$ for the ribbed model and a small change in C_{pb}.

The meridional rib roughness is described by the following parameters:

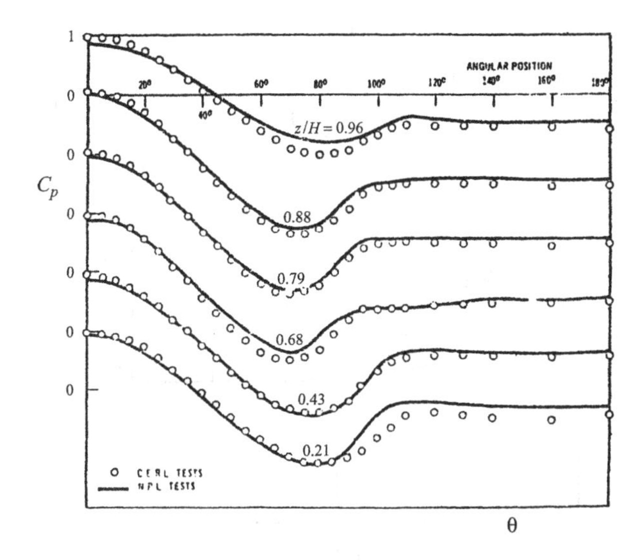

FIG. 22.73. Comparison of measured pressure distribution around and along cooling tower model at Re = 7.1M, and 185k, Ti = 3.5%, K/D = 0.09%, Armitt (1968P)

(i) K/D_m, the relative roughness is defined as the ratio of rib height K and mean diameter D_m. The actual relative roughness K/D varies along the height: $(K/D)_{\max}$ occurs at the throat and $(K/D)_{\min}$ at the root of the tower;

(ii) b/K, ratio of the width of the rib b and height K;

(iii) p/K, the circumferential pitch p is divided by the height K;

(iv) N, number of ribs is related to p as

$$N = \frac{D\pi}{p} \tag{22.10}$$

Farell *et al.* (1976J) carried out tests on a 1:250 model of the Weisweiler cooling tower, as shown in Fig. 22.74(a), in the range 150k < Re < 470k, D/B = 0.07, and Ti = 0.2%. The effect of N and K/D on $C_{p\min}$ was investigated. Figure 22.75(a,b) shows the $C_{p\min}$ variation in terms of Re and K/D, respectively. Figure 22.75(a) shows that only three sizes of ribs are used K/D = 0.15% and 0.40%, but there are eight different $C_{p\min}$ curves related to the different number of ribs.

The effects of K/D on $C_{p\min}$ are compiled in Fig. 22.75(b) from the model tests by Armitt (1968P), Cowdrey and O'Neill (1956P), Davenport and Isyumov (1966P), Ebner (1968P), Farell *et al.* (1976J), Golubovic (1957J), Hayn

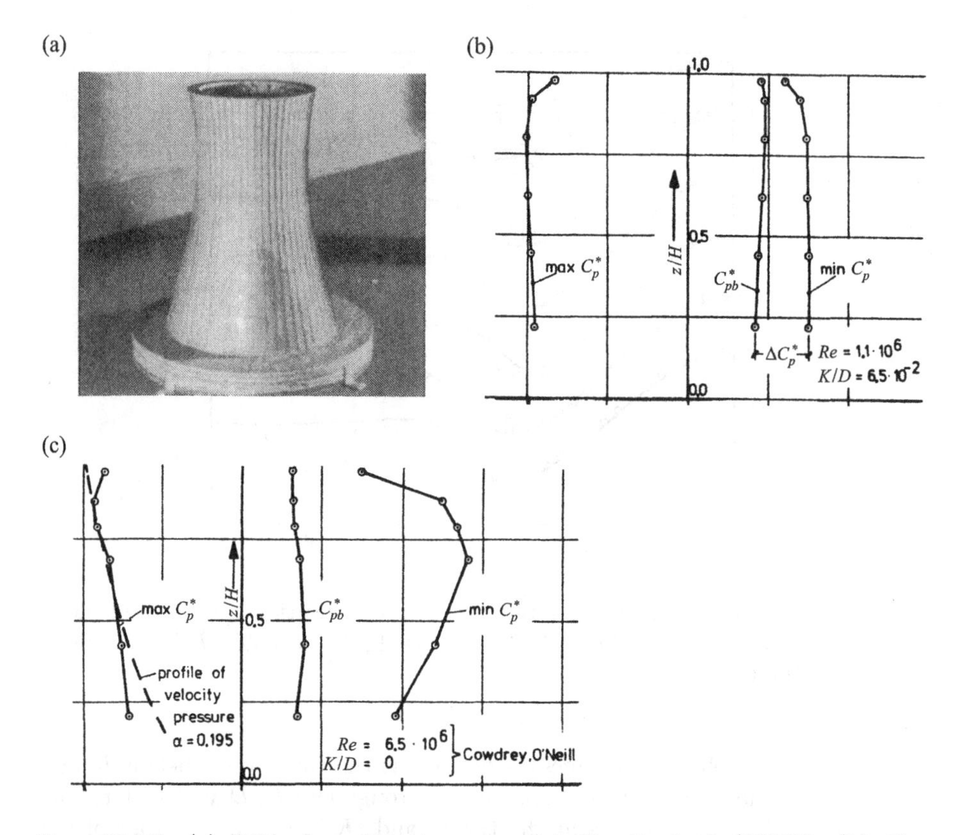

FIG. 22.74. (a) Ribbed cooling tower model, Farell *et al.* (1976J), (b) C_{po}, $C_{p\min}$, C_{pb} with ribs, (c) C_{po}, $C_{p\min}$, C_{pb} without ribs, Niemann (1971P)

(1976P), Niemann (1971P), and Pris (1959P), and full-scale data in wind by Ebner (1968P) and Niemann (1971P). Each experimental point is accompanied by a number indicating the p/K value. A decrease in $C_{p\min}$ seems to level off for $K/D > 0.5\%$, and the agreement of points from various sources is good. The most notable is a good agreement between the model and full-scale tests for $K/D = 0.2\%$ (open circle) and for $K/D = 0.35\%$ (cross and circle).

Figure 22.76 shows a compilation of the adverse pressure rise, $\Delta p = C_{p\min} - C_{pb}$ in terms of K/D. The decrease in Δp seems to level off at $K/D > 0.4\%$. As the boundary layer becomes retarded and thicker with increased K/D, the flow separates more easily when subjected to an adverse pressure rise. This causes a monotonous decrease in Δp up to $K/D \approx 0.4\%$. This argument was confirmed by Farell *et al.* (1976J).

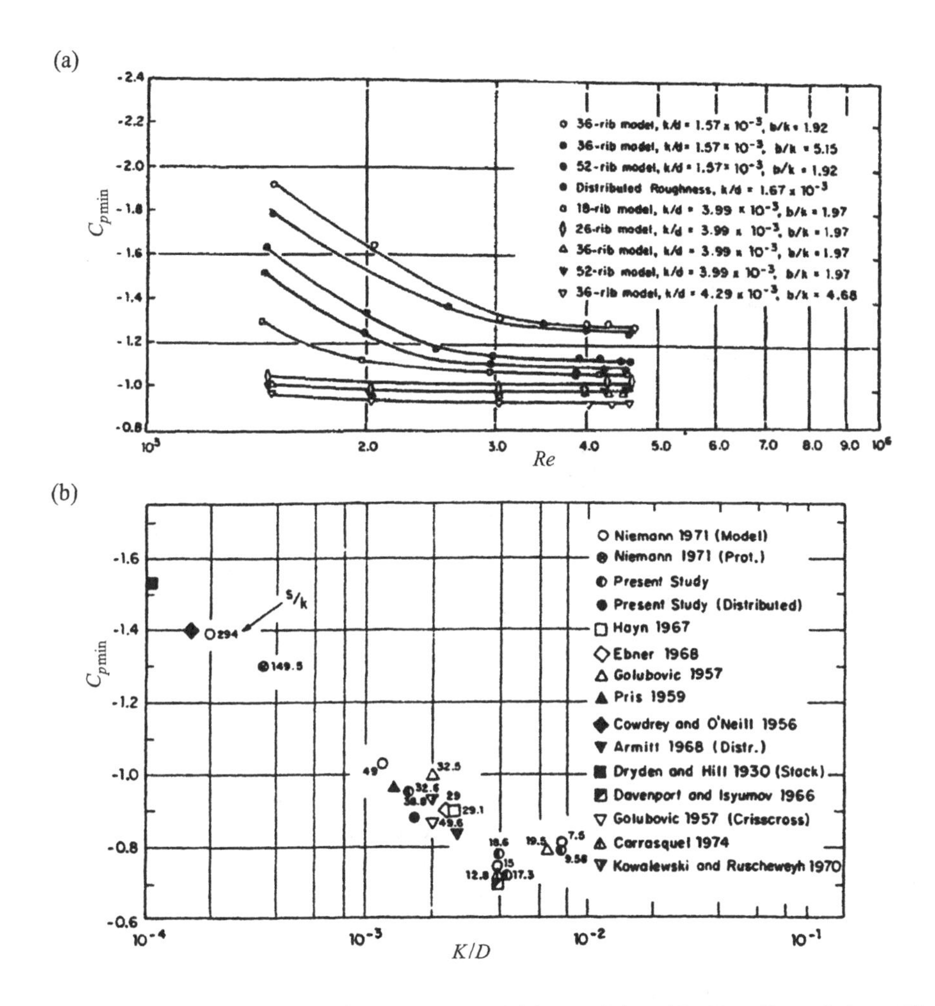

FIG. 22.75. Minimum pressure in terms of (a) Re, (b) K/D, Farell *et al.* (1976J)

22.8.7 *Cooling tower model in a gust*

Sawyer (1973P) measured C_p' on a 1:375 model of the Ferrybridge cooling tower at $Re = 120\text{k}$ and seven elevations. The wind profile was simulated by an array of cylindrical rods spaced according to Owen and Zienkiewicz's (1957J) method. The unique feature of the open jet wind tunnel was the production of a transverse gust[125] by flapping an array of aerofoils placed across the jet at variable frequency and amplitude. The electro-hydraulic actuators were driven by a tape-recorded wind signal.

[125] A gust is defined as a sudden change in the direction of the wind velocity.

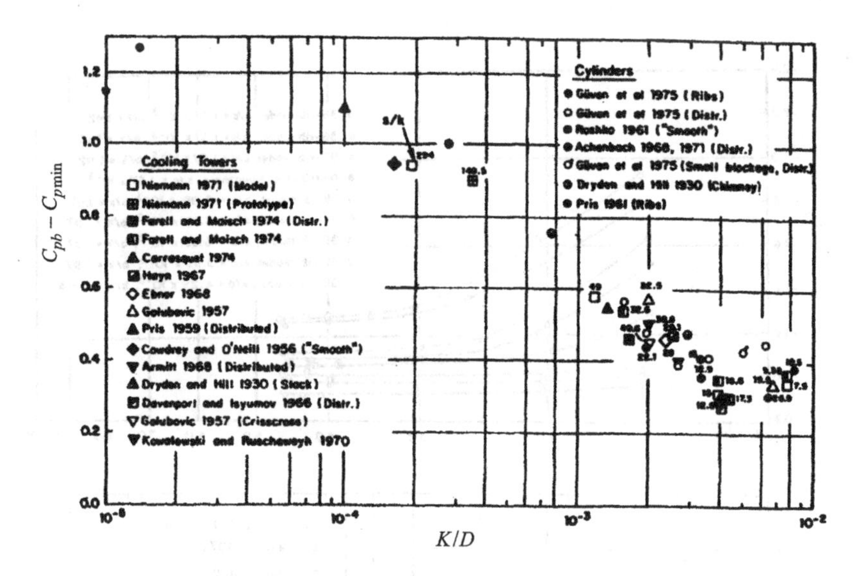

FIG. 22.76. Compilation of adverse pressure gradients in terms of K/D, Farell *et al.* (1976J)

A survey of measured fluctuating pressure distribution on a single and 15° staggered tower showed that $C'_{p\max}$ occurred slightly below the throat at $z/H = 0.735$. Figure 22.77(a) shows a typical C'_p distribution around the single tower for smooth, $Ti = 2.4\%$ and 4.7%, gust turbulence at $z/H = 0.735$. Figure 22.77(b,c) shows the measured frequency power spectra at $\theta = 60°$ and $120°$, respectively. There is a pronounced peak for $Ti = 4.7\%$ at $\theta = 60°$, and a notable displacement of the power spectra to higher frequencies for $\theta = 120°$. These measurements show the effect of a gust on the laminar boundary layer at medium *Re*.

22.8.8 *Full-scale tests in natural wind*

Full-scale tests of cooling towers in natural wind appear deceptively attractive as a method of reaching very high *Re*. However, natural wind brings a series of uncontrollable and unrepeatable parameters, such as random magnitude, direction, gustiness of the shear velocity, and intensity, scale, and frequency spectra of turbulence. An advantage of full-scale tests is the presence of other buildings and structures on the site, so that the interference effects are included. The limitations of full-scale tests are of three kinds:

(i) the fixed geometry of the cooling tower and the single K/D value for surface roughness;

(ii) random variation in *Re*, Ti, and Ts;

(iii) the difficulty in measuring the free stream static pressure, which varies

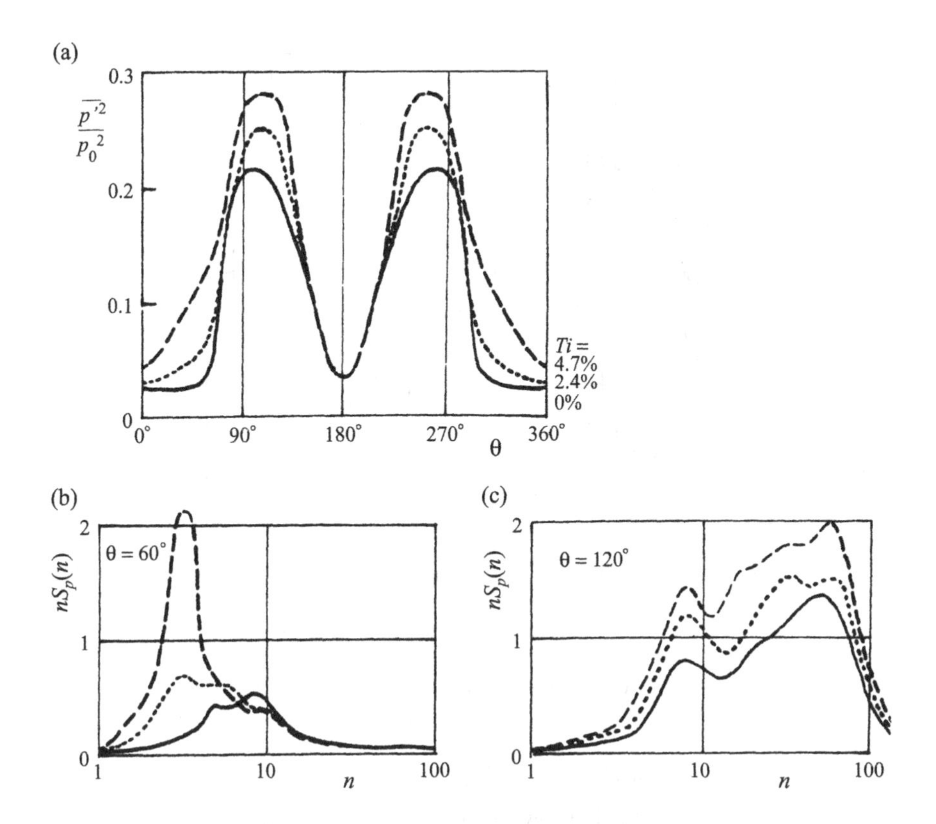

FIG. 22.77. Cooling tower model in a gust, (a) fluctuating pressure distribution, (b) frequency power spectra at $\theta = 60°$, (c) at $\theta = 120°$, Sawyer (1973P)

with time, and the dynamic pressure, which varies with the height of the cooling tower and time.

Figure 22.78 illustrates the extrapolation gap between the wind tunnel and full-scale tests. Niemann (1971P) measured the C_p distribution at elevation $z/H = 0.59$ on two cooling towers in Germany: Weisweiler, $K_r/D = 0.65\%$, and Schmehausen, $K_r/D = 2.3\%$. Figure 22.79(a) shows the $C_{p\min}$ and θ_s lines on the sketched cooling tower. The C_p distribution on the Weisweiler and Schmehausen towers at $Re = 65$M and 61M are shown in Fig. 22.79(b). The two C_p curves are in good agreement except around $C_{p\min}$. It seems that despite different K_r/D the adverse pressure rise is in the same place and of the same magnitude.

Niemann and Pröpper (1975/76J) also measured C_p' on the same cooling towers. His data were strongly affected by the adjacent 60 m building. Ruscheweyh (1975/76J) reported measurements on a single tower at $Re = 60$M. The complementary model tests at $Re = 300$k and 600k considerably underestimated

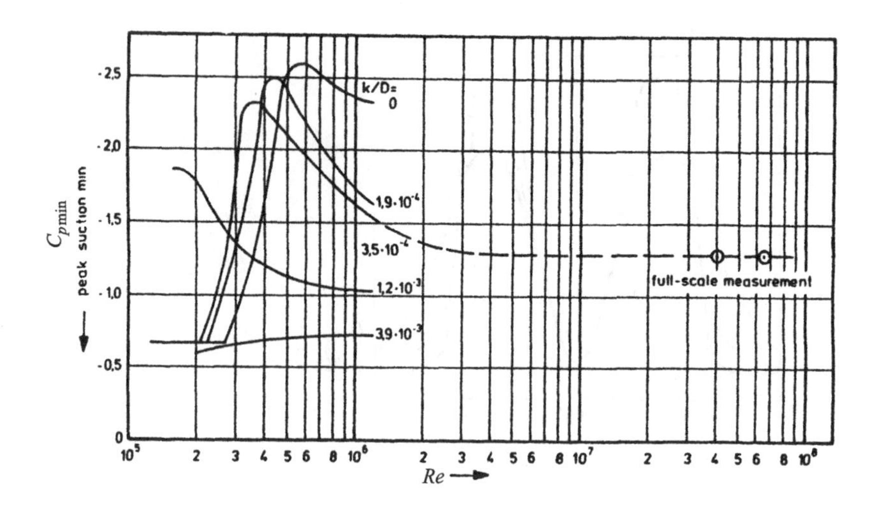

FIG. 22.78. Effect of relative roughness and Re on minimum pressure, Niemann (1971P)

the C_p' distribution. However, Davenport and Isyumov's (1966P) model tests at $Re = 300$k, and added turbulence, were in good agreement with full-scale tests.

22.8.9 *Possible causes of the Ferrybridge failure*

The collapse of the Ferrybridge towers prompted an official inquiry. The official cause of the collapse of the Ferrybridge cooling towers was found to be an excessive wind loading around $\theta = 0°$. A possible 'channelling effect' of the wind produced by the first row of cooling towers was proposed. However, other findings did not support this hypothesis. The local C_d was evaluated from the local C_p distribution at $Re = 2.9$M. Figure 22.80 shows three C_d curves:

(1) single tower, $C_D = 0.47$ (dotted line);

(2) tower 1A in the group of eight, $C_D = 0.34$ (full line);

(3) tower 3A in the group of eight, $C_D = 0.54$ (dashed line).

Paradoxically, the wind loading on tower 1A, which collapsed first, was the lowest. This might indicate that either a different loading occurred at $Re = 130$M or that some other effect increased the wind loading.

Owen (1967P) proposed a theoretical explanation for flow around grouped cooling towers. He noted the tendency of the group to slow down the approaching wind, and to ameliorate the so-called 'channelling effect'.[126] The group diverts

[126] The 'channelling' hypothesis was popular at that time. It assumed that the front row towers channelled the wind towards the second row towers.

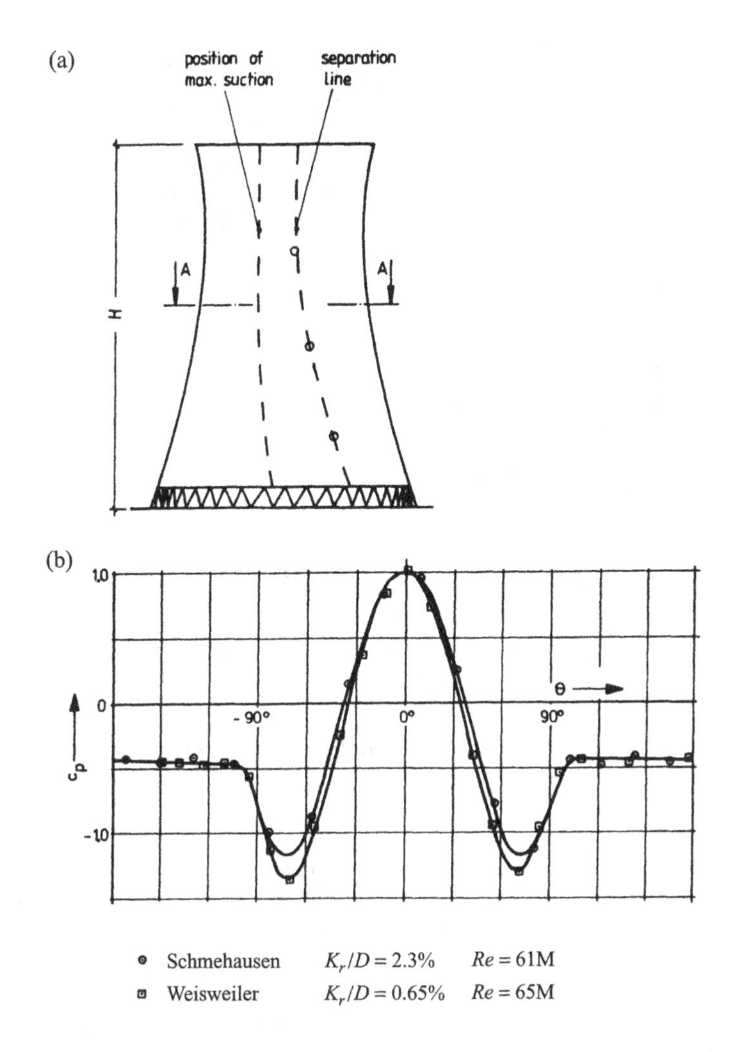

FIG. 22.79. Cooling towers in wind, (a) $C_{p\min}$ and θ_s lines, (b) mean pressure, Niemann (1971P)

part of the wind, and the approaching wind is slowed down in the vicinity of the front row. This was confirmed by a flow visualization at low *Re*. Figure 22.81 shows streamlines through a two-row array of two-dimensional cylinders arranged in the same way as the Ferrybridge towers. In order to simulate the observed flow pattern, Owen (1967P) replaced the cylinders by sources and vortices. This yields

$$V_1/V_0 = \tfrac{1}{2}\left[1 + (1 - 2C_{D0})^{1/2}\right] \tag{22.11}$$

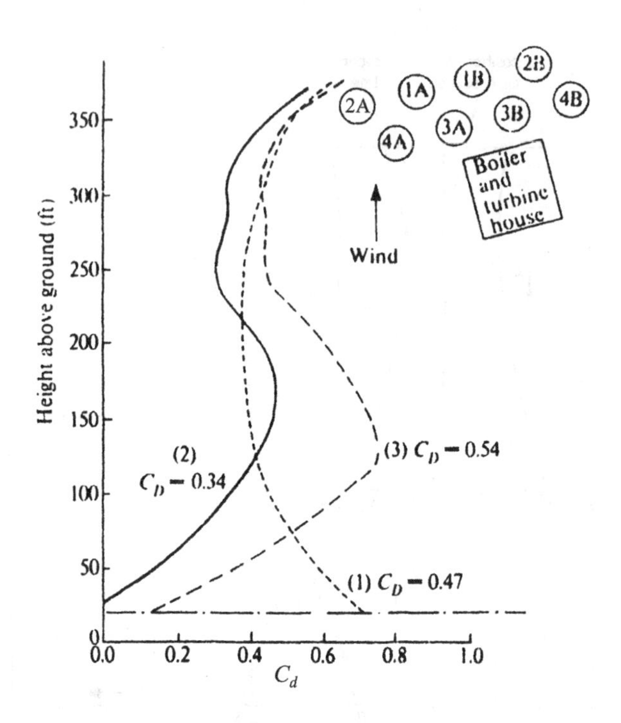

FIG. 22.80. Local drag force acting on cooling tower, (1) single, (2) tower 1A, (3) 3A, Scruton (1981B)

where C_{D0} is the overall drag coefficient of the entire array; V_0 is the free stream velocity; and V_1 is the velocity just upstream of the array. The measured $C_D =$ 0.59 and 0.39 for the upstream and downstream rows, respectively, yield an average $C_D = 0.44$ and from eqn (22.11) $V_1/V_0 = 0.67$. Hence, the wind velocity decreases 33% before reaching the front row. This explains the survival of the front row towers but does not account for the collapse of the rear row towers.

Maull (1966J) tested a group of five finite non-tapered cylinders having $H/D = 1.82$ and arranged in two rows, $T/D = S/D = 2$ at $Re = 150$k. The resultant force C_F, and its direction, were evaluated from the measured C_p distribution on one cylinder located in the rear row. Figure 22.82 shows that over a small range of wind direction around $\alpha = 26°$ the C_F can double in magnitude. If the wind direction changes by $\pm 2°$ around $\alpha = 26°$, then the C_F varies from 0.6 to 1.2. Note that the gale wind at Ferrybridge was in the range $5° < \alpha < 10°$, and so the proposed mechanism cannot be applied to the Ferrybridge towers.

Sun and Zhou (1983J) measured the external C_p and internal C_{pi} distributions around a model and full-scale Moomin cooling tower at $Re = 900$k and 5.4M, respectively. Figure 22.83 shows the internal pressure (dashed line) and C_{pi} (full line) distribution for various wind speeds at full, half, and no draught

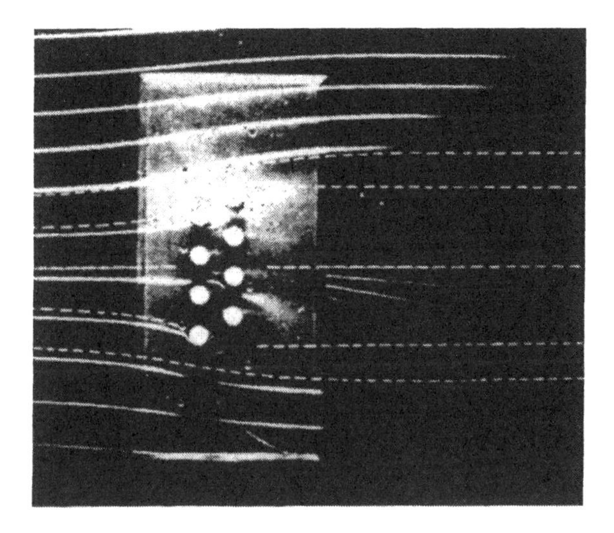

FIG. 22.81. Visualization of streamlines approaching two rows of two-dimensional cylinders, Owen (1967P)

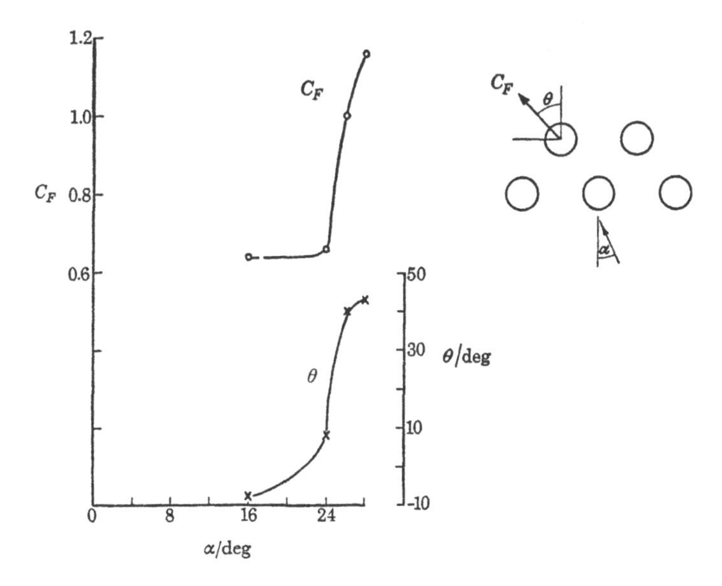

FIG. 22.82. Resultant force magnitude and direction on the rear row cylinder in terms of wind direction, Maull (1966J)

inside the cooling tower. At low wind, an almost constant $C_{pi} = -0.15$ was found around the inner circumference. However, at high wind, the flow over the open

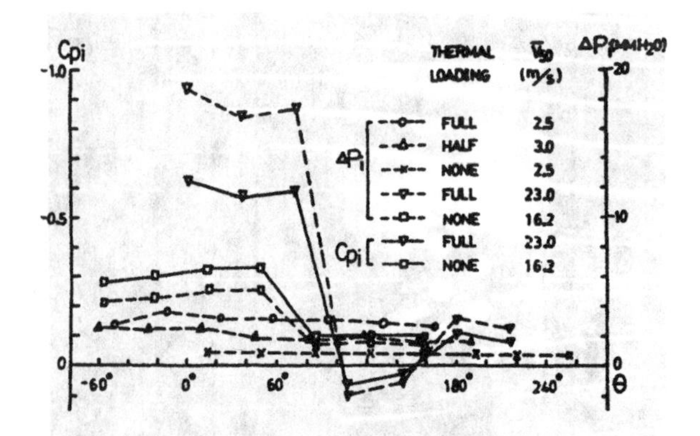

FIG. 22.83. Internal pressure distribution in a cooling tower with full, half, and without draught, Sun and Zhou (1983J)

end induced additional draught, and $C_{pi} = -0.6$ in the range $0° < \theta < 60°$ (the cooling tower was in full operation). The external $C_{po} = 1$ and internal $C_{pi} = -0.6$ add up to $C_p = 1.6$ at $\theta = 0°$. Hence, it might be that an enhancement of the wind action on the inner draught was responsible for the collapse of the Ferrybridge cooling towers.[127]

[127]This is a hindsight speculation by the author 30 years after the 1966 disaster.

23

BLOCKAGE AND WALL PROXIMITY

23.1 Introduction

The walls in wind and water tunnels induce a confining effect on the model that has been termed *blockage*. The blockage ratio is defined as the ratio of the cylinder diameter to the breadth of the test section, D/B. A convenient classification of blockage effects on circular cylinders is as follows:

(i) for $D/B < 0.1$, the blockage effect is small, and may be ignored;

(ii) for $0.1 < D/B < 0.6$, the blockage modifies the flow, and correction of the measured data is necessary;

(iii) for $D/B > 0.6$, the blockage radically alters the flow around the cylinder, and the correction of data is meaningless.

It should be pointed out that the above rough classification is applicable only for the TrSL, TrBL, and T states[128] of flow. For very low Re in the L1 regime, the blockage is important, even for $D/B < 0.001$.

An asymmetric kind of blockage takes place when the cylinder is placed away from the test section axis of symmetry. The side walls are at a different distance from the cylinder, and the effect of the nearer wall is dominant. This kind of asymmetric blockage rarely occurs in wind tunnel testing, and consequently few data are available. However, the extreme case when the farther wall is at infinity corresponds to the cylinder near a boundary. This kind of asymmetric blockage can be found in a wide range of engineering applications, such as submarine pipelines on or near the sea bed subjected to currents and waves in offshore engineering; pipelines above sand in a desert or snow in Alaska are subjected to wind in chemical engineering, etc.

All kinds of blockage effects are discussed in this chapter in terms of Re starting from the laminar state of flow L, followed by TrW and TrSL, and concluding with the TrBL state of flow. Theoretical models that yield correction factors necessary to remove the effect of blockage are described in the second part of this chapter. The final, third part is dedicated to the important asymmetric kind of blockage in engineering when the cylinder is in the proximity of the plane boundary. Flow regimes and related force coefficients are discussed in detail.

[128]The adopted classification and abbreviation of flow regimes is given in Section A3 in the Appendix.

23.2 Laminar, L, state of flow

23.2.1 *Creeping flow, the L1 regime*

White (1946J) carried out simple experiments by dropping 11 wires, $0.19\,\text{mm} < D < 1.44\,\text{mm}$, $4.9 < L/D < 50$, and $0.0001 < Re < 0.13$, into a jar filled with two glycerine solutions. The range of blockage ratios was $0.0015 < D/B < 0.11$, and by timing the free fall, the resistance force R was evaluated. The resistance coefficient C_R was used in the form

$$C_R = \frac{R}{\mu V L} \tag{23.1}$$

where μ is the viscosity of glycerine solutions, V is the free fall velocity, and L is the length of the wire.

Figure 23.1 shows C_R in terms of D/B for the creeping flow, L1 regime. There is an upward trend in C_R for rising D/B. For example, C_R increases three-fold as D/B rises from 0.0017 to 0.12. A logarithmic approximation with an arbitrary origin taken at $D/B = 0.8$ is in the form

$$C_R = \frac{6.3}{\log 0.8B/D} \tag{23.2}$$

Note that eqn (23.2) is independent of Re, i.e., D/B is the governing parameter even at such a low blockage. This becomes more evident in Fig. 23.2 where C_R is plotted versus Re. Separate horizontal lines designate each D/B value in region A.

'As the Re increases, the inertia term begins to shield the wire from the distant boundaries', wrote White (1946J). Curve B corresponds to Lamb's equation (given in Fig. 23.2), and experimental data conform to it in region C. For example, for $D/B = 0.02$ and 0.002, Lamb's equation is valid for $Re > 1$ and 0.05, respectively.

White (1946J) also examined the shape effect of the confining walls. He found that the wire fell a little faster between plane walls than within a circular jar of

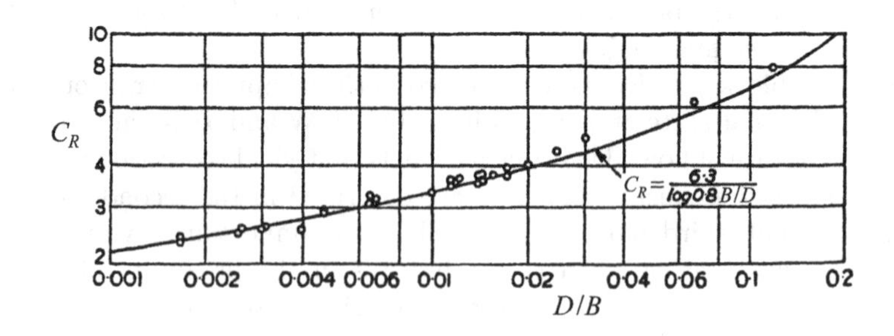

FIG. 23.1. Resistance coefficient in terms of blockage D/B in creeping flow, White (1946J)

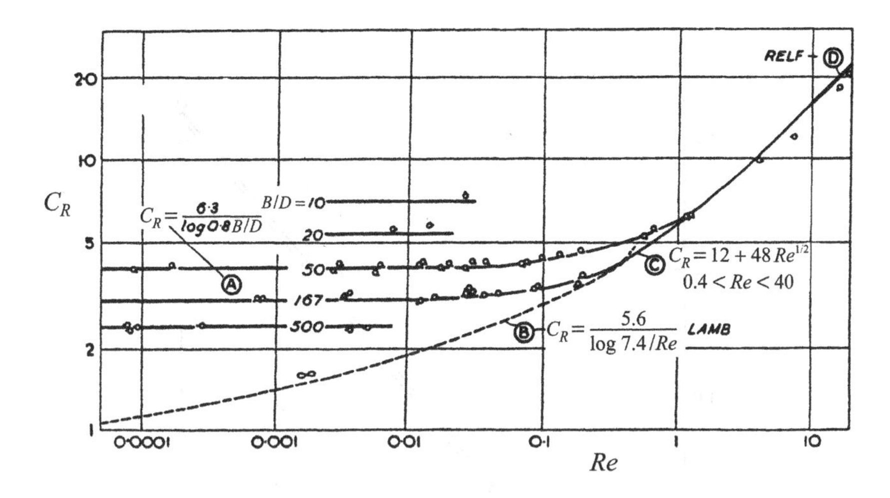

FIG. 23.2. Resistance coefficient in terms of Re for various B/D, White (1946J)

diameter D_j. The constant 6.3 in eqn (23.2) for the circular jar changes to 6.4 for the square jar.

Taneda (1964J) measured the resistance force R on a towed cylinder in the range $10^{-5} < Re < 10^{-1}$ for $0.001 < D/B < 0.1$. He noted that the product $C_D Re$ = constant. This product actually equals $2C_R$, and his experimental points are included in Fig. 23.2.

23.2.2 *Closed near-wake, the L2 regime*

The end of the L1 regime is marked by the initiation of separation and the formation of the closed near-wake in the L2 regime. The precise value of Re_s when separation starts is difficult to measure because the size of the separated region is almost zero. Coutanceau and Bouard (1977J) found $Re_s = 5.2$, 7.2, and 9.6 for $D/B = 0.024$, 0.67, and 0.12, respectively. The three Re_s values showed a significant effect of relatively small blockage on the beginning of separation. An extrapolation to $D/B = 0$ gave $Re_s = 4.4$ for the blockage-free flow.

A typical effect of blockage on the length of the closed near-wake L_w/D in terms of Re is shown in Fig. 23.3, where L_w is measured from the flow visualization photographs. Taneda's (1956J) and Homann's (1936J) data for a small blockage are in good agreement. Grove *et al.* (1964J) found that the D/B rise decreases the slope of the L_w/D curves in Fig. 23.3. However, Coutanceau and Bouard's (1977J) results do not show this in free fall tests.

As soon as the closed near-wake has been formed, a reverse velocity appears along the near-wake axis. The end of the near-wake, designated by L_w, represents a stationary confluence point with zero velocity. Figure 23.4 shows the effect

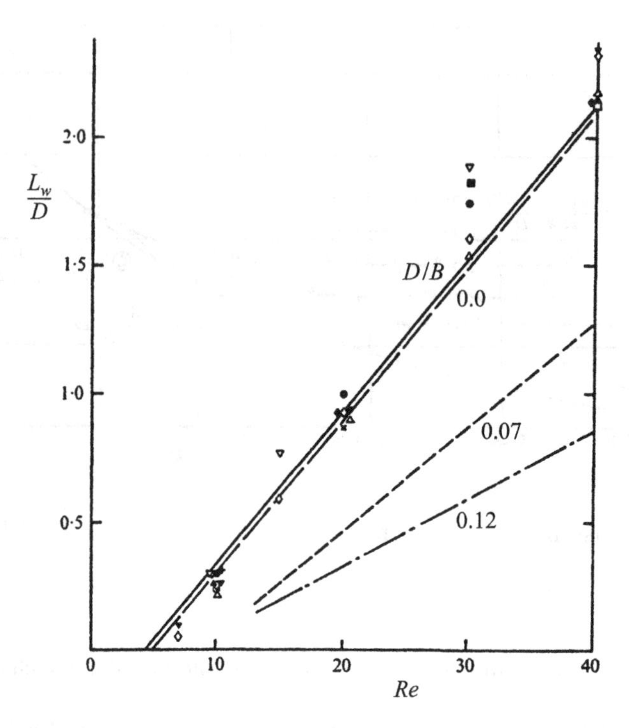

FIG. 23.3. Effect of blockage D/B on relative length of near-wake, Coutanceau and Bouard (1977J)

of blockage D/B on the shape and size of the near-wake at three Re values. The increase in D/B at the same Re shortens and narrows the near-wake. The location of twin circulation centres in the near-wake has already been shown in Vol. 1, p. 28, Fig. 2.10. The variation in the streamwise direction was linear while that in the transverse direction was parabolic. The angular separation θ_s in terms of Re was also shown in Vol. 1, p. 24, Fig. 2.5 for the three blockages. Note that experimental points in the range $150° < \theta < 180°$ were missing, which may suggest that the separation initiates at around $\theta_s = 150°$ rather than at $\theta_s = 180°$, as predicted by the boundary layer theory, Schlichting (1979B).

Coutanceau and Bouard (1977J) measured the relative reverse velocity u/V along the near-wake axis in the range $5 < Re < 40.5$ for $D/B = 0.02$, 0.07, and 0.12. Figure 23.5 shows velocity distributions at $Re = 20$ and D/B. The crossings at the $u/V = 0$ axis designate the length of the closed near-wake L_w. The universal reverse velocity curve can be obtained by scaling $u/u_{\min}$ versus $(x - D/2)/L_w$, Vol. 1, p. 30, Fig. 2.11. The magnitude of the reverse velocity is very small; for $Re = 20$, it is only 5% of the free stream velocity. It was for this reason that the term 'weak recirculating region' was adopted instead of 'twin

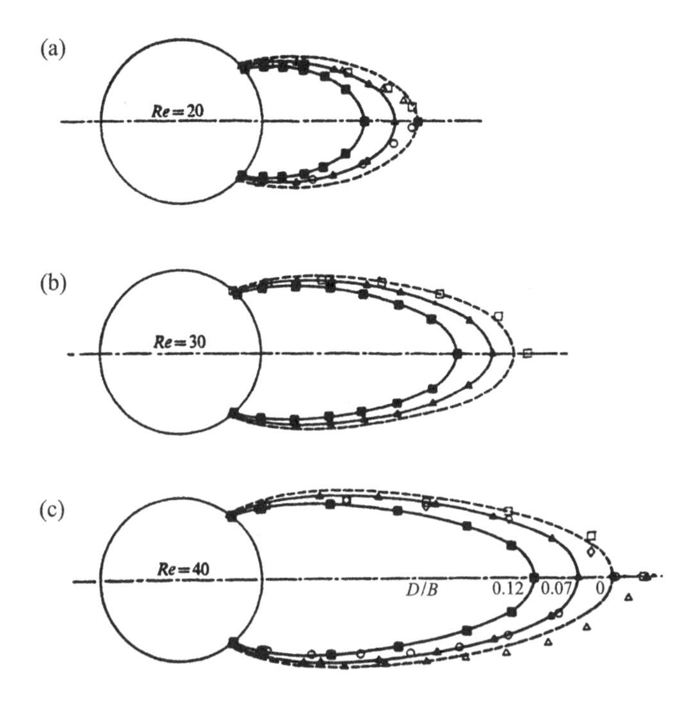

FIG. 23.4. Shape of near-wake for various blockages, (a) $Re = 20$, (b) $Re = 30$, (c) $Re = 40$, Coutanceau and Bouard (1977J)

eddies' in Vol. 1, pp. 329–30.

Rosenhead and Schwabe (1930J) carried out flow visualization in a water tank. Figure 23.6(a,b) shows the near- and far-wake behind a towed cylinder at $Re = 25$ and $D/B = 0.19$, and $Re = 386$ and $D/B = 0.67$. A straight and steady trail is seen at both Re. The weak symmetric recirculating regions are seen adjacent to the cylinder for $Re = 25$ and several diffusing symmetric eddy pairs at $Re = 386$. The near-wake and trail instabilities are fully suppressed for $Re = 386$ and $D/B = 0.67$, so that the near-wake is in the L2 regime.

23.2.3 *Instability of the near-wake*

It has been discussed in Vol. 1, Chapter 3, pp. 33–4 that the termination of the steady L2 regime is marked by instability of the near-wake end. It is manifested by several features such as the formation of gathers (spikes) along the free shear layers, the transverse oscillation of the confluence point, and the circulation centres. Taneda (1956J) stated that the trail emanating from the confluence point began to oscillate before the gathers were formed, while Gerrard (1978J) observed that the appearance of gathers resulted in a wavy trail. Presumably, each triggered the other thereby preventing any accurate estimate of the initiation of instability Re_{osc}.

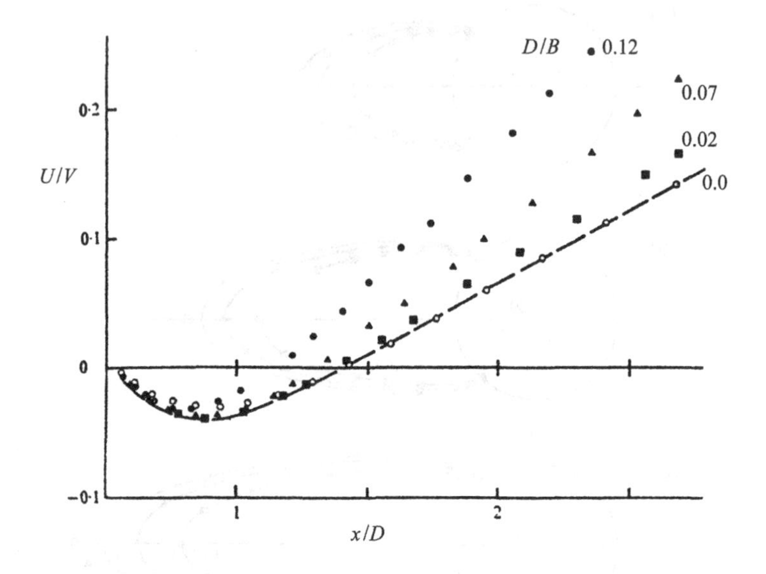

FIG. 23.5. Velocity distribution along near-wake axis at $Re = 20$, extrapolation for $D/B = 0$ (dashed line), ∘, computation Nieuwstadt and Keller (1973J), Coutanceau and Bouard (1977J)

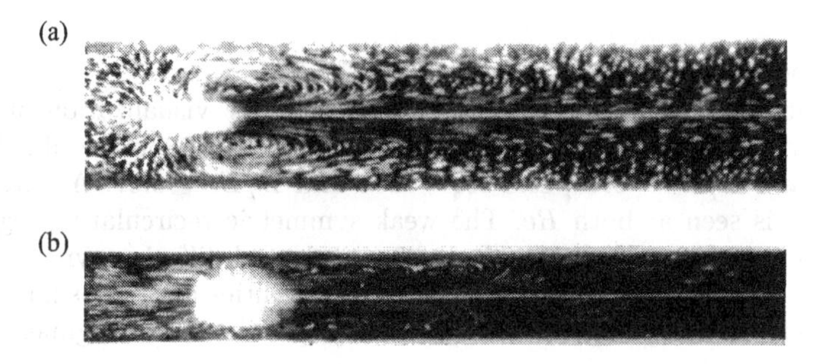

FIG. 23.6. Water surface visualization, (a) $Re = 25$, $D/B = 0.192$, (b) $Re = 386$, $D/B = 0.667$, Rosenhead and Schwabe (1930J)

Figure 23.7 is a compilation of various data, and shows Re_{osc} in terms of D/B. Coutanceau and Bouard (1977J) and Mitry (1977P) determined Re_{osc} from the displacement of the near-wake maximum width. The latter increased on one side and decreased on the other during oscillation. Taneda's (1956J) estimate was made by observing the wavy trail. The two different methods may have caused

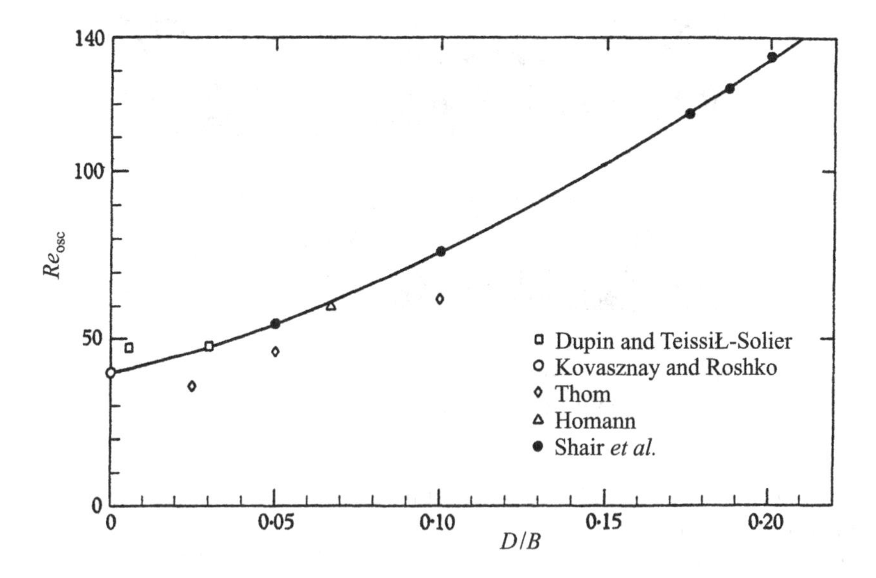

FIG. 23.7. Oscillation Re_{osc} in terms of blockage, Shair *et al.* (1963J)

a consistent discrepancy. All the data in Fig. 23.7 demonstrate that the stability of the steady near-wake is considerably enhanced by the blockage. Coutanceau and Bouard (1977J) noted that at Re_{osc} the maximum near-wake width became nearly equal to the cylinder diameter.

Shair *et al.* (1963J) explained this as follows:

> The near-wake instability is caused primarily by disturbances which generate in a direction perpendicular to both that of the undisturbed flow and the near-wake axis. Thus, it would appear reasonable to expect that the propagation of such disturbances should be inhibited by the presence of the walls which confine the streamlines and near-wake.

The laminar wake instability, which produces a wavy trail and transverse flapping of the rear of the near-wake, will certainly be subdued by the blockage, and postponed to higher Re_{osc}. For example, if $D/B = 0.2$, then Re_{osc} rises 300% more than any other parameter or coefficient.

23.2.4 *Laminar periodic wake, the L3 regime*

Systematic flow visualization of periodic laminar wakes in the L3 regime were carried out by Rosenhead and Schwabe (1930J). Figure 23.8(a,b) shows Kármán–Bénard eddy streets at $Re = 57$ and 95, respectively, for $D/B = 0.19$. Shair *et al.* (1963J) could not observe a near-wake oscillation for $D/B = 0.2$ before $Re_{osc} = 120$. The reason is evident in Fig. 23.8(a): the near-wake does not start

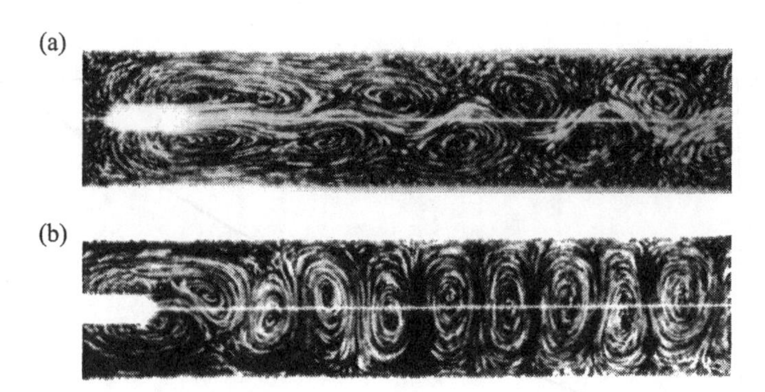

FIG. 23.8. Water surface visualization for $D/B = 0.192$, (a) $Re = 57$, (b) $Re = 95$, Rosenhead and Schwabe (1930J)

to oscillate until $x/D > 15$. The instability is not suppressed, but delayed to occur in the far-wake.

Another important aspect of blockage is the distortion of the spacing ratio $b/a = Ka$ (where b is the distance between the rows and a the distance between consecutive eddies[129]). The wall confinement reduces Ka to zero when two rows become aligned at $x/D > 50$. The eddies are then distorted and become elliptically shaped, as seen in Fig. 23.8(b). The blockage affects the eddy street more strongly in the far-wake than in the near-wake.

One of the features of the laminar eddy street is that the streamwise distance a between eddies in the same row is remarkably constant along the wake. Rosenhead and Schwabe (1930J) noted that in almost every experiment, no matter how irregular the configuration of eddies, the horizontal distance between consecutive eddies remained nearly constant. However, the value of a/D depends on D/B, as seen in Fig. 23.9(a). There is an orderly decrease in a/D with rising D/B.

The velocity perpendicular to the wake axis manifests itself in the form of an eddy motion towards the confining walls. When x/D becomes large, the eddies cling to the walls. Figure 23.9(b) shows the estimated b/D in terms of Re and D/B. There is a sharp decrease in b/D with rising D/B. The values correspond to the configuration observed immediately behind the cylinder. Note that in Fig. 23.8(b) $b/a = 0$, the single row, and in Fig. 23.9(b) $b/a = 1.2$ instead. The widening of the two rows of eddies depends not only on Re and D/B but also on x/D.

The spacing ratio was around $b/a = 0.32$ at the start of the eddy street. Taking into account the experimental error, this value holds for all D/B up to

[129] Kármán (1956B) theoretically determined the spacing ratio b/a and this ratio was called the Kármán number, Ka, in Vol. 1.

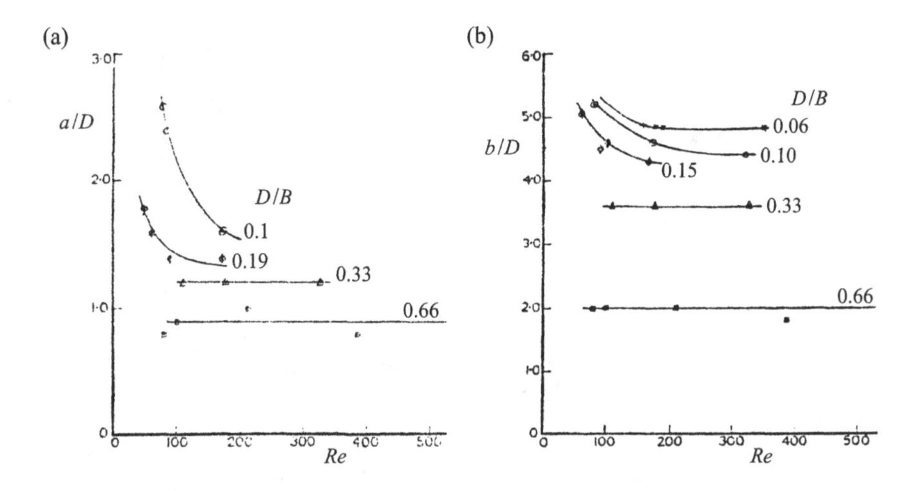

FIG. 23.9. Spacing of eddies in the Kármán–Bénard street, (a) streamwise, (b) transverse, Rosenhead and Schwabe (1930J)

0.3. The results agree with the experimental evidence by Bénard (1926J) in that the b/a values are greater than the Kármán theory, $Ka = 0.28$.

The effect of blockage on the mean pressure distribution around the cylinder at $Re = 173$ is shown in Fig. 23.10. It is evident that the combined effect of an increase in C_{po} and a decrease in C_{pb} produce a considerable increase in $C_D = 0.68,\ 0.91$, and 1.06 for $D/B = 0.05,\ 0.10$, and 0.19, respectively.

The complementary flow visualization at $Re = 186$ by Taneda (1964J) for high blockages $D/B = 0.33,\ 0.50$, and 0.67 is shown in Fig. 23.11(a,b,c), respectively. For $D/B = 0.33$, the Kármán–Bénard eddy street is visible all along the wake, for $D/B = 0.50$ only half-way downstream, and for $D/B = 0.67$, it is fully suppressed. The last is similar to Fig. 23.7(b) for $Re = 386$ and $D/B = 0.67$.

Rosenhead and Schwabe (1930J) also estimated the convection velocities of eddies from the flow visualization. Figure 23.12 shows the eddy velocity scaled by the cylinder velocity V_C/V in terms of Re for various blockages D/B. There is a small effect of D/B on V_C/V and a large effect on Re. Note that for $D/B = 0.67$ the Kármán–Bénard street is suppressed, and the experimental points correspond to the velocity in the wake.

Finally, Fig. 23.13 shows the estimated angular separation θ_s in terms of Re and D/B. Mitry (1977P) measured θ_s in the wide range $0.06 < D/B < 0.50$ and $40 < Re < 10\text{k}$. There is a particularly strong effect of D/B on the laminar and transitional wakes below $Re = 1\text{k}$.

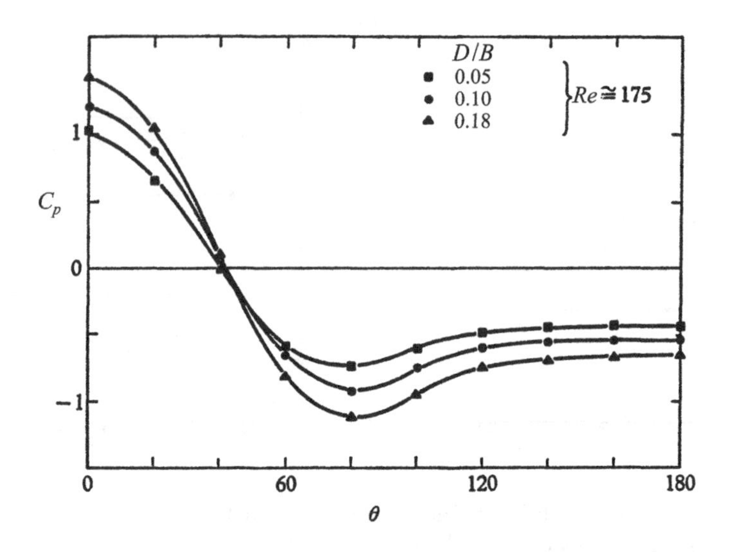

FIG. 23.10. Pressure distribution for Re = 175 and three blockages, Grove *et al.* (1964J)

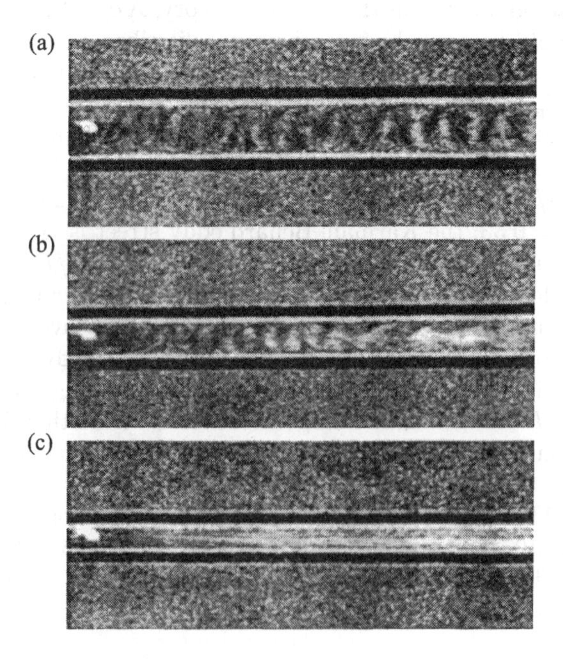

FIG. 23.11. Water surface visualization $Re = 186$, (a) $D/B = 0.33$, (b) $D/B = 0.50$, (c) $D/B = 0.67$, Taneda (1965J)

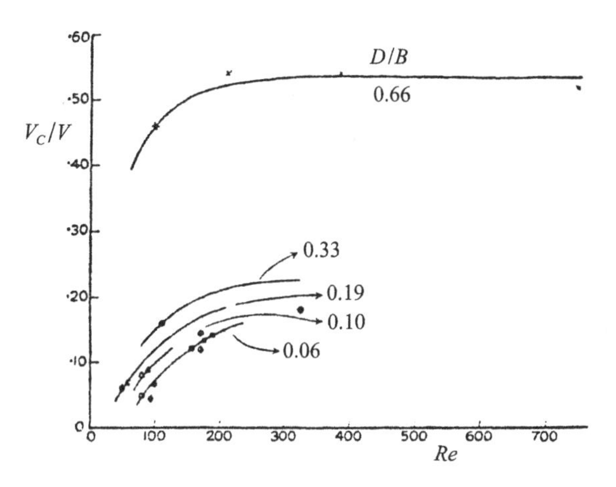

FIG. 23.12. Relative convective velocity of eddies in terms of Re for various blockages, Rosenhead and Schwabe (1930J)

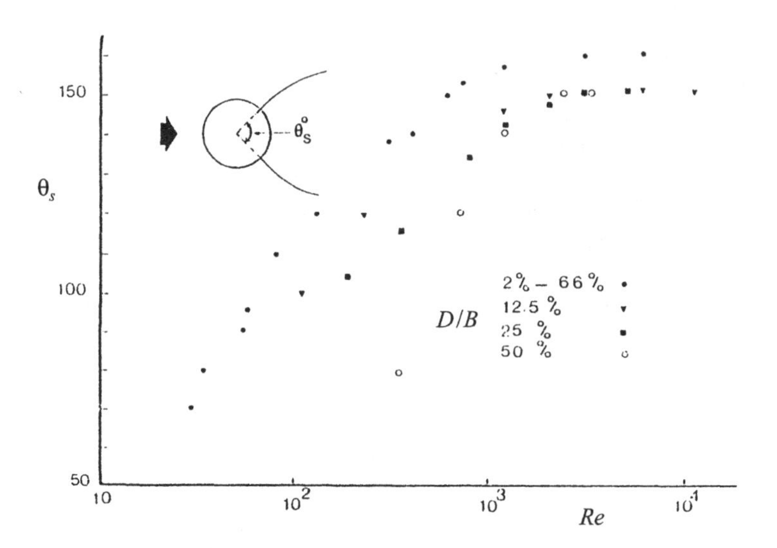

FIG. 23.13. Position of angular separation in terms of Re for various blockages, Mitry (1977P)

23.3 Transition in shear layers, the TrSL state of flow

23.3.1 *Mechanics of blockage*

Fage (1929P) wrote:

> It is well known that when a cylindrical body of infinite length is immersed in airstream flowing with a uniform velocity in a channel (tunnel) of finite breadth the presence of the walls influences the flow around the body, so that its drag differs from that which would be experienced in an infinite stream flowing with the same general velocity. In most wind tunnel investigations, an endeavour is made to keep this difference small, but occasions arise, for example, when it is necessary to use a body whose dimensions are considerably larger than those normally used, even though it is known that the interference of the walls cannot be neglected.

The blockage affects the flow around cylinders in several ways:

(i) The cylinder in a closed test section reduces the cross-sectional area locally. This 'solid' blockage causes an increase in velocity around the cylinder.

(ii) The side walls affect the widening of the wake, and cause *wake blockage*. The increased velocity outside the wake reduces the pressure in it.

(iii) The increase in local velocity along the test section is associated with an increase in the streamwise pressure gradient. The latter contributes to the drag increase.

The complexity of the blockage mechanics can be appreciated from the flow visualization. Figure 23.14(a,b,c,d) shows a water surface visualization by Okamoto and Takeuchi (1975J) at $Re = 5.4\text{k}$ for $D/B = 0.09$, 0.18, 0.25, and 0.34, respectively. The blockage effect is small for $D/B = 0.09$, visible for $D/B = 0.18$, and excessive for $D/B = 0.25$ and 0.34. The strength of the eddies is related to their visibility. They are weak for $D/B = 0.09$, medium for $D/B = 0.18$, and strong for $D/B = 0.25$ and 0.34, where the eddies appear compressed between the walls. Okamoto and Takeuchi (1975J) observed an increase in K/VD from 4.13 to 5.11 at $Re = 32\text{k}$, when D/B increased from 0.05 to 0.34, respectively.

23.3.2 *Mean pressure distribution and drag*

The increase in local velocity around the cylinder by blockage has been associated with a corresponding lowering of the C_p curves. Figure 23.15(a,b) illustrates this trend at $Re = 10\text{k}$ and 45k for $0.06 < D/B < 0.35$ and $0.60 < D/B < 0.8$, respectively. Hiwada and Mabuchi (1981J) added arrows at the estimated θ_s to emphasize how small the effect of blockage on θ_s is. For very high blockage $D/B = 0.8$, the value of $\theta_s = 90°$ at the minimum gap.

The excessive reduction in C_{pb} with the increasing D/B has a direct effect on the rise in C_D. Modi and El-Sherbiny (1971P) measured $1.2 < C_D < 2.4$ for $0.06 < D/B < 0.38$ at $10\text{k} < Re < 120\text{k}$. The extremely wide range $0.2 <$

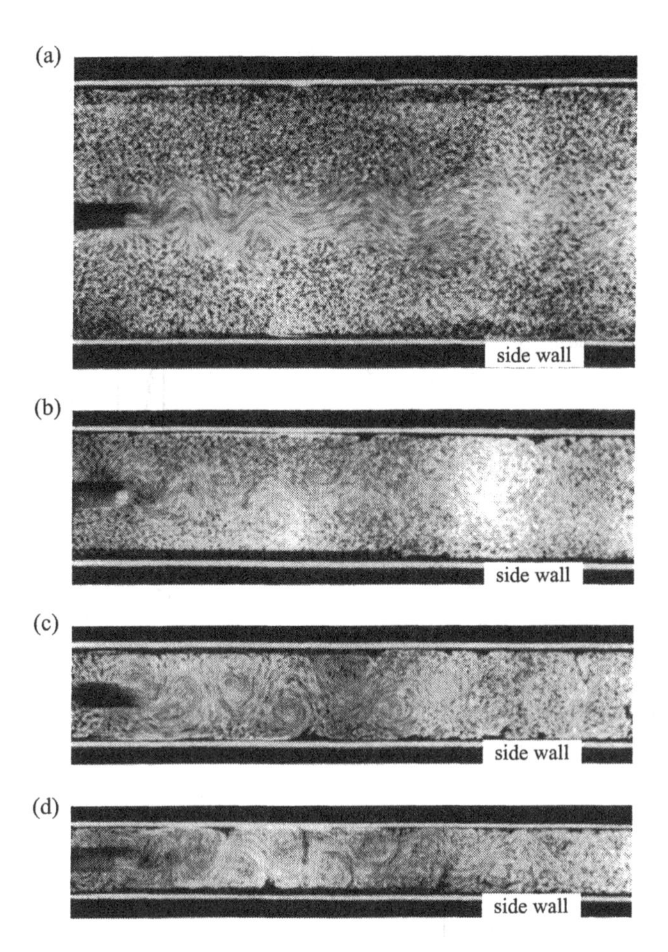

FIG. 23.14. Surface flow visualization for Re = 5.4k, (a) $D/B = 0.083$, (b) $D/B = 0.181$, (c) $D/B = 0.25$, (d) $D/B = 0.34$, Okamoto and Takeuchi (1975J)

$D/B < 0.8$ was covered by Hiwada and Mabuchi (1981J) and $C_D = 20$ was reached for $D/B = 0.8$, see Fig. 23.16. It is interesting to note that an empirical approximation up to $D/B = 0.5$ cannot be extrapolated to higher D/B values.

The excessive increase in C_D for high values of D/B prompted some authors to replace the free stream velocity by some other reference velocity. Ramamurthy and Lee (1973J) carried out tests in the range $0.07 < D/B < 0.31$ and 69k $<$ $Re <$ 204k. They proposed the separation velocity $V_s = kV$, where $1 - k^2 = C_{pb}$, and the jet velocity $V_j = C_c V_s$, where C_c is the contraction coefficient for the gap flow between the cylinder and wall. Figure 23.17 shows that the parabolic increase in conventional C_D with D/B is replaced by the constants C_{Ds} and C_{Dj}

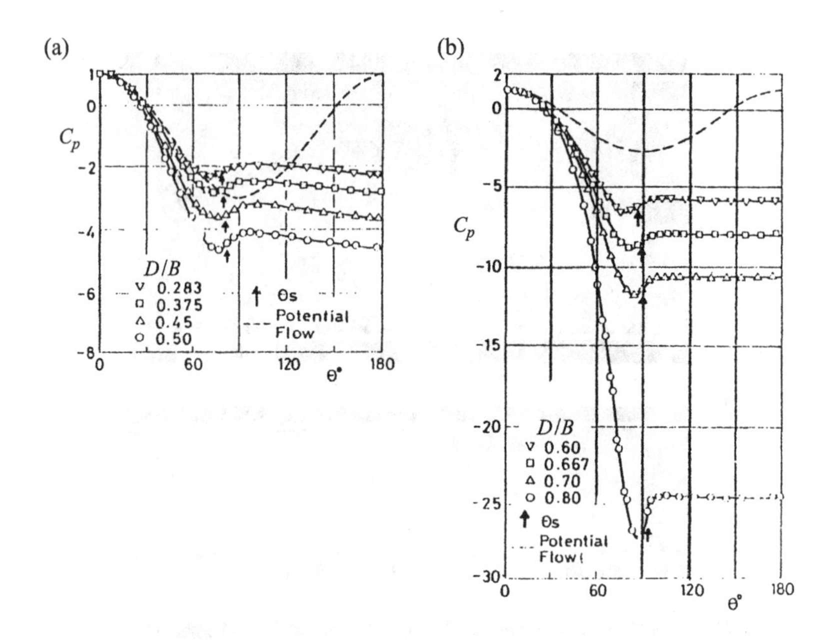

FIG. 23.15. Pressure distribution as affected by: (a) low blockage, Re = 10k, (b) high blockage, Re = 45k (arrows show position of separation), Hiwada and Mabuchi (1981J)

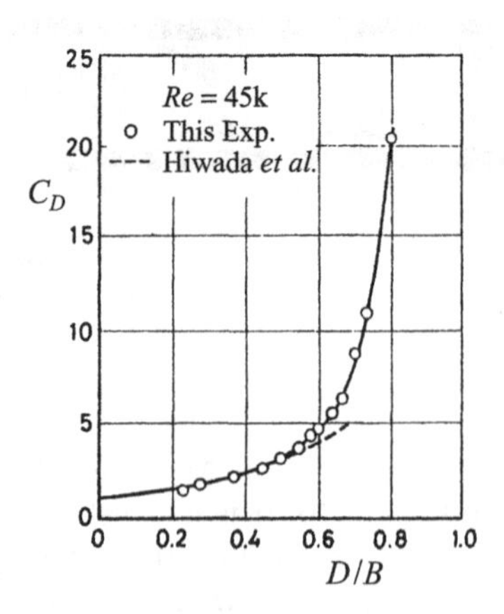

FIG. 23.16. Drag coefficient in terms of blockage at Re = 45k, Hiwada and Mabuchi (1981J)

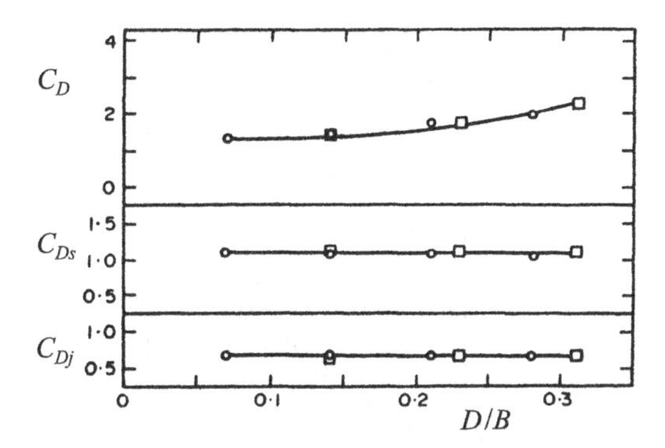

FIG. 23.17. Conventional C_D, separation velocity C_{Ds}, and jet velocity C_{Dj} in terms of blockage, Ramamurthy and Lee (1973J)

for V_s and V_j, respectively.

23.3.3 *Strouhal number and fluctuating pressure*

It has been discussed in Vol. 1, Chapter 5.3, p. 120, and Chapter 6.4, p. 172, that the St value is closely related to the near-wake width. The wider the wake, the lower the St, and vice versa. The effect of blockage is expected to narrow the near-wake, and consequently to increase the St value.

Hiwada and Mabuchi (1981J) covered a wide range $0.2 < D/B < 0.8$ at $Re = 45$k, and Fig. 23.18 shows a steep rise in St. Beyond $D/B = 0.6$, the peak in the frequency spectra disappeared indicating that eddy shedding ceased. The dashed line corresponds to the empirical correlation derived for small D/B. The experimental proof that St does not depend on Re was given by Modi and El-Sherbiny (1971P, 1973P). Figure 23.19 shows St in terms of Re for $0.07 \leqslant D/B \leqslant 0.42$.

An attempt to replace free stream velocity by the separation velocity is shown in Fig. 23.20. Toebes and Ramamurthy (1970P) succeeded in transforming a parabolic rise in St into a constant St_s in the range $0.04 < D/B < 0.44$ and $10\text{k} < Re < 67\text{k}$.

Richter and Naudascher (1976J) pointed out that St may also depend on the aspect ratio L/D. Figure 23.21 shows a compilation of St in terms of D/B. Toebes (1971P) varied the aspect ratio in the range $1 < L/D < 3.8$, Shaw (1971P) in the range $2 < L/D < 6$, while Richter and Naudascher (1976J) used $L/D = 8.6$ throughout. The effect of L/D on St is not clear.

The increase in C_D and strength of eddies K with D/B are likely to affect C_p'. Modi and El-Sherbiny (1971P) measured C_p' at $10\text{k} < Re < 105\text{k}$ for $0.03 < D/B < 0.38$. Figure 23.22 shows $C'_{p\max}$ in terms of Re for a series of D/B. The rise in $C'_{p\max}$ in the range $10\text{k} < Re < 40\text{k}$ is in the TrSL2 regime, and is associated with a shortening of the eddy formation regime. It becomes

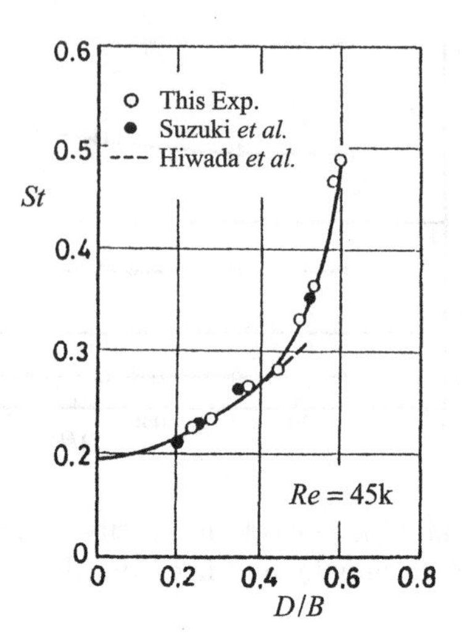

FIG. 23.18. Strouhal number variation in terms of blockage for Re = 45k, Hiwada and Mabuchi (1981J)

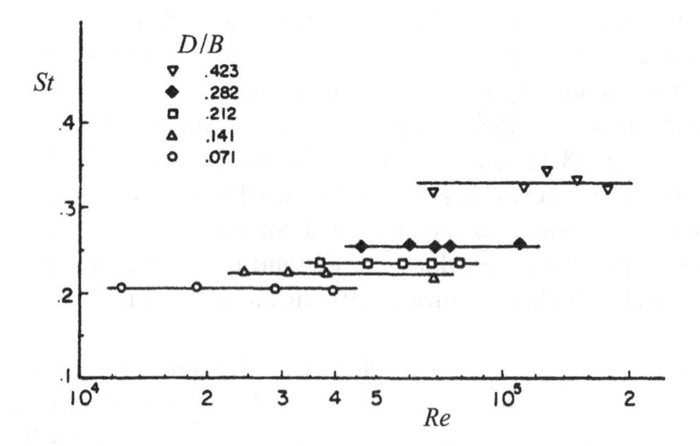

FIG. 23.19. Strouhal number in terms of Re for various blockages, Ramamurthy and Ng (1973J)

independent of Re in the TrSL3 regime.

23.3.4 *Suppression of eddy shedding*

It was mentioned in the previous section that at about $D/B = 0.6$ the peak in the frequency spectra disappeared. This meant that a different flow regime started at high D/B values.

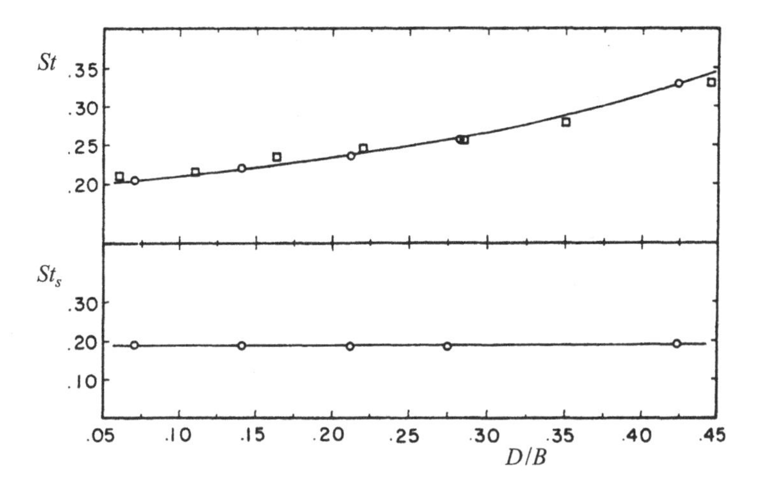

FIG. 23.20. Conventional St and that based on separation velocity St_s in terms of blockage, Toebes and Ramamurthy (1970P)

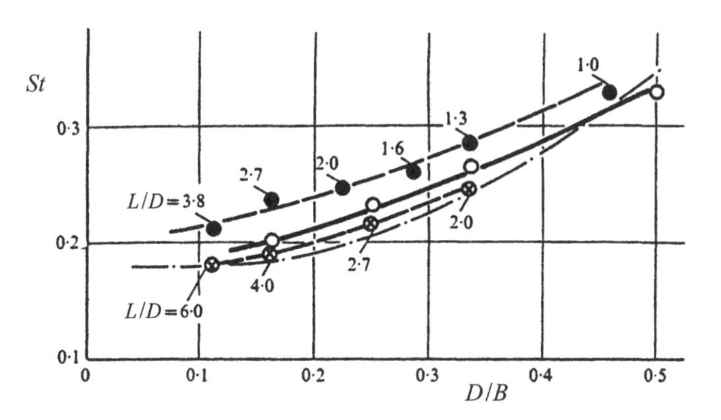

FIG. 23.21. Strouhal number in terms of blockage for various aspect ratios, Richter and Naudascher (1976J)

Hiwada and Mabuchi (1981J) measured velocity fluctuations at $x/D = 1$, $y/D = 0.56$, and $Re = 40$k. Figure 23.23 shows the frequency spectra for $D/B = 0.45$ and 0.8 with and without a peak, respectively. Hiwada and Mabuchi (1981J) also measured the local distribution of the mass transfer, Sh. Figure 23.24(a,b) shows the Sh distribution for $D/B = 0.375$ and 0.8, respectively. There are three differences in the local mass transfer distributions:

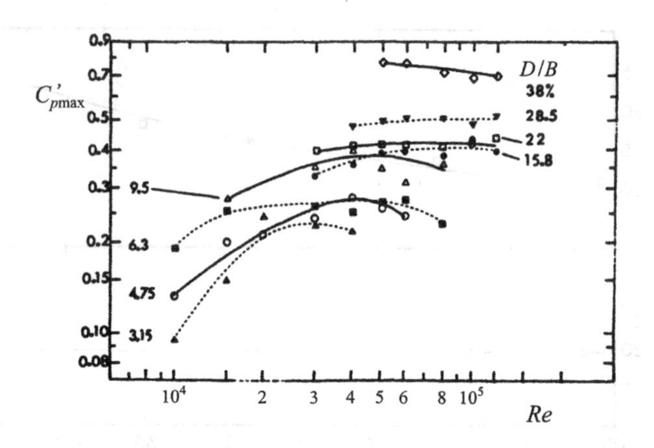

FIG. 23.22. Maximum fluctuating pressure coefficient in terms of *Re* for various blockages, Modi and El-Sherbiny (1971P)

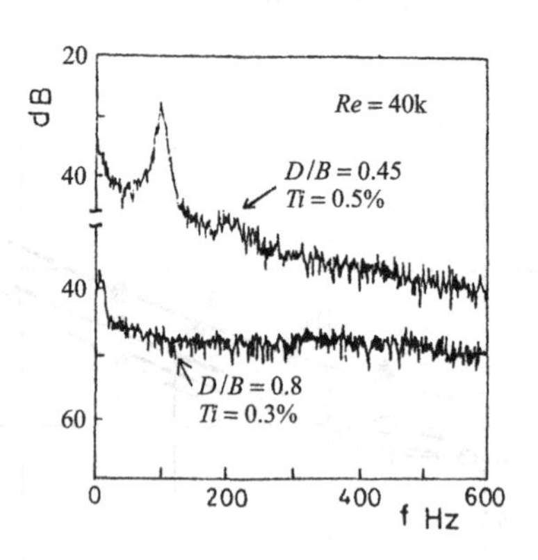

FIG. 23.23. Frequency spectra at $Re = 40\text{k}$ for $D/B = 0.45$ and 0.8, Hiwada and Mabuchi (1981J)

(i) an increase in Sh in the range $60° < \theta < 80°$ for $D/B = 0.8$;

(ii) a displacement of $Sh_{\min}$, which corresponds to θ_s from 85° to 95° for $D/B = 0.375$ and 0.8, respectively;

(iii) a new $Sh_{\min}$ appears on all Sh curves for $D/B = 0.8$ at $\theta = 120°$ to 130°.

The origin of the two $Sh_{\max}$ seen in Fig. 23.24(b) for $D/B = 0.8$ may be attributed to the wall separation downstream of the gap. The first $Sh_{\max}$ at $\theta \approx 70°$ does not appear up to $D/B = 0.6$ and is related to the suppression of eddy shedding. The second $Sh_{\max}$ at $\theta = 100°$ is displaced at $D/B = 0.6$.

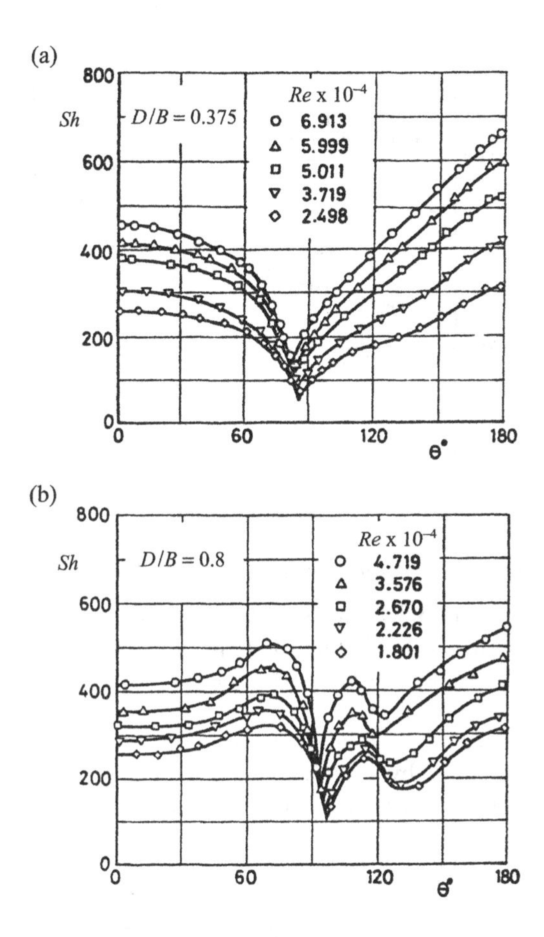

FIG. 23.24. Local mass transfer distribution, (a) $D/B = 0.375$, (b) $D/B = 0.8$, Hiwada and Mabuchi (1981J)

Hiwada *et al.* (1979J) observed the formation of a wall eddy on the side where the eddy rolls up at $D/B \approx 0.4$. These alternating eddies on both walls are swept by the main eddy street. However, when the eddy street is absent the intermittent separation becomes permanent, and the wall separation forms a separation bubble. The walls and cylinder act as a wide angle diffuser with separation on both sides. The deflection of flow towards the cylinder increases the heat transfer and Sh peaks at $\theta = 115°$.

Hiwada and Mabuchi (1981J) demonstrated the existence of two blockage flow regimes. Figure 23.25 shows the variation of mean $\overline{Sh}$ in terms of D/B for a constant $Re = 40$k. There is a marked discontinuity in the $\overline{Sh}$ curve near $D/B = 0.6$ with a sudden drop in $\overline{Sh}$.

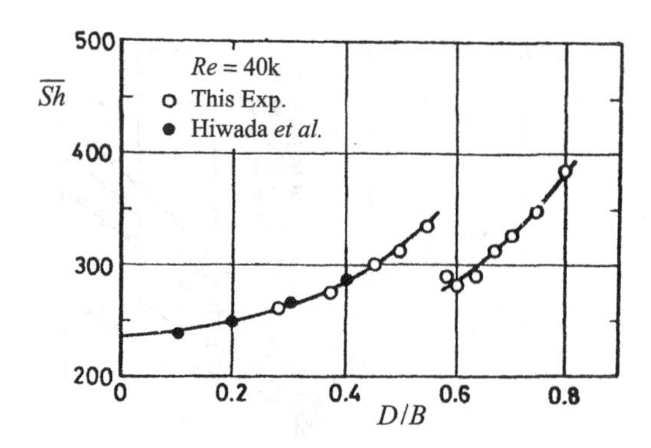

FIG. 23.25. Mean mass transfer coefficient in terms of blockage for Re = 40k, Hiwada and Mabuchi (1981J)

23.3.5 *Strength and correlation of eddies*

It was shown in Section 23.3.1 that eddies appear stronger, and last longer, as D/B increases, see Fig. 23.14. It may be expected that the increase in velocity around the cylinder due to blockage will generate more vorticity and stronger eddies.

Hiwada *et al.* (1979Ja) calculated the velocity distribution along the outer edge of the boundary layer by using three methods (see Fig. 23.26): experiment (open circles), potential theory (dashed line), and analytic series expression (full

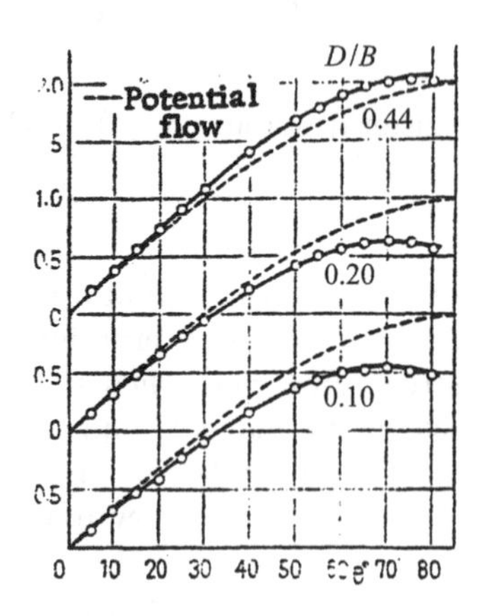

FIG. 23.26. Distribution of velocity along the outer edge of the boundary layer, experiment (open circles), potential flow (dashed line), analytic approximation (full line), Hiwada *et al.* (1979J)

line). Figure 23.26 shows good agreement between the evaluation of $V/V_{\max}$ from the measured C_p distribution and analytical approximation. The maximum V/V_m rises from 1.51 to 2 for $D/B = 0.10$ and 0.44, respectively.

The stronger eddies are likely to induce a better spanwise correlation. Blackburn (1994J) carried out tests on the cylinder $L/D = 11.75$, with end plates $D_e/D = 7$, and $D/B = 0.056$, 0.20, and 0.40 at $Re = 4$k in the TrSL2 regime. Figure 23.27 shows significant improvement in the spanwise correlation. The correlation length was found to be $7D$, $13D$, and $20D$ for the three respective blockages. It should be pointed out that these results are valid for the TrSL2 regime characterized by a long eddy formation region. The increase in Re leads to a shortening of the eddy formation region, reaching a minimum in the TrSL3 regime, and correlation length of $3D$ for $D/B = 0$. Thus these results should not be extrapolated to the TrSL3 regime before verification.

The increase in eddy strength and spanwise correlation are likely to affect the velocity profiles in the far-wake. Okamoto and Takeuchi (1975J) traversed the far-wake, and Fig. 23.28 shows the similar velocity profiles. The blockage data are consistently above the Schlichting's (1930J) theoretical curve.

The final effect of blockage is on the wake width. Figure 23.29 shows the wake half-width development along the far-wake up to $x/D = 40$. For small $D/B = 0.05$, the b/D continues to widen all along the near-wake. The widening stops at $x/D = 30$, 10, and 6 for $D/B = 0.10$, 0.20, and 0.30, respectively.

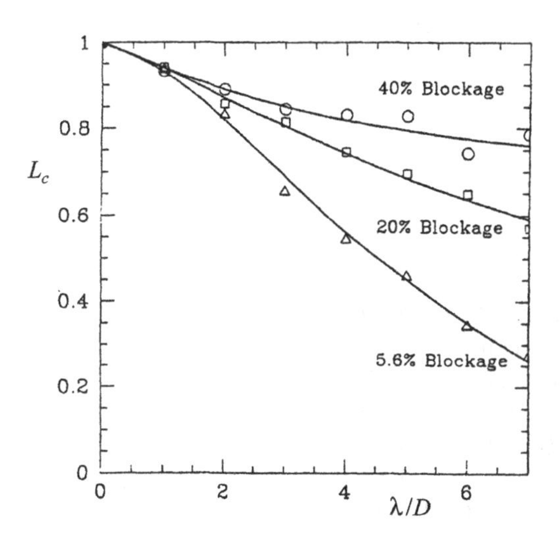

FIG. 23.27. Correlation of spanwise pressure fluctuation for $Re = 40$k, Blackburn (1994J)

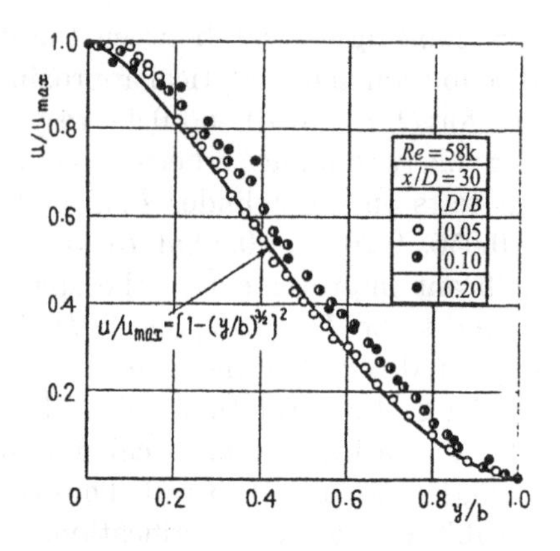

FIG. 23.28. Similarity velocity profiles across the wake at $x/D = 30$ for $Re = 58$k, Okamoto and Takeuchi (1975J)

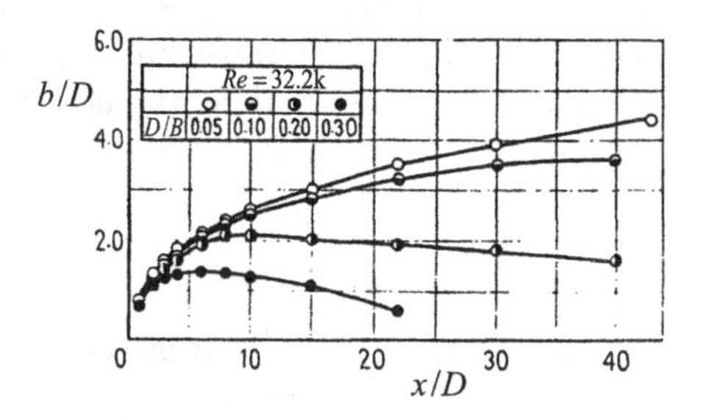

FIG. 23.29. Development of wake width for $Re = 32.2$k, Okamoto and Takeuchi (1975J)

23.3.6 *Effect of free stream turbulence*

The effect of free stream turbulence on flow around circular cylinders has been discussed in Vol. 1, Chapter 14. The combined effect of blockage and free stream turbulence was investigated by Modi and El-Sherbiny (1971P, 1973P, 1975P). The cylinders $L/D = 18$, 9, 3.9, and 2.2 corresponded to $D/B = 0.045$, 0.09, 0.20, and 0.35, respectively. The last two L/D had a significant effect on flow, as discussed in Chapter 21. The turbulence grid positioned upstream at varying distances produced a free stream turbulence of $Ti = 6.7\%$, 9.2%, and 12%, and for the last value two scales $Ts = 0.2$ and 0.6.

Figure 23.30 shows the C_p distribution for $D/B = 0.20$ as affected by Ti and Ts at $Re = 50$k. The effect of four turbulence grids is to disturb the boundary layer and postpone separation to higher Re. The effect is significant only for $Ti = 12\%$, curves d and e, at $Re = 50$k.

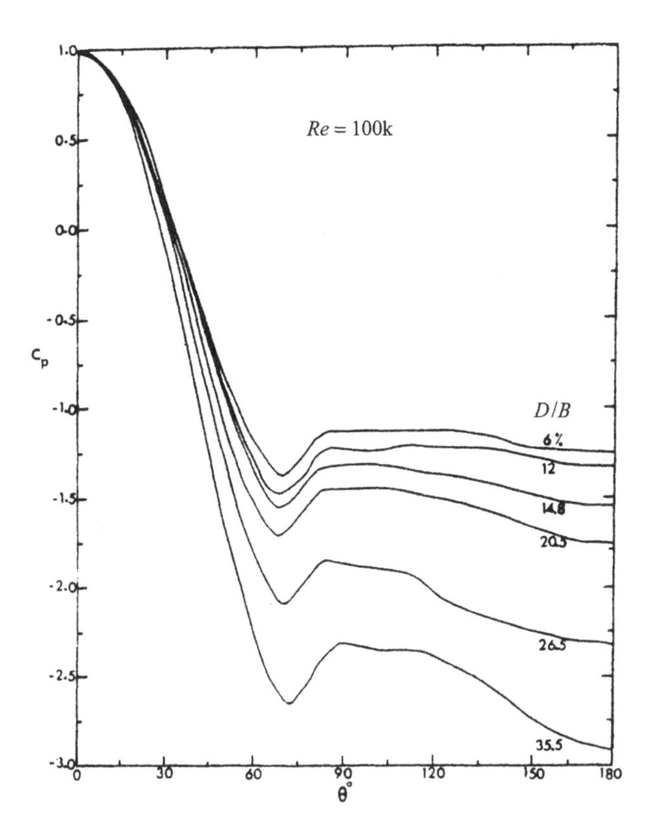

FIG. 23.30. Pressure distribution for various D/B at Re = 100k, Modi and El-Sherbiny (1971P)

Figure 23.31 shows the variation in C_d evaluated from the C_p distribution in terms of Re for $D/B = 0.20$ and 0.35, and five Ti values. There is a considerable departure of the C_d curve for low Ti from the others. The features evident in Fig. 23.31 are:

(i) the start of the fall in C_d occurs at about the same Re for both D/B values;

(ii) there is a small effect of Ti on the C_d curves, which are all close together;

(iii) the flat part of the $C_{d\min}$ curves for the two blockages is similar to that found for a low blockage, as shown in Vol. 1, Fig. 14.16, p. 445.

The mechanism of the increase in C_d with rising D/B is not related to wake widening because the angular separation remains the same, see Fig. 23.15. The low C_{pb} is associated with an increase in the separation velocity leading to increased suction in the near-wake.

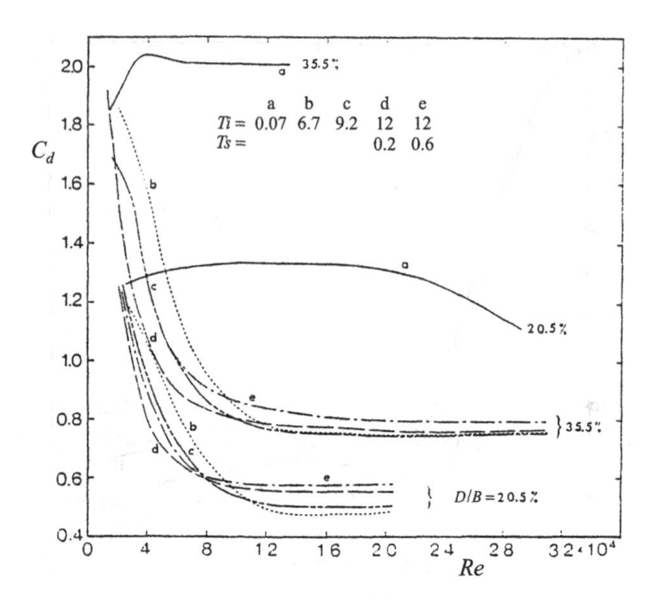

FIG. 23.31. Drag coefficient in terms of *Re* for $D/B = 0.20$ and 0.33, Modi and El-Sherbiny (1975P)

23.4 Transition in the boundary layer, the TrBL state

23.4.1 *Introduction*

The TrBL state of flow has been discussed in Vol. 1, Chapter 6. It encompasses five flow regimes:

(i) the TrBL0, precritical regime, where transition reaches separation;

(ii) the TrBL1 and 2, with one and two separation bubbles;

(iii) the TrBL3, supercritical regime, where eddy shedding ceases;

(iv) the TrBL4, postcritical regime, where eddy shedding re-starts.

23.4.2 *Drag variation with blockage*

The effect of blockage on C_D is expected to be different for each TrBL regime. The increased favourable pressure gradient as induced by the confining walls should have a stabilizing effect, and postpone the TrBL0 regime to the higher *Re* values. However, the separation velocity also increases, and related Re_s should destabilize the boundary layer to lower *Re* values. These two opposing tendencies might cancel each other out, as will be shown shortly.

Richter and Naudascher (1976J) carried out experiments to examine the blockage effect on C_D in the TrBL state. Unlike Toebes and Ramamurthy (1970P), Ramamurthy and Lee (1973J), and Modi and El-Sherbiny (1971P,

1973P, 1975P), who decreased L/D in order to increase D/B, Richter and Naudascher (1976J) kept $L/D = 8.6$ constant in the range $0.16 < D/B < 0.50$, and $20\text{k} < Re < 400\text{k}$. Figure 23.32 shows the measured mean drag force in terms of Re for four D/B values. There is a very small (almost negligible) displacement of the C_D fall towards low Re with increasing D/B. Wieselsberger's curve for small blockage shows the same shape as the others, which means that the TrBL0 regime is followed by the formation of one bubble and two bubbles in the TrBL1

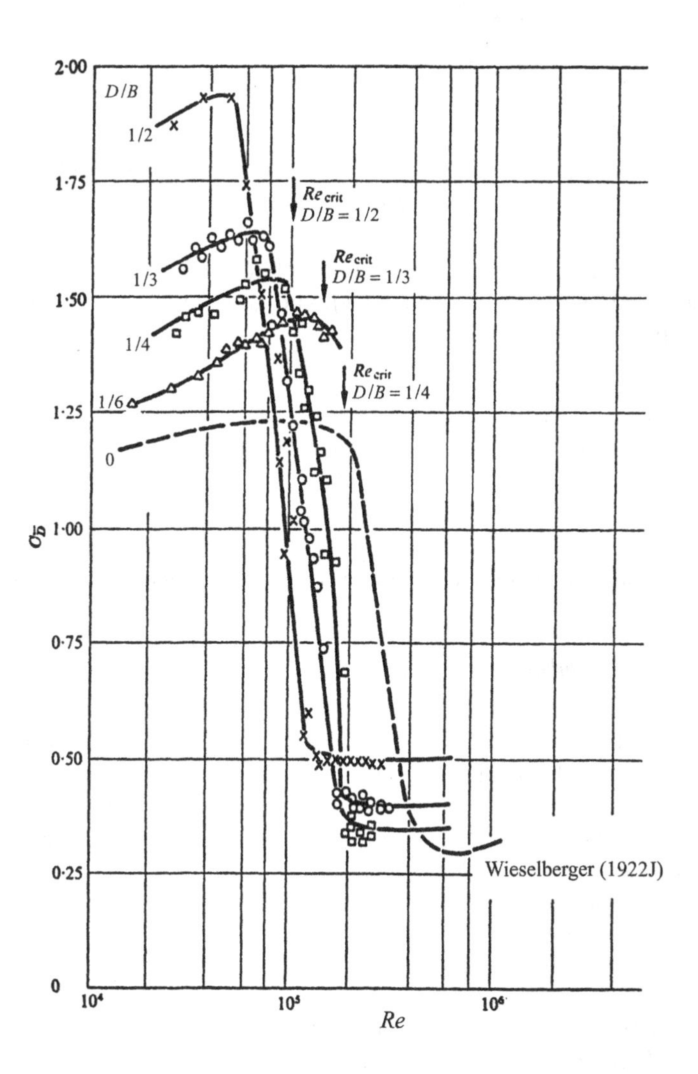

FIG. 23.32. Mean drag coefficient in terms of Re and D/B, Richter and Naudascher (1976J)

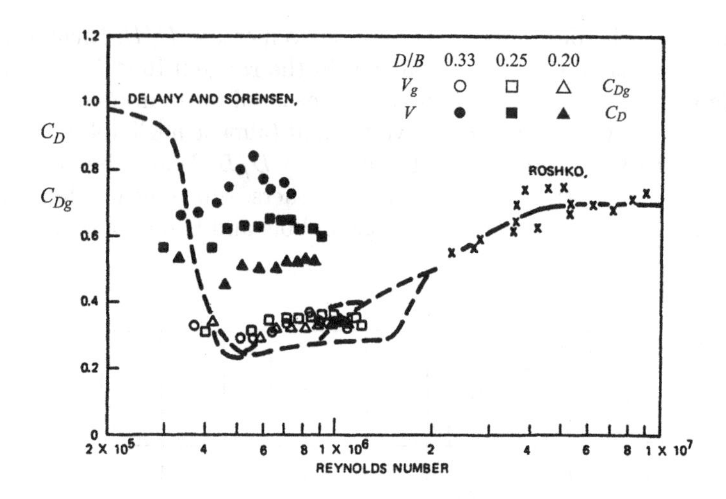

FIG. 23.33. Mean drag coefficient in terms of Re and D/B expressed through the free stream velocity (closed circles), and gap velocity (open circles), Chen and Doepker (1975J)

and 2 regimes, respectively.

Chen and Doepker (1975J) extended tests to 350k $< Re <$ 1.2M for $D/B =$ 0.20, 0.25, and 0.33 ($L/D =$ constant). Figure 23.33 shows the conventional C_D and C_{Dg} (based on the gap velocity) in terms of Re and three D/B tested. The increase in C_D with D/B is replaced by the collapse of data onto a single curve when C_{Dg} is used. It also appears that the TrBL2 regime extends unabated to $Re =$ 1.2M.

23.4.3 *Strouhal number and fluctuating force*

Richter and Naudascher (1976J) evaluated St from the frequency spectra of C'_L. Figure 23.34(a,b) shows a distinct jump in St when separation bubbles are formed. The re-plot of data in Fig. 23.34(b) in terms of the universal St_R proposed by Roshko (1955J)[130] produced a remarkable collapse of data into $St =$ 0.16 and 0.28 before and beyond the St jump, respectively. The $D/B =$ 0.5 was an exception in the TrBL2 regime.

Richter and Naudascher (1976J) measured C'_L and C'_D on the cylinder segment, $L/D =$ 6, in the range 20k $< Re <$ 400k. Figure 23.35(a) shows a significant increase in conventional C'_L by blockage in the TrBL0 regime; for example, there is a seven-fold increase in C'_L for D/B = 0.5. The high C'_L values are followed by extremely small C'_L in the TrBL2 regime. The C'_L curves by Fung (1960J) and Schmidt (1965J) are in good agreement, as can be seen in Fig.

[130]The diameter of the cylinder is replaced by the wake width and the free stream velocity by the separation velocity, Vol. 1, Chapter 13.

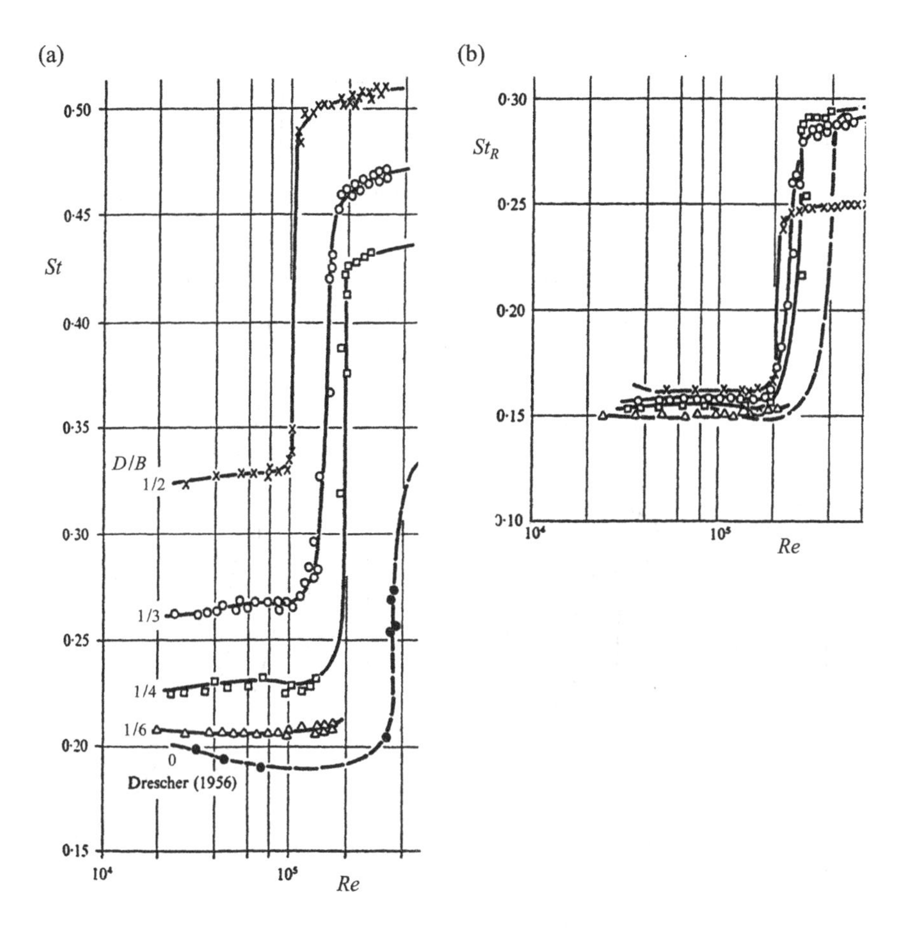

FIG. 23.34. Strouhal number in terms of Re for various D/B, (a) conventional, (b) universal, Richter and Naudascher (1976J)

23.35(a). The low values of C'_L may be attributed to a considerable lengthening of the eddy formation region, as has been found for small blockages.

There is an unexpected trend in measured C'_D shown in Fig. 23.35(b). In the TrBL0 regime C'_D, the curves are in the opposite order, i.e., the highest C'_D curve corresponds the smallest $D/B = 0.16$. There are another two peculiar features:

(i) There is a sharp increase in C'_D with decreasing Re even in the TrSL3 regime. It has been shown in Vol. 1, Fig. 5.53, p. 138 that C'_D is constant in the TrSL3 regime, and the values are small $0.02 < C'_D < 0.04$ in the range $20\text{k} < Re < 200\text{k}$.

(ii) The sharp drop in the C'_D curves precedes a similar drop in the C'_L curves. Richter and Naudascher (1976J) argued that the three-dimensional flow in

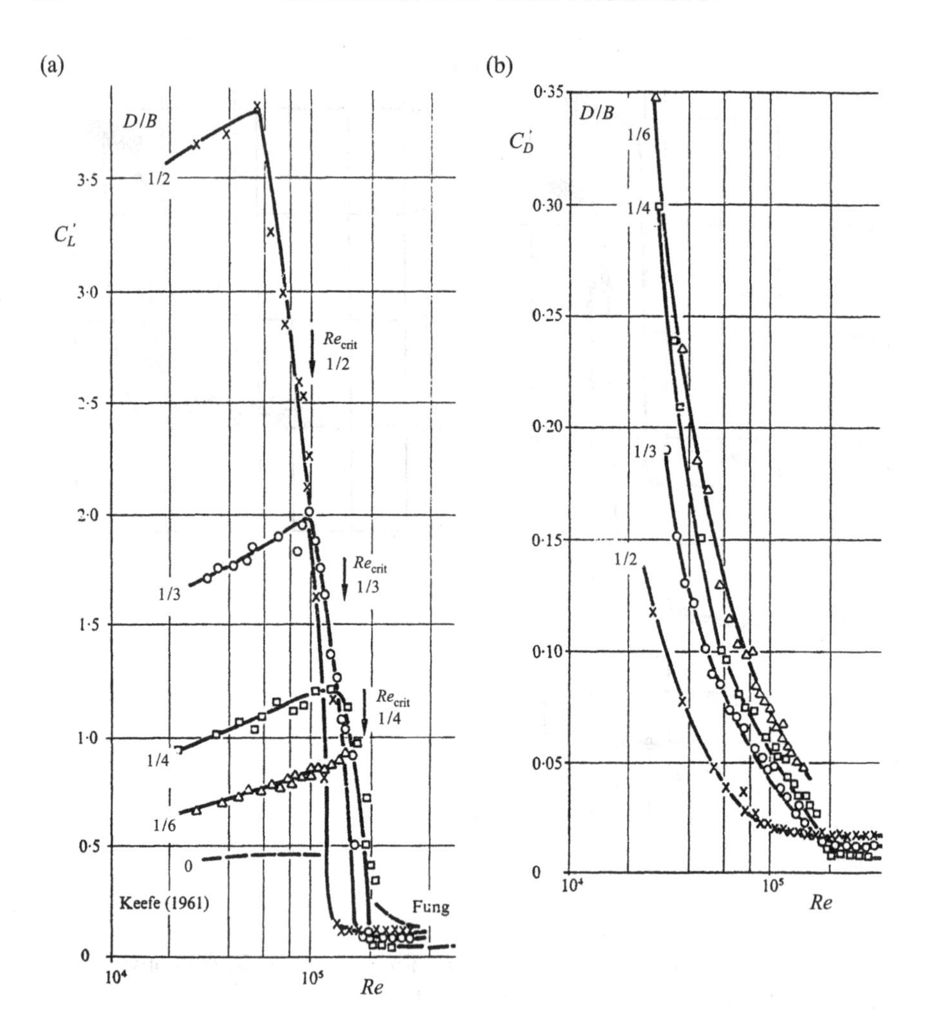

FIG. 23.35. Fluctuating (a) lift, (b) drag coefficients in terms of Re and D/B, Richter and Naudascher (1976J)

transitional boundary layers should be first felt by C'_D.

Chen and Doepker (1975J) extended tests in the TrBL3 regime, $350\text{k} < Re < 1.2\text{M}$, $20 < D/H < 0.33$, and $L/D = 3$. They found that C'_D was about five times smaller than C'_L. Figure 23.36(a,b) shows the measured C'_L and C'_D, respectively. Both C'_L and C'_D exhibit a monotonous increase with rising Re in the TrBL3 regime similar to that found for the small blockage.

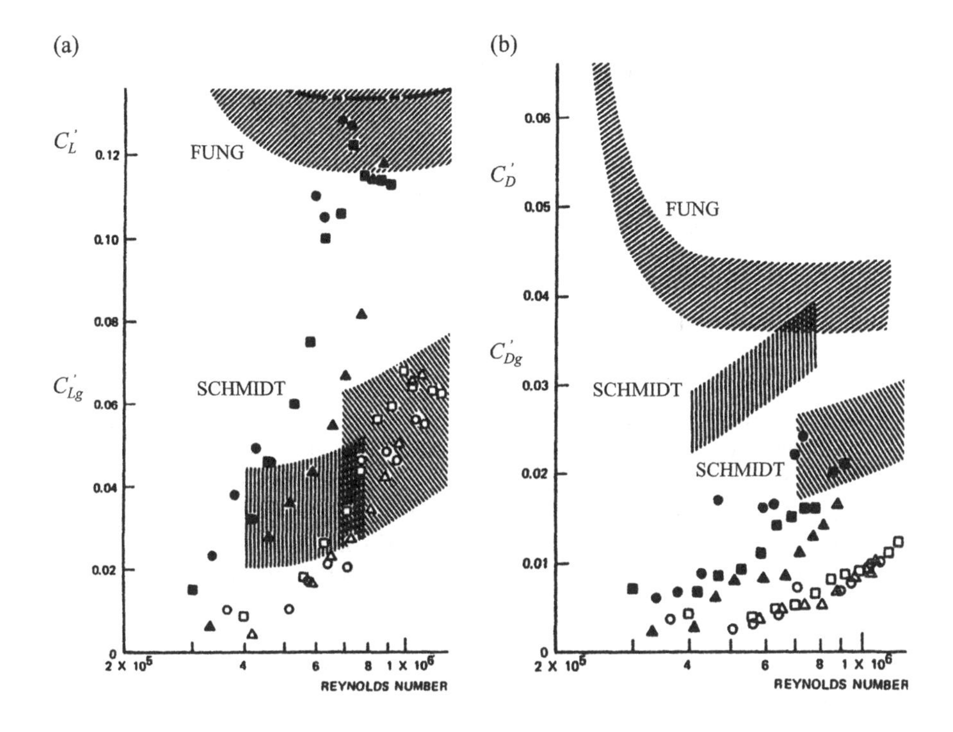

FIG. 23.36. Fluctuating (a) lift, (b) drag coefficients in terms of Re expressed through free stream V and gap velocity V_g, Chen and Doepker (1975J)

23.5 Theoretical correction models

23.5.1 *Introduction*

Blockage effects are always present in closed test sections of wind and water tunnels. Corrections of measured data are required in order to obtain blockage-free values relevant to most practical applications. This was recognized in the early days of aviation and wind tunnel testing when the first theoretical models for blockage correction were derived by Fage (1929P), Lock (1929P), and Glauert (1933P). These early correction models were derived for bodies of small drag, such as Rankine's ovals, aerofoils, etc. and not for high drag bodies like circular cylinders.

The blockage effect depends not only on D/B and Re, but also on the shape of the body and test section. The latter may be rectangular, square, elliptical, octagonal, or circular. The theoretical correction models are derived for two-dimensional flows between straight and parallel walls. It has been discussed in Chapter 21 that a nominally two-dimensional cylinder spanning a rectangular test section is always subjected near its ends to horseshoe-swirls formed by the roll-up of wall boundary layers. Hence, the flow near the cylinder ends is not two-dimensional.

It has been discussed in Section 23.1 that the cylinder presence in the test section induces three kinds of blockage: solid blockage, wake blockage, and a streamwise pressure gradient. The first correction models dealt with the solid blockage only; subsequent models dealt with the wake blockage, and only recently was the streamwise pressure gradient included.

A mathematical prediction of blockage effects on flow past cylinders in the closed test section involves a complete solution of the Navier–Stokes equations. However, most theoretical models are based on the inviscid potential flow theory or general momentum balance.

23.5.2 *Fage's blockage correction*

A simple method for predicting the blockage effects on the drag of symmetric bodies was derived by Fage (1929P). The prediction involved D/B, the shape of the test section, and an empirical factor. The method is based on the potential two-dimensional flow around Rankine's ovals, and the circular cylinder appears as a special case when the chord is equal to the thickness of the oval.

Rankine (1881B) derived a complex potential function for an oval by using a source and sink of equal strengths in a uniform free stream. The source followed by the sink produced an oval streamline, which could be regarded as a rigid and frictionless boundary. When the origin was taken at the oval centre the stream function was given by

$$\psi = V_0 y - \frac{V_0 t}{4\cot^{-1}(t/s)} \tan^{-1}\left[\frac{sy}{x^2 + y^2 - 0.25 s^2}\right] \tag{23.3}$$

where s is the distance between the source and sink and t is the thickness of the oval. For $\psi = 0$, the length of the oval chord is

$$a = s\left[1 + \frac{t}{s\cot^{-1}(t/s)}\right]^2 \tag{23.4}$$

Note that eqns (23.3) and (23.4) were derived for ovals in unconfined flow.

A modified equation was obtained for a Rankine oval between walls as

$$\psi = V_0 y + A\tan^{-1}\left(\frac{\sinh\dfrac{\pi s}{B}\sin\dfrac{2\pi y}{B}}{\cosh\dfrac{2\pi x}{B} - \cosh\dfrac{\pi s}{B}\cos 2\pi y B}\right) \tag{23.5}$$

where

$$A = \frac{V_0}{2}\tan^{-1}\left(\frac{\sinh\dfrac{\pi s}{B}\sin\dfrac{\pi t}{B}}{1 - \cosh\dfrac{\pi s}{B}\cos\dfrac{\pi s}{B}}\right) \tag{23.6}$$

where the angles must be taken to lie between 0 and π.

If the velocity at any point on the surface of an oval in an unconfined stream is V_0, then the velocity at the same point when the oval is in the test section is increased to mV. It can be calculated at any point x, y on the oval from

$$\left(\frac{mV}{V_0}\right)^2 = \left\{1 - \frac{2\pi y}{BC}\left[\frac{\sin 2C}{2\tan\dfrac{2\pi y}{B}} - \frac{\sin^2 C}{\tanh\dfrac{\pi s}{B}}\right]\right\} + \left[\frac{2\pi y^2 C \sinh\dfrac{2\pi x}{B}}{BC\sinh\dfrac{\pi s}{B}\sin\dfrac{2\pi y}{B}}\right]^2 \tag{23.7}$$

where

$$C = \frac{2y}{t}\tan^{-1}\left[\frac{\sinh\dfrac{\pi s}{B}\sin\dfrac{\pi t}{B}}{1-\cosh\dfrac{\pi s}{B}\cos\dfrac{\pi t}{B}}\right]$$

As an illustration of m, Fage (1929P) calculated m at $x = 0$ and $y/t = 0.5$ on the surface following: $s/t = 1$ (circle), 2.5, 5.0, 9.5, and 18 in the range $0.05 < D/B < 0.25$. Figure 23.37 shows the variation of m in terms of B/t for four ovals and a circle ($s/b = 1$).

The correction factor n represents the ratio of the measured drag force between the walls nF_D and without walls $F_D(B \to \infty)$. Fage (1929P) suggested the following relationship

$$n = 1 + K(m^2 - 1) \tag{23.8}$$

where $K = 2.33$ for the circle can be regarded as an empirical factor related principally to the body shape and $m^2 - 1$ is a factor dependent on c and t (chord and thickness) of the oval, and on the breadth of the test section B. Figure 23.38 shows the correction factors n and m^2 in terms of B/D. Few experimental points (crosses) are added to show how the constant 2.33 was obtained to fit the curve. By using the plot, it is possible to predict the drag in an unconfined free stream from the drag measured in the confined stream between walls B apart.

23.5.3 *Lock's method of images*

A two-dimensional inviscid theoretical model for circles was derived by Lock (1929P). He considered the flow past a circle in a closed test section with straight parallel walls of breadth B. The straight streamlines along the walls were simulated by an infinite series of images, as shown in Fig. 23.39. The images of the doublets lie on a straight line normal to the test section, all are of equal strength, and all are a distance B apart. The images induce an additional velocity u_i in the neighbourhood of the original doublet in the test section

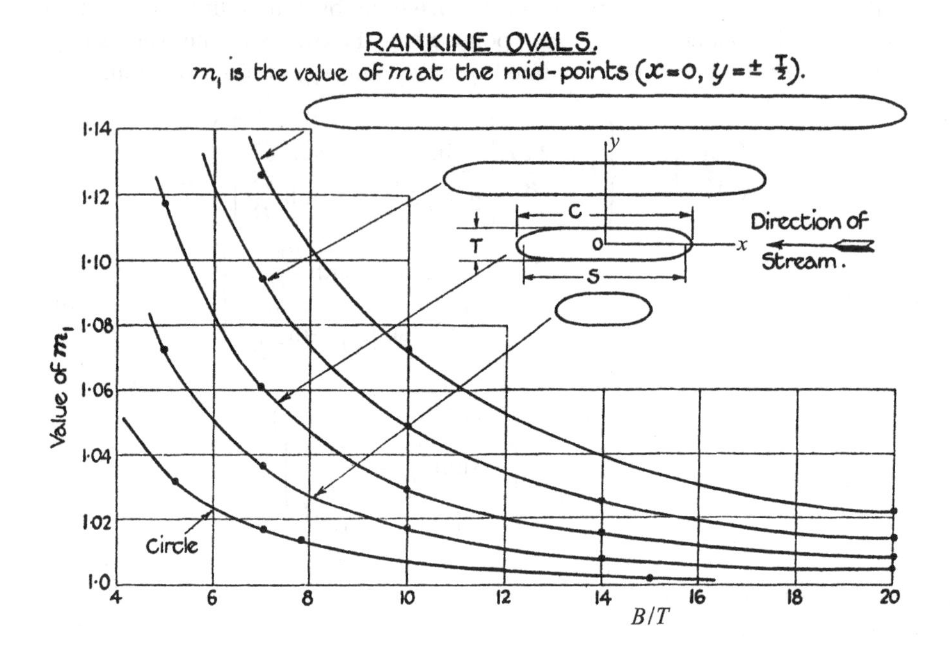

FIG. 23.37. Velocity factor, m, in terms of B/D for ovals and circle, Fage (1929P)

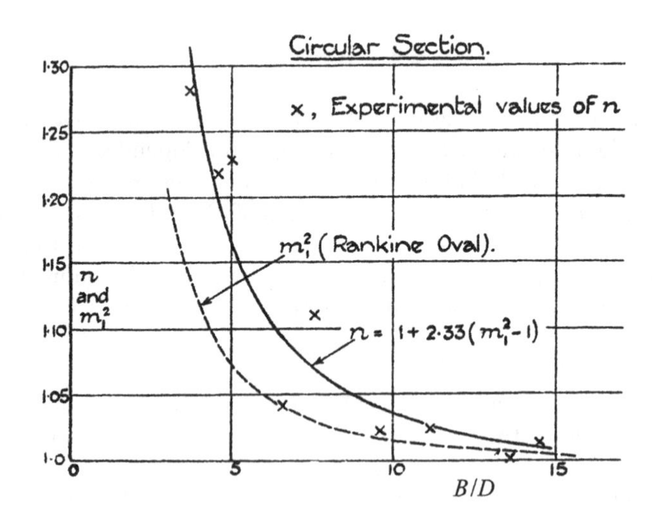

FIG. 23.38. Drag correction factor, n, in terms of B/D for circle, Fage (1929P)

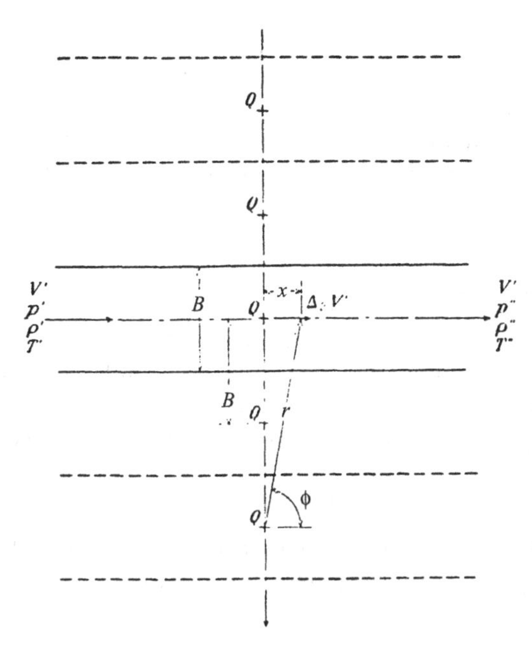

FIG. 23.39. Layout of image doublet or source system adapted from Allen and Vincenti (1948P)

$$u_i = \left(\frac{2Q_0}{B^2}\right) \sum_{n=1}^{\infty} \frac{1}{n^2} \tag{23.9}$$

where summation yields $\pi^3 Q_0/3B^2$. Under these conditions, the shape of the original wall streamline will be altered, and in order to recover the straight streamline the strength Q_0 must be replaced by Q such that

$$\frac{Q}{Q_0} = \frac{V + u_i}{V} \tag{23.10}$$

It follows that the magnitude of the velocity at any point in the neighbourhood of the body will be altered in this same ratio. Hence, it is expected that the pressure drag of the body will be increased in the ratio $(1 + u_i/V)^2$.

Lock (1929P) applied his theory to elliptic cylinders and Zhukovski aerofoils. Only the first will be discussed because the circle is a special case of the ellipse having the same semi-axes, $a = b$. The complex potential for the ellipse is given by

$$\omega = V\left(z + c \sinh \xi_0 e^{\xi_0 - \xi}\right) \tag{23.11}$$

where $z = c \cosh \xi$ and semi-axes: $a = c \cosh \xi_0$ and $b = c \sinh \xi_0$. For large values of z

$$\frac{Q}{VB^2} = \frac{1}{2}\left(1 + \frac{a}{b}\right) \tag{23.12}$$

where for a circle $a = b$ and $Q = VB^2$.

Lock (1929P) compared his model with Fage's (1929P) semi-empirical formula, eqn (23.8). He suggested that the empirical factor $K = 2.33$ should be replaced by $K_1 = 3.13$.

23.5.4 *Glauert's semi-empirical formula*

Glauert (1933P) wrote:

> On reviewing the problem it would appear that Lock's (1929P) analysis is based on sound principles for a body in a perfect fluid, but in such a fluid, body has no drag. On comparison of Lock's (1929P) prediction with experiments, it is found that the theoretical induced velocity must be altered by an empirical factor whose value is 3.13 for a circular cylinder. The magnitude of this factor casts doubt on the adequacy of the theoretical model.

Glauert (1933P) re-examined the blockage effect and argued that the governing parameter is $(D/B)^2$. He suggested a semi-empirical formula

$$\frac{1}{n} = \left[\frac{1 - C(D/B)^2}{1 + 0.822(D/B)^2}\right]^2 \tag{23.13}$$

where C is an empirical factor determined from experiment ($C = 0.3$ for a circular cylinder). Figure 23.40 shows Glauert's comparison of his correction formula (full line) with Fage's (1929P) experiments (crosses). There is an agreement for higher values of D/B and the discrepancy for the lower D/B, which he attributed to measuring errors. The dashed line represents the correction anticipated in a free jet.

Modi and El-Sherbiny (1973P) applied Glauert's (1933P) formula to their C_D measurements in the range $0.03 < D/B < 0.35$. Figure 23.41(a) shows a wide scatter for $D/B < 0.2$, and above that the C_D values are not reduced. In an attempt to improve the correction formula, they changed the empirical factor to $C = 0.6$. Figure 23.41(b) shows a considerable improvement and reduced scatter of points.

23.5.5 *Allen and Vincenti's source model*

23.5.5.1 *Simulation of wake blockage*

Since the local velocity within the wake is less than in the external flow, the mass flow rate per unit area $\dot{m}$ is less inside the wake than outside. In the closed test section, the requirement of continuity of flow between a station upstream of the cylinder, and a station across the wake requires that $\dot{m}$ outside the wake is greater than upstream of the cylinder. This implies that as the flow proceeds down the tunnel the velocity in the main portion of the stream undergoes an increase. The wake blockage gives rise to a velocity increment coupled with the velocity and pressure gradients, which are not present in the unconfined stream.

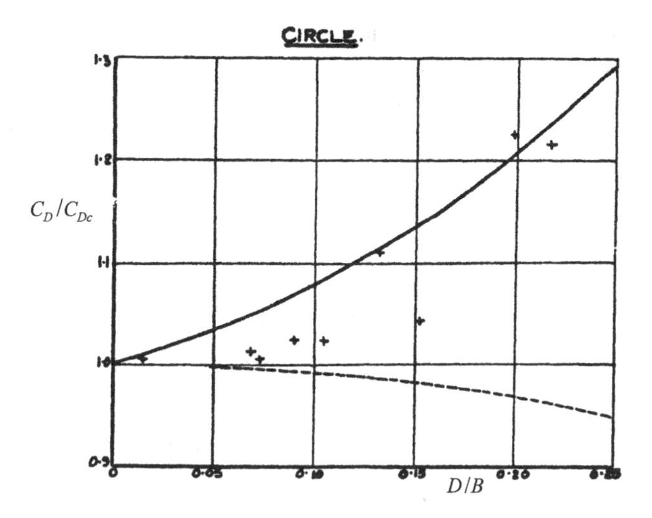

FIG. 23.40. Ratio of wind tunnel and corrected drag in terms of D/B, Glauert (1933P)

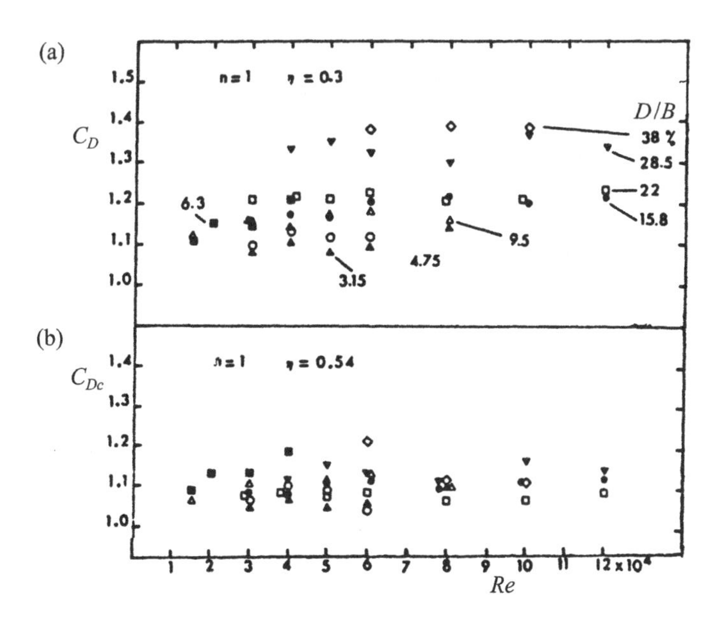

FIG. 23.41. (a) Experimental data, (b) corrected, using Glauert's formula, Modi and El-Sherbiny (1971P)

Allen and Vincenti (1948P) derived a theoretical model for the wake blockage. Two stations in the tunnel are considered, far upstream from the cylinder and far enough downstream so that the wake has spread to the walls, and the velocity is again uniform across the tunnel. The difference in static pressure between the two stations is evaluated as a function of the measured cylinder drag.

The cylinder and its wake are replaced by a source located at the position of the cylinder. The source strength is then related to the cylinder drag by ensuring that the static pressure difference between the two stations in the tunnel is the same as that when the cylinder and wake are present.

The tunnel walls can then be simulated by an infinite system of sources placed directly above and below the position of the cylinder at intervals equal to the breadth of the tunnel, as shown in Fig. 23.39. The system of image sources would induce a small finite negative velocity upstream, so that it is necessary to superimpose on the flow field an additional uniform flow of equal velocity in the opposite direction in order to satisfy the original condition that the upstream velocity be unchanged. The velocity of this flow expressed in terms of the source strength and cylinder drag gives a velocity increment caused at the cylinder by the interference between the wake and the walls. It is evident that this method fails to satisfy conditions as regards the velocity at infinity downstream. This discrepancy arises through the fundamental difference between the actual flow in the wake and the source flow by which it is represented, and is unavoidable as long as this representation is used.

23.5.5.2 *Derivation of the source model* Allen and Vincenti's (1948P) theoretical model considers an inviscid, compressible,[131] and two-dimensional flow in a closed wind tunnel, as sketched in Fig. 23.39. At a station far upstream in the undisturbed region, the velocity is denoted by a dot and far downstream by double dots. The equation of continuity is

$$\rho B V'' = \rho B V' + Q \quad \text{or} \quad V''/V' = 1 + Q/\rho B V' \tag{23.14}$$

where the cylinder and its wake are replaced by the source of strength Q. The latter is required to produce the same pressure difference as drag, hence

$$Q = \tfrac{1}{2}\rho C_D D V' \tag{23.15}$$

The streamwise velocity increment ΔV induced by the infinite system of sources necessary to simulate wind tunnel walls is

$$\Delta V = \sum_{n=1}^{\infty} Q/\pi\rho r_n \cos\theta_n \tag{23.16}$$

where r_n and θ_n are the radial distance and the polar angle of the point relative to the source at distance nB above and below the centre line, respectively, Fig. 23.39. The final result after summation is

[131] In all equations presented here an incompressible flow is considered, $Ma = 0$. For compressible flow, see Allen and Vincenti (1948P).

$$\Delta V = -Q/2\rho B \tag{23.17}$$

In order to keep the original upstream velocity V' unchanged, the $-\Delta V$ must be counterbalanced by another uniform flow of equal magnitude but of opposite sign. The addition of this flow at all points in the field will result in a speeding up by

$$\Delta V = \frac{1}{4} C_D \frac{D}{B} V' \tag{23.18}$$

The pressure–drag coefficient correction terms are all negative and tend to reduce the measured pressure–drag coefficient C_d. The final expression is

$$C_{dc} = C_d \left[1 - \frac{\pi^2}{4} \left(\frac{D}{B} \right)^2 - \frac{C_d}{2} \left(\frac{D}{B} \right) \right] \tag{23.19}$$

where the second term in the bracket represents both the solid blockage and the effect of the streamwise pressure gradient, and the third term is due to the wake blockage. The velocity and pressure gradients are obtained from

$$\frac{dV'}{dx} = \frac{\pi}{12} \frac{C_d D}{B^2} V' \quad \text{and} \quad \frac{dp}{dx} = \frac{\pi}{6} \frac{C_d q D}{B^2} \tag{23.20}$$

23.5.5.3 *Correction model validation* Allen and Vincenti (1948P) used Fage's (1929P) data to check their correction model. They found good agreement between the corrected C_d and the straight line corresponding to C_d at $D/B = 0$, see Fig. 23.42. Dalton (1971J) repeated calculations and found a different value for the corrected C_D by using eqn (23.19). Figure 23.42 shows C_D in terms of D/B according to Fage (1929P), open circles, Allen and Vincenti (1948P), crosses, and Dalton (1971J), open squares. The good agreement with $D/B = 0$ (dashed curve) reported by Allen and Vincenti (1948P) turned out to be an overcorrection when calculated by Dalton (1971J).[132] Dalton (1971J) suggested that Allen and Vincenti's (1948P) equation should not be used for $D/B > 0.1$, and proposed instead that the coefficient $\pi^2/4$ should be replaced by an empirical 1/4 in eqn (23.19).

Another validation of Allen and Vincenti's (1948P) correction formula was reported by Farell *et al.* (1977J). The cylinder has $L/D = 8$, $0.07 < D/B < 0.16$, 72k $<$ *Re* $<$ 150k, for a small wind tunnel, and 150k $<$ *Re* $<$ 250k, for a large wind tunnel. These two *Re* ranges correspond to the TrSL3 and TrBL0 regimes, respectively. Figure 23.43(a,b) shows C_d, $|C_{p\min}|$, $|C_{pb}|$, and $C_{pb} - C_{p\min}$ in terms of D/B. The uncorrected experimental values (full lines) increase with D/B, while the corrected ones, Allen and Vincenti's (1948P) (dashed line), and Maskell's (1963P) are on $D/B = 0$, blockage-free line. Allen and Vincenti's (1948P) correction gives better agreement along the blockage-free line, particularly for $D/B > 0.1$.

[132] Dalton (1971J) stated that a numerical error crept in in Allen and Vincenti's (1948P) calculations.

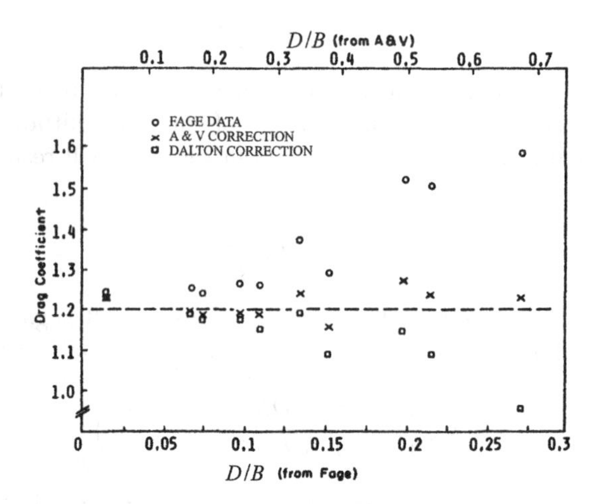

FIG. 23.42. Uncorrected C_D (circles), corrected C_{Dc} (crosses), Allen and Vincenti (1948P) corrected C_{Dc} (squares). Dalton (1971J)

23.5.6 *Maskell's correction model*

Maskell (1963P) developed a semi-theoretical model, which was intended for axisymmetric wakes and stalled wings of any aspect ratio. He stated:

> An implied assumption in the theory is that the origin of the wake, i.e. boundary layer separation on the body is independent of blockage. And so it may be necessary to exclude well-rounded bluff bodies (like the circular cylinder), for which a small change in pressure distribution might lead to a significant movement of the separation point.

Maskell (1963P) proposed to simulate the wake by the stream surface, as illustrated in Fig. 23.44. This extends downstream to station 2, where the cross-sectional area of the wake is at a maximum. At the stream surface there is a constant pressure p_b and a corresponding velocity kV, where V is the free stream velocity. The further development of the wake is of no immediate interest. The shape of the constant pressure surface is unknown.

The fundamental hypothesis of the model is that the pressure distribution over the body remains self-similar for various D/B values

$$\frac{p - p_b}{p_0 - p_b} = \text{independent of blockage} \tag{23.21}$$

where p is the pressure at any point on the body surface, and p_0 is the stagnation pressure of the free stream. Since $p_0 - p_b = k^2 q$, where q is the dynamic pressure, it follows that

$$C_D/k^2 = \text{constant} \tag{23.22}$$

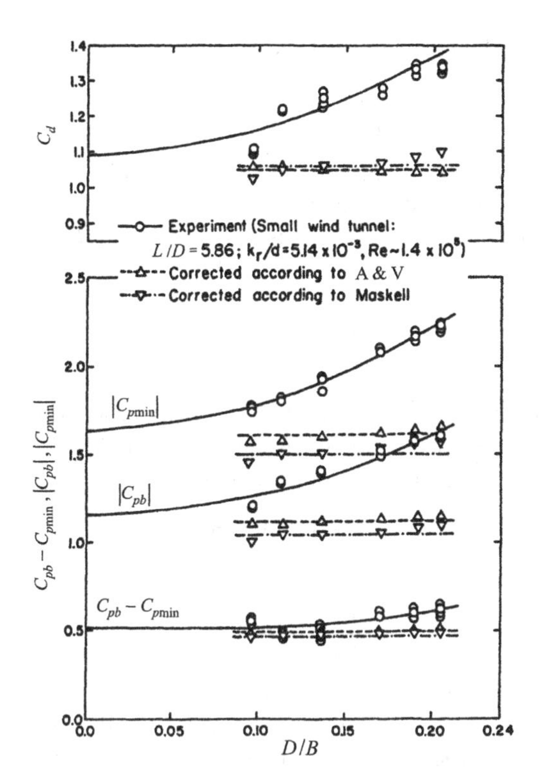

FIG. 23.43. Blockage effect uncorrected (full line), corrected (dashed line) according to Allen and Vincenti (1948P), Maskell (1963P), and Farell *et al.* (1977J)

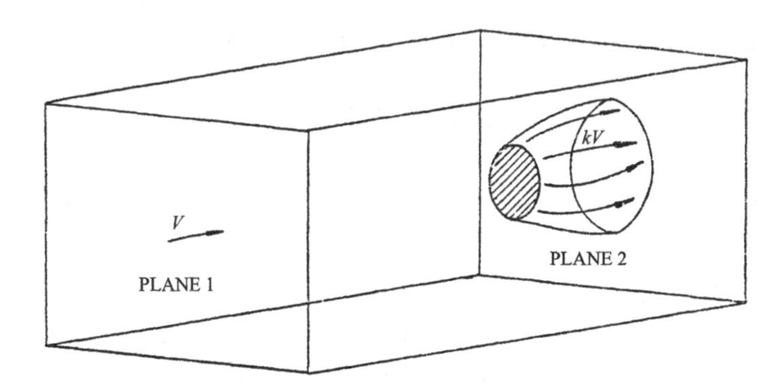

FIG. 23.44. Definition sketch of disc and near-wake, Maskell (1963P)

where C_D is the drag coefficient. It also follows that $k_c V_c = kV$, where the subscript c refers to the corrected values. Hence

$$\frac{V_c^2}{V^2} = \frac{k^2}{k_c^2} = \frac{C_D}{C_{Dc}} \tag{23.23}$$

The validity of the hypothesis in eqn (23.22) was confirmed by experiments on square plates in the range $1.38 < k^2 < 1.6$, and it was found that $C_D/k^2 = 0.84$.

Another equation is required because both V_c and k_c are unknown. The momentum equation for the control surface shown in Fig. 23.44 is

$$F_D + p_b A_w = \iint_A \left(p_1 + \rho V_c^2\right) \, dy \, dz - \iint_{A-A_w} \left(p_2 + \tfrac{1}{2}\rho V_c^2\right) \, dy \, dz \tag{23.24}$$

where F_D is the drag force, and A and A_w are the cross-sectional areas of the test section and wake, respectively. Since the flow is outside the wake, Bernoulli's equation is applicable and the continuity equation is

$$V_1 A = V_2 \left(A - A_w\right) \tag{23.25}$$

which gives the final expression

$$C_D = \frac{A_w}{A_B} \left(k^2 - 1 - \frac{A_w}{A}\right) \tag{23.26}$$

where A_B is the cross-sectional area of the body.

A further relationship is required to account for the distortion of the wake by the blockage. Maskell (1963P) wrote: 'To do this theoretically would involve a greater understanding of the internal mechanics of the wake than was available at present.' So he suggested an empirical formula

$$\frac{k^2}{k_c^2} = 1 + \frac{C_D}{k^2 - 1} \frac{A_B}{A} + O\left\{\left(\frac{A_B}{A}\right)^2\right\} \tag{23.27}$$

Note that $A_B/A = D/B$ for a circular cylinder. The effect of the wake distortion is to replace C_{Dc} in the correction term by the measured C_D. Alternatively, by using q it yields

$$\frac{\Delta q}{q} = \varepsilon \frac{C_d D}{B} \tag{23.28}$$

where Δq is the effective increase in dynamic pressure for the free stream q, and $\varepsilon = (k^2 - 1)^{-1}$, the blockage factor for the bluff body flow ($\varepsilon = 0.96$ for two-dimensional bluff bodies).

It has been shown that Farell *et al.* (1977J) found good agreement by using Maskell's (1963P) correction formula. Another check in a wider range $0.03 < D/B < 0.35$ was done by Modi and El-Sherbiny (1973P). Figure 23.45(a) shows that the correction formula is not valid beyond $D/B = 0.2$. Modi and El-Sherbiny (1973P) attempted to improve the formula by using higher order terms. This led to a limited improvement.

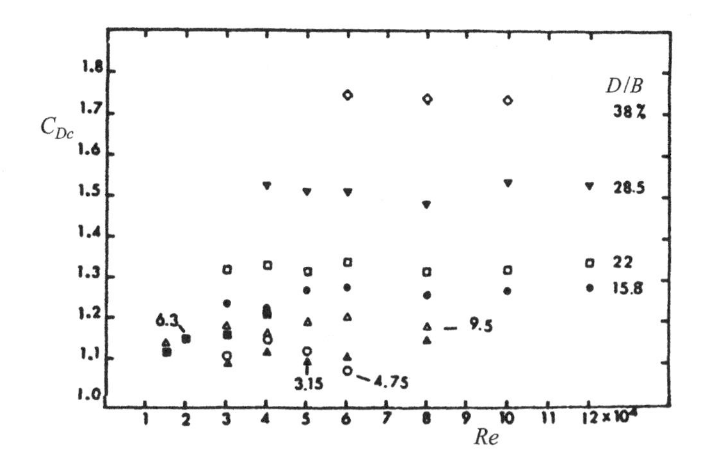

FIG. 23.45. Correction of experimental C_D using Maskell's model, Modi and El-Sherbiny (1971P)

23.5.7 *Modi and El-Sherbiny's streamline model*

Modi and El-Sherbiny (1977J) extended Parkinson and Jandali's (1970J) model developed for a cylinder in an infinite stream to flow between parallel walls. Parkinson and Jandali's (1970J) model was discussed in Vol. 1, Chapter 10, pp. 304–6. The near-wake was simulated by two sources attached to the downstream side of the circle and two sinks placed at the circle centre. The source strength was determined by the location of the separation points.

Modi and El-Sherbiny's blockage derivation is based on the flow around a circle between the parallel walls separated by the distance B. The physical plane z is transformed to a quasi-arc in the plane ω, as depicted in Fig. 23.46(a,b), respectively. Arc S_1AS_2 is mapped in the ω plane by the transformation

$$\omega = \exp\left[\frac{2\pi(z+iB)/2}{B}\right] \tag{23.29}$$

and two walls are mapped on the u-axis. The transformation also reduces the uniform flow by the sink $-2Q$ of strength VB at the origin. It is possible to map the arc into the circle in the plane ζ by using the transformation

$$z = \frac{B}{2\pi}\left[\frac{1}{2}\sin\left(\frac{\pi b}{B}\right)\left(\zeta + \cot B - \frac{1}{\zeta + \cot B}\right) + \cos\left(\frac{\pi b}{B}\right)\right] - \frac{B}{2} \tag{23.30}$$

By using Parkinson and Jandali's (1970J) transformation the arc becomes a circle of radius r_1 in the plane ζ. Points S_1 and S_2 become critical points and free streamlines become normal to the circle. The source $2Q$ appears at a distance ζ_m from the origin. It is necessary to add an equal source Q' at the image point

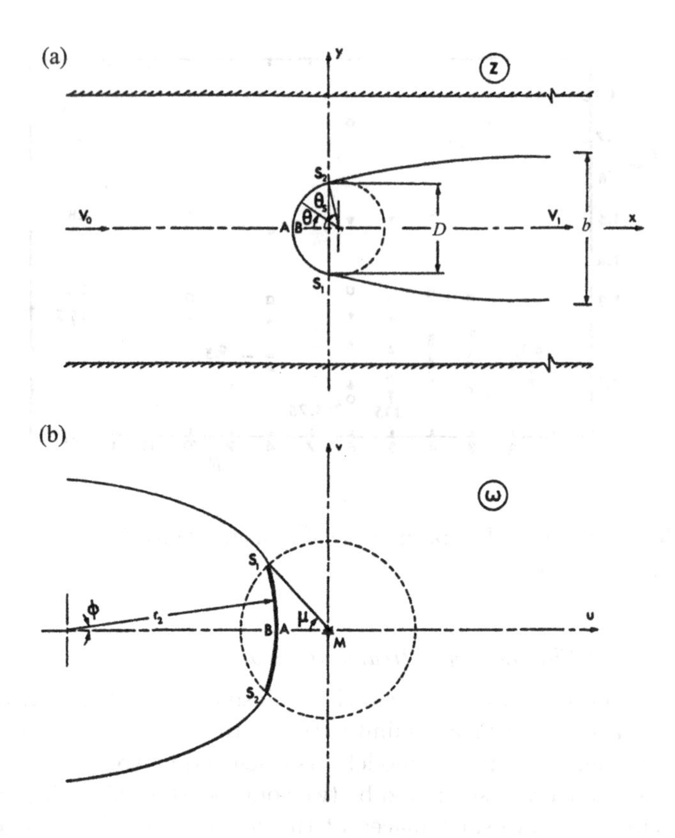

FIG. 23.46. Transformation of circle from the physical plane z to ω, Modi and El-Sherbiny (1977J)

ζ'_m, and at the corresponding sink at the origin. A pair of double sources of strength $2Q$ are located symmetrically on the circumference at angles $\pm\delta$ and their image sinks at the origin.[133] The complex potential is

$$F(\zeta) = \frac{VB}{2\pi}\left[\ln(\zeta - \zeta_m) + \ln(\zeta - \zeta'_m) - \ln\zeta\right] + \frac{Q}{\pi}\left[\ln\left(\zeta - r_1 e^{i\delta}\right) + \ln\left(\zeta - r_1 e^{-i\delta}\right) - \ln\zeta\right] \tag{23.31}$$

To make the free streamlines normal to the surface at S_1 and S_2, the sources of strength Q and the angles $\pm\delta$ are adjusted in such a way as to fix the S_1 and S_2 stagnation points, $V_\zeta = 0$.

[133] Note that the strength of the sink is $-2Q$ and two sources located externally at the circumference $+4Q$ are not cancelled. Thus the model is similar to that of Allen and Vincenti (1948P).

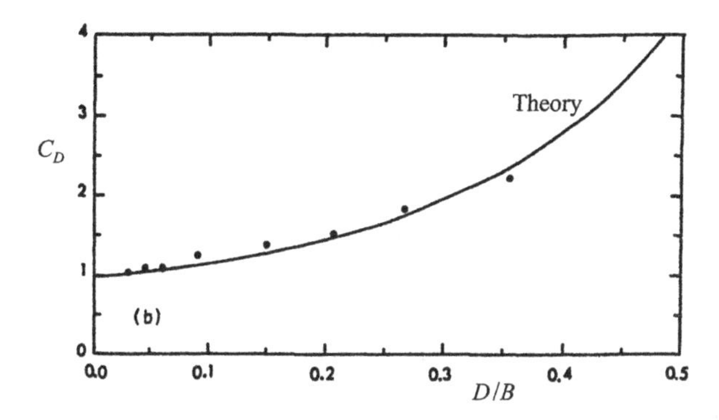

FIG. 23.47. Comparison of theory and experiment for the rise in C_D in terms of D/B, Modi and El-Sherbiny (1977J)

The experiments carried out to validate the model covered a wide range, $0.03 < D/B < 0.35$, $10\text{k} < Re < 120\text{k}$, and $2.3 < L/D < 26.6$. Figure 23.47 shows the measured C_D (full circles) and model prediction (full line) in terms of D/B. Even for high values of D/B, the theoretical model predicted C_D quite well. The only limitation of the model is the concept of the wake without eddy street.

23.6 Asymmetric blockage

It has been stated in the introduction in Section 23.1 that an asymmetric blockage would arise if the model was placed away from the tunnel axis. Most wind tunnels are designed in such a way that the model is always placed half way between the walls. Limited research has been carried out into the occurrence of asymmetric blockage.

23.6.1 *Laminar wake*

Taneda (1964J) examined the effect of cylinder displacement away from the flume axis for $D/B = 0.007$, 0.01, and 0.04. Figure 23.48 shows measured C_R across the water tank at $Re = 4.5 \times 10^{-5}$, 1.7×10^{-4}, and 3.5×10^{-4}. The $C_{R\min}$ occurs about half way between the tank axis and side walls. The $C_{R\min}$ is around 10% lower than C_R at the axis. The rapid increase in C_R towards the wall is caused by the proximity to the wall.

Figure 23.49(a,b) shows surface visualization at $Re = 0.02$ and 170 for $G/D = 0.5$ and 0.1, respectively. Most of the flow approaching the cylinder is directed away from the wall. Such a gross asymmetry creates C_L. Figure 23.49(b) shows a unique single row eddy street, which is discussed shortly.

23.6.2 *Turbulent wake*

Ramamurthy and Lee (1973J) carried out tests for $D/B = 0.31$ and $Re = 70\text{k}$, 110k, and 150k. Figure 23.50(a) shows the C_p distribution for the cylinder $e/D =$

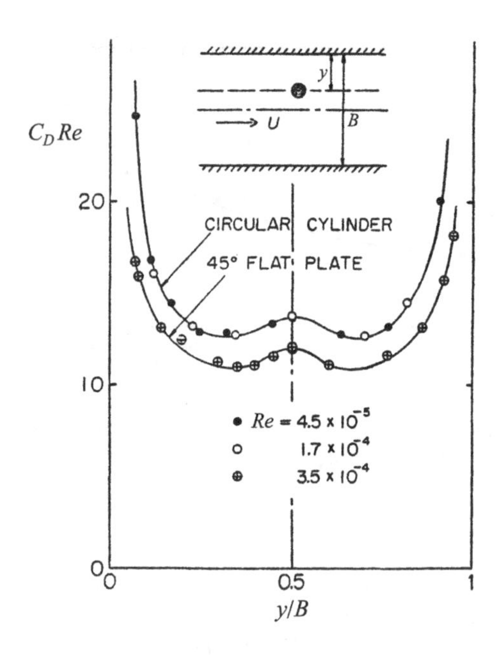

FIG. 23.48. Resistance coefficient variation across test section, Taneda (1964J)

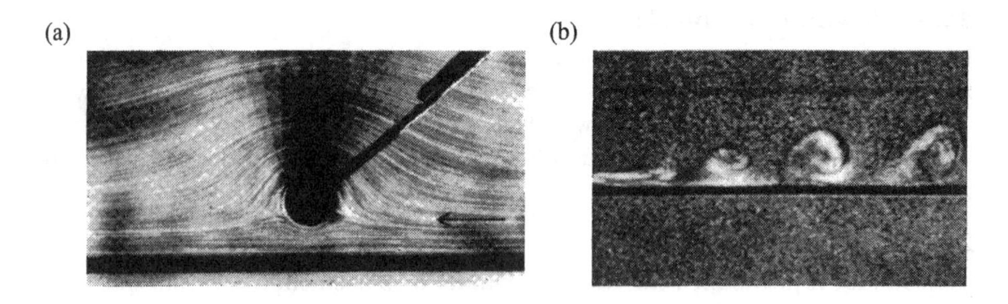

FIG. 23.49. Flow visualization, (a) Re = 0.02, G/D = 0.5, (b) Re = 175, $G/D = 0.1$, Taneda (1964J)

0.82 away from the tunnel axis. The two C_p curves that correspond to the TrSL3 and the start of the TrBL0 regime are close to each other. There is no asymmetry in the C_p distribution, and $C_{p\min}$ appears the same on both sides of the cylinder despite the proximity of one wall.

Figure 23.50(b) shows the conventional C_D, $C_{D1} = (1 - D/B)^2 C_D$, and $C_{Dj} = C_D/k^2$ in terms of G/D. The measured C_D shows a decreasing trend with rising G/D. However, when C_D is multiplied by $(1 - D/B)^2$ the corrected C_{D1}

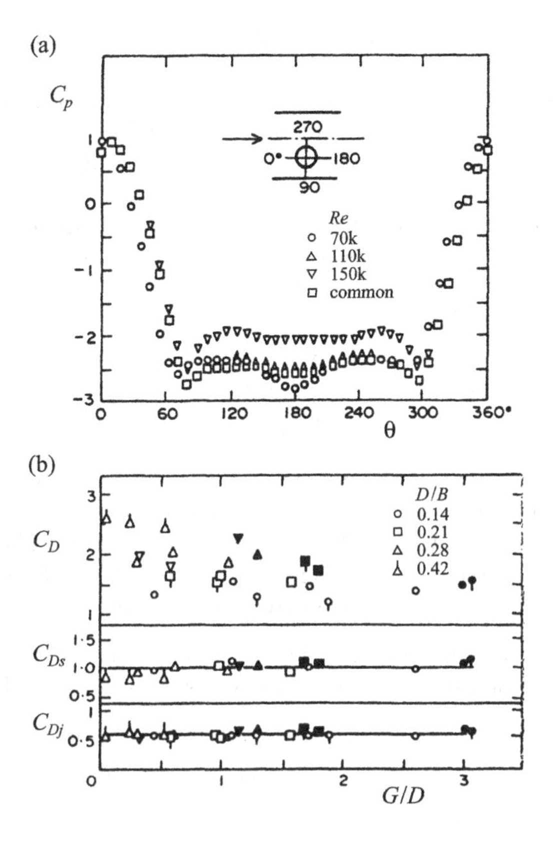

FIG. 23.50. Cylinder away from the test section axis, (a) pressure distribution for $D/B = 0.31$ and $E/D = 0.82$, (b) drag in terms of G/D, Ramamurthy and Lee (1973J)

remains nearly constant for the tested range of gap ratios. When the separation velocity was taken as the reference one, the straight line was lowered to $C_{Dj} = 0.55$.

23.7 Proximity to a boundary

23.7.1 *Introduction*

Pipelines have been widely employed in various branches of engineering, such as chemical, civil, offshore, etc. They are subjected to wind, tide, currents, or waves, and are usually laid in the proximity of the ground or seabed, termed the *boundary* for short. The boundary may be smooth or rough, flat or undulated, compact or erodible, etc. When pipelines are laid on a sandy boundary wind or waves move the sand away. The boundary is *scoured*, i.e. eroded and changed in shape underneath the pipeline. The phenomenon of scour is important for subsea pipelines.

The proximity of a cylinder to a boundary affects the flow around *both*, and consequently alters the magnitude of forces exerted on the cylinder. The interference is complex and depends not only on Re, and the other influencing parameters, but also on the gap to diameter ratio G/D, the relative boundary layer thickness δ_B/D, whether the boundary layer is laminar, transitional, or turbulent, etc. When the cylinder is in contact with the boundary, $G/D = 0$, the entire flow is forced to pass around the other side of the cylinder. This leads to an asymmetric flow resulting in a significant mean lift force C_L. When G/D is small, a weak gap flow is established. The asymmetry of the flow and C_L decrease as G/D increases. The interference ceases beyond a certain G/D value when the flow becomes symmetric around the cylinder, and the same pressure is exerted on both sides.

The straight and rigid boundary imposes parallel flow along itself, and prevents near-wake diffusion on its side. This one-sided confinement is further enhanced by the boundary layer developing along the boundary. The effect depends not only on the relative boundary layer thickness δ_B/D but also on δ_B/G. There are three possibilities for the relative size of the boundary layer and gap:

(i) $\delta_B/G = 0$, which can be achieved either when the cylinder is towed through water at rest, or when the boundary is moved at a flow velocity;

(ii) $0 < \delta_B/G < 1$, the cylinder is partly submerged in the boundary layer;

(iii) $\delta_B/G > 1$, the cylinder is fully submerged in the boundary layer.

The first case is not relevant for most engineering applications, and the last is very important in wind and offshore engineering. For the latter, the decrease in drag force is caused not only by the small G/D but also by the large δ_B/G, which brings the reduced shear velocity near to the cylinder.

23.7.2 *Classification of flow regimes*

The classification is related to the observed flow regimes caused by the interference between the cylinder and the boundary. The governing parameter is G/D, and the proposed flow regimes are:

(i) The contact regime, $G/D = 0$; the cylinder lies on the boundary. Figure 23.51(a) shows side-by-side the sketch and flow visualization. The cylinder retards the upstream flow and gives rise to flow separation from the boundary. This results in the formation of the upstream separation region as depicted in Fig. 23.51(a). The wake behind the cylinder and the boundary form the downstream separation region. The free shear layer separated from the cylinder reattaches to the boundary downstream at $x/D = 6$–10 depending on Re. Periodic eddy shedding is not found, although occasionally the random roll-up of the free shear layer into eddies may produce large eddies in the wake.

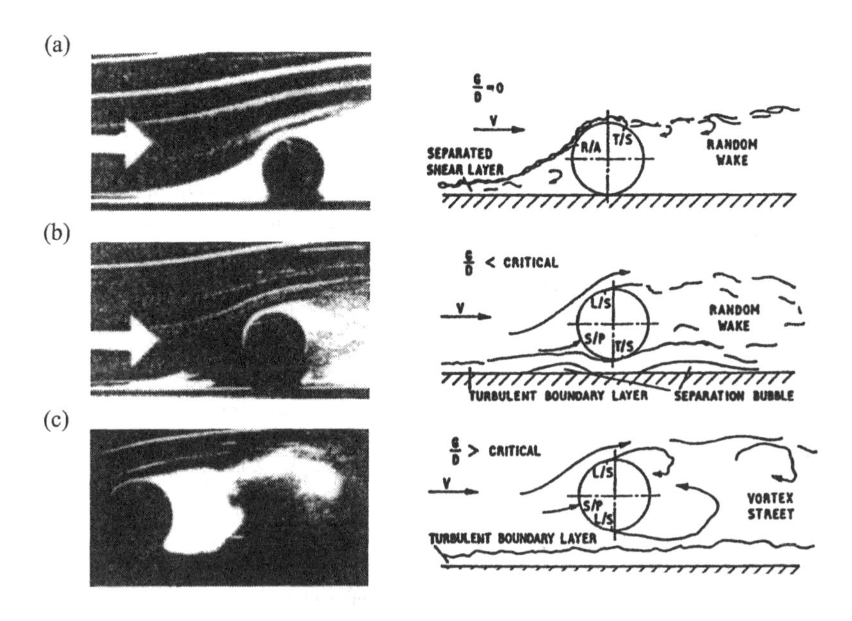

FIG. 23.51. Flow visualization and sketched topology, (a) $G/D = 0$, (b) $G/D < (G/D)_c$, (c) $G/D > (G/D)_c$, Zdravkovich (1982P, 1985P)

(ii) The narrow-gap regime, $0 < G/D \leqslant (G/D)_c$, where the subscript c stands for critical. The narrow gap may induce a jet-like flow along the boundary. The separated flow reattaches to the boundary due to the favourable pressure gradient produced by the contraction between the cylinder and the boundary. The former separation region is reduced to a separation bubble attached to the boundary, as seen in Fig. 23.51(b). The second separation from the boundary is caused by the wide-angle 'diffuser' downstream of the minimum gap inducing an adverse pressure gradient. No periodic eddy shedding is found in this regime until $G/D = (G/D)_c$.

(iii) The wide-gap regime, $G/D > (G/D)_c$. Regular periodic shedding is established in the wake, as seen in Fig. 23.51(c). The retardation of the upstream flow is considerably reduced due to the wide gap, and consequently the upstream separation bubble is obliterated. The wide gap also reduces the adverse pressure gradient along the gap. The periodic formation of strong eddies in the vicinity of the boundary induces the moving separation bubbles along the boundary between the consecutive eddies. At a fixed point along the boundary, the separation bubbles appear periodically in the rhythm of eddy shedding. Regular periodic eddy shedding is found for all $G/D > (G/D)_c$. As G/D increases, the flow asymmetry around the cylinder diminishes until at about $G/D = 2$ the flow becomes almost symmetric. For $G/D > 2$, the effect of the boundary is negligible on C_D and $C_L = 0$.

23.7.3 *Contact regime*

The cylinder appears like an obstacle or fence laid on the boundary, and as such forms the upstream and downstream separation regions. Among others, Bearman and Zdravkovich (1978J) measured the C_p distribution around the cylinder and along the boundary at $Re = 48$k, for $L/D = 32$, $\delta_B/D = 0.8$ (turbulent), and $Ti = 0.2\%$. Figure 23.52 shows the C_p distributions around the cylinder in polar coordinates and along the boundary in the range $-3 < x/D < 3$. The upstream separation region displays an almost constant positive C_p, and the downstream separation region a similar negative C_p. They are separated by a discontinuous jump at the contact between the cylinder and the boundary.

The C_p distribution around the cylinder is markedly asymmetric relative to the horizontal axis passing through the cylinder centre. It is notable that medium $C_{pb} = -0.5$ and low $C_{pb\min} = -1.2$ at $\theta = 45°$ contribute to a low C_D value. It will be shown that a minimum value of C_D is always found in this flow regime. Another feature of the C_p distribution is the low negative C_p above and the positive C_p below the bottom side. These contribute to a high value of mean C_L. A maximum C_L is always found in this flow regime, as will be discussed shortly.

An alarming discrepancy in the evaluated C_l and C_d has been found in the contact regime by various authors. Table 23.1 is a compilation of data for a wide range of Re and influencing parameters. The data are grouped in terms of Re and based on an inferred state of the cylinder boundary layer. The latter is affected by the separated boundary layer which reattaches onto the cylinder. The effect is analogous to free stream turbulence and triggers an early transition of the cylinder boundary layer. The state of the cylinder boundary layer is estimated from the magnitude of the adverse pressure recovery in C_p distributions. Table 23.1 shows that the low and high C_l are associated with the laminar and turbulent

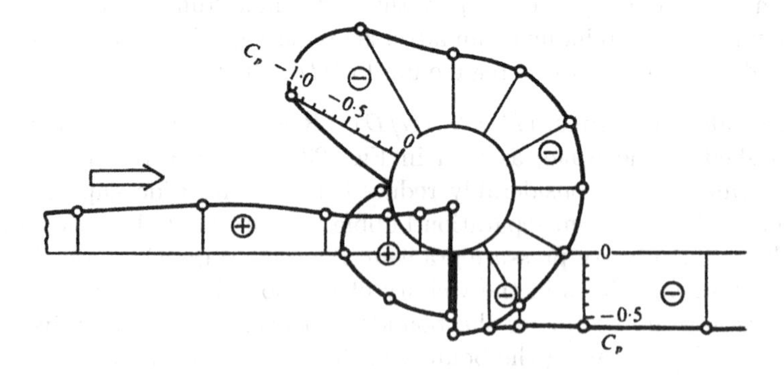

FIG. 23.52. Pressure distribution around the cylinder and along the boundary, $Re = 48$k, $G/D = 0$, Bearman and Zdravkovich (1978J)

Table 23.1 *Compilation of data for $G/D = 0$*

Cylinder boundary layer	Author(s)	Re	δ_B	δ_B/D	L/D	D/B	C_l	C_d
Laminar	Roshko *et al.* (1975P)	20k	T	0.5	10	0.10	0.6	0.8
	Hiwada *et al.* (1986J)	20k	T	0.2 2.8	20	0.05	0.7	0.45
Transitional	Bearman and Zdravkovich (1978J)	48k	T	0.8	32	0.04		
	Brown (1967J)	53k	T	?	5.5	0.19	1.1	0.90
	Brown (1967J)	60k	T	?	4.0	0.25	0.8	0.90
Turbulent	Brown (1967J)	97k	T	?	5.5	0.19	1.3	0.85
	Brown (1967J)	150k	T	?	4.8	0.25	1.3	0.55
	Göktun (1975P)	90k	L	?	6.7	0.20	1.3	0.85
	Göktun (1975P)	150k	L	?	6.7	0.20	1.4	0.65
	Göktun (1975P)	250k	L	?	6.7	0.20	1.4	0.61

cylinder boundary layers, respectively.

The origin of C_l is the asymmetric pressure distribution. A compilation of C_p distributions is reproduced in Fig. 23.53, as a polar plot with positive C_p inside the circle. The potential theory (curve 1) is inadmissible because $C_l = 4.49$ and $C_d = 0$. Font's curves (2 and 3) are obtained for $Re = 97$k and 53k, respectively. The adverse pressure recovery ($\Delta C_p = C_{pb} - C_{p\min}$) is 1.2 and 0.4 and indicate turbulent and laminar separation on the cylinder, respectively. Brown's (1967J) curve (4) for $Re = 330$k, yields $\Delta C_p = 0.6$. Göktun's (1975P) curves (not shown) yield $\Delta C_p = 1.4$, 1.6, and 1.9 at $Re = 90$k, 150k, and 250k, respectively.

A strong effect of the upstream length of the boundary on C_d and C_l was noted by Göktun (1975P). When the cylinder was located at $x/D = 2$, 4, and 8 from the boundary leading edge $C_d = 0.75$, 0.64, and 0.60, and $C_l = 1.50$, 1.44, and 1.20, respectively, at $Re = 53$k. The upstream separation region was affected when the cylinder was located at $x/D = 2$ and 4 downstream.

The effect of turbulent boundary layer thickness was examined by Hiwada *et al.* (1986J) in the range $0.23 < \delta_B/D < 2.82$ at $Re = 20$k. They found $C_d = 0.80$ and 0.56 for $\delta_B/D = 0.23$ and 2.82, respectively. The decrease in C_d was related to the cylinder immersion in the boundary layer for $\delta_B/D = 2.82$. However, C_l was not affected by δ_B/D, and remained unchanged at 0.45.

An attempt to reduce the high C_l was made by Brown (1967J). He examined the effect of a spoiler attached to the cylinder in the form of a fence $h/D = 0.25$. Figure 23.54 shows the C_p distribution around a plain cylinder and with the spoiler attached at 0°, 45°, 90°, and 135°. The reduction in C_l is seen for all angular locations of the spoiler. The effect of θ_{sp} on C_d is different. It increases C_d

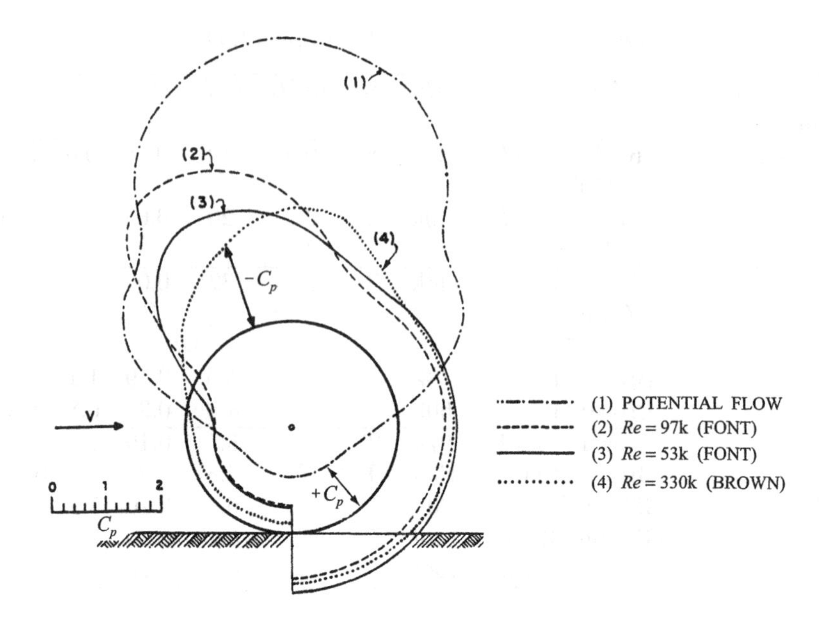

FIG. 23.53. Compilation of pressure distribution around cylinder for $G/D = 0$, (1) theoretical, (2) $Re = 97$k, Font, (3) 53k, Font, (4) 330k, Brown (1967J)

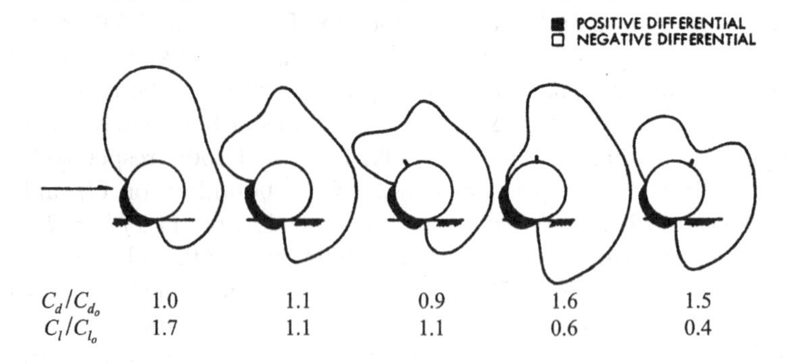

FIG. 23.54. Effect of spoiler on C_p, C_d/C_{d_o}, and C_l/C_{l_o} for $G/D = 0$, $Re = 150$k, Brown (1967J)

for all spoiler locations except at $\theta_{sp} = 45°$. The latter seems to be the optimal location of the spoiler for the resultant force reduction.

23.7.4 *Potential flow for a circle on a boundary*

Milne-Thomson (1967B) gave a full derivation of the velocity potential for an inviscid flow around a circle touching a flat boundary. The complex potential–stream function was found to be in the form

$$\omega = \pi V \frac{D}{2} \coth \frac{\pi D}{2z} \tag{23.32}$$

where the properties of eqn (23.32) are as follows

$$\phi + i\psi = \frac{\pi D V}{2} \coth\left(\frac{\pi D}{2r^2} - i\frac{\pi y D}{2r^2}\right) \tag{23.33}$$

where $r^2 = x^2 + y^2$. If $y = 0$, eqn (23.33) reduces to the real term only, and represents a streamline $\psi = 0$. The function ω also becomes real if $\pi i y D/2r^2 = \pi/2$. Thus, the zero streamline consists of $y = 0$ (the boundary) and $x^2+y^2 = Dy$ (the circle touching the boundary). For large values of $|z|$, the complex potential reduces to

$$\omega = \frac{\pi D V}{2}\frac{2z}{D\pi} \quad \text{or } Vz \tag{23.34}$$

i.e. a uniform flow parallel to the real axis, and directed from left to right.
The velocity potential was obtained by differentiating ω

$$\frac{\mathrm{d}\omega}{\mathrm{d}z} = \frac{D^2\pi^2 V}{4r^2}\left[\frac{2}{\cosh\dfrac{D\pi x}{r^2} - \cos\dfrac{\pi D y}{r^2}}\right] \tag{23.35}$$

where $r^2 = Dy$ on the surface of the circle (subscript c)

$$V_c = \frac{\pi^2 D V}{8y^2 \cosh^2 \dfrac{D\pi x}{r^2}} \tag{23.36}$$

and along the boundary $y = 0$ (subscript B)

$$V_B = \frac{\pi^2 D^2 V}{4x^2 \sinh^2 \dfrac{\pi D}{2x}} \tag{23.37}$$

The calculated potential flow pressure distribution around a circle, $p = \frac{1}{2}\rho V_c^2$, was shown in Fig. 23.53. The lift and drag coefficients evaluated from the theoretical C_p distribution were 4.49 and 0, respectively. Both values are physically inadmissible.

23.7.5 *Narrow-gap regime*

Once the cylinder is detached from the boundary, $G/D > 0$, a fraction of flow passes through the gap. This strongly affects both the upstream and downstream separation regions. The former is reduced to a separation bubble on the boundary, and the latter separates the wake from the boundary. Zdravkovich (1982P) carried out a combined smoke and surface flow visualization at $Re = 25$k. Figure 23.55(a,b) shows two kinds of visualization in two wind tunnels:

(a) Smoke is introduced upstream and downstream of the cylinder through the boundary. The leading edge of the boundary was $6D$ from the cylinder, and the boundary layer was laminar. The shape of the upstream separation bubble is seen in Fig. 23.55(a,b) for $G/D = 0.2$.

(b) For the surface flow visualization, the boundary layer was turbulent (tripping wire $d = 1\,\text{mm}$). The wire and the cylinder are located $7D$ and $32D$ from the boundary leading edge, respectively. Figure 23.55(b) shows the pattern around the upstream separation bubble reattachment and secondary separation downstream behind the cylinder. The horizontal line with arrows shows the cylinder axis position during the tests. The same surface flow visualization on the cylinder itself revealed straight separation lines.

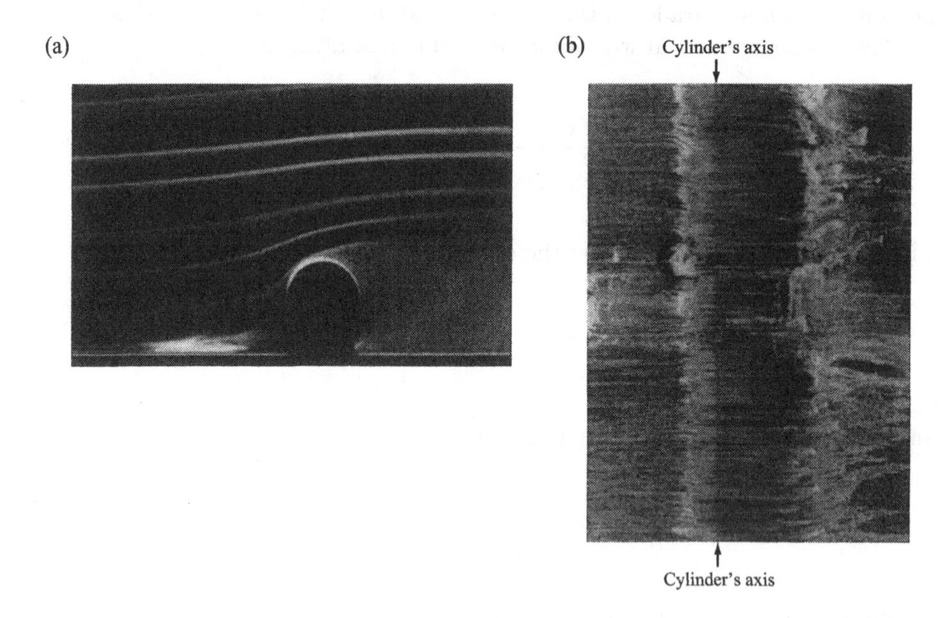

FIG. 23.55. Flow visualization for $G/D = 0.2$, (a) $L/D = 1$, laminar boundary layer, (b) $L/D = 32$, turbulent boundary layer $\delta_B/D = 0.8$, Zdravkovich (1982P)

Bearman and Zdravkovich (1978J) measured the mean C_p distribution for $G/D = 0.1$, $L/D = 32$, $Re = 48$k, and $\delta_B/D = 0.8$. Figure 23.56 shows that the discontinuous jump in pressure at $\theta = 270°$ has been replaced by a steep favourable pressure gradient up to $(G/D)_{\min}$. A similar favourable pressure gradient along the cylinder continues up to $\theta = 315°$. This may indicate that the gap flow remained attached to the cylinder and detached from the boundary.

Another feature of the narrow-gap regime is the formation of the stagnation region with $C_{p0} = +1$. The angular position of C_{p0} is displaced towards the boundary. This means that the incoming flow is deflected around the upper part of the cylinder. As G/D increases, C_{p0} is displaced away from the boundary, and more of the incoming flow is diverted towards the gap.

The most notable feature of the narrow-gap regime is the absence of periodic eddy shedding. There might be several reasons for this:

(i) The gross asymmetry of the flow around the cylinder produces different velocities at the upper and lower sides. This is analogous to the cylinder being submerged in shear flow where eddy shedding is suppressed, Vol. 1, Chapter 15.

(ii) The vorticity generated along the cylinder side facing the boundary is opposite in sense to that generated along the boundary. For small G/D, the mixing of the vorticity of the opposite sense reduces the circulation.

(iii) The small flow rate through the gap may be insufficient for the entrainment in the near-wake required for eddy formation, Vol. 1, Chapter 5.3.8.

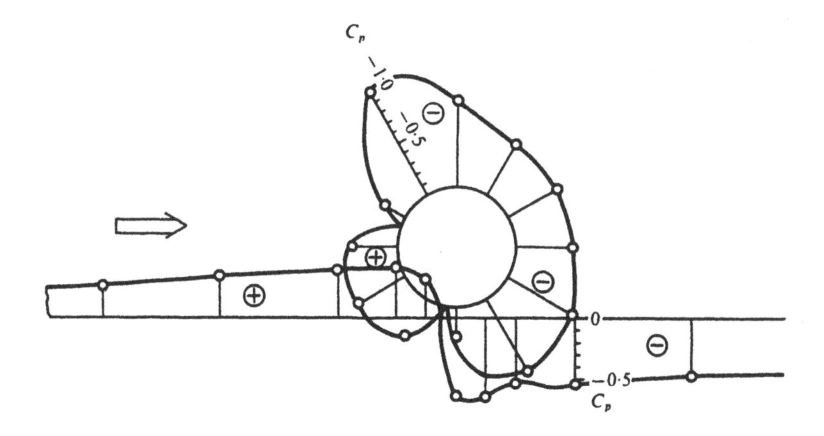

FIG. 23.56. Pressure distribution around the cylinder and along the boundary $Re = 48$k, $G/D = 0.1$, $\delta_B/D = 0.8$, Bearman and Zdravkovich (1978J)

The main feature of the narrow-gap regime is a rapid decrease in flow asymmetry and associated C_L. Hiwada *et al.* (1986J) measured the location of separation points on the gap side (closed circle) and outer side (open circle). Figure 23.57(a) shows that both separation points rapidly converge towards a symmetric position at $\theta_s = \pm 85°$. These rapid displacements of θ_s cause an equally rapid decrease in C_L in Fig. 23.57(b). Note that a single C_L curve corresponds to a wide range $0.82 \leqslant \delta_B/D \leqslant 2.82$.

Zdravkovich (unpublished) measured drag C_D in the range $72\text{k} < Re < 300\text{k}$, $L/D = 21$, $\delta_B/D = 0.12$, and $D/B = 0.6$ in a wind tunnel with a fixed and moving floor. Figure 23.58 shows C_L and C_D variation in terms of G/D. The widening of the gap rapidly enhances the gap flow rate, and reduces the flow asymmetry around the cylinder. It is evident that C_L is independent of Re except for $G/D = 0$. For all practical purposes, C_L becomes negligible for $G/D > 2$. Figure 23.58 also shows an unexpected kink in the C_D curve in the range $0.12 < G/D < 0.25$ for all higher Re. The smooth increase in C_D is found at $Re = 20\text{k}$ by Roshko *et al.* (1975P), Hiwada *et al.* (1986J), and at $Re = 48\text{k}$ by Bearman and Zdravkovich (1978J).

The measurements of C_D and C_L were also carried out by running the moving floor at the flow velocity, $\delta_B/D = 0$. Figure 23.59 shows the C_D and C_L curves for the fixed and moving floor. For the latter, the decrease in C_D does not occur, and C_L is reduced slightly. These features demonstrate the importance of δ_B/D.

23.7.6 *Wide-gap regime*

The precise value of the start of eddy shedding, $(G/D)_c$, depends on Re, δ_B/D, the state of the boundary layer, and other influencing parameters. As soon as the cylinder is placed beyond $(G/D)_c$, the change-over of the flow regime takes place. The sequences of the change-over are shown in Fig. 23.60(a–d). The narrow-gap, $G/D = 0.3$, and wide-gap, $G/D = 0.8$, regimes behind a towed cylinder in a water tank are two extremes for $\delta_B/D = 0$ and $Re = 3.5\text{k}$. For the

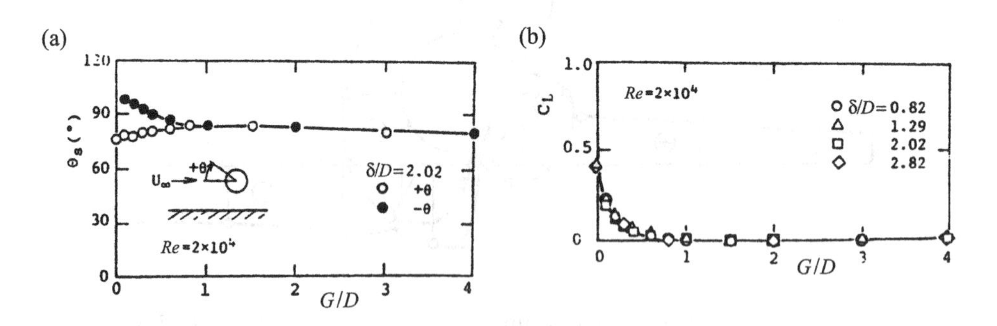

FIG. 23.57. (a) Separation on two sides of the cylinder, (b) lift for $Re = 20\text{k}$, Hiwada *et al.* (1986J)

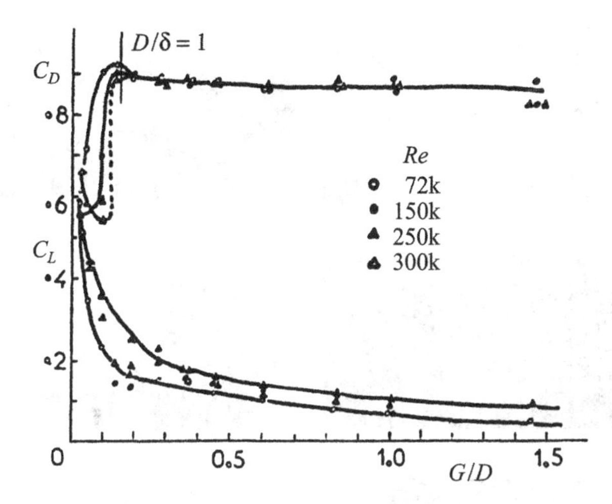

FIG. 23.58. Drag and lift coefficients in terms of G/D, Zdravkovich (unpublished)

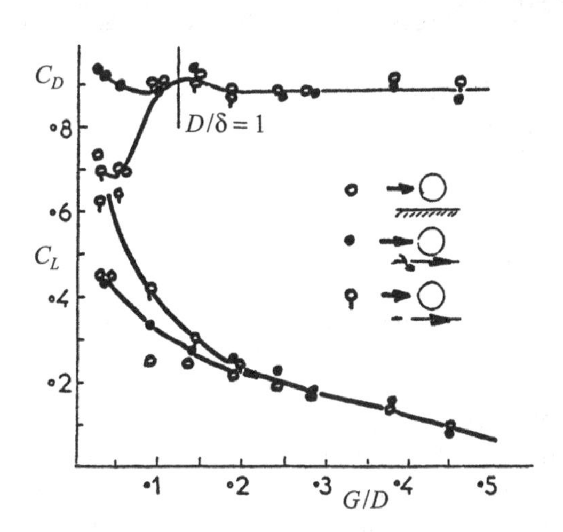

FIG. 23.59. Drag and lift coefficients in terms of G/D for the $Re = 250\text{k}$ boundary, stationary (open circles), moving (closed circles), without suction (tail), Zdravkovich (unpublished)

narrow-gap regime, several randomly spaced eddies of different size emanate from the cylinder side away from the boundary, Fig. 23.60(a). For the wide-gap regime, the vigorous and equally spaced eddies emanate from both sides of the

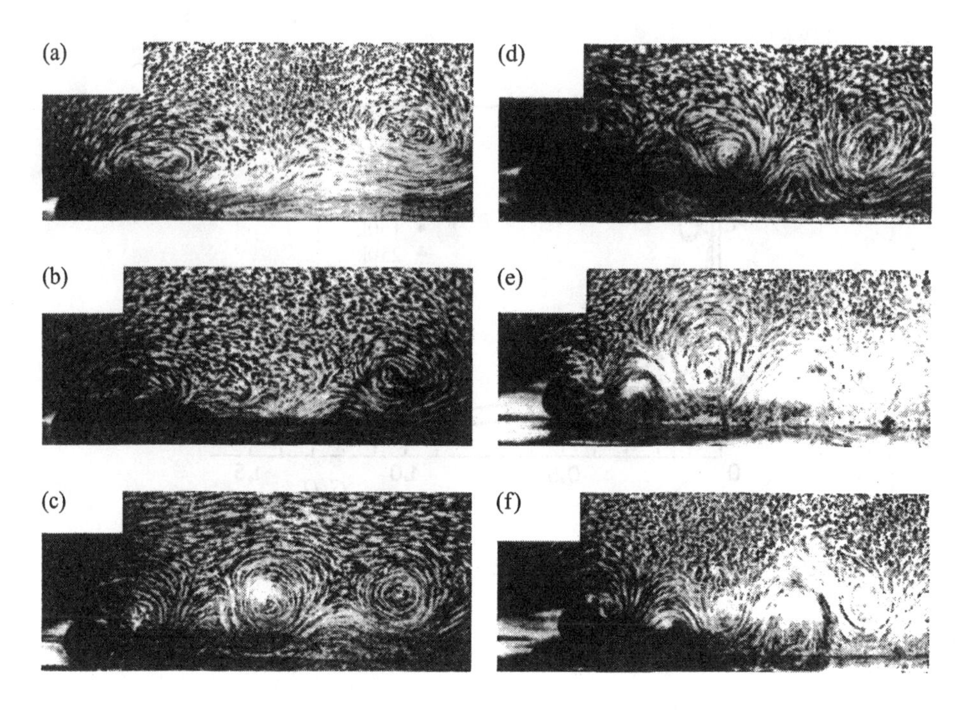

FIG. 23.60. Flow visualization at Re = 3.55k, $\delta_B/D = 0$, (a) $G/D = 0.1$, (b) $G/D = 0.3$, (c) $G/D = 0.4$, (d) $G/D = 0.6$, (e) $G/D = 0.8$, (f) $G/D = 1.0$, Zdravkovich (1985P)

cylinder, Fig. 23.60(d).

The change-over from no eddy shedding to fully developed eddy shedding is separated by a unique appearance of a single-row eddy street. Figure 23.60(b,c) shows the single-row eddy streets for G/D = 0.4 and 0.6, respectively. The second row of eddies adjacent to the boundary is absent, and only the undulated streamline is seen. Note that single-row eddy shedding was reported by Taneda (1965J) in the L3 regime, Shaw (1971J) in the TrSL2 regime, Göktun (1975P) in the TrBL0 regime, Bearman and Zdravkovich (1978J) in the TrSL3 regime, and others. The phenomenon is an intrinsic feature of the interference between the cylinder and the boundary and not of any particular flow regime.

Figure 23.61 shows the C_p distribution at Re = 48k for $G/D = 0.4$ (Note that $(G/D)_c = 0.3$ for these tests.) There are three new features in comparison with $G/D = 0.1$ shown in Fig. 23.56:

(i) the position of $C_{p\max}$ is now nearer to $\theta = 0°$;

(ii) there is an increase in $-C_{pb}$ induced by eddy shedding;

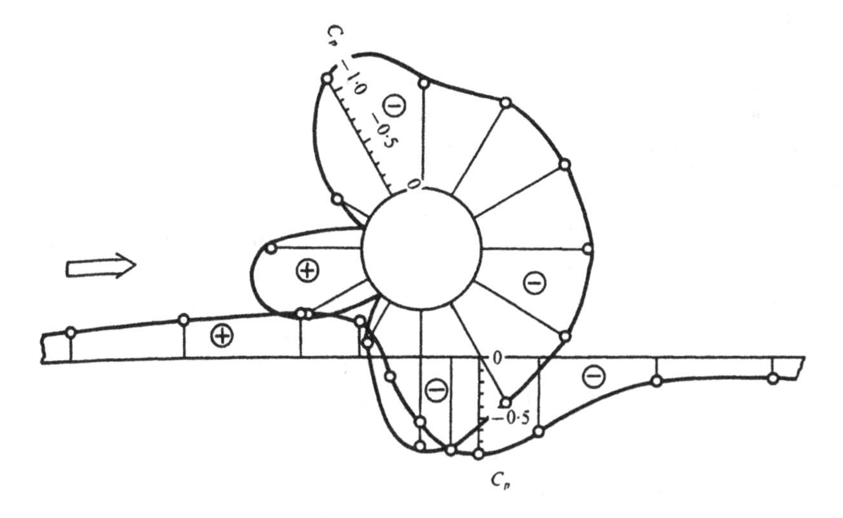

FIG. 23.61. Pressure distribution around the cylinder and along the boundary, $Re = 48k$, $\delta_B/D = 0.8$, $G/D = 0.4$, Bearman and Zdravkovich (1978J)

(iii) there is a significant decrease in C_p along the boundary upstream and downstream of the gap.

A typical example of frequency spectra for single-row eddy shedding is given in Fig. 23.62 for $G/D = 0.2$. The hot-wire locations were at $(x/D, y/D) = (1.0, 1.5)$ and $(0.4, 0.6)$. Figure 23.62(a,b) shows the frequency peak correspond-

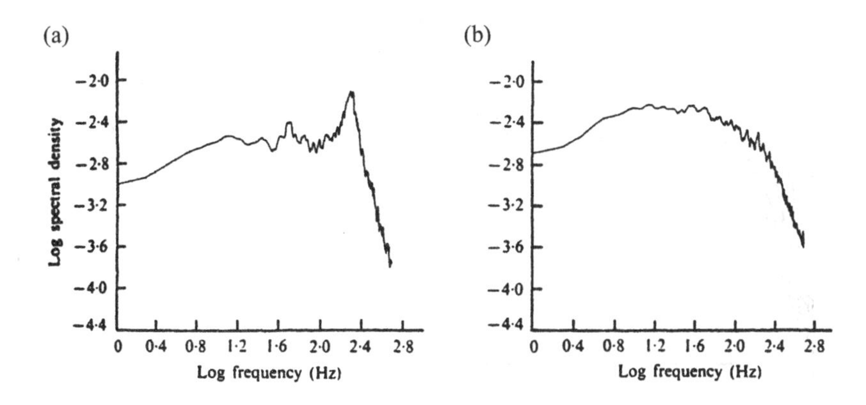

FIG. 23.62. Frequency spectra of hot-wire signals for $G/D = 0.2$, $Re = 48k$, (a) away from the boundary, (b) near the boundary, Bearman and Zdravkovich (1978J)

ing to the single row away from the boundary, and no peak adjacent to the boundary, respectively. It is interesting to point out that the repeat of the experiment at $G/D = 0.2$ without the tripping wire attached to the boundary (laminar boundary layer) showed a small frequency peak near the lower boundary. Thus the appearance of two-row eddy shedding depends on the state of the boundary layer in addition to G/D and δ_B/D.

Regular two-row eddy shedding was found for all $G/D > 0.3$. The remarkable feature was that the St remained constant at 0.21 and independent of G/D. This indicates that the width of the near-wake was not affected by the boundary proximity. Another feature of eddy shedding is the gradual reduction in the frequency modulation typical of the TrSL3 regime, see Vol. 1, Chapter 5.5. The modulation disappears at $G/D = 0.3$, and velocity fluctuations appear more highly tuned than those recorded at large gaps.

23.7.7 *Effect of wall boundary layer*

Hiwada *et al.* (1986J) carried out tests using several tripping wires of different diameters to render the wall boundary layer thick and turbulent in the range $0.23 < \delta_B/D < 2.82$ at $Re = 20$k (other details are given in Table 23.1). Figure 23.63(a,b) shows the variation in C_d (evaluated from C_p) in terms of G/D for various δ_B/D. When C_d is expressed through the free stream velocity there are separate C_d curves for each δ_B/D in Fig. 23.63(a). The start of the decrease in C_d is closely related to $\delta_B/D = G/D$. The thicker the boundary layer the earlier is the start in the C_d decrease. However, when C_d is expressed through the local velocity at the cylinder centreline all C_d curves collapse into one (except $\delta_B/D = 0.23$), see Fig. 23.63(b). This demonstrates that the main cause of the C_d decrease is the lower velocity in the boundary layer.

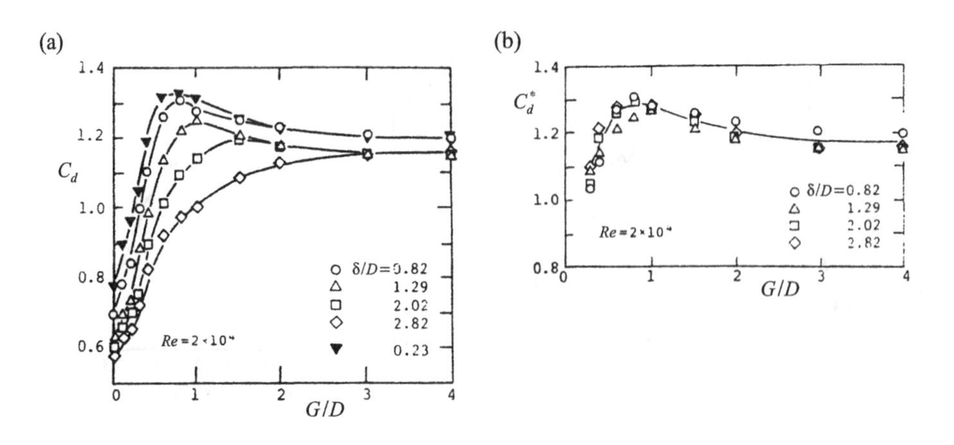

FIG. 23.63. Drag coefficient at $Re = 20$k based on (a) the velocity of the free stream and (b) the local velocity at the cylinder centre, Hiwada *et al.* (1986J)

Hiwada *et al.* (1986J) also measured St by using the hot-wire and frequency spectra. Figure 23.64(a,b) shows two G/D ranges where St was constant:

(i) At the start of the wide-gap regime, $G/D = (G/D)_c$, $St < 0.2$ and the value depended on δ_B/D. Presumably, this constant St range corresponded to single-row eddy shedding.

(ii) When the cylinder was outside the boundary layer, $\delta_B/G > 1$, another constant $St = 0.2$ was found, which corresponded to fully developed two-row eddy shedding.

(iii) There was a transitional range between (i) and (ii) where St gradually increased from the low to the high value.

An attempt to replace V by the local velocity did not produce a collapse of St curves, see Fig. 23.64. Hence, the observed variation in St could not be attributed to the velocity defect in the boundary layer.

There was another unexpected feature in C_L variation for Re = 60k caused by shear in the thick wall boundary layer. Zdravkovich (1985Jb) found that C_L became negative, directed towards the boundary, in the range $0.3 < G/D < 1.2$. Figure 23.65(a,b) shows the unexpected trend. This anomalous trend was confirmed by Lei *et al.* (1999J) for other thick boundary layers in Fig. 23.65(b).

Taniguchi and Miyakoshi (1990J) carried out tests at Re = 94k in the range $0.34 < \delta_B/D < 1.05$ on a cylinder with $L/D = 14$, $D/B = 0.07$, and $Ti = 0.2\%$. They measured fluctuating C'_D and C'_L by using a load cell mounted inside the cylinder segment $1.2D$ long. Figure 23.66(a,b) shows the rms variation in C'_D and C'_L in terms of G/D (abscissa) and δ_B/D (superimposed plots), respectively. The prominent kinks in the curves corresponded to the kinks observed in the C_D curves, see Fig. 23.58. The kinks transformed to troughs for thick boundary

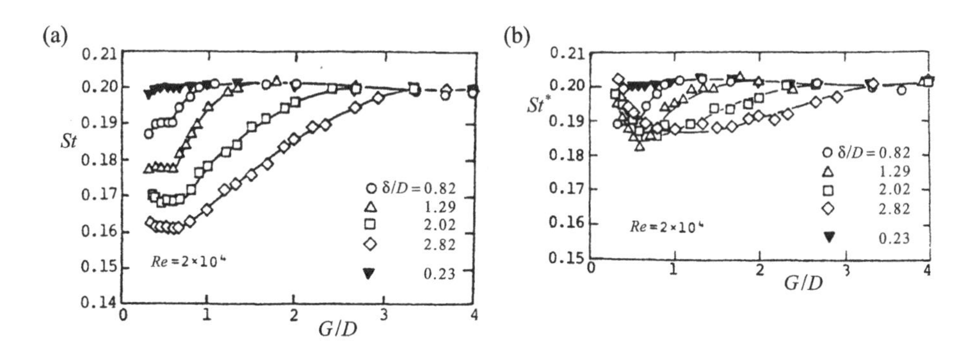

FIG. 23.64. Strouhal number in terms of G/D for Re = 20k based on (a) the velocity of the free stream and (b) the local velocity at the cylinder centre, Hiwada *et al.* (1986J)

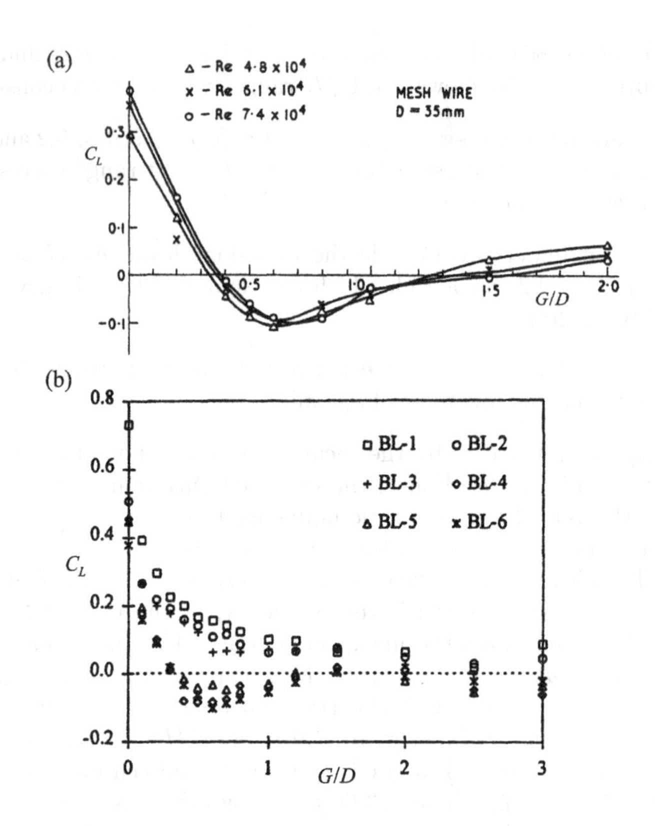

FIG. 23.65. Mean lift coefficient in terms of G/D, (a) $\delta_B/D = 0.97$, Zdravkovich (1985Jb), (b) $0.14 \leqslant \delta_B/D \leqslant 2.89$, Lei *et al.* (1999J)

layers, $\delta_B/D > 0.68$.

The arrows in Fig. 23.66(a,b) designate the location of $(G/D)_c$ estimated from the first appearance of the peak in the frequency spectra. The hot wire is located at $x/D = 4.5$ and $y/D = 2.5$ on the side away from the boundary. Thus, the estimated $(G/D)_c$ corresponds to the appearance of single-row eddy shedding. The initial increase in the C'_D and C'_L curves up to the arrows is associated with the narrow-gap regime. The second increase from the arrows up to the maximum corresponds to the transition from one-row eddy shedding to two-row eddy shedding.

Figure 23.67 shows the variation of $(G/D)_c$ in terms of δ_B/D for $Re = 93$k. Note that it is not a universal curve, and $(G/D)_c$ strongly depends on Re, L/D, type of wall boundary layer, etc. For example, Hiwada *et al.* (1986J) found variations in $(G/D)_c$ in terms of δ_B/D.

Further insight into the origin of the kinks might be gained from the variation and correlation in C'_p on two sides of the cylinder at 90° and 270°. Figure

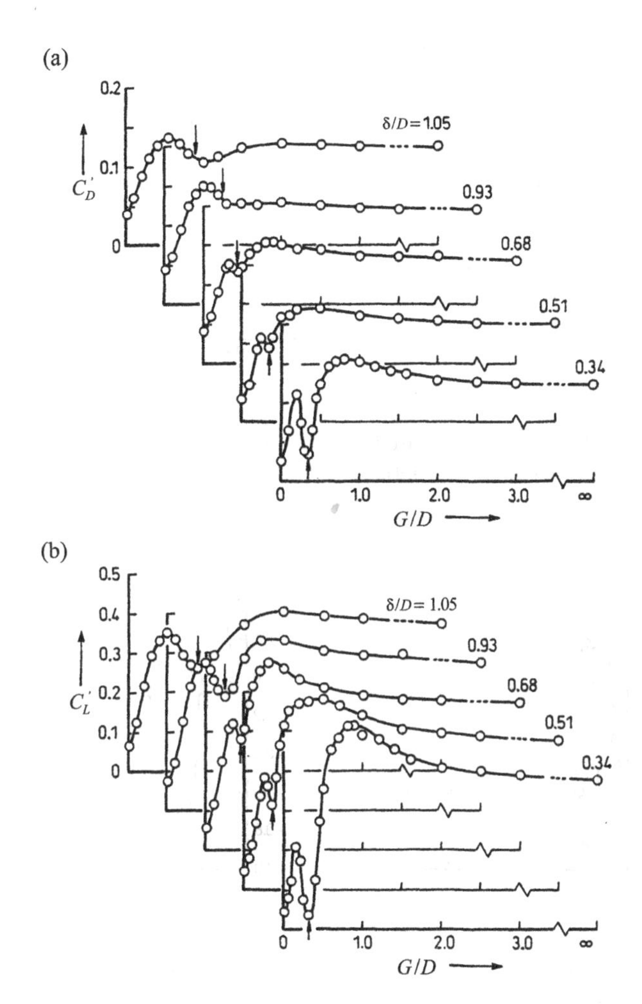

FIG. 23.66. Fluctuating (a) drag, (b) lift, for Re = 94k in terms of G/D for various δ_B/D, Taniguchi and Miyakoshi (1990J)

23.68(a,b) shows C_p' and R_{12} in terms of G/D for $\delta_B/D = 0.34$ and 0.93. Positive peaks in R_{12} are found, i.e. the fluctuating pressures on two sides of the cylinder act in phase and are due to one-row eddy shedding. The peaks in the C_L' and C_D' curves are reached at the same G/D where R_{12} attained a minimum. The fluctuations on two sides of the cylinder are out-of-phase and relate to two-row eddy shedding.

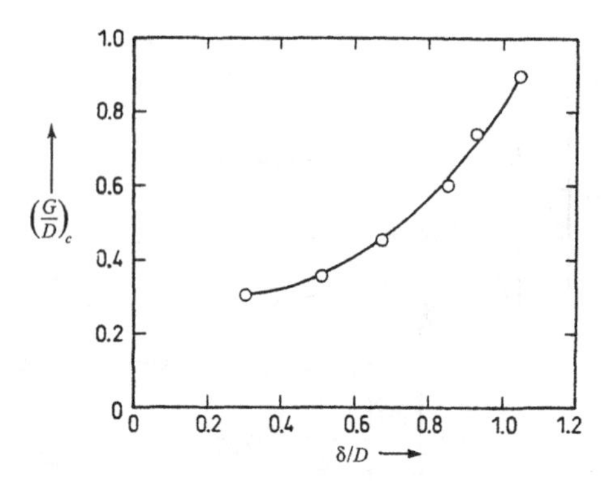

FIG. 23.67. Critical gap-to-diameter ratio in terms of δ_B/D for Re = 94k, Taniguchi and Miyakoshi (1990J)

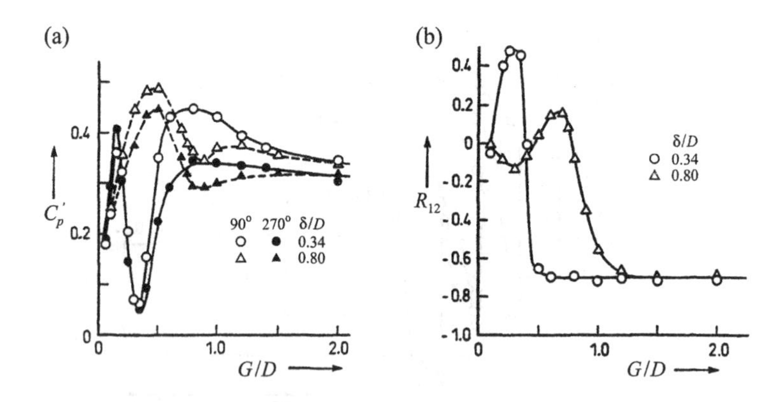

FIG. 23.68. (a) Pressure fluctuations at $\theta = 90°$ and $270°$ of the cylinder at Re = 94k for δ_B/D = 0.34 and 0.80, (b) cross-correlation, Taniguchi and Miyakoshi (1990J)

23.8 Erodible boundary, scour

23.8.1 *Scouring mechanism*

Submarine pipelines have been laid near to or on an erodible sandy seabed. When exposed to currents and/or waves a *scour* develops. The scour is a result of the erosion and removal of sand by the current underneath the pipeline. The scour depth gradually increases, and the shape of the boundary becomes similar to a streamline around the pipeline. The scour depth and shape have a considerable effect on flow around and forces on the pipeline.

If a pipeline is rigid, then the formation of the scour is expected to develop evenly along the span, and the scour underneath is expected to be nominally two-dimensional. However, due to the irregularities of a seabed surface and related non-uniform flow along the span, the scour develops unevenly.

When a flexible pipeline is initially laid on the seabed the scour gradually develops, and sequences are sketched in vertical columns in Fig. 23.69(A–A, B–B). The uneven scour starts to develop at station B—B. The spanwise spread of the scour with time lengthens the unsupported span of the pipeline, which starts to sag. The latter enhances the increase in the scour depth, Fig. 23.69(B–B) This process terminates when the sagged pipeline eventually reaches the bottom of the scour. In that position, the pipeline is shielded from the current. The sediment transport deposits sand in the trench, and the pipeline gradually becomes buried, as depicted in Fig. 23.69(d).

Fredsøe *et al.* (1988J) carried out two-dimensional scouring tests on $L/D = 20$ without end plates at $Re = 60\text{k}$. The sagging was simulated by moving the pipe downwards at a slow speed. Figure 23.70(a–d) shows the shape of the scoured boundary after a time given in minutes (dotted line). The initial burial of the sagged pipeline takes place on the downstream side first, after 150 min (dotted line).

23.8.2 *Forces and Strouhal number*

Jensen *et al.* (1990J) carried out tests in a water flume using a strain-gauged cylinder $L/D = 10$, $D/B = 0.13$, $\delta_B/D = 2$, and $Re = 10\text{k}$. They measured the

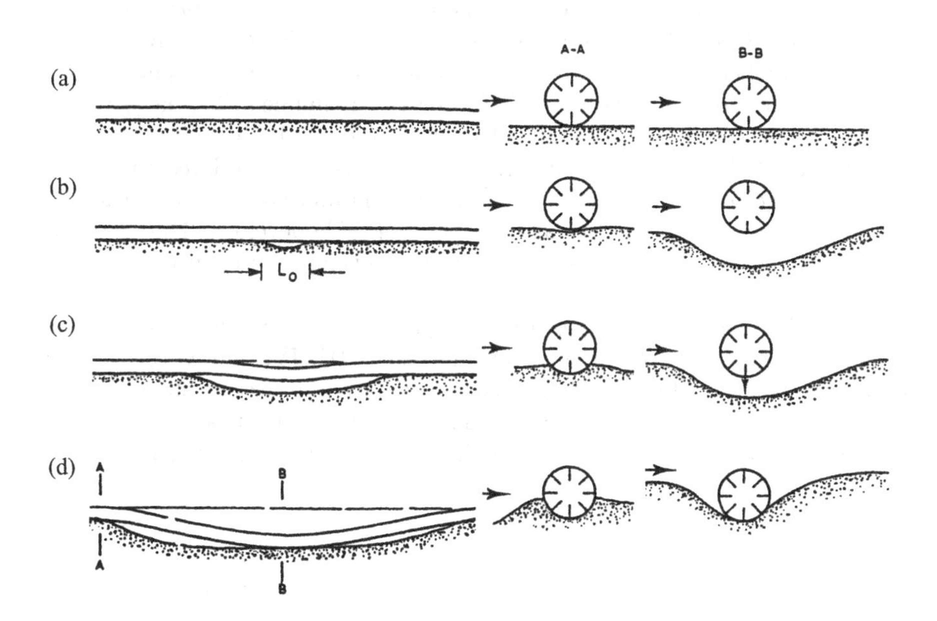

FIG. 23.69. Sketch of a three-dimensional scour, upstream view (first column), (A–A) side cross section, (B–B) central cross section, Fredsøe *et al.* (1988J)

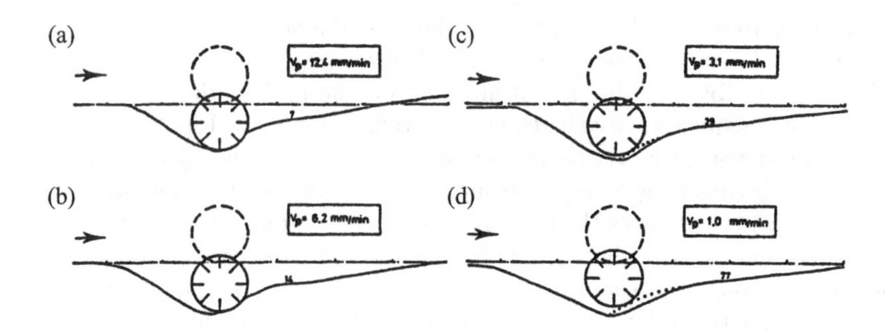

FIG. 23.70. Development of scour profiles in terms of sagging velocity, (a) 7 min, (b) 14 min, (c) 29 min, (d) 77 min, Fredsøe *et al.* (1988J)

mean velocity u, v, C_D, C_L as well as the fluctuating u', v', C'_L, and C'_D. They chose five profiles starting with I, the flat boundary at $G/D = 0$, and finishing with the final scour profile V, when the hump disappeared.

Figure 23.71(a–d) shows C_D, C_L, $2C'_L$, and St in terms of time of scouring. The dashed curves correspond to the flat boundary, and full curves to the scoured boundary. The values of C_D are less for the scoured profiles in comparison with the flat boundary except for $G/D = 0.1$ at the early stage of scouring, see Fig. 23.71(a). The difference in C_L between the flat and the scoured profiles is even greater in Fig. 23.71(b). It is remarkable that the positive C_L (away from the flat boundary) becomes a negative C_L (towards the scoured boundary). The $C_{L\min}$ is not found when the scoured profile is streamlined. The negative C_L indicates that the incoming flow is deflected more towards the gap than to the opposite side of the cylinder. The variation in C'_L is closely related to St, and cannot be found for $G/D < 0.3$. Figure 23.71(d) shows that St is higher for the scoured boundary, and decreases as the scoured profile develops. This indicates that the near-wake widens as the scour profile develops, as confirmed by flow visualization.

Figure 23.72 shows sketches of the flow pattern from the flow visualization together with three photos of the scouring stages III, IV, and V (in stages I and II there was no eddy shedding). The regularity improves in stage IV, and becomes two-row eddy street in stage V. The initially narrow near-wake in stage III becomes wide in stage V, and results in a decreasing St in Fig. 23.71(d).

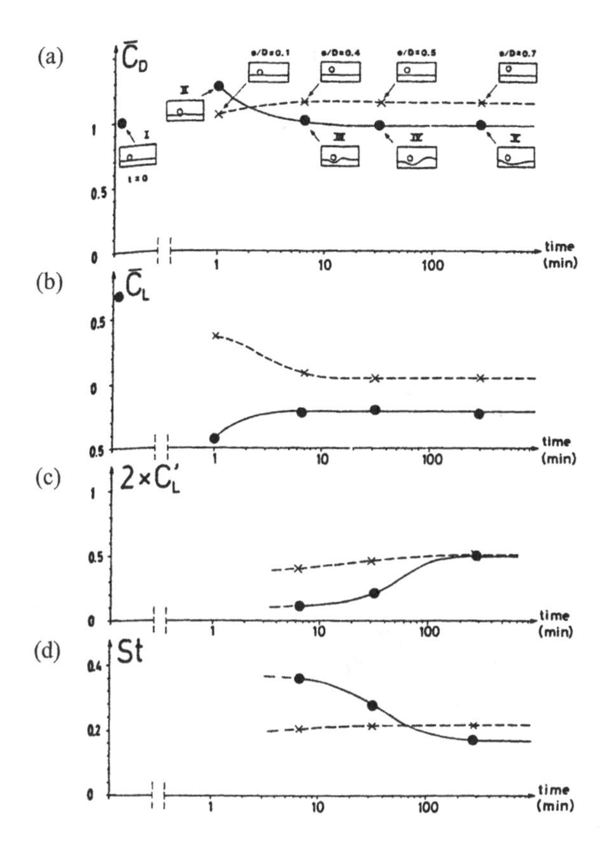

FIG. 23.71. Comparison of parameters over a scoured and plane boundary, (a) mean drag, (b) mean lift, (c) fluctuating lift, (d) Strouhal number, Jensen *et al.* (1990J)

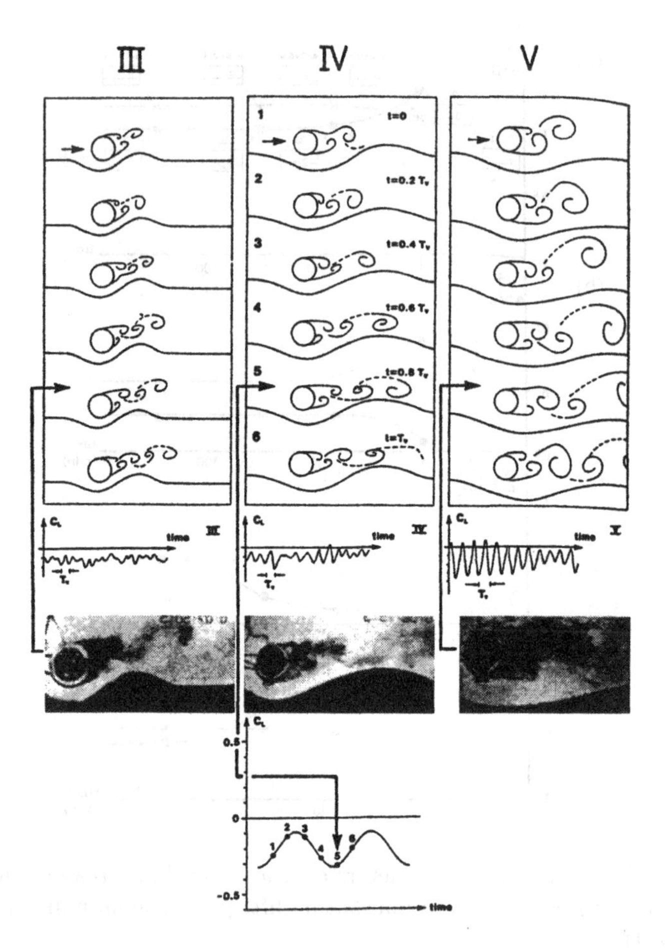

FIG. 23.72. Sketch of sequences of eddy shedding 1 to 6 (vertical), lift, and flow visualization for stages III, IV, and V (horizontal), Jensen *et al.* (1990J)

24

BOUNDARY LAYER CONTROL

Prandtl's epoch-making lecture on the importance of the 'thin layer' adjacent to the solid boundary on flow of fluid of very low viscosity was given in 1904. This lecture put the entire theory of Fluid Mechanics on a new basis. At least one could understand the cause of the large energy losses, which were found in practice and which, up to then, had been so mysterious.

A. Betz (1961P)

24.1 Introduction

Prandtl's (1961B) historical lecture on the concept of boundary layer was given in the same year as the concept of its control was initiated. He started his renowned flow visualization tests and filming of a circular cylinder with a slot on one side where suction was applied. The flow remained attached up to the suction slot at $\theta_s = 140°$, see Fig. 24.54. The second method of boundary layer control, a rotating cylinder, followed in 1906, see Fig. 24.1. The surface moving in the same direction as the flow postponed separation in proportion to the rotational velocity. A flow pattern analogous to that produced by suction was also produced by the rotating cylinder.

In this chapter the boundary layer control methods will be restricted to circular cylinders. Rotating cylinders will be treated first because much research has been carried out in the past in this area. This will be followed by two relatively short sections on boundary layer control by suction and blowing.

24.2 Rotating cylinder

24.2.1 *Magnus effect*

It has been known for a long time that missiles and bullets, when spinning around the long axis, always depart from their ballistic paths. The cause of that departure is an additional aerodynamic force in the direction perpendicular to the missile axis and the velocity vector, the so-called Magnus (1853J) effect. The term 'effect' instead of force might give the impression that a mysterious origin was behind it, which presumably had been the case before the time of Magnus (1853J).

Gustav Magnus (1853J) carried out early experiments and established the existence of a side force on the rotating cylinders. The direction of the side force was from the cylinder centre towards the side where the rotational velocity was in the same direction as the free stream velocity. Lafay (1910J, 1912J) discovered the inversion of the Magnus effect within a small range of the rotational velocities and high *Re*. The inversion meant that the side force was directed in the opposite direction to that found by Magnus (1853J). It took some time before a correct explanation for the inversion was established by Krahn (1956J), and proved by Kelly and Van Aken (1956J), Swanson (1961R), and others.

Prandtl (1961B) noted that the flow past rotating cylinders represented an example of the boundary layer control. He wrote:

> Boundary layers may also be influenced artificially so as to prevent separation of the flow taking place. If, for example, a cylinder immersed in flow is made to rotate so that its circumferential velocity is equal to or greater than the greatest velocity of flow occurring at the surface of the cylinder, the boundary layer on the side where the fluid and the surface of the cylinder are moving in the same direction will not be retarded at all, but will, on the contrary, be accelerated by the motion of the cylinder. This layer will then have a more powerful effect than the external current in overcoming the rises in pressure causing retardation, so that on this side backward flow does not set in at all, and the flow therefore does not become separated from the boundary. On the other side, the surface of the cylinder is moving in the direction opposite to that of the fluid, and subsequently the separation of a strong eddy takes place there. Together with the eddy, there arises a circulation about the cylinder in the direction opposite to that of the eddy. This gives rise to the transverse force known as the Magnus effect.

24.2.2 *Classification of flow patterns*

Prandtl (1961B) carried out a revealing flow visualization around a rotating cylinder at $Re \approx 4\text{k}$. Figure 24.1 shows a variety of flow patterns in terms of the non-dimensional ratio V_r/V, where the cylinder rotational velocity V_r is divided by the free stream velocity V. A selection of six Prandtl and Tietjens (1957B) frames from the cine film are reproduced in Fig. 24.1. The first ratio, $V_r/V = 0$, corresponds to a non-rotating cylinder in the free stream, and the last ratio, $V_r/V = \infty$, represents the rotating cylinder in water at rest. There are gradual changes in the flow patterns with rising V_r/V as follows:

(i) The eddy street is seen in Fig. 24.1(a,b) only for $V_r/V = 0$ and 1. The eddies are formed and shed alternately on two sides of the cylinder. The long eddy formation region typical for $Re = 4\text{k}$ at $V_r/V = 0$ is considerably shortened for $V_r/V = 1$.

(ii) Eddy formation and shedding can no longer be seen for $V_r/V = 2$ and beyond in Fig. 24.1(c–e). The stagnation point is displaced towards the side where V and V_r are in opposite directions. The near-wake is considerably reduced in length and width, and biased towards the side where V and V_r are in the opposite direction.

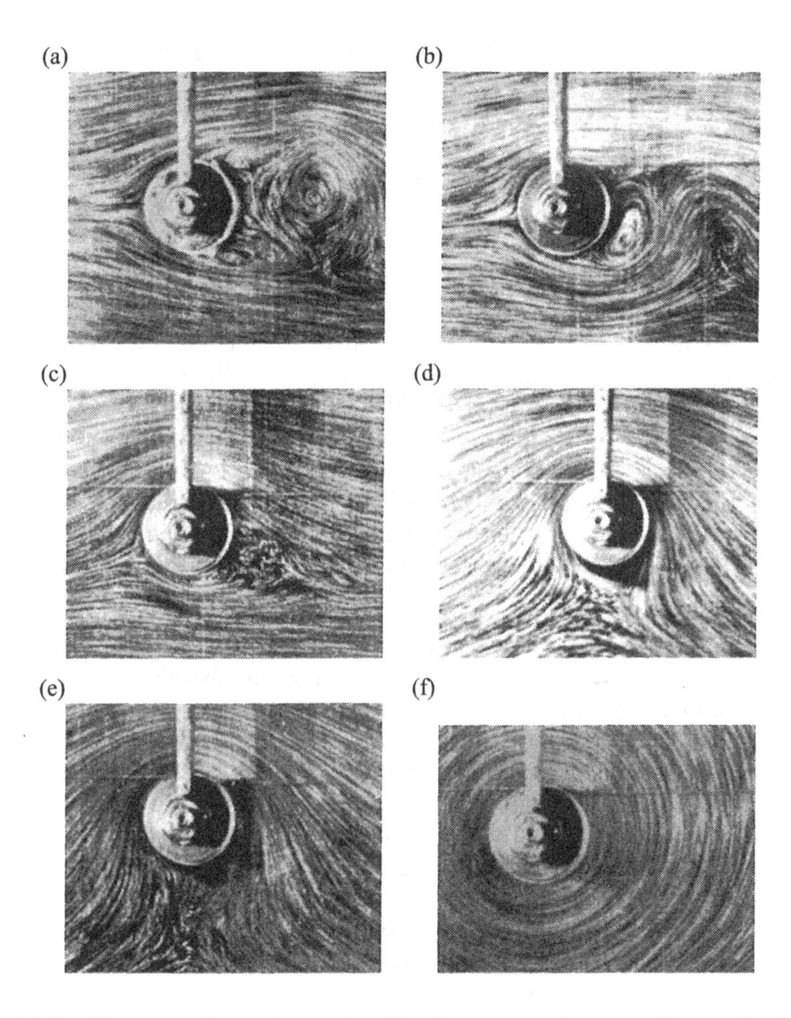

FIG. 24.1. Water surface flow visualization around a rotating cylinder at Re = 4k, (a) $V_r/V = 0$, (b) $V_r/V = 1$, (c) $V_r/V = 2$, (d) $V_r/V = 4$, (e) $V_r/V = 6$, (f) $V_r/V = \infty$, Prandtl (1961B)

(iii) A further increase in V_r/V towards 4 brings the small near-wake at the side where V and V_r are in opposite directions. The flow pattern appears almost symmetric to the axis perpendicular to the free stream velocity, as seen in Fig. 24.1(d,e) for $V_r/V = 4$ and 6.

(iv) The extreme case of $V_r/V = \infty$ represents a single large eddy induced by the rotating cylinder in water at rest. The eddy flow is confined along concentric circles, as seen in Fig. 24.1(f).

Two distinct flow regimes can be recognized in the flow past rotating cylinders in Fig. 24.1:

(i) $V_r/V \leqslant 2$, the periodic eddy shedding persists;

(ii) $V_r/V > 2$, the periodic eddy shedding ceases.

24.2.3 *Prandtl's concept of circulation*

Prandtl (1961B) proposed the following mechanism of circulation production:

> Whether the cylinder is rotating or not the initial motion of fluid, when the stream is started from rest is irrotational without circulation. On one side, the surface of the rotating cylinder is moving with the current, and there is no separation and no formation of eddies. On the other side, however, where the surface is moving against the current, separation takes place. Since there is now a greater production of vorticity on that side than if the surface is at rest, there is a greater transport of vorticity into the wake. This vorticity is practically concentrated into a single eddy, which is carried downstream leaving the circulation around the cylinder of the opposite sense, the same as the rotation.

Prandtl (1961B) provided experimental proof of his circulation concept. He carried out tests on a rotating cylinder starting impulsively from rest. Flow visualization of the temporal development of the flow pattern is shown in Fig. 24.2. Only one eddy is formed on the side where V and V_r are in the opposite direction, Fig. 24.2(a). As the flow develops over time, the eddy grows in size, Fig. 24.2(b), and eventually becomes detached from the rotating cylinder, Fig. 24.2(c). Finally, when the steady flow for $V_r/V = 4$ is reached, Fig. 24.4(d) shows that the anti-clockwise eddy is cast off downstream, and the clockwise circulation is left around the rotating cylinder.

Coutanceau and Menard (1983P) carried out flow visualization around impulsively started rotating cylinders at $Re = 200$. They deduced from the frames of cine films the path of the eddy centre during $0 < t^* < 4$, where the non-dimensional $t^* = tV/D$. Figure 24.3 shows that for $V_r/V = 0$ the path of the developing eddy is closest to the near-wake axis. The path for $0.28 \leqslant V_r/V \leqslant 1$ remains within $y/D < 0.5$. As soon as the eddy ceased to be formed on the opposite side for $V_r/V = 2.07$ and 3.25, there is a significant change in the first eddy path, $y/D > 0.5$. When the steady flow is reached, eddy shedding will not take place for $V_r/V > 2$.

24.2.4 *Potential flow theory*

> An enormous simplification of the mathematical theory has been achieved by ignoring viscosity and vorticity in an ideal fluid. This idealized flow is termed the potential flow. Goldstein (1965B).

A combination of a uniform potential flow, source, and sink of equal strength simulates a potential flow past a 'circle'. Note that the term 'circle' is not asso-

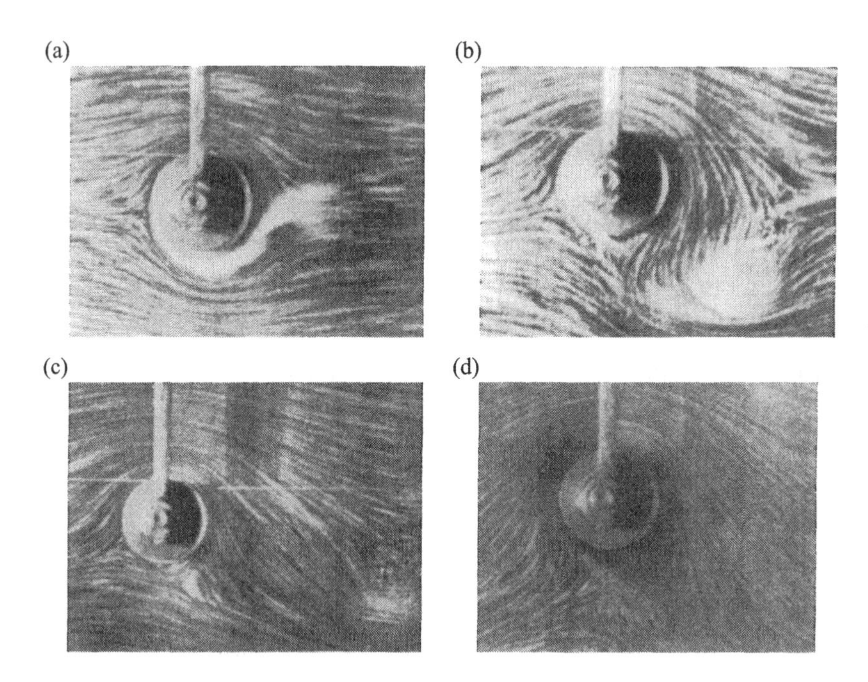

FIG. 24.2. Water surface visualization of flow development from rest for $V_r/V = 4$, (a) initial eddy formation, (b) detachment, (c) shedding, (d) carried downstream, Prandtl and Tietjens (1957B)

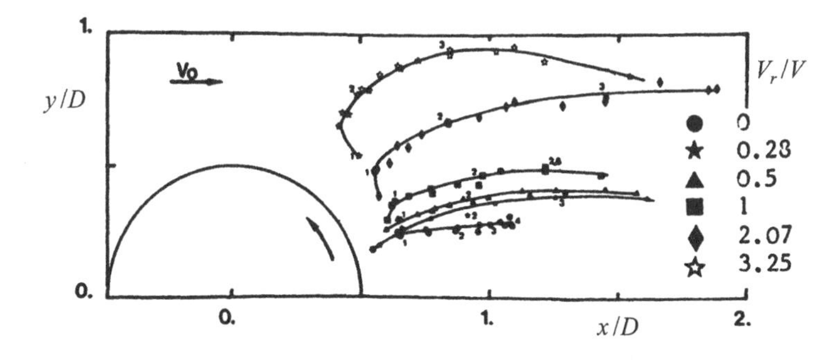

FIG. 24.3. Path line of eddy centre for $Re = 200$ for various V_r/V, Coutanceau and Menard (1983P)

ciated with a solid boundary but represents a single circular streamline.[134] The potential flow past a circular streamline, or circle for short, predicts the following features:

(i) the pressure distribution is doubly symmetric relative to the velocity vector and perpendicular to it;

(ii) the separation from the circle-streamline is mathematically impossible;

(iii) the rear stagnation point, $C_{p180} = 1$, indicates a full pressure recovery;

(iv) the zero drag results from the symmetrical pressure distribution relative to the vertical axis, and is known as the D'Alembert paradox.

All four features predicted by the potential flow theory are inadmissible for real flows, where:

(i) the pressure distribution is symmetrical only relative to the horizontal axis;

(ii) there is always flow separation from the cylinder;

(iii) the rear stagnation point does not exist because the flow is separated;

(iv) as a result of (i), (ii), and (iii) there is always a considerable drag force.

Some improvement in the potential flow theory was achieved when the theory was extended to a rotating cylinder. The rotation is simulated by adding circulation K in the centre of the circle. The imposed circulation leads to:

(i) the asymmetric pressure distribution relative to the horizontal axis is predicted;

(ii) the separation from a circle-streamline remains impossible;

(iii) the stagnation points are displaced from $\theta = 0°$ and $180°$;

(iv) the lift is predicted, but zero drag remains.

Typical streamline patterns for three values of circulation are shown in Fig. 24.4. There are three possible locations for the stagnation points:

(i) for $0 < V_r/V < 4$ or $K < 2\pi DV$, there are two symmetrically arranged stagnation points in Fig. 24.4(a);

(ii) for $V_r/V = 4$ or $K = 2\pi DV$, there is only one stagnation point at $\theta_0 = 270°$ in Fig. 24.4(b);

(iii) for $V_r/V > 4$ or $K > 2\pi DV$, the stagnation point moves away from the circle, and becomes a confluence point, see Fig. 24.4(c).

Prandtl (1961B) suggested that a maximum lift $C_L = 4\pi$ should be reached when a single stagnation point occurred at $\theta_0 = 270°$ for $V_r/V = 4$. He argued

[134]The more precise term might be the circle-streamline.

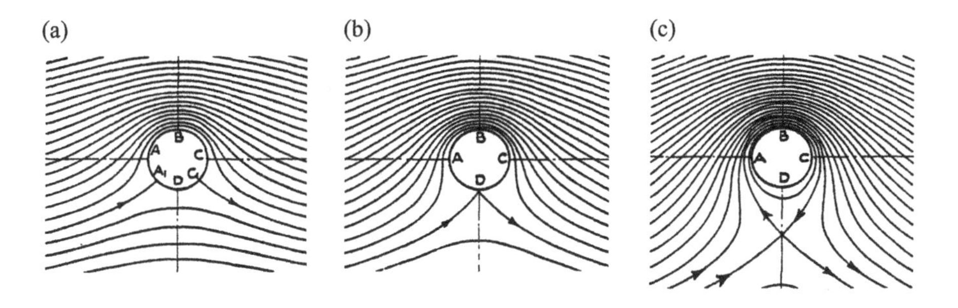

FIG. 24.4. Potential theory streamlines, (a) $V_r/V = 2$, (b) $V_r/V = 4$, (c) $V_r/V = 6$, Prandtl and Tietjens (1957B)

that there should be neither further change in circulation nor vorticity generation. Early tests by Betz (1925J) using a cylinder $L/D = 4.7$ at $Re = 52$k without end plates yielded $C_L = 4$ at $V_r/V = 3$, see Fig. 24.25. By adding end plates $D_e/D = 1.7$, he found $C_L = 9.5$ at $V_r/V = 4$. Further tests by Busemann (1932P) covered $L/D = 4.5$ and $D_e/D = 1.12$ and extended up to $V_r/V = 12.5$. He found $C_L = 11.6$ at $V_r/V = 4$ but C_L continued to rise up to $C_L = 16$ at $V_r/V = 12$. This was confirmed by Reid (1924P) and Swanson (1961R). Betz (1961J) commented that beyond $V_r/V = 4$ the cylinder was surrounded by a rotating boundary layer, which increases the effective diameter of the cylinder.

24.2.5 *Bickley's potential model*

A more sophisticated potential model was constructed by Bickley (1928J). He added a single potential vortex downstream of a circle in order to simulate the net vorticity shed into the 'wake'. According to Prandtl's concept, the circulation around the circle K should be in the opposite circulation to the shed vortex[135] $-k$, as depicted in Fig. 24.5. The circulation K acts at the centre of the circle $z = 0$, and vortex A at $z_A = ce^{i\theta}$, where c is the distance from the circle centre and θ is the polar angle. This arrangement of K and $-k$ will not maintain the circle as a streamline in potential flow. Bickley (1928J) added an image vortex B of opposite circulation k inside the circle at the inverse distance $z_B = D^2/(4c)\mathrm{e}^{i\theta}$. The latter cancelled the additional circulation of vortex B, and maintained the total circulation within the circle to K.

The vortex at A (c, θ) is a free vortex, that at B $\left(D^2/4c, \theta\right)$ is the inverse image of vortex A, and both vortices in the centre $z = 0$ are bound vortices. Only with this combination of three vortices inside the circle is the circle $c = D/2$ preserved as the streamline. The complex potential–stream function, $\omega = \phi + i\psi$, for such a model is

[135]Bickley (1928J) assumed that the circulation around the circle has the same circulation as the shed vortex.

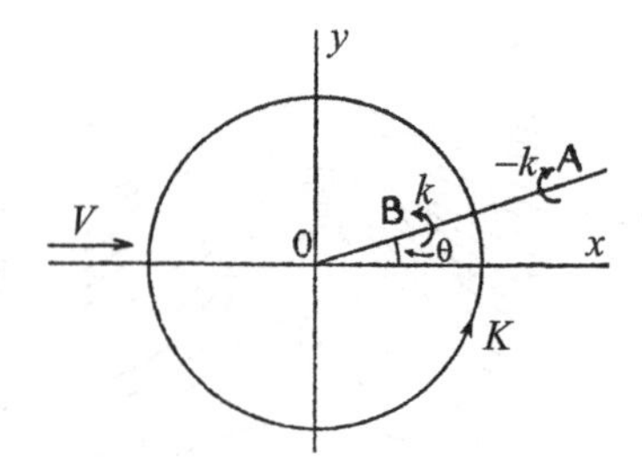

FIG. 24.5. Definition sketch of circle and vortices, Bickley (1928J)

$$\omega = -V\left(z + \frac{D^2}{4z}\right) - \frac{-i(K-k)}{2\pi}\ln z + \frac{ik}{2\pi}\ln\frac{z - z_A}{z - z_B},$$
$$u - iv = V\left(1 + \frac{D^2}{4z^2}\right) + \frac{i(K-k)}{2\pi z} - \frac{ik}{2\pi}\left(\frac{1}{z - z_A} - \frac{1}{z - z_B}\right) \tag{24.1}$$

Bickley (1928J) also calculated the velocity components of vortex A for the arbitrary magnitude of K and k as

$$u_A = V\left(1 - \frac{D^2}{4c^2}\cos 2\theta_A\right) + \left[\frac{K-k}{2\pi c} - \frac{kc}{2\pi(c^2 - D^2/4)}\right]\sin\theta_A,$$
$$v_A = -V\frac{D^2}{4c^2}\sin 2\theta_A - \left[\frac{K-k}{2\pi c} - \frac{kc}{2\pi(c^2 - D^2/4)}\right]\cos\theta_A \tag{24.2}$$

and a complex force expressed through the components $\boldsymbol{D}$ (drag) and $\boldsymbol{L}$ (lift) is

$$\boldsymbol{D} + i\boldsymbol{L} = i\rho[(K-k)V + k(u_A + iv_A)] \tag{24.3}$$

The unique feature of eqn (24.3) is that the drag force component $\boldsymbol{D}$ is predicted by this potential model if vortex A moves downstream, $u_A > 0$ and $v_A > 0$. If free vortex A is stationary, $u_A = 0$ and $v_A = 0$, then $\boldsymbol{D} = 0$; the D'Alembert paradox becomes valid. The potential model predicts the lift force component $\boldsymbol{L}$, which is proportional to $K - k$ and k according to eqn (24.3).

Bickley (1928J) applied Prandtl's concept that the overall circulation around any closed circuit embracing both the circle and free vortex A should be zero, $K - k = 0$. This meant that the circulation $-k$ of the free vortex A induced the circulation K in the circle equal in magnitude and of the opposite circulation. Introducing the coefficients of drag C_D and lift C_L, and applying the $K - k = 0$ condition yielded

$$C_D = \frac{\cos(\theta - \pi)}{2\pi} \frac{D}{2c} \left(\frac{2k}{DV} \right)^2 \tag{24.4}$$

$$C_L = \left(1 - \frac{D^4}{16c^4} \right) \left(\frac{2k}{DV} \right) - \frac{\sin(\theta - \pi)}{2\pi} \frac{D}{2c} \left(\frac{2k}{DV} \right)^2 \tag{24.5}$$

The C_L and C_D curves obtained from eqns (24.4) and (24.5) are all parabolas. Bickley (1928J) calculated curves for $\theta - \pi = 0$ (wake axis) and $c/D = 1,\ 2,$ and 3. Figure 24.6(a) shows that the curves for three values of c/D are close to Flettner's curves with and without end plates. Bickley (1928J) also calculated C_L and C_D curves for $c/D = 3$ and $\theta - \pi = -30°,\ -15°,\ 0°,\ 15°,$ and $30°$.

Swanson (1961R) used Bickley's (1928J) potential model, and compared the calculated C_L and C_D curves with his experimental curve. Figure 24.7(a) shows C_L and C_D curves for $c/D,\ \theta = 2,\ 210°$ (dotted), $2,\ 224°$ (dashed), $1.7,\ 242°$ (dash-dot), and experimental (full line). The agreement is good between the last two curves in $4 < C_L < 8$, medium in $0 < C_L < 2$, and poor for $C_L > 8$. A particularly large disagreement between the experiments and potential model is seen in Fig. 24.7(b), where C_L is plotted in terms of V_r/V for $c/D,\ \theta = 1.7,\ 242°$ (dotted), $2,\ 224°$ (dashed), and the experimental (full line).

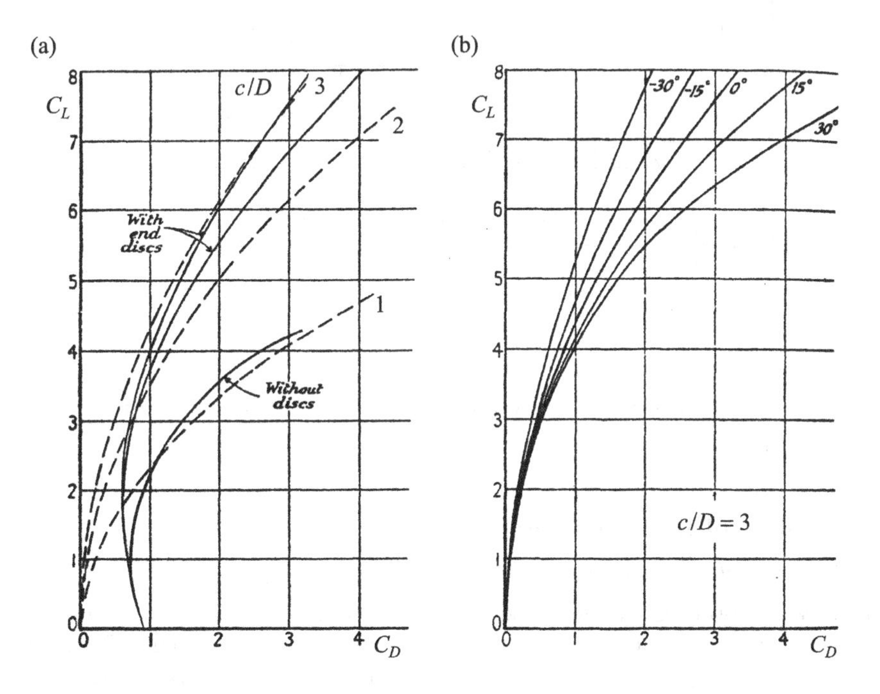

FIG. 24.6. Comparison of calculated lift and experiments (a) various $2c/D$, (b) $2c/D = 6$, Bickley (1928J)

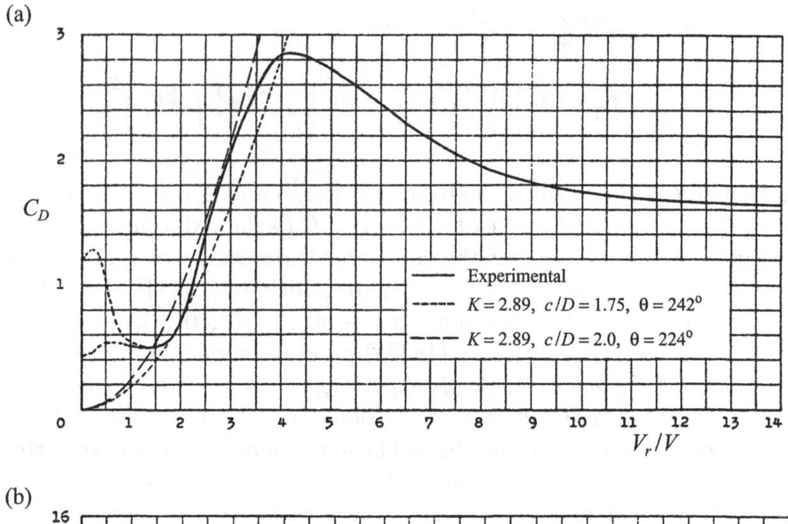

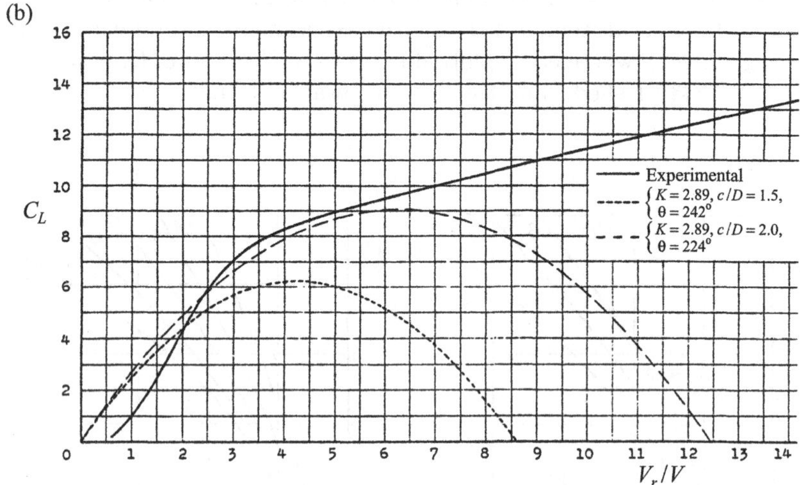

FIG. 24.7. Comparison of theory and experiment, (a) drag, (b) lift, Swanson (1961R)

24.3 Effect of Reynolds number

24.3.1 *Laminar, L3, and transitional, TrW, wakes*

A flow past a rotating cylinder is likely to be affected by the state of the boundary layers and wakes, which are laminar, transitional, and turbulent. Each state spreads over a range of *Re*, and is affected by the influencing parameters as well. The presentation in this chapter will be in the order of rising *Re* as in Vol. 1.

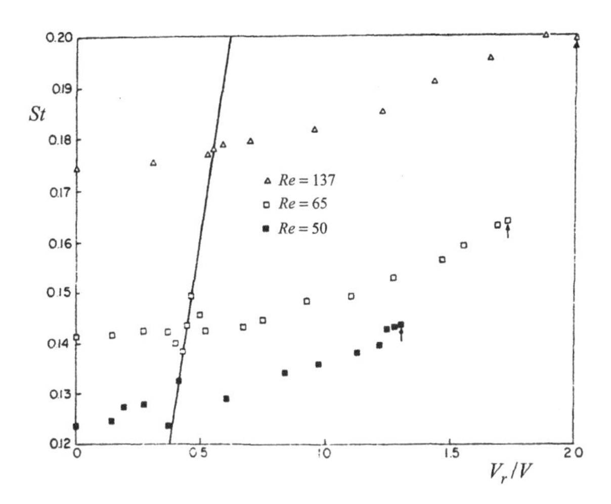

FIG. 24.8. Strouhal number in terms of V_r/V for laminar wakes, Jaminet and Van Atta (1969J)

Jaminet and Van Atta (1969J) carried out tests at low Re on rotating cylinders (wires) having $L/D = 40$, 52.5, and 64.6, and $D/B = 0.002$ to 0.005 fitted with square and stationary[136] end plates $L_e/D = 140$, 105, and 68, respectively. The eddy shedding frequency was measured with a hot wire at $x/D = 11$ and y/D where the maximum amplitude of the velocity fluctuations was detected. The cylinder rotation caused y/D to be displaced transversely due to the biased wake, as seen in Fig. 24.1.

Figure 24.8 shows that the eddy shedding frequency and related St increase slowly with the increasing rate of the cylinder rotation until a point is reached when eddy shedding ceases, marked by arrows. The solid straight line corresponds to St_r based on the rotation frequency of the cylinder. As the rotation frequency increases and approaches the more slowly increasing shedding frequency, eddy shedding becomes disturbed and difficult to measure. A further increase in rotation results in synchronization of the shedding frequency with the rotational frequency. In the synchronization range, eddy shedding is very stable and regular. When the rotation rate is increased further, the shedding frequency no longer follows the rotation frequency, see Fig. 24.8. There is an analogy with the synchronization caused by sound, Vol. 1, Chapter 18, and by transverse oscillations of a flexible cylinder, Vol. 3, Chapter 33.

For non-rotating cylinders at $Re < 48$, eddy shedding is not found. However, eddy shedding can always be induced in the range $32 < Re < 48$ behind the

[136]The cylinders (wires) pass through the holes in end plates and tunnel walls. This arrangement avoids additional motion induced by the rotation of end plates.

rotating cylinder. The forced eddy shedding frequency is initially equal to twice the rotation frequency followed by a complete disintegration of eddy shedding at somewhat higher rotation rates. The initial eddy shedding triggered by the rotation is probably caused by local disturbances resulting from very small eccentric movements (whipping) of the cylinder (wire) during the rotation.

Jaminet and Van Atta (1969J) also examined the critical velocity ratio, $(V_r/V)_c$, required to suppress eddy shedding in the range $50 < Re < 190$. Figure 24.9 shows $(V_r/V)_c$ in terms of Re. The forced eddy shedding for $Re < 48$ is at a low St. As the $(V_r/V)_c$ is approached, the hot-wire signal decreases, which indicates that eddies are weak. The hot-wire wake traverse shows that eddies become aligned in a single row. Finally, as $(V_r/V)_c$ is reached, the eddies in the single row become weak and suppressed simultaneously. Only intermittent random fluctuations are found in the wake then. Figure 24.9 shows that for $Re > 80, (V_r/V)_c = 2$. The same value was found by Prandtl (1961B) for $Re = 4$k, as seen Fig. 24.1.

There is yet another kind of eddy induced by the rotating cylinder in the fluid at rest after an impulsive start. Taneda (1980P) demonstrated the appearance of counter-rotating pairs of eddies of toroidal shape. Figure 24.10 shows the toroidal eddy pairs along the span of the rotating cylinder at an early stage of formation after the impulsive start. The flow is three-dimensional but the spanwise pitch of the eddy pairs appears to be constant.

Matsui (1977P) carried out flow visualization in water for $L/D = 6$, $D/B = 0.012$, $150 < Re < 400$, and $2 < V_r/V < 2.7$. The suppression of eddy shedding seems to be associated with the interaction of toroidal eddies. Figure 24.11 shows a remarkable 'net' pattern for $Re = 214$ and $V_r/V = 2.4$. The main feature of the net is the periodic arrangement of eddy filaments both in the spanwise and streamwise directions.

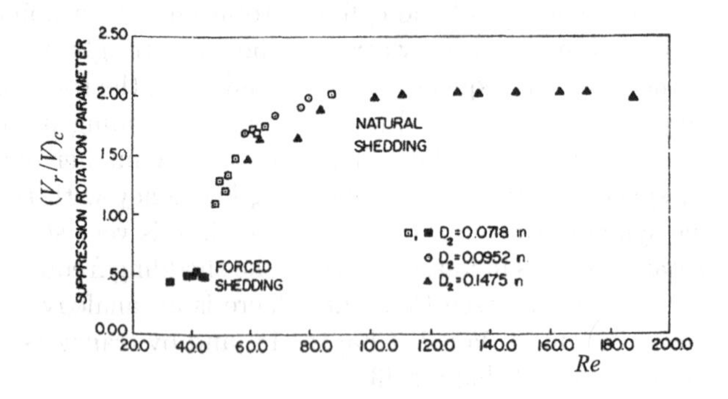

FIG. 24.9. Critical velocity ratio for suppressing eddy shedding in terms of Re, Jaminet and Van Atta (1969J)

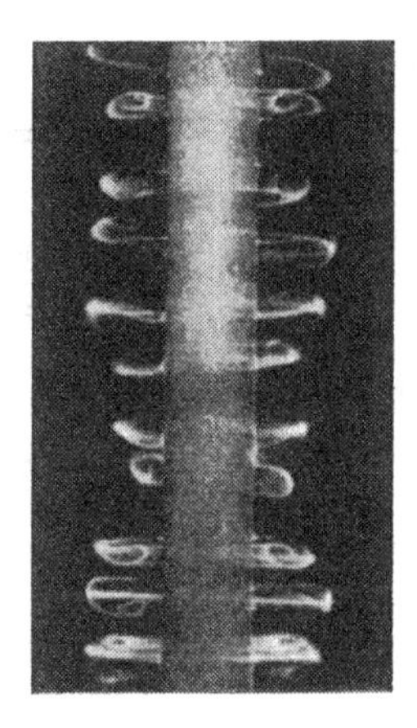

FIG. 24.10. Dye visualization of ring eddies around a rotating cylinder after 4.7 s, Taneda (1980P)

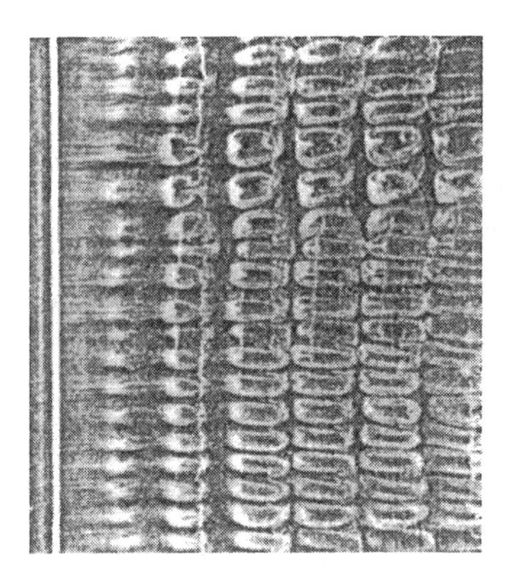

FIG. 24.11. Smoke visualization of wake behind a rotating cylinder at $Re = 214$, $V_r/V = 2.4$, Matsui (1977P)

24.3.2 *Pressure distribution in the TrSL state*

The simplest of all measurements has been the mean (time-averaged) pressure distribution around a stationary cylinder immersed in a free stream. This measurement becomes a daunting task when the cylinder rotates. Lafay (1912J) measured the pressure distribution by using a stationary pressure probe placed adjacent to the cylinder surface.

Thom (1926Pb) designed a special apparatus for measuring the mean pressure

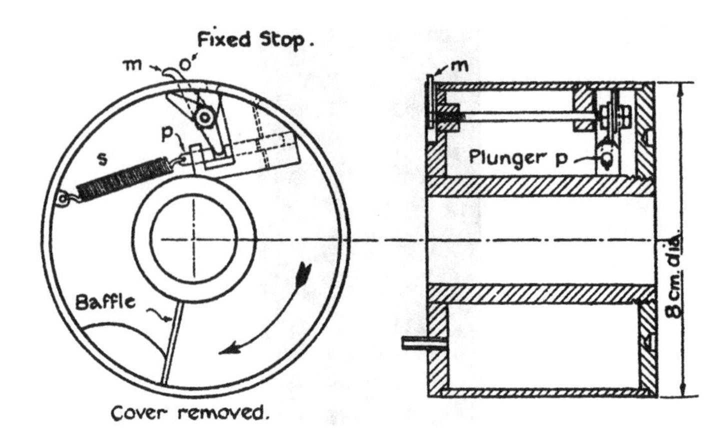

FIG. 24.12. Apparatus for pressure measurement around a rotating cylinder, Thom (1926Pb)

on the rotating cylinder. Figure 24.12 shows two views of the apparatus. The pressure tapping on the cylinder surface continues through a plunger p. The pressure tapping may be opened or closed by the movement of the plunger. The latter is moved by a lever on the spindle, which passes through the pressure chamber formed by a small aspect ratio central section of the cylinder, Fig. 24.12(b). The other side of the lever protrudes through the cylinder surface in the form of a cam m. The plunger opens the pressure tapping each time the fixed stop hits the cam, and closes afterwards by the action of a spring s. The pressure distribution is measured by adjusting the fixed stop at the required angular position θ.

Thom's (1926Pb) rotating cylinder had $L/D = 7.7$, $D/B = 0.13$, no end plates, and was tested for $V_r/V = 0,\ 1,\ 2,\ 3,$ and 4 in the range 7.7k $< Re <$ 30.7k. Figure 24.13 shows the C_p distribution for stationary $V_r/V = 0$ and rotating $V_r/V = 1,\ 3,$ and 4 cylinders. The potential flow curves are added for comparison (dashed lines). The potential flow theory (circle with circulation) is in better agreement with experiments as V_r/V increases. There are important changes in the C_p distribution brought about by the rotation:

(i) The stagnation point C_{p0} gradually becomes displaced for $V_r/V = 1,\ 3,$ and 4 to $\theta_0 = 355°,\ 310°$, and $290°$, respectively.

(ii) A drastic decrease in $C_{p\min}$ takes place for $V_r/V = 0$, 1, 3, and 4, $C_{p\min} = -1.1,\ -2.9,\ -9$, and -10.5, on the side where V and V_r are in the same direction. However, on the other side, where V and V_r are in the opposite direction, $C_{p\min}$ is obliterated for $V_r/V > 1$.

(iii) The adverse pressure recovery is significantly increased. For example, $\Delta C_p = 2,\ 7$, and 8 at $\theta = 100°,\ 125°$, and $130°$ for $V_r/V = 1,\ 3$, and 4, respectively.

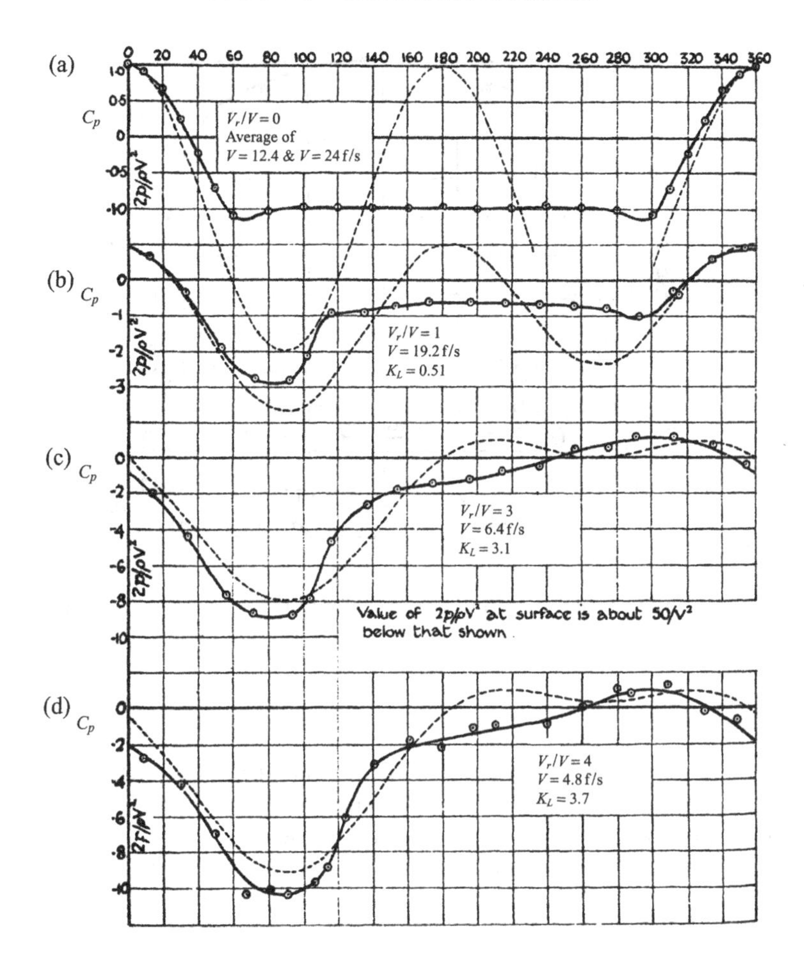

FIG. 24.13. Mean pressure distribution around a rotating cylinder, (a) $V_r/V = 0$, (b) $V_r/V = 1$, (c) $V_r/V = 3$, (d) $V_r/V = 4$, Thom (1926Pb)

(iv) The large asymmetry in pressure distribution on the two sides results in a high side/lift[137] force. For example, $C_L = 1.02,\ 6.2$, and 7.5 for $V_r/V = 1, 3$, and 4, respectively.

Thom (1926Pb) also measured the C_p distribution at different spanwise locations for $V_r/V = 2$. Figure 24.14 shows the spanwise variation in the local k_l and k_d[138] evaluated from C_p for $V_r/V = 2$ and $Re = 15.4$k. Thom (1926P)

[137] The side force can be called the lift force because it is perpendicular to the free stream velocity.

[138] Note that since the 1930s k_d has been replaced by $\frac{1}{2}c_d$ and k_l has been replaced by $\frac{1}{2}c_l$ because the old dynamic pressure used, ρV^2, has been replaced by $\frac{1}{2}\rho V^2$, respectively.

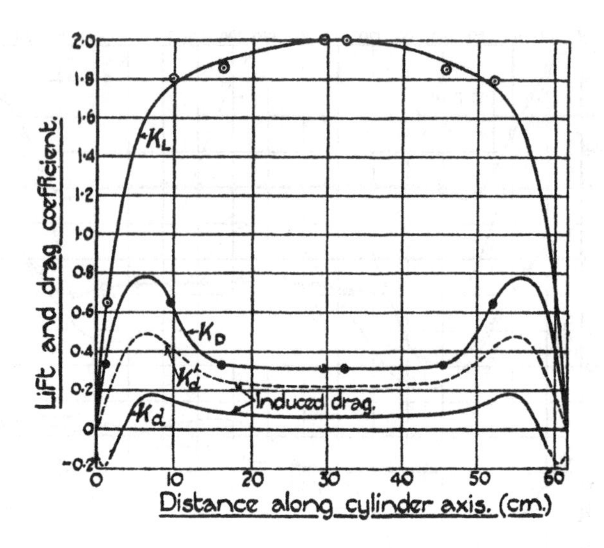

FIG. 24.14. Spanwise distribution of local drag and lift at $Re = 15\text{k}$, $V_r/V = 2$, Thom (1926Pb)

commented:

> It will be seen that the distribution of lift and drag along the span is not uniform except possibly for a short distance of the midspan. The lift would appear to drop to zero at the end of the cylinder. That this drop is not the result of a fall in velocity towards the end is shown by the comparatively normal value of $C_{po} = 0.8$ on the end section indicating a fall in velocity of only about 10%.

It was possible to calculate the approximate distribution of induced drag k_{di} from the known $\mathrm{d}k_l/\mathrm{d}z$ in Fig. 24.14. If z is measured along the span and downwash $\mathrm{d}w$ is produced by the trailing vortices at a position z', then

$$\mathrm{d}w = \frac{DV}{4\pi}\frac{\mathrm{d}C_l}{\mathrm{d}z}\frac{\mathrm{d}z}{z' - z} \tag{24.6}$$

Thom (1926Pb) performed a graphical integration and avoided infinite values for $z' = z$ by using Rankine's type of vortex. The curve for k_{di} obtained in this manner is added in Fig. 24.14. The above method ignores the presence of tunnel walls. The effect of the tunnel walls is taken into account by Lock's modified equation

$$\mathrm{d}w = \frac{DV}{4B}\frac{\mathrm{d}C_l}{\mathrm{d}z}\cot\frac{\pi(z' - z)}{B}\,\mathrm{d}z \tag{24.7}$$

where B is the distance between the tunnel walls. The lowest k_d curve in Fig. 24.14 shows the induced drag calculated from Lock's equation.

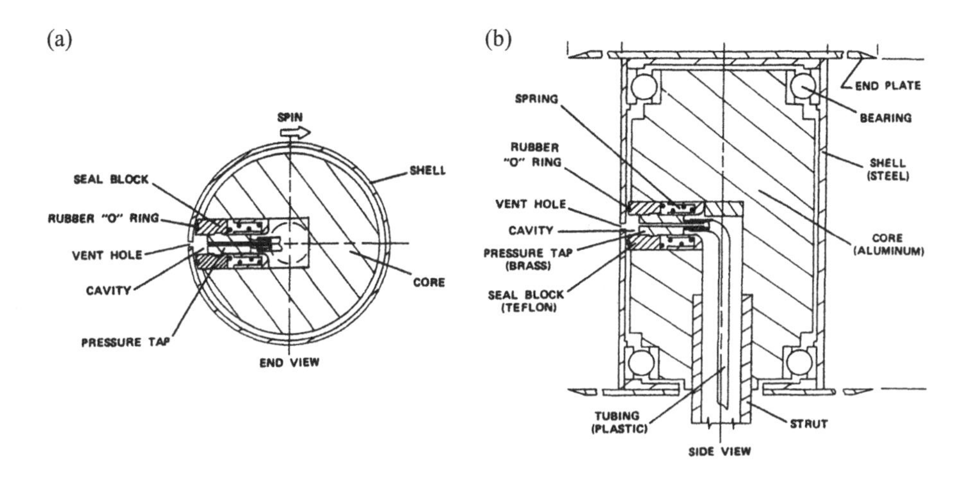

FIG. 24.15. Apparatus for measuring pressure on a rotating cylinder, (a) cross-section, (b) spanwise, Miller (1976J)

Thom's pioneering pressure measurements around rotating cylinders has been neither repeated nor superseded for almost half a century. Miller (1976J) used a different apparatus, as shown in Fig. 24.15. Only the cylinder with the pressure tapping rotates while the inner cylindrical block is stationary. The pressure tapping is sealed by the block with the cavity, and is attached to the cylinder surface by the spring. The pressure is measured over an angle which corresponds to the size of the cavity, and not to the pressure tapping size.

Miller (1976J) measured C_p on a cylinder $L/D = 1.64$ and $D/B = 0.04$, with end plates $D_e/D = 1.93$. Three C_p distributions for $V_r/V = 0.17,\ 0.77$, and 2.05 at Re = 340k, 440k, and 220k, respectively, are shown in Fig. 24.16. The local lift and drag coefficients are evaluated from C_p, $C_l = -0.39,\ 1.16$, and 3.07, and $C_d = 0.69,\ 0.49$, and 1.67, respectively. Note that the negative value of C_l for $V_r/V = 0.17$ and Re = 340k represents inversion of the Magnus effect, and will be discussed in the next section.

24.3.3 *Inversion of the Magnus effect*

Lafay (1910J, 1912J) was among the first researchers to find the inverse Magnus effect, i.e. the side force directed in the opposite direction to that predicted by Magnus (1853J). He carried out tests on a rotating cylinder $L/D = 3.5$ without end plates in an open jet in the range $0 < V_r/V < 1.3$ for 57k $< Re <$ 198k. Figure 24.17 shows the measured resultant force inclination relative to the free stream direction for Re = 62k, 88k, 128k, 168k, and 198k. Except for the lowest Re, all the other curves initially show a negative inclination α at the low speed of rotation. This means that initially the side force component is negative, i.e. inverse to the Magnus effect.

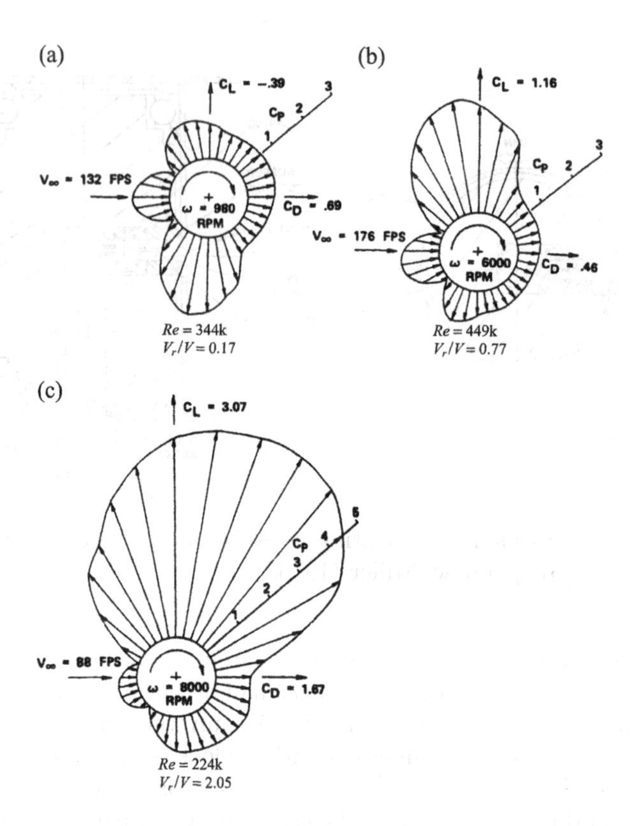

FIG. 24.16. Polar presentation of mean pressure distribution, (a) Re = 344k, $V_r/V = 0.17$, (b) Re = 449k, $V_r/V = 0.77$, (c) Re = 224k, $V_r/V = 2.05$, Miller (1976J)

Lafay (1912J) verified that for $V_r/V = 0.21$, Re = 128k by measuring the pressure distribution. The minimum pressure on the side where V_r and V were in the opposite direction was lower than on the side where V_r and V were in the same direction. This resulted in the side force being directed towards the side where V_r and V were in the opposite direction. At higher $V_r/V = 0.65$ and 1.30, Re = 128k, the minimum pressure was lower on the side where V_r and V were in the same direction, confirming the Magnus effect.

The fact that the inversion of the Magnus effect occurred in the TrBL0 regime prompted Krahn (1956J)[139] to attribute the origin of the inverse Magnus effect to a different start of transition on two sides of the rotating cylinder. He introduced the effective Re, based on the relative velocity of the rotating surface as $Re(1 \pm V_r/V)$, where the positive sign referred to the side where V_r and V were in the

[139] Krahn (1956J) did not carry out tests to verify his hypothesis.

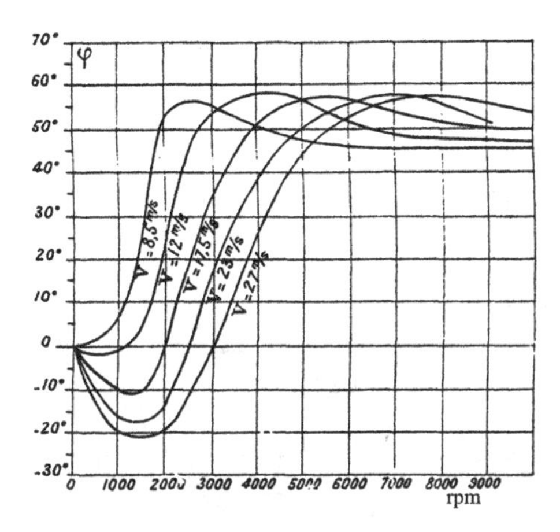

FIG. 24.17. Inclination of resultant force relative to wind velocity in terms of speed of rotation, Lafay (1912J)

opposite direction, while the negative sign referred to the side where V and V_r were in the same direction. The boundary layer became transitional first for the $Re(1+V_r/V)$ side causing the displacement of separation and lowering of $C_{p\min}$. This in turn induced a negative Magnus effect. A further increase in V triggered transition on the opposite side where V and V_r are in the same direction, and the Magnus effect took over.

Krahn's (1956J) explanation was partly proven experimentally by Kelly and Van Aken (1956J). They re-plotted Lafay's (1912J) data and proved that the inverse Magnus effect did not show before $Re = 98$k, and according to their tests it was not found for $Re > 605$k. The minimum value of the negative side force, $C_{L\min} = -0.7$, was found for $Re = 302$k. Hence, the appearance of the Magnus effect was confined to the TrBL0, TrBL1, and TrBL3 regimes, as proposed by Krahn (1956J).

The most complete and detailed experiments on rotating cylinders were carried out by Swanson (1961R), who also gave a thorough review of the phenomena. He covered $35\text{k} < Re < 500\text{k}$ and $0 < V_r/V < 1$. The rotating cylinder protruded through the side holes of the tunnel walls, so Swanson (1961R) assumed that $L/D = \infty$. The measured variation in C_L in terms of V_r/V is given in Fig. 24.18(a,b) up to $Re = 300$k and beyond, respectively. For $34\text{k} < Re < 99\text{k}$, the inverse Magnus effect does not take place. It starts to develop for $Re > 99$k, and the minimum value of $C_{L\min} = -0.6$ is reached for $Re = 325$k. This coincides with the TrBL1 regime where the separation bubble is formed on the stationary cylinder. A further increase in Re up to 500k produced a gradual decrease in the negative C_L until the inversion of the Magnus effect disappears for $Re = 501$k.

All C_L curves in Fig. 24.18(a,b) are bound by three envelopes:

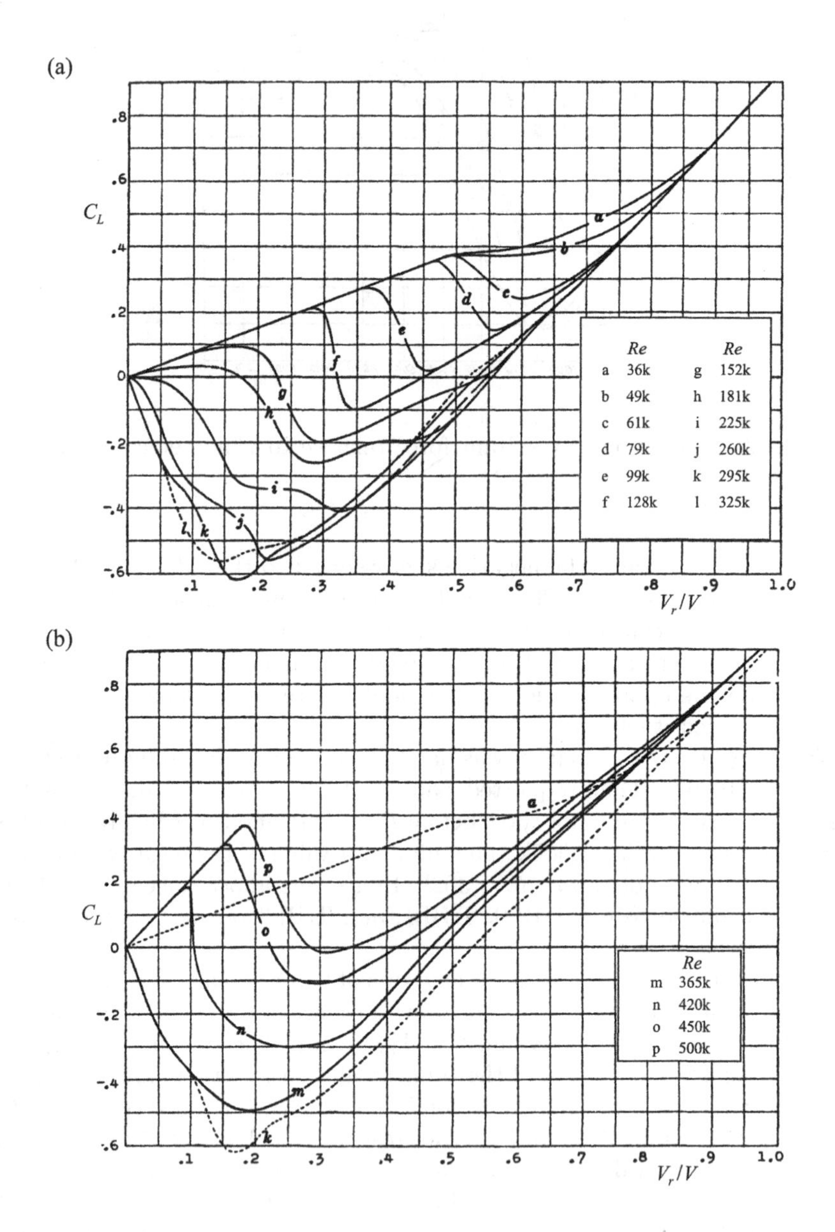

FIG. 24.18. Measured force on a rotating cylinder in terms of V_r/V, (a) lift 36k < *Re* < 325k, (b) lift 365k < *Re* < 501k, Swanson (1961R)

(i) for the laminar boundary layer in Fig. 24.18(a), the slope is $dC_L/d(V_r/V) = 0.8$;

(ii) for the transitional boundary layer in Fig. 24.18(b), the slope is 2;

(iii) when the pressure difference is dominated by the side where V_r and V are in the same direction the envelopes in Fig. 24.18(a,b) are not straight, and the slope is $0.9 < dC_L/d(V_r/V) < 1$.

The measured C_D is shown in Fig. 24.19, and displays remarkably less variation in comparison with C_L. There is a lower envelope at $C_D = 0.55$ for all C_D curves. The decrease in C_D for $V_r/V = 0$ mimics the TrBL0 regime.

24.3.4 *Boundary layer*

Early measurements of the boundary layer adjacent to a surface of the rotating cylinder where reported by Thom (1931P). He tested a cylinder spanning a test section of $L/D = 5.33$ and $D/B = 0.187$ at $Re = 18\text{k}$ and $V_r/V = 2$. The small Pitot and static tubes were used separately in different runs. Hence, the calculated velocity profiles might be of limited accuracy.

Swanson (1961R) also measured the boundary layer on the rotating cylinder at $Re = 40\text{k}$ for $V_r/V = 1$ and 2. He argued that the origin of the boundary layer on the rotating cylinder should not coincide with the stagnation point. For the rotating cylinder, the stagnation point is displaced in a direction opposite to the direction of rotation. Swanson (1961R) suggested that the origin of the boundary layer was where the surface velocity is equal to the free stream velocity,

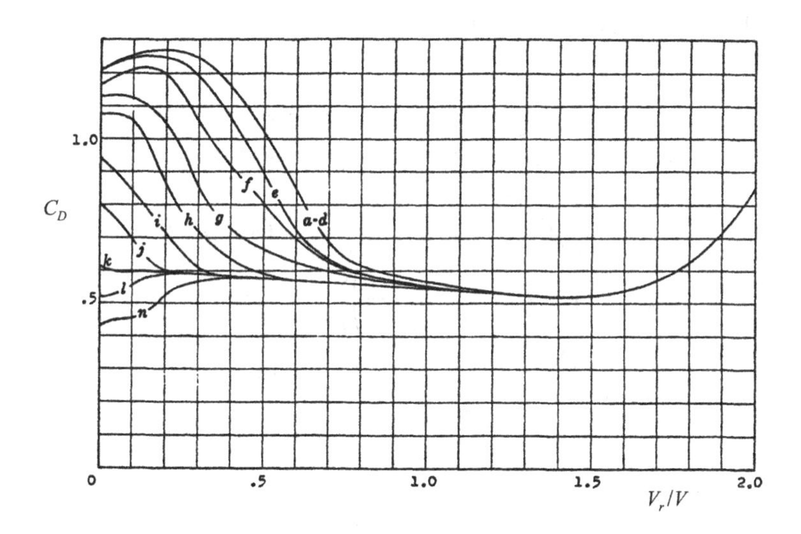

FIG. 24.19. Measured drag $36\text{k} < Re < 420\text{k}$ in terms of V_r/V, Swanson (1961R)

i.e. at the zero relative velocity. This means that the boundary layer origin is displaced in the same direction as the moving surface and opposite to that of the stagnation point, as depicted in Fig. 24.20.

Peller (1986J) tested a cylinder $L/D = 1.25$ spanning test section without end plates, $D/B = 0.3$ (blockage corrected) at $Re = 48\text{k}$ by using LDV. Figure 24.21(a,b) shows mean velocity profiles around the rotating cylinder at $V_r/V = 0.5$ and 2. The thin line tangential to the cylinder surface designates the circumferential surface velocity. There is a considerable increase in the high-velocity region for $V_r/V = 2$ up to 240°.

Peller (1986J) also confirmed Swanson's (1961J) argument about the displacement of the origin of the boundary layer on the rotating cylinder. Figure 24.22 shows the measured thickness distribution of the boundary layer along the upstream and downstream moving surfaces of the cylinder at $Re = 48\text{k}$ for $0 \leqslant V_r/V \leqslant 2$. The separation angles measured from the fixed $\theta = 0°$ are written as the vertical numbers on both sides. The separation angles θ_{so} measured from the approximate location of the boundary layer origin are written as horizontal numbers. Only for $V_r/V = 0$ are both angles equal to 80° because the stagnation point and the location of zero velocity are the same. However, for $V_r/V = 0.5$, the location of zero velocity is at $\theta_s - \theta_{so} = 10°$. The separation angles measured from $\theta_0 = 10°$ became $\theta_{so} = 98°$ and 99°, as given by the horizontal numbers. The displacement of the zero relative velocity location towards the downstream moving side shortens the boundary layer length on that side, and lengthens it on the opposite side.

Tanaka and Nagano (1973J) determined the angular location of the fluctuating separation by using a hot-wire probe as close to the surface as $y/D = 0.007$ (0.5 mm). The rotating cylinder spanned a test section, $L/D = 2.4$, $D/B = 0.18$

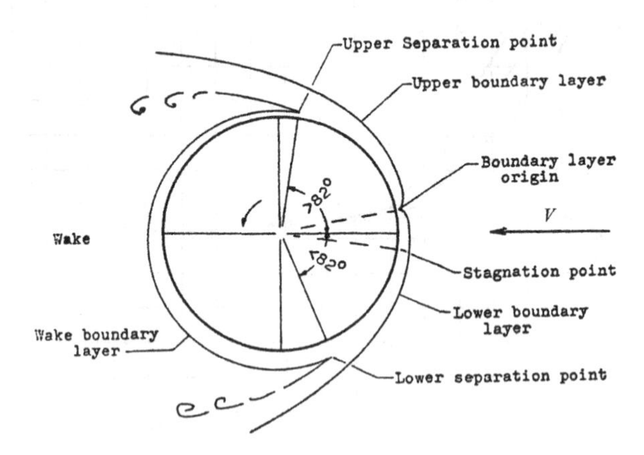

FIG. 24.20. Definition sketch of zero relative velocity at the origin of the boundary layer on a rotating cylinder, Swanson (1961R)

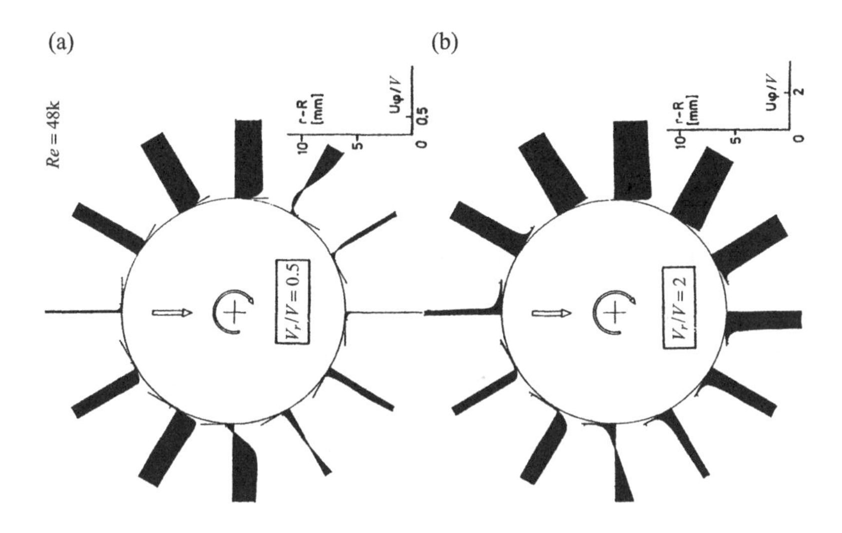

FIG. 24.21. Velocity profiles around a rotating cylinder at Re = 48k, (a) $V_r/V = 0.5$, (b) $V_r/V = 2$, Peller (1986J)

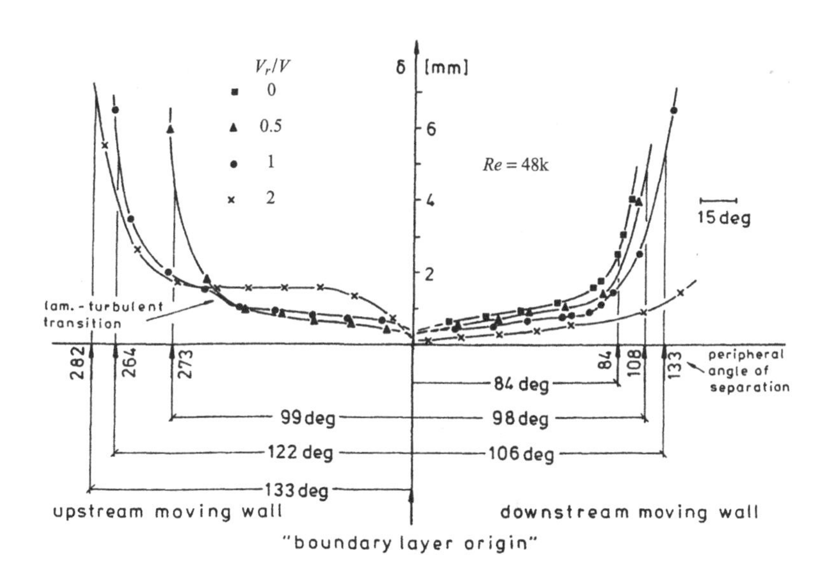

FIG. 24.22. Distribution of boundary layer thickness at Re = 48k, Peller (1986J)

(uncorrected), and $Ti = 0.2\%$. The test range was $0 \leqslant V_r/V < 1.1$ and $48\text{k} < Re < 371\text{k}$. They found that the free shear layer near the separation oscillates in the rhythm of eddy shedding. The local velocity reduces to a minimum when the separation point reaches the upstream end, and increases to a maximum when the separation point reaches the downstream end of the oscillation.

Figure 24.23 shows the effect of V_r/V on the estimated extent of the separation fluctuation on the top and bottom halves of the downstream and upstream moving surfaces, respectively. The origin of the coordinate system is at $\theta = 90°$, while $\theta = 0°$ and $180°$ correspond to the front and the rear of the cylinder, respectively. The shaded area designates the V_r/V range where eddy shedding is found in the wake. The increase in V_r/V causes a monotonous increase in θ_s on the downstream moving surface as expected. However, there is a pronounced kink around $(V_r/V)_c$ in the range $65° < \theta_s < 85°$ and a sudden jump to $90° < \theta_s < 100°$ on the opposite side. Tanaka and Nagano (1973J) related the kink in θ_s on the upstream moving surface to the dip in the C_L curve, which in turn may lead to the inverse Magnus effect at higher Re.

24.3.5 *Strouhal number*

It has been shown by Prandtl (1906J) in Fig. 24.1 that for $Re = 4.1\text{k}$ eddy shedding was suppressed behind a rotating cylinder for $V_r/V \geqslant 2$. The disappearance of eddy shedding was associated with the gradual narrowing, lengthening, and biasing of the near-wake.

Tanaka and Nagano (1973J) found that eddy shedding ceased at a considerably lower value, V_r/V, in the range $0.67 < (V_r/V)_c < 0.72$ in the tested Re range. This might have been affected by the high Re, low L/D, and partly by the distant location of the hot wire at $x/D = 2.89$ and $y/D = -0.8$. Figure 24.24

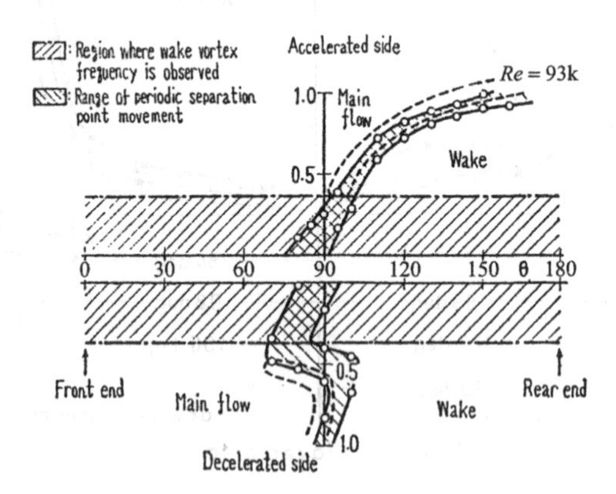

FIG. 24.23. Angular extent of fluctuating separation in terms of θ at $Re = 93\text{k}$, downstream moving surface (top), upstream moving surface (bottom), Tanaka and Nagano (1973J)

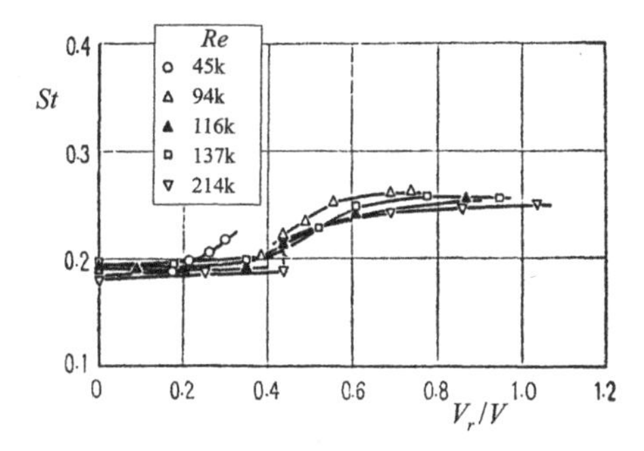

FIG. 24.24. Strouhal number in terms of V_r/V, Tanaka and Nagano (1973J)

shows the variation in St in terms of the product[140] $Re \times V_r/V$ for different Re. For small V_r/V values, St is constant within a band 0.18–0.195 as found behind the non-rotating cylinder. As V_r/V is increased, there is a small rise in St to the 0.24–0.26 band.

Diaz *et al.* (1982P, 1983J, 1985J) repeated the St measurements using a cylinder $L/D = 30$ and $D/B = 0.03$ at $Re = 9$k and $0 \leqslant V_r/V < 2.5$. The frequency power spectra showed the St peak up to $V_r/V = 1.5$, and beyond it became wide-banded.

24.3.6 *Effect of end plates*

Early measurements of the resultant force by Lafay (1912J), and the mean lift and drag forces by Betz (1925J), were carried out on short aspect ratio cylinders without end plates. According to Betz (1961P), Prandtl argued that two-dimensional flow could not be achieved without end plates. So Betz (1925J) repeated experiments on the $L/D = 4.7$ cylinder with end plates $D_e/D = 1.7$, and measured $C_L = 9$ almost twice that without the end plates, Fig. 24.25(a). Another set of measurements by Reid (1924P) in the range 39k $< Re <$ 116k used a long cylinder, $L/D = 13.3$, which passed through the side walls of the test section and without end plates. He also measured $C_L = 9$ at $V_r/V = 4$, as seen in Fig. 24.25(a).

Figure 24.25(b) shows the measured C_D in terms of V_r/V. Note that the C_D scale is magnified 10× in comparison with C_L in Fig. 24.25(a). Betz's (1925J) C_D values for $V_r/V = 0$ and $Re = 52$k are below 1.2, which indicates that the flow is still not two-dimensional with the end plates used. Reid (1924P) found $C_d = 0.94$ for $V_r/V = 0$ and $Re = 116$k. It is hard to explain such a large discrepancy in C_D compared with the close agreement of the C_L curves.

An early criticism of the use of the *rotating end plates* for rotating cylinders came from Ahlborn (1929J). He suggested that two entirely different flow topolo-

[140]This unusual parameter, which Tanaka and Nagano (1973J) called the rotational Re_r, is not appropriate because eddy shedding is produced by V and not V_r.

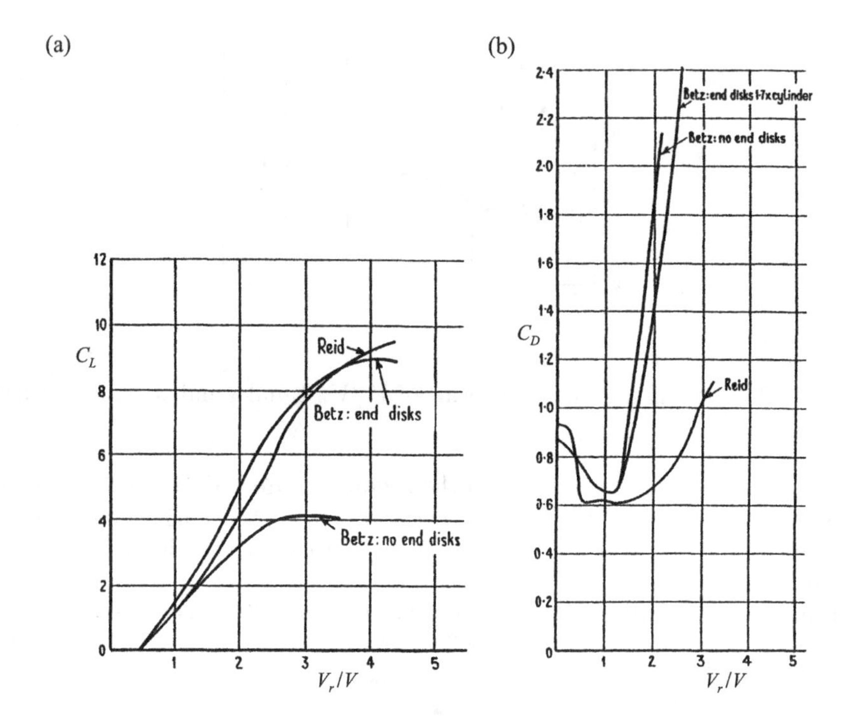

FIG. 24.25. Measured force on a rotating cylinder in terms of V_r/V, (a) lift, (b) drag, Betz (1925J), Reid (1924P)

gies were produced with and without end plates on the rotating cylinder. Figure 24.26(a) shows that without end plates the secondary axial flow is induced by the rotation of the cylinder ends. The result is a pair of counter-rotating toroidal eddies. When the end plates rotate, the secondary radial flow is induced along both sides of the end plates. The system of two counter-rotating eddies is formed as depicted in Fig. 24.26(b) arranged symmetrically relative to the midspan axis. Ahlborn (1929J) also added a third pair of counter-rotating toroidal eddies induced by the secondary toroidal eddy. Hence, the rotating end plates induced additional secondary flow, and do not improve the two-dimensionality of the flow.

The other option for rotating cylinders was to protrude through the test section, as used by Reid (1924P) and Swanson (1961R). This arrangement could not be considered as $L/D = \infty$ as suggested by Swanson (1961R). Presumably, the stationary end plates, as used by Jaminet and Van Atta (1969J), might be the best option.

Thom (1934P) carried out tests on two cylinders $L/D = 13$ and 26, $D/B = 0.04$, 0.08, respectively, in the range 5k $< Re <$ 12.5k, without and with end

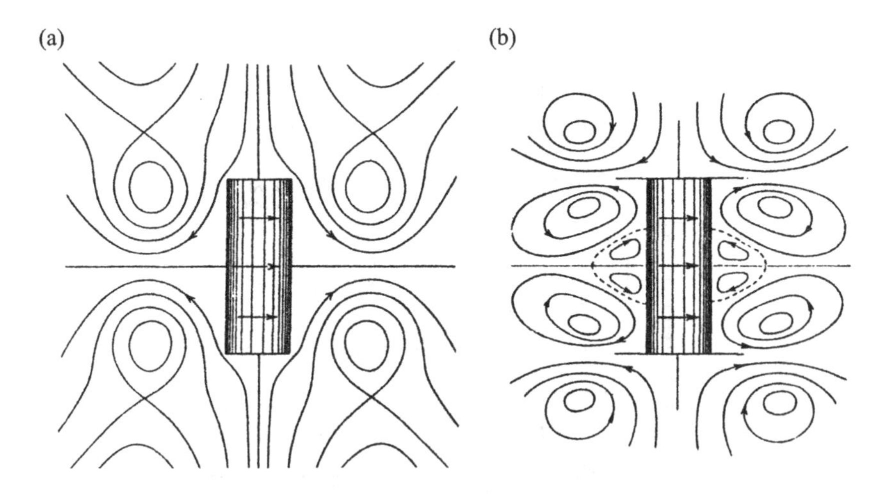

FIG. 24.26. Tentative topology of secondary flow induced by a rotating cylinder, (a) without end plates, (b) with end plates, Ahlborn (1929J)

plates $D_e/D = 3$. Figure 24.27(a,b) shows the variation in k_L ($\frac{1}{2}C_L$) and k_D ($\frac{1}{2}C_D$) in terms of V_r/V. The experimental points are not shown, but the *Re* effect is negligible. However, there is a marked effect of adding end plates on both the k_L and k_D curves (marked plain and end discs in Fig. 24.27). It is notable that the increase in k_L obtained by fitting end plates is associated with a decrease in k_D. This means that the k_L/k_D ratio is considerably increased.

24.3.7 *Effect of surface roughness and fins*

Thom (1934P) also examined the effect of surface roughness on the lift and drag forces. He did not specify the relative roughness and mentioned that he glued sand onto the cylinder surface. The range 5k < *Re* < 12.5k corresponds to the TrSL2 regime where the laminar boundary layer is very stable. Presumably, this is the reason why both k_L and k_D are close to the plain cylinder in Fig. 24.27(a,b).

Thom (1934P) also added equally spaced fins along the span of the rotating cylinder $L/D = 13$ and 26, $D/B = 0.04$ and 0.08, fin ratio $D_f/D = 3$, and $p/D = 0.75$ and 1.5, where p is the pitch between fins. Figure 24.27 shows a significant rise in k_L and a drop in k_D for $V_r/V = 4$. Particularly unexpected is $C_D < 0$ for $V_r/V < 7$, which indicates that the inclination of the resultant force is tilted upstream. The significant rise in k_L ($C_L > 30$) may be attributed to the secondary flow around 17 fins. The overall lift is produced by the combined action of the rotating cylinder and fins.

Another important parameter is the shape of the cylinder ends. Thom (1926Pa) examined the effect of flat (square) and elliptical ends on the flow around the rotating cylinders. All k_D curves exceeded those measured on the

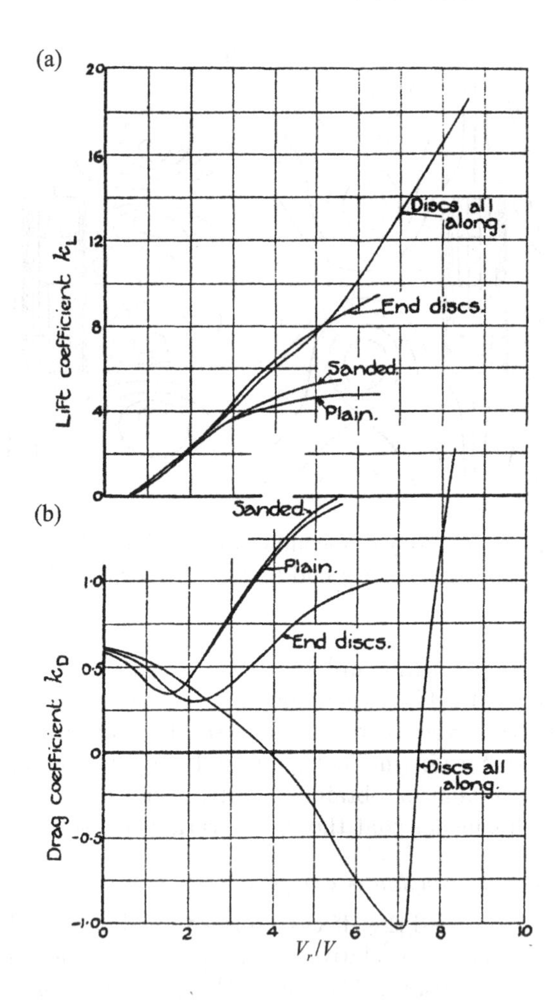

FIG. 24.27. Measured force in terms of V_r/V, (a) lift, (b) drag, Thom (1934P)

long aspect ratio cylinder with flat ends in the range $1 < V_r/V < 3$.

24.3.8 *Far-wake development*

Diaz *et al.* (1985J) traversed the wake behind a rotating cylinder, $L/D = 30$, $D/B = 0.03$, $Re = 9\text{k}$, $0 \leqslant V_r/V < 2.5$, and $1.5 < x/D < 100$. Figure 24.28 shows the decay of the velocity defect along the wake for $V_r/V = 0$ and 2.5, respectively. The biased near-wake displaces the maximum velocity defect y_m away from the wake axis y_c. The displacement rate is large up to $x/D = 7.5$, and an asymptotic value is reached for $x/D > 30$.

The normalized velocity defect in terms of $(y - y_m)/b$ is shown in Fig. 24.29 for $V_r/V = 0,\ 0.5,\ 1, 2$, and 2.5. All experimental points collapse onto a single

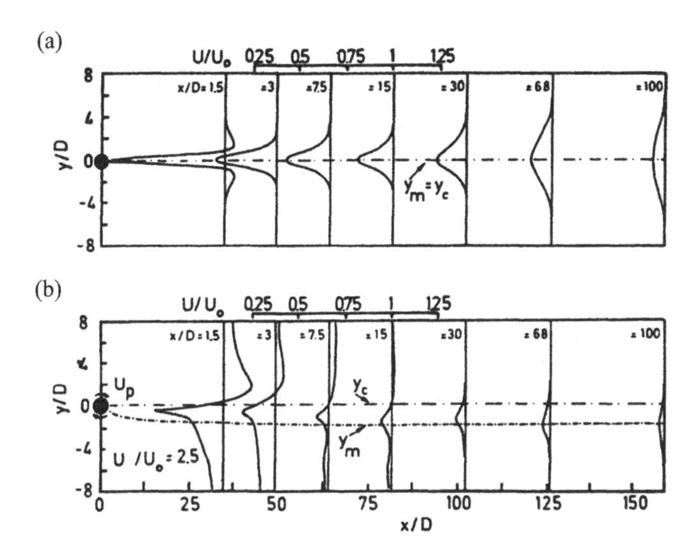

FIG. 24.28. Mean velocity profiles, Re = 9k, (a) $V_r/V = 0$, (b) $V_r/V = 2.5$, Diaz *et al.*, © (1982P) AIAA – reprinted with permission

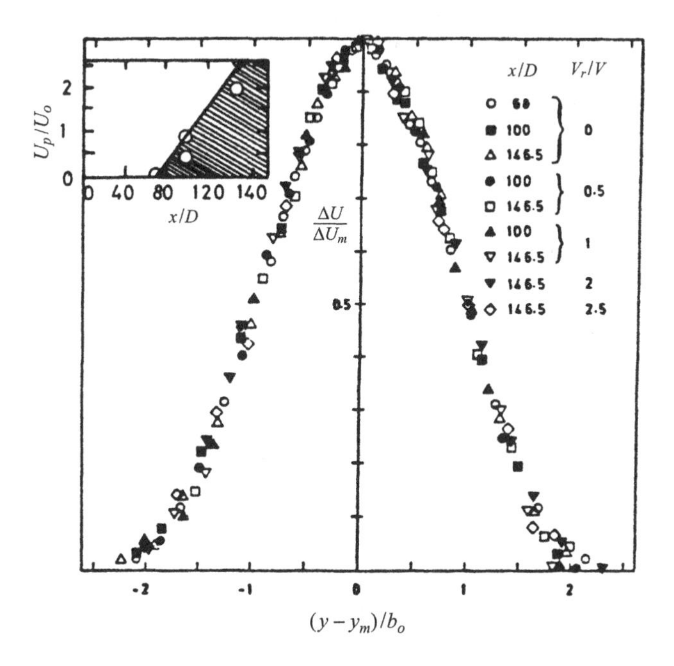

FIG. 24.29. Normalized defect velocity profiles at Re = 9k, Diaz *et al.*, © (1985J) AIAA – reprinted with permission

curve. This indicates that the diffusion and decay of biased asymmetric wakes is self-similar as for non-rotating cylinders.

Variations in u'/V and v'/V for $V_r/V = 1$ and $x/D = 10$ are shown in Fig. 24.30(a,b). A large peak occurs on the side of the cylinder rotating against the free stream velocity. Figure 24.31(a,b) shows the turbulence intensity variations

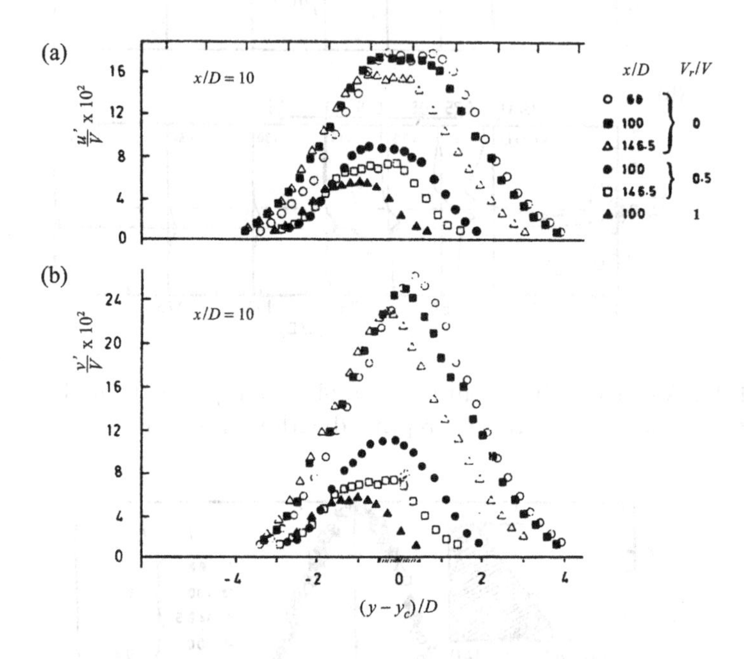

FIG. 24.30. Fluctuating velocity profiles at $x/D = 10$, (a) u_{rms}/V, (b) v_{rms}/V, Diaz *et al.*, © (1985J) AIAA – reprinted with permission

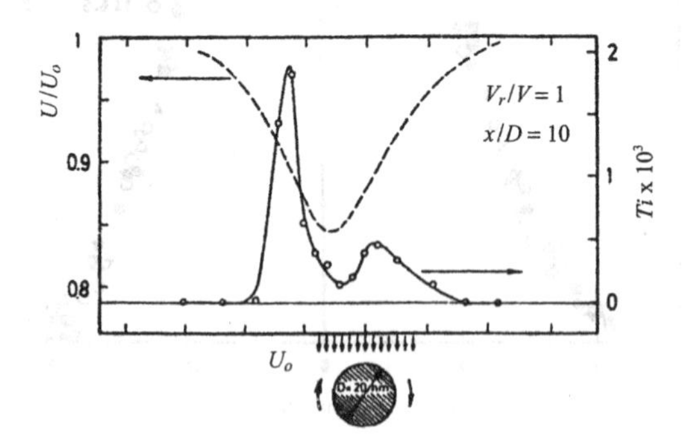

FIG. 24.31. Total kinetic energy of turbulent intensity at $x/D = 10$, Diaz *et al.*, © (1985J) AIAA – reprinted with permission

u' and v' across the wake at $x/D = 10$ for V_r/V from 0 to 2.5. The streamwise component u' is less than the transverse v' turbulence intensity. There is a notable grouping of curves for V_r/V with strong eddy shedding, and with weak or no eddy shedding (low curves). It can be concluded that the absence of eddy shedding for $V_r/V > 2$ results in a four-fold reduction in turbulence intensity, and a three-fold decrease in the maximum defect velocity in comparison with $V_r/V = 0$.

24.4 Applications

24.4.1 *Flettner's rotor ship*

Rotating cylinders, called *rotors* in applications, were used for ship propulsion by Flettner (1925R). He converted the sailship *Buckau*, later named *Baden Baden*, by fitting two cylindrical rotors with small end plates, Fig. 24.32. The rotorship's maiden voyage in the North Sea in midwinter was followed by two crossings of the Atlantic. She passed through terrific storms but the rotors proved to be entirely reliable and efficient.

Flettner (1925R) wrote:

> As far as power is concerned, the rotors accomplished as much as sails of about ten times the area of the former, and under certain circumstances are superior to them. In place of the crew, which handled the sails before the ship is converted, there is an electric controller on the bridge for each of the rotors.

Figure 24.33 shows the wind tunnel models of the rotorship and sailship which convert the same wind force into propulsion.

Figure 24.34 shows the overall loading on two rotors of the rotorship *Baden Baden* in terms of wind velocity. The upper curve represents the constant speed of rotation and the lower for one and both rotors when stationary. For the particular conditions selected, the wind loading on the rotors ceases to increase after the

FIG. 24.32. Converted rotor ship *Buckau*, Flettner (1925R)

FIG. 24.33. Model of sail ship and rotor ship of same effective sail area, Flettner (1925R)

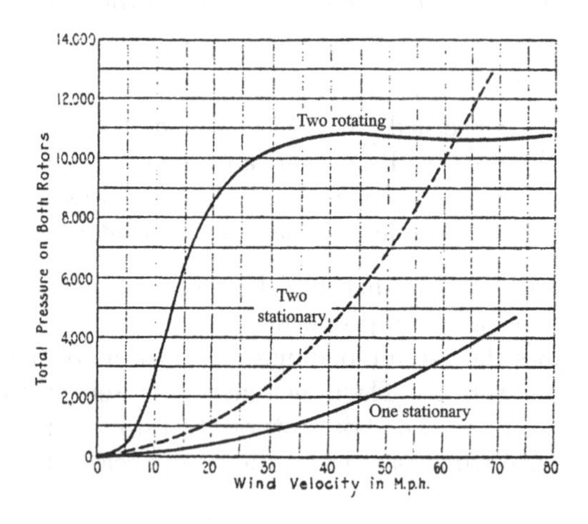

FIG. 24.34. Wind loading on two rotors of *Baden Baden*, Flettner (1925R)

wind has reached a velocity of 48 km/h. The power required to revolve the rotor at the velocity giving maximum C_L is less than 10% of the energy absorbed from the wind.

The second rotorship, the 3 000 tonne *Barbara* (the first ship actually built as a rotorship) was in regular commercial service, and had already made two trans-Atlantic voyages in 1927. According to the reports received, 'her performance is everything which her designers anticipated'. Regrettably, neither sailships nor rotorships could beat propeller-driven ships.

24.4.2 *Rotor windmill*

Flettner also studied windmills and carried out extensive tests on a rotor windmill, see Willhofft (1927R). Figure 24.35 shows the prototype of Flettner's windmill having a four-rotor propeller, $D = 19.77\,\text{m}$, on the top of $H = 32.4\,\text{m}$ tower. The rotors were tapered, $D_1 = 0.85\,\text{m}$ and $D_2 = 0.69\,\text{m}$, having $L = 4.92\,\text{m}$. The housing at the top of the tower contained the electric generator, which was driven by the rotor propeller through 1:100 transmission. The generator was of a special design giving constant voltage at widely varying speeds and fed either into a line or into a storage battery.

Flettner (1925R) intended to build a full-scale rotor windmill power plant generating 136 kW. He planned a 30 m diameter four-rotor propeller with small Savonius rotors attached at the tips of rotors for self-starting. The rotor propeller has important advantages in comparison with conventional propellers:

(i) At high wind velocities, the rotors need not be turned off even when the wind exceeds the value for which the propeller has been designed (40 km/h). The rotors will continue to absorb the same amount of energy from the wind as they absorbed at 40 km/h.

(ii) The rotor maintains a uniform lift force along the whole span at all wind velocities. The latter could be controlled independently from the wind by adjusting V_r.

Föttinger, as discussed in Flettner (1925R), proposed the same concept of rotor propeller for ship propulsion.

FIG. 24.35. Flettner's rotor windmill, Willhofft (1927R)

24.4.3 *Madaras Power Plant Project*

Another interesting attempt to use the Magnus effect was the Madaras Power Plant Project in the USA. Several rotors driven by the wind were patented by Julius Madaras in 1912, and project planning started in the 1920s.

The basic principle of the plant is the Magnus effect produced by the wind acting on rotating cylinders. The plant consists of a large number of rotating cylinders, 33.5 m high ($L/D = 4.2$), Fig. 24.36, mounted vertically on four-wheel trucks in such a way that the Magnus effect drives them along a circular track. The rotation of cylinders upon their vertical axes is secured by means of small electric motors. Upon each track is mounted an electric generator, which is turning as the truck moves forward. By means of a trolley, see Fig. 24.36, the current generated is taken from the generator and transmitted to the power line. Madaras estimated that a 5 MW plant can be built for about \$50/kW (1928), and started to plan a 10 MW plant. The low investment combined with the possibilities of a high use factor makes this type of plant particularly promising.

A pilot model of a single full-scale Madaras rotor is shown in Fig. 24.37, and was actually erected at Burlington, NJ, in 1933.

> Before it had had a chance to prove itself, it blew down in a high wind. With that failure, the attempt to generate commercial power from the wind was quietly put at rest by Detroit Edison Co and its partners, and has not been

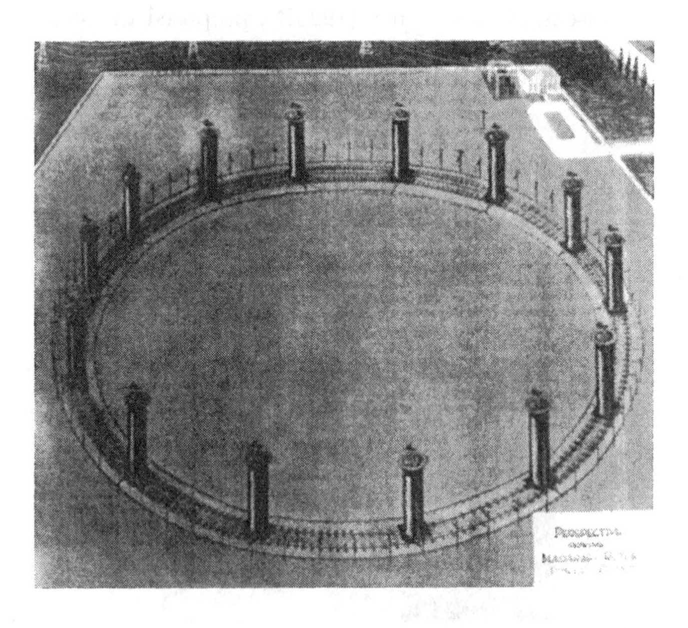

FIG. 24.36. Artist's conception of the Madaras rotor power plant, Hirschfeld (1977R)

FIG. 24.37. Full-scale rotor erected at Burlington, NJ, in 1933, Hirschfeld (1977R)

heard of since.

The wind was and remains a very hazardous, treacherous, and unpredictable element.

24.4.4 *Wallis's 'dam-buster'*

An ingenious device was designed and tested by Sir Barnes Wallis[141] in 1942–3 to breach large gravity dams, and stop electricity generation.

The colossal size and enormous strength of concrete dams made it impossible for conventional bombing to inflict any damage. Moreover, the dams were protected by dual screens of torpedo netting suspended from two rows of large cylindrical steel buoys, Fig. 24.38, effectively preventing any form of underwater attack. Thus, with all conventional methods of attack eliminated, there remained the possibility of using a special weapon delivered from a low altitude flight so that it could bounce on the surface of the water in a similar fashion to the ricochet cannon-balls used in Naval gunnery centuries ago. Wallis (1964P).

Extensive experiments in 1942–3 showed that the number of bounces, and hence the distance travelled over the surface of the water by such a weapon if released from a low height and at high speed, could be very greatly increased by imparting to it a high rate 'backspin' about a horizontal axis when compared with the number of bounces (one or two at most), and distance covered by the

[141]For 20 years (1943–1963) the details of the weapon and aircraft installation had remained a secret. The background notes were written by Sir Barnes Wallis in 1964 as a chief of BAC, Weybridge Division, England.

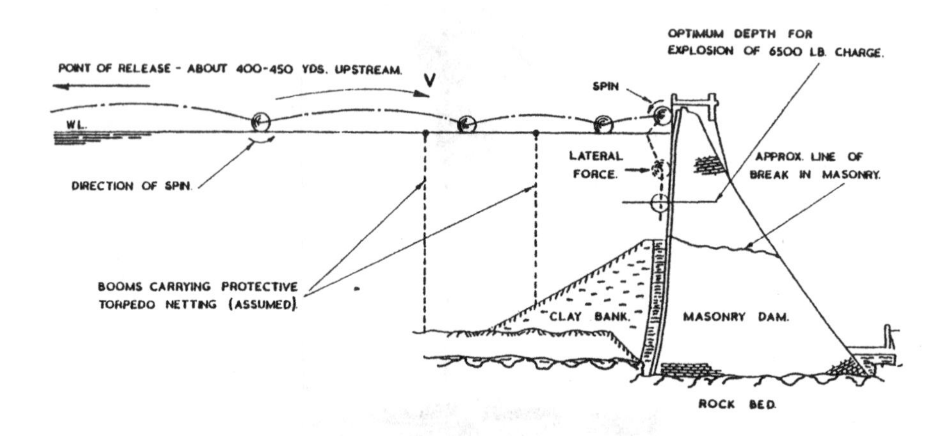

FIG. 24.38. Dam buster, Wallis (1943/1963P), Reproduced by Courtesy of Science Museum, London

sphere that was not spinning. Moreover, the great momentum of the weapon when travelling at high speed ensured that if it did not jump over the lines of buoys, it would easily burst through them, and would reach the dam while still spinning at a high rate. After the bounce back, due to a tremendous impact with the dam, the spin, acting in conjunction with a vertical velocity of sink, then caused the weapon to continue forward in the water until it reached the face of the dam, down which it rolled in close contact, until the critical depth for explosion is reached. Wallis's original (1964P) description.

The original design of the weapon was spherical, but this had to be given up, and a cylindrical container substituted owing to the impossibility of obtaining steel billets for the dyes that would be required to make spherical casings in the time available. The final weapon was 1.5 m long and 1.25 m in diameter. Each end of the cylindrical casing carried a hollow circular track 0.5 m in diameter into which disc-like wheels mounted on the supporting calliper arms were fitted. The total weight of the weapon was 4200 kg of which 3000 kg was charge weight. The rotation opposite to rolling of 500 rpm was imparted by means of a VSG hydraulic motor and belt drive to one of the discs. The height of release was 18 m estimated by the pilot when two spot lights mounted fore and aft of the fuselage came together on the water surface. The correct distance for the release of 360–400 m from the target was judged with the aid of a simple triangular sight using the known distance between the dam towers for a base. Finally, the speed of release of 106–111 m/s was determined by the Lancaster's speed at low diving.

The characteristics of this kind of rotating cylinder were $L/D = 1.2$ and $V_r/V = 0.35$ at the launch. Very low L/D and no end plates combined with the low V_r/V induced a small $C_L \approx 0.3$, and reduced the drag to $C_D = 0.5$–0.6.

However, after impact with the dam and during sinking the value of $V_r/V = 5$–10 induced significant lift $C_L = 3$–4. The Magnus effect forced the bouncing weapon to move back towards the dam during sinking, as sketched in Fig. 24.38 by Wallis (1964P).

24.5 Rotary angular oscillation of a surface

24.5.1 *Physical background*

It has already been pointed out that the distinct feature of flow past rotating cylinders is the upstream and downstream moving surfaces on two opposite sides. If a cylinder is subjected to a rotary finite amplitude angular oscillation around the long axis, then the moving surfaces perform an alternate periodic motion, which consists of an alternate acceleration and deceleration between the fixed angular amplitude. This is another way of boundary layer control of flow separation.

Taneda (1980P) discussed the intrinsic difference between the instantaneous streamlines and time-integrated streaklines in unsteady flows in periodic cylinder wakes. Figure 24.39 shows two methods: aluminium powder and electrolytic precipitation visualization in water around a stationary cylinder at $Re = 100$. The streamlines and streaklines do not coincide because the wake is unsteady. The eddy centres and confluence points are not seen in the streakline pattern. This is because the streamline pattern depends on the reference frame, while the integrated streaklines are invariant.

Taneda (1980P) suggested a tentative topology for an oscillating moving surface. Figure 24.40(a,b,c) shows both streamlines and streaklines for the stationary, accelerating, and decelerating surfaces, respectively. Outside the thin secondary boundary layer, $\delta_s = \sqrt{\nu t}$, where ν is the kinematic viscosity and t is the time, the primary streakline is moved downstream or upstream by the main

FIG. 24.39. Combined aluminium powder and electrolytic precipitation visualization methods, $Re = 100$, Taneda (1980P)

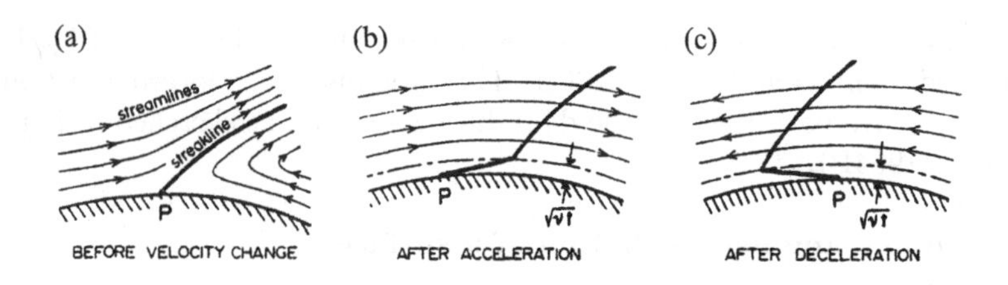

FIG. 24.40. Sketch of streamlines and streaklines boundary, (a) stationary, (b) moving downstream, (c) moving upstream, Taneda (1980P)

flow. Since the secondary boundary layer is very thin at small time t, it is very difficult to determine the integrated streakline separation point. Taneda (1980P) also argued that the integrated streakline separation does not exist on the rotary cylinder.

24.5.2 *Laminar L2 and L3 regimes*

Two parameters are required to describe the angular rotary oscillation of a cylinder: the angular amplitude of oscillation $\Delta\theta$ and the forced non-dimensional frequency parameter[142]

$$N_{\text{osc}} = nD/V \quad \text{or} \quad D\dot{\theta}/2V \tag{24.8}$$

where n is the forced frequency of the rotary oscillation, and $\dot{\theta}$ is the peak rotational rate related to $V_{r\max}/V$.

Taneda (1978J) carried out tests on the rotary oscillating cylinder by using simultaneously aluminium powder and an electrolytic precipitation for flow visualization. Figure 24.41 shows the evolution of the streamline and streakline patterns with time after an impulsive start of the rotary oscillation at $Re = 35$, $N_{\text{osc}} = 6$, and $\Delta\theta = \pm 45°$. Immediately after the onset of oscillation, the two streakline separations from the cylinder begin to move towards $\theta = 180°$, Fig. 24.41(a,b). They reach $\theta = 180°$ and remain there, Fig. 24.41(c,d). The closed near-wake is gradually swept downstream, and L1, the attached flow regime, is established at 35 s.

Another example of the rotary oscillation will demonstrate the effect of N_{osc} on the flow at $Re = 61$ and $\Delta\theta = \pm 45°$. Figure 24.42(a) shows the original eddy shedding pattern for a stationary cylinder $N_{\text{osc}} = 0$. The forced rotary oscillation at $N_{\text{osc}} = 2$ and 2.5 suppresses the near-wake instability, and leads to the stable closed near-wake in Fig. 24.42(b,c). An increase in $(N_{\text{osc}})_c$ to 2.8 transforms the L2 regime to the L1 regime, as in Fig. 24.42(d). Taneda (1978J) observed a similar stabilization for $Re = 111, (N_{\text{osc}})_c = 3.4$, and $\Delta\theta = \pm 45°$. The blockage affected the near-wake stabilization as well as a decrease in the aspect ratio. These effects are summarized in Table 24.1.

[142] Some authors called N_{osc} the Strouhal number based on the forced frequency. This might be misleading because St is related to the eddy shedding frequency.

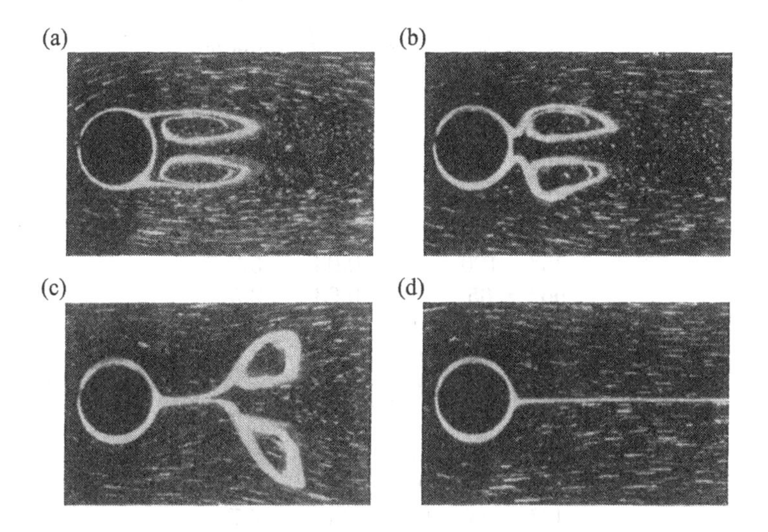

FIG. 24.41. Evolution of streamline and streakline pattern after starting rotary oscillation, $Re = 35$, $\Delta\theta = \pm 45°$, $N_{\rm osc} = 6$, (a) $t = 0\,$s, (b) $t = 4.4\,$s, (c) $t = 10\,$s, (d) $t = 35\,$s, Taneda (1978J)

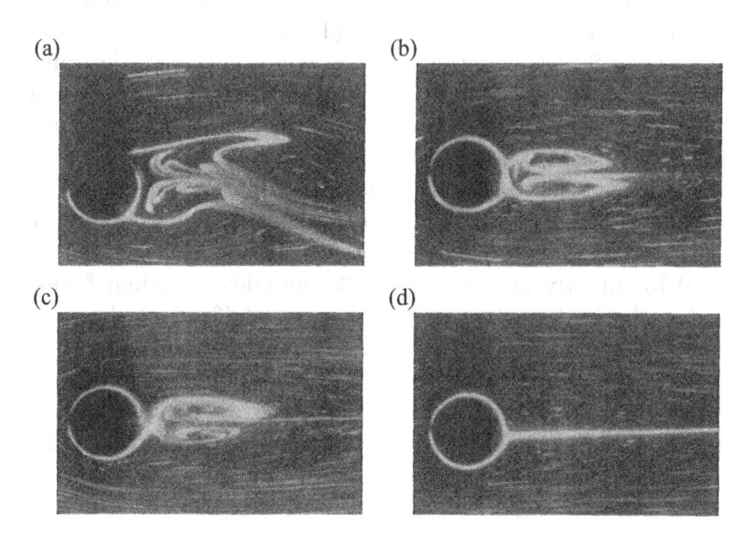

FIG. 24.42. Evolution of streamline and streakline pattern after starting rotary oscillation, $Re = 61$, $\Delta\theta = \pm 45°$, (a) $N_{\rm osc} = 0$, (b) $N_{\rm osc} = 2$, (c) $N_{\rm osc} = 2.5$, (d) $N_{\rm osc} = 2.8$, Taneda (1978J)

Table 24.1 *Observed critical forced oscillation number*

$\pm\Delta\theta$	Re	L/D	D/B	$(N_{osc})_c$
45°	36	30	0.02	2.8
45°	62	30	0.02	2.8
45°	60	25	0.024	3.2
30°	60	25	0.024	4.8
45°	36	14	0.03	5.4
45°	65	14	0.03	5.2
45°	120	14	0.03	5.2
90°	65	14	0.03	2.5
90°	115	14	0.03	2.5
45°	36	14	0.087	7.5
45°	110	14	0.087	8.5
45°	32	6.8	0.087	11.0
45°	111	6.8	0.087	11.0
45°	110	9.7	0.062	7.5
45°	300	9.7	0.062	7.7

24.5.3 *A solution of Navier–Stokes equations*

A pioneering numerical simulation of a laminar flow past a rotary oscillating cylinder was reported by Okajima *et al.* (1975P). They used unsteady two-dimensional Navier–Stokes equations for $Re = 40$ and 80 in the range $0.02 < N_{osc} < 0.30$ and for amplitude $0.2 < V_{r\max}/V < 1$. The periodic wake was not found for $Re = 40$ behind a stationary cylinder. However, the forced rotary oscillation triggered a periodic wake for $Re = 40$ at $St = 0.11$.

Eddy shedding at $St = 0.13$ was calculated behind a stationary cylinder for $Re = 80$. For a small $N_{osc} = 0.02$, eddy shedding continued at $St = 0.13$, and the fluctuating C_L' was superimposed upon the long period of the forced oscillation. For $N_{osc} = 0.15$, slightly above $St = 0.13$, the eddy shedding frequency was synchronized with the forced frequency. Figure 24.43(a,b,c) shows calculated streamlines, surface velocity oscillation, and fluctuating C_L', respectively. The streamlines were calculated for C_L' near the maximum ($t = 28$ s) and at the minimum ($t = 36$ s). Note that the formation of a new eddy on the top side was associated with the former C_L', and a vigorous small one at the bottom side for $C_{L\min}$. The same topology was found behind the stationary cylinder at $Re = 113$k, see Vol. 1, Chapter 5.13, Fig. 5.58. The forced surface and C_L' oscillations are shown in Fig. 24.43(b,c). Note that C_{Lp}' and C_{Ls}' correspond to the pressure and friction components of the C_L'.

Figure 24.44(a–c) shows the flow at $Re = 80$ with the rotary oscillation at $N_{osc} = 0.3$, well beyond synchronization. The calculated streamline patterns were for $t = 10.5$ s far from $C_{L\min}'$, and for 17.56 s near the $C_{L\min}'$. The new eddy on the

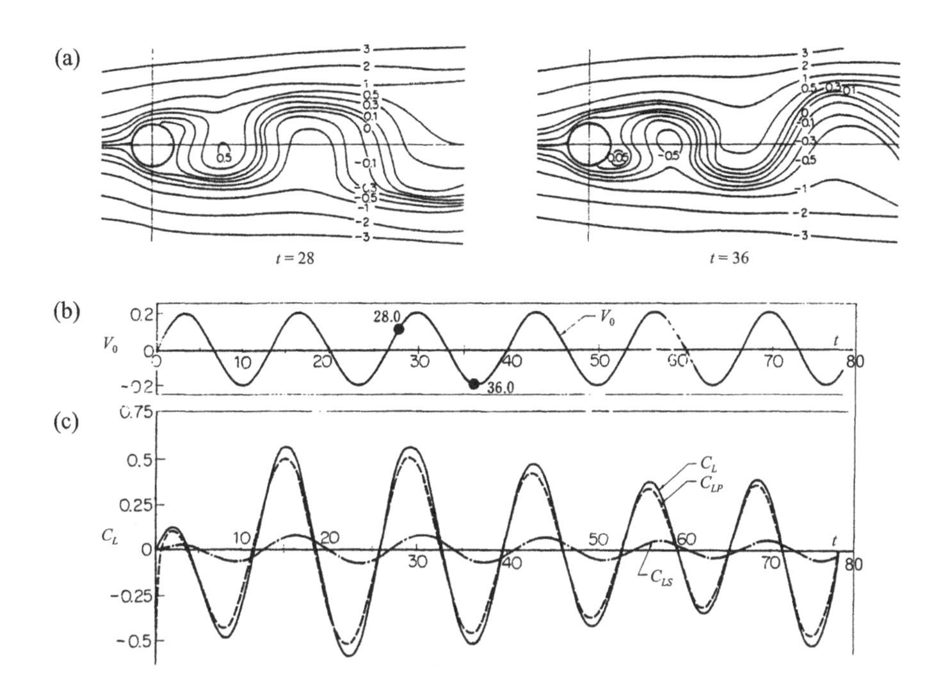

FIG. 24.43. Computation of rotary oscillation for $Re = 80$, $N_{\text{osc}} = 0.15$, $\Delta V = 0.2$, (a) streamlines, (b) surface velocity, (c) lift, Okajima *et al.* (1975P)

lower side was not formed for the former and was in an initial stage of formation for the latter. Figure 24.44(c) shows that the beyond the synchronization range C'_L is irregular both in frequency and amplitude.

Okajima *et al.* (1975P) also carried out experiments in an oil/water towing tank for low/high Re, respectively. A three-section cylinder, $L/D = 12.3$ and $D/B = 0.04$ was forced to perform rotary oscillations in the range $0.05 < N_{\text{osc}} < 0.3$ and $40 < Re < 160$ (oil), $3\text{k} < Re < 6.1\text{k}$ (water). The calculated and measured C'_L and ϕ (phase-lag) were compared for $Re = 40$ and 80 and found to be in good agreement. Figure 24.45(a) shows that the C'_L is unaffected by N_{osc} for $Re = 40$. For all other low Re, C'_L rises gradually, and $C'_{L\max}$ is reached at different values of N_{osc}. This reflects the linear rise in St for the stationary cylinder in the laminar periodic L3 regime. Note that $N_{\text{osc}} > St$, for example, for $Re = 80$, we have $St = 0.13$ and $N_{\text{osc}} = 0.15$ for $C'_{L\max}$. This feature of synchronization will be discussed in Vol. 3, Chapter 33. Another feature is the close relationship between the C'_L and ϕ variations. The maximum in C'_L and the subsequent decrease are matched by the start of a high rate of decrease in ϕ. The latter causes the end of the synchronization range, as will be discussed in Vol. 3, Chapter 33.

For high $Re = 3\text{k}$, 4.56k, and 6.1k, Fig. 24.45(a,b) shows a single curve for

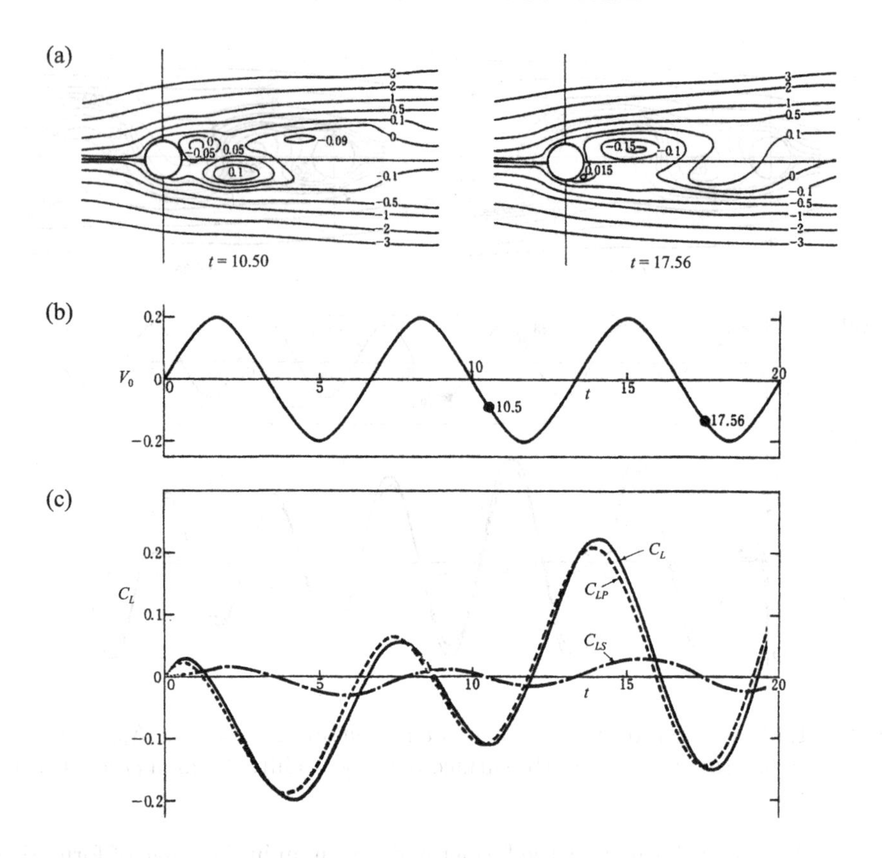

FIG. 24.44. Computation of rotary oscillation for $Re = 80$, $N_{\rm osc} = 0.3$, $\Delta V = 0.2$, (a) streamlines, (b) surface velocity, (c) lift, Okajima *et al.* (1975P)

C'_L and ϕ. The common $C'_{L\max}$ occurs at $N_{\rm osc} = 0.19$. Further development of decreasing C'_L is different from that for laminar wakes. The experimental scatter increases, as shown by the vertical bars covering the fluctuating range of the measured C'_L. Some kind of bistable flow develops, the dash-dot line in Fig. 24.45(a). A beating wave-form is composed of the forced and eddy shedding frequencies for $N_{\rm osc} > 0.21$.

Finally, Fig. 24.45(c) shows the measured variation in C_D and C'_D in terms of $N_{\rm osc}$. For low Re, $C_{D\max}$ appears at $N_{\rm osc}$ where $C'_{L\max}$ is found. Note that a large increase in C'_L in the synchronization range is associated with an only 20% increase in C_D. For high Re, C_D curves are similar to the laminar curves except that $C_{D\max}$ is at $N_{\rm osc} = 0.19$. The fluctuating C'_D is an order of magnitude lower than C_D, and seems to be unaffected by the synchronization.

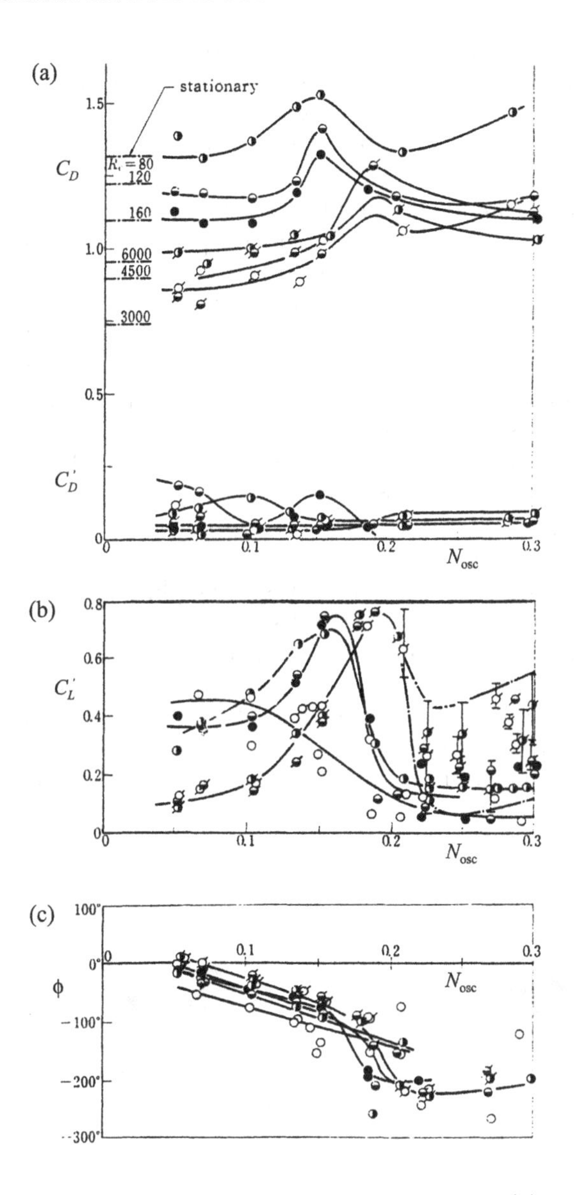

FIG. 24.45. Comparison of computation and experiment, (a) mean and fluctuating drag, (b) lift, (c) phase angle, Okajima *et al.* (1975P)

24.5.4 *Forced rotary oscillation eddy shedding*

It has been discussed in Vol. 1, Chapter 3, that the low-speed mode of eddy shedding is induced by the laminar wake instability, and in Chapter 5, that the high-speed mode of eddy shedding is governed by the distance between the free shear layers. For the forced rotary oscillation, it is possible to induce eddy shedding by the alternate variation in the strength of free shear layers. The synchronization of the eddy shedding frequency to the forced frequency can occur in all three modes, but the extent of the synchronization range is by far the greatest in the third mode.

Tokumaru and Dimotakis (1991J) carried out tests in water at $Re = 15\text{k}$, $L/D = 4.5$, $D/B = 0.22$, and without end plates. The range of the forced frequency of the rotary oscillation was $0.17 < N_{\text{osc}} < 3.3$, and the ratio of the maximum rotational circumferential velocity to the free stream velocity was up to 16. Dye visualization of the eddy street is shown in Fig. 24.46(a,b,c,d) for $Re = 15\text{k}$, $V_{r\max}/V = 8$, and $N_{\text{osc}} = 0,\ 0.3,\ 0.9$, and 3.5, respectively. There is a rapid dye diffusion in the wake behind a stationary $N_{\text{osc}} = 0$ cylinder in comparison with the rotary oscillating cylinder. The length of the eddy formation region L_f is rather long for $N_{\text{osc}} = 0$, and typical for the TrSL2 regime. These two features, i.e. the rate of diffusion and L_f, are drastically reduced for the forced rotary oscillation at $N_{\text{osc}} = 0.3$ in Fig. 24.46(b). The streamwise and transverse spacing of eddies remains similar.

The narrow eddy street with closely spaced eddies is seen in Fig. 24.46(c) for $N_{\text{osc}} = 0.9$. The eddy street is still synchronized at $St = 0.9$. The extent of the synchronization range exceeds by far the observed synchronization at the low- and high-speed modes. It is a new shedding mode induced by the angular rotary oscillation of separation points, periodic variation, and cut-off of vorticity in the free shear layers. Note that as N_{osc} increases, the period of time for the eddy formation is reduced resulting in smaller and weaker eddies. Finally, the

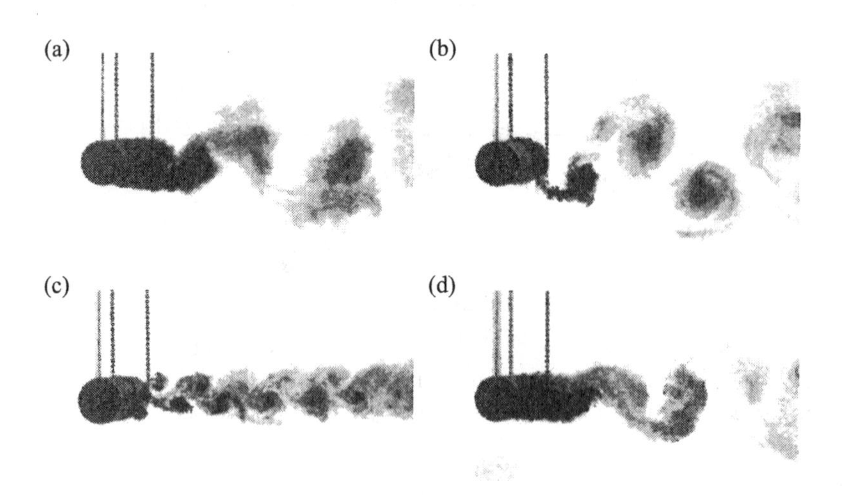

FIG. 24.46. Dye visualization for $Re = 15\text{k}$, $V_r/V = 8$, (a) $N_{\text{osc}} = 0$, (b) $N_{\text{osc}} = 0.3$, (c) $N_{\text{osc}} = 0.9$, (d) $N_{\text{osc}} = 3.3$, Tokumaru and Dimotakis (1991J)

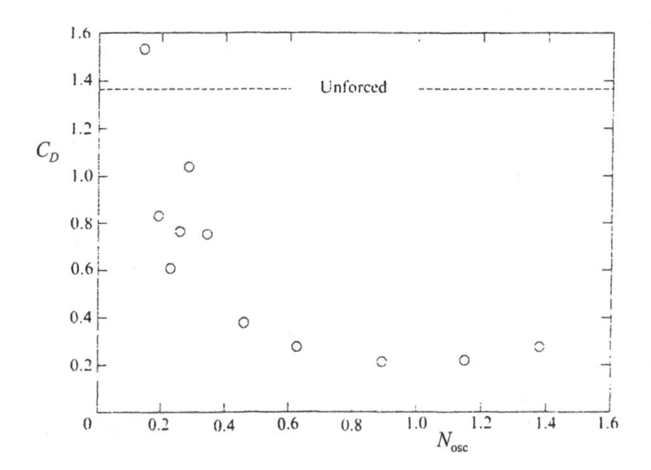

FIG. 24.47. Mean drag coefficient in terms of N_{osc} at $Re = 15$k, Tokumaru and Dimotakis (1991J)

synchronization ends in Fig. 24.46(d) where the eddy street becomes similar to $N_{osc} = 0$ having $St = 0.2$ again.

The rotary oscillation of the cylinder surface affects vorticity generation in the boundary layers. The downstream motion of the surface weakens the vorticity generation, while the upstream surface motion enhances the generation of vorticity. The periodic weakening and enhancement of the boundary layers governs the periodic eddy formation and shedding in the rhythm of the forced rotary oscillation frequency. Hence, this mode is a combination of the forced synchronized vorticity generation and eddy shedding.

The unusual feature of the forced rotary oscillation is seen in Fig. 24.47. The measured C_D for $V_{r\max}/V = 2$ is given in terms of N_{osc}. For the stationary (unforced) cylinder $C_D = 1.35$ (high blockage). The start of synchronization around $N_{osc} = 0.2$ produces a medium C_D value. However, for $0.4 < N_{osc} < 1.4$, where forced oscillations synchronize and generate a narrow wake, the value of C_D is reduced to a low $C_D = 0.2$. Tokumaru and Dimotakis (1991J) suggested that this *active* boundary layer control leading to low C_D may be more effective in reducing loading on structures than the *passive* control capable of suppressing eddy shedding but subject to a high C_D loading.

24.6 Concentric rotating cylinders

24.6.1 *Introduction*

The steady motion of a viscous flow between two concentric rotating cylinders has been referred to as a *Couette flow*. The arrangement of one fixed cylinder and the other rotating was widely used in viscometry in the last century, see

Dryden *et al.* (1956B). The torque to rotate one of the cylinders was linearly related to the viscosity of the fluid when the flow was laminar. In some early experiments by Couette (1890J), the viscosity was measured while the inner cylinder was stationary and the outer cylinder rotated, while Mallock (1896J) used the opposite combination. They both found that the linear relationship between the torque T and the speed of rotation ω changed at higher speed when T started to increase rapidly in a non-linear manner. The departure from linearity was attributed to a transition from laminar to turbulent flow. The critical speed ω_c for transition was much lower when the inner cylinder D_i was rotating, and the outer D_o was at rest.

The instability of Couette flows was treated theoretically by Rayleigh (1916J). He examined a basic swirling flow of an inviscid fluid, and by a simple physical argument derived a criterion for stability. An inviscid Couette flow was unstable whenever the square of the circulation decreased outwards, $\mathrm{d}(r^2v^2)/\mathrm{d}r < 0$. Rayleigh's criterion did not hold when cylinders rotated in the opposite directions. A full derivation of Rayleigh's theory, and subsequent developments, were given by Drazin and Reid (1981B).

An outstanding theoretical and experimental paper on concentric rotating cylinders was published by Taylor (1923J). The excellent agreement between the theoretical and experimental points was remarkable. He took into account viscosity to improve Rayleigh's stability criterion, verified his calculations experimentally, and observed the secondary laminar flow regime in the form of a column of toroidal eddies, now referred to as *Taylor eddies.* The steady and regular secondary laminar flow was triggered by the onset of instability.

Coles (1965J) discovered a series of other transitional and turbulent flow regimes in the form of secondary and tertiary instabilities. Particularly fascinating was the alternation of laminar and turbulent 'spirals' in the form of the 'barber pole', now referred to as *Coles' spirals.*

24.6.2 *Taylor's theory and experiment*

The remarkable feature of Taylor's (1923J) paper is a complete theoretical derivation and analysis supported by thoroughly designed and executed experiments complemented with flow visualization. The paper was divided into two parts, theory and experiment. Full derivation of the theoretical model was given in the paper and subsequently in textbooks, for example Drazin an Reid (1981B). Here, only an outline of the fundamental assumptions and physical arguments will be given.

Taylor (1923J) reconsidered Rayleigh's criterion of stability derived for the inviscid fluid. If the flow was unstable according to Rayleigh's criterion, then it might be expected to continue to be stable due to the viscous effect. Taylor's starting argument was that the effect of viscosity could not be neglected in a Couette flow. He considered the full unsteady Navier–Stokes equation in a polar coordinate system with boundary conditions imposed by the rotating concentric cylinders. These implied that the mean axial and radial velocity components would be zero. This greatly simplified the Navier–Stokes equations to

$$\frac{1}{\rho}\frac{\mathrm{d}p}{\mathrm{d}r} = \frac{v^2}{r} \quad \text{and} \quad \nu\left(\frac{\partial v}{\partial t} + \frac{v}{r}\frac{\partial v}{\partial \theta}\right)\left(\frac{\partial v}{\partial r} + \frac{v}{r}\right) = 0 \tag{24.9}$$

This implied that a viscous Couette flow is described by $\log v = Ar + B/r$.

The next argument was that the perturbed flow was axisymmetric. This was based on the observation of a steady secondary flow in the form of alternating toroidal Taylor eddies spaced at an equal pitch along the cylinders. This was an important observation, which allowed a significant simplification

$$\begin{gathered}\left(\frac{\partial u'}{\partial r} + \frac{u'}{r}\right)\left[\frac{\partial u'}{\partial t} - \nu\left(\frac{\partial u'}{\partial r} + \frac{u'}{r}\right)\right] = 2\frac{\partial^2 v'}{\partial z^2} \\ \frac{\partial v'}{\partial t} - \nu\left(\frac{\partial^2 v'}{\partial r^2} + \frac{1}{r}\frac{\partial v'}{\partial r} + \frac{\partial^2 v'}{\partial z^2}\right) = -\left(\frac{\partial^2 v'}{\partial r^2} + \frac{1}{r}\frac{\partial v'}{\partial r} + \frac{\partial v'}{\partial z}\right)u'\end{gathered} \tag{24.10}$$

These equations could be further simplified by analysing the normal mode disturbances.

$$u', v' = (u, v)\exp(st + ikz) \tag{24.11}$$

where k was the axial flow wave number. A further important physical argument was the adoption of length scale as the gap between the rotating cylinders. This allowed a narrow-gap approximation and simplification. The final argument used the principle of the stability exchange. This was restricted to $\omega_i/\omega_o \geqslant 0$.

The second part of Taylor's (1923J) paper was the experiment. Taylor (1923J) used a precision cut glass for the outer cylinder so that the flow inside was visible by injecting dye at six levels. He used various inner cylinders in the range $0.73 < D_i/D_o < 0.90$. To avoid end effects the cylinder height-to-gap ratio was $81 < H/G < 382$. The ratio of angular velocities ω_i/ω_o was constant during each run by using the free-fall weight and electric motor. The angular velocities were gradually increased in magnitude until the onset of instability was observed. The angular velocities ω_o and ω_i at the onset of instability were distinct and repeatable on different runs with an accuracy of 1%. The wavelength of the axisymmetric mode of instability was also measured.

Figure 24.48(a) shows the flow visualization of toroidal eddies formed in the gap between the rotating cylinders. Figure 24.48(b) shows calculated streamlines in terms of the stream function. There is full agreement between experiment and theory.

Figure 24.49 shows the original Taylor stability plot of the ω_i/ν vertical axis and the ω_o/ν horizontal axis with negative values corresponding to the counter-rotating cylinders. The straight dashed line designates Rayleigh's stability criterion derived for the inviscid fluid. The theoretical calculated points (open circles) and experimental points (closed circles) are virtually on the same

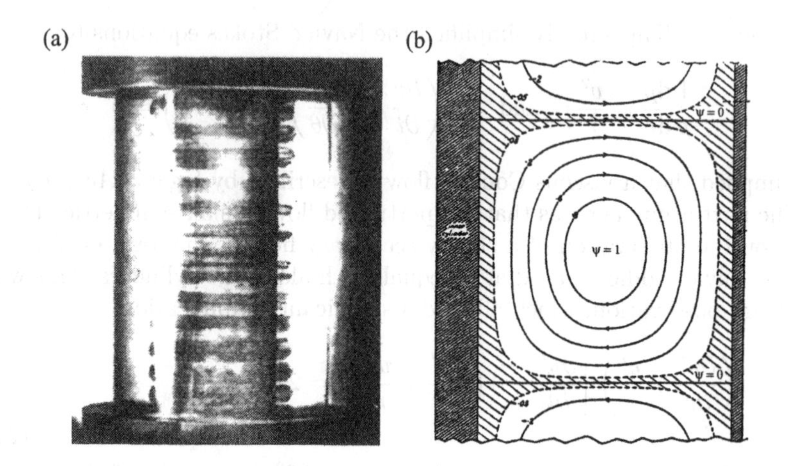

FIG. 24.48. (a) Flow visualization of ring eddies, (b) theoretical streamlines, Taylor (1923J)

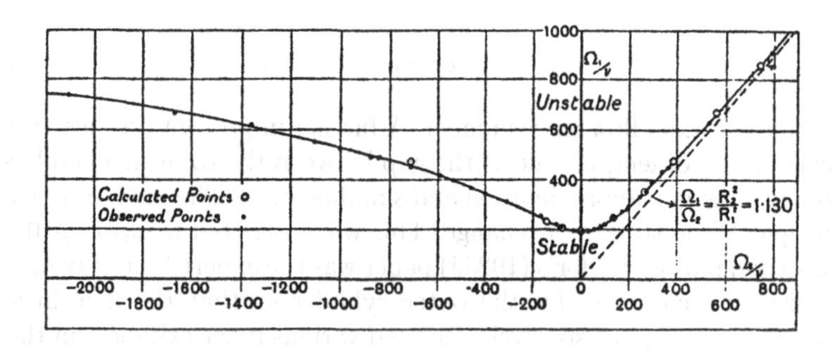

FIG. 24.49. Observation and narrow-gap calculation of flow stability, Taylor (1923J)

curve (full line). The curve separates the stable (below) and unstable (above) flow regions. Taylor (1923J) wrote: 'The accuracy with which these points fall on the curve appears remarkable when it is remembered how complicated was the analysis employed in obtaining them.'

24.6.3 *Coles' further transitions*

Coles (1965J) carried out extensive experiments in the range $0 < Re_i < 20\text{k}$ and $-80\text{k} < Re_o < 10\text{k}$, where $Re_{i,o} = \omega_{i,o} R_{i,o}^2/\nu$, and the subscripts i and o refer to the inner and outer cylinder, respectively. The Re range is presented in three scales in Fig. 24.50(a,b,c), the full 1:1, 1:3, and 1:20, respectively. Note that the last scale corresponds to Taylor's range of variables. The open circles correspond

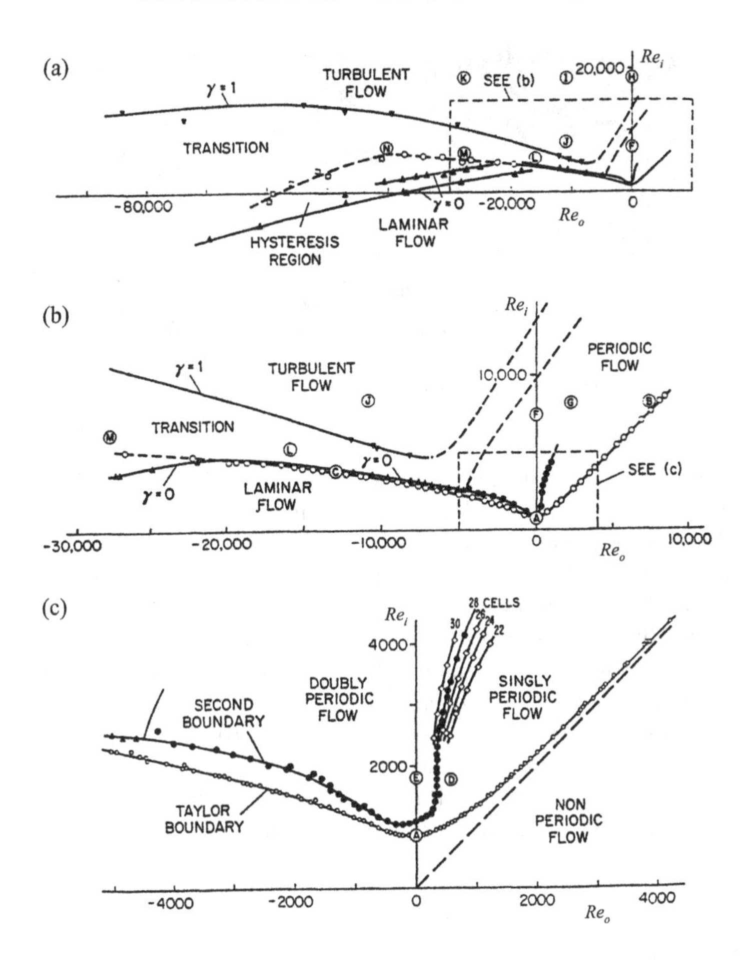

FIG. 24.50. Different flow regimes in Couette flow, (a) large-scale plot, (b) medium, (c) small, Coles (1965J)

to visual observation of the onset of Taylor's (1923J) instability and the curve to Taylor's theoretical model. The second set of points (full circles) corresponds to a rotating pattern of tangential waves superimposed on the original Taylor eddies. This is the double periodic flow which was first described qualitatively by Taylor (1923J) and later by Schultz-Grünow and Hein (1956J). Coles (1965J) established quantitatively the boundary between the single and double periodic flow. As seen in Fig. 24.51, an important feature of the double periodic flow is a phase shift between eddies in adjacent rows. The tertiary flow is essentially a combined rotation and translation in a stream tube of periodically varying area.

The end of the transition region in Fig. 24.50(a–c) is estimated by the disappearance of a smooth laminar part of the hot-wire signal, described by the

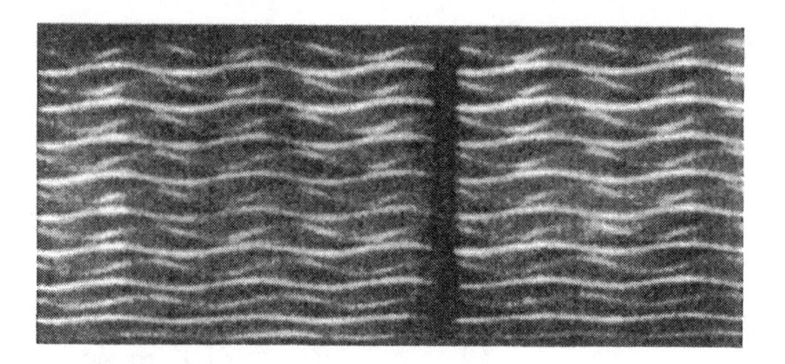

FIG. 24.51. Doubly periodic secondary instability, Coles (1965J)

intermittency factor $\gamma = 1$. The laminar range extends through the single and double periodic flow up to the lower-dashed line (tentative). There is a considerable delay between the appearance of the double periodic flow for $\omega_i > 0$, and a small delay for $\omega_i < 0$. There is another distinction, namely a hysteresis in the appearance and disappearance of double waviness.

Figure 24.50(c) shows that the boundary separating the single and double periodic flow regimes in the right quadrant is not unique, but depends on the axial wave number, which can itself take on several discrete values. When the waves first appear their angular velocity is close to that of a rotating inner cylinder, in agreement with the theoretical estimate by Di Prima (1961J). As the speed increases, however, the dimensionless wave velocity approaches a value of approximately 0.34. This value is virtually constant, and does not depend on Re_i.

24.6.4 *Classification of flow regimes*

Coles (1965J) defined transitional regimes using two integers denoting the number of Taylor eddies, and the number of tangential waves superimposed on them. A change in flow regime implies a discrete change in one or both integers. The non-uniqueness is exhibited by the existence of a number of hysteresis loops, in which the changes from one regime to another take place as the speed ω_i is slowly increased or decreased. A total of 74 such regimes are identified by Coles (1965J) (each observed at least three times). Figure 24.52 shows the classification of flow regimes. The base is a rectangular grid in terms of the number of Taylor eddies and the number of tangential waves. The height above the base is taken to be proportional to ω_i. Each change in flow regime can be represented by a horizontal bridge from one column to another.

For flows dominated by the rotation of the outer cylinder, Fig. 24.50(a) shows that the laminar region is extensive. The opposite trend is seen for $\omega_o < 0$; the transition region is between the curves $\gamma = 0$ and 1, and is fully laminar and fully turbulent, respectively. For $\omega_o < 0$, the secondary flow boundary $\gamma = 0$ turns downwards to cross the Taylor boundary, and beyond this intersection there is

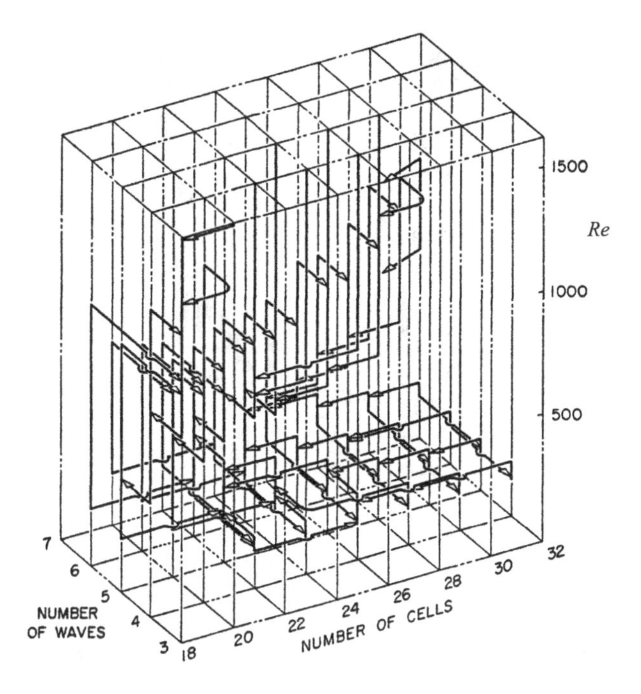

FIG. 24.52. Classification of Couette flow regimes, Coles (1965J)

a region of hysteresis and non-uniqueness.

Two different fluids are used to obtain the data in Fig. 24.50(a). Thus there are two $\gamma = 0$ curves. Below the $\gamma = 0$ line the turbulence, if present, will always decay. Above the $\gamma = 0$ line the turbulence, if present, will persist indefinitely. If the flow is not disturbed, the Taylor boundary can still be observed if it is approached from the laminar side. Once turbulence has appeared it takes control over the flow. This is designated as the hysteresis region in Fig. 24.50(a).

Among various transitional flow regimes there is a special configuration which Coles (1965J) called *spiral turbulence.* Figure 24.53 shows a spiral band of turbulence (either right-handed or left-handed) which rotates steadily at the mean angular velocity of the two cylinders without changing its shape or losing its identity.

The spiral moves in nearly circular paths so that the fluid element near either wall repeatedly traverses the spiral pattern, and participates alternately in the laminar and turbulent regions. It follows that there have to be two kinds of interfaces separating regions of laminar and turbulent flow. One kind represents the transition from laminar to turbulent flow, and the other kind represents the inverse transition with turbulent regions becoming laminar. The two kinds of interface are sharply defined, and propagate at comparable speeds with respect to the fluid velocity.

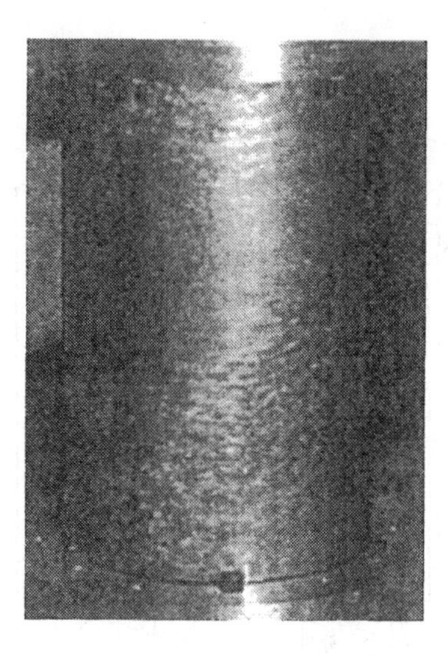

FIG. 24.53. Spiral turbulence region in Couette flow, Coles (1965J)

24.7 Boundary layer control by suction and blowing

24.7.1 *Suction*

In the same year, 1904, Prandtl delivered a lecture on the 'thin layer' along a boundary, and demonstrated the effect of boundary layer removal by suction on the flow around a circular cylinder. Figure 24.54 shows the water surface flow visualization past a cylinder with a slot on one side. The small suction through the slot has a great influence on the flow past a cylinder. The separation is forced to take place at the slot, $\theta = 140°$, eddy shedding is absent, and asymmetry results in a lift force at right angles to the free stream. The reduced size of the near-wake indicates a decrease in drag force.

It should be noted that the flow pattern around a cylinder with suction is analogous to the flow around a rotating cylinder, see Fig. 24.1. The slow flow in the boundary layer is removed by suction, and energized by rotation for the latter. The same effect could be achieved on a non-rotating cylinder by blowing a jet tangential to the cylinder surface. Hence, boundary layer control can be achieved by three apparently dissimilar methods: a rotating cylinder, suction, and blowing the boundary layer.

The early experiments by Prandtl on suction in 1904 were not taken up by subsequent researchers. The revival of interest in suction and blowing came about forty years later from works by Thwaites (1948J), Cheeseman (1968J), and others. The main attraction of boundary layer control was the scope to increase considerably the lift on streamlined and bluff bodies independently of the angle of incidence.

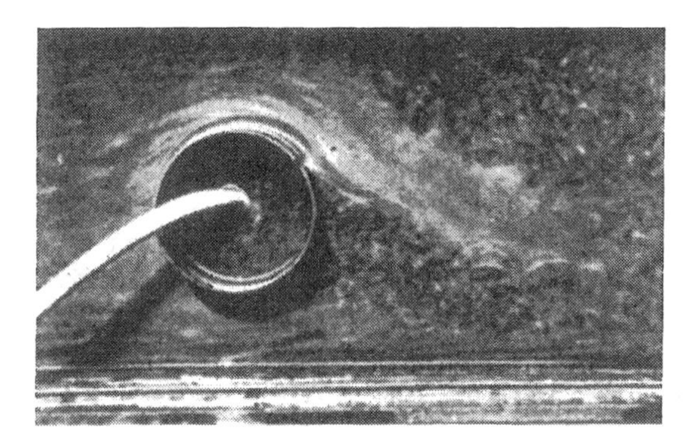

FIG. 24.54. Flow visualization of a cylinder with a suction slot at $\theta = 140°$, $Re = 4.1$k, Prandtl (1961B)

24.7.2 *Porous surface suction*

It is well known that the potential flow past circles does not agree with real flow because the viscosity generates a boundary layer, and fundamentally alters the flow pattern around a circular cylinder. However, it is less well known that the potential flow can be simulated in a real flow by removing the boundary layer by surface suction.

Thwaites (1948J), Pankhurst and Thwaites (1950P), and Hurley and Thwaites (1951P) carried out a series of experiments on a porous cylinder with and without suction. The cylinder was made of sintered bronze and had $L/D = 5$[143] and $D/B = 0.06$. The suction coefficient $C_Q = Q/VDL$, where Q is the suction flow rate, V is the free stream velocity, and DL is the projected area of the cylinder. The C_Q was combined with Re to form a range $15 < C_Q Re^{1/2} < 45$.

Figure 24.55 shows the C_p distribution at $Re = 101$k with and without suction. For the latter, a typical C_p distribution for a small aspect ratio is obtained ($C_{po} = 1$, $C_{p\min} = -1.2$, $C_{pb} = -1$). When a suction of $C_Q Re^{1/2} = 25.2$ is applied, $C_{p\min} = -3.2$ and $C_{pb\max} = +0.3$ positioned at $\theta = 160°$. The C_p distribution is found to be unsteady. The inability of the suction to prevent separation suggests that some type of instability occurs. Thwaites (1948J) tried to stabilize the observed oscillation by placing a flap[144] in the form of a short splitter plate at $\theta = 180°$. Figure 24.55 shows that with the flap, $0.04 < f/D < 0.25$ and sufficient suction ($C_Q Re^{1/2} > 30$) the separation is entirely prevented, and a complete pressure recovery is obtained at $C_{p180} = 1$. A remarkably close approximation to the potential flow is achieved. Two discrepancies are $C_{p\min} = -4$ on one side and $C_D = 0.088$ may be due to non-uniform porosity.

The wake traverses confirm the other features of the potential flow. Figure

[143]The overall span was restricted because of the limited volume flow rate of the fan.

[144]Now referred to as Thwaites' flap.

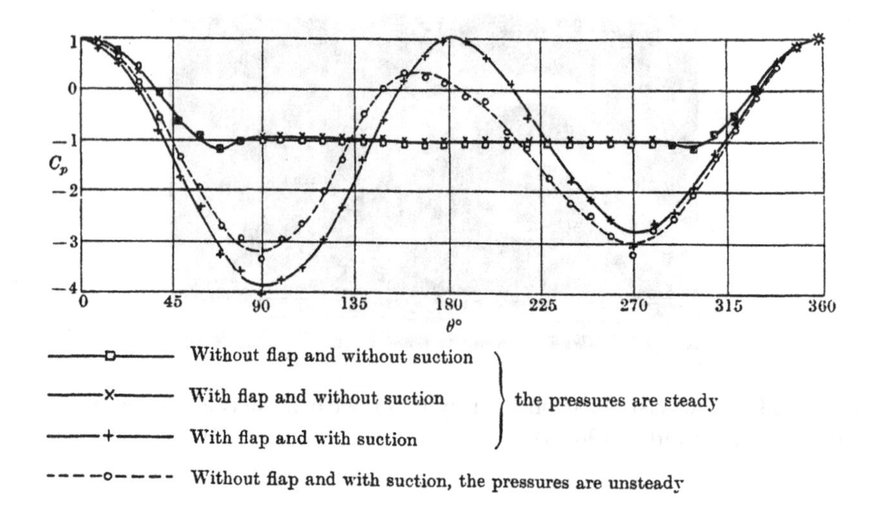

FIG. 24.55. Mean pressure distribution around a porous cylinder with and without flap, Pankhurst and Thwaites (1950P)

24.56 shows the highest velocity defect for the cylinder without suction and flap. The flap without suction slightly decreases the peak in the velocity defect. A considerable reduction in the velocity defect is achieved by suction without the flap, and a complete annihilation of the velocity defect is achieved with the flap and suction. The D'Alembert paradox is confirmed that without the boundary layer and wake the drag on the cylinder is actually zero.

Pankhurst and Thwaites (1950P) also examined the effect of the suction flow quantity C_Q on the C_p distribution and C_D. Figure 24.57 shows the C_p distribution for four values of $C_Q Re^{1/2}$ at two Re with the flap $f/D = 0.17$. Without the suction, $C_D = 0.96$ at $Re = 101\text{k}$, $L/D = 5$, and $C_Q Re^{1/2} = 14$, 17, and 42, with values of $C_D = 0.69$, 0.47, and 0.01, respectively. The value $C_Q Re^{1/2} = 30$ is quoted as optimal by the authors.

24.7.3 *Thwaites' flap*

Thwaites (1948J) fitted a thin flap in contact with and perpendicular to the cylinder surface. The purpose of the flap was to fix the position of the rear dividing streamlines, as sketched and visualized in Fig. 24.58(a,b), respectively. The flap provided some form of separation control. The circulation and lift were determined by the flap's angular position behind the cylinder. The short flap experienced no force as it coincided with the dividing streamline.

The measured pressure distributions over the midspan of the porous cylinder at $Re = 67.6\text{k}$ with the suction and flap $f/D = 0.25$ are shown in Fig. 24.59. The flap angular position α is measured from $\theta = 180°$ ($\alpha = \theta - 180°$). For $\alpha = 0$,

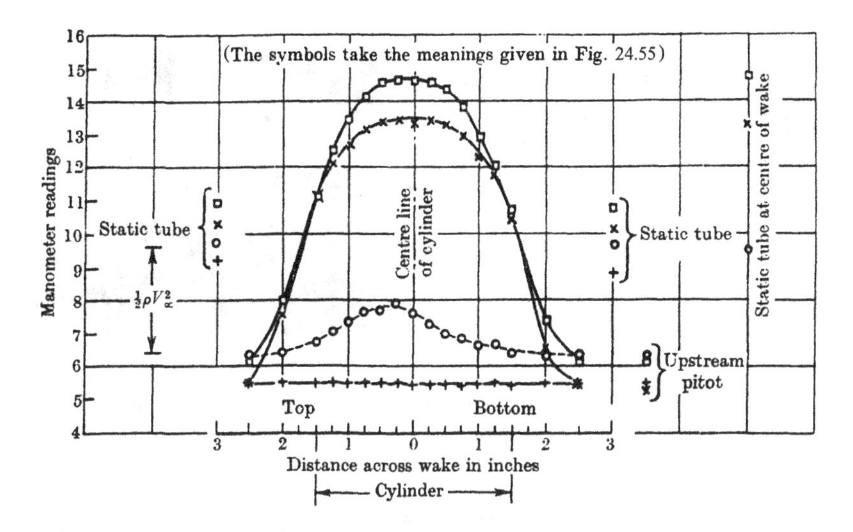

FIG. 24.56. Static pressure profiles across the wake with and without suction, Pankhurst and Thwaites (1950P)

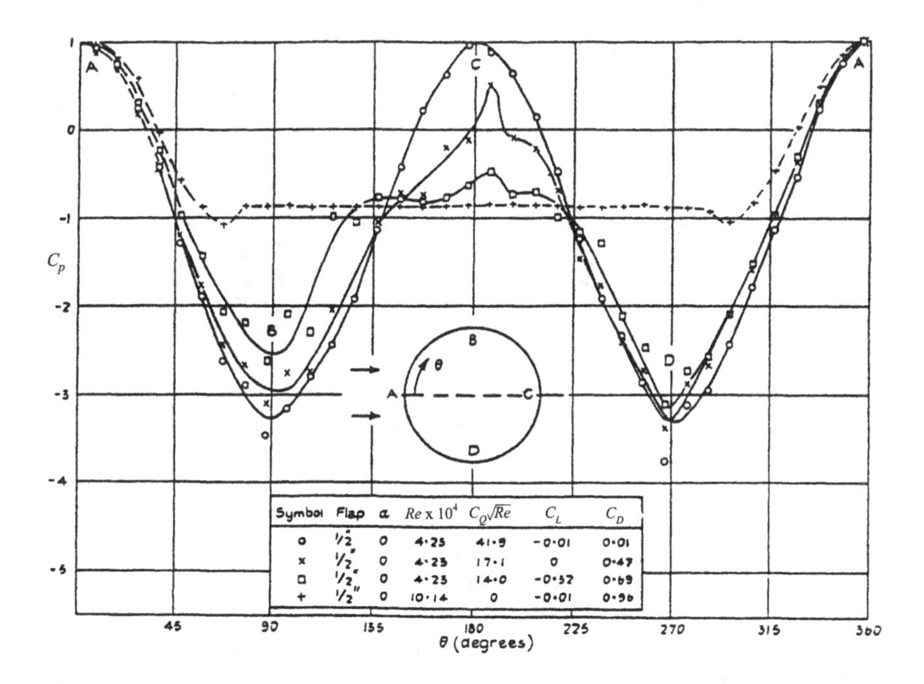

Symbol	Flap	α	$Re \times 10^4$	$C_Q\sqrt{Re}$	C_L	C_D
o	1/2"	0	4·25	41·9	-0·01	0·01
×	1/2"	0	4·25	17·1	0	0·47
□	1/2"	0	4·25	14·0	-0·37	0·69
+	1/2"	0	10·14	0	-0·01	0·96

FIG. 24.57. Mean pressure distribution around a porous cylinder for $Re = 4.2\text{k}$ to 10.1k, $0 < C_Q Re^{1/2} < 41.9$, Pankhurst and Thwaites (1950P)

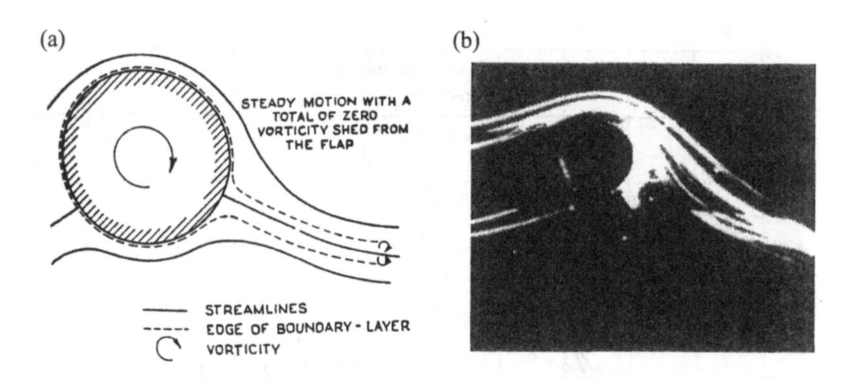

FIG. 24.58. Thwaites' flap, (a) streamlines, (b) smoke visualization, Thwaites (1948J)

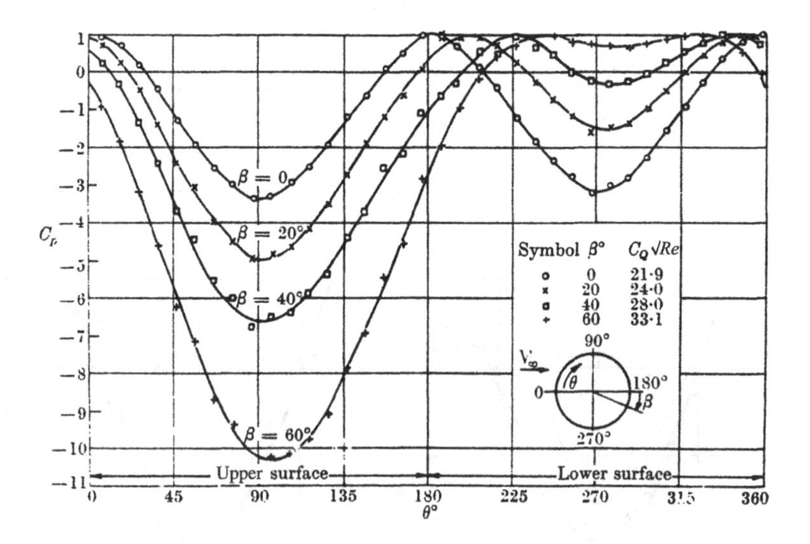

FIG. 24.59. Mean pressure distribution around a cylinder with suction and Thwaites' flap, $Re = 6.7$k, Pankhurst and Thwaites (1950P)

an almost potential C_p distribution is obtained for $C_Q Re^{1/2} = 21.9$ suction. For $\alpha = 20°,\ 40°$, and $60°$, $C_{p\min} = -5,\ -6.5$, and -10.4, and the rear stagnation point coincides with the flap location.

The experiments on porous cylinders were extended by Hurley and Thwaites (1951P) to the measurement of the boundary layer characteristics, which were then compared with the results of theoretical analysis. Fair agreement was obtained for the distribution of momentum thickness, see Thwaites (1960B).

24.7.4 *Jet-blowing*

Boundary layer control by blowing a jet of air through a narrow tangential slot dates back to 1921, see Lachmann (1961B). The high-speed jet introduced through the slot re-energizes the boundary layer and delays its separation. The latter lowers the pressure on the slot side and introduces a lift force on the cylinder. From a practical point of view, boundary-layer blowing is more convenient than suction because the compressed air can be tapped from any engine. Another reason is that blowing tends to keep the injection slot clear of dirt, in contrast to suction, which can clog the slots.

The number of parameters that affect the flow past a cylinder with jet-blowing is very large. They can be grouped as follows:

(i) The slot position, number, design, size, etc.

(ii) The jet velocity, momentum, uniformity along the span, pressure, temperature, etc.

(iii) The boundary layer state (laminar, transitional, turbulent), δ, Re, Ma, surface roughness, etc.

The most important parameter in the first group is the angular position of the slot, α_s, measured from $\theta = 90°$. When two slots are used the angular separation between them, $\Delta\alpha_s$, is an additional parameter. A wide variety of slot designs has been used, and some are displayed in Fig. 24.60. The geometry of nozzles is important for a stable and uniform jet. Another parameter is the slot width.

The most important parameter in the second group is the jet momentum. It is expressed through the momentum coefficient:

$$C_J = \frac{\text{jet momentum per unit span}}{\text{free stream dynamic head} \times \text{diameter}} \tag{24.12}$$

The uniformity of the jet is adversely affected by the sharp-cornered nozzles. The ratio of the jet to the free stream velocity is analogous to V_r/V, but this parameter is not used for the jet-blowing tests. The third group attracted proper attention, as will be described shortly.

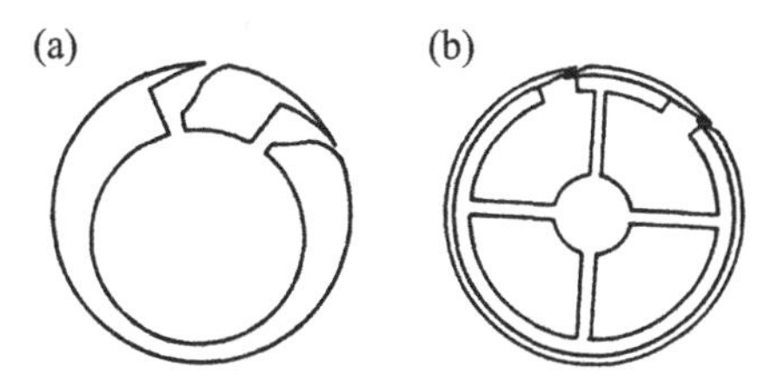

FIG. 24.60. Slot design, (a) Dunham (1968J), (b) Cheeseman (1967J)

24.7.5 *Lift and drag forces*

The effect of single-jet blowing depends on the position of the slot. It was found that the best position is slightly behind the separation for the no-blowing slot. The jet re-energizes the boundary layer, and postpones separation. The suction is increased on the jet-side surface, and decreased on the opposite side. Therefore, the lift force is generated, and the drag force is reduced.

If the slot is positioned upstream of the separation, less lift is generated by a given jet momentum. If the slot is too far beyond the separation, it cannot prevent separation, and the lift drops sharply (stalls). The angle of incidence of the slot at which the lift reaches its maximum value for a given momentum is about $\alpha_s = 5°$ for the subcritical regime, and $\alpha_s = 25°$ to $30°$ for the supercritical regime, according to Dunham (1968J).

Even the presence of the unblown slot influences the cylinder's lift by affecting the upper surface boundary layer. Thus, the lift is not zero in the no-blowing condition. Even a small amount of slot-blowing suppresses the periodic eddy shedding, and improves the spanwise uniformity of the flow.

A second slot re-energizes the boundary layer again, and delays separation further. The second slot is best placed where the boundary layer produced by the first slot would separate, or slightly downstream. As with the single slot, if it is too far back, jet-blowing becomes ineffective at inducing reattachment.

24.7.6 *Dunham's theoretical model*

Dunham's (1968J) jet-blowing model was a 'conglomerate' of several models, and semi-empirical relationships dealing separately with the pressure distribution, development of the laminar boundary layer, transition, the turbulent boundary layer, separation, and superimposed jet-blowing. The main objective of the model was to predict the lift and drag coefficients on a circular cylinder with jet-blowing. Further sub-models included the modified potential flow, the semi-empirical calculation of laminar and turbulent boundary layers, semi-empirical criteria for the transition, and separation of the compressible inviscid jet-blowing.

The potential flow around a circle with circulation as simulated by the doublet, bounded vortex, and the free stream has been known to be inadequate. As an improvement, the displacement effect of the real wake was simulated by adding a source S placed at $\theta = 180°$ and $z = d/2$. This also required a sink of the source strength $-S/2$ placed at the centre of the circle in order to maintain the circle-streamline. The complex velocity potential ω was given by

$$\frac{\omega}{V} = z + \frac{D^2}{4z} + \frac{iDC_K}{4\pi} \ln z + \frac{DC_Q}{4\pi} \ln(z - Z) - \frac{DC_Q}{8\pi} \ln z \tag{24.13}$$

where $Z = -\frac{1}{2}D \exp(-i\pi)$, K is the vortex strength, and S is the source or sink strength. The tangential and normal velocity components are

$$v_r = 0, \qquad v_t = V\left[2\sin\theta + \frac{K}{2\pi} - \frac{S}{4\pi}\cot\tfrac{1}{2}(\pi - \theta)\right] \tag{24.14}$$

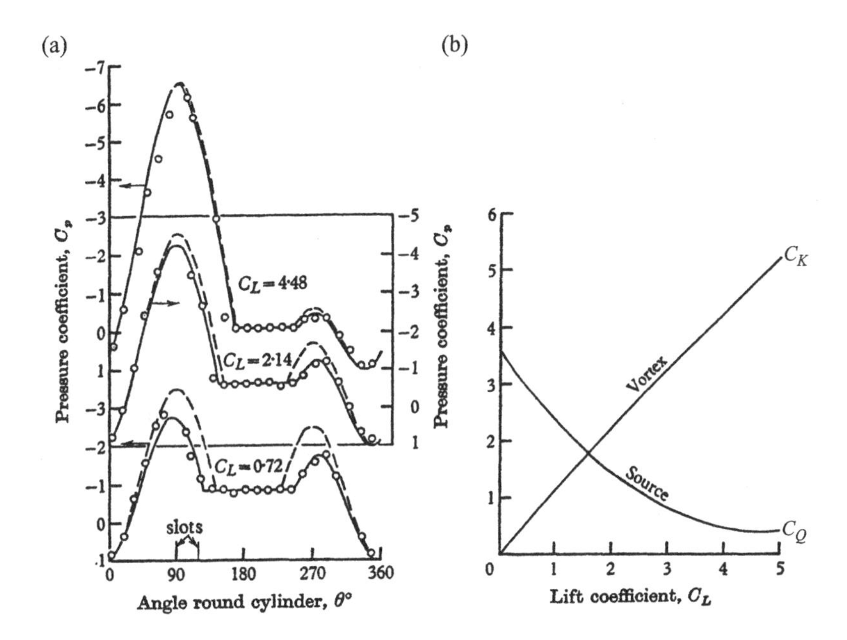

FIG. 24.61. Experiment and theory, (a) mean pressure distribution, (b) required match of vortex and source strength, Dunham (1968J)

The pressure coefficient C_p calculated from this equation is shown in Fig. 24.61(a). The agreement with the measured C_p is achieved by adjusting the source S and vortex K strengths. As C_L increases, the wake contracts, and the value of S diminishes, as shown in Fig. 24.61(b). Eventually, for $C_L > 5$, there is no need for the source, and the simple potential model of a circle with circulation becomes adequate.

The laminar boundary layer development is calculated using Thwaites's (1960B) model step by step from the stagnation point to the transition point

$$\frac{\delta}{D} = \left[\frac{0.45}{u^6 Re} \int_0^{s/D} u^5 \, \mathrm{d}\left(\frac{s}{D}\right)\right]^{1/2} \tag{24.15}$$

where s is the surface distance.

Transition is assumed to occur either at the first slot, or when the velocity gradient becomes adverse, whichever occurs first.[145] Spalding's (1964P) semi-empirical model is used to calculate the development of the turbulent boundary layer step by step from transition to any slot(s) and beyond them to separation. The combined velocity profile of the boundary layer and the tangential jet gradually converges to a boundary layer profile far downstream.

[145]Note that this assumption is valid only for the TrBL0 regime.

Dunham (1968J) assumed that the momentum thickness is that given by Thwaites' model at the end of the laminar portion. The entrainment relationship for jet flow is modified to take account of the increase in entrainment due to wall curvature. Both modifications considerably affect the assessment of separation pressure.

The tangential jet may be subsonic or sonic (choked flow) and the one-dimensional compressible flow theory is adopted. The wall-jet strength is expressed by its momentum coefficient C_J

$$C_J = \frac{\gamma p (Ma)^2 s}{1/2 \rho V^2 D} \tag{24.16}$$

where $\gamma = 1.4$ for air (ratio of specific heats), p is the static pressure at the slot exit, Ma is the wall-jet Mach number, and s is the slot width. At the slot, the cylinder boundary layer meets the wall-jet, and along some of its length the downstream velocity profile cannot conform to Spalding's profile. There is a strong entrainment in this region, and a transverse pressure gradient must be present to align the wall jet with the main stream and wall. A typical example of the calculated development of the turbulent boundary layer is shown in Fig. 24.62(a,b) for $C_L = 4.48$ along the lower and the upper surface, respectively. The separations are predicted at $\theta_{sl} = 243°$ and $\theta_{su} = 170°$ for $C_J = 0.36$.

Dunham (1967J) summarized all the available experimental evidence in terms of the following parameters: Re, C_J, Ma, the number of slots, their position θ_j, slot width s/D, and design. For a single value of $Re = 360$k, there is reasonable agreement for the theoretical and experimental C_L versus C_J curves in Fig. 24.63. The C_L curve can stall like aerofoils when slots become ineffective behind the separation point. The effect of jet Ma is small on the C_L and C_J curves. However, the effect of the slot width is considerable, and produced the biggest disagreement between theory and experiment.

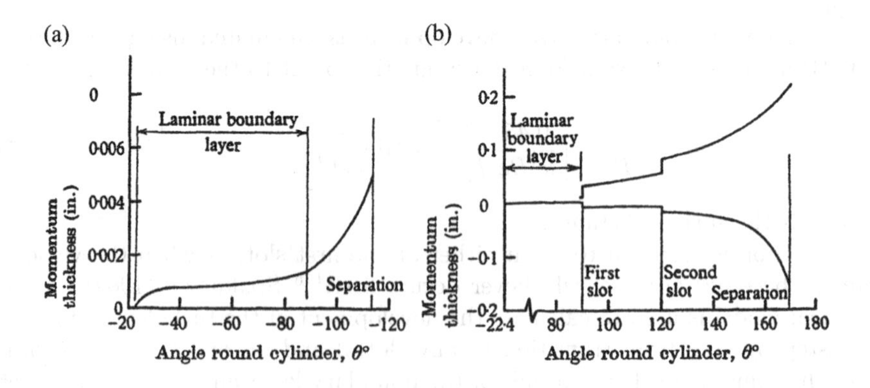

FIG. 24.62. Calculation of boundary layer development, (a) lower surface, (b) upper, Dunham (1968J)

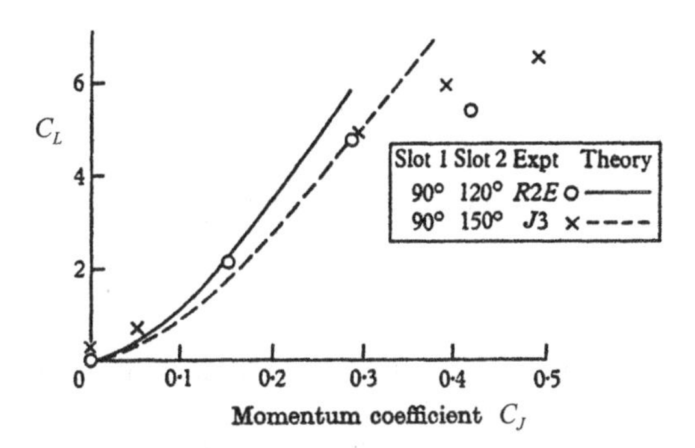

FIG. 24.63. Comparison of theory and experiment, lift versus the wall's jet coefficient, Dunham (1968J)

Yoshino and Furuya (1974J) improved the theoretical model by taking into account the effect of the surface curvature on the tangential wall jet. Three types of differential equations were derived for the boundary layer, mixing region, and the effect of thickness on the mixing region.

25

YAWED CYLINDERS

25.1 Introduction

In some engineering applications, circular cylinders have not been positioned perpendicular to the free stream, but rather at a certain *yaw angle*. For example, electrical transmission lines may be exposed to an oblique wind; offshore structural members submerged in slanted waves and currents; tubes in heat exchangers fitted with baffle plates; and missiles flying at high angle of incidence.

The yaw angle has been defined as that between the cylinder axis and the normal to the free stream velocity.[146] A wide variety of symbols has been used to designate the yaw angle, and in this book the uppercase Greek letter Λ was chosen because it had already been adopted in Aerodynamics for swept-back wings. The yaw angle appeared to be a complement to the angle of incidence, α, for missiles,

$$\Lambda + \alpha = 90^\circ \tag{25.1}$$

as can be deduced from Fig. 25.1.

There are several distinct flow features related to yawed cylinders, as depicted in Fig. 25.1:

(i) When a circular cylinder is yawed the cross-section becomes elliptical. The ratio of major-to-minor axis of the ellipse is proportional to Λ.

(ii) The free stream velocity has two components:

(a) one normal to the cylinder axis, $V_n, = V \cos \Lambda$;

(b) the other parallel, i.e. along the span, $V_t = V \sin \alpha$.

It may be argued that V_n has a major effect on the flow, while V_t should have a minor effect. For $\Lambda = 45^\circ$, both components are equal in magnitude, $V_n = V_t$, while for $\Lambda > 45^\circ, V_t > V_n$.

25.1.1 *Independence principle*

An early attempt to replace the free stream velocity by V_n was reported by Relf and Powell (1917P). It was only partially successful in correlating the measured force data. A theoretical confirmation for $V_n = V \cos \Lambda$ was provided by Sears (1948J). He simplified the Navier–Stokes equations of motion for a yawed circular

[146]The same yaw angle is formed by the free stream velocity and the normal to the cylinder axis, see Fig. 25.1.

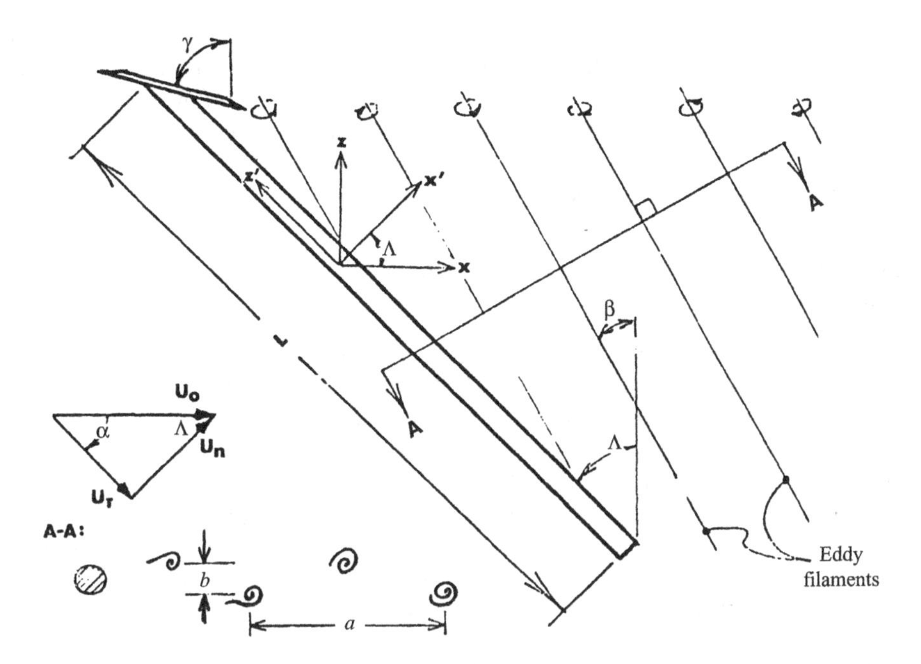

FIG. 25.1. Definition sketch and nomenclature, Ramberg (1978P)

cylinder of infinite span by applying Prandtl's arguments regarding the order of magnitude of terms for small viscosity (large Re) in a laminar boundary layer. He showed that the projected flow in the normal plane, based on $V \cos \Lambda$, was described by the same equations as for the flow around an unyawed cylinder. The spanwise boundary layer flow could be calculated by the integration of a linear second-order differential equation. As the governing equations of motion were uncoupled, the normal and spanwise flows were independent, hence the term the *Independence Principle.*

Several important limitations are expected to be applicable to the Sears (1949J) theory, such as:

(i) The laminar boundary layer theory becomes invalid beyond separation. This restricts the validity of application of the Independence Principle to within the stagnation and separation lines. It implies that the base pressure, and drag force related to it, might not be correlated by the Independence Principle.

(ii) The mathematical idealization of two-dimensional flow past cylinders of infinite aspect ratio cannot be realized in practice. Flow around cylinders of finite aspect ratio are subjected to end effects, which have not been accounted for in the theory.

(iii) The transition around separation for unyawed cylinders occurs at a certain Re_c due to the instability of the laminar boundary layers. By assuming that the flow past yawed cylinders is subjected to the same flow disturbances, it may be expected that the instability would appear at the same Re_c. However, the spanwise flow is likely to modify the disturbances, and the value of Re_{nc} for flow past yawed cylinders.

25.2 Laminar wakes in the L2 and L3 regimes

25.2.1 *Strouhal number*

An early experimental attempt to verify the Independence Principle for laminar wakes was carried out by Hanson (1966J). He measured the frequency of the wake trail oscillations, and subsequently eddy shedding in the range $40 < Re < 150$ for the yaw angles $0° \leqslant \Lambda \leqslant 72°$. Two piano wires were tested in a circular wind tunnel, $\phi = 0.1\,\text{m}$, having $L/D = 125$ and 1500.

The effect of Λ upon the onset of trail oscillation, Re_{osc}, is shown in Fig. 25.2. The Re_{osc}, based on the free stream velocity, rises with increasing Λ. However, when $V_n = V \cos \Lambda$, i.e. Re_n, is used the Re_{osc} = constant up to $\Lambda = 50°$. The Independence Principle ceases to be valid for $\Lambda > 50°$. Hanson also noted a hysteretic effect for $\Lambda > 45°$.

Systematic measurements of the eddy shedding frequency in terms of Λ are shown in Fig. 25.3(a,b). Hanson (1966J) used Roshko's number ($Ro = fD^2/\nu$), and obtained different straight lines for various Λ. However, when the same data were scaled by Re_n the linear relationship derived for unyawed cylinders could be extended to the yawed cylinders

$$Ro_n = 0.212Re_n - 4.5, \qquad \text{for } \Lambda < 50° \tag{25.2}$$

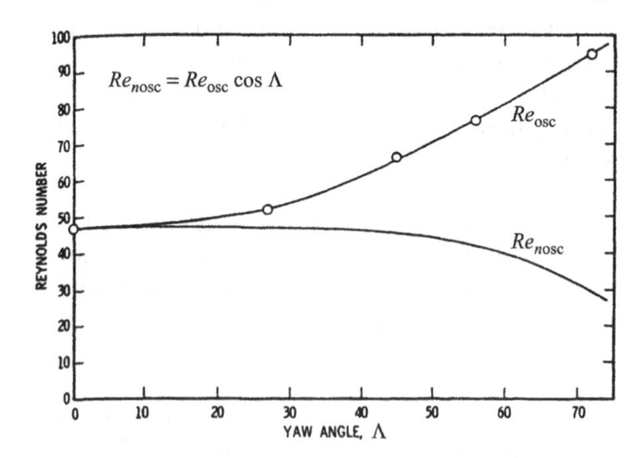

FIG. 25.2. Effect of yaw upon Re_{osc} in terms of yaw angle Λ, Hanson (1966J)

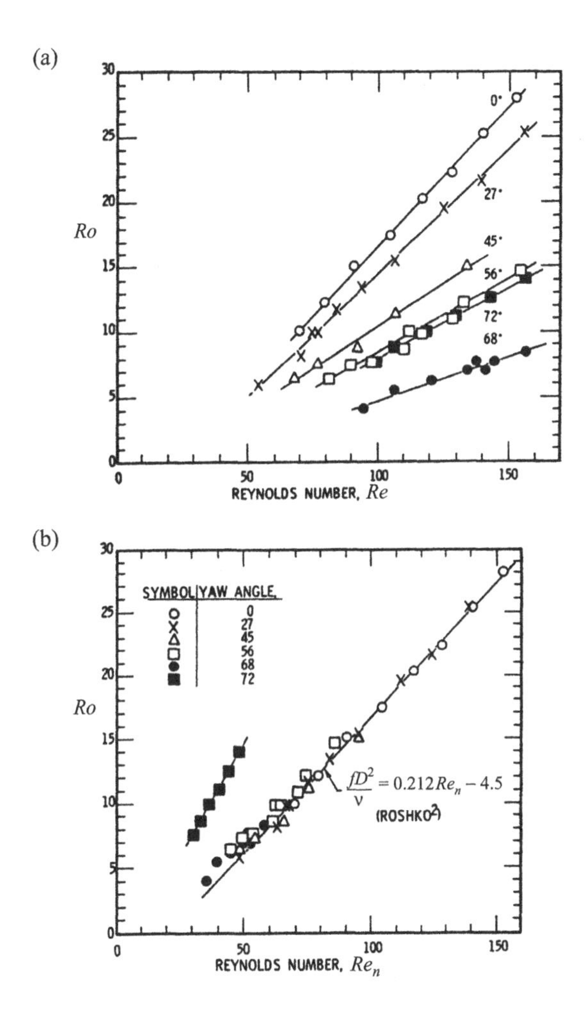

FIG. 25.3. Roshko's number in terms of (a) Re, (b) Re_n, Hanson (1966J)

Another careful measurement of the eddy shedding frequency was reported by Van Atta (1968J). He pointed out that the piano wires used as cylinders may be excited to oscillate at higher harmonics when synchronized with the eddy shedding frequency. The tension of the wire appeared as an influencing parameter. Within the synchronization range, the hot-wire signal was regular, while outside that range it was always modulated. All Van Atta's data were taken outside the synchronization range. The aspect ratio was in the range $240 < L/D < 1875, 50 < Re < 150$, and $0° \leqslant \Lambda \leqslant 78°$.

Figure 25.4(a) shows the conventional St in terms of Λ for three constant $Re_n = 50, 80$, and 150. For each Re_n, the St decreases steadily as Λ increases,

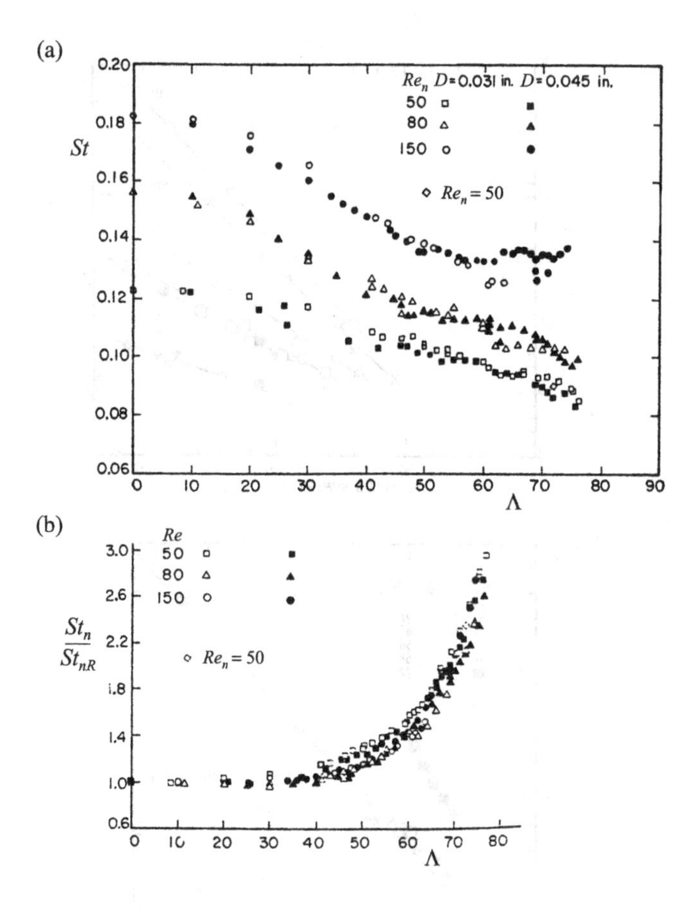

FIG. 25.4. (a) Strouhal number in terms of Λ, (b) St_n/St_{nR} in terms of Λ, Van Atta, © (1968J) AIAA – reprinted with permission

except for $Re_n = 50$. Figure 25.4(b) re-plots the same data for St_n, based on $V \cos \Lambda$ and normalized by St_{nR}. Roshko's (1955J) empirical correlation of data for non-yawed cylinders was extended to yawed cylinders, as follows

$$St_{nR} = 0.212 - 4.5/Re_n \tag{25.3}$$

If the Independence Principle is valid, then the measured St_n, and that calculated from eqn (25.3), should be the same, i.e. $St_n/St_{nR} = 1$. Figure 25.4(b) shows that data for cylinders of different diameters collapse fairly well into a single curve. St_n is virtually constant for $0° \leqslant \Lambda \leqslant 40°$, and increases for $\Lambda > 40°$. Hanson (1966J) and Van Atta (1968J) found different Λ ranges for the validity of the Independence Principle within the laminar Re values. Van Atta (1968J) suggested that different tension of wires, and related oscillation in the synchronization range might be the reason.

Ramberg (1978P, 1983J) demonstrated convincingly that St_n could be influenced by the aspect ratio, and conditions at the cylinder ends. Van Atta's (1968J) relationship (obtained for lower Re) was reproduced by moving the upstream end of the yawed cylinder away from the wind tunnel wall. Conversely, when the same cylinder touched the wall, experimental points were below the $V \cos \Lambda$ curve, and close to Hanson's (1966J) data. It appeared that the range of the Independence Principle validity strongly depended on the end condition.

25.2.2 *Effect of end plate*

End plates have frequently been used on unyawed cylinders in wind tunnel testing, as discussed in Chapter 21.2. The aims were to reduce the end effects, prevent three-dimensional flow, and render the flow *nominally* two-dimensional. The use of end plates on yawed cylinders could neither reduce end effects nor prevent three-dimensional flow. The spanwise flow, $V \sin \Lambda$, along the yawed cylinder is directed away from the wall at the upstream end, and towards the wall at the downstream end. These inherently different end conditions remain when the end plates are fitted.

Ramberg (1978P, 1983J) carried out experiments in an open jet, $160 < Re < 1.1\text{k}$, $10° < \Lambda < 60°$, and $20 < L/D < 100$ with an end plate (truncated disc) $5 < D_e/D < 8$. He noted that the condition at the upstream end of the yawed cylinder dominated the flow. So he fitted the end plate only onto the upstream end, and varied the end plate angle, γ, measured between the plate plane and normal to the free stream velocity, see Fig. 25.1. Some of the flow visualization photographs obtained for $Re = 160$ and $\Lambda = 40°$ are shown in Fig. 25.5(a–d). The prominent feature is that for fixed Re and Λ, there is a significant variation in wake patterns affected solely by the end plate angle γ. For $\gamma < 90°$, Fig. 25.5(a,b), the end plate 'opens' upstream, and the eddy filaments are slanted $\beta > \Lambda$. For $\gamma > 90°$, Fig. 25.5(c,d), the end plate 'closes' upstream, and the eddy filaments eventually become normal to the free stream velocity.

The observed variation in the eddy filaments' slanted angle β is related to the variation in the eddy shedding frequency and Strouhal number. The relationships of St/St_n and β in terms of γ are shown in Fig. 25.6 for $Re = 160$ and $\Lambda = 30°$. For fixed Re and Λ, there is a wide range $0.6 < St/St_n < 1.1$ depending on the end plate angle γ. However, according to the Independence Principle $St/St_n = \cos \Lambda$ should be 0.86 for $\Lambda = 30°$ (arrows in Fig. 25.6). It appears that eddy shedding is not unique and may assume two modes for $\gamma \lesssim 90°$.

Another feature evident in Fig. 25.6 is the change-over in St/St_n in the narrow range $87° < \gamma < 110°$. This may account for many discrepancies in measured shedding frequencies and St mode by various authors.

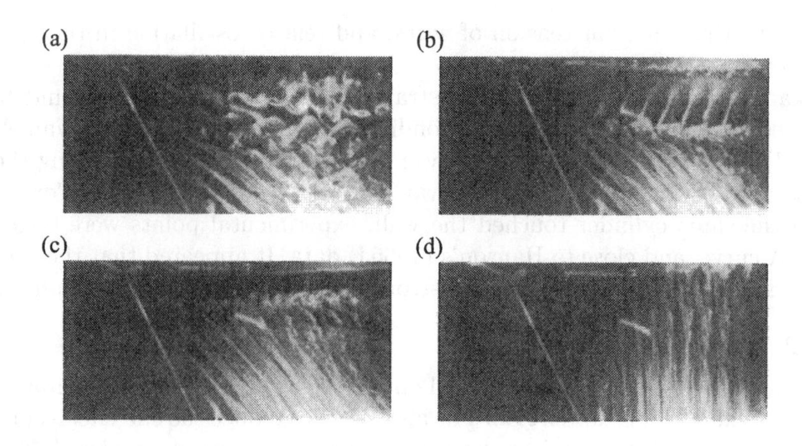

FIG. 25.5. Smoke visualization at $Re = 160$ for $\Lambda = 40°$, (a) $\gamma = 70°$, (b) $\gamma = 80°$, (c) $\gamma = 110°$, (d) $\gamma = 120°$, Ramberg (1978P)

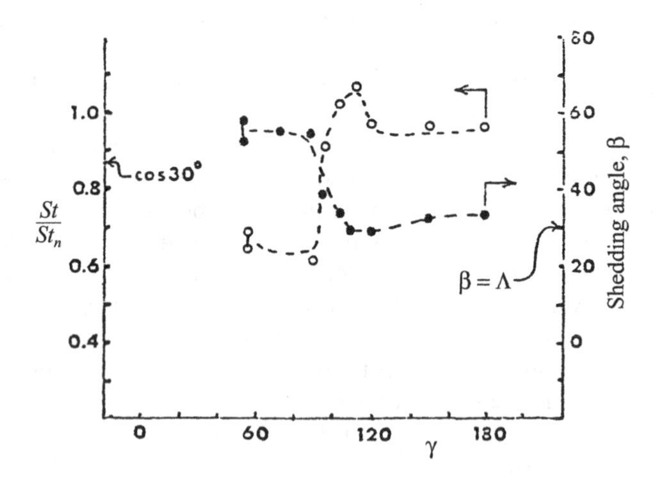

FIG. 25.6. Normalized Strouhal number and shedding angle in terms of end plate angle γ for $\Lambda = 30°$, $Re = 160$, Ramberg (1978P)

25.2.3 *Free-ended and yawed cylinders*

The effect of the free end on the flow past unyawed cylinders has been discussed in Chapter 21.8. The affected region was limited to the vicinity of the free end. For yawed cylinders, the affected region was proportional to the yaw angle Λ, and for high Λ it spread along the entire span.

Ramberg (1978P, 1983J) also examined the free-end case in detail in addition to the effect of end plates, as described previously. Flow visualization was carried out at $Re = 160$ for ten values of Λ. However, Fig. 25.9 shows only $\Lambda = 0°, 5°, 20°, 30°, 40°$, and $50°$, respectively. The evolving wake patterns will be described in three phases:

(i) the appearance and development of the free-end cell;

(ii) the weakening and disappearance of the free-end cell;

(iii) the appearance and spreading of the streamwise trailing pattern.

Figure 25.7(a) shows the wake pattern for the unyawed cylinders, $\Lambda = 0°$, which is also similar to all small $\Lambda < 5°$. The eddy filaments are curved, and the wake pattern is unsteady. As Λ is increased from 5° to 20°, Fig. 25.7(b,c), the free-end effect is seen first as the cross-link of eddy ends in loops. When the two cells are

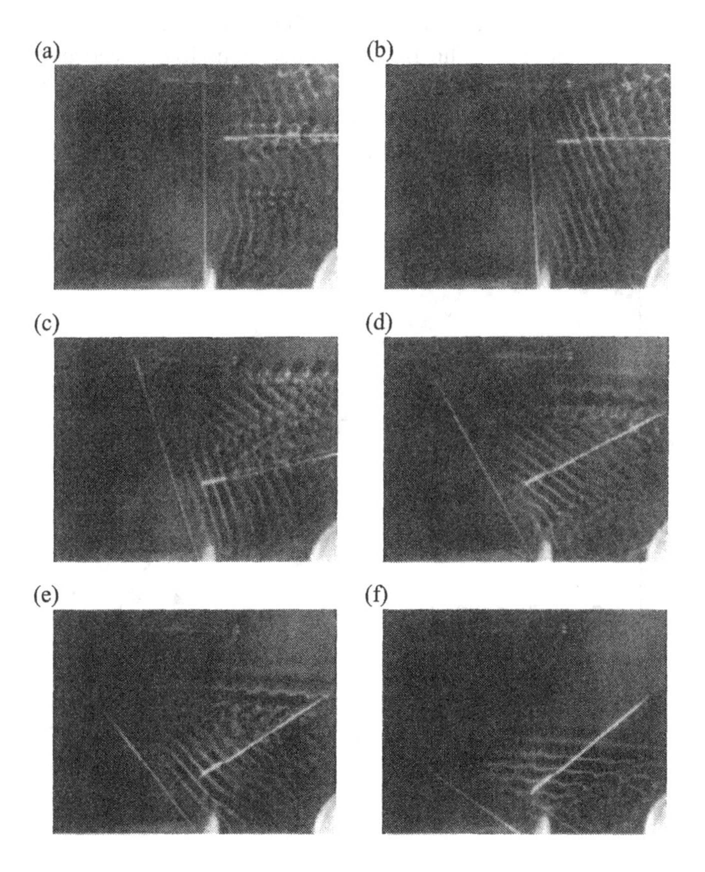

FIG. 25.7. Smoke visualization at $Re = 160$ in terms of Λ, (a) $\Lambda = 0°$, (b) $\Lambda = 5°$, (c) $\Lambda = 20°$, (d) $\Lambda = 30°$, (e) $\Lambda = 40°$, (f) $\Lambda = 60°$, Ramberg (1978P)

formed along the span, the strobe light could be made to 'freeze' one cell but not both, which indicates a difference in eddy shedding between the cells. The St for the free-end cell (high β) is lower than for the root cell (small β). The two frequencies cause beating in the hot-wire signal, as observed by Van Atta (1968J).

The second phase of the wake pattern development is shown in Fig. 25.7(d,e) for $\Lambda = 30°$ and $40°$. The eddy linking loops move away from the free end to be replaced by a wavy trail in the streamwise direction. The free-end cell spreads along the entire span, and β decreases considerably. This phase represents transition from the slanted eddy filament pattern to the streamwise trailing pattern.

The third phase is shown in Fig. 25.7(f) for $\Lambda = 60°$. The streamwise trailing pattern is spread across the entire wake, and the slanted eddy filaments are obliterated.

Ramberg (1978P, 1983J) also measured St in a wide range $160 < Re < 1.1\text{k}$, $20 < L/D < 90$, for yawed cylinders with the upstream free end. Figure 25.8 shows St/St_n in terms of Λ. The Independence Principle is represented by the $\cos\Lambda$ curve (dashed). The points above the $\cos\Lambda$ curve (open symbols) are taken in the main cell, and below the $\cos\Lambda$ curve (closed symbols) are detected in the free-end cell. The relationship between β and St/St_n follows the same inverse

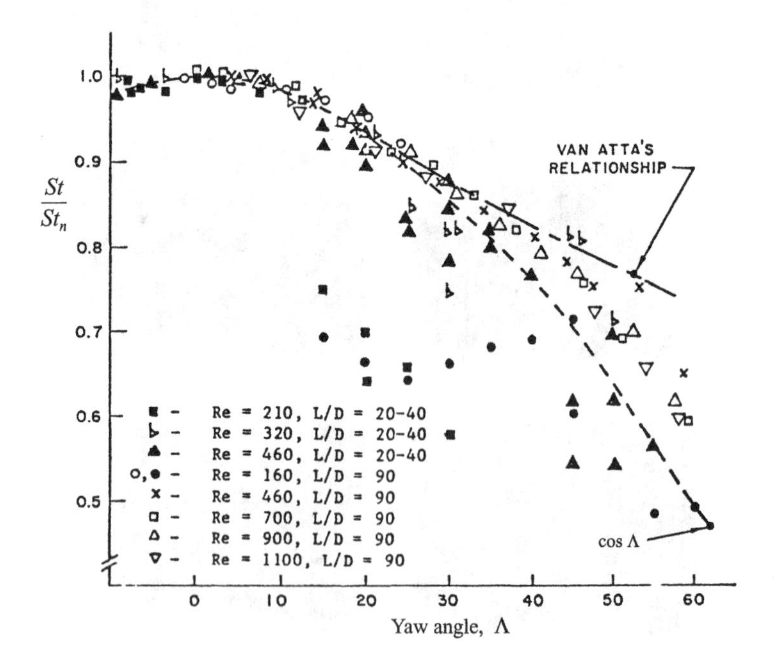

FIG. 25.8. Ratio of conventional and normal velocity Strouhal number in terms of Λ for $160 < Re < 1.1\text{k}$, Ramberg (1978P)

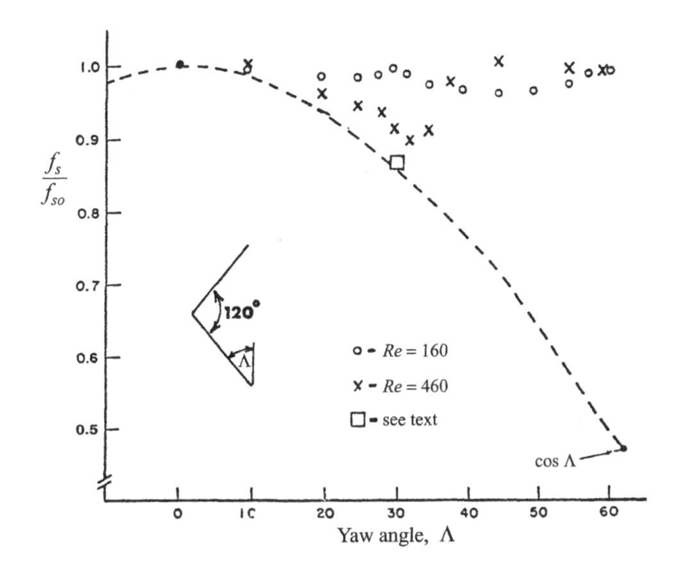

FIG. 25.9. V-shaped cylinder's St/St_n ratio in terms of Λ, Ramberg (1978P)

trend as found for yawed cylinders with end plates.

Ramberg (1978P, 1983J) also tested two yawed cylinders forming a letter V with an opening angle of 120°, see sketch in Fig. 25.9. The measurements St/St_n for $Re = 160$ and 460 depart consistently from the $\cos \Lambda$ curve. The eddy filaments are fixed at $\beta = 0$.

25.3 Transition-in-shear layers, TrSL, state

25.3.1 *Eddy formation region and base pressure*

The length of the eddy formation region L_F has been defined as the distance from the cylinder to the position of a fully formed eddy. Ramberg (1978P, 1983J) measured the location of the maximum of the second harmonic of velocity fluctuation along the wake axis. The transverse distance between the two maxima, $u_{\max}$, is taken as the wake width B_F at the L_F station.

The Independence Principle is not valid beyond separation, i.e. it is not expected to correlate B_F and L_F in terms of Λ. The measured L_F and B_F are shown in Fig. 25.10 in terms of Re_n. The L_F/D does not scale results onto the $\Lambda = 0°$ curve.

The base pressure measurements for yawed cylinders are taken at $Re = 550$ and 750. Figure 25.11 shows C_{pbn}/C_{pb} versus Λ. Again, the Independence Principle fails to correlate C_{pbn}/C_{pb}.

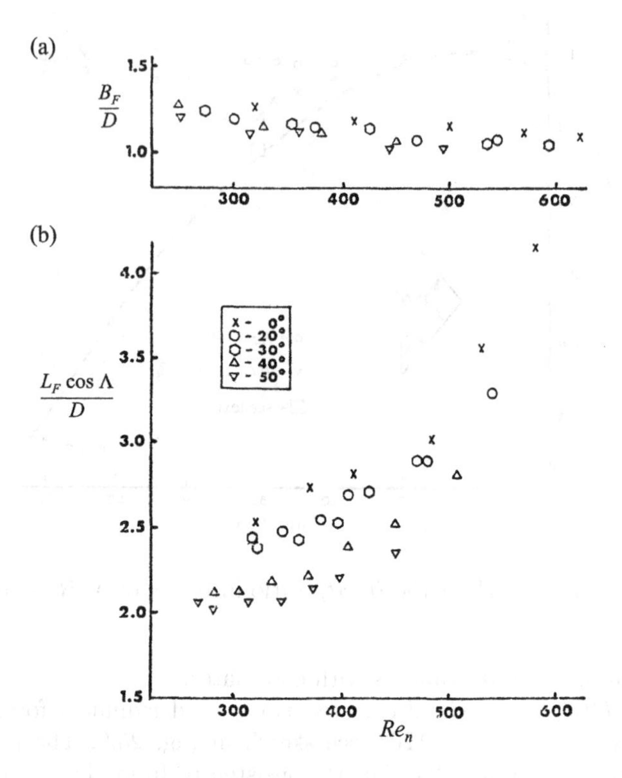

FIG. 25.10. (a) Width of near-wake and (b) length of eddy formation region along normal velocity in terms of Re_n, Ramberg (1978P)

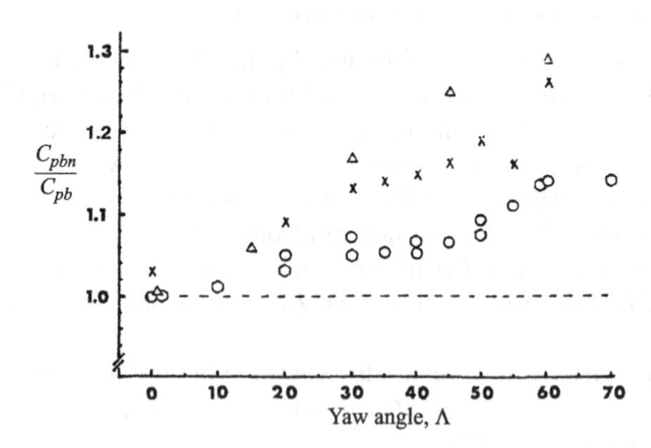

FIG. 25.11. Ratio of normal velocity and conventional base coefficient in terms of Λ, Ramberg (1978P)

25.3.2 *Elliptic cross-section*

It has already been mentioned that yawed cylinders alter the circular cross-section in the plane of the free stream velocity to an elliptic section. This raises the question as to whether there is an analogy between flows around yawed cylinders, and unyawed elliptic cylinders of equivalent section.

It is known that elliptic cylinders are less bluff than circular cylinders. This implies a narrower wake and higher St than for the circular cylinder. As the St for the yawed cylinder is less than for the unyawed cylinder, it is expected that the cross-section analogy does not hold.

The experimental verification was provided by Shirakashi *et al.* (1986J). The circular cylinders were tested in the range $1\text{k} < Re < 52\text{k}$ for $\Lambda = 0°$, $30°$, and $45°$, and for the equivalent elliptic cylinders at $\Lambda = 0°$ (both without end plates). Additional tests were carried out with partition plates placed behind the yawed cylinders at $x/D = 0.9$. Figure 25.12 shows the measured eddy shedding frequency in terms of the free stream velocity (St is proportional to the slope). The data for the circular cylinder at $\Lambda = 0°$, $30°$, and $45°$ show the decreasing trend as Λ rises. The data for the elliptic cylinders show an opposite trend: St increases for the elliptic cylinders. The analogy between the yawed circular and unyawed elliptic cylinders does not hold. However, by fitting the partition plates,

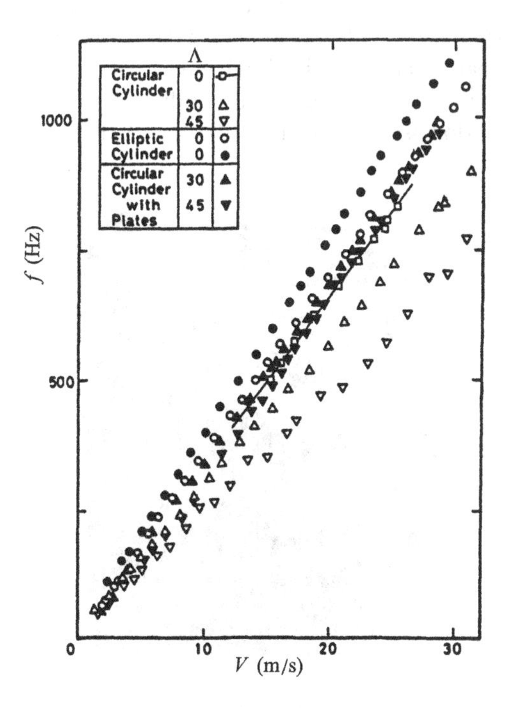

FIG. 25.12. Frequency of eddy shedding in terms of V, Shirakashi *et al.* (1986J)

$x/D = 0.9$, the departure of points from $\Lambda = 0°$ is prevented; the reason for this will be discussed in the next section.

Flow visualization around yawed cylinders at $Re = 3.1$k for $\Lambda = 30°$ without end plates is shown in Fig. 25.13. The streaklines gradually bend as they approach the stagnation line, and pass the cylinder at almost a right angle to the cylinder axis. The turbulence conceals the flow direction in the near-wake.

25.3.3 *Effect of end plates*

It has been argued in Section 25.2.2 that end plates fitted on yawed cylinders are subjected to different types of three-dimensional flow at the upstream and downstream ends. The spanwise flow, $V \sin\Lambda$, along the yawed cylinder is directed away from the upstream end plate, and towards the downstream end plate. The mechanism of interference between the flow and two end plates is inherently different.

Hayashi *et al.* (1992J) carried out tests in an open jet at $Re = 10$k, $0° \leqslant \Lambda \leqslant 30°$, with end plates, $5 \leqslant L/D \leqslant 15$. They also tested yawed cylinders placed across the jet without end plates. The latter allowed an uninterrupted spanwise flow along the yawed cylinder. The pressure distribution, C_p, was measured at the mid-section, and at two other stations at $\pm 4.7D$.

Typical C_p distributions for $\Lambda = 30°$ are shown in Fig. 25.14(a,b). The paradoxical feature is that C_{pb} distributions are more uniform along the span of the yawed cylinder without the end plates than with them. The lower value of C_{pb} occurs on the upstream section for both with and without end plates. This pressure difference is expected to induce a secondary flow in the near-wake. The direction is from the downstream to the upstream end of the yawed cylinder, i.e.

FIG. 25.13. Smoke visualization of stagnation streamlines for $Re = 3.1$k at $\Lambda = 30°$, Shirakashi *et al.* (1986J)

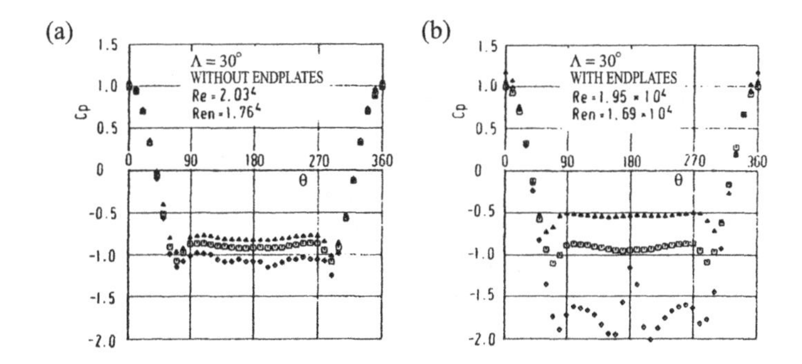

FIG. 25.14. Mean pressure distribution at Re = 20k for $\Lambda = 30°$, (a) without end plates, (b) with end plates, Hayashi *et al.* (1992J)

the opposite to the streamwise velocity.

The secondary flow in the near-wake of a yawed cylinder was examined by Smith *et al.* (1972J). They used a hot-film probe to determine the flow direction by traversing the near-wake in the range $0.5 \leqslant x/D \leqslant 10$ for Re = 7.4k. The measured flow directions are presented in Fig. 25.15 for $\Lambda = 30°$. The flow direction is denoted by α and represents an angle between the local and the free stream velocity. Figure 25.15 shows that close to the cylinder ($x/D = 0.5, 1, 2$) the flow direction may be $\alpha = (\pi/2) - \Lambda$, i.e. the flow is along the span. Further towards the wake axis, $\alpha > (\pi/2) - \Lambda$ and $V \cos\Lambda < 0$, a back flow region exists.

Hayashi *et al.* (1992J) checked the Independence Principle regarding C_{po} and C_{pb} for yawed cylinders with and without end plates. Figure 25.16 shows the spanwise distribution of C_{pon} and C_{pbn} (expressed through $V \cos\Lambda$) at Re = 20k and $\Lambda = 0°, 10°, 20°$, and $30°$. The Independence Principle is valid for C_{pon} except near the jet. The decrease in C_{pon} occurs at both ends for $\Lambda = 0°$, and

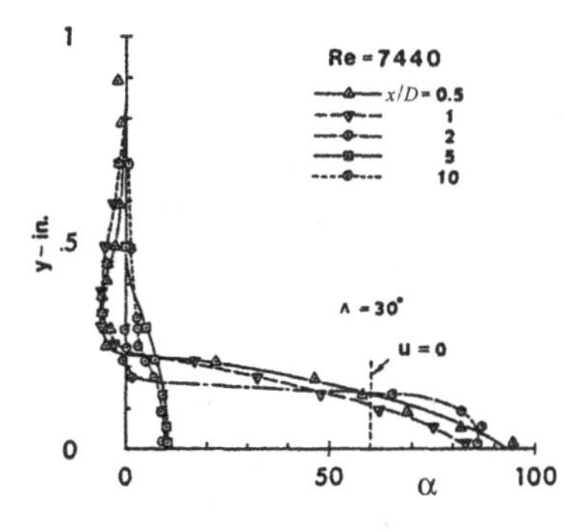

FIG. 25.15. Profile of velocity incidence angle for Re = 7.4k at $\Lambda = 30°$, Smith *et al.* (1972J)

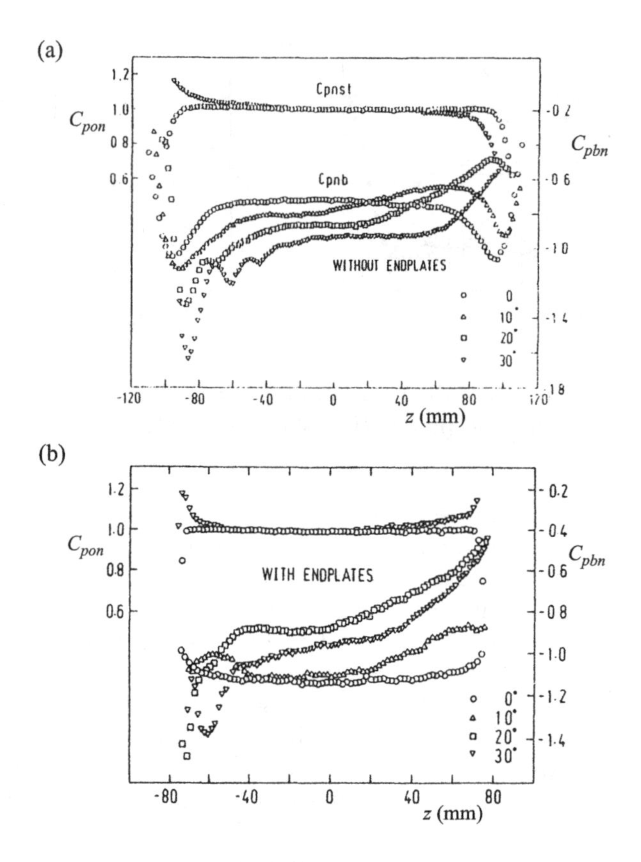

FIG. 25.16. Mean stagnation and base pressure distribution along span for $Re = 20$k, at $\Lambda = 30°$, (a) without end plates, (b) with end plates, Hayashi *et al.* (1992J)

increases at the upstream end for $\Lambda = 30°$. The increase in $C_{pon} > 1$ indicates that the approach velocity is greater than $V \cos \Lambda$. The flow in the end region changes direction from V parallel to the end plate to $V \cos \Lambda$ away from the end plate. Figure 25.16(a) also shows that the Independence Principle is invalid for C_{pbn} all along the span and for all Λ. The value of C_{pbn} at the upstream end is lower than at the downstream end. The flow interference between the jet boundaries and yawed cylinder is particularly pronounced at the upstream end. The typical feature is the minimum in C_{pbn} near the jet boundaries for yawed and unyawed cylinders (for unyawed cylinders, see Chapter 21).

The addition of end plates separates the jet boundaries from the flow around yawed cylinders, and interrupts the spanwise flow along the yawed cylinders. Figure 25.16(b) shows that C_{pon} remains mostly unaffected; however, $C_{pon} > 1$ is seen at both ends. This is because both end plates are aligned with the free stream velocity, and impose that velocity on both cylinder ends. Figure 25.16(b) also shows the C_{pbn} distribution along the span between the end plates. There are several important differences in comparison with cylinders without end plates:

(i) all C_{pbn} curves are above the curve for the unyawed cylinder, i.e. precisely opposite to Fig. 25.16(a) (without end plates);

(ii) the pressure gradient is greater at the downstream end of the yawed cylinder;

(iii) there is an anomaly, namely the $\Lambda = 20°$ curve appears above the $\Lambda = 30°$ curve;

(iv) the end plates fully suppress the flow interference between the jet boundaries.

It may be concluded that the end plates substantially change the flow past yawed cylinders, and at present the mechanism of the end plate interference is unresolved. For example, Kawamura and Hayashi (1992J) carried out computer simulations of flow around infinite and finite yawed cylinders at $Re = 2\text{k}$ for $\Lambda = 30°$. They calculated pressure and time-averaged streamlines at six stations along the yawed cylinder for $L/D = 9$ with end plates. The wake is narrowest near the upstream end and widest near the downstream end of the yawed cylinder.

25.3.4 *Effect of the aspect ratio*

Hayashi *et al.* (1992J) studied the effect of the aspect ratio variation between end plates in the range $5 \leqslant L/D \leqslant 15$, $Re = 20\text{k}$, and $\Lambda = 30°$. Figure 25.17 shows C_{pon} and C_{pbn} distributions along the span. As before, $C_{pon} = 1$ everywhere except near the end plates where $C_{pon} > 1$. The end plates align the flow to the free stream velocity.

The C_{pbn} curves show an orderly change in the pressure gradient, which increases as L/D decreases. It is notable that the minimum C_{pbn} remains at around -1.4 irrespective of the aspect ratio. Another feature is that the highest C_{pbn} value is at the downstream end and remains constant at around -0.5. Thus, the only effect of reducing L/D is to increase the C_{pbn} gradient, and consequently the intensity of the secondary flow along the span.

25.3.5 *Skin friction*

Detailed measurements of the fluctuating and time-averaged surface velocity gradient[147] were carried out by Tournier and Py (1978J). They used the electrochemical method developed by Dimopoulos and Hanratty (1968J). Tournier and Py (1978J) used a series of split-film electrochemical probes arranged around the circumference and along the span. The cylinder, $L/D = 10$ and $D/B = 0.1$, was tested at a single $Re = 13.9\text{k}$ for $\Lambda = 0°, 10°, 20°$, and $30°$. The signals from

[147] The skin friction is proportional to the velocity gradient at the wall and the constant of proportionality is the viscosity of the fluid.

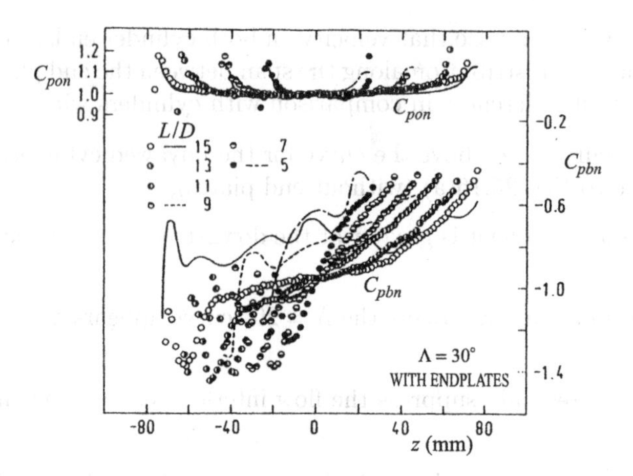

FIG. 25.17. Mean stagnation and base pressure along the span for Re = 20k, at $\Lambda = 30°$, and different L/D, Hayashi *et al.* (1992J)

the split-film probes were processed so as to separate the normal and spanwise velocity gradients as well as the fluctuating and time-averaged contributions.

Figure 25.18(a,b) shows the distribution of time-averaged velocity gradients projected normal to the cylinder axis and spanwise along the axis, respectively. The split-film signals are normalized by using the value at $\theta = 50°$. The shape of the curve in Fig. 25.18(a) is identical to the C_f curve for the unyawed cylinder, see Vol. 1, Fig. 3.46, p. 71, and Fig. 9.6, p. 288. The collapse of points obtained for $\Lambda = 0°, 10°, 20°$, and $30°$ proves the validity of the Independence Principle. The distribution of the spanwise velocity gradient, which is zero for the unyawed cylinder, shows three separate curves for each tested yaw angle Λ. The scaling by $V \cos \Lambda$ is inappropriate. The notable feature is that the maximum relative velocity gradient occurs along the 'stagnation' line. Evidently, the term 'stagnation' cannot be used because $V \sin \Lambda$ flows along $\theta_o = 0°$. Another unusual feature is the finite value of the velocity gradient at separation, $\theta_s = 80°$. This indicates a complex three-dimensional separation.

Figure 25.19 shows the magnitude of the fluctuating velocity gradient (rms) in the normal (top curves) and spanwise (bottom curves) directions, respectively. The fluctuations underneath the laminar boundary layer up to the separation were due to periodic eddy formation and shedding. The maximum occurs at θ_s as for the unyawed cylinder, see Vol. 1, Fig. 17.17, p. 537. The second peak at $\theta = 180°$ is specific for the yawed cylinders and will be discussed later. The spanwise component in Fig. 25.19 is much smaller in the boundary layer region but exceeds the normal component in the separated region. This is caused by the coherent surface eddy flow, mostly in the spanwise direction.

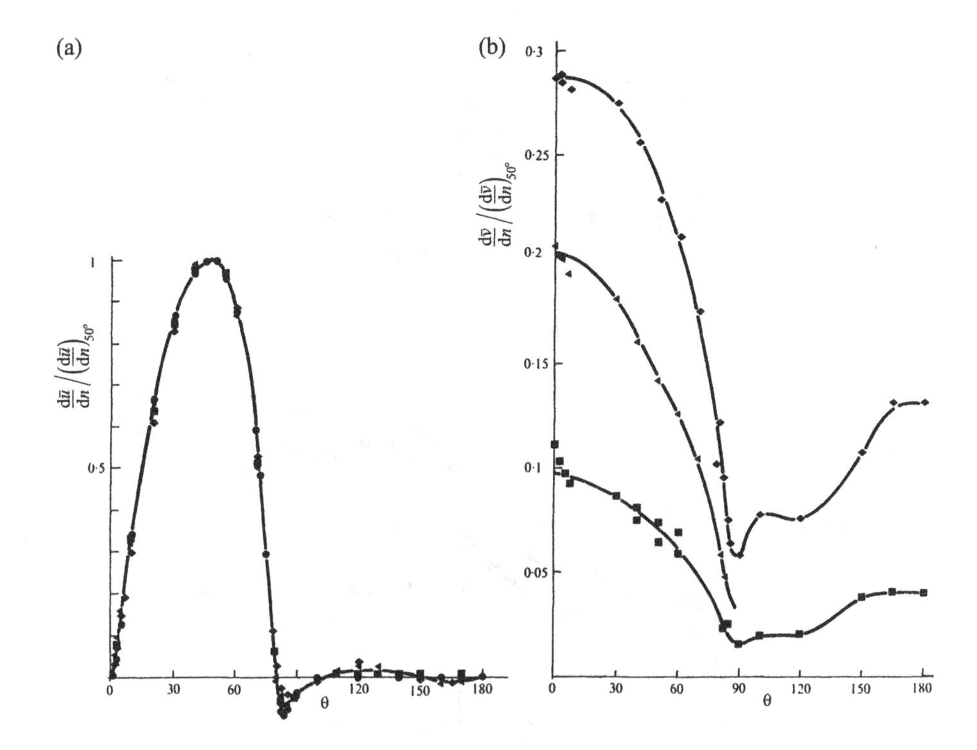

FIG. 25.18. Mean surface velocity gradient distribution at $Re = 39$k, (a) streamwise, (b) spanwise, Tournier and Py (1978J)

Conditional phase-averaging reconstructs an instantaneous surface flow. Figure 25.20 shows six sequences out of 24 during one shedding cycle for $\Lambda = 10°$. There are three distinct regions:

(i) the surface flow underneath the laminar boundary layer up to separation;

(ii) the separation is followed by an intermittent reattachment line;

(iii) the separated region is marked with the spanwise surface flow alternating towards and away from the separation line during the eddy formation cycle.

The first region exhibits a small oscillation of the stagnation line. A similar oscillation was observed on the unyawed cylinder, see Vol. 1, Fig. 5.48, p. 135. However, beyond the stagnation line the surface streamlines are yawed in the direction of the spanwise velocity component. Farther along the circumference the surface streamlines appear aligned with the free stream velocity. Finally, just before separation, the surface streamlines are turned again in the spanwise direction.

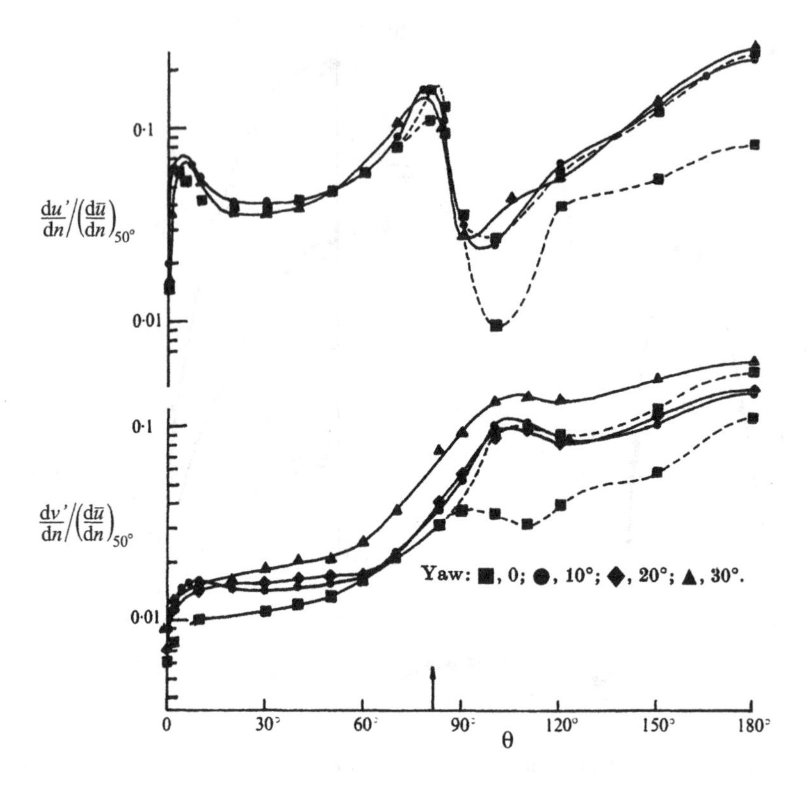

FIG. 25.19. Fluctuating surface velocity gradient distribution for Re = 39k, Tournier and Py (1978J)

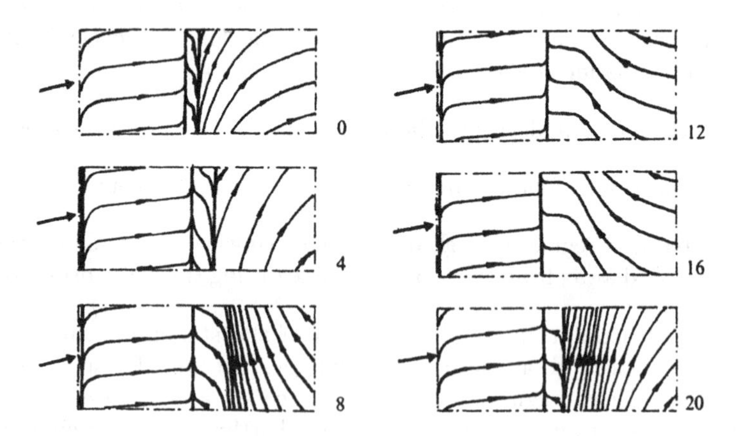

FIG. 25.20. Conditionally-averaged instantaneous surface velocity sequences during half a shedding cycle, Tournier and Py (1978J)

The second region may be interpreted as the surface flows underneath the swirling[148] eddy filament. The rolled-up shear layer has a large spanwise velocity component. The second line is the reattachment line induced by the swirling eddy. The growing swirling eddy is gradually carried downstream and upwards, sequences 4 and 8. This causes both separation and reattachment lines to be displaced downstream, and the distance between them increases in sequences 4 and 8. The farther the eddy is from the surface, the farther is the reattachment, which finally disappears in sequences 12 and 16 when the reverse flow reaches the separation line. Note that the latter is now displaced upstream again to reach the most upstream position in sequence 6. The final sequence 20 shows the reappearance of the reattachment, and a somewhat wider separation bubble than that in sequence 0.

The third region displays the biggest variation in the surface flow pattern, although it is underneath a fully separated flow. As the separation bubble moves downstream, the surface streamlines become compressed and change direction from outwards to inwards, sequences 0, 4, and 8. As the separation bubble disappears, the surface streamlines form a reverse flow within small spanwise component in sequences 12 and 16. When the new separation bubble appears, the surface streamlines are compressed, and directed mostly in the spanwise direction, sequence 20. It is evident that there is a strong correlation between eddy formation, shedding, and spanwise velocity gradient fluctuations.

25.3.6 *Strouhal number*

Early measurements of the eddy shedding frequency in the turbulent wake of yawed cylinders were reported by Smith *et al.* (1972J). The tests were carried out in an open jet by using a cylinder without end plates, $L/D = 39.3$, $2\text{k} < Re < 10\text{k}$, $\Lambda = 0°, 15°, 30°, 45°$, and $60°$, and $Ti = 0.22\%$. For unyawed cylinders, the frequency peaks were sharp and narrow-banded. For yawed cylinders, the peaks became less pronounced, and wide-banded. For $\Lambda = 60°$, the lack of coherence resembled the disorganized turbulent wake.

Another measurement of eddy shedding frequency in a wider range $50 < Re < 12\text{k}$ for $\Lambda = 39°$ was reported by Shirakashi *et al.* (1984J). Three cylinders, $L/D = 2070, 415$, and 69 spanned a closed test section through large slots $10D \times 20D$ ($D = 5.98\,\text{mm}$). The collapse of experimental points onto the empirical curve was found to be good in the L3 and TrSL1 regimes. The notable scatter in the TrW and TrSL2 regimes indicated the interferences between the yawed cylinder with the transitional three-dimensional wakes.

Both sets of measured shedding frequency mentioned previously were confined to a single point behind the mid-span of the yawed cylinder. Hayashi *et al.* (1992J) extended measurements of the eddy shedding frequency along the span of yawed cylinders with and without end plates, $L/D = 15$ and $Re = 20\text{k}$. Figure 25.21(a,b,c,d) shows the frequency spectra for $\Lambda = 10°$ and $20°$ without and with end plates, respectively. There are three features to be noted:

[148]The term ‘swirl’ describes an eddy having a velocity component in the direction of its axis.

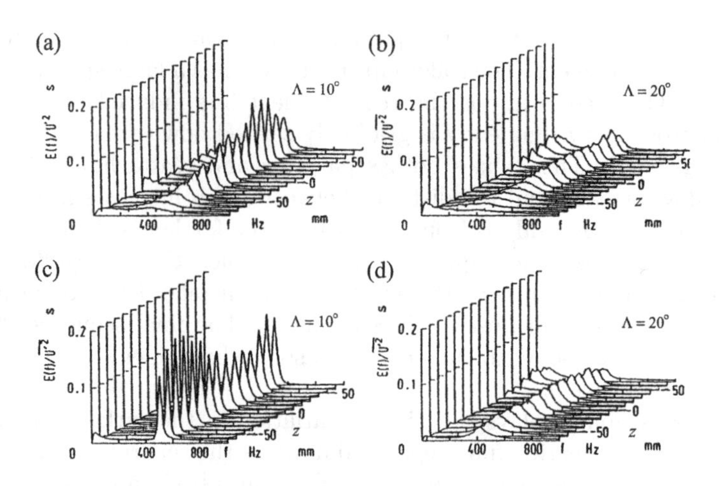

FIG. 25.21. Frequency power spectra along the span for $Re = 20$k, (a) without end plates, $\Lambda = 10°$, (b) $\Lambda = 20°$, (c) with end plates, $\Lambda = 10°$, (d) $\Lambda = 20°$, Hayashi *et al.* (1992J)

(i) eddy shedding is not uniform along the span;

(ii) the addition of end plates for $\Lambda = 10°$ markedly improves the uniformity of eddy shedding, which becomes very regular at the upstream end of the yawed cylinders;

(iii) there is a notable collapse of strong eddy shedding for $\Lambda = 20°$ with and without end plates. The sharp- and narrow-banded peaks along the span are replaced by an order of magnitude weaker and wide-banded peaks. This trend continues for $\Lambda = 30°$ (not shown in Fig. 25.21).

Figure 25.22 shows a compilation of all shedding frequencies and St (based on V) along the span. For yawed cylinders, $\Lambda = 10°$ without end plates, and the single value of St is still preserved. However, when the end plates are fitted, the points are above the $\Lambda = 0°$ curve for the downstream end and below for the upstream end. This may indicate two eddy shedding cells. This trend further develops for $\Lambda = 20°$ where, despite the large scatter and wide-banded peaks in the energy spectra, several shedding cells may be recognized. For example, for $\Lambda = 20°$ with end plates, three cells are seen at the upstream end, central span, and along the downstream end. It appears that the weakening in eddy shedding is associated with the appearance of shedding cells at different frequencies.

An indirect piece of evidence for the cause of erratic eddy shedding in turbulent wakes behind yawed cylinders was also provided by Shirakashi *et al.* (1986J). They used partition plates, $04D \times 4D$, and placed them behind the yawed cylinders at $\theta = 180°$. The partition plates affected neither the upstream nor side

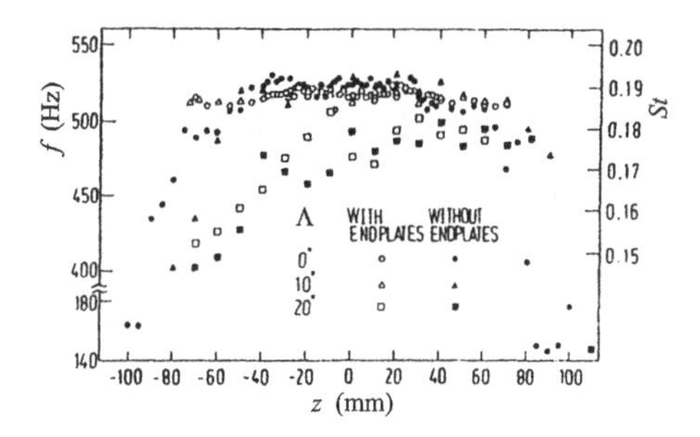

FIG. 25.22. Compilation of eddy shedding frequency and St variation along the span with and without end plates, Hayashi *et al.* (1992J)

flows. They prevented secondary flow in the spanwise direction. The remarkable recovery of strong eddy shedding between the plates was marked by sharp and narrow-banded peaks. These occurred in the range $0.2 < L/D < 0.8$. This piece of evidence indicated that the spanwise velocity component interfered with the eddy shedding mechanism.

25.3.7 *Drag coefficient*

The value of the drag coefficient for unyawed cylinders is related to the base pressure coefficient. It has been argued that the latter should not be correlated by using the Independence Principle. As an example, Fig. 25.23 shows the pressure distribution, C_{pn} (based on $V \cos \Lambda$), for $0° \leqslant \Lambda \leqslant 60°$ at $Re_n = 6.5$k. There is an orderly departure in C_{pbn} with rising Λ values from $\Lambda = 0°$, and the Independence Principle is invalid for all values of Λ.

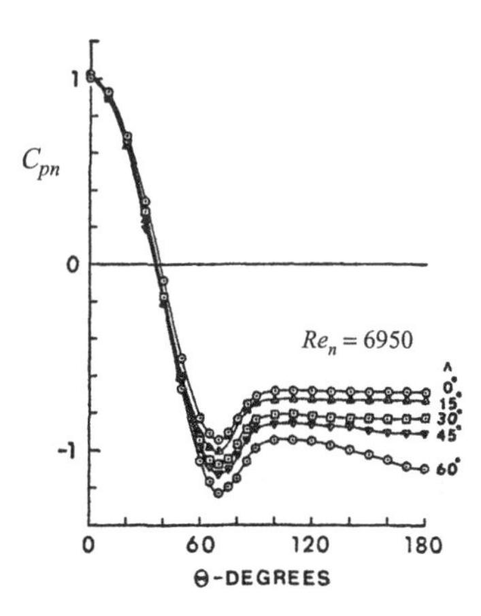

FIG. 25.23. Mean pressure distribution for $Re_n = 6.95$k, Smith *et al.* (1972J)

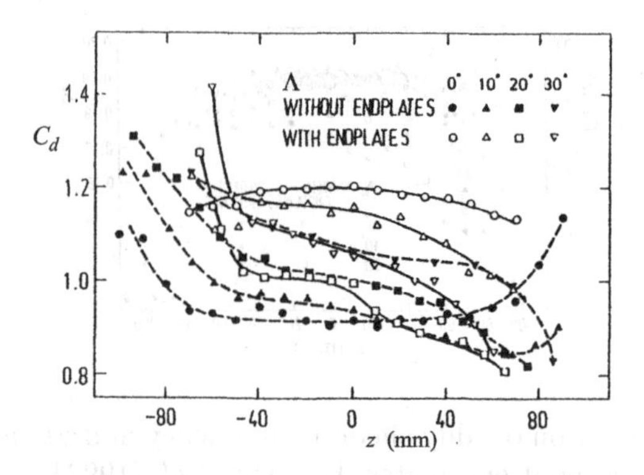

FIG. 25.24. Local drag variation along the span for Re = 20k, Hayashi *et al.* (1992J)

The local drag coefficient evaluated from C_{pn} measured at the mid-span by Smith *et al.* (1972J) showed a similar departure of curves for different Λ. The variation of local C_d along the span was evaluated from local C_p by Hayashi *et al.* (1992J). Figure 25.24 shows C_d variation along the span for Re = 20k and $\Lambda = 0°, 10°, 20°$, and 30° for yawed cylinders with and without end plates. Without end plates there is a rise in C_d with rising Λ in an orderly fashion. Adding the end plates changes the effect of the yaw angle Λ. The values of C_d decrease below the $\Lambda = 0°$ curve. The high C_d remains at the upstream end and low C_d at the downstream end. This inexplicable effect of end plates underlines yet again that the flow around yawed cylinders without end plates is dissimilar to that with end plates.

25.4 Turbulent wakes in the TrBL state

Bursnall and Loftin (1951P) carried out tests on yawed cylinders in the range 102k $< Re <$ 596k, $0 \leqslant \Lambda \leqslant 60°$, $18 < L/D < 36$, and $D/B = 0.02$. The spanwise flow uniformity was checked by measuring the pressure distribution at three stations $3D$ apart. The inverted pressure distributions at the start of the TrBL0 regime, expressed as C_{pn} in terms of Re_n, is shown in Fig. 25.25. The Independence Principle does hold along the favourable pressure gradient for $\Lambda \leqslant 45°$. The onset of transition in the TrBL2 regime occurred progressively at lower Re_n as Λ increased. For example, the fall in C_{dn} was found at Re_n = 370k, 331k, 301k, 230k, and 106k for $\Lambda = 0°, 15°, 30°, 45°$, and 60°, respectively. These values indicate that the onset of the TrBL state cannot be correlated by the Independence Principle.

The local drag coefficient, C_{dn} (based on $V \cos \Lambda$), was evaluated from $C_{pn}0$ measured at the mid-span.[149] Figure 25.26 shows C_{dn} curves in terms of Re_n

[149]The flow was found to be non-uniform along the span.

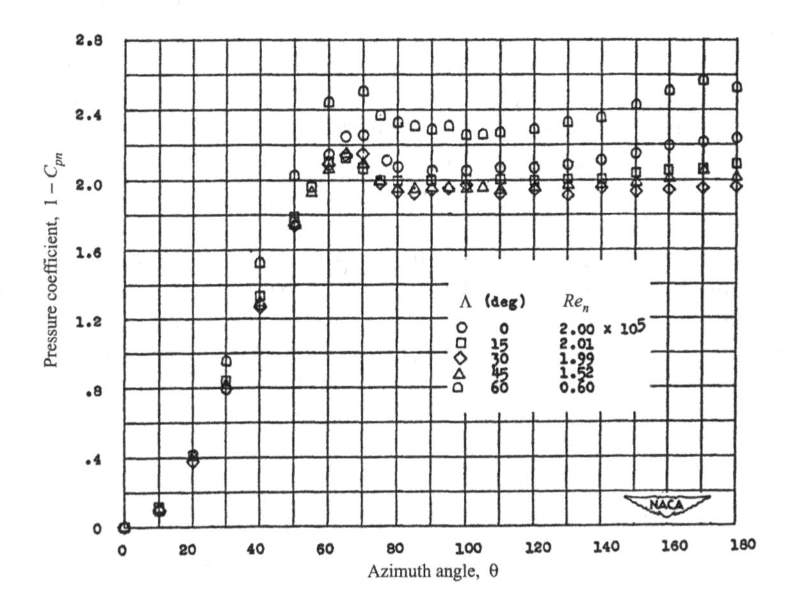

FIG. 25.25. Inverse mean pressure distribution C_{pn} for various Λ, Bursnall and Loftin (1951P)

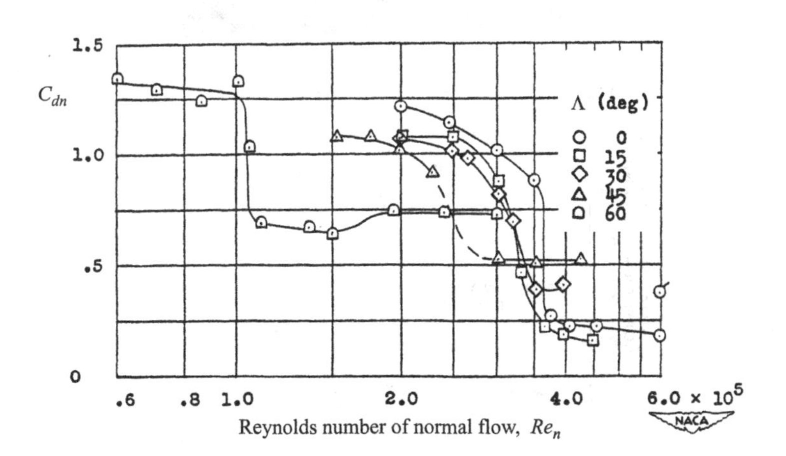

FIG. 25.26. Variation of local drag C_{dn} in terms of Re_n, Bursnall and Loftin (1951P)

for tested Λ. The Re_c has been defined in Vol. 1, Chapter 6, as the Re at which C_D falls rapidly. For $\Lambda = 0°$, $Re = 365$k and it decreases to $Re_{cn} = 106$k for $\Lambda = 60°$. The decrease in C_{dn} becomes more gradual as Λ is increased. However, for $\Lambda = 60°$, the fall in drag is abrupt at three times lower Re_n than for the unyawed cylinder. Another important feature in C_{dn} curves is the gradual and monotonous rise in C_{dnmin} with rising Λ.

25.4.1 *Marine surface roughness*

The effect of marine growth on the hydrodynamic forces exerted on unyawed cylinders has been discussed in Chapter 22. Norton *et al.* (1981P) extended measurements to yawed cylinders at $\Lambda = 0°, 10°, 20°, 30°$, and $40°$ for $Re = 2$M, $10 < L/D < 13, D/B = 0.143$, and $K/D = 3.5\%$. Two force components were measured, normal to the cylinder axis C_{Da} and spanwise along the axis C_{Da}. Figure 25.27 shows that the C_{Da} can be neglected up to $\Lambda = 30°$.

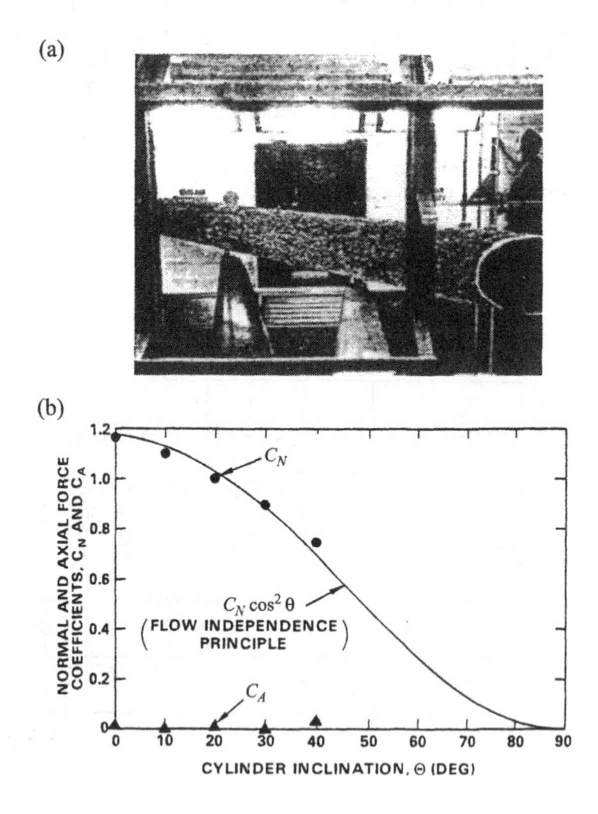

FIG. 25.27. (a) Yawed rough cylinder $K/D = 3.5\%$ in the test section, (b) drag components in normal velocity and span directions for $Re = 2$M, Norton *et al.* (1981P)

Norton *et al.* (1981P) also measured velocity fluctuations at the mid-span, $x/D = 2$ and $y/D = 1$, by using a hot-film probe. At $Re = 2$M, the narrow-band peak, as detected at $\Lambda = 0°$ and $10°$, rapidly decreased in magnitude. The frequency peak disappeared into the background turbulence for $\Lambda = 40°$. This trend is remarkably similar to that observed by Hayashi *et al.* (1992J) at $Re = 20$k.

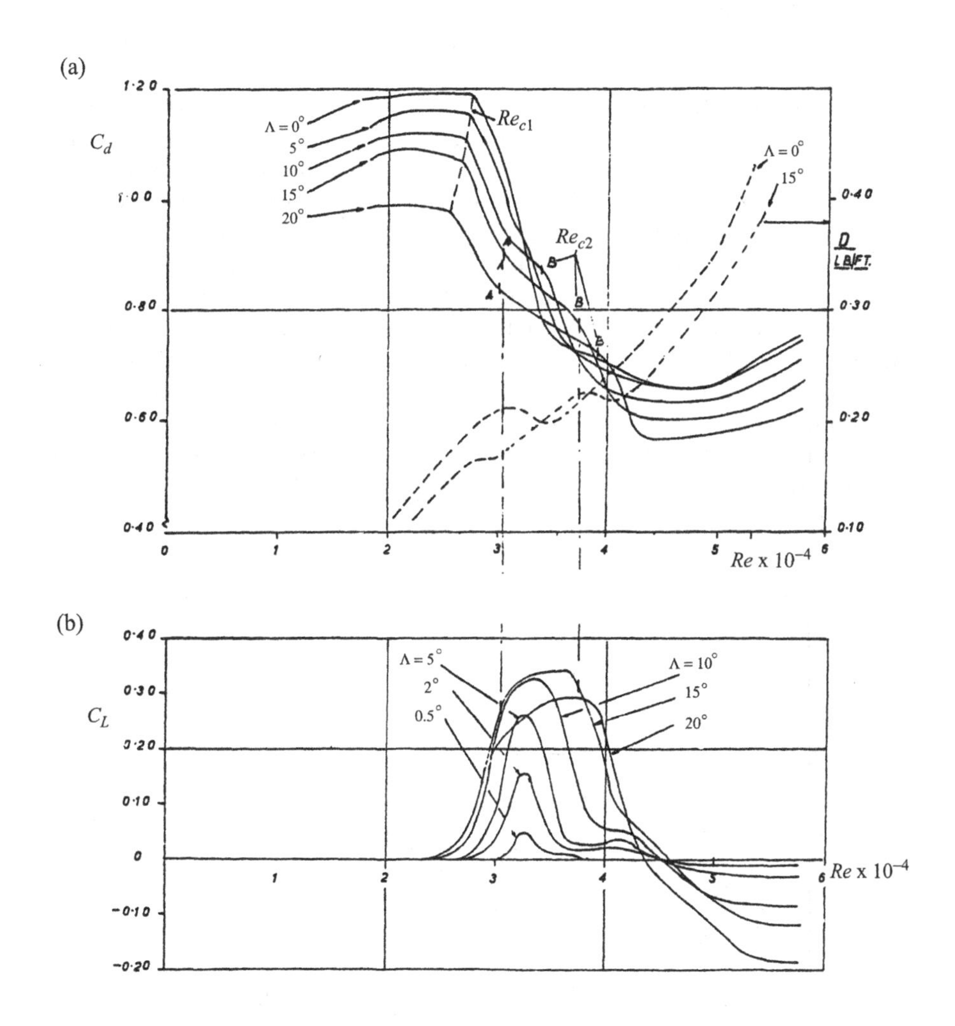

FIG. 25.28. Force components on a stranded conductor for $N = 42$ in terms of *Re*, (a) drag, (b) lift, Counihan (1963P)

25.4.2 *Stranded cables and conductors*

Stranded cables have already been discussed in Chapter 22, Section 22.3.8 as a special type of surface roughness. The strands appear differently orientated on the opposite side of the cable, having the helix pitch angle, $\pm\gamma$. For yawed stranded cables, however, the relative strand angle changes to $\Lambda \pm \gamma$. The wind may pass across the strands on one side, and along strands on the other side. Although the relative roughness, K/D, remains unchanged, the texture changes considerably. Hence, the yawed stranded cable appears 'smooth' on one side, and 'rough' on the other side.

Counihan (1963P) carried out drag and lift measurements on the stranded cables having $N = 18, 24$, and 42 strands, $L/D = 66$, and $D/B = 0.02$ fitted with end plates in the range 18k $< Re <$ 56k. The variation in C_D with Re is shown in Fig. 25.28(a). For $\Lambda > 5°$, the shape of the C_D curves begins to change.

As the yaw angle Λ is increased, the strands on one surface become more out of line to the wind direction so that the surface becomes effectively rougher, and the opposite effect is obtained on the other surface. These trigger the TrBL0 regimes at different Re on two sides of the yawed cylinder. The first one decreases Re from 28k to 24k at $\Lambda = 22.5°$, and the second increases to approximately Re = 40k. At any Re greater than Re_c, the asymmetry of flow around two surfaces produces a side or lift force. Figure 25.28(b) shows the variation in C_L with Re. Wrapping the conductor with PVC tape reduces the surface roughness, this in turn raises Re_c = 60k at $\Lambda = 0°$, to Re_c = 40k at $\Lambda = 15°$. The measured

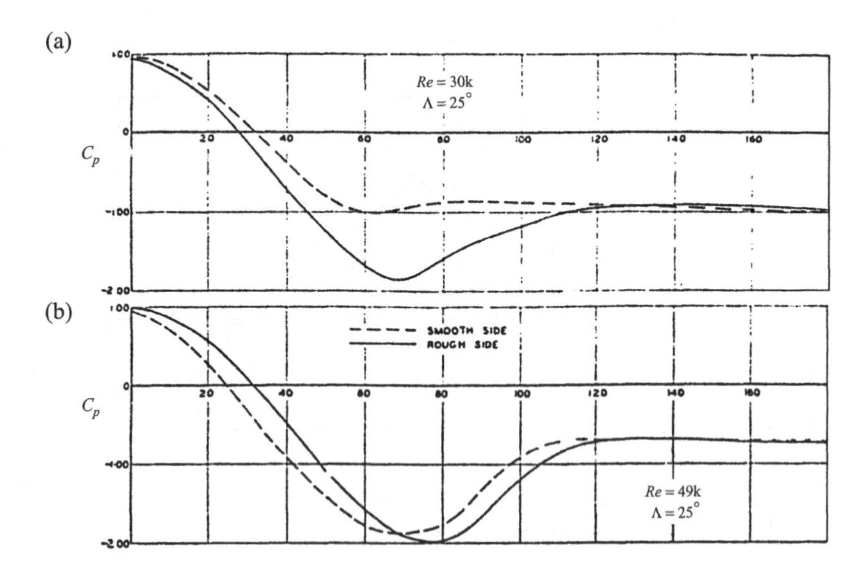

FIG. 25.29. Mean pressure distribution on a stranded conductor for $\Lambda = 25°$ and $N = 42$, (a) Re = 30k, (b) Re = 49k, Counihan (1963P)

C_L becomes negligible in comparison with the untaped stranded cable.

The pressure distribution measured on a 42-strand cable confirmed the existence of a flow asymmetry. Figure 25.29(a) shows the asymmetric C_p distribution for $\Lambda = 25°$ at $Re = 30$k. The 'rough' surface separation moves downstream, and there is an indication that the 'smooth' surface separation moves slightly upstream. With a further increase in Re, the smooth surface becomes displaced downstream. The result is a pressure difference, shown in Fig. 25.29(b), which produces C_L in the opposite direction to that produced at the lower Re.

25.5 High angle of incidence

25.5.1 *Classification of flow regimes*

In missile aerodynamics, the angle of incidence α has been defined as that between the free stream velocity and the missile longitudinal axis. This angle is precisely a complement to the yaw angle, $\alpha + \Lambda = 90°$. It means that the flow at a small α corresponds to high Λ and vice versa. Enormous research has been carried out on missiles at small and high Λ, as reviewed by Wardlaw (1978R), Nielsen (1979R), Ericsson and Reding (1980R), etc. The separated three-dimensional flow is dominated by swirling eddies.

An excellent flow visualization in water was carried out by Werle (1979P). He used coloured dye released from the upstream side of the cylinder, $L/D = 16$, with an ogive $4D$ long at $\alpha = 38°$ and $Re = 2$k. Figure 25.30(b) reveals two symmetrically disposed swirling flows along the lee side of the cylinder. The swirling flows consist of the roll-up free shear layers separated from the cylinder, and the entrained neighbouring flow marked by the dye filaments. Farther along the cylinder, first one and then the other swirl detaches, and forms a steady trail of swirls, Fig. 25.30(a).

The air bubble visualization in Fig. 25.30(c,d,d',e) shows the structures of the swirl cores at $x/D = 3.9, 7.75$, and 11.6, respectively. At first, two symmetric small eddies are seen in Fig. 25.30(c). They increase in size, become asymmetric, and one is eventually detached from the cylinder, Fig. 25.30(d). The detached eddy may become distorted occasionally, as seen in Fig. 25.30(d'). Finally, one eddy detaches from the cylinder, and only the attached one is seen in Fig. 25.30(e). For a long cylinder at a high angle of incidence α a series of detached swirls along the cylinder form a swirl trail.

The change-over from the symmetric to asymmetric arrangement of eddies depends on a series of parameters: $\alpha, x/D$, nose shape, $L/D, Re, Ma$, etc. Fiechter (1966P) established the change-over boundary from his flow visualization. Figure 25.31 shows the boundary between symmetric and asymmetric eddies α in terms of x/D for a 4.2D ogive followed by an $L/D = 16$ cylinder.

Based on flow visualizations over a whole range of α, the flow past cylinders at small, medium, and large α can be classified as follows:

(i) small α: $0 \leqslant \alpha \leqslant 1/2\alpha_c$ or $1/2\alpha_o$, where α_c and α_o are the inclusive cone and ogive angles, respectively. The onset of separation is observed when α

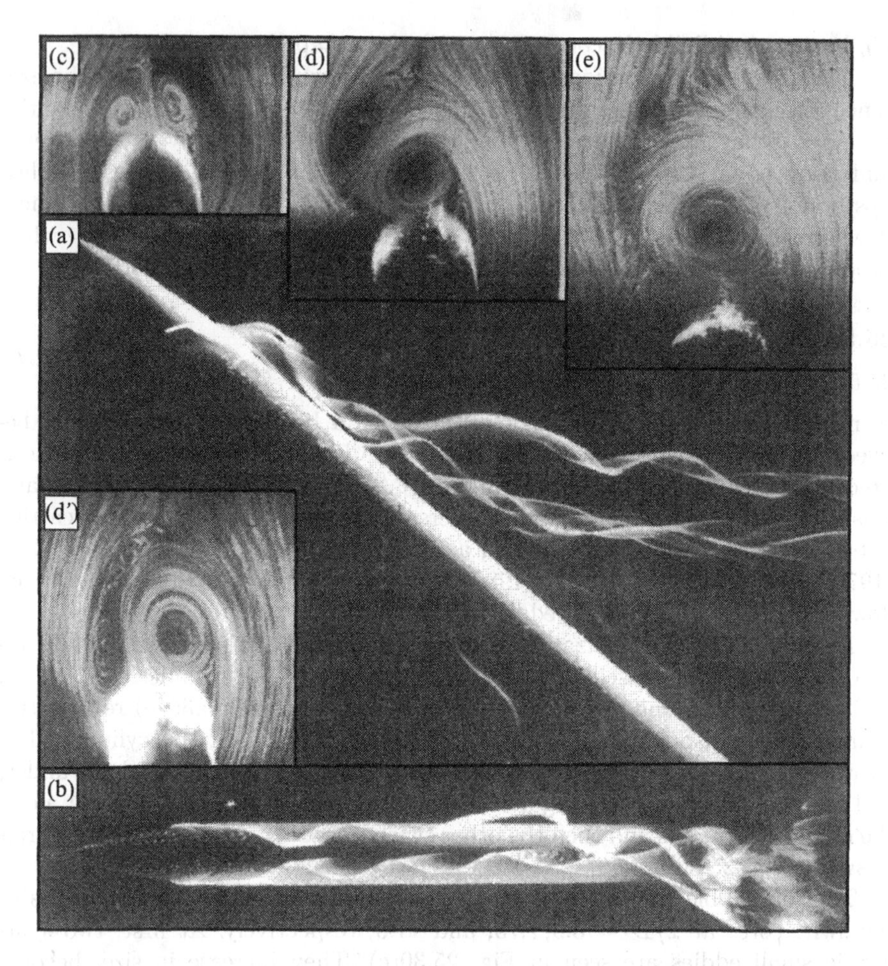

FIG. 25.30. Flow visualization past a missile at $\alpha = 38°$ for $Re = 2k$, (a) side view, (b) top view, (c) $x/D = 3.9$, (d,d') $x/D = 7.75$, (e) $x/D = 11.6$, Werle (1979P)

exceeds approximately $1/2\alpha_c$;

(ii) low medium α: $1/2\alpha_c < \alpha < 10°$ to $50°$, depending upon x/D; the stable and steady symmetric eddies remain attached to the cylinder surface inside a closed near-wake, see sketches in Fig. 25.32(a);

(iii) high medium α: $10°$ to $50° < \alpha < 60°$ to $75°$ depending on x/D, Fig. 25.32(b). The quasi-steady asymmetric eddies are alternately detached along and downstream of the cylinder. After each detachment, a new eddy is formed by the rolled-up free-shear layers. The detached streamwise eddy

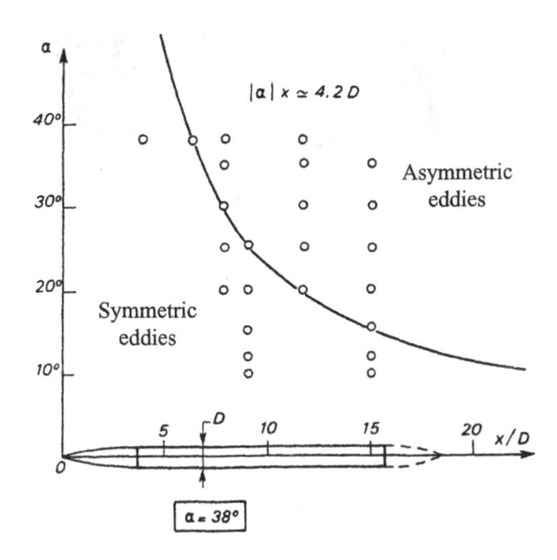

FIG. 25.31. Boundary between symmetric and asymmetric eddies for $2\text{k} < Re < 4\text{k}$ in terms of x/D, Fiechter (1966P)

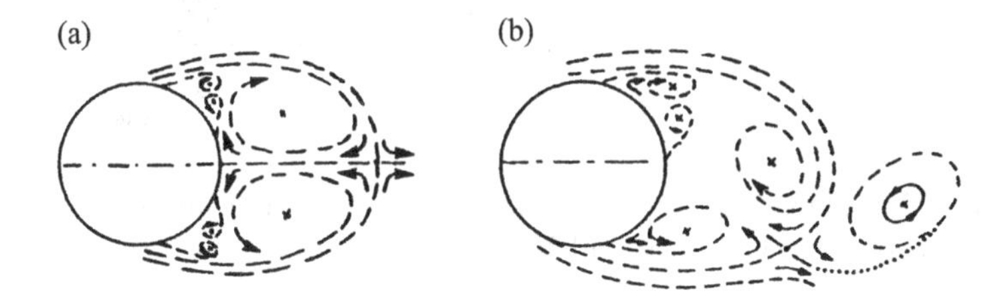

FIG. 25.32. Sketch of flow topology, (a) symmetric eddies, (b) asymmetric eddies, Werle (1979P)

filaments appear as a stationary alternate eddy street. Figure 25.33(a,b) shows Thomson and Morrison's (1971J) schlieren visualization for $\alpha = 30°$, $M_n = 0.25$, and the topology sketch, respectively;

(iv) large α: 60° to $75° < \alpha \leqslant 90°$, depending upon L/D; the asymmetric and alternate arrangement of eddies is established in the form of the Kármán–Bénard eddy street in the vertical cross-section, see Fig. 25.33(a,b).

The overlapping of the flow regimes is strongly affected by the shape, extent, and orientation of the nose, and x/D location. The force exerted on the cylinder depends on α, L/D, Re_n, Ma_n, and the other influencing parameters.

(a)

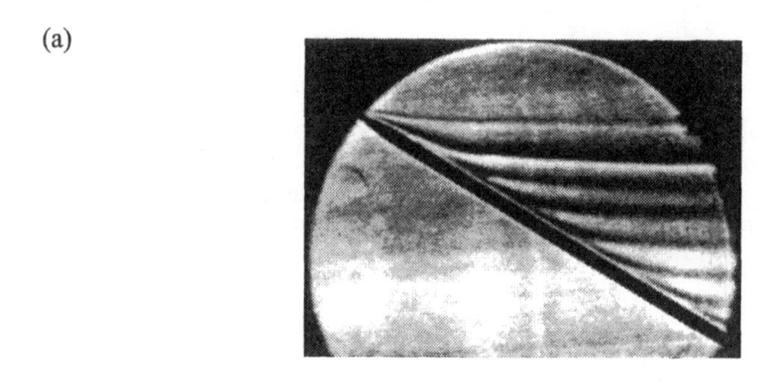

(b)

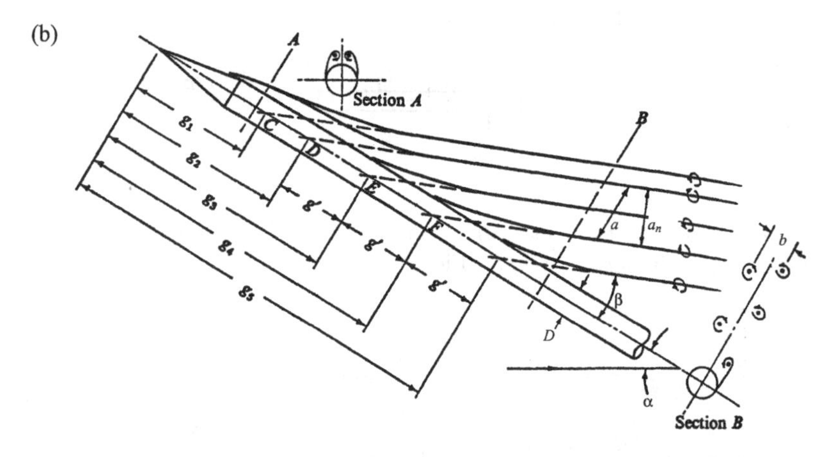

FIG. 25.33. Missile at $\alpha = 30°, Ma = 0.5$, $Re = 75$k, (a) schlieren visualization, (b) flow topology, Thomson and Morrison (1971J)

25.5.2 *Impulsive cross-flow analogy*

The cross-flow analogy was suggested by Allen and Perkins (1951P). It was established by Prandtl (1961B) that flow sequences for an impulsively started unyawed cylinder are always in the following order:

(i) the potential-like flow without separation at the start;

(ii) later, the effect of viscosity results in separation and the formation of two symmetrical counter-rotating eddies;

(iii) the two eddies become unstable and different in size;

(iv) finally, an alternate periodic eddy shedding is established.

All these sequences of flow develop gradually along the yawed cylinder at the medium α. The separation initiates at the nose followed by the attached symmetric eddies. Farther along the yawed cylinder they become unequal in size at a certain x/D distance, and eventually one eddy becomes detached. The progressive development of eddies along the yawed cylinder when viewed in cross-flow planes is analogous to the development of the impulsive flow behind the unyawed cylinder.

25.5.3 *Strength of detached vortices*

Kármán's vortex street theory, as discussed in Vol. 1, Chapter 13.7, p. 333, was used[150] by Thomson and Morrison (1971J) to evaluate the strength of vortices K:

$$\frac{K}{VD\sin\alpha} = 2\frac{a}{D}\left(1 - \frac{\tan\beta}{\tan\alpha}\right)\coth\left(\frac{\pi b}{a}\right) \qquad (25.4)$$

This equation is derived for an incompressible and potential (inviscid fluid) flow. However, Thomson and Morrison extended the use of eqn (25.4) to subsonic and supersonic Ma_n flows. They determined α, β, and b/a from the flow visualization and calculated $K_n/VD\sin\alpha$. Figure 25.34 shows the wide range of

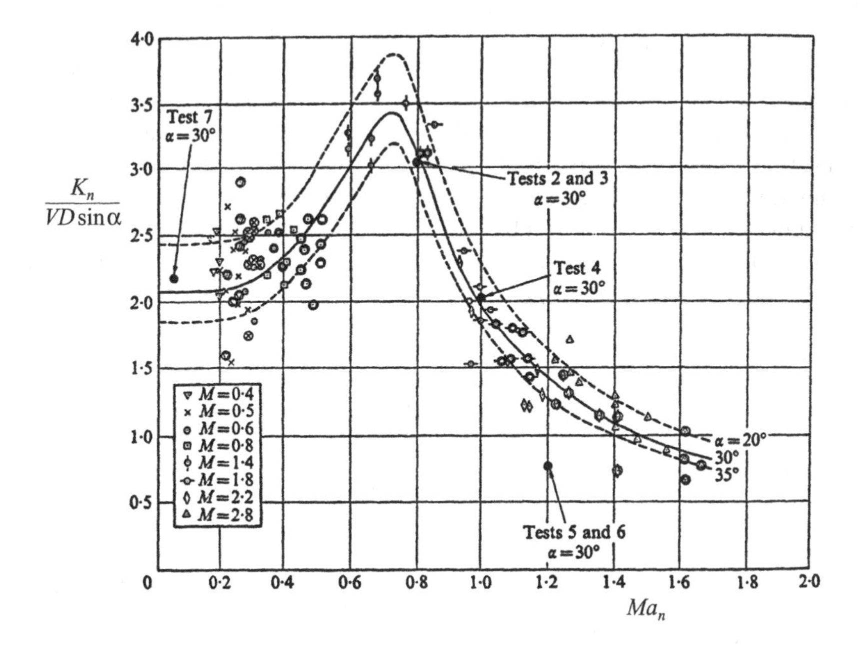

FIG. 25.34. Non-dimensional strength of eddies in terms of Ma, Thomson and Morrison (1971J)

[150]Morkovin (1964R) suggested this approach when a cross-flow analogy is applicable.

the free stream, $0.4 \leqslant Ma_n \leqslant 1.8$ and $20° < \alpha < 35°$. The three fitted curves show that K decreases with increasing α. All points correspond to the strength of the first vortex for a spacing ratio of $b/a = 0.19$.

The distinct peak in the $K/VD \sin\alpha$ curve arises from the minimum in $\tan\beta/\tan\alpha$ at $Ma_n = 0.7$. It has been shown in Vol. 1, Chapter 16, p. 500 that there was a peak in C_D at around $Ma = 0.7$ caused by the appearance of permanent shock waves before the separation. The shock waves were not incorporated into the incompressible theory.

25.5.4 *Normal and side force components*

When symmetric eddies are formed, the resultant drag force lies in the plane formed by the free stream velocity and cylinder axis. However, the asymmetric eddies induce an asymmetric pressure distribution on two sides of the cylinder. The resultant force has an out-of-plane component at right angle to the free stream, which is called the *side* force. The flow topology shown in Fig. 25.32(a,b) indicates that the side force varies along the cylinder depending upon the strength and arrangement of the asymmetric eddies.

There have been numerous measurements of the side force with large discrepancies and inconsistencies. The flow was highly sensitive to shape, length, and eccentricity of the nose, downstream station x/D, overall $L/D, Re_n, Ma_n$, and other influencing parameters. Lamont and Hunt (1976J) carried out tests at $Re = 110\text{k}$, $L/D = 11, D/B = 0.02$, and $30° < \alpha < 75°$. The pressure was measured simultaneously at 10 stations approximately $1/2D$ apart with 36 tappings at each one. The normal drag, C_{dn}, and side force, C_{sn}, were evaluated along the cylinder from the local C_p distributions for a $4D$ long ogive. Figure 25.35 shows that the variation in C_{dn} is similar to that found for the impulsively started unyawed cylinder, see Sarpkaya (1966J). The variation in C_{sn} is also sinusoidal, and eventually becomes zero for $\alpha > 75°$. The initial maximum and minimum

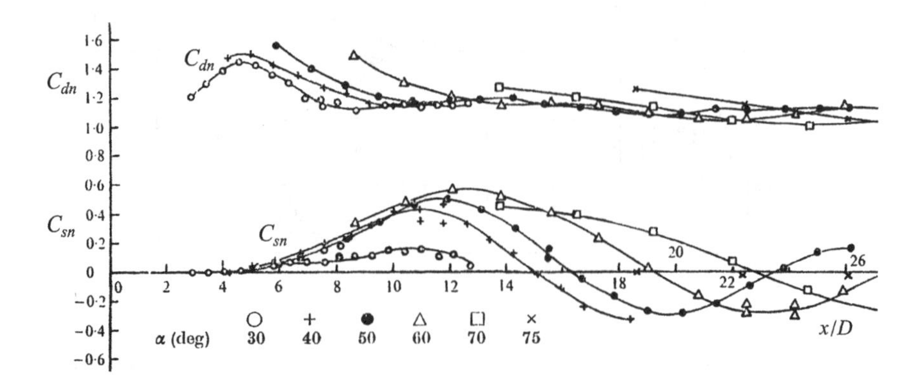

FIG. 25.35. Local normal drag and side force for a $2D$ ogive at $Re = 110\text{k}$, Lamont and Hunt (1976J)

amplitudes are produced by the first eddies, and the wavelength decreases after each detachment. This is related to the position and strength of eddies from the surface.

Lamont and Hunt (1976J) also found a strong effect of the nose shape and length on C_{dn} and C_{sn}. For example, the ogives $1D, 2D$, and $3D$ long, and a cone $2D$, produced $C_{sn\max} = 0.5, 0.6, 1.4$, and 2.0, respectively. The cone $2D$ long produces 42% higher $C_{sn\max}$ than the $2D$ ogive fitted to the same cylinder. Another source of a huge discrepancy was the eccentricity of the apex. Figure 25.36 shows a large variation in C_{sn} with roll angle for the $2D$ ogive at $\alpha = 50°$ and Re = 110k. The roll angle was varied in steps of 30° from 0° to 180°. The values of C_{dn} are slightly affected by γ but C_{sn} is greatly influenced. Another test for $L/D = 7.5$ cylinders with a $2D$ ogive at $\alpha = 55°$ and Re = 4M shows a change in direction of C_{sn}. Figure 25.37 shows C_{sn} based on area $= D^2\pi/4$ in terms of roll angle ϕ. For the turbulent separation, a switch in eddy detachment took place, leading to a different sign for C_{sn}.

The origin of the side force and its variation is related to the asymmetric pressure distribution. Figure 25.38 shows the C_p distribution at the station where a maximum side force occurs for the $2D$ ogive cylinders at $\alpha = 50°$ and Re = 112k. The difference in the base pressure produces $C_{sn} = 0.71,\ 0.69$, and 0.6 for

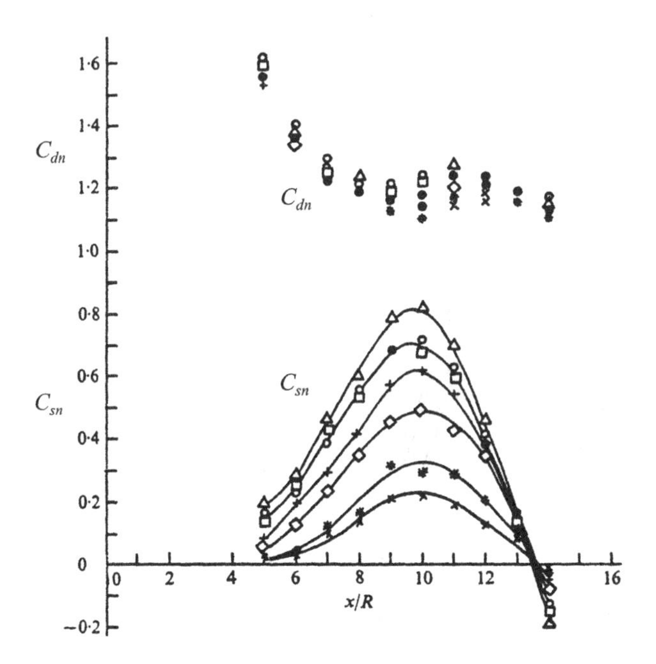

FIG. 25.36. Effect of roll on side force for $\alpha = 55°$, Re = 110k, 2D ogive, Lamont and Hunt (1976J)

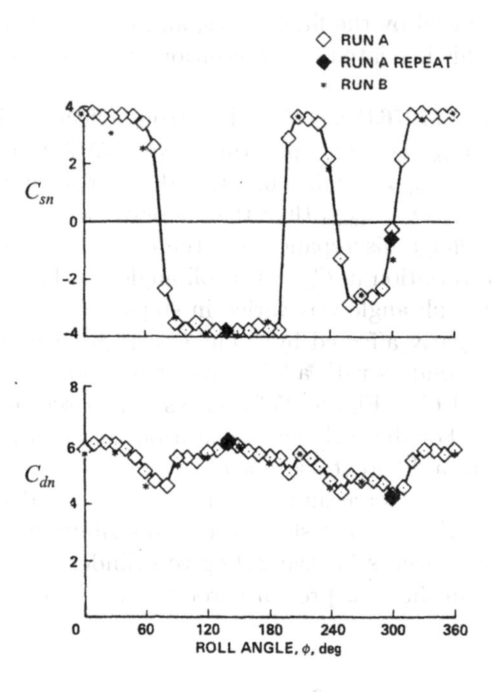

FIG. 25.37. Effect of roll on drag and side force for $\alpha = 50°$, $Re = 4$M Lamont (1980P)

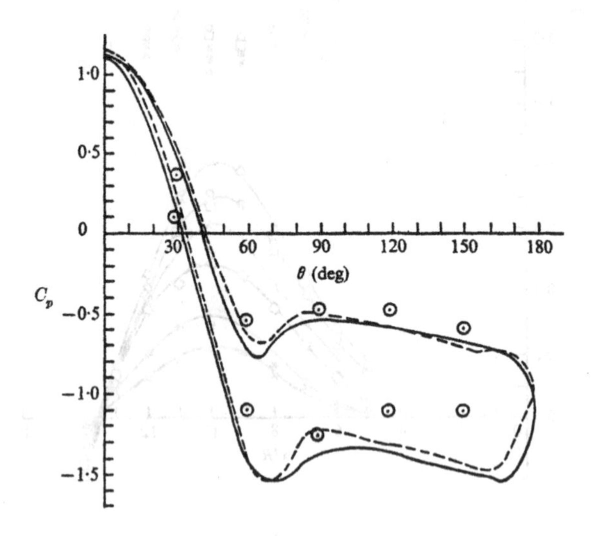

FIG. 25.38. Pressure distribution for $\alpha = 50°$, $Re = 110$k, at x/D for maximum local side force, Lamont and Hunt (1976J)

the $2D$ ogive (full line), the $3D$ ogive (dotted line), and unyawed cylinder $C'_{p\max}$ (points), Drescher (1956J).

25.5.5 *Effect of Reynolds number*

The swirling eddies emanate from the separated boundary layers. It is expected that the transition from laminar to turbulent separation may disrupt the formation and detachment of swirling eddies. Lamont (1980P) carried out tests on a $L/D = 7.5$ cylinder with a $2D$ ogive nose in the range $20° < \alpha < 90°$ and 200k $< Re <$ 4M. Mean pressure was measured at each of the 12 stations with 36 tappings and fluctuating pressure at four stations with 12 transducers each.

Figure 25.39 shows the C_{sn} (based on $D^2\pi/4$) variation along the cylinder x/D for $\alpha = 55°$ in a wide range 200k $< Re <$ 4M. As Re increases, the peak measured at $Re =$ 200k rapidly decreases at $Re =$ 400k in the TrBL2 regime. At $Re =$ 800k, the supercritical TrBL3 regime brings a fragmented separation line, and C_{sn} is virtually annihilated. As soon as straight turbulent separation lines are re-established at $Re =$ 4M, the critical TrBL4 regime, large C_{sn}, re-appears.

The origin of the maximum C_{sn} for laminar and turbulent separation can be traced to the asymmetric C_p distribution. Figure 25.40 shows the C_{pn} distribution for $\alpha = 55°$ at $x/D = 3$ for laminar $Re =$ 200k, and turbulent $Re =$ 4M. Both distributions on one side have only one $C_{p\min}$ at $\theta = 65°$ and $85°$, with the end of the pressure recovery ending at $\theta = 85°$ and $100°$ for laminar and turbulent separation, respectively. There is a dominant second $C_{p\min}$ peak at $\theta = 150°$ and $160°$ on the other side caused by the vicinity of laminar and turbulent eddies,

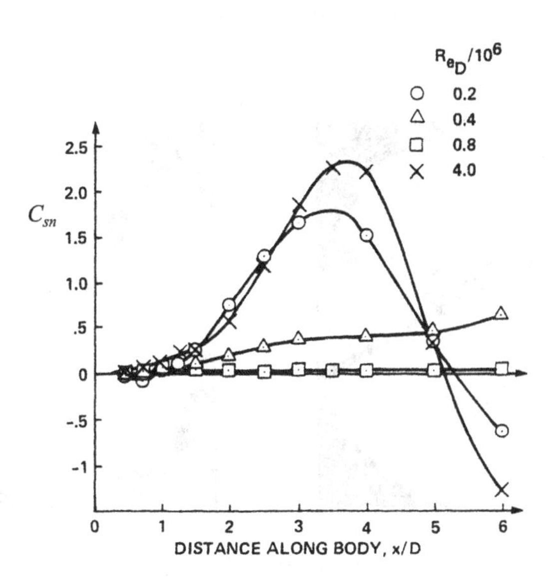

FIG. 25.39. Local side force along the missile for various Re, Lamont (1980P)

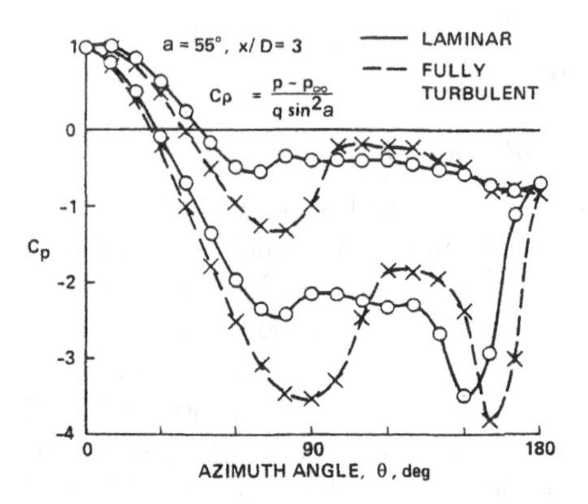

FIG. 25.40. Mean pressure distribution for $\alpha = 55°$, $x/D = 3$, $Re = 200$k, and 4M, Lamont (1980P)

respectively.

25.5.6 *Effect of Mach number*

The compressibility effects on flow past circular cylinders have been discussed in Vol. 1, Chapter 16. The shock waves periodically alternate on two sides of the unyawed cylinder at $Ma > 0.4$, and become steady and permanent for $Ma > 0.7$. When $Ma > 1$ a detached bow shock wave is formed followed by a subsonic flow. The periodic eddy shedding ceases for all supersonic free streams, $Ma > 1$.

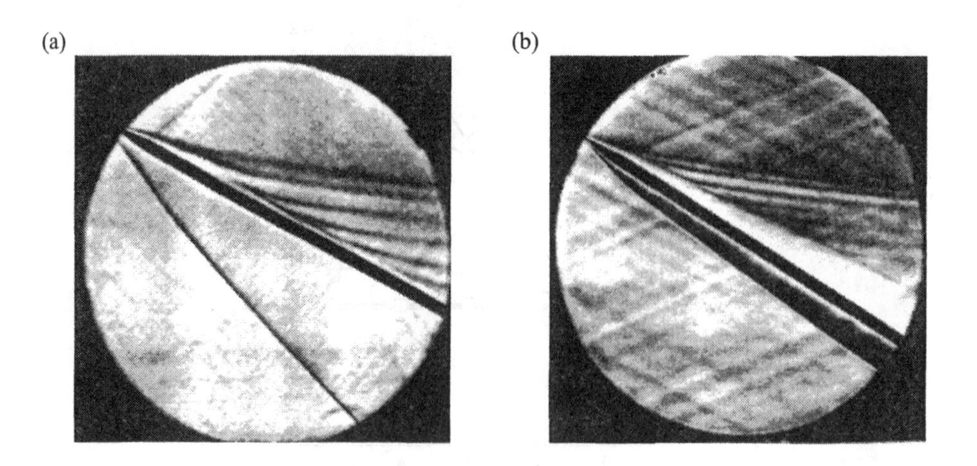

FIG. 25.41. Schlieren visualization at $\alpha = 30°$, (a) $Ma = 1.4$, $Ma_n = 0.7$, (b) $Ma = 2.4$, $Ma_n = 1.2$, Thomson and Morrison (1971J)

Thomson and Morrison (1971J) carried out a systematic schlieren visualization behind yawed cylinders at $Ma > 1$. Figure 25.41(a) shows the cylinder at $\alpha = 30°$ and $Ma = 1.4$, having a normal component, $Ma_n = 0.7$. The oblique shock wave is seen below the cylinder, and detached eddy filaments appear similar to those obtained for the subsonic free stream. Six equally spaced trailing eddy filaments are seen along the cylinder's upperside. Once the normal $Ma_n > 1$, the pattern of the trailing eddy filaments changes. Figure 25.41(b) shows a cylinder at $\alpha = 30°$ and $Ma = 2.4$, $Ma_n = 1.2$. The oblique shock wave is a few D below the cylinder and the Ma behind the oblique shock wave is supersonic. The weak trailing eddies are formed only behind the ogive section and they are close to each other.

Thomson and Morrison (1972J) also measured the distribution of total head, p/p_0, in the wake of a cone-cylinder at $\alpha = 30°$. Figure 25.42 shows isobars

FIG. 25.42. Total head isobars at ten x/D stations for $Ma = 1.6$, $Ma_n = 0.8$, $Re_n = 50\text{k}$, and $\alpha = 30°$, Thomson and Morrison (1971J)

at various stations along the cylinder for $Ma = 1.6$, $Ma_n = 0.8$, $Re = 50\text{k}$, and cone half-angle 10°. The symmetric eddies persist at the first three stations followed by asymmetric eddies up to $x/D = 14$. Beyond $x/D = 15$, detached eddies form a cluster of alternate eddies up to the last station $x/D = 18.6$. This $Ma_n = 0.8$ is near the $K_{\max}$ evaluated by using Kármán's incompressible potential flow theory.

25.5.7 *Detachment instability*

Thomson and Morrison (1971J) observed that the detached eddy filaments may become unstable in a certain α range. For $\alpha < 20°$, the eddy pattern was generally stable. Beyond that, in the range $25° < \alpha < 40°$, the eddy pattern became unstable within a fraction of α. The stable eddy pattern before and after an unstable region always appeared similar in schlieren photographs, but the position of the eddy trails in relation to the apex sometimes changed. When $Ma_n < 1$, the eddy pattern was displaced at a downstream position after the instability without a change in the eddy spacing. When $Ma_n > 1$, there is a flow displacement coupled with a marked decrease in the eddy spacing. It was suggested that the detachment instability originated at the apex initially affected the formation of the symmetric eddies. The disturbed eddy detachment eventually produced a new alternate detached pattern.

25.5.8 *Suppression of eddy detachment*

It has been discussed in some detail that the origin of the side force on the yawed cylinder stems from the asymmetric eddies and their alternate detachment along the cylinder. It appears that the way to annihilate the side force is either to keep the symmetric eddies attached at all α, or to weaken the asymmetric eddies to such an extent that the side force becomes negligible.

A wide variety of means has been tried in the past with no, little, or limited success. All means were attached or applied near the apex and might be grouped into three categories:

(i) surface roughness, tripping wires, and strakes;

(ii) jet-blowing, sidewise, upstream and/or downstream;

(iii) ogive or cone spinning and/or entire cylinder spinning.

The surface roughness is expected to be effective in reducing C_{sn} only for the transitional boundary layer. The reduction in C_{sn} is found in the same Re range level as for the smooth ogive in TrBL regimes.

Rao (1978P) carried out tests with symmetrical straight, helical, and truncated wires. Figure 25.43 shows that helical strakes are used as separation wires to guide a symmetric separation. The straight, inversed, and insufficiently thick wires have little or no effect. However, the properly sized helical wires showed suppression in C_{sn} for $Re = 470\text{k}$, in the transitional range where the side force is low. The second method that attempted to alleviate the side force on yawed

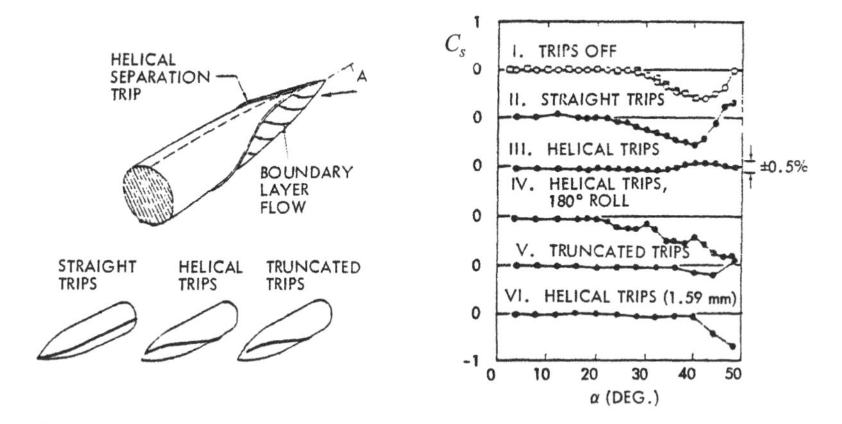

FIG. 25.43. Effect of helical and straight wires upon side force, Rao in Ericsson and Reding, © (1978P) AIAA – reprinted with permission

cylinders was tangential surface blowing. The method proved to be ineffective in most cases, and partly effective in some.

The third method introduced the spinning of a part or a whole yawed cylinder around its axis. Krause *et al.* (1979P) designed the model $L/D = 11.5$ in three separate parts: ogive tip $L/D = 1.5$, the rest of the ogive $L_o/D = 3$, and cylindrical body $L/D = 7$. The tests covered four roll positions: $\gamma = 0°, 90°, 180°$, and $220°$. The tip rotation did not have any effect on C_{sn}. The effect of the whole ogive or body rotation reduced C_{sn} only for one γ roll position.

26

TWO CYLINDERS

26.1 Introduction

Two parallel circular cylinders placed close to each other have been employed in many areas of engineering, such as in aeronautical engineering (twin struts to support biplane wings), in hydronautical engineering (periscope and its guide), in space engineering (twin booster rockets), in civil engineering (twin chimney stacks and jetties), in offshore engineering (platform structural members), in electrical engineering (transmission lines), in chemical engineering (pipe racks), etc. However, despite the wide-spread applications, systematic investigations of flow around two cylinders started rather late in comparison with investigations into a single cylinder.

One reason was the a priori assumption that the flow around and forces on two cylinders should be similar, or even identical to those on a single cylinder. This implied that the flow interference between the two cylinders would be weak or negligible. The early *ad hoc* tests showed that the interference effects might cause considerable changes in flow patterns, magnitude of forces, and eddy shedding. The proliferation of the *ad hoc* testing produced a lot of undigested, uncorrelated, and sometimes inconsistent data.

The interference effects strongly depend on the arrangement of the two cylinders and their orientation to the free stream. There are an infinite number of possible arrangements of two parallel cylinders positioned perpendicularly to the free stream velocity. A simple categorization adopted in Zdravkovich (1977R) is sketched in Fig. 26.1. The categories are:

(i) T, the tandem arrangement; one cylinder behind the other relative to the free stream;

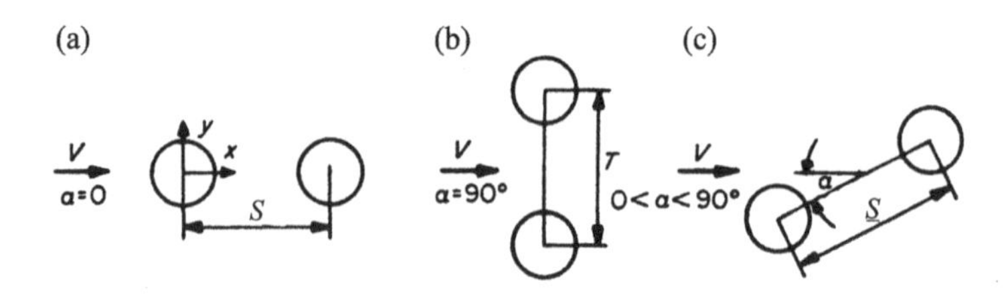

FIG. 26.1. Definition sketch, (a) tandem, T, (b) side-by-side, SS, (c) staggered, S, arrangements

(ii) SS, the side-by-side arrangement; both cylinders face the free stream;

(iii) S, the staggered arrangement; the cylinders are at any combination of streamwise S/D and transverse T/D spacing ratio components.

The governing interference parameters are the spacing ratios: streamwise S/D and transverse T/D for categories (i) and (ii), respectively. For category (iii), either the combination of $S/D, T/D$ or $\underline{S}/D, \alpha$ are taken as shown in Fig. 26.1. Note that $\underline{S}/D$ is the actual spacing ratio, and α is the stagger angle formed by the cylinders' axes plane and the free stream direction. The T/D value gives an immediate insight into whether the downstream cylinder is immersed in the upstream cylinder wake or not. The second option, $\underline{S}/D, \alpha$ does not reveal this important feature. However, it is useful for some applications such as chimney stacks exposed to wind, where $\underline{S}/D = \text{const}$, and α is the variable wind direction.

The adopted categorization of the two-cylinder arrangements served as a framework for this chapter. The tandem, side-by-side, and staggered arrangements are discussed in separate sections in terms of rising Re. This is followed by two cylinders of different diameters for all three categories. Also, a brief discussion is included of the effect of finite height cylinders, free stream turbulence, and surface roughness.

26.1.1 *Basic interference flow regimes*

The three categories of two-cylinder arrangements are associated with different interference flow regions, as sketched in Fig. 26.2. The approximate borders (dash-dot lines) may serve as a rough guide. The hatched areas indicate special bistable flows, and the dashed semi-circle line designates the loci of the cylinders' centres when they are in contact.

The reference cylinder at $S/D = 0$ and $T/D = 0$ is fixed, and the other

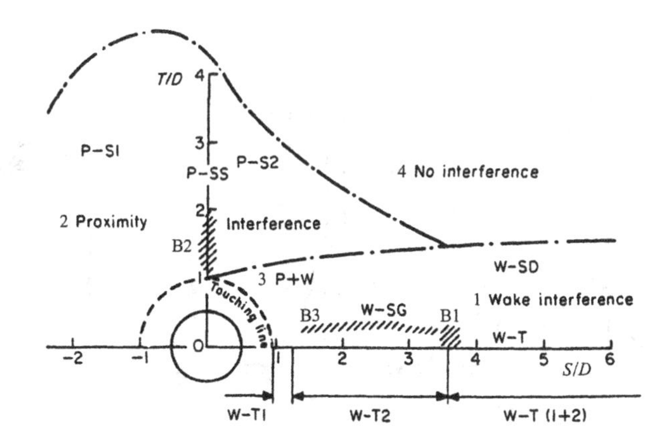

FIG. 26.2. Interference flow regions, Zdravkovich (1987R)

cylinder may assume any upstream or downstream position in the S/D, T/D plane. Four main interference regions may be identified:

(1) W, the wake interference; the downstream cylinder is near to or submerged in the upstream cylinder wake, but is not in close proximity to the reference cylinder;

(2) P, the proximity interference; the cylinders are close to each other;

(3) P+W, the combined proximity and wake interference represents an overlap of P and W, as shown in Fig. 26.2;

(4) no interference anywhere outside the P, W, and P+W regions. The flow around and forces on both cylinders are the same as for the single cylinder.

Figure 26.3 shows the S/D, T/D plot with an approximate location and extent of the observed flow regimes. Category (i), tandem cylinders, exhibits two basically different kinds of wake interference: with and without eddy shedding behind the upstream cylinder. Further subdivision of the wake interference regimes is as follows:

(a) W-T1, the regime without the reattachment, $1 < S/D < 1.1$–1.3, depending on Re. The free shear layers separated from the upstream cylinder do not reattach onto the downstream cylinder. The eddy street behind the latter is formed by the free shear layers separated from the upstream cylinder, as depicted in Fig. 26.3(a);

(b,c,d) W-T2, the three reattachment regimes, 1.1–$1.3 < S/D < 3.5$–3.8. The free shear layers separated from the upstream cylinder may reattach alternately (b), permanently (c), or intermittently (d) onto the downstream cylinder, as depicted in Fig. 26.3(b,c,d), respectively. The common fea-

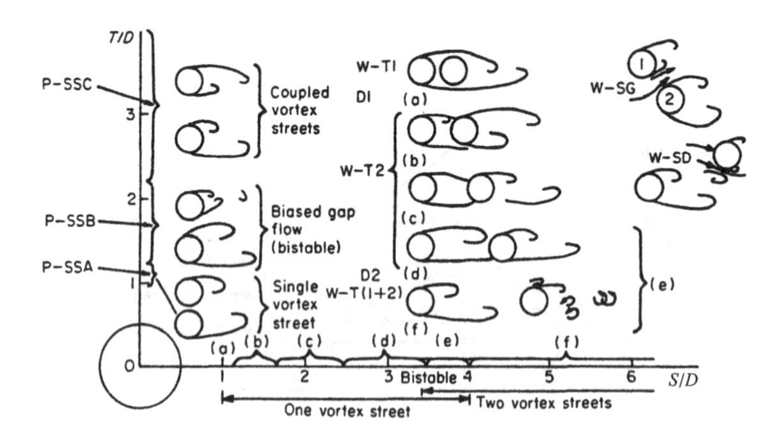

FIG. 26.3. Classification of interference flow regimes, Zdravkovich (1987R)

ture of all three regimes is that eddy shedding takes place only behind the downstream cylinder. Also note that all three flow regimes, (b), (c), and (d), are in the P+W interference region.

(e) W-T (1 or 2), the bistable regime, B1, $3.0 < S/D < 4.0$, depending on *Re*. An intermittent change-over takes place with and without eddy shedding behind the upstream cylinder, either (d) or (f) in Fig. 26.3.

(f) W-T (1+2) the coupled eddy shedding regime, $3.8 < S/D <$ 5–6. Two eddy streets are synchronized in frequency and phase. This means that the eddies from the upstream cylinder pair with the eddies behind the downstream cylinder and form the *binary*[151] eddy street, see sketch in Fig. 26.3(f) and Fig. 26.14.

(g) W-T (1,2) the uncoupled eddy shedding regime behind both cylinders for $S/D >$ 5–6. Two different shedding frequencies are found up to $S/D =$ 60, Okajima (1977P).

Category (ii) encompasses side-by-side arrangements SS. Three distinct interference flow regimes have been observed depending on the transverse spacing ratio T/D between the two cylinders:

(A) P-SSA, the single eddy street regime, $1 < T/D <$ 1.1–1.2; the single eddy street is formed behind both cylinders, which appear as a single bluff body with a weak flow through the gap, as depicted in Fig. 26.3;

(B) P-SSB, the biased flow regime, 1.1–1.2 $< T/D <$ 2–2.2; narrow and wide wakes are formed behind two *identical* cylinders. The gap flow forms a jet *biased* towards the narrow wake. The biased gap flow is bistable, and may intermittently switch to either side, see B2 in Figs 26.2 and 26.3;

(C) P-SSC, the coupled wakes regime, 2–2.2 $< T/D <$ 4–5; both wakes are equal in size, and eddy shedding is synchronized in frequency and phase. The predominant out-of-phase coupling produces two eddy streets, which mirror each other relative to the gap axis, Fig. 26.3.

Category (iii) covers staggered arrangements S in regions 1, 2, and 3, as sketched in Fig. 26.2. For the upstream cylinder, the interference is restricted to the P-S1 region, while for the downstream cylinder, it covers the P-S2, P+W, and W regions, as shown in Fig. 26.2. The common feature for all staggered arrangements S is the narrow wake behind the upstream cylinder and the wide wake behind the downstream cylinder. Note that there is an anomaly for staggered cylinders. The narrow wakes are associated with a high drag and wide wakes with a low drag. This is an inversion of the rule established for single cylinders. The reason is the different origin of the reduction in base pressure. For a single cylinder, the shape and size of the near-wake are free to adjust according to *Re* variations, while for the staggered cylinders the near-wake is forced to adapt to

[151]The term binary was proposed by Williamson (1985J) as an analogy with the binary stars which rotate around each other.

the interference effects. This intrinsic dissimilarity between the two cases has not been sufficiently appreciated.

Two different interference flow regimes are observed in the P+W and W regions for staggered arrangements, S:

(i) (P+W)-SG, the gap flow regime, $1.1 < S/D < 3.5$ and $T/D > 0.2$; the strong gap flow induces a high lift force directed towards the upstream cylinder wake axis. The gap flow may intermittently cease, and result in a zero lift force. The gap flow is bistable, as depicted in Fig. 26.2 by B3 (hatched).

(ii) W-SD, the wake displacement regime, $S/D > 2.8$ and $T/D > 0.4$; the downstream cylinder partly displaces the upstream cylinder wake. The streamlines are squeezed, and this produces a lift force directed towards the upstream cylinder wake axis.

The 12 interference flow regimes represent a tentative overview and may serve as a guide restricted to the transition-in-shear-layers, TrSL, state of flow. It will be shown that a considerable modification occurs for the laminar, L, transitional, TrW, and fully turbulent, T, states. The location and extent of interference regimes in the S/D, T/D plane is governed by Re, and affected by free stream turbulence, surface roughness, finite height, and other influencing parameters.

26.2 Tandem arrangements

26.2.1 *Creeping flow regime, L1*

It has been discussed in Vol. 1, Chapter 1, that there are three flow regimes in the laminar state of flow:[152] no-separation L1, closed steady near-wake L2, and periodic Kármán–Bénard eddy street L3.

Tatsuno and Ishii (1983P) and Tatsuno *et al.* (1990J) towed two cylinders, $L/D = 20$ and $D/B = 0.025$, in a glycerine tank in the range $0.02 < Re < 0.20$. The flow visualization and long-exposure (3 min) photographs are shown in Fig. 26.4(a–d). When the spacing ratio is large the separation of flow does not occur, as seen in Fig. 26.4(a) for $S/D = 3$. As S/D is reduced the local separation region occurs in Fig. 26.4(b) for $S/D = 2.25$ behind the upstream and in front of the downstream cylinder. For $S/D = 2$, the separation regions are linked between the cylinders, as seen in Fig. 26.4(c). A further decrease in S/D to 1.7 leads to a widening of the gap separation, as seen in Fig. 26.4(d).

Figure 26.5 shows the measured separation angle θ_s at $Re = 0.02$ from the photographs compared with Miyazaki and Hasimoto's (1980J)[153] Stokes flow calculation in terms of S/D. The separated region starts at $\theta_s = 140°$ when two cylinders are in contact $S/D = 1$, and disappears for $S/D = 2.6$. Tatsuno and

[152]The classification of flow regimes for a single cylinder is given in the Appendix.

[153]Hasimoto should be spelt as Hashimoto according to the current Japanese spelling convention.

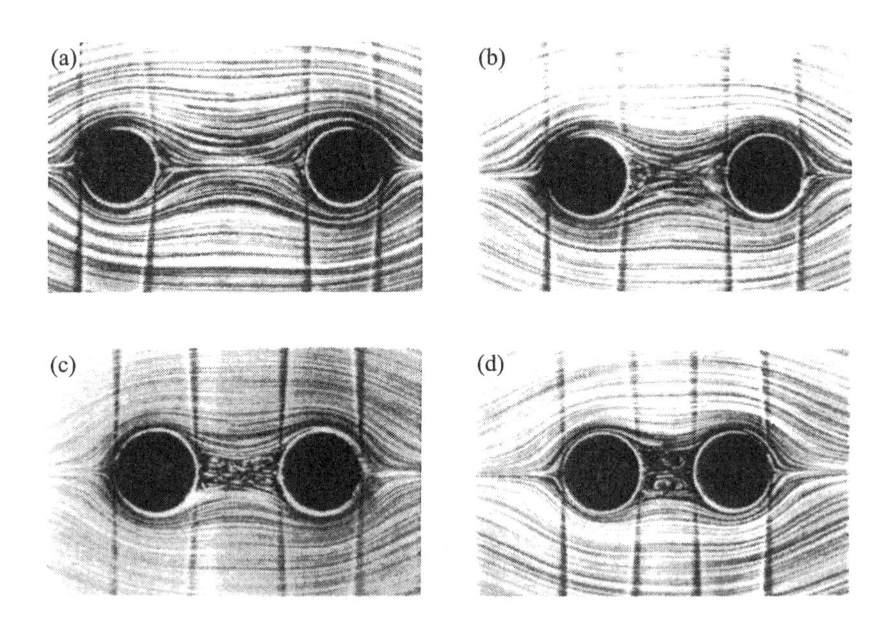

FIG. 26.4. Creeping flow visualization at Re = 0.02, (a) S/D = 3, (b) $S/D = 2.25$, (c) $S/D = 2$, (d) $S/D = 1.7$, Tatsuno (1989J)

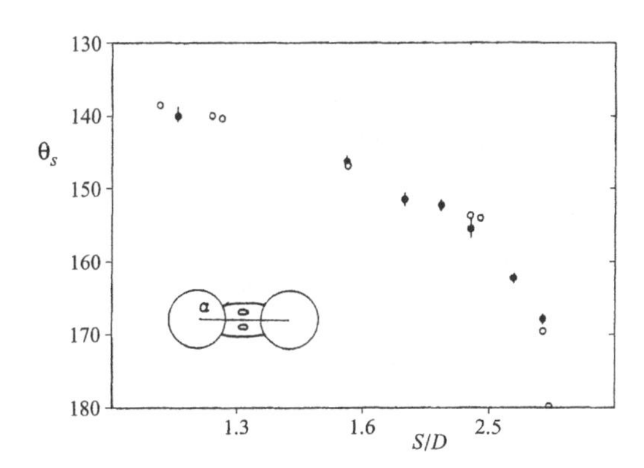

FIG. 26.5. Separation angle versus S/D for Re = 0.02, Tatsuno and Ishii (1983P)

Ishii (1983P) also measured the resistance force and expressed it through a drag coefficient C_D. Figure 26.6(a,b) shows measured C_{D1} and C_{D2} for the upstream and downstream cylinders, respectively. The interference effect is remarkably similar for both cylinders inasmuch as it reduces C_{D1} and C_{D2} values as S/D

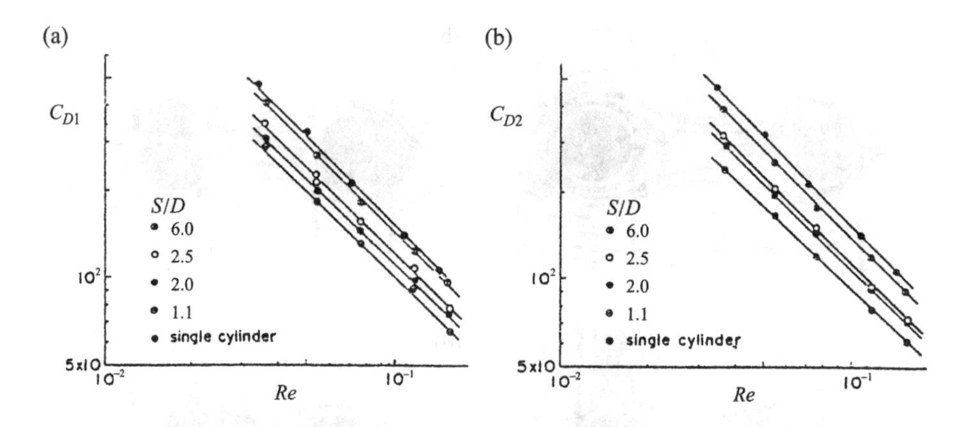

FIG. 26.6. Two cylinders' drag in terms of *Re*, (a) front cylinder, (b) rear, Tatsuno and Ishii (1983P)

decreases. Note that high values of $C_D = 200$–500 may be scaled down by using the resistance coefficient C_R as described in Vol. 1, Chapter 2, pp. 22–3.

26.2.2 *Kármán–Bénard street, L3 regime*

A laminar periodic Kármán–Bénard eddy street generated by a single nominally two-dimensional cylinder first appears for $Re_{\text{osc}} > 40$, and the wake becomes transitional for $Re_{Tr} > 220$. The interference between tandem cylinders modifies the eddy streets considerably, and changes both Re_{osc} and Re_{Tr}.

Thomas and Kraus (1964J) tested two rods $L/D = 32$ and $D/B = 0.08$ at $Re = 62$ in the range $3.6 < S/D < 16$. They observed that eddies generated by the upstream cylinder coalesced with those behind the downstream cylinder, which resulted in the expansion of the eddy street. As S/D decreased, a contraction took place. The width of the eddy street becomes less than behind the single cylinder at the same *Re*. For $S/D \leqslant 8.5$ 'beat patterns'[154] occurred in the form of alternating contractions and expansions of the eddy street at a frequency much lower than the eddy shedding frequency. Finally, for $S/D = 3.6$, the eddy street was completely suppressed behind the upstream cylinder. Thomas and Kraus (1964J) suggested that the beating phenomenon was related not only to S/D but also to $a/2D$, the relative streamwise distance between the consecutive eddies. The maximum interference occurred when S/D was an odd multiple of $a/2D$.

Zdravkovich (1972J) extended tests in the range $40 < Re < 250$ reaching the transition-in-wake state TrW. Two wires $L/D = 140$ and $D/B = 0.0003$ were arranged $1 < S/D < 12$. The smoke visualization used a single filament $D_f/D = 1.9$. Figure 26.7(a,b,c,d) shows the wakes behind the tandem $S/D = 12$, 3, 2,

[154] Similar beating patterns were also photographed by Zdravkovich (1967J).

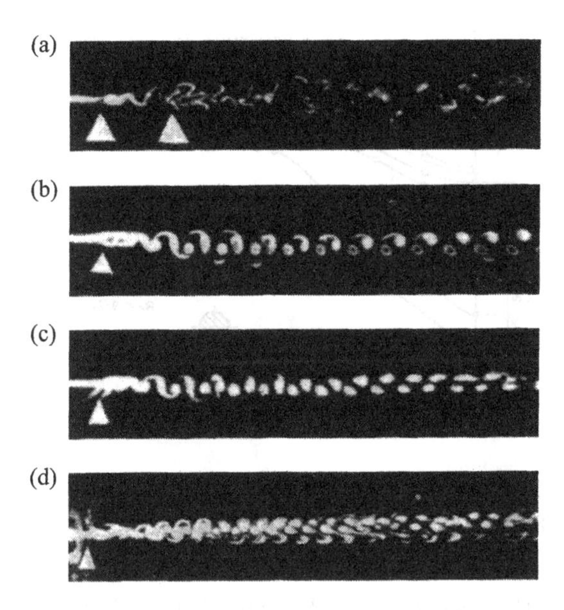

FIG. 26.7. Smoke visualization of tandem cylinders, (a) $Re = 156$, $S/D = 12$, (b) $Re = 156$, $S/D = 3$, (c) $Re = 205$, $S/D = 2$, (d) $Re = 246$, $S/D = 1$, Zdravkovich (1972J)

and 1, and $Re = 156$, 156, 205, and 246, respectively. For the large $S/D = 12$, the wake is transitional at $Re = 156$ though the single cylinder wake is laminar. For $S/D = 3$ and $Re = 205$, the tandem wake is laminar up to $x/D = 40$ while the single cylinder wake is transitional. The tandem wake also remains laminar for $S/D = 2$ and 1 at $Re = 205$ and 246, respectively. Hence, it appears that $S/D > 4$ disturbs the tandem wake, but $S/D < 4$ stabilizes the tandem wake, and preserves the laminar Kármán–Bénard eddy street up to $Re = 250$.

26.2.3 *Early research in the TrSL state of flow*

The first interest in flow around two circular cylinders began in aeronautical engineering in the early era of biplanes. The common use of cylindrical and later streamlined struts to join two wings stimulated research.

Pannell *et al.* (1915P) measured the overall force on two parallel wires and streamlined struts. They varied the spacing ratio S/D from 1 (wires in contact) to 6, at $Re = 9.7$k for tandem and staggered arrangements. Figure 26.8 shows the overall drag force (full line) and twice the drag of the single wire (dashed lines). They wrote: 'It is interesting to notice that, in the case of circular wires, the minimum drag on two wires in contact is only 40% of the drag on one wire alone.'

Another feature of the drag force is also evident in Fig. 26.8. The increments

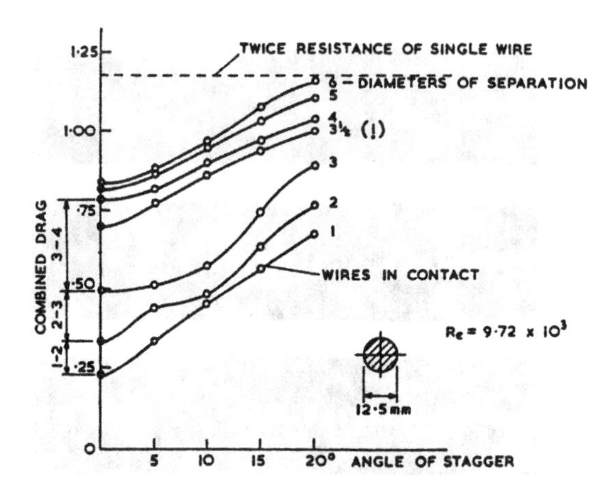

FIG. 26.8. Tandem and staggered cylinders for Re = 9.7k, combined drag in terms of stagger angle, Pannell *et al.* (1915P)

in drag force for the spacings 1–2, 2–3, and 3–4 are large in comparison with 4–5 and 5–6. There is a particularly large increment between spacings 3–4. The authors added an intermediate spacing 3.5, which joined the upper group of curves.

Biermann and Herrnstein (1933P) measured drag separately on two tandem cylinders in the range 65k < Re < 163k. They introduced an *interference drag* coefficient C_{Di} defined as the difference between the drag coefficient measured on one of the cylinders in tandem, and the drag coefficient of the single cylinder at the same Re. Figure 26.9 shows the measured interference drag coefficient for upstream A, downstream B, and both cylinders C, in terms of spacing ratio S/D and Re. The negative value of C_{Di1} for the upstream cylinder is only in the range $2 < S/D < 4$, while C_{Di2} is negative for all S/D values tested. The minimum C_{Di2} value takes place for $S/D = 1$, and corresponds to $C_{D2} = -0.4$, i.e. the thrust force. The kink around $S/D = 3$ coincides with the minimum in C_{Di1}.

The peculiar behaviour of the C_D curves for tandem cylinders around the spacing $S/D = 4$ was resolved by flow visualization. Ishigai *et al.* (1972J) used the schlieren method and heated cylinders between 50°C and 120°C. The cylinders had $L/D = 11$, $D/B = 0.09$, and 2k < Re < 4k in the TrSL2 regime. This flow regime was characterized by the longest eddy formation region and elongated laminar free shear layers. Thus only qualitative similarities of flows might be expected between the flow visualization in the TrSL2 regime and tests in the TrSL3 regime.

Figure 26.10 shows a selection of flow visualization frames from a cine film. For $S/D = 1.5$, there is an alternating reattachment on both sides of the downstream cylinder in Fig. 26.10(a). As the spacing increases further to $S/D = 2$ and

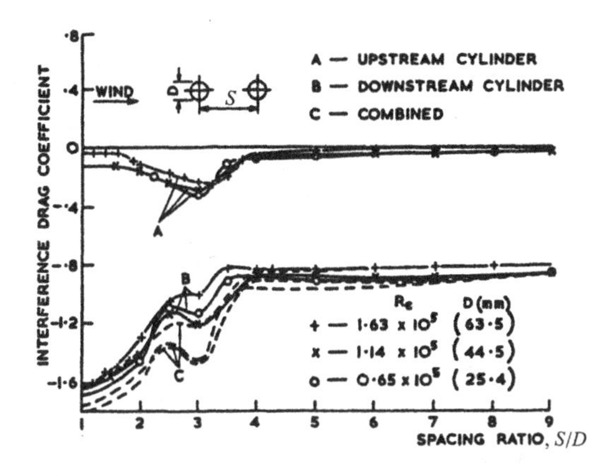

FIG. 26.9. Interference drag coefficient in terms of S/D, Biermann and Herrnstein (1933P)

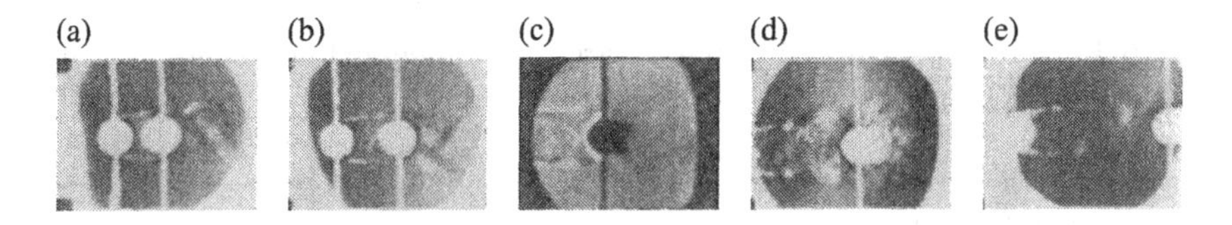

FIG. 26.10. Schlieren visualization of tandem cylinders, (a) $S/D = 1.5$, (b) $S/D = 2$, (c) $S/D = 3$, (d) $S/D = 4$, (e) $S/D = 5$, Ishigai *et al.* (1972J)

3, the reattachment becomes quasi-stationary and symmetric in Fig. 26.10(c,d). For $S/D = 4$, the upstream cylinder free shear layers start to roll up symmetrically on both sides before impinging onto the downstream cylinder, as can be seen in Fig. 26.10(e). Finally, in Fig. 26.10(f) there is a fully developed eddy street behind the upstream cylinder. The spacing ratio $(S/D)_c$ at which alternate eddy shedding commences behind the upstream cylinder is termed *critical.* Note that $(S/D)_c > 4$ is in the TrSL2 regime, and is reduced to $(S/D)_c < 3.5$ in the TrSL3 regime.

26.2.4 *Modification of pressure distribution*

The different variations in C_{D1} and C_{D2} were produced by two kinds of C_p distribution. The early measurements of C_p by Hori (1959P) are shown in Fig. 26.11 in polar plots. The upstream cylinder shows the first kind of C_p distribution:

(i) positive C_{p1} in the stagnation region with $C_{p01} = 1$;

(ii) favourable and adverse pressure gradients;

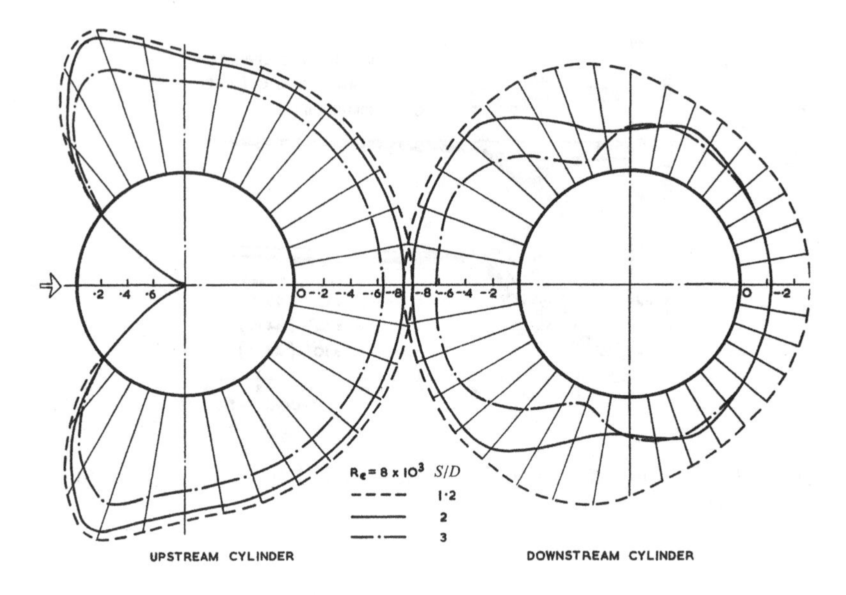

FIG. 26.11. Mean pressure distribution around tandem cylinders, adapted from Hori (1959P)

(iii) almost constant C_{pb1}.

The second kind of C_p distribution is seen around the downstream cylinder:

(i) the base pressure behind the upstream cylinder C_{pb1} faces the upstream side of the gap, and the gap pressure C_{pg2} is exerted along the front side of the downstream cylinder. Note that $C_{pb1} = C_{pg2}$ indicates the lack of flow in the gap;

(ii) the gap pressure C_{pg2} is lower than the base pressure C_{pb2}. This induces the negative C_{D2};

(iii) the two symmetrically arranged peaks at the sides of the downstream cylinder for $S/D = 2$ and 3 are produced by the reattachment of the free shear layers separated from the upstream cylinder. For $S/D = 1.2$, the peaks are absent because the downstream cylinder is wrapped inside the upstream cylinder free shear layers.

Igarashi (1981J, 1984J) carried out extensive C_p and C'_p measurements around tandem cylinders, and classified flow regimes in terms of S/D and Re. The cylinders had $L/D = 4.4$, $D/B = 0.06$, 8.7k $< Re <$ 52k, and $1.03 \leqslant S/D \leqslant 5.00$. Figure 26.12(a,b) shows mean C_{p1} and C_{p2} distributions at Re = 35k in terms of S/D. The interference affects only $C_{p1\min}$ and C_{pb1} on the upstream cylinder. For the downstream cylinder, the C_{p2} curves in Fig. 26.12(b) are of the second

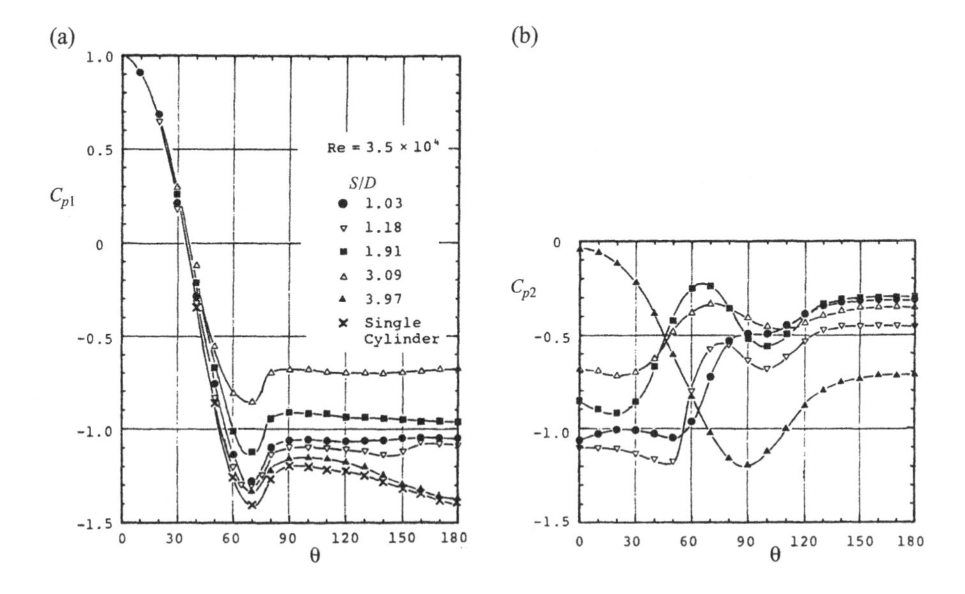

FIG. 26.12. Mean pressure distribution for Re = 35k, (a) front cylinder, (b) rear cylinder, Igarashi (1981J)

kind, as are with the reattachment peaks, except for $S/D = 1.03$ and 3.97.

Figure 26.13(a,b) shows complementary C'_{p1} and C'_{p2} distributions, respectively. The absence of eddy shedding behind the upstream cylinder in the range $1.03 \leqslant S/D \leqslant 3.09$ results in the low separation peak. However, when eddy shedding commences for $S/D = 3.82$ the separation peak exceeds that for the single cylinder. The second peak at $\theta = 160°$ is caused by the vicinity of the fully grown eddy to the cylinder surface, see Igarashi (1981J).

Figure 26.13(b) shows an exceedingly complex C'_{p2} distribution around the downstream cylinder for various S/D values. The description is facilitated by grouping the C'_{p2} curves according to the interference flow regimes as outlined in Fig. 26.3. The main features of the C'_{p2} curves are the reattachment and separation peaks:

(a) W-T1, no reattachment of free shear layers for $S/D = 1.03$; thus only the separation peak in Fig. 26.13(b). The downstream cylinder is wrapped in the free shear layers separated from the upstream cylinder.

(b) W-T2, the alternate reattachment for $S/D = 1.18$; the low reattachment peak is followed by the separation peak. The third peak at $\theta = 45°$ is caused by the secondary separation of the reattached flow.

(c) W-T2, the permanent reattachment for $S/D = 1.62$ and 1.91; both the reattachment and separation peaks are less than that of (b).

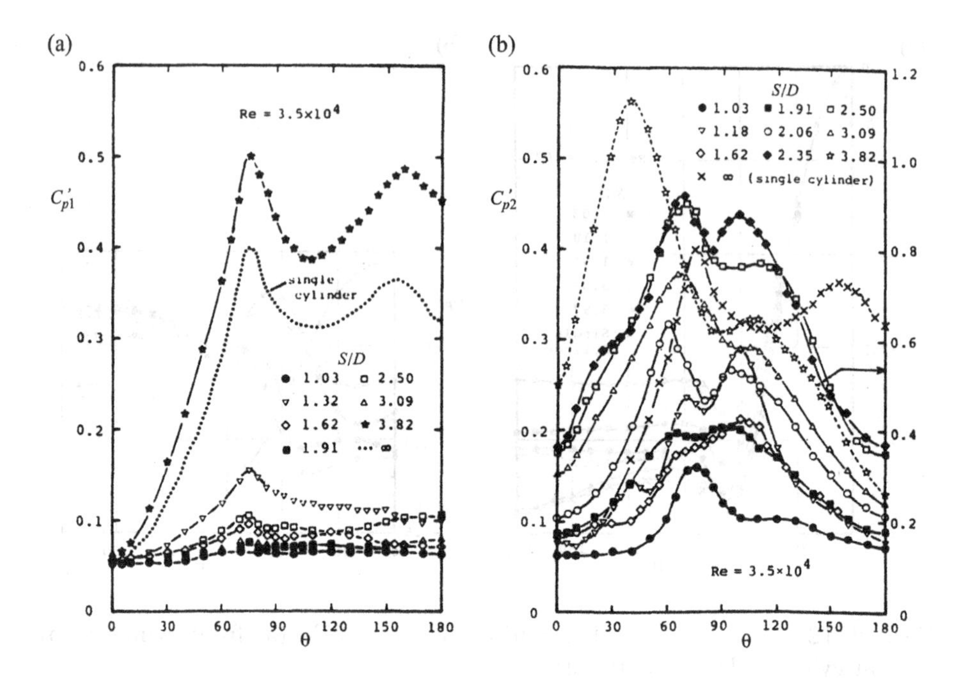

FIG. 26.13. Fluctuating pressure distribution for $Re = 35$k, (a) front cylinder, (b) rear cylinder, Igarashi (1981J)

(d) W-T2, the quasi-stationary reattachment of free shear layers for $S/D =$ 2.06, 2.35, and 2.50; the high reattachment and low separation peaks are seen in Fig. 26.13(b). The magnitude of peaks exceeds those found for the single cylinder.

(e) W-T(1 or 2), the bistable flow with one or two eddy streets for $S/D = 3.09$; the reattachment peak is lower than that of (d).

(f) W-T(1+2), the two eddy streets regime is for $S/D = 3.82$. The extreme peak at $\theta = 45°$ (right scale) is caused by the impingement of eddies shed from the upstream cylinder. The separation peak is low because the upstream cylinder eddies are carried around the downstream cylinder sides, and pair with the eddies formed behind the downstream cylinder. The *binary* eddy street is shown in Fig. 26.14 for $S/D = 4.5$ and $Re = 700$. Kobayashi (in Nakayama, 1984B), used red dye behind the upstream cylinder and blue dye behind the downstream cylinder, denoted by R and B in Fig. 26.14. This made binary eddies visible as a combination of B and R.

The systematic measurements of C'_{p1} and C'_{p2} enabled Igarashi (1981J) to classify in more detail the interference flow regimes for tandem cylinders in terms

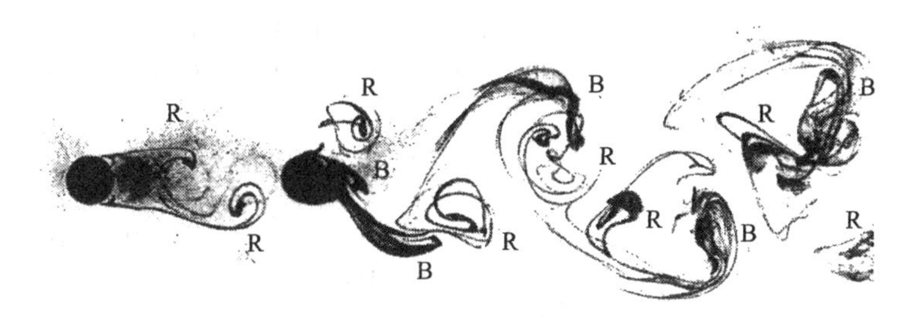

FIG. 26.14. Dye visualization of eddy streets behind tandem cylinders for $Re = 700$, $S/D = 4.5$, Kobayashi, in Nakayama (1984B)

of S/D and Re, as shown in Fig. 26.3. His experiments covered the upper TrSL2 regime and the lower TrSL3 regime. The boundaries between the interference regimes were independent of Re in the TrSL3 regime (Re >20k), and displaced towards higher S/D in the TrSL2 regime.

26.2.5 *Strouhal number*

The basic features of St for a single cylinder will be reiterated. For flow past a single cylinder, a linear increase in the free stream velocity results in a linear rise in the eddy shedding frequency, so the St remains constant throughout the TrSL state of flow. In some special cases, such as shedding cells along a yawed cylinder, and for an oscillating cylinder in the synchronization range, a constant shedding frequency persists and the St value decreases with rising Re. Both types of variation are found behind the stationary tandem cylinders.

Hiwada *et al.* (1982J) measured the eddy shedding frequency behind tandem cylinders, $L/D = 15$, in an open test section at $Re = 50$k. Figure 26.15 shows St measured behind the upstream cylinder (full circles) and downstream cylinder (open circles) in terms of S/D. There are three distinct ranges:

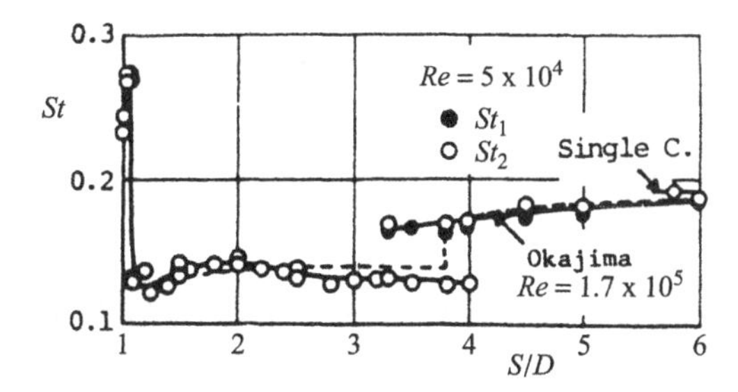

FIG. 26.15. Strouhal number in terms of S/D for Re $=$ 50k, Hiwada *et al.* (1982J)

(i) The high St values (0.24–0.28) are related to the no-reattachment regime. Two closely spaced cylinders appear 'streamlined' to the oncoming flow, and produce a narrow wake, and thus high St.

(ii) The low St values (0.12–0.15) are found in the reattachment regime. A constant St value is found only behind the downstream cylinder up to $S/D = 4$. However, in the bistable range $3.3 < S/D < 4.0$, eddy shedding behind the upstream cylinder emerges intermittently at the higher St value, as can be seen in Fig. 26.15.

(iii) The slowly rising St (0.17–0.19) takes place in the two-eddy-street regime. There is a good agreement with Okajima's (1979J) values (dashed line) obtained at the higher $Re = 170$k.

The change from the no-reattachment to the reattachment regime results in a discontinuous St_2 drop. Igarashi (1981J, 1984J) examined this St_2 drop in detail by restricting tests to the narrow range $1.20 < S/D < 1.33$ across 15k $< Re <$ 100k.[155] Figure 26.16 shows measured St_2 in terms of Re. The frequency spectra exhibit either high St_2 (full circles) or low St_2 (open circles) in the range 20k $< Re <$ 40k. This is the transitional range of bistable flow regimes where a constant shedding frequency results in decreasing St_2. Beyond that for $Re >$ 45k, there is only a constant $St_2 = 0.1$ typical of the reattachment regime. Igarashi (1984J) also examined the effect of a short splitter plate placed in the gap between the cylinders. As soon as the gap flow was stopped the bistable[156] flow ceased.

Okajima (1977P, 1979J) and Okajima and Sugitani (1980P, 1984J) carried out extensive experiments on flow around tandem cylinders. The cylinders had

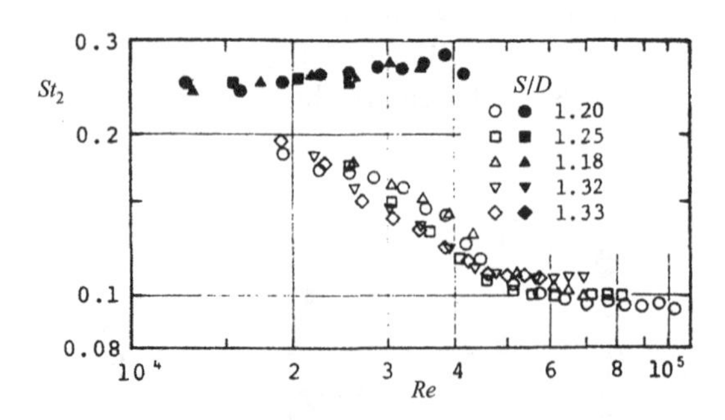

FIG. 26.16. Strouhal number in terms of Re, Igarashi (1984J)

[155] The experimental details of his cylinders have already been given in Section 26.2.2.

[156] Igarashi (1981J) used the term 'unstable flow' due to the quick change-over of two regimes.

$L/D = 6.7$, $D/B = 0.075$, $5 < S/D < 66$, and $16\text{k} < Re < 236\text{k}$. The important feature of these tests was a simultaneous measurement of eddy shedding frequency behind the upstream and downstream cylinders. The hot wires were located at $(x/D, y/D) = (3.5, 2)$; (1.5, 1.5). Figure 26.17 shows St_1 (horizontal) and St_2 (rising curve) in terms of S/D and Re.

There is a fundamental difference between the flow approaching two identical tandem cylinders. The upstream cylinder is exposed to a uniform and low turbulence-free stream, while the downstream cylinder is submerged in the wake of the upstream cylinder, i.e. a non-uniform and turbulent stream. The different flow conditions result in a different St_1 and St_2 as shown in Fig. 26.17 in the range $6 < S/D < 65$. For the upstream cylinder, St_1 is constant up to $Re = 94\text{k}$, while for the downstream cylinder St_2 rises from 0.12 to 0.18.

The downstream cylinder is exposed to the velocity defect of the upstream cylinder wake. The calculated St_2 is based on the free stream velocity, hence $St_2 < 0.2$. As S/D increases, the velocity defect decreases, and St_2 rises towards 0.18.

26.2.6 *Drag coefficients*

The pioneering force measurements on tandem cylinders by Pannell *et al.* (1915P), Fig. 26.8, and Biermann and Herrnstein (1933P), Fig. 26.9, were followed by Hori (1959P), Counihan (1963P), Wardlaw *et al.* (1974P), Suzuki *et al.* (1971P), Wardlaw and Cooper (1973P), Taneda *et al.* (1973J), Cooper (1974P), and others.

Zdravkovich (1977R) compiled all these data into a single plot of C_{D1} and

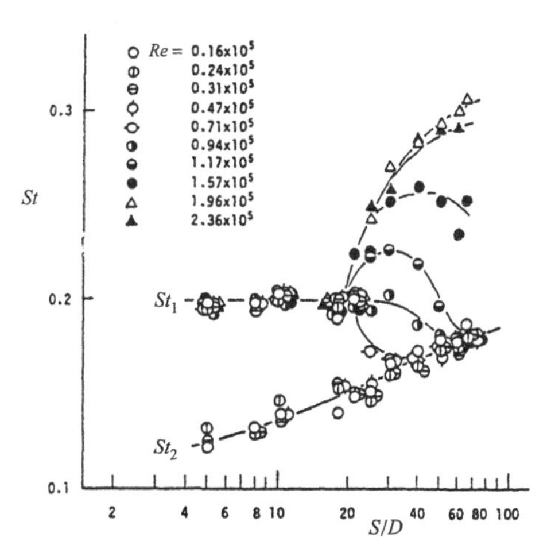

FIG. 26.17. Strouhal number in terms of Re at high Re, Okajima and Sugitani (1984J)

C_{D2} in terms of S/D and Re. Figure 26.18 shows C_{D1} (closed symbols) and C_{D2} (open symbols) in the range $1 < S/D < 30$ and $80 < Re < 230$k. The most prominent feature is a discontinuous jump in C_{D1} and C_{D2} at $3 < S/D < 4$. A notable exception is the laminar flow at $Re = 80$ where the jump takes place at $S/D = 5$. The latter was subsequently confirmed by Aude *et al.* (1985P). Another anomaly of the laminar flow is that C_{D2} remains positive for $S/D < 5$ and $Re = 80$. The other C_{D1} and C_{D2} curves show the following trend:

(i) There is a negligible Re effect on C_{D1} curves. The common feature is the decrease in C_{D1} up to the critical $(S/D)_c$.

(ii) There is a strong Re effect on C_{D2} curves. For $S/D < 2$, C_{D2} is negative, and acts as a thrust force. For $S/D > 3.5$, C_{D2} becomes positive for high Re.

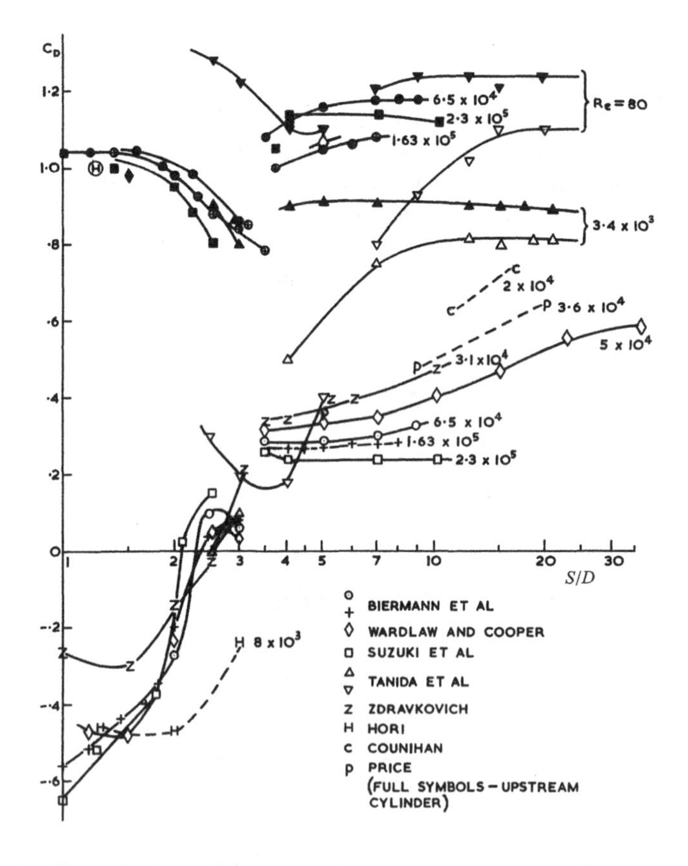

FIG. 26.18. Compilation of drag coefficients on tandem cylinders in terms of S/D and Re, Zdravkovich (1977R)

(iii) The increase in C_{D2} beyond $(S/D)_c$ relates to the decreasing *Re* in an orderly manner. The variation in C_{D2} reflects the effect of the upstream wake turbulence on the flow past the downstream cylinder.

26.2.7 *Transition-in-boundary-layer, TrBL, state*

It has been discussed in Vol. 1, Chapter 6, that for a single cylinder, several distinct flow regimes take place due to the interaction between the transition in boundary layer and separation. These flow regimes are relevant only to the upstream cylinder of the tandem. The downstream cylinder is submerged into the turbulent wake, which affects transition and separation. It has been discussed in Vol. 1, Chapter 14, that the free stream turbulence has a particularly strong effect on the TrBL state of flow. It can obliterate one- and two-bubble, TrBL1 and TrBL2 regimes, as well as the supercritical TrBL3 regime. Hence, it may be expected that the flow around the downstream cylinder would be governed by the upstream wake turbulence.

Okajima (1977P, 1979J) carried out tests on tandem cylinders, $L/D = 6.7$, $D/B = 0.075$, $1 < S/D < 6.5$, and $40\text{k} < Re < 630\text{k}$. As a representative of the reattachment regime, Okajima (1977P) chose $S/D = 3$, and for the two-eddy-street regime, $S/D = 5$. Figure 26.19(a,b) shows measured C_{D1}, C_{D2} and St_1, St_2 in terms of *Re*. The variation in C_{D1} shows a similar trend as for the single cylinder. There is an opposite trend for C_{D2} with a wide difference in magnitude for $S/D = 3$ and 5. For $Re > 450\text{k}$, $C_{D1} < C_{D2}$, because the former is in the TrBL1 regime and the latter in the TrBL3 regime. This is confirmed by St_1 and St_2 variations where St_2 is absent for $Re > 450\text{k}$. Okajima (1977P)

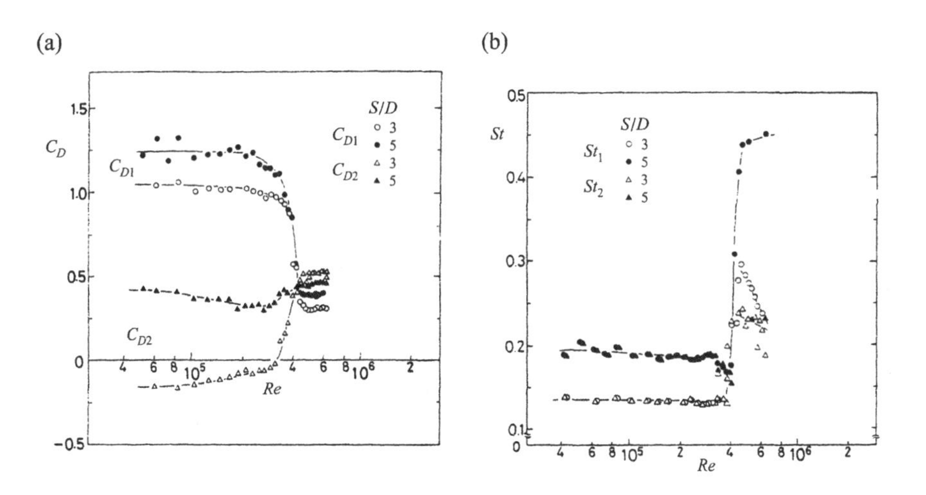

FIG. 26.19. Tandem cylinder for $S/D = 3$ and 5, (a) drag, (b) Strouhal number, Okajima (1979J)

carried out surface flow visualization and observed separation bubbles on the upstream cylinder, but not on the downstream one.

Okajima (1977P) also gave an alternative presentation of C_{D1}, C_{D2} and St_1, St_2 in terms of S/D. The most important feature was the disappearance of the jump for $(S/D)_c$ for $Re > 500\text{k}$ due to the absence of eddy shedding in the TrBL3 regime.

26.2.8 *Effect of free stream turbulence*

The effect of free stream turbulence on the boundary and free shear layers of a single cylinder has already been discussed in Vol. 1, Chapter 14. The free stream turbulence acts as a random and three-dimensional unsteady perturbation. The transition in the boundary and free shear layers is promoted at lower Re than in the smooth flow. If the boundary and free shear layers are turbulent, then the free stream turbulence enhances mixing, entrainment, and turbulent diffusion. These general remarks are pertinent to the upstream cylinder, while the downstream cylinder is already submerged in the turbulent wake.

Tandem cylinders are separated by a gap which is bounded by the free shear layers. The gap can be thought of as a 'wake' of restricted length, and the free stream turbulence triggers an early transition in the free shear layers. This leads to a more rapid thickening and widening of the free shear layers. The perturbed gap flow is expected to instigate an early start of eddy shedding behind the upstream cylinder at smaller $(S/D)_c$.

Most past research was restricted to low free stream turbulence. An exception was Bokaian and Geoola's (1984J) tests in a water flume, having $Ti = 6.5\%$. The measured St_1 was on cylinders $L/D = 18.7$, $D/B = 0.08$, $1.1\text{k} < Re < 5.6\text{k}$, and $1 < S/D < 5$. The eddy shedding behind the upstream cylinder was detected at $(S/D)_c = 2$. This was well below the $(S/D)_c$ reported by other authors for $Ti < 0.5\%$.

Ljungkrona *et al.* (1991J) examined the effect of Ti on C_{D1}, C_{D2}, St_1, and St_2. They used tandem cylinders $L/D = 8$, $D/B = 0.08$, and $1.1 < S/D < 5$, in the range $3.3\text{k} < Re < 20\text{k}$. Figure 26.20(a,b) shows C_{D1}, C_{D2}, St_1, and St_2 in terms of S/D for smooth and turbulent flows. There is a significant reduction in $(S/D)_c$ from 3.2 to 2.6 and 2.2 for $Ti = 0.4\%$, 1.4%, and 3.2%, respectively. It is also evident that the jump in C_{D1} and drop in St_2 disappeared. Another effect of turbulence is the disappearance of high St in the no-reattachment regime, $S/D < 1.2$. The free stream turbulence disturbs the free shear layers, and induces an early reattachment.

Zhang and Melbourne (1992J) carried out measurements for $Re = 110\text{k}$, $Ti = 4.5\%$ and 11%, $L/D = 8$, $D/B = 0.05$, and $1 < S/D < 10$. They found $(S/D)_c = 2.3$ and 2.0 for $Ti = 4.5\%$ and 11%, respectively. The change-over from the reattachment to two-eddy-street regimes was gradual, and the bistable regime was absent. Figure 26.21 shows the interference drag coefficient C_{Di} in terms of S/D and Ti. The free stream turbulence considerably increased C_{Di2} for $Re = 110\text{k}$.

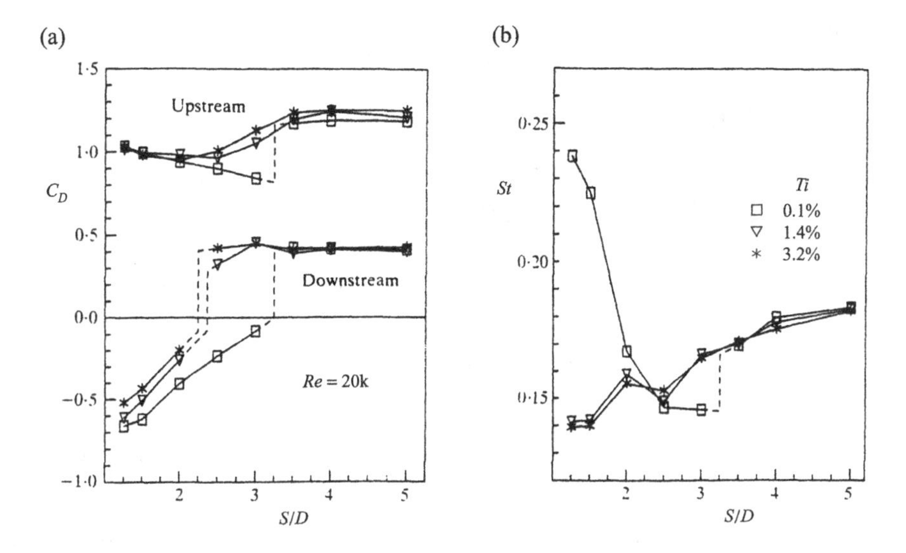

FIG. 26.20. Effect of free stream turbulence on tandem cylinders, (a) drag, (b) Strouhal number, Ljungkrona *et al.* (1991J)

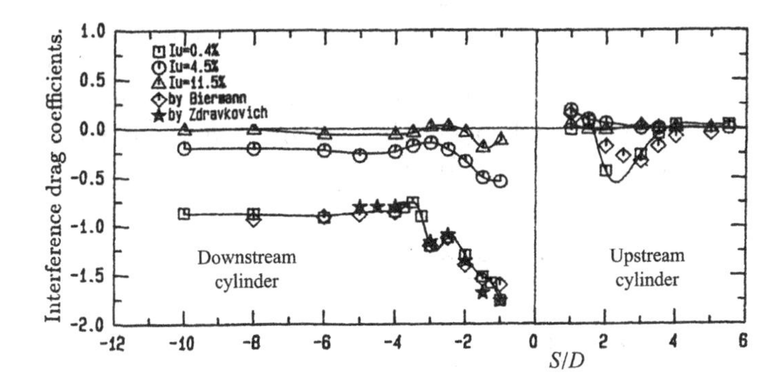

FIG. 26.21. Effect of turbulence on the interference drag coefficient for tandem cylinders, Zhang and Melbourne (1992J)

The effect of free stream turbulence $Ti = 10\%$ on high $Re = 650\text{k}$ for tandem cylinders was examined by Gu *et al.* (1993J). The tandem cylinders had $L/D = 15$, $D/B = 0.05$, $195\text{k} < Re < 650\text{k}$, and $1 < S/D < 7$. They first established that at $Re = 650\text{k}$ and $Ti = 10\%$ there was eddy shedding behind a single cylinder at $St = 0.26$. Figure 26.22(a,b) shows the C_{p2} and C'_{p2} distributions for different S/D values. The reattachment peak in the C_{p2} distribution is seen

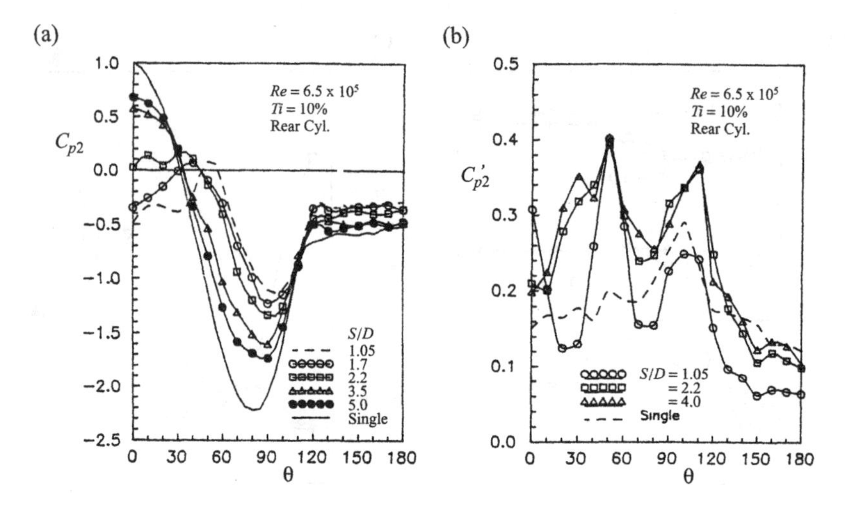

FIG. 26.22. Rear cylinder at Re = 650k, Ti = 10%, (a) mean pressure distribution, (b) fluctuating pressure distribution, Gu *et al.* (1993J)

for $S/D = 1.05$, and $(S/D)_c$ occurred around 2.2. The two peaks in the C'_{p2} distribution correspond to the impingement of the eddy street from the upstream cylinder at $\theta = 45°$ and separation at $\theta_s = 110°$. The C'_{p2} fluctuations are above those found behind the single cylinder (dashed line). The variation in C_{D1} and C_{D2} in terms of S/D is given in Fig. 26.23 for Re = 650k. There is a distinct jump in C_{D2} at $S/D = 2.2$ and no jump in the C_{D1} curve. Note that $C_{D2} > C_{D1}$ beyond $(S/D)_c$, and $C_{D2} = -0.15$ for $S/D = 1$ and $Ti = 10\%$.

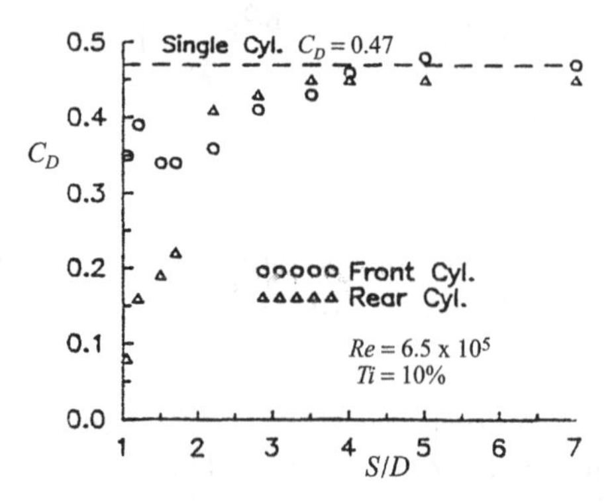

FIG. 26.23. Drag on tandem cylinders at Re = 650k, Ti = 10% in terms of S/D, Gu *et al.* (1993J)

26.2.9 *Effect of surface roughness*

The effect of surface roughness on flow past a single cylinder has been discussed in Chapter 22. The surface roughness triggers the TrBL state of flow at a lower *Re*, obliterates separation bubbles in the TrBL1 and TrBL2 regimes, as well as inhibiting the supercritical TrBL3 regime. The disappearance of the last three flow regimes means that the precritical TrBL0 regime is followed by the post-critical TrBL4 regime with uninterrupted eddy shedding.

Okajima (1977P, 1979J) modified his smooth cylinders by glueing polystyrene beads onto the surface (relative roughness $K/D = 0.9\%$). Two combinations were tested: the rough cylinder was followed by the smooth cylinder and two rough cylinders.

Figure 26.24(a,b) shows C_{D1}, C_{D2}, and St_1, St_2 in terms of *Re* for both combinations of tandem cylinders at $S/D = 3$ and 5. The added dashed lines show the respective curves obtained for the rough and smooth tandem cylinders. The effect of surface roughness induces $C_{D1\min}$ at $Re = 80$k. The curves for the rough upstream cylinder are virtually identical to those expected. The rough downstream cylinder shows a considerable increase in C_{D2} in comparison with the smooth cylinder only for $S/D = 5$. It is notable that the St_1 and St_2 curves remained the same for both combinations.

26.2.10 *Effect of finite height*

Flow past a finite single cylinder has been discussed in Chapter 21.8. There are two distinct flow features: a horseshoe-swirl near the boundary and the free-end

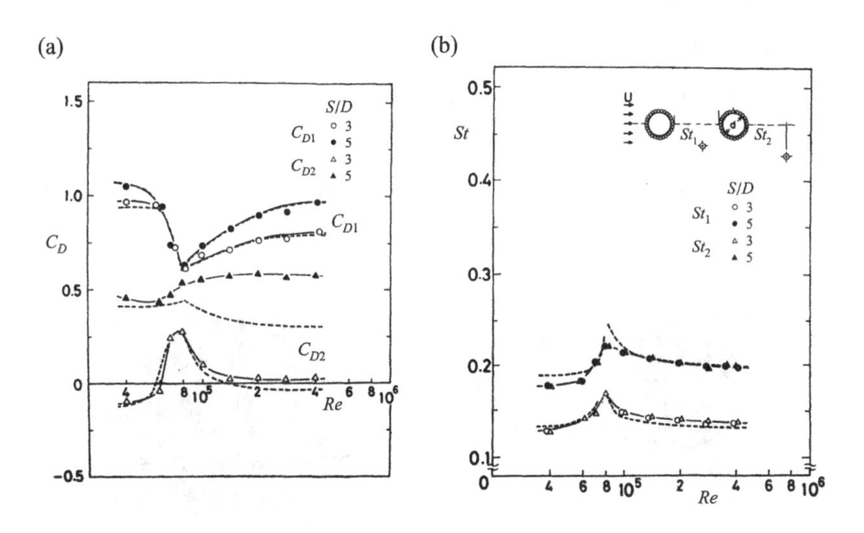

FIG. 26.24. Effect of surface roughness $K/D = 0.9\%$ in terms of *Re*, on (a) drag, (b) Strouhal number, Okajima (1979J)

effect at the tip. The effect of the interference between the tandem cylinders on these two flow features is considered below.

Taniguchi *et al.* (1982J) carried out tests on short $H/D = 3$ tandem cylinders partly submerged in a turbulent boundary layer, $\delta/H = 0.87$ at $Re = 15.5$k, in the range $1.2 < S/D < 4.0$. For $H/D = 3$, there is no eddy shedding along the cylinder span. There are two regions: the horseshoe-swirl, and the free end. Smoke wire visualization was employed at various cylinder elevations. Figure 26.25(a,b) shows the flow pattern for $S/D = 1.5$ and 4 at $z/H = 0.11$ and $Re = 620$. A single horseshoe-swirl is seen for $S/D = 1.5$ and two separate ones for $S/D = 4$. The flow pattern at the midheight $z/H = 0.49$ and $S/D = 2.5$ shows neither eddy street nor reattachment of the free shear layers. The effect of the free end can be deduced from the measured C_p distribution at $z/H = 0.22$, 0.58, and 0.89 for $S/D = 2$ and $Re = 15.5$k. The reattachment type of C_p distribution occurred at the midheight elevation, while the stagnation type of C_p distribution was found near the free end. Near the free end, the flow from the upstream cylinder impinges on the downstream cylinder.

Another example is tandem cylinders of $H/D = 8$ and $D/B = 0.02$ at $Re =$ 33k. Luo *et al.* (1996J) carried out C_p measurements at eight elevations, and evaluated local C_{d1} and C_{d2}. Figure 26.26(a,b) shows the variation in local C_{d1} and C_{d2} in terms of z/H for $S/D = 1$ and 3, respectively. The free-end flow increases C_{d2} considerably. The effect is noticeable when the tandem cylinders are in contact, as seen in Fig. 26.26(a), and it spreads down the span for $S/D = 3$ in Fig. 26.26(b). The free-end effect is important for the design of twin chimney stacks because the rise in the local C_{d1} and C_{d2} near the tip significantly increases the overturning moment caused by the wind loading.

26.2.11 *Effect of heat transfer*

Heat transfer from a single cylinder has been discussed in Vol. 1, Chapter 17. The maximum in the local heat transfer $Nu/Re^{1/2}$ was found at the stagnation point, while a minimum occurred near the separation θ_s. However, an excessive $(Nu/Re^{1/2})_{\max}$ was found in the TrBL1 and TrBL2 regimes at the end of the

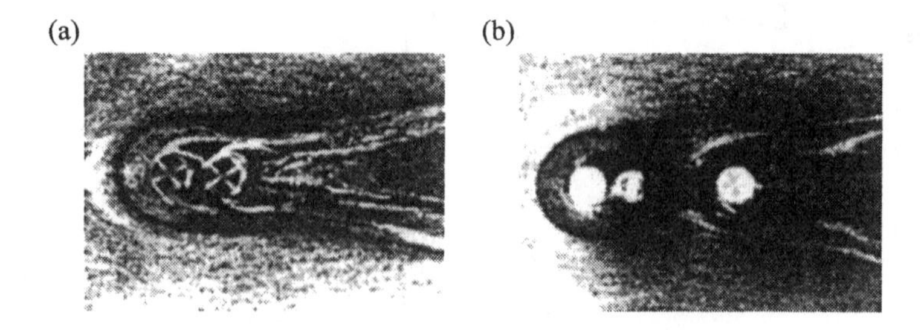

FIG. 26.25. End-plate flow visualization behind tandem cylinders, (a) $S/D = 1.5$, (b) $S/D = 4$, Taniguchi *et al.* (1982J)

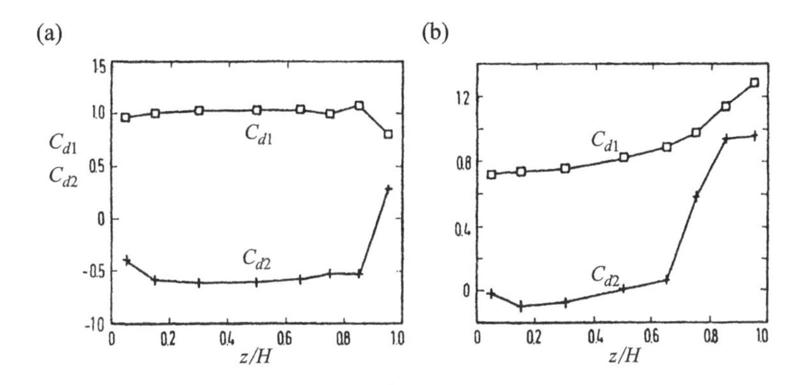

FIG. 26.26. Local drag on tandem cylinders in terms of z/H, (a) $S/D = 1$, (b) $S/D = 3$, Luo *et al.* (1996J)

separation bubble. This showed that the reattachment was the most effective mechanism in enhancing the local heat transfer.

Igarashi and Yamasaki (1992P) carried out heat transfer tests on tandem cylinders for $S/D = 1.3$, $L/D = 5$, $D/B = 0.075$, and $39\text{k} < Re < 62\text{k}$. It has been shown in Fig. 26.16 that for $S/D = 1.3$ no reattachment was found for $Re < 20\text{k}$, a permanent reattachment for $Re > 30\text{k}$, and a bistable regime was observed in between the two Re.

Local heat transfer measurements from tandem cylinders are presented in Fig. 26.27(a,b) showing $Nu/Re^{1/2}$ in terms of various Re. The reattachment peaks on

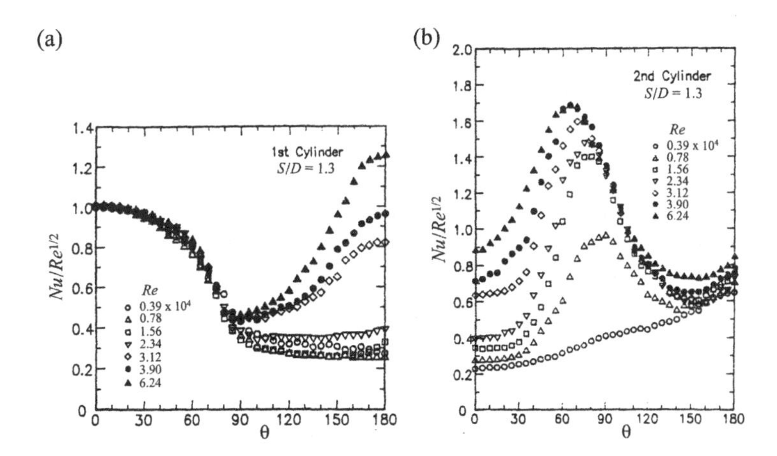

FIG. 26.27. Local heat transfer from tandem cylinders, (a) front cylinder, (b) rear cylinder, Igarashi and Yamasaki (1992J)

the downstream cylinder appear from $Re = 7.8\text{k}$, and grow in strength with rising Re. The increase in $Nu/Re^{1/2}$ for the upstream cylinder starts for $Re > 23.4\text{k}$. The delay in the enhancement of $Nu/Re^{1/2}$ for the downstream cylinder may be attributed to the laminar conditions in the free shear layers emanating from the upstream cylinder. The local heat transfer from the downstream cylinder is significantly enhanced by the reattachment so that an overall $\overline{Nu}$ appears to be almost twice[157] that for the upstream cylinder.

26.3 Side-by-side arrangements

26.3.1 *Classification of interference regimes*

It has been shown in Fig. 26.2 that two parallel cylinders in the side-by-side arrangement SS are subjected to the proximity interference P. Three distinct flow regimes are sketched in Fig. 26.3 depending on the transverse spacing ratio T/D:

(A) P-SSA, $1 < T/D < 1.1$–1.2; the single eddy street is formed behind both cylinders with a weak gap flow between them.

(B) P-SSB, 1.1–$1.2 < T/D < 2$–2.2; narrow and wide wakes are formed behind two identical cylinders, and the gap flow is biased towards the narrow wake. The biased flow is bistable, and can switch to either side. This leads to an interchange of the narrow and wide near-wakes behind the cylinders.

(C) P-SSC, 2–$2.2 < T/D < 5$–6; the coupled eddy streets are synchronized not only in frequency but also in phase. The eddies are formed simultaneously behind the gap so that the two eddy streets mirror each other relative to the gap axis.

The visualization of flow past two side-by-side cylinders has revealed the general nature and minute details of the interference flow regimes. An early flow visualization by Ishigai *et al.* (1972J) used the schlieren method.[158] They heated cylinders from 50 °C to 120 °C so that only the hot free shear layers were made visible. Figure 26.28(a–e) shows cine film frames as T/D decreased from 2.5 to 1.25 in the range $3.4\text{k} < Re < 4\text{k}$ (the TrSL2 regime). Figure 26.28(a) shows two perfectly synchronized and phased eddies formed in the gap for $T/D = 2.5$, representing the coupled eddy street in the P-SSC regime. Next, Fig. 26.28(b,c) shows the initiation and development of the biased gap flow for $T/D = 2$ and 1.5, respectively. Figure 26.28(d,e) shows two biased flows for the same $T/D = 1.25$ observed at different times. The intermittent changes of two biased flows are intrinsic to the bistable flow regime.

26.3.2 *Laminar wakes*

It has been discussed in Vol. 1, Chapter 2, that for $Re < 5$, in the creeping L1 regime, the flow remained attached all around a single cylinder. Taneda

[157] Igarashi and Yamasaki (1992P) calculated $\overline{Nu_1} = 70$ and $\overline{Nu_2} = 120$ at $Re = 20\text{k}$.

[158] The details of the experimental arrangement are given in Section 26.2.

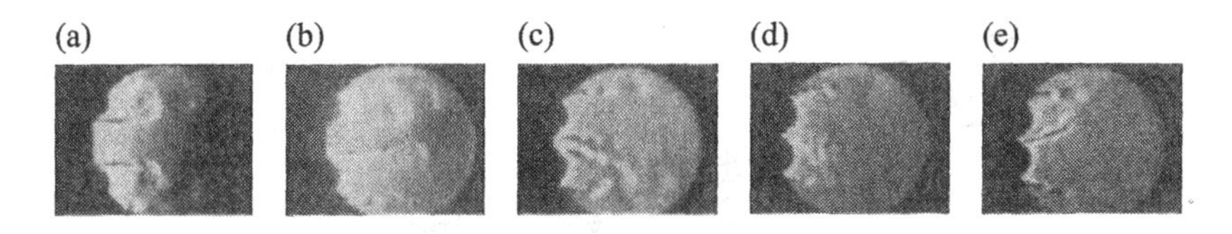

FIG. 26.28. Schlieren visualization of side-by-side cylinders, (a) $T/D = 2.5$, (b) $T/D = 2$, (c) $T/D = 1.5$, (d,e) $T/D = 1.25$, Ishigai *et al.* (1972J)

(1979J) and Tatsuno (1989J) carried out flow visualization around towed side-by-side cylinders. Figure 26.29(a,b,c) shows streamlines at $Re = 0.02$ and $T/D =$ 1.1, 1.5, and 2.0, respectively. The proximity interference induces local separation regions on the upstream and downstream sides, which gradually diminish with rising T/D.

Taneda (1956J) measured the repulsive force acting on the side-by-side cylinders in the range $0.01 < Re < 1.6$ and $4.6 < T/D < 50$. Figure 26.30 shows the measured C_L in terms of Re, and the theoretical curves are calculated for $T/D = 10$ and 50. The theory is based on the Oseen approximation, and derived for the side-by-side cylinders. The agreement between the measured and predicted C_L is good and shows a monotonous decrease in C_L with rising Re. It

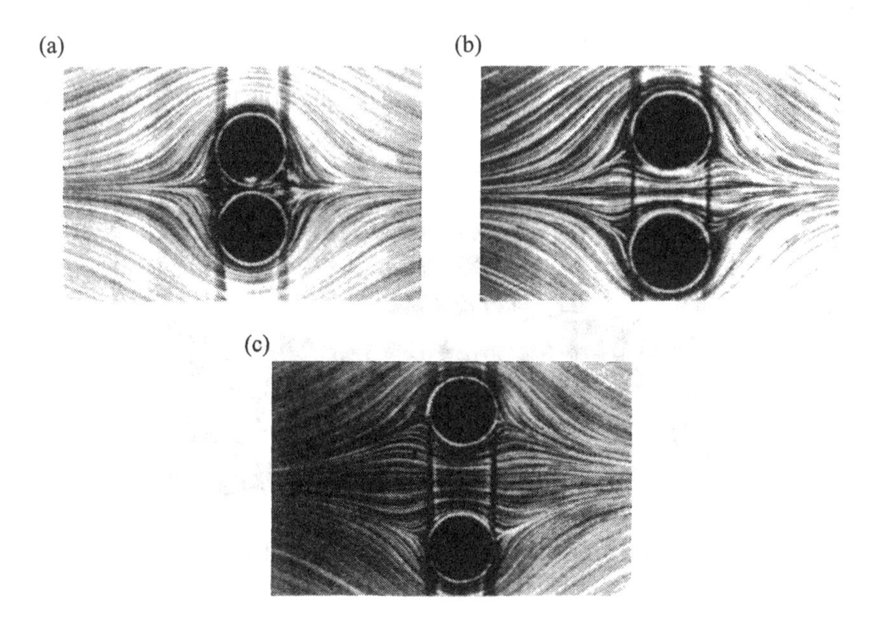

FIG. 26.29. Creeping flow visualization around side-by-side cylinders, (a) $T/D = 1.1$, (b) $T/D = 1.5$, (c) $T/D = 2$, Tatsuno (1985J)

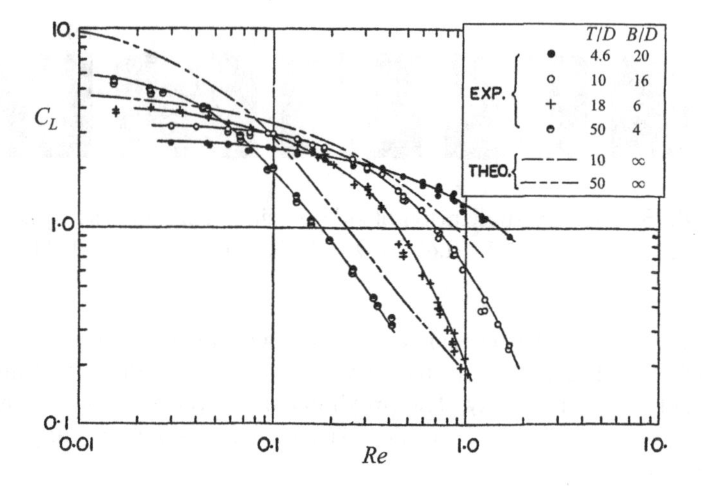

FIG. 26.30. Lift (repulsive) in terms of Re, Taneda (1956J)

is notable that for the creeping flow at low Re a significant repulsive force exists even at $T/D = 50$.

The separation develops and forms a steady closed near-wake behind a single cylinder in the L2 regime. Williamson (1985J) carried out flow visualization around two side-by-side cylinders at $Re = 55$[159] and $T/D = 2$. Figure 26.31 shows that slightly biased gap flow results in a large and small closed near-wake. The length ratio of the closed near-wakes (dark) behind the two cylinders

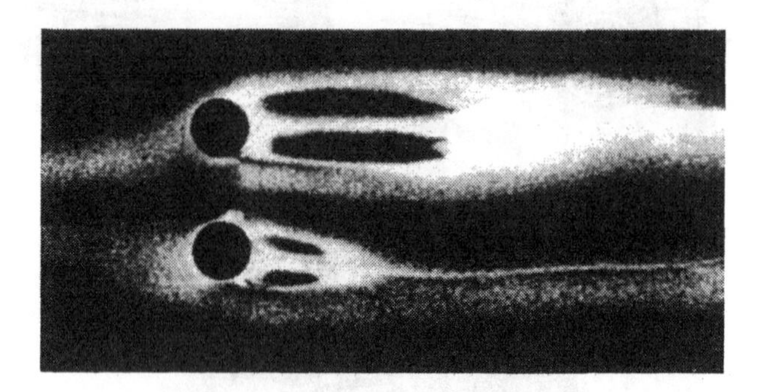

FIG. 26.31. Smoke visualization past side-by-side cylinders at $Re = 55$ and $T/D = 2$, Williamson (1985J)

[159] The oscillating trail starts at $Re \approx 40$, but the Kármán–Bénard eddy street is not formed before $Re = 60$–65.

is around three to one. The important conclusion is that the biased gap flow regime exists when eddy shedding is absent.

The laminar periodic eddy street in the L3 regime was also investigated by Williamson (1985J). The coupled eddy street regime was found in the range $2 < T/D < 6$ for the out-of-phase mode. The eddies are shed simultaneously in the gap and later at the outer sides, as illustrated in Fig. 26.32(a) for $Re = 200$ and $T/D = 6$. The two out-of-phase eddy streets remained unchanged far downstream. However, Williamson (1985J) occasionally observed the in-phase mode in Fig. 26.32(b) at $Re = 100$ and $T/D = 4$. The two eddy streets are not stable, and form a single binary eddy street far downstream. The gap side eddies move out and pair with the eddies of the same rotation formed at the outer sides of the cylinders. The final result is a single binary eddy street.

The biased gap flow regime produces narrow and wide near-wakes, and consequently high and low eddy shedding frequencies, respectively. Williamson (1985J) also carried out dye visualization in a water channel at $Re = 200$ and $T/D = 1.85$. Figure 26.33 shows that initially two eddy streets transform into a large-scale single Kármán–Bénard eddy street. This time, the eddies are comprised of one or three separate eddies. The dominant outer side eddies are convected downstream at a higher speed than gap eddies. The disparity in the convection speed facilitates a swift pairing process. Williamson (1985J) suggested that eddy shedding takes place at the harmonic modes, i.e. a multiple of the fundamental mode. However, the monotonous increase in the biased flow angle and near-wake widening with reducing T/D do not support the step-like change in St.

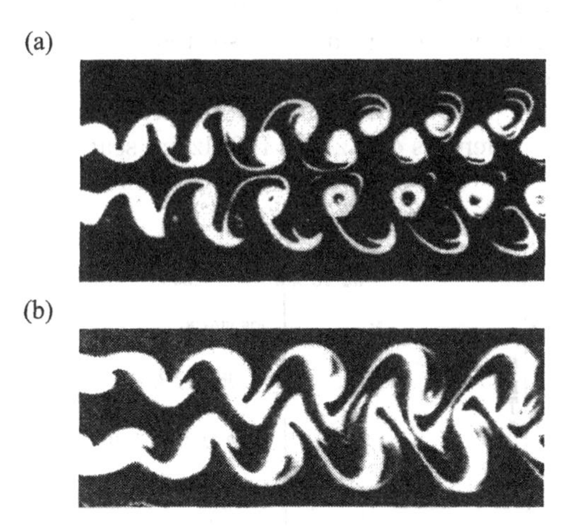

FIG. 26.32. Smoke visualization past side-by-side cylinders, (a) $Re = 200$, $T/D = 6$, (b) $Re = 100$, $T/D = 4$, Williamson (1985J)

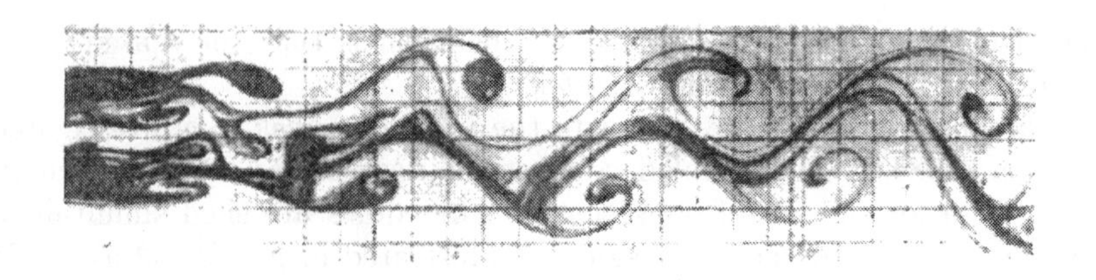

FIG. 26.33. Dye visualization past side-by-side cylinders at $Re = 200$, $T/D = 1.85$, Williamson (1985J)

26.3.3 *Strouhal number*

It might be instructive and appropriate to start with an early milestone paper. Spivack (1946J) carried out a meticulous hot-wire exploration of a flow field behind two parallel cylinders in side-by-side arrangements. The experimental details were $L/D = 39$, $D/B = 0.025$, and $5\text{k} < Re < 93\text{k}$. Twenty-one spacings were examined within the range $1 \leqslant T/D < 6$: four in the P-SSA, ten in the P-SSB, and seven in the P-SSC regimes.

A single eddy shedding frequency was found everywhere in the flow field at $Re = 28\text{k}$ in the range $1 \leqslant T/D < 1.09$ (P-SSA regime) except along the wake axes. A constant $St = 0.1$ corresponded to $St = 0.2$ if D was replaced by $2D$ as the reference projected diameter. A single eddy shedding frequency was also found in the range $2 < T/D < 6$ (P-SSC regime) leading to a constant $St = 0.2$.

Additional frequencies were found behind the gap and outside the two wakes in the range $1.18 < T/D \leqslant 2.00$ (P-SSB regime). The most puzzling feature was, as Spivack (1946J) wrote, 'usually, these frequencies occurred at different positions in the field of flow, but different frequencies were also found simultaneously in the same place'. A typical example is shown in Fig. 26.34 where mutually unrelated frequencies 47 Hz and 108 Hz are scattered everywhere in the

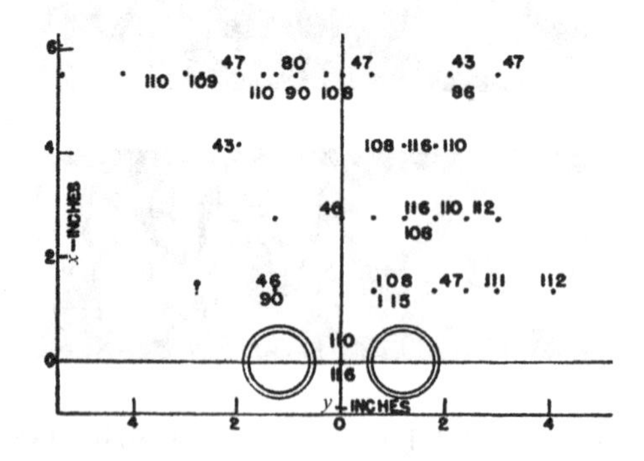

FIG. 26.34. Measured frequencies behind side-by-side cylinders for $T/D = 1.73$, Spivack (1946J)

flow field. The two frequencies may indicate two eddy streets, but their simultaneous occurrence behind both cylinders could not be explained. Spivack (1946J) offered a plausible but incorrect explanation: 'There may be two modes of eddy formation at the outside of the two cylinders and in the gap.'

The phenomenon embodied two paradoxes seemingly absurd but actually taking place. The preconceived notion was that the flow around symmetrically arranged and identical cylinders should be symmetric. In the P-SSB regime, the flow was asymmetric with narrow and wide wakes separated by the biased gap flow. Another preconceived notion was that only one stable flow would be possible. The existence of two quasi-stable flows produces the bistable biased flow, which intermittently switches in the P-SSB regime. The nature of the bistable biased flow was resolved by flow visualization by Ishigai *et al.* (1972J), and Bearman and Wadcock (1973J).

Spivack (1946J) concluded that both St and $(T/D)_c$ are not dependent on Re in the range tested. The low St branch continued unabated in Fig. 26.35 and the high St branch suddenly terminated at $T/D = 1.5$. Ishigai *et al.* (1972J) found high St down to $T/D = 1.3$, while Okajima (1979J) measured the high St down to $T/D = 1.15$ (see insert in Fig. 26.35). The reason for these discrepancies was the short length of the narrow wake. The high-frequency eddy street merged fast into the low-frequency eddy street, and could not be detected if the hot wire was placed too far downstream.

26.3.4 *Drag and lift forces*

The drag force exerted on a single cylinder has been related to the width of the near-wake. The existence of the narrow and wide wakes behind side-by-side cylinders results in different drag forces on the two cylinders. Biermann and Herrnstein (1933P) measured the drag force in the range $1 < T/D < 5$

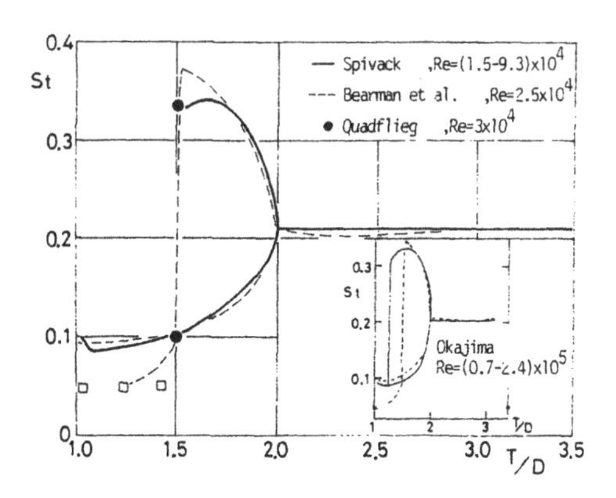

FIG. 26.35. Strouhal number in terms of T/D, Okajima *et al.* (1986J)

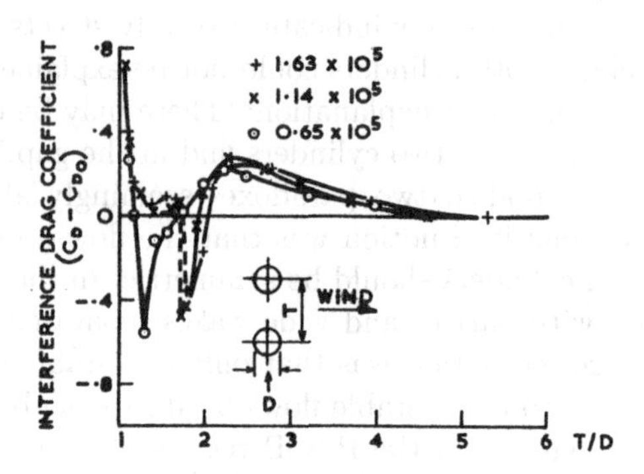

FIG. 26.36. Interference drag in terms of T/D, Biermann and Herrnstein (1933P)

and 65k $\leqslant Re \leqslant$ 163k.[160] Figure 26.36 shows the variation in the interference coefficients C_{Di} in terms of T/D. For $T/D < 5$, C_{Di} starts to increase, and reaches a maximum at $T/D = 2.2$ in the P-SSC regime. For $T/D < 2$, the C_{Di} variation becomes erratic between the positive and negative values. The authors made a remarkable observation: 'Apparently the type of flow changes rapidly with a change in spacing; it may even change while the spacing is held constant.' This was the very first hint that some kind of bistable flow had occurred.

Bearman and Wadcock (1973J) were the first to measure simultaneously C_{pb} on both side-by-side cylinders. A single C_{pb} value is found for $T/D < 1.1$ (P-SSA regime) and $T/D > 2.2$ (P-SSC regime), while two C_{pb} values are found in the P-SSB regime. The authors wrote: 'The base pressure changed from one steady value to another, or simply fluctuated between the two extremes. Stopping or starting the tunnel could cause pressure to change over.'

Two C_{pb} values produced two C_D values in the P-SSB regime. A compilation of measured C_D is given in Fig. 26.37. The early measurements by Biermann and Herrnstein (1933P) at Re = 160k are re-plotted, but the experimental points are connected by two curves: high and low C_D. It is believed that this rectified the effect of the bistable change-over observed during the tests. Similar high and low C_D values were evaluated by Hori (1959P) for Re = 8k, and Zdravkovich and Pridden (1977J) at Re = 60k. Only high C_D values are quoted by Bearman and Wadcock (1973J) at Re = 25k.

Some analogy exists between flows past side-by-side cylinders and a single cylinder in the proximity of the plane boundary. A lift force component emerges at small T/D for the former and at small G/D for the latter. In both cases, the stagnation points are displaced towards the gap, and a maximum C_L occurs when the $T/D = 1$ and $G/D = 0$ cylinders are in contact and touching the

[160]The experimental details are given in Section 26.3.

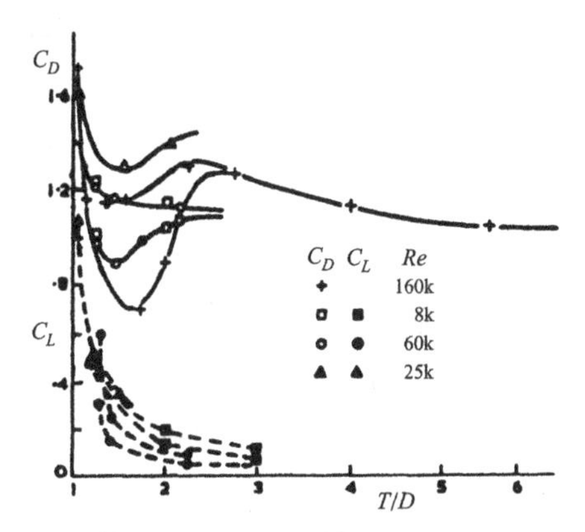

FIG. 26.37. Drag and lift in terms of T/D, Zdravkovich and Pridden (1977J)

boundary, respectively.

Bearman and Wadcock (1973J) noted: 'The movement of stagnation point due to the proximity of the other cylinder rotates the resultant force vector (towards the gap), thus producing the component in the lift direction.' To test this hypothesis the angle $\theta_0' = \tan^{-1} C_L/C_D$ was calculated, and compared with θ_0 from the C_p distributions. The agreement was good because the C_p distribution was symmetrical relative to the displaced stagnation point.

Figure 26.37 also shows the variation in C_L in terms of T/D. The variation in C_L resembles closely that for a cylinder near the plane boundary, see section 24.8. The biased gap flow causes a greater deflection of the narrow wake, and this produces a higher C_L value. Unlike the C_{pb} variation with T/D, the difference between the high and low C_L is small. Figure 26.37 shows a monotonous decrease in C_L from the maximum value at $T/D = 1$. Note that the positive C_L designates a repulsive force between the cylinders.

Quadflieg (1977J) carried out extensive tests on fixed and flexible side-by-side cylinders having $L/D = 10$, $D/B = 0.28$, and $113\text{k} < Re < 157\text{k}$. He measured the angular location of the stagnation θ_0 and separation θ_s. For $T/D > 1.5$, the deflection of θ_0 towards the gap is the same for both cylinders. However, in the range $1.1 < T/D < 1.5$, a difference of about 2° appears.

Figure 26.38(a–c) shows stagnation, θ_0, and separation on the gap side, θ_{sg}, and on the outer side, θ_{so}. There is about an 8° deflection in θ_{sg} on the two cylinders at the gap side in the range $1.1 < T/D < 1.5$ due to the wide and narrow wakes. The maximum deflection in θ_{so} at $T/D = 1.1$ on the gap side does not exceed 14°, while θ_0 exceeds 33°. Figure 26.38(c) shows a single θ_s curve without branching as measured on the outer side of the cylinders. The lack of branching may suggest that the biased gap flow phenomenon is confined to the gap only.

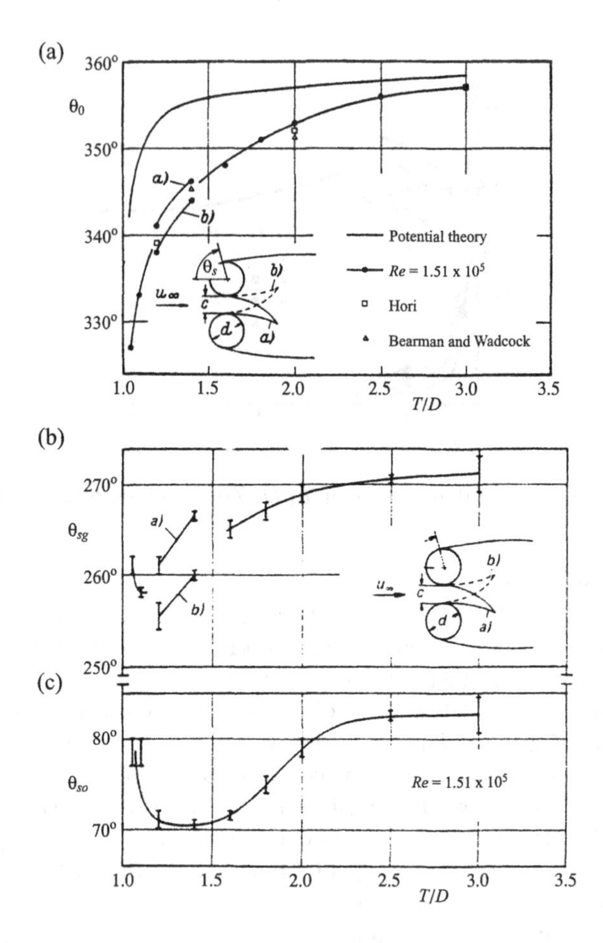

FIG. 26.38. (a) Angular position of stagnation point for $Re = 151\text{k}$, (b) separation angle gap side, (c) outside, Quadflieg (1977J)

26.3.5 *Origin of biased gap flow*

It has already been discussed that the bistable biased gap flow has two paradoxical features. The first is that an entirely symmetrical oncoming flow leads to the asymmetric narrow and wide wakes behind the two identical side-by-side cylinders. The second is that a uniform and stable flow induces a non-uniform and random bistable flow. The origin of bistable biased flow has been attributed to various causes, but still remains unresolved.

Ishigai *et al.* (1972J) suggested that the Coanda effect is responsible for the biased gap flow. The Coanda effect is observed when a jet tangentially attached to a curved surface becomes deflected by following the surface. According to this definition, it is essential to have a rounded surface and displacement of the separation point.

Bearman and Wadcock (1973J) carried out preliminary tests to verify the Coanda hypothesis by using side-by-side flat plates. The biased gap flow was found despite the absence of a rounded surface and the presence of fixed separation points. This proved that the Coanda effect could not be the main cause for the biased gap flow. Bearman and Wadcock (1973J) suggested that the biased gap flow was due to a wake interaction rather than to separation displacement.

Zdravkovich (1977R, 1987R) noted that the stable narrow and wide wakes were common features for the upstream and downstream cylinders in staggered arrangements, respectively. As the side-by-side arrangement was approached, the above distinction persisted so that one cylinder remained 'upstream', with a narrow wake, and the other 'downstream', with a wide wake. The asymmetric flow structure was preserved but became bistable because neither of the side-by-side cylinders was 'upstream' nor 'downstream'. The flow structure consisting of two identical wakes appears to be intrinsically unstable and hence impossible.

Zdravkovich (1995P) attempted to unriddle the phenomenological mechanism of the biased flow. He suggested two stages in the interference leading to the bias and its random change-over:

(i) The initial stage of the interference takes place in the gap between the two shear layers of the vorticity of opposite sign. As the shear layers are in the vicinity of the closely spaced side-by-side cylinders in the P-SSB regime, they mutually inhibit the roll-up into the eddies. However, a small deflection in the gap flow to either side disturbs the balance of the induced velocity, and one of the shear layers starts to roll up. This triggers a further increase in the bias towards the narrow wake until the finite bias is reached for a particular value of G/D.

(ii) The mechanism of the intermittent change-over of the biased direction may also be related to the initial deflection in the free shear layers. The separated shear layers may be deflected to the opposite side, and the initial roll-up changes to that side. Further development is the same as described in (i).

The two basic features of the phenomenon, i.e. the bias and the bistable gap flow, might be eliminated either by using a partition plate between the cylinders, or an external acoustic excitation of transition eddies in the free shear layers.

26.3.6 *Effect of partition plate and sound*

Kim and Durbin (1988J) examined the effect of a partition plate inserted between the side-by-side cylinders on the bistable biased flow. The experimental details were: $L/D = 26.8$, $D/B = 0.076$, with square end plates, $D_e/D = 10.5$, $Re =$ 3.3k, and $T/D = 1.75$. A typical variation of time histories in C_{pb1}, C_{pb2}, and the biased angle α_b are shown in Fig. 26.39 without the partition plate. The random and simultaneous change-over in C_{pb1} and C_{pb2} is triggered by the gap flow switch α_b. Although the time between the change-over is random, the magnitudes of

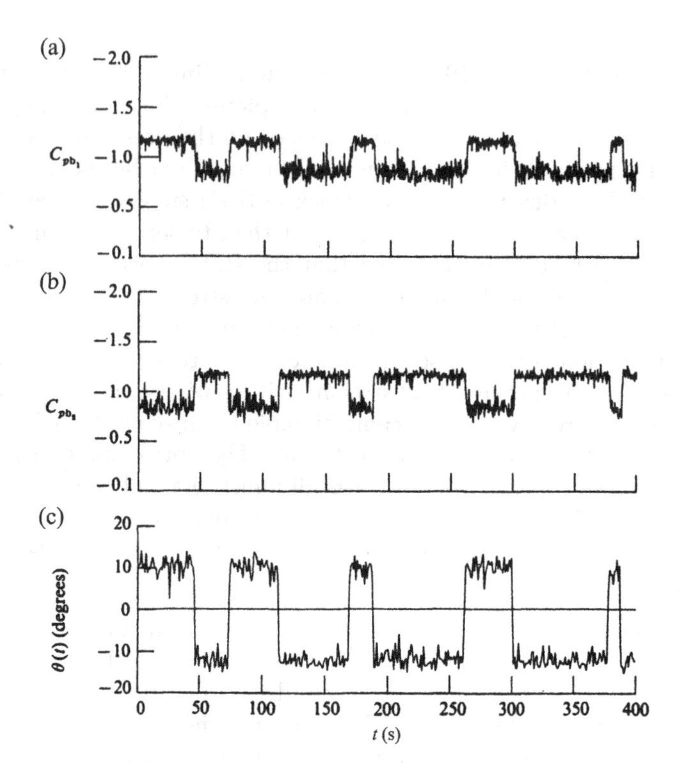

FIG. 26.39. Time variation in base pressure, (a) cylinder 1, (b) cylinder 2, (c) bias angle, Kim and Durbin (1988J)

C_{pb1}, C_{pb2}, and α_b are essentially constant. This confirms that the two quasi-stable flows exist, and that either of the two is possible. The mean period between the change-over decreases exponentially with rising *Re*. The period of the change-over could be described by a Poisson stochastic process (see Kiya *et al.*, 1992P).

The partition plate, 4.3D long, used had a sharp leading edge, LE, and a trailing edge, TE. The interference between the partition plate and the gap flow strongly depended on the relative locations of TE and LE along the x-axis. Three flow regimes were observed:

(i) The steady biased regime, $1.3 < (x/D)_{\mathrm{TE}} < 2.7$, shows the suppression of the bias change-over. The same can be achieved with the partition plate farther downstream, $1.2 < (x/D)_{\mathrm{LE}} < 7$, Fig. 26.40(a).

(ii) The steady symmetrical regime, $2.7 < (x/D)_{\mathrm{TE}} < 4.3$, shows the suppression of both bistable and biased flow. The most effective location is shown in Fig. 26.40(b).

(iii) The periodic biased flow, $4.3 < (x/D)_{\mathrm{TE}} < 5.8$ and $0 < (x/D)_{\mathrm{LE}} < 1.5$, is

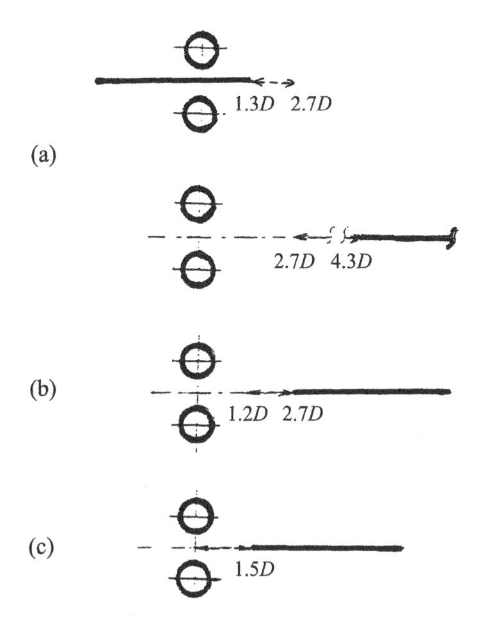

FIG. 26.40. Partition plate position, (a) steady bias, (b) no bias, (c) oscillating bias, adapted from Kim and Durbin (1988J)

shown in Fig. 26.40(c). The alternate formation and shedding of separation bubbles at LE from the two sides of the partition plate is synchronized with the biased flow oscillation. The non-dimensional frequency of oscillation, $St_{\mathrm{osc}} \simeq 0.002$, is two orders of magnitude lower than St for eddy shedding.

The partition plate was removed and a loudspeaker was positioned in front of the wind tunnel entrance. The sound pressure was 100 dB, and the sound frequency was chosen at $St_a = 1.45$ to match the synchronization of the transition eddies in the free shear layers at $Re = 3.6$k. The wake traverse revealed two symmetric near-wakes without bias. It appeared that the increased regularity and periodicity of transition eddies in the free shear layers stabilized the gap flow and prevented bias.

26.3.7 *Landweber's theoretical model*

It has been discussed in Vol. 1, Chapter 11, that the convection velocity in a two-row potential vortex street is induced by vortices. The velocity components along the x and y axes are:

$$u_c = \frac{K\pi}{a} \tanh \frac{\pi b}{a}, \qquad v_c = 0 \tag{26.1}$$

where u_c and v_c are the streamwise and transverse convection velocity components, respectively, K is the strength of each vortex, and a and b are the streamwise distances between vortices and two rows, respectively. The streamwise velocity u in row 1 is induced only by the vortices in the opposite row 2, and vice versa.

The addition of another two rows produced by two cylinders in a side-by-side arrangement changes the kinematics so that the streamwise velocity in row 1 is induced by the other three rows, u_2, u_3, and u_4. This was recognized by Landweber (1942P) who extended Kármán's theory derived for a two-row vortex street stretching to infinity in the upstream and downstream direction to two parallel vortex streets. He stipulated the following conditions:

(i) The streamwise-induced velocity components have to be the same in all rows. This condition is arbitrary because the observed pairing of eddies by Williamson (1985J) can occur only if $u_1 \neq u_2$ and $u_3 \neq u_4$.

(ii) The transverse-induced velocity component $v_i = 0$, for $i = 1, 2, 3, 4$, so that the vortex rows remain aligned. The above condition is possible only for the out-of-phase and in-phase staggered four rows as sketched in Fig. 26.41(a,b).

(iii) The distance between the rows is assumed to be $b/D = 1$ for the TrSL state and $b/D = \frac{1}{2}$ for the TrBL state, which is acceptable as a rough approximation.

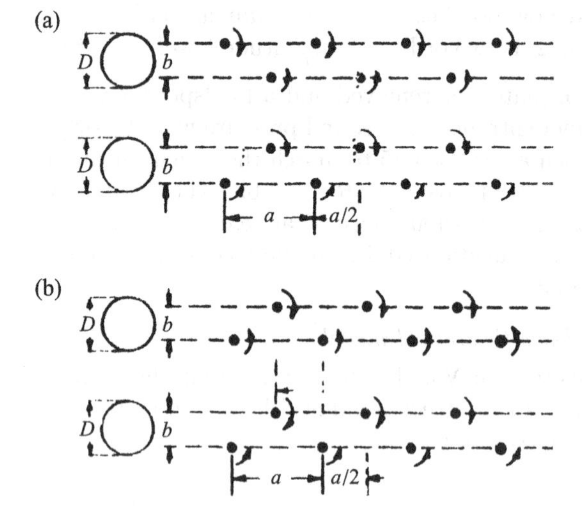

FIG. 26.41. Modes of vortex street coupling, (a) out-of-phase, (b) in-phase, Landweber (1942P)

The simple eqn (26.1) derived for a single vortex street is extended for an out-of-phase configuration, Fig. 26.41(a):

$$\left.\begin{aligned} u_1 &= \frac{1}{2a}\left[K_2 \tanh\frac{\pi b}{a} - K_3 \tanh\frac{\pi(b+k)}{a} + K_4 \coth\frac{\pi(2b+k)}{a}\right] \\ u_2 &= \frac{1}{2a}\left[K_1 \tanh\frac{\pi b}{a} - K_3 \coth\frac{\pi k}{a} + K_4 \tanh\frac{\pi(b+k)}{a}\right] \end{aligned}\right\} \tag{26.2}$$

where k is the distance between rows 2 and 3.

Landweber (1942P) assumed $K_1 = K_4$ and $K_2 = K_3$ and imposed $u_1 = u_2$ owing to the symmetry of two side-by-side cylinders

$$\frac{K_1}{K_2} = \frac{\tanh\frac{\pi b}{a} - \tanh\frac{\pi(b+k)}{a} + \coth\frac{\pi k}{a}}{\tanh\frac{\pi b}{a} + \tanh\frac{\pi(b+k)}{a} - \coth\frac{\pi(2b+k)}{a}} \tag{26.3}$$

where K_1 is the circulation of vortices in rows 1 and 4, and K_2 that in rows 2 and 3. The distance between rows 1 and 4 is $2b+k$. Since the hyperbolic cotangent is greater than the hyperbolic tangent (except when the argument is infinite when both are equal to unity) the numerator in eqn (26.3) is greater than $\tanh \pi b/a$ and the denominator is less than $\tanh \pi b/a$, hence $K_1 > K_2$.

A similar derivation yields for the in-phase vortex configurations, Fig. 26.41(b)

$$\frac{K_1}{K_2} = \frac{\tanh\frac{\pi b}{a} - \coth\frac{\pi(b+k)}{a} + \tanh\frac{\pi k}{a}}{\tanh\frac{\pi b}{a} + \coth\frac{\pi(b+k)}{a} - \tanh\frac{\pi(2b+k)}{a}} \tag{26.4}$$

Applying the same reasoning as before, the numerator is less than, and the denominator greater than, $\tanh \pi b/a$. Hence, $K_1 < K_2$. For $b/a = 0.3$, there is a limit $k/a = 0.45$ ($T/D = 2.8$), which is not found in real flows.

Okajima and Sugitani (1980P) traversed the wakes at $x/D = 2.5$ behind side-by-side cylinders for $Re = 240$k. Figure 26.42(a) shows that the outer and gap wake velocities are the same for $T/D = 3$. However, there is a marked increase in the outer velocity and decrease in the gap velocity for $T/D = 1.125$. According to Landweber's (1942P) theory, this means $K_3 < K_4$ in Fig. 26.42 for the out-of-phase vortex configuration.

26.3.8 *Other theoretical models*

In Vol. 1, Chapter 12, it was discussed how Rosenhead (1931J) replaced a continuous free shear layer with a sheet of equally spaced discrete potential vortices, the so-called vortex sheet. The small initial sinusoidal displacement of the vortex sheet induced a velocity field, which in time led to the concentration of vortices into clusters. The same method was extended to two vortex sheets of the opposite

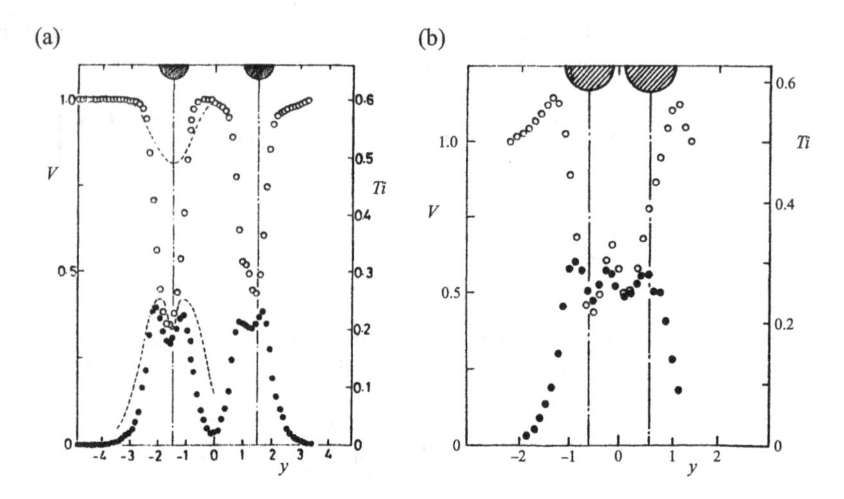

FIG. 26.42. Velocity profiles behind side-by-side cylinders at $x/D = 2.5$, *Re* = 240k, (a) $T/D = 3$, (b) $T/D = 1.125$, Okajima and Sugitani (1980P)

rotation by Abernathy and Kronauer (1962J). The concentration of the individual vortices of the same rotation formed two rows of clusters. The interaction of two vortex sheets resulted in the attraction of some vortices of the opposite rotation into clusters, effectively reducing the cluster circulation by mixing.

Kamemoto (1976J) applied the vortex sheet method to four rows of potential vortices. He distinguished four growing and four decaying periodic modes being in-phase and out-of-phase. The time step and number of vortices were the same as in Kronauer's computation. A typical development of clusters after the initial periodic perturbation (amplitude $A/a = 0.0125$, $b/a = 0.28$, and $T/D = 2$) is shown in Fig. 26.43(a,b), the out-of-phase and in-phase modes, respectively. For the former, four rows of clusters are formed downstream as observed by Williamson (1965J), Fig. 26.32(a). For the in-line mode, the initial four vortex sheets gradually rearranged to form a single cluster street as observed by Williamson (1985J), and shown in Fig. 26.32(b).

26.3.9 *Transition-in-boundary-layer, TrBL, state*

It has been discussed in Vol. 1, Chapter 6, that for a smooth cylinder in a low *Ti* free stream the separation bubbles are formed in the range 340k < *Re* < 450k. At the lower *Re*, one separation bubble is formed on either side of the cylinder, the TrBL1 regime, followed by the second bubble formed on the opposite side at higher *Re*, the TrBL2 regime.

The interference between two side-by-side cylinders considerably changes the sequence of events and the *Re* range of the bubble appearance. The formation of the first separation bubble always takes place on the outer side, and is associated

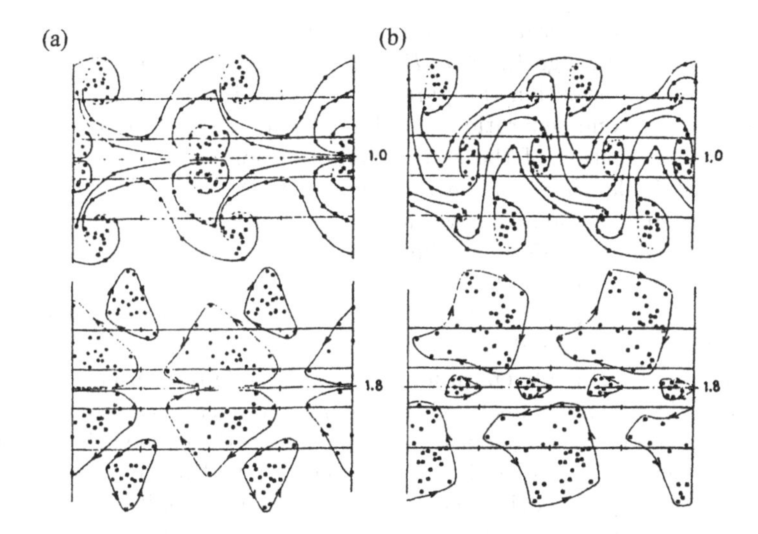

FIG. 26.43. Computation of vortex clusters after initial disturbance, (a) symmetric, (b) anti-symmetric, Kamemoto (1976J)

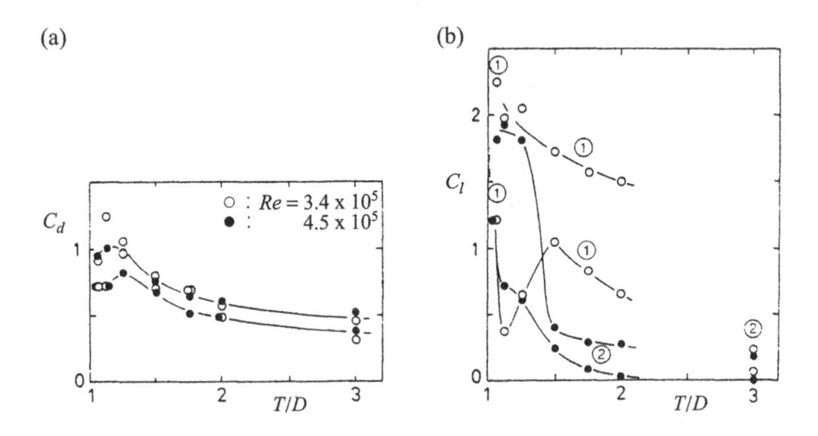

FIG. 26.44. (a) Local drag, (b) local lift for side-by-side cylinders at $Re = 340$k and 450k, Okajima *et al.* (1986J)

with an extremely high C_L exceeding 2. The appearance of the second bubble makes the C_p distribution almost symmetric and leads to a collapse of high C_L.

Okajima *et al.* (1986J) carried out tests in the range $1 < T/D < 3$ at $Re =$ 340k and 450k.[161] Figure 26.44(a,b) shows C_{d1}, C_{d2}, and C_{l1}, C_{l2} evaluated from

[161] The experimental details are given in Section 26.2.

the C_p distribution in terms of T/D, respectively. Note that the C_{d1} and C_{d2} differ slightly in the range $1.2 < T/D < 2.3$, presumably due to a weak biased flow caused by the formation of separation bubbles. The sequence of appearance of the first and second separation bubbles, ① and ②, may be followed in Fig. 26.44(b). The formation of the second bubble leads to the collapse of C_l for $T/D > 2.4$. A similar development takes place for $Re = 450$k at $T/D = 1.3$.

Another unique phenomenon was found for the SS arrangements by Okajima *et al.* (1986J). Figure 26.45 shows the evaluated C_{d1}, C_{d2}, C_{l1}, and C_{l2} in terms of T/D. The repulsive C_l suddenly becomes attractive (negative) for $T/D = 1.2$, and appears so only at $Re = 120$k and 240k. Okajima *et al.* (1986J) carried out smoke visualization, and observed that a highly biased gap flow triggered a significant displacement of separation on the gap side of the narrow wake. At $Re = 120$k and 240k, the transition in the boundary layer is triggered by a biased gap flow resulting in the unique attractive (negative) C_l on one cylinder only.[162]

An even higher Re was reached by Sun *et al.* (1992Ja). They measured C_p' around both side-by-side cylinders at $Re = 325$k and 650k[163] for $T/D = 2.2$. Figure 26.46(a,b) shows the C_{p1}' and C_{p2}' distributions and C_p' around a single cylinder for comparison. Figure 26.46(a) shows that the separation peaks of different magnitude appear on two side-by-side cylinders at $Re = 325$k. The separation peaks are considerably reduced and are all equal in magnitude for

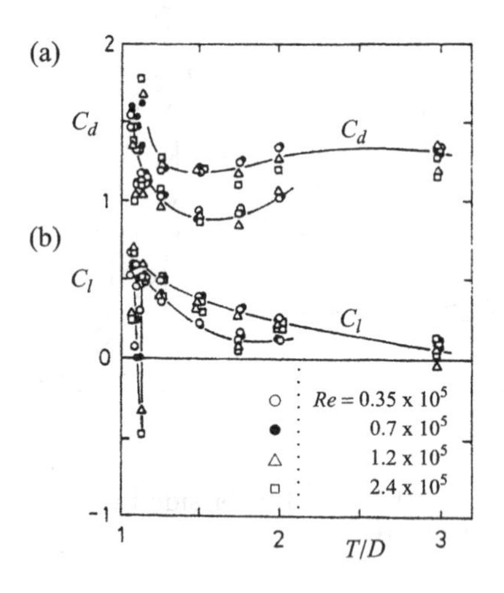

FIG. 26.45. (a) Local drag, (b) local lift in terms of T/D for side-by-side cylinders at $Re = 120$k and 240k, Okajima *et al.* (1986J)

[162]The phenomenon is analogous to the inversion of the Magnus effect and blowing slot disturbance, as discussed in Chapter 24.

[163]The experimental details are given in Section 26.2.

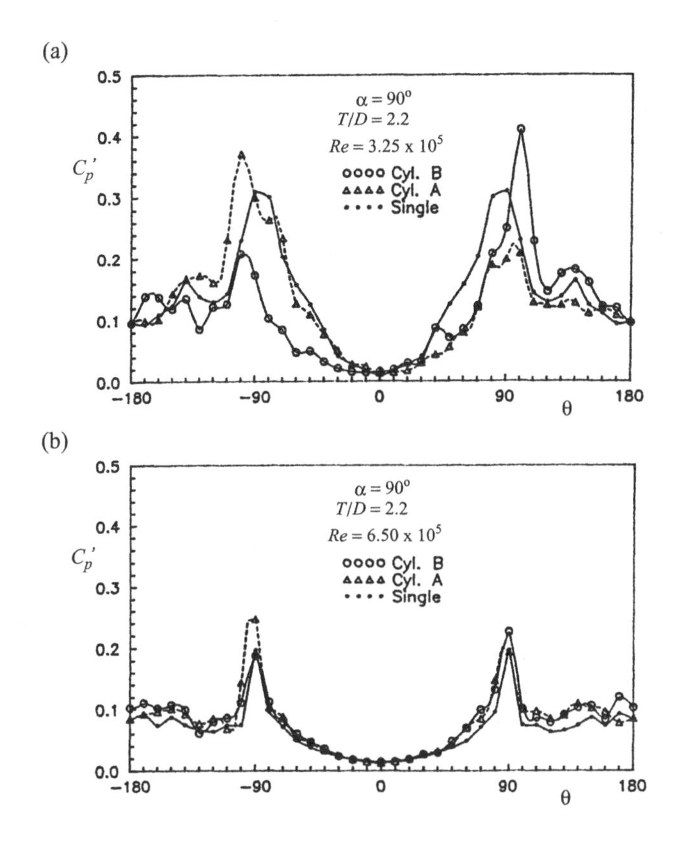

FIG. 26.46. Fluctuating pressure distribution around side-by-side cylinders for $T/D = 2.2$, (a) $Re = 325\text{k}$, (b) $Re = 650\text{k}$, Sun *et al.* (1992Ja)

$Re = 650\text{k}$ in Fig. 26.46(b). This may indicate that the biased gap flow is inhibited at $Re = 650\text{k}$ for $T/D = 2.2$.

26.3.10 *Effect of free stream turbulence*

The effect of free stream turbulence has been found to be particularly strong in the biased flow regime. Zhang (1993P) carried out tests on side-by-side cylinders subjected to $Ti = 0.4\%$, 4.5%, and 11.5%. He measured C_D, C_D', C_L' only on one of the cylinders in the range $1 < T/D < 4$. The C_{Di} value is reduced with increasing T. The disruption of the biased flow regime is seen in Fig. 26.47.

26.4 Staggered arrangements

26.4.1 *Classification of interference flows*

Staggered arrangements have often occurred in many engineering applications but have attracted less research in comparison with tandem and side-by-side

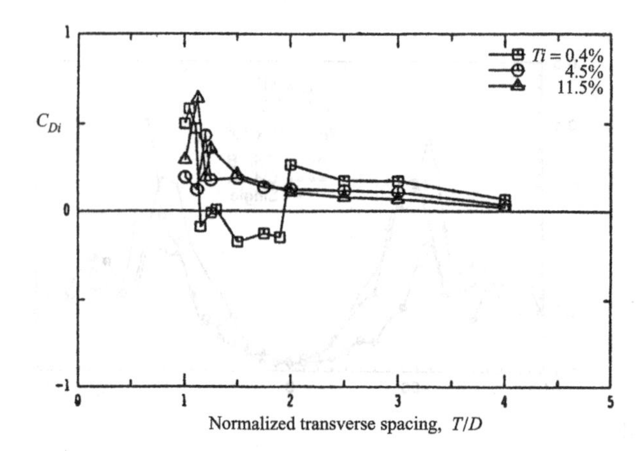

FIG. 26.47. Interference drag coefficient for side-by-side cylinders in terms of T/D for $Ti = 0.4\%$, 4.5%, and 11.5%, Zhang (1993P)

arrangements. The early measurements of forces on two cylinders in staggered arrangements were carried out by Pannell *et al.* (1915P), see Fig. 26.8.

Ishigai *et al.* (1972J) also carried out the schlieren visualization around heated cylinders in a wide range of staggered arrangements in the range $2k < Re < 4k$. Figure 26.48 shows a selection of some frames in order of the decrease in spacing ratio between the staggered cylinders. The common feature of all spacings is the narrow and wide wakes behind the upstream and downstream cylinders, respectively. Figure 26.48 shows that the bigger the spacing, the less is the bias of the gap flow and the difference in size of the two wakes. The extreme bias is seen for $S/D = 0.68$ and $T/D = 1$ when the bias angle exceeds 45°. The bias becomes bistable for $S/D < 0.1$–0.15.

Hori (1959P) evaluated C_d and C_l from the C_p distribution and presented their results in a table. These data were converted into the resultant interference force coefficient C_{ri} by Zdravkovich (1977R), and are displayed in the T/D, S/D

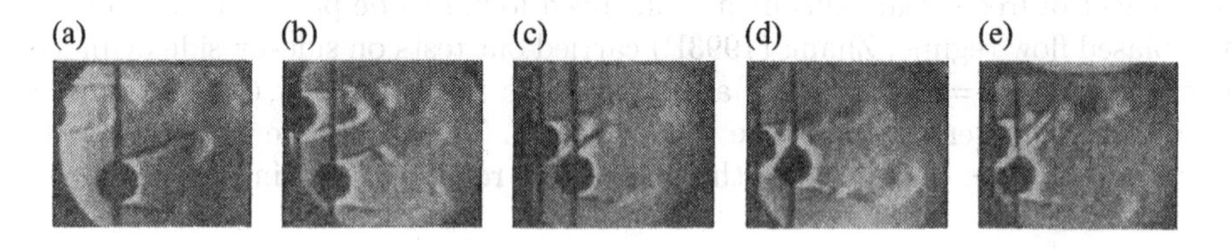

FIG. 26.48. Schlieren visualization of staggered cylinders, (a) $S/D = 2$, $T/D = 2$, (b) $S/D = 1$, $T/D = 2$, (c) $S/D = 1$, $T/D = 1$, (d) $S/D = 1$, $T/D = 0.5$, (e) $S/D = 0.68$, $T/D = 0.5$, Ishigai *et al.* (1972J)

plane. The arrows are related to the monitored cylinder location. Figure 26.49 shows a wide variation in magnitude and direction of C_{ri}, and its components C_{li} and C_{di}. The lift component may be directed towards or away from the upstream cylinder wake. The boundary between the attractive and repulsive forces is $C_L = 0$ (dashed-dot line). Zdravkovich (1977R) suggested that the following magnitude and sign of C_l should be used as a criterion for the classification of staggered arrangements:

(1) $C_l \approx 0$, the upstream cylinder is in the upstream proximity region in tandem arrangements or in the no-interference region;

(2) $C_l > 0$, the side proximity region; the repulsive lift force and the increased or decreased drag force;

(3) $C_l < 0$, the wake interference region; the attractive lift force and the increased or decreased drag force.

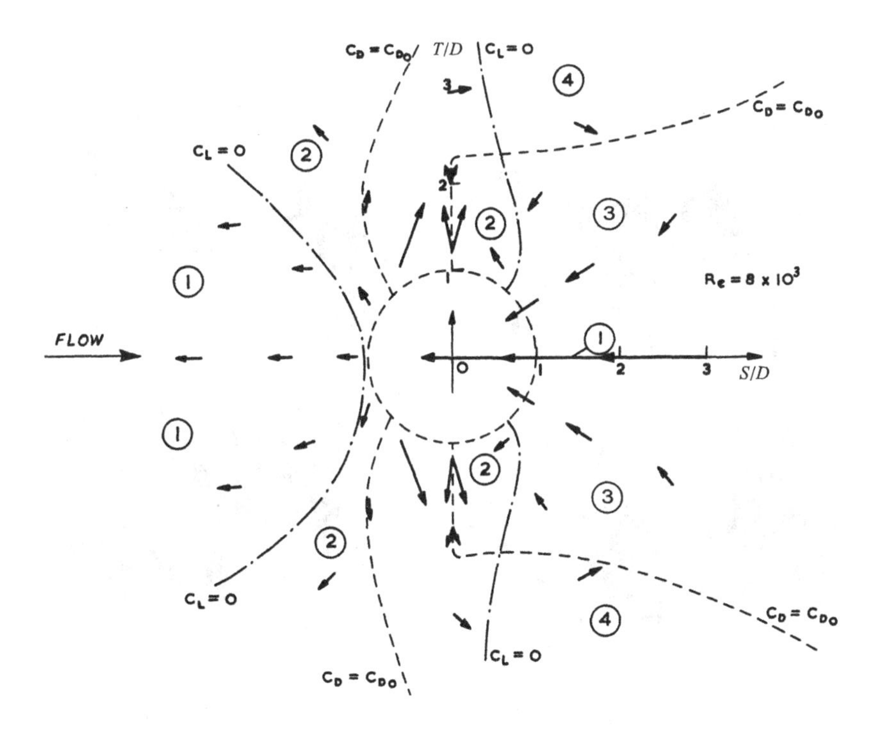

FIG. 26.49. Force vectors in the T/D, S/D plane at $Re = 8$k, adapted from Hori (1959P)

26.4.2 *Laminar, L, state of flow*

Tatsuno and Ishii (1983P) and Tatsuno (1989J) carried out flow visualization around tandem, side-by-side, and staggered cylinders at $Re = 0.02$. Figure 26.50(a–d) shows staggered arrangements for $\underline{S}/D = 1.5$ and 2.5, $\alpha = 5°$, $15°$, and $45°$. The weak gap flow is established for $\underline{S}/D = 1.5$ as can be seen in Fig. 26.50(a,c) for $\alpha = 5°$ and $15°$, respectively. A fully developed gap flow is seen in Fig. 26.50(b,d).

Zdravkovich (1972J) carried out flow visualization in the L3 regime, where the laminar Kármán–Bénard eddy street is formed behind a single cylinder. Figure 26.51(a,b) shows eddy streets behind staggered cylinders $\underline{S}/D = 6$ and $\alpha = 45°$ at $Re = 100$. Two eddy streets are formed with a different scale and of the same frequency. The eddy street behind the upstream cylinder persists along the wake, while the other behind the downstream cylinder diffuses, and loses its identity. The angle of stagger promotes the transition to turbulence to a considerably lower Re.

26.4.3 *Mean pressure distribution in the TrSL state*

The detailed measurements of the mean pressure distribution around staggered cylinders were reported by Hori (1959P), Suzuki *et al.* (1971P), Wardlaw and

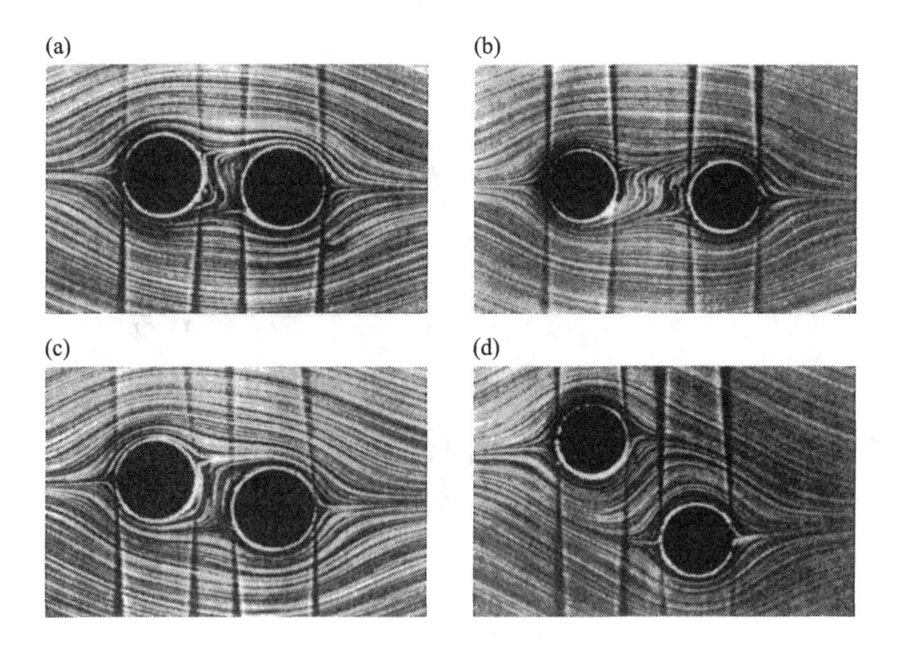

FIG. 26.50. Creeping flow visualization for staggered cylinders, $Re = 0.2$, (a) $\underline{S}/D = 1.5$, $\alpha = 5°$, (b) $\underline{S}/D = 2$, $\alpha = 5°$, (c) $\underline{S}/D = 1.5$, $\alpha = 15°$, (d) $\underline{S}/D = 2$, $\alpha = 45°$, Tatsuno and Ishii (1983P)

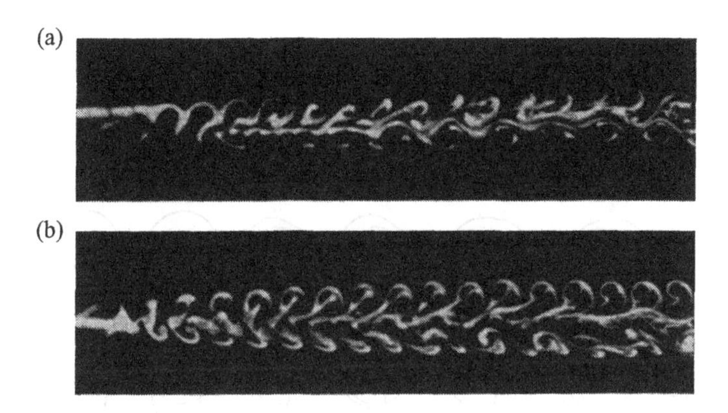

FIG. 26.51. Smoke visualization of staggered cylinders $\underline{S}/D = 6$, $\alpha = 45°$, $Re = 100$, (a) filament above cylinders, (b) between cylinders, Zdravkovich (1972J)

Cooper (1973P), Price (1976J), Zdravkovich and Pridden (1977J), Zdravkovich (1977R), Dayoub (1982J), Price and Paidoussis (1984J), Bokaian and Geoola (1985P), Ohya *et al.* (1988R), Zhang and Melbourne (1989P), Sun and Gu (1995J), etc.

Suzuki *et al.* (1971P) presented the C_p distribution in the form of polar plots around the downstream cylinder and evaluated the resultant force for each cylinder location in Fig. 26.52(a,b), respectively. The top row corresponds to tandem arrangements $\alpha = 0°$, and the other three rows represent an increase in the staggered angle from $\alpha = 5°$ to $15°$. Three features are evident in Fig. 26.52:

(i) The stagnation point (the maximum positive C_p) is displaced away from the upstream wake axis.

(ii) A significant $C_{p\min}$ is developed along the gap side of the downstream cylinder. This results in a large lift force directed towards the upstream cylinder wake.

(iii) The asymmetry in the C_p distribution diminishes for $S/D > 2.5$, and becomes symmetric relative to the axis passing through θ_0 and the cylinder centre. The value of C_l is related to the inclination of the resultant force.

26.4.4 *Lift and drag in the TrSL state*

Systematic measurements of the lift and drag components of the resultant force exerted on the downstream cylinder in the range $0.1 < T/D < 2$ and $0 \leqslant S/D \leqslant 4$ at $Re = 60$k were carried out by Zdravkovich (1977R). Two identical cylinders had $L/D = 15.7$, $D/B = 0.06$, spanned the end panels, and $Ti = 0.1\%$. The drag and lift forces were measured separately on a central cylinder section attached to the six components' strain gauge balance.

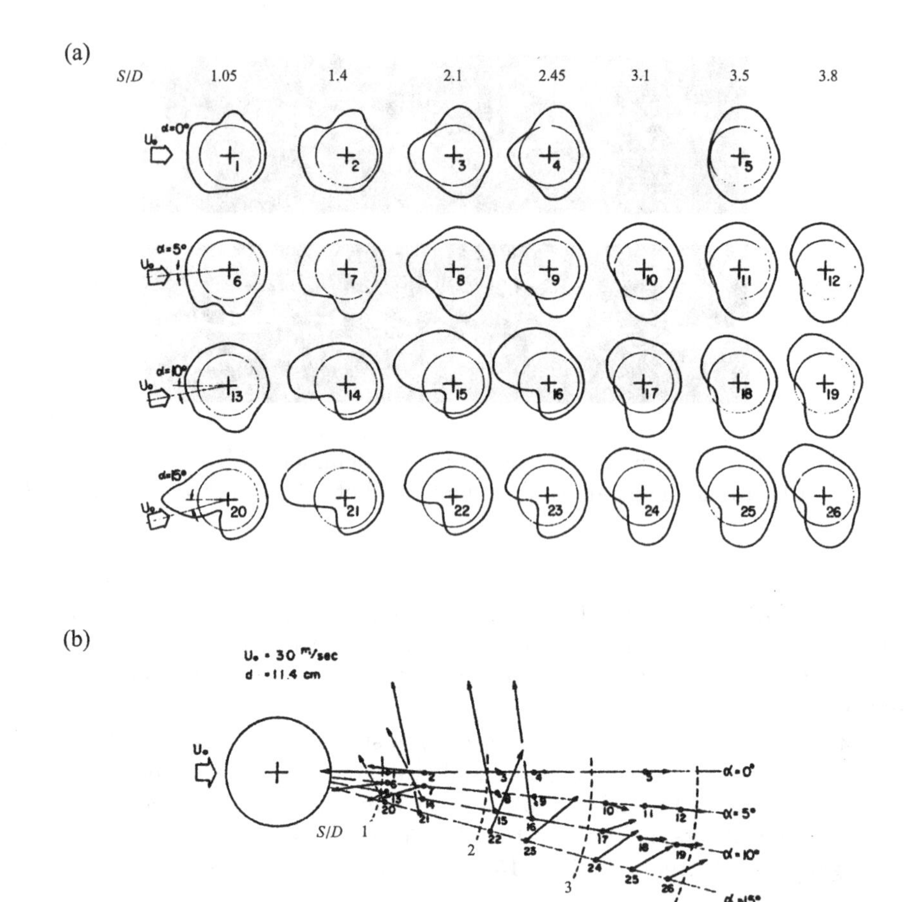

FIG. 26.52. (a) Polar plots of mean pressure distribution around the rear cylinder, (b) force vectors, Suzuki *et al.* (1971P)

The novel way of data presentation is displayed in Fig. 26.53(a,b). A reference cylinder is fixed in the origin of the T/L, S/L coordinate system. The dashed semi-circle designates the loci of the monitored cylinder centre when two cylinders are in contact. The positive quadrant represents forces on the downstream cylinder and the negative (left) quadrant on the upstream cylinder. The measured values of C_L and C_D are interpolated in such a way that the position and magnitude of equi-C_L and equi-C_D are traced. Figure 26.53(a,b) shows the curves of the constant equi-C_D and equi-C_L, and estimated $C_{L\min}$ (dashed-dot line). Additional semi-circles designate the constant spacing ratios between the cylinders.

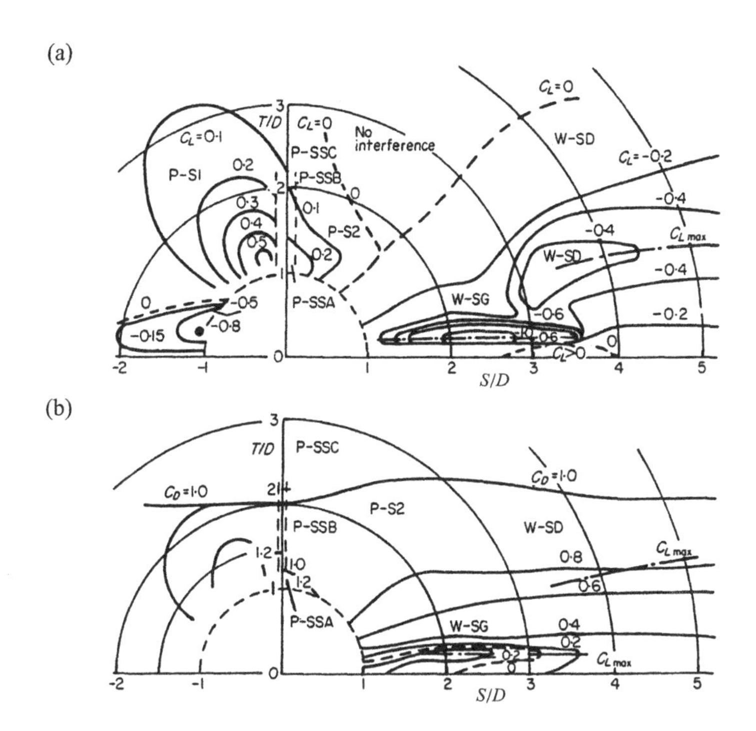

FIG. 26.53. Measured and interpolated, (a) equi-drag, (b) equi-lift, in the T/D, S/D plane, Zdravkovich (1977R)

There are three distinct interference flow regions displayed in Fig. 26.53(a):

(i) the attractive force $C_L < 0$ for the upstream cylinder, $C_{L\min} = -0.8$;

(ii) the repulsive force $C_L > 0$ for the upstream and downstream cylinders, $C_{L\max} > 0.5$;

(iii) the attractive force $C_L < 0$ exhibits two subregions:

(a) W-SG, gap flow having $C_{L\min} < -1$ at about $T/D = 0.25$ and $L/D < 3$;

(b) W-SD, displacement flow, $C_{L\min} \approx -0.5$ near the wake border for $L/D > 2.5$.

Okajima and Sugitani (1976P) examined the lift and drag forces on the upstream cylinder in the proximity of the downstream cylinder. They traversed the upstream cylinder across $S/D = 1.25$ at $Re = 100$k in the range $0 \leqslant T/D \leqslant 0.5$. Figure 26.54(a,b) shows the variation in C_d and C_l in terms of T/D. There is a rapid rise in C_l beyond $T/D = 0.2$ to a maximum at $T/D = 0.25$. Price and Paidoussis (1984J) increased the traversing distance to $S/D = 1.5$, and found a

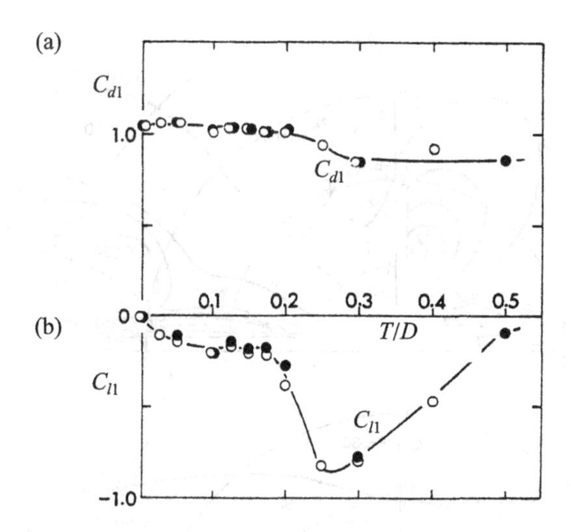

FIG. 26.54. Variation in local (a) drag, (b) lift in terms of T/D for $S/D = 1.25$ at $Re = 100\text{k}$, Okajima and Sugitani (1976P)

reduced $C_{L\min} = -0.19$. The high magnitude of $C_{L\min}$ appears to be restricted to small S/D gaps between the cylinders, as depicted in Fig. 26.53(a).

The upstream side of the interference region produced the repulsive $C_L > 0$, as has already been discussed in side-by-side arrangements. The bistable biased flow regime yielded two different C_L values not only for $S/D = 0$ but also within the gap $S/D = \pm 0.15$, as depicted in Fig. 26.53(a). The maximum C_L value was found for $S/D = 1$ when the cylinders were in contact.

The downstream wake was subjected to two different flow regimes:

(i) W-SG, dominated by the staggered gap flow;

(ii) W-SD, the staggered wake displacement regime.

Zdravkovich (1977R) termed $C_{L\min}$ in the gap and wake displacement regime as 'inner' and 'outer', because they appeared at $T/D = 0.2$ and 0.6, respectively.

26.4.5 *Gap flow interference regime*

The gap flow regime, W-SG, is confined to the proximity and wake interference region in the range $1.2 < S/D < 3.5$. The attractive force, $C_L < 0$, becomes more negative as T/D is reduced and reaches $C_{L\min} = -1$ at $T/D = 0.25$. However, a further reduction in T/D brings a sudden fall in C_L, as depicted in Fig. 26.53(a) in the range $1.5 < S/D < 2.5$.

The reason for the sudden fall in C_L is due to the drastic change in the pressure distributions. Figure 26.55(a) shows two kinds of C_p distributions for $S/D = 3$, $T/D = 0.25$, and $Re = 60\text{k}$. The gap flow regime is marked by the

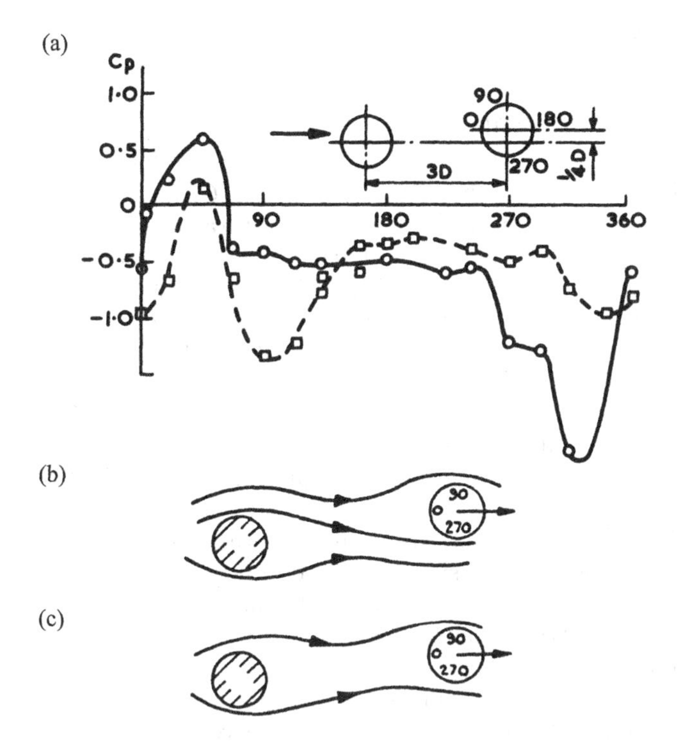

FIG. 26.55. (a) Two mean pressure distributions for $S/D = 3$, $T/D = 0.25$, $Re = 60\text{k}$, (b) gap flow topology, (c) reattachment flow, Zdravkovich (1977R)

low $C_{p\min}$ at $\theta = 330°$ and $C_{p\max}$ at $\theta = 45°$, as sketched in Fig. 26.55(b). An intermittent change in the C_p distribution shows the disappearance of $C_{p\min}$, and the C_p distribution resembles that behind the downstream tandem cylinder with the reattachment at $\theta = 45°$. The C_L collapse is associated with the cessation of the gap flow, as sketched in Fig. 26.55(c).

Sun and Gu (1995J) extended measurements to $Re = 330\text{k}$, $S/D = 1.7$, and $Ti = 2\%$. They found a bistable range in the C_p distribution for $\alpha = 9.75°$ and $\underline{S}/D = 1.7$, which corresponded to $T/D = 0.28$. This confirmed the existence of the gap flow regime at high Re.

Figure 26.53(b) showed less variation in C_D in comparison with C_L. The only exception is between $C_{L\min}$ (dashed-dot line) and the wake axis where $C_D < 0$. The $C_D < 0$ is induced by the low pressure in the gap, as shown in Fig. 26.53(b). It is notable that the lowest $C_D < 0$ is around $T/D = 0.2$ and not for the tandem arrangements, $T/D = 0$.

Another unexpected feature of the drag and lift variations is the different extent of the interference region. The wake boundary is a line along which the velocity becomes the same as in the free stream. The interference line, however, is the line along which $C_L = 0$. The latter always exceeds the wake boundary by at least $D/2$.

26.4.6 *Wake displacement interference regime*

The 'inner' $C_{L\min}$ at $T/D = 0.25$ was replaced by the 'outer' $C_{L\min}$ along the wake border for $S/D > 3$. The origin of the 'outer' $C_{L\min} = -0.3$ was attributed to the upstream cylinder wake displacement by Zdravkovich (1977R). He argued that the downstream cylinder was not *immersed* in the upstream cylinder wake but *displaced* it instead. Figure 26.55(b) shows the 'squeezed' streamline pattern by the downstream cylinder. The displaced stagnation point deflected more flow towards the wake, and induced $C_{L\min} < 0$ near the former upstream wake boundary.

There was a plethora of other explanations for the 'outer' $C_{L\min}$. Maekawa (1964J) suggested that the static pressure gradient across the upstream cylinder wake could induce a 'buoyancy' force towards the wake axis. Cooper (1973P) showed that less than half of $C_{L\min}$ could be attributed to the pressure gradient. Another of Maekawa's (1964J) arguments was that the wake turbulence could affect the separation on the sides facing the upstream wake. This might lead to an asymmetric C_p and C_L. However, this argument was not confirmed by experiments in the TrSL2 and the lower TrSL3 regimes.

Mair and Maull (1971R) proposed a wake entrainment mechanism. It was related to the inclined free stream velocity and to the resulting tilted force. Price (1976J) pointed out that the tilt itself could not account for the full magnitude of $C_{L\min}$. Price (1976J) also showed that when the downstream cylinder was twice the diameter of the upstream one, the $C_{L\min}$ increased. This was in line with the displacement argument as well as the opposite effect for the small cylinder in the wake of the large one.

The variation in C_D across and along the upstream cylinder wake was scaled by Dayoub (1982J). He suggested the following empirical relationship:

$$\frac{C_D - C_{D\min}}{C_{D\max} - C_{D\min}} = \frac{1}{2}\left[1 - \cos\frac{\pi y/D}{(y/D)_{\max}}\right] \tag{26.5}$$

where y/D and $(y/D)_{\max}$ are the transverse distance and half wake width, respectively. Figure 26.56 shows good agreement of experimental points in the range $7 < S/D < 20$. The method is analogous to the velocity defect profiles as derived by Schlichting (1979B).

26.4.7 *Strouhal number*

Ishigai *et al.* (1972J) were among the first to measure St in the range $1 < S/D < 6$ and $1 < T/D < 4$. The cylinders had $L/D = 11$ in an open jet, Ti was not stated, and $1.5\text{k} < Re < 15\text{k}$. Figure 26.57(a,b) shows St in terms of S/D and T/D, respectively. There is a distinct grouping of curves at $St > 0.2$ for the upstream cylinder (narrow wakes). Similarly, all St curves for the downstream cylinder are below $St = 0.2$.

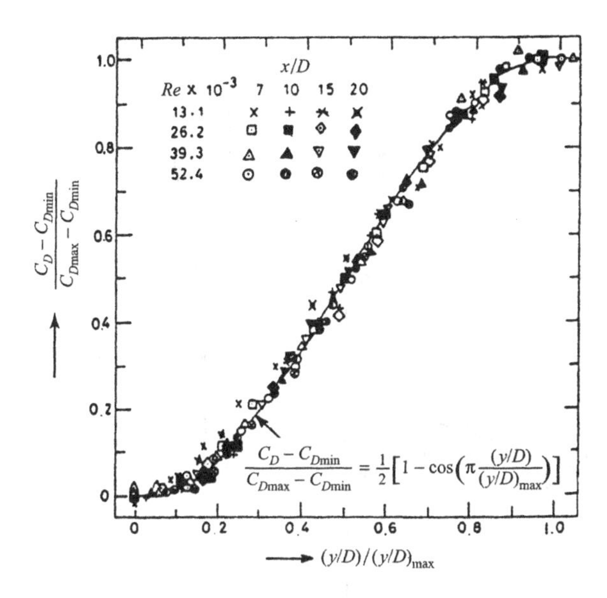

FIG. 26.56. Normalized drag in terms of non-dimensional wake width, Dayoub (1982J)

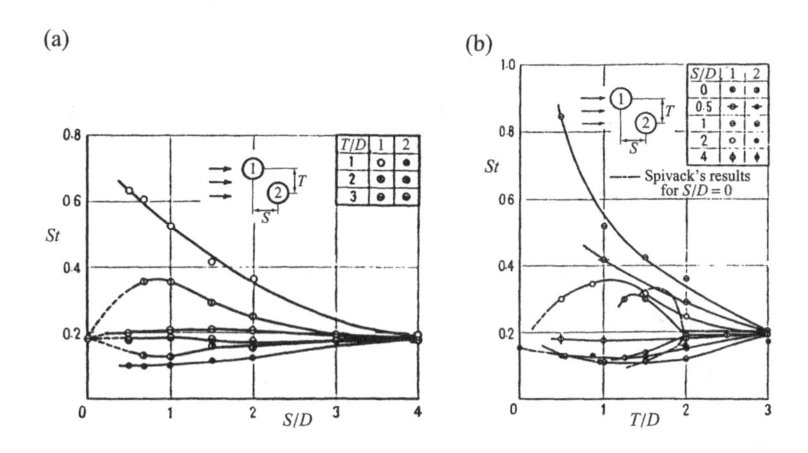

FIG. 26.57. Strouhal number in terms of (a) S/D, (b) T/D, Ishigai *et al.* (1972J)

Subsequent more detailed measurements were carried out by Kiya *et al.* (1980J). The cylinders had $L/D = 20.9$, $D/B = 0.042$, $Ti = 0.8\%$, $Re = 15.8\text{k}$, $1 \leqslant \underline{S}/D \leqslant 5.5$, and $0 < \alpha < 360°$. Figure 26.58(a,b) shows St in terms of $\underline{S}/D$ and α for the upstream cylinder in tandem, $\alpha = 0°$, and staggered, $15° \leqslant$

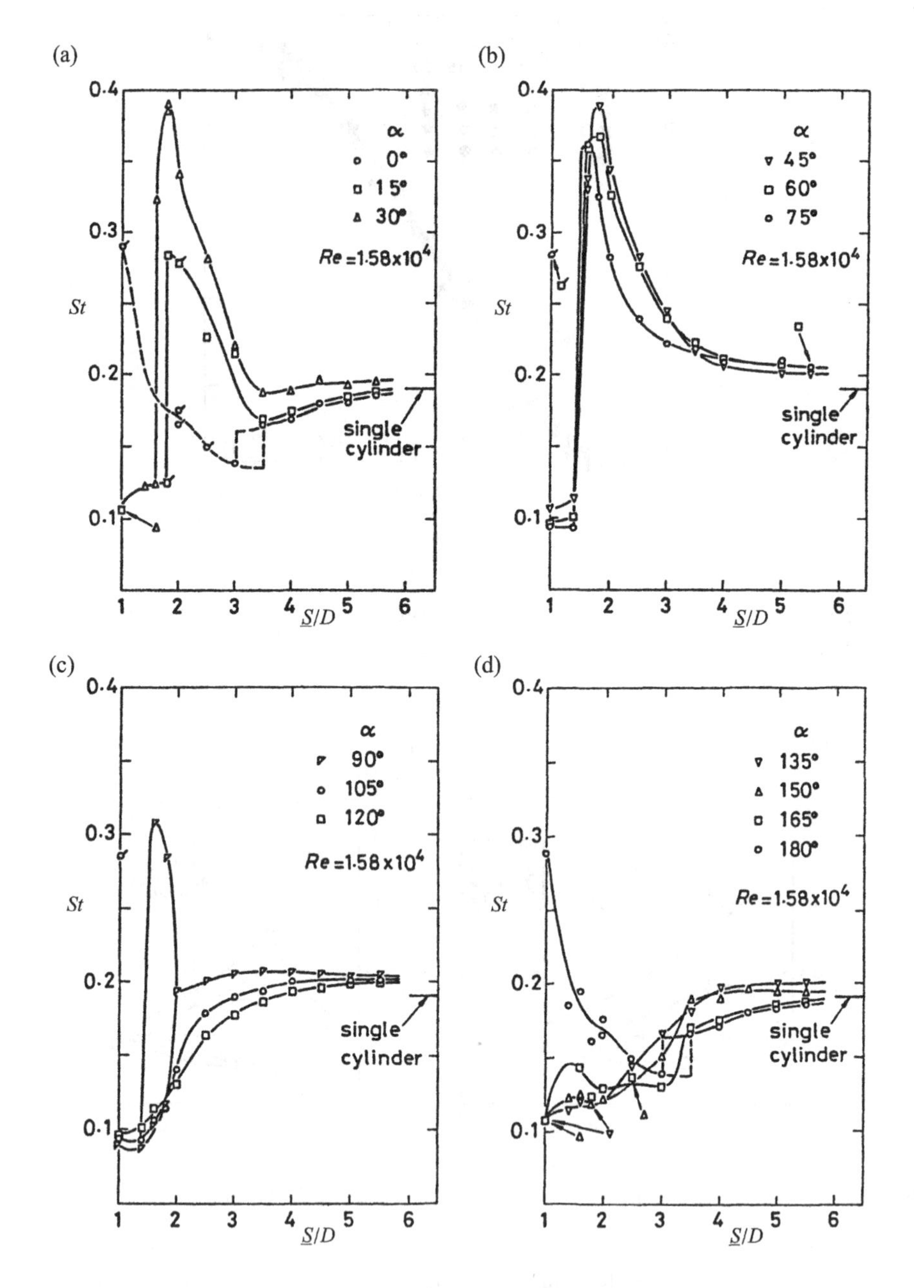

FIG. 26.58. Strouhal number in terms of $\underline{S}/D$ for Re = 15.8k, (a) 0° to 30°, (b) 45° to 75°, (c) 90° to 120°, (d) 135° to 180°, Kiya *et al.* (1980J)

$\alpha \leqslant 75°$, arrangements. Figure 26.58(c,d) exhibits the side-by-side, $\alpha = 90°$, staggered, $105° < \alpha < 165°$, and tandem, $\alpha = 180°$, arrangements.

The notable feature in Fig. 26.58(a,c,d) is that the St curves for the tandem $\alpha = 0°$ and $180°$, and the side-by-side, $\alpha = 90°$, arrangements are entirely different from the St curves for staggered arrangements. All St curves for the upstream cylinder in the staggered arrangements attain $St_{\max}$ for $S/D > 1.5$. The downstream cylinder forms a wide wake from the start, $S/D = 1$, and St gradually increases with rising S/D.

26.4.8 *Transition-in-boundary-layer, TrBL, state*

The near-wake narrowing in the TrBL state for a single cylinder was caused by the significant displacement of the turbulent boundary layer separation. It is expected that the narrow wake will reduce the wake interference region, and considerably modify the drag and lift forces exerted on the two cylinders.

Sun *et al.* (1992Ja) measured C_p and C_p', and evaluated C_l and C_d for staggered cylinders at $Re = 220$k, 325k, and 650k. The cylinders had $L/D = 15.1$, $D/B = 0.05$, $Ti = 0.12\%$, and $1.05 \leqslant \underline{S}/D \leqslant 7$. The measured C_p distributions for $Re = 325$k were of two kinds around the downstream cylinder:

(i) the stagnation C_{p0} is at $\theta = 0$ for $S/D > 4$;

(ii) the reattachment $C_{p\max}$ is in the range $30° < \theta < 60°$ for $S/D = 2.2$ and 3.5.

For $Re = 650$k, all C_p curves for the downstream cylinder have C_{p0} at $\theta = 0°$. This indicates that a gap flow is established even for $S/D = 2.2$ due to the narrow wake behind the upstream cylinder.

Figure 26.59 shows the C_p' distribution on two staggered cylinders for $\underline{S}/D = 2.2$ and $\alpha = 12.5°$ at $Re = 325$k and 650k. A drastic reduction in C_{p2}' occurs for $Re = 650$k. The shedding frequency peaks are absent behind the upstream and downstream cylinders at $Re = 650$k.

26.4.9 *Effect of free stream turbulence*

It has been shown that free stream turbulence had a profound effect on the interference flow regimes for tandem and side-by-side cylinders. Gu *et al.* (1993J) carried out tests at $Re = 650$k, $Ti = 10\%$, $L/D = 15.1, D/B = 0.05$, without end plates, $1 < S/D < 4$, and $1 < T/D < 4$. The C_d and C_l were obtained by integrating the C_p distribution. The shedding frequency peak for a single cylinder was absent for $Ti = 0.12\%$ and present for $Ti = 10\%$ ($St = 0.26$).

An overall variation in C_l and C_d for the downstream cylinder in the S/D, T/D plane is given in Fig. 26.60(a,b), respectively . The plot is remarkably dissimilar to the $S/D, T/D$ plot for the TrSL state as shown in Fig. 26.53(a,b). The free stream turbulence of $Ti = 10\%$ obliterated both the gap flow and wake displacement regimes so that the wake interference becomes negligible. The local C_d is always less than that for the single cylinder ($C_d = 0.5$).

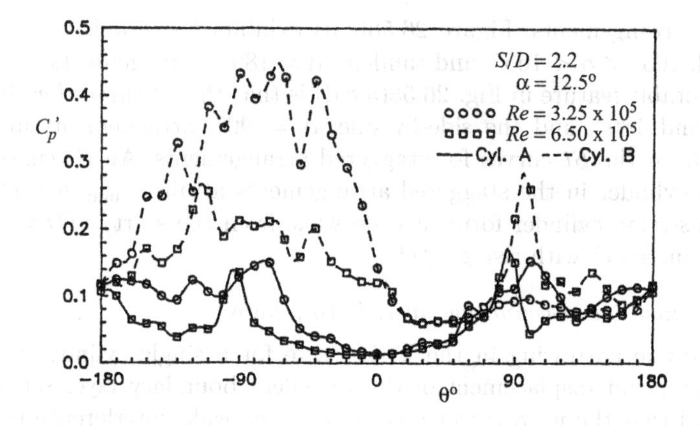

FIG. 26.59. Fluctuating pressure distribution on staggered cylinders for $\underline{S}/D = 2.2$, $\alpha = 12.5°$, $Re = 325$k and 650k, Sun *et al.* (1992Ja)

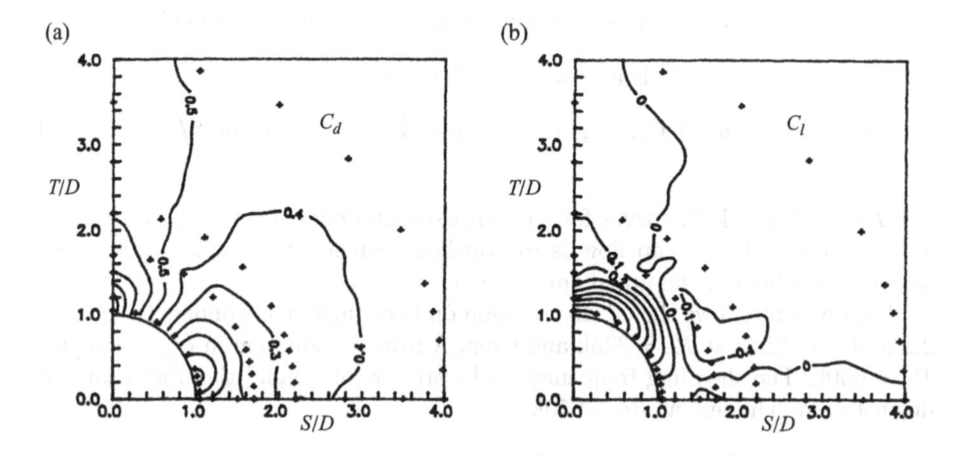

FIG. 26.60. (a) Equi-drag, (b) equi-lift in the T/D, S/D plane at $Re = 650$k, Gu *et al.* (1993J)

Zhang (1993P) measured C_L and C_D at the lower $Re = 110$k and the higher $Ti = 11.5\%$. The $S/D, T/D$ plot of equi-C_L and equi-C_D was found to be similar to Sun *et al.* (1992Ja). This indicated that Ti was the governing parameter and not Re. Both results confirmed that the free stream turbulence changes the mechanism of interference in staggered arrangements.

26.4.10 *Stranded conductors*

Wardlaw and Cooper (1973P) and Cooper (1974P) carried out a wind tunnel investigation of steady aerodynamic forces on smooth and stranded conductors. The compilation of their data in the S/D, T/D plane and the interpolation of measured C_L and C_D by the equi-lines is shown in Fig. 26.61(a,b) for a 'chukar' stranded conductor. It has been shown in Section 22.3.8 that a twisted stranded cable displaces $C_{D\min}$ towards low Re. The tested $Re = 150$k is in the postcritical TrBL4 regime for the single conductor. This means a narrow wake and a reduced interference region in comparison with the smooth cylinder. This is evident in the reduced extent of both C_L and C_D in Fig. 26.61(a,b), respectively. It is also notable that $C_{L\min}$ in the gap flow regime is halved in comparison with the smooth cylinders. Counihan (1963P) investigated spacings $S/D = 10.7$ and 16, which was important for the transmission lines.

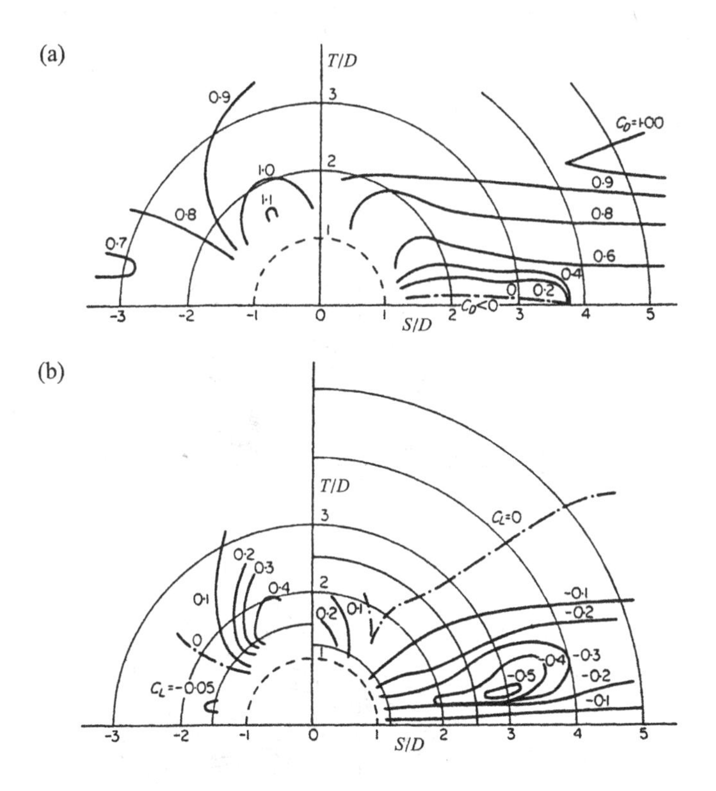

FIG. 26.61. (a) Equi-drag, (b) equi-lift in the T/D, S/D plane for a stranded conductor at $Re = 150$k, adapted from Wardlaw and Cooper (1973P)

26.4.11 *Effect of the finite aspect ratio*

It has been discussed in Chapter 21.8 that a cylinder free end induces a strong three-dimensional flow. If the aspect ratio H/D is less than 5–6, then eddy shedding is suppressed along the cylinder. The effect of the free end on the interference regions of two finite cylinders has only occasionally attracted research.

Taniguchi *et al.* (1982J) carried out tests on short staggered cylinders $H/D = 3$, $D/B = 0.004$, $\delta/H = 0.87$, $Re = 15.5$k, and $Ti = 0.3\%$. The interpolated equi-C_l and equi-C_d from the measured C_p distribution are presented in the S/D, T/D plane. Figure 26.62(a,b) shows a qualitative similarity to Fig. 26.53(a,b) for the two-dimensional cylinder. The attractive $-C_l$ regions are separated by the repulsive $+C_l$ around the side-by-side arrangement. The extreme values $C_{L\min} = -0.2$ and -0.46, and $C_{L\max} = 0.25$ are considerably less: $C_{L\min} = -0.8$ and -1.0, and $C_{L\max} = 0.5$ for the two-dimensional cylinder.

Zdravkovich (1980J) carried out tests for a single spacing $\underline{S}/D = 1.32$, $H/D = 4.6$, $D/B = 0.034$, and $Re = 200$k, and surface roughness $K/D = 0.017\%$. The pressure tappings were at $z/H = 0.25$, 0.50, 0.75, and 0.87 elevations. The local lift and drag were calculated at these elevations from the C_p distributions. Figure 26.63(a,b) shows the local C_l and C_d at the corresponding

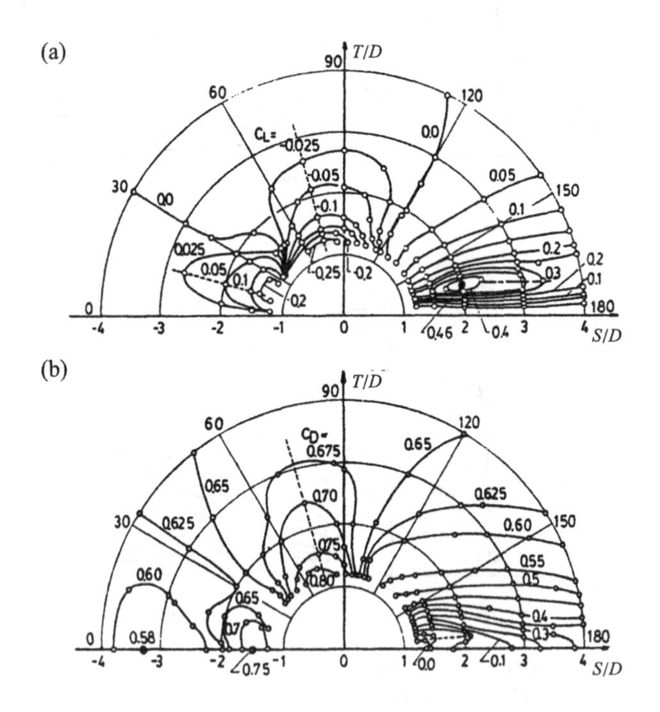

FIG. 26.62. (a) Equi-lift lines, (b) equi-drag lines for $H/D = 3$, $Re = 15.5$k, Taniguchi *et al.* (1982J)

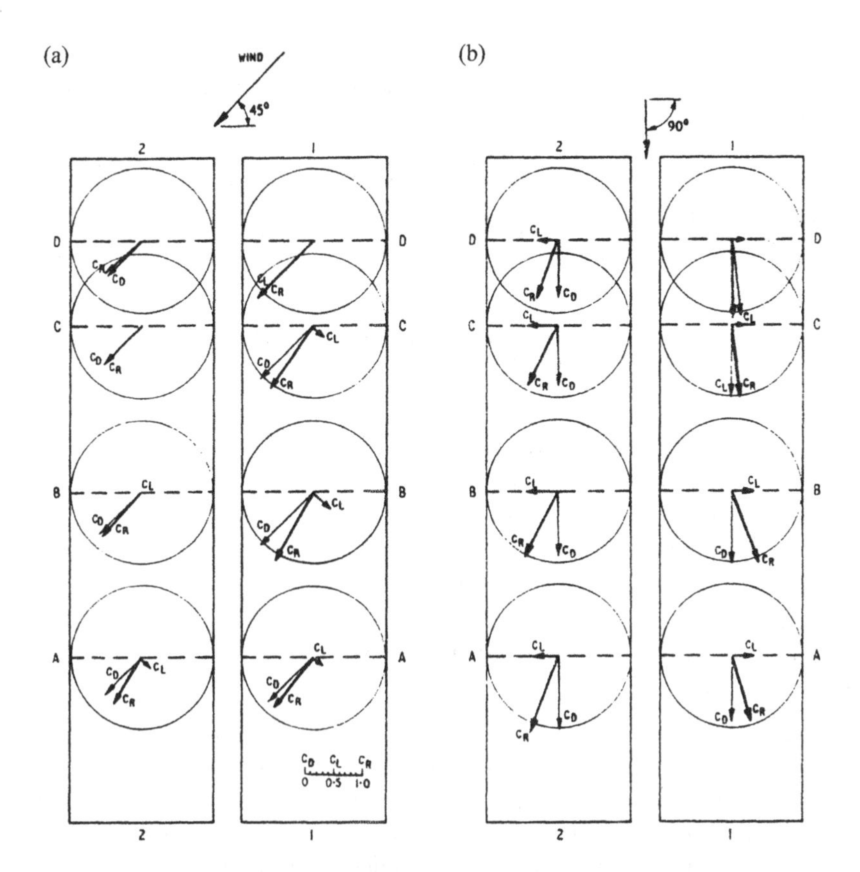

FIG. 26.63. Local lift, drag, and resultant force for $\underline{S}/D = 1.32$, $H/D = 4.6$, $Re = 200$k, (a) $\alpha = 15°$, (b) $\alpha = 45°$, Zdravkovich (1980J)

elevations for the staggered angles $\alpha = 15°$ and 45°, respectively. For $\alpha = 15°$, only small variations in C_l and C_d are seen along the upstream cylinder 1, while for downstream cylinder 2, a considerable variation occurs. For example, the negative C_d at A–A becomes positive near the free end at D–D. For $\alpha = 45°$, small variations in C_l and C_d are seen along both cylinders.

Zhang and Melbourne (1992P) carried out tests on finite staggered cylinders of $H/D = 8$, $D/B = 0.04$, $Ti = 4.5\%$ and 11.5%, and $Re = 110$k. They measured the base moment of the cantilever cylinders. The interference was suppressed for $Ti = 11.5\%$.

26.4.12 *Twin cooling towers*

Sun and Gu (1995J) carried out full-scale tests on two and four cooling towers at $Re = 75$M and wind tunnel models at $Re = 142$k, respectively. The full-scale towers had ribs equally distributed around their circumference, and the

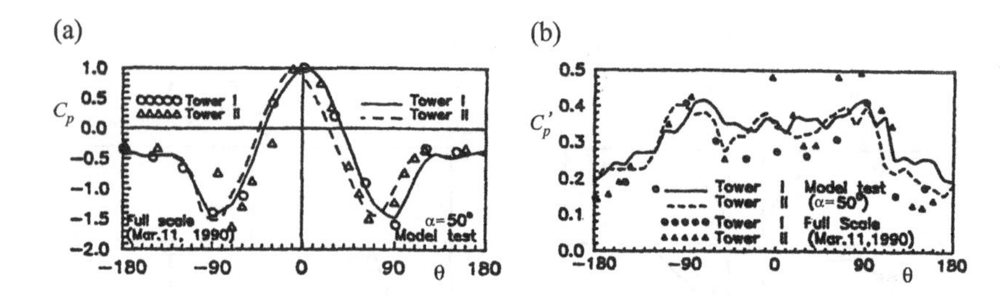

FIG. 26.64. Pressure distribution around cooling towers for $\alpha = 50°$ and $Re = 142$k and 75M, (a) mean, (b) fluctuating, Sun and Gu (1995J)

model was roughened with scotch tape. The spacing between cooling towers was $\underline{S}/D_b = 1.6$, where D_b was the base diameter. The measured variation in C_p and C_p' on the model and cooling towers is shown in Fig. 26.64(a,b), respectively. There is a good agreement between the model measurements at $Re = 142$k and full scale at $Re = 75$M for the angle of stagger $\alpha = 50°$.

26.5 Two cylinders of unequal diameter

26.5.1 *Categorization of arrangements*

Flow past two parallel cylinders of equal diameter has been discussed in previous sections. However, two parallel cylinders of unequal diameter may also be employed in some applications such as in chemical, civil, offshore, wind engineering, etc. The observed interference effects for two equal cylinders are likely to be modified by the unequal size of two cylinders. For two unequal cylinders, an additional governing parameter is the ratio of the diameters D_1/D_2.

The well-established categories for two equal cylinders have to be subdivided in terms of D_1/D_2 as follows:

(i) $0 < D_1/D_2 < 1$; the small cylinder D_1 is placed upstream from the large cylinder D_2. The small cylinder D_1 should be taken as the reference diameter used to describe the spacing ratio S/D_1.

(ii) $D_1/D_2 > 1$; the large diameter cylinder D_1 is positioned upstream of the small downstream cylinder D_2. The large diameter D_1 should be taken as the reference diameter for describing the spacing ratio S/D_1.

(iii) $0.01 < D_c/D < 0.15$; a special category is the case when a very small cylinder D_c is used to control the flow around the large cylinder.

26.5.2 *Tandem cylinders, $D_1/D_2 < 1$*

Hiwada *et al.* (1979Jb) carried out tests on flow around and heat transfer from two unequal cylinders. The ratio of cylinder diameters was $0.132 \leqslant D_1/D_2 \leqslant 0.526$, $L/D_1 = 59.8$, $L/D_2 = 7.9$, $Ti = 0.7\%$, $15\text{k} < Re_2 < 80\text{k}$, and $1 < S/D_1 <$

9. The two basic interference regimes have been specified for equal tandem cylinders as being without and with eddy shedding behind the upstream cylinder. Figure 26.65 shows the boundary between the two flow regimes for unequal tandem cylinders as $(S/D_1)_c$ in terms of D_2/D_1. Surprisingly, the relationship is linear

$$(S/D_1)_c = 1.5[1 + 0.67D_2/D_1] \tag{26.6}$$

Hiwada *et al.* (1979Jb) evaluated C_{d1} and C_{d2} from the C_p distributions. Figure 26.66 shows C_{d1} and C_{d2} in terms of S/D_1 at $Re_2 = 50$k. There are discontinuous jumps in C_{d1} at $(S/D_1)_c$ when eddy shedding behind the small upstream cylinder commences. The C_{d2} is not affected except for $D_1/D_2 = 0.526$. For small S/D_1, two kinds of unstable flow have been observed around the big downstream cylinder:

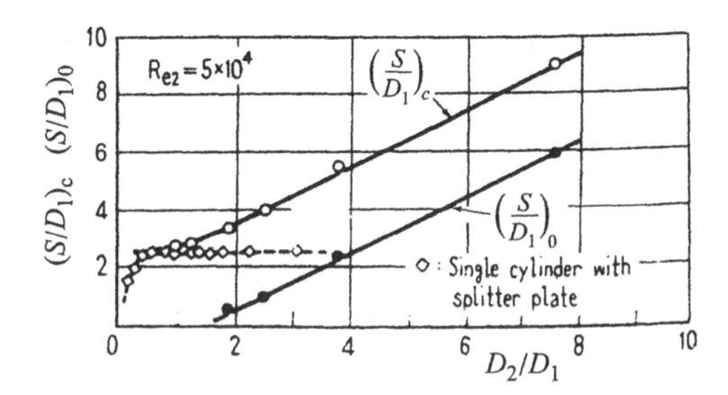

FIG. 26.65. Critical and optimal spacing ratios in terms of D_2/D_1, Hiwada *et al.* (1979Jb)

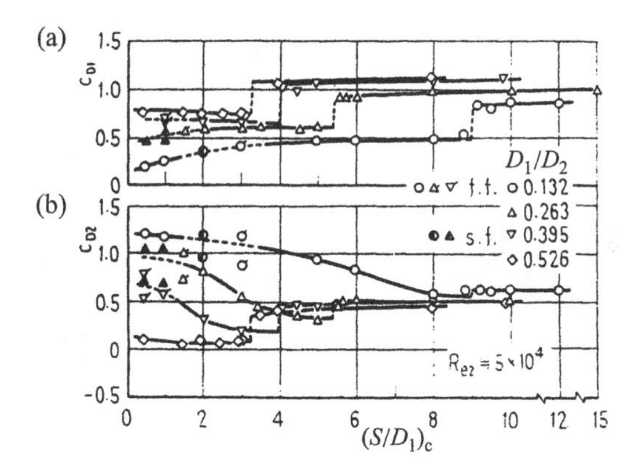

FIG. 26.66. Drag of unequal tandem cylinders in terms of S/D for $Re = 50$k, Hiwada *et al.* (1979Jb)

(i) when disturbed, externally a stable biased gap flow to one side may occur (half-closed symbols in Fig. 26.66);

(ii) the alternate biased flow to both sides occurs intermittently without the external disturbance.

Hiwada *et al.* (1979Jb) stated that both biased flows were not due to misalignment of the tandem cylinders. They also measured heat transfer from both cylinders. Figure 26.67(a,b) shows the Sh distribution for various D_1/D_2. Figure 26.67(a) shows the similarity of Sh curves for $S/D_1 < (S/D_1)_c$. The first $Sh_{\max}$ coincides with the reattachment of shear layers. The second $Sh_{\max}$ is due to the vicinity of the transitional shear layers to the surface as shown in Vol. 1, Figs 17.19 and 17.20, pp. 539–40. Figure 26.67(b) shows the similarity of Sh curves for $S/D_1 > (S/D_1)_c$. The reattachment is absent, the first $Sh_{\max}$ is small, and the second $Sh_{\max}$ is more prominent due to the additional disturbance produced by the upstream cylinder eddy shedding. Note that the small upstream cylinder enhanced the heat transfer from the big downstream cylinder. Hiwada *et al.* (1979Jb) found the optimal $(S/D_1)_0 = (S/D_1)_c - 3$, which appears as a lower curve in Fig. 26.65.

26.5.3 *Tandem cylinders, $D_1/D_2 > 1$*

The interference between the downstream small cylinder and the big upstream one is expected to be greater than for the inverse tandem arrangements. This means that the critical spacing $(S/D)_c$, where eddy shedding behind the upstream cylinder commences, should be reduced with decreasing D_2/D_1. Another feature is that the eddy shedding behind the small downstream cylinder will be affected more than for the inverse tandem arrangement.

Igarashi (1982J) tested tandem cylinders $D_1/D_2 = 1.47$, $L/D_{1,2} = 3.0$ and

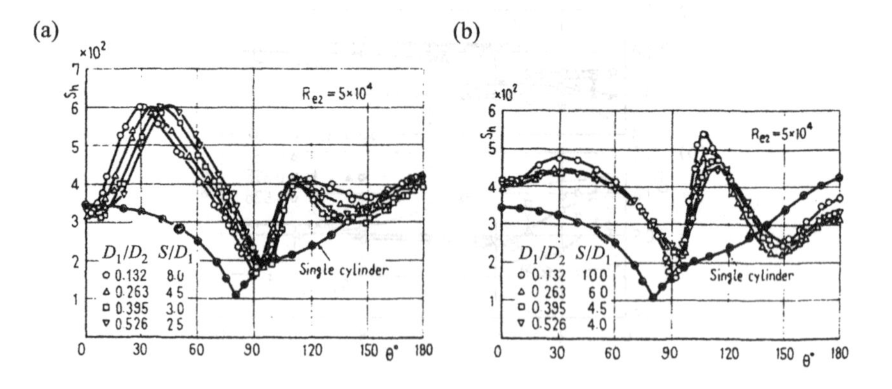

FIG. 26.67. Local heat transfer from rear cylinder at $Re_2 = 50$k, (a) $S/D_1 < (S/D_1)_c$, (b) $S/D_1 > (S/D_1)_c$, Hiwada *et al.* (1979Jb)

4.0, $D_{1,2}/B = 0.08$ and 0.06, and $Ti = 0.5\%$ in the range $13\text{k} < Re_1 < 58\text{k}$ and $0.9 < S/D_1 < 4.0$. Figure 26.68 shows the variation in C_{d1} and C_{d2} evaluated from the C_p distribution in terms of S/D_1 and Re_1 for $D_2/D_1 = 0.68$ (open symbols) and 1 (closed circles), respectively. The critical $(S/D_1)_c$ jump in C_{d1} is displaced to $(S/D_1)_c = 2.2$–2.4 in comparison with $(S/D)_c = 3.5$ for equal cylinders. However, there is also a discontinuous fall in C_{d1} at $S/D_1 = 1.0$–1.1, which is absent for $D_2/D_1 = 1$. The measured C_p distribution around the small downstream cylinder for $S/D_1 = 0.9$ and 1.0 shows an almost constant $C_p = -0.9$ all around the circumference. Beyond that, the reattachment regime brings $C_{pg} < C_{pb}$ and negative $C_{d2} = -0.5$.

The mechanism of interference flow regimes is reflected in the C'_p distributions. Figure 26.69(a,b) shows the C'_{p1} and C'_{p2} distributions for $D_2/D_1 = 0.68$ at $Re = 32\text{k}$ for S/D_1 from 0.9 to 4.0. The negligible and flat C'_{p1} and C'_{p2} for $S/D = 0.9$ and 1.0 indicate the absence of eddy shedding behind both cylinders. The three-peak curves for $S/D_1 = 1.1$, 1.4, and 1.8 represent the reattachment regime, where $\theta_r = 70°$–$80°$ is surrounded by the separation peaks. Note that the magnitude of the reattachment peak is relatively small because only the low velocity part of the free shear layer hits the small cylinder. Finally, the last two-peak C'_{p2} curves correspond to the two-eddy-street regime. The prominent peak at $\theta_e = 40°$–$50°$ is produced by the impinging eddies from the large upstream cylinder.

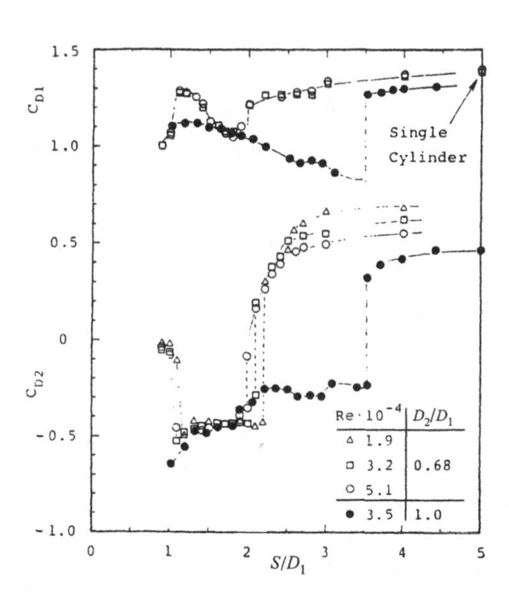

FIG. 26.68. Drag of $D_2/D_1 = 0.68$ tandem cylinders in terms of S/D_1, Igarashi (1982J)

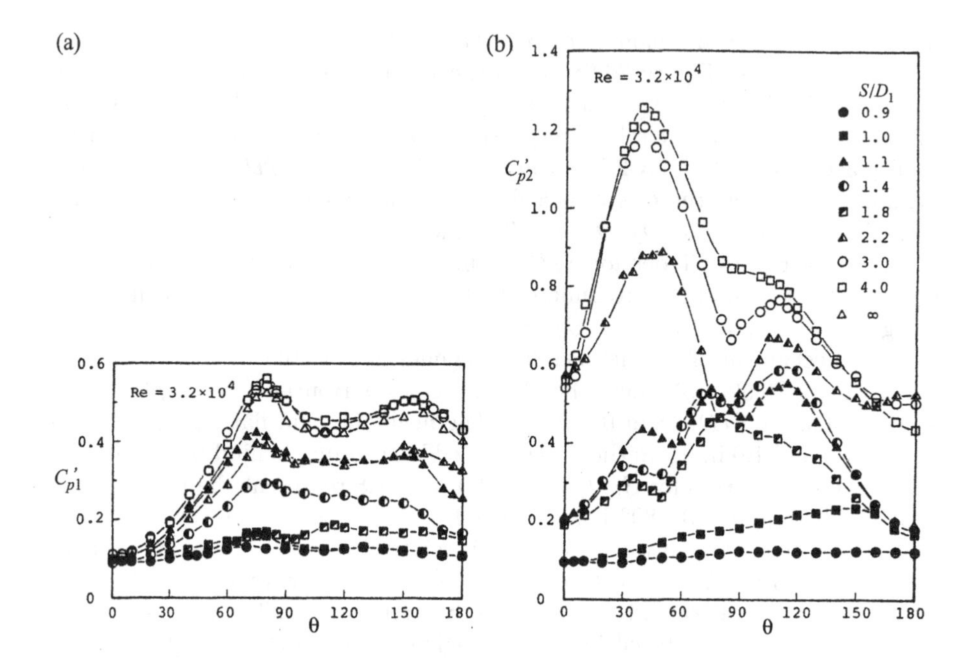

FIG. 26.69. Fluctuating pressure distribution for $Re = 32$k, $D_2/D_1 = 0.68$, (a) front cylinder, (b) rear cylinder, Igarashi (1982J)

26.5.4 *Strouhal number for tandem cylinders*

Igarashi (1982J) measured eddy shedding frequencies behind the small downstream cylinders for $D_1/D_2 = 1.47$ in the range $19\text{k} \leqslant Re_1 \leqslant 51\text{k}$. Figure 26.70 shows a compilation of St_2 in terms of S/D_1 for $D_2/D_1 = 0.5$, Novak (1975P), 0.68, Igarashi (1982J), and 1.0, Igarashi (1981J). The prominent discontinuous jump in St_2 at $S/D_1 = 3.3$, as found for equal cylinders, is barely noticeable at $S/D_1 = 2.2$ for $D_2/D_1 = 0.68$, and absent for $D_2/D_1 = 0.5$. The measured frequency is due to eddy shedding behind the upstream cylinder as shown in Igarashi's flow visualization.

26.5.5 *Synchronization of eddy shedding*

It has been discussed in Section 26.2 that for two cylinders of equal diameter synchronization of eddy frequencies occurs in the range $3.5\text{–}4 < S/D < 6\text{–}7$. In that S/D range, $St_1 = St_2$, and a binary eddy street is formed behind the downstream cylinder. The alternate eddies shed from the upstream cylinder govern the phase of formation and shedding frequency of the downstream cylinder eddies. The subsequent pairing of the eddies of the same rotation leads to the binary eddy street as shown in Fig 26.14. A similar kind of synchronization may take place behind unequal tandem cylinders.

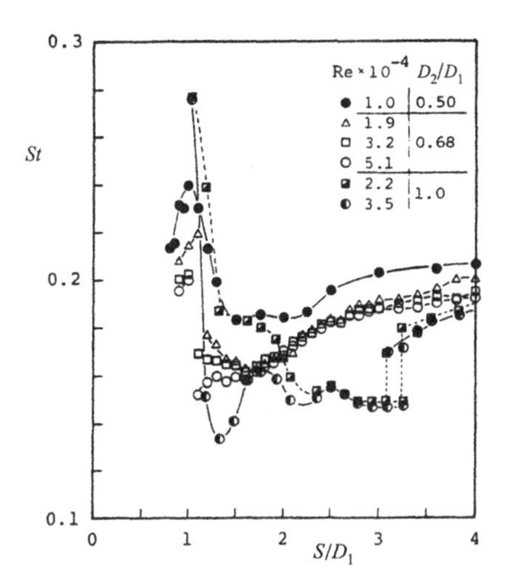

FIG. 26.70. Strouhal number behind rear cylinder in terms of S/D_1, Igarashi (1982J)

Baxendale *et al.* (1985J) found the synchronization for staggered cylinders, $D_1/D_2 = 0.5$. Figure 26.71 shows St in terms of the stagger angle α and $\underline{S}/D = 3.6$, where $St_1 = nD_1/V$ for the upstream cylinder, and $St_2 = ND_2/V$ for the downstream cylinder. It was found that for $S/D \geqslant 3$, there is a range of α for which $St_1 = St_2$. The shedding frequency is synchronized to the shedding frequency f_1 in the ratio 1:2. The synchronization ceases for $\alpha > 30°$–$35°$ when the downstream cylinder is outside the upstream cylinder wake. The corresponding C_{pb} distribution is shown in Fig. 26.72.

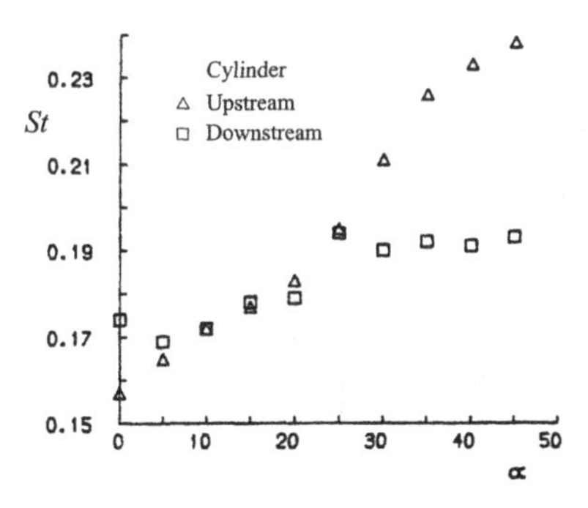

FIG. 26.71. Strouhal numbers for staggered cylinders in terms of α for $S/D = 3.5$, Baxendale *et al.* (1985J)

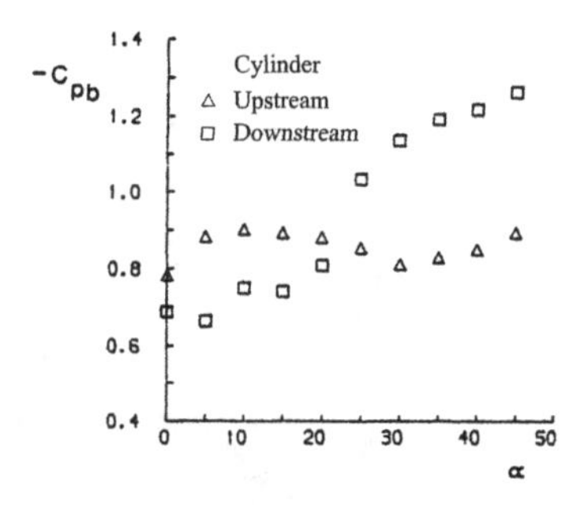

FIG. 26.72. Base pressure in terms of stagger for two cylinders, Baxendale *et al.* (1985J)

Barnes *et al.* (1986J) carried out more tests to find the conditions required for synchronization. Seven cylinders were used in turn for the upstream cylinder in the range $0.42 < D_1/D_2 < 0.57$, $40.7 < L/D_1 < 50.1$, $L/D_2 = 22.8$, $D_e/D = 4$, $D_2/B = 0.05$, $15.5\text{k} < Re_2 < 49.2\text{k}$, and $Ti = 0.1\%$. The synchronization $f_1 = 2f_2$ takes place only for $D_1/D_2 > 0.5$. Figure 26.73 shows the C_{pb2} variation in terms of D_1/D_2 for $Re_2 = 15.5\text{k}$, 31.8k, and 49.2k. The value $-C_{pb2} = 1.3$ for $T/D_2 = 1.8$ in the no-synchronization range is followed by a steep rise in the synchronization range.

The inverse staggered arrangement, large upstream cylinder D_1 and small downstream cylinder D_2 was tested by Sayers and Saban (1994J). The five upstream cylinders and a single downstream cylinder were in the range $1.25 < D_1/D_2 < 2.4$, $3.4 < S/D_1 < 6.6$, $7.2 < L/D_1 < 13.8$, and $D_1/B = 0.21$ in an open jet. Figure 26.74 shows the frequency $f_1 = 0.5f_2$ for the upstream and downstream cylinders, respectively. For all upstream cylinders, $D_1/D_2 > 2$, the f_2/f_1 ratio is synchronized at 2:1. At the small $D_1/D_2 < 2$ for all arrangements,

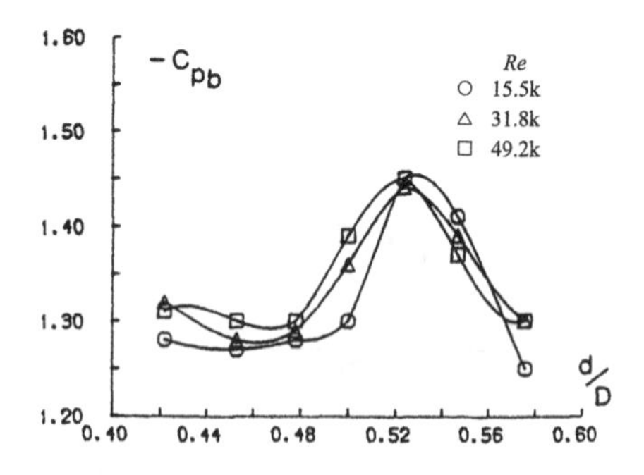

FIG. 26.73. Base pressure behind downstream cylinder inside upstream wake in terms of D_1/D_2, Barnes *et al.* (1986J)

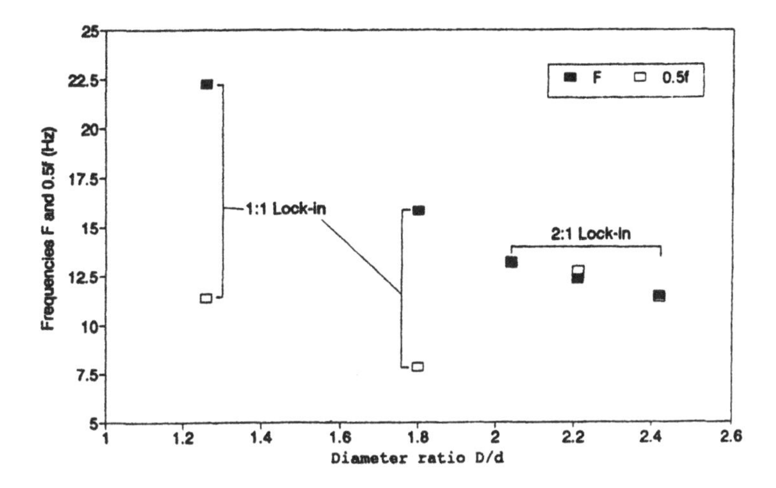

FIG. 26.74. Eddy shedding frequency in terms of D_1/D_2, Sayers and Saban (1994J)

the eddy shedding frequency ratio was 1:1. Sayers and Saban (1994J) suggested that a critical D_1/D_2 ratio exists where a switch from 1:1 to 2:1 synchronization takes place.

26.5.6 *Unequal side-by-side cylinders*

It has been discussed in previous sections that two equal cylinders arranged side by side produced a bistable biased flow in the range $1.2 < T/D < 2$. Two unequal cylinders arranged side-by-side are likely to stabilize the biased flow. However, it is not clear which of the unequal wakes is dominant.

Palmer and Keffer (1972J) carried out tests on two side-by-side cylinders of $D_1/D_2 = 0.6$ and 0.8 separated by $T/D_1 = 1.5$. The aspect ratio was $L/D_1 = 60$, $Ti = 0.004\%$, and $Re = 5\text{k}$. The mean and fluctuating velocity profiles were measured at $x/D = 14$ and 80. For the two unequal cylinders, the wake width was expressed as the sum of the small and large cylinder half-widths. The location of the maximum velocity defect was selected as the origin of the coordinate system $x - x_0$.

All measured mean velocity profiles for $D_1/D_2 = 0.6$ and 0.8 showed that the half-wake width continued to be greater on the small cylinder side from $x/D = 14$ up to 80. The lateral extent of the velocity defect zone was larger on the small-cylinder side of the combined wake. The turbulent velocity profiles for $\overline{u'^2}$ across the wake for $x/D = 14$ and 80 are shown in Fig. 26.75 for $D_1/D_2 = 0.6$ and $T/D_1 = 1.3$. The $\overline{u'^2}/V^2$ peak for $x/D = 14$ is on the large-cylinder side. The maximum absolute value of the Reynolds stress $u'v'$ occurs on the small-cylinder side of the combined wake.

Palmer and Keffer (1972J) made another interesting observation. For the

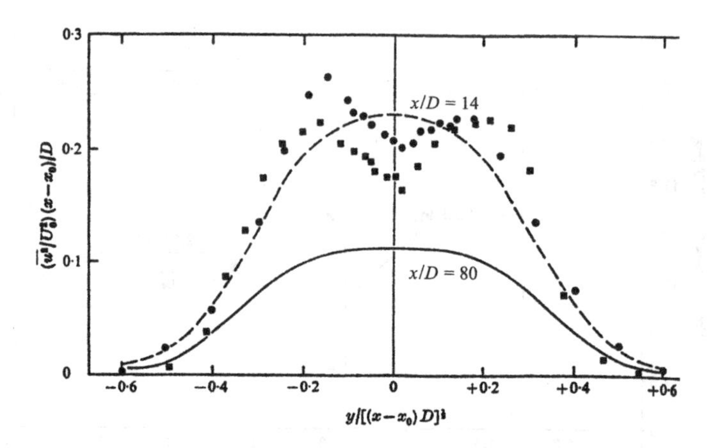

FIG. 26.75. Fluctuating velocity profile across the wake, $D_1/D_2 = 0.6$, $T/D_1 = 1.3$, $x/D = 14$ and 80, $Re = 5\text{k}$, Palmer and Keffer (1972J)

unequal cylinder pair, the point of zero Reynolds stress did not coincide with the maximum velocity defect. They termed it the *turbulent energy reversal.* The flow visualization revealed well-defined eddy shedding behind both cylinders. The eddies remained persistent and easily identifiable downstream of the cylinders at $St = 0.18$ and 0.206 behind the small and large cylinders, respectively. Synchronization of shedding frequencies did not occur.

26.5.7 *Control cylinder upstream*

It has been discussed in Vol. 1, Chapter 13, that free stream turbulence may considerably reduce C_D for a single cylinder in the Re range where the boundary layers are transitional. The reduction in C_D is not produced by the entire turbulence field but rather restricted to a turbulence band interacting with the cylinder boundary layer. Hence, it appears that the same results would be obtained by introducing a small control cylinder upstream in order to generate the turbulence band, and reduce the overall drag of the downstream cylinder.

Lesage and Gartshore (1987J) carried out tests by using control cylinders of $D_1/D_2 = 0.17$, 0.33, and 0.50 at $Re = 33\text{k}$ and 65k, $L/D_2 = 18$, and $D_2/B = 0.04$. The measured C_D for a single cylinder was 1.19. Figure 26.76 shows the ratio C_{D2}/C_D in terms of S/D_2 for $Re = 33\text{k}$ (crosses) and 65k (open circles). The drag reduction is more effective at higher Re and for the larger control cylinder. This is expected because the boundary layer on the downstream cylinder is more transitional at a higher Re. The minimum at C_{D2}/C_D occurs in the range $1.2 < S/D_2 < 1.8$.

Lesage and Gartshore (1987J) found that for $Re = 65\text{k}$ the optimal diameter ratio $D_1/D_2 = 0.33$ gives a minimum $C_{D2}/C_D = 0.41$. This was confirmed by Prasad and Williamson (1997J), who replaced the control cylinder with a

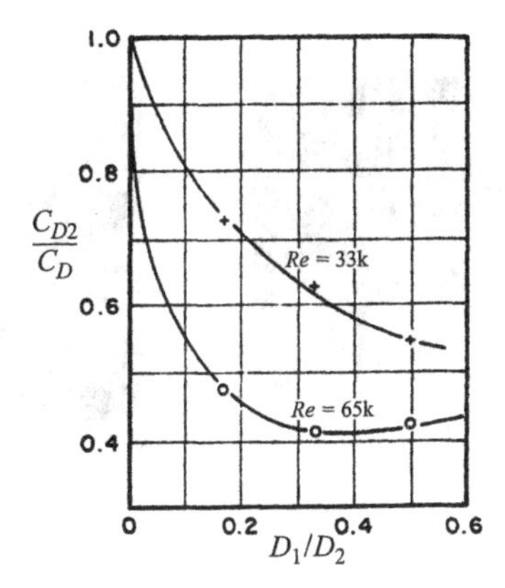

FIG. 26.76. Ratio of rear cylinder to single cylinder drag in terms of D_1/D_2, Lesage and Gartshore (1987J)

flat plate and tested in the range $0.03 < D_p/D_2 < 1$ at $Re = 50$k. They also showed that the spacing between the control plate and downstream cylinder was $S/D_p = 5.9$. This spacing corresponded to $(S/D_1)_c$ where eddy shedding behind the upstream cylinder became bistable.

26.5.8 *Control cylinder outside laminar wake*

The mechanics of laminar near-wake instability have been discussed in Vol. 1, Chapter 3. The steady and closed near-wake becomes unstable beyond $Re = 30$–48, depending on the experimental conditions. The transverse oscillation of the near-wake end initiates a wavy trail along the far-wake. At a higher $Re > 55$–70, the roll-up of free shear layers takes place at the crests and troughs of the wavy trail. The ensuing Kármán–Bénard eddy street is not shed from the cylinder, but gradually evolves along the near-wake end. The length of the near-wake reduces from $L_w/D = 3.5$–4 to 1.5–2 with rising Re. These essential facts are pertinent for the effect of a small control cylinder placed outside the large cylinder wake.

Strykovski and Sreenivasan (1990J) carried out tests on a large cylinder D_1 followed by a small control cylinder D_2 in the range $3 < D_1/D_2 < 20$, $14 < L/D_1 < 60$, between end plates, $45 < Re_1 < 120$, and $10 < Re_2 < 25$.[164] They found that a small control cylinder was ineffective in influencing the large cylinder wake except when placed around $x/D_1 = 1.2$ and $y/D_1 = 0.8$, where a complete suppression of the Kármán–Bénard eddy street took place. Figure 26.77(a,b) shows this unexpected full suppression of the eddy street by the control cylinder $D_1/D_2 = 7$ at $Re = 70$. Note that the control cylinder becomes ineffective at higher Re_1.

[164]Note that flow past small control cylinder was in the steady closed near-wake regime.

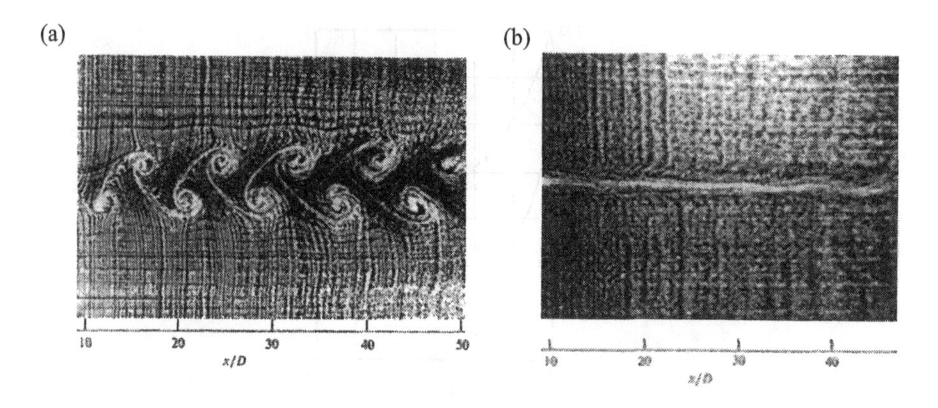

FIG. 26.77. Kármán–Bénard eddy street at $Re = 70$, (a) without control cylinder, (b) with $D_1/D_2 = 7$, Strykovski and Sreenivasan (1990J)

The effectiveness of the control cylinder at suppressing the large cylinder eddy street depends on its size D_1/D_2 and Re_1. Figure 26.78(a) shows the location and extent of the effective regions for various D_1/D_2 at $Re_1 = 80$. The largest control cylinder $D_1/D_2 = 3$ is effective in the wide region while the smallest $D_1/D_2 = 20$ is effective at a single point. Figure 26.78(b) shows the effective regions for $D_1/D_2 = 10$ in terms of Re_1. There are two different kinds of effective regions:

(i) $46 < Re_1 < 50$; the border spreads from $-0.3 < x/D_1 < 4$ and $0 \leqslant y/D_1 < 2$. This Re range corresponds to the onset of the wavy trail regime. The effective regions engulf the origin of the wavy trail and stabilize the near-wake.

(ii) $55 < Re_1 < 80$; a significant reduction in the effective regions to around $x/D_1 = 1.2$ and $y/D_1 = 1$. This Re_1 range corresponds to the Kármán–Bénard eddy street gradually evolving into a roll-up along the far-wake.

Strykovski and Sreenivasan (1990J) stated that the addition of another control cylinder on the other side of the large cylinder wake prolongs the effectiveness of the eddy street suppression from $Re_1 = 80$ to 100. The effectiveness of the control cylinder disappears at high Re. For example, for $D_1/D_2 = 20$ the eddy street reappears beyond $Re_1 = 80$. They argued that the control cylinder may divert part of the outside flow into the near-wake. This was not found in the instantaneous streamline pattern computed for $Re = 55$. Another argument was the concept of absolute stability. It was impossible to explain the suppression of wake instability by the control cylinder located outside the wake. The possible re-distribution of the vorticity by the control cylinder was ruled out because the locus of the maximum vorticity lay away from the effective region.

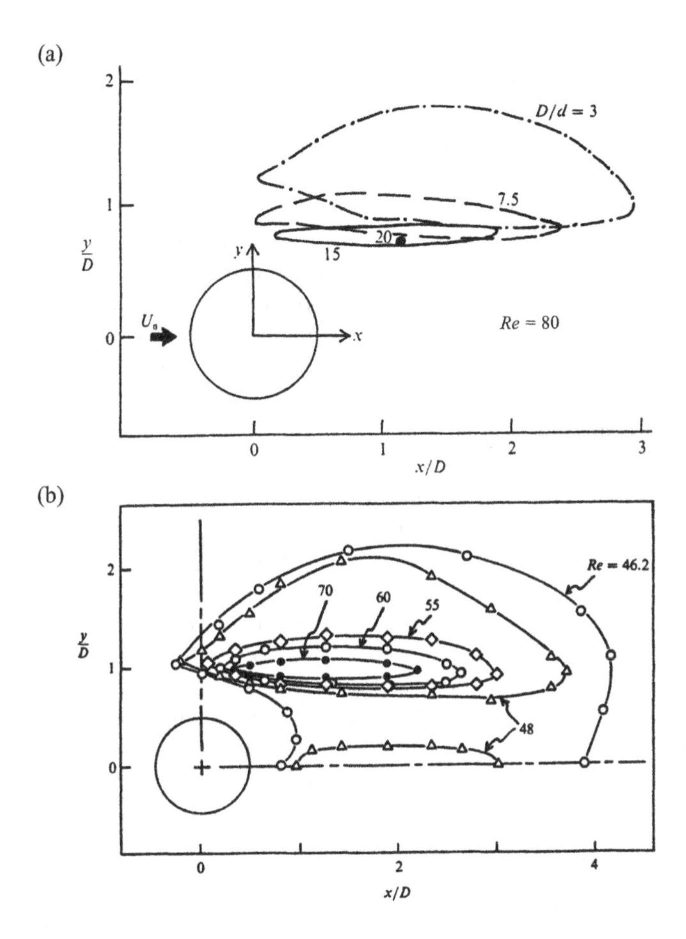

FIG. 26.78. Location and extent of effective control cylinder region in terms of (a) x/D and D/D_c, (b) x/D and Re, Strykovski and Sreenivasan (1990J)

The steady and closed near-wake behind the control cylinder may significantly contribute to the damping of transverse oscillation of the trail behind the upstream cylinder. For higher Re, L_w/D_2 shrinks and x/D_1 decreases accordingly. The loci of the effective location are closely related to the confluence end of the main wake and the lateral effective location was outside the thick free shear layers. When the control cylinder was immersed in the free shear layer the Re_2 would decrease, and the closed near-wake would shrink. Hence, the phenomenon seemed to be associated with the steady and stable closed near-wake behind the control cylinder, which dampens the wavy trail instability behind the main cylinder.

26.5.9 *Boundary layer control*

The flow control has been achieved by placing a small cylinder D_c in close proximity to the large main cylinder D. Sakamoto and Haniu (1994J) carried out experiments on $L/D = 8.2$, $D/B = 0.08$, $Ti = 0.2\%$, $D/D_c = 16.3$, and $1 < G/D_c < 2.6$ in the range $0° < \theta_c < 180°$. The mean and fluctuating C_D and C_L were measured on a cylinder segment of $L_s/D = 0.92$. Figure 26.79 shows the C_D variation in terms of θ_c at $Re = 60\text{k}$. A significant reduction in C_D is achieved by the control cylinder being placed in the range $10° < \theta_c < 60°$. The effectiveness becomes markedly reduced for $G/D_c = 2.66$, and the optimal range is confined to $1.3 \leqslant G/D_c \leqslant 2$. When the control cylinder was in the range $70° < G/D_c < 110°$ the value of C_D exceeded that for a single cylinder. The reduction in C_D is recovered for $110° < \theta_c < 135°$. Beyond that the control cylinder is fully submerged in the wake of the large cylinder, and has no effect on C_D.

The effect of the control cylinder on the mean C_p distribution is seen in Fig. 26.80 for $G/D_c = 2$. The curves for the plain cylinder and $\theta_c = 40°$ are close together. For $\theta_c = 63°$, there is not only an increase in C_{pb} but also two stable C_{pmin} states.

The overall variation in C_D and C_L is given in the $G/D_c, \theta_c$ plane in Fig. 26.81. There is a marked reduction in C_D around $\theta_c = 60°$ and less so for $\theta_c = 120°$. The C_L also peaked to 1.1 in the same range. The maximum in C_L and minimum in C_D occurred for $G/D_c = 1.66$ and $\theta_c = 60°$. Remarkably similar plots were found for C_D' and C_L'. Flow visualization revealed that the control cylinder triggered transition of the boundary layer on the main cylinder,

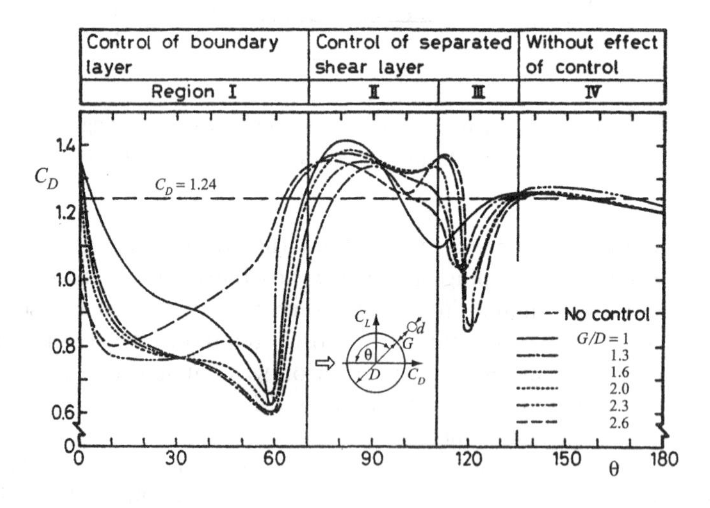

FIG. 26.79. Drag in terms of control cylinder angular location, $D/D_c = 16.1$, $Re = 60\text{k}$, Sakamoto and Haniu (1994J)

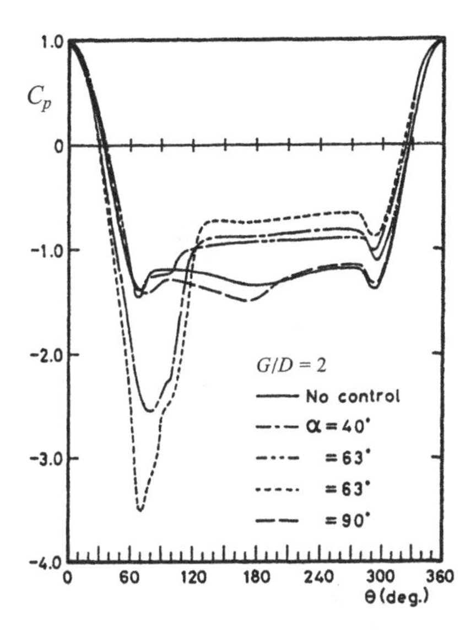

FIG. 26.80. Mean pressure distribution as affected by control cylinder, $D/D_c = 16$, $G/D_c = 2$, Sakamoto and Haniu (1994J)

and postponed separation. The asymmetric separation produced C_L.

26.5.10 *Free shear layer control*

A complex interference takes place between the control cylinder and the main cylinder when the former is partly or fully submerged in the free shear layer of the latter. Igarashi and Tsutsui (1994P) and Igarashi *et al.* (1997P) carried out detailed tests for the small control cylinder $D/D_c = 20$. The main cylinder had $L/D = 3.75$, $D/B = 0.1$, $Ti = 0.5\%$, $1 < G/D_c < 2.3$, and $4.5\text{k} < Re < 41\text{k}$. The location of the control cylinder D_c was varied in the range $100° < \theta_c < 140°$. The control cylinder was ineffective for $\theta_c < 118°$, $\theta_c \geqslant 130°$, and was the most effective for $\theta_c = 122°$.

The variations in C_d and C_l were evaluated from the C_p distribution and are shown in Fig. 26.82(a,b). The drop in C_d in the range $118° < \theta_c < 123°$ is associated with a jump in C_l. The variation in measured St at $Re = 41\text{k}$ is shown in Fig. 26.83 in terms of θ_c. The increase in St coincides with the fall in C_D.

Igarashi and Tsutsui (1994P) carried out smoke visualization, Fig 26.84(a,b), and observed the following flow regimes:

(a) $C_{l\max}$ and $C_{d\min}$ occur when the control cylinder induces reattachment, and the separation is delayed beyond $\theta_s = 180°$;

(b) the control cylinder disturbs the free shear layer, and the roll-up is delayed.

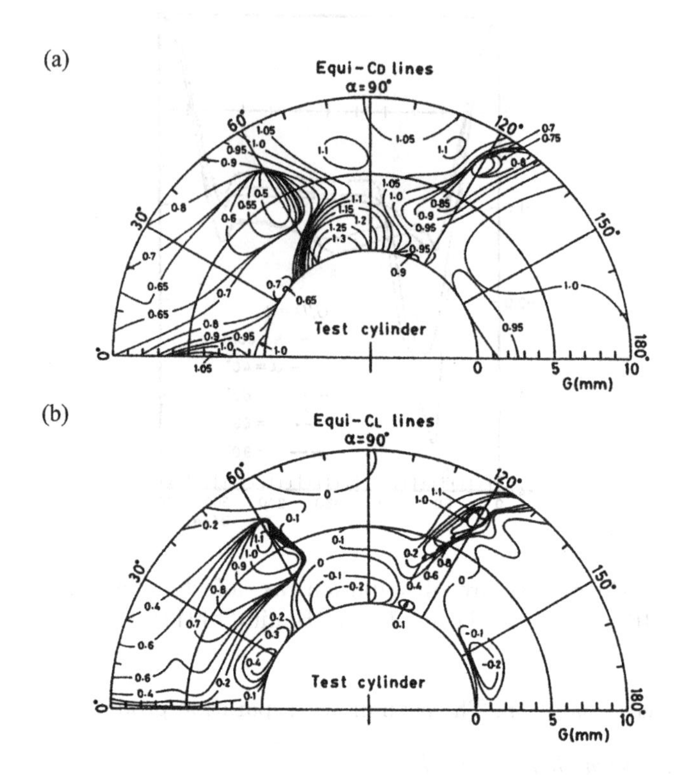

FIG. 26.81. Control cylinder in the G, α, plane, $Re = 60\text{k}$, (a) equi-drag lines, (b) equi-lift lines, Sakamoto and Haniu (1994J)

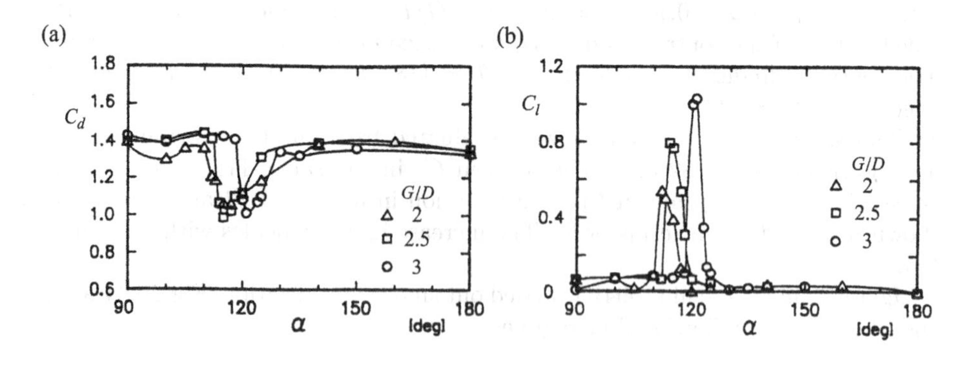

FIG. 26.82. Main cylinder at $Re = 41\text{k}$, $D/D_c = 20$, (a) drag, (b) lift, Igarashi and Tsutsui (1994P)

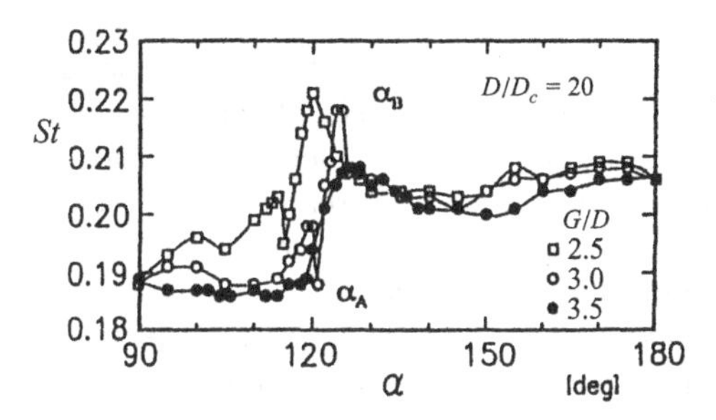

FIG. 26.83. Main cylinder at $Re = 41\text{k}$, $D/D_c = 20$, Strouhal number, Igarashi and Tsutsui (1994P)

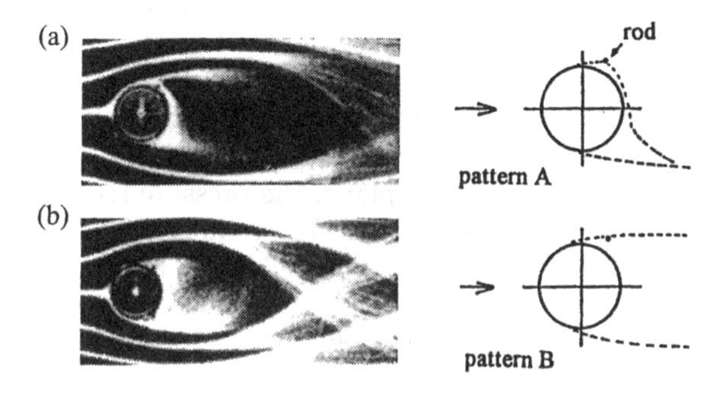

FIG. 26.84. Smoke visualization around the main and control cylinders, (a) reattachment at $\theta_c = 122°$, (b) $\theta_c = 130°$, Igarashi and Tsutsui (1994P)

26.6 Two cylinders crossing at right angles

26.6.1 *Introduction*

The interference of two parallel circular cylinders in close proximity has attracted a great deal of research. However, the related interference of two cylinders with perpendicular axes was ignored until the early 1980s. The practical significance of the interference between two cylinders forming a cross may be found in various fields of fluids engineering. For example, all offshore structures consist of a large number of intersecting circular members. Screens formed of interwoven fine wires were extensively used in wind and water tunnels to reduce the free stream turbulence by breaking down large eddies. The opposite effect was achieved by using biplanar grids to generate isotropic turbulence. The crossing of two cylinders could be considered as an element of offshore platforms, wire gauzes, and biplanar grids. The multiple crossing would be simulated by using the low aspect ratio cylinders and high blockage in test sections.

The three-dimensional structure of the interference between the two crossed cylinders is sketched in Fig. 26.85(a,b). The top view shows the upstream cylinder U and downstream cylinder D. The streamlines passing around cylinder U are forced to converge along cylinder D, which acts as an obstacle. The stagnation region and streamlines up to the separation line are slightly affected by cylinder D. The presence of cylinder D creates a high local pressure, and induces a secondary flow SF (arrows) away from point C. Figure 26.85(b) shows the side view of cylinder D placed across the flow. The stagnation pressure along cylinder D is gradually reduced as the latter is submerged in the near-wake of cylinder U. The pressure difference along the upstream side of cylinder D induces the secondary flow SF towards the point of contact C between the two cylinders.

26.6.2 *Local pressure and forces along cylinders*

Zdravkovich (1983J) measured local mean pressure distributions C_p around two cylinders forming a cross, $L/D = 9.18$, $D/B = 0.056$ and 0.113, and 21.2k < Re < 107.4. Figure 26.86(a,b) shows C_{pU} and C_{pD} at $Re = 42$k. The strongest interference is found at the plane of contact, which is also a plane of symmetry. The $C_{p\min}$ is reached at around $\theta = 90°$, and the pressure recovery continues unabated up to 160°. The curve resembles the potential flow pressure distribution without separation around a single circle. As the monitored plane is displaced farther and farther from the contact plane, the pressure recovery collapses, and $C_{p\min}$ is reduced. The latter reached the lowest value at $z/D = 0.75$. The base pressure C_{pb} becomes flat for $z/D > 2$, but well below that measured for a single cylinder at $Re = 42$k.

Figure 26.86(b) shows C_{pD} at $Re = 42$k. Contrary to the considerable variation in C_{pb} along cylinder U, the C_{pb} is almost unaffected along the span of cylinder D. The pressure distribution measured near the contact plane $z/D = 0$ differs considerably from the rest. The kink and peak at $\theta = 60°$ for $y/D = 0.1$ is similar in shape to that found on the rear of the tandem cylinders in close proximity. The peak in the latter case is attributed to the reattachment.

Surface flow visualization was carried out on the same models at the same

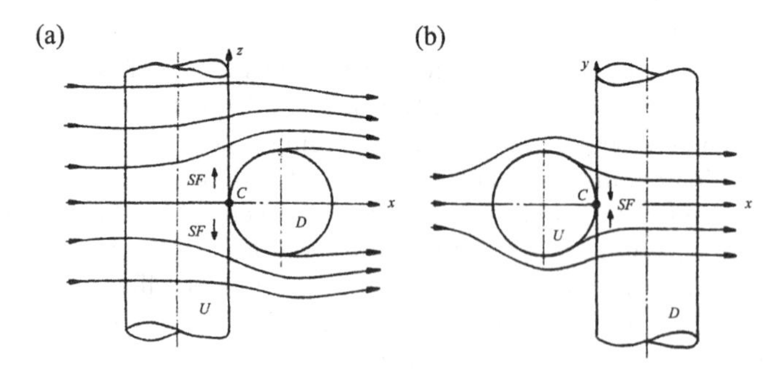

FIG. 26.85. Three-dimensional flow past crossed cylinders, C–contact, U–upstream, D–downstream, SF–secondary flow, Zdravkovich (1983J)

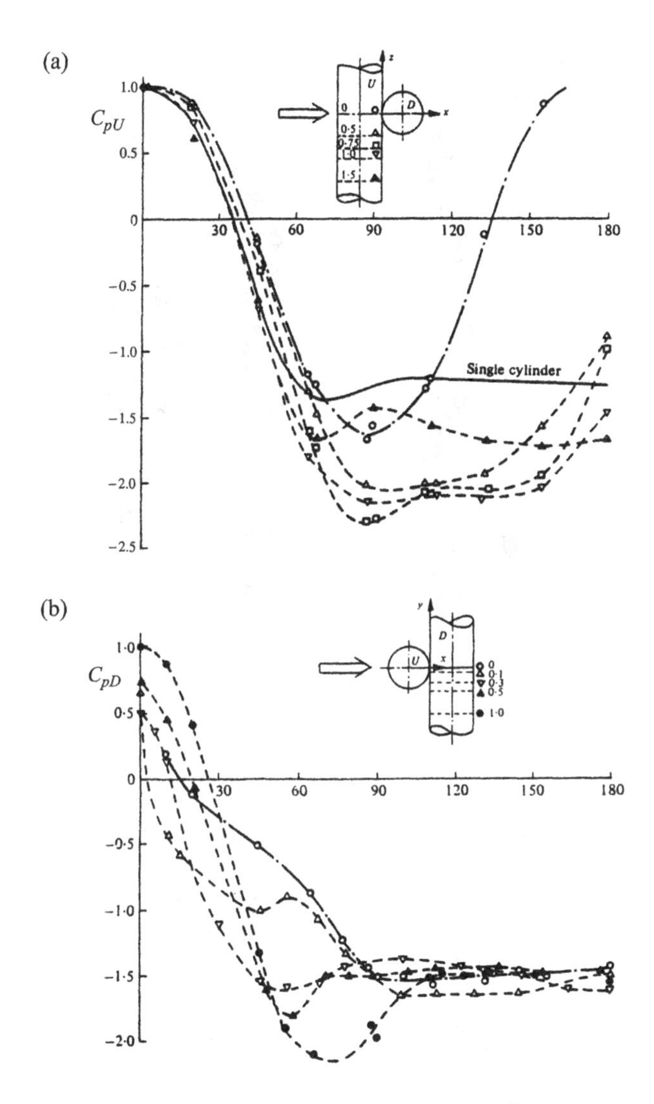

FIG. 26.86. Mean pressure distribution at Re = 42k, (a) upstream cylinder, (b) downstream cylinder, Zdravkovich (1983J)

Re. It revealed separation lines and secondary flows. Figure 26.87(a–c) shows the same surface flow pattern on cylinder U as viewed from the top, $\theta = 90°$, oblique $\theta = 135°$, and rear $\theta = 180°$. The bow-shaped separation line is displaced upstream in front of cylinder D. The secondary flow is seen in Fig. 26.87(b,c) behind the bowed separation line in the form of 'blobs' located symmetrically to the contact plane. The latter coincided with the dip in $C_{p\min} = -2.5$ at

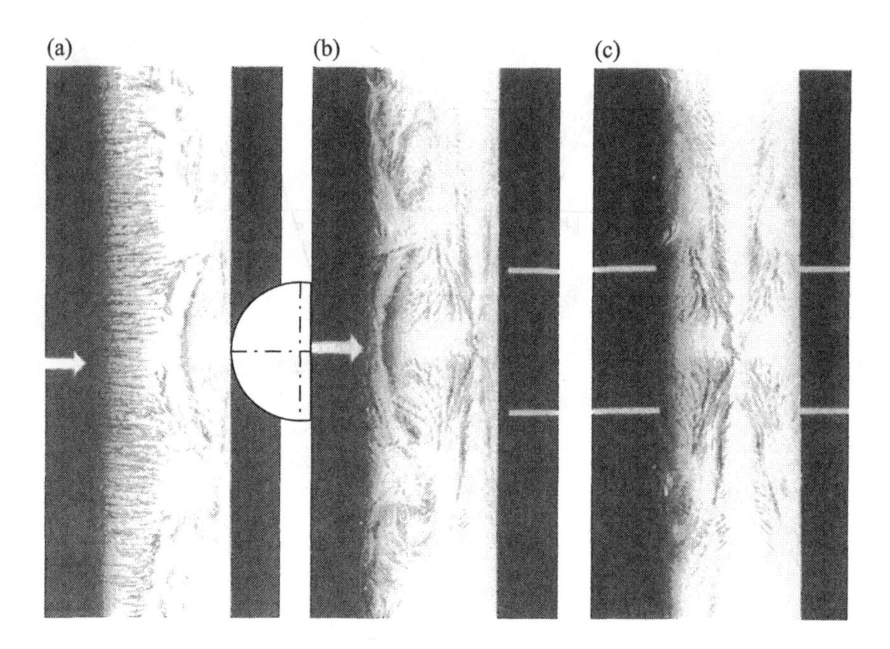

FIG. 26.87. Surface flow visualization on the upstream cylinder at Re = 42k, (a) side view, (b) oblique view, (c) rear view, Zdravkovich (1983J)

$z/D = \pm1$ and $\theta = 85°–90°$. Zdravkovich (1983J) suggested that the blobs represented the imprint on the surface of the swirling flow by two streamwise eddies.

Figure 26.88(a,b) shows the surface flow pattern on cylinder D, as photographed from the top and the rear, $\theta = 75°$ and $165°$, respectively. All streamlines up to the separation line are deflected towards the plane of contact. The straight separation line, however, is not affected by that deflection. The separation line undergoes significant distortion in the narrow wedge-like region within $y/D = \pm0.25$. Two surprising features are:

(i) cylinder U promotes separation almost from the point of contact C;

(ii) the deep intrusion of the secondary flow along the rear side within $y/D = \pm0.25$. A distinct wedge-like pattern is seen in Fig. 26.88(b).

26.6.3 *Tentative topology*

The bulk of the secondary flow behind the cylinders cannot be visualized except for the time-averaged imprint in the immediate vicinity of the surface. However, the blob pattern appears remarkably similar to that observed by Gould *et al.* (1968P) near the free end of the finite height cylinder. They interpreted the blobs as being the foot of the detached streamwise eddies.

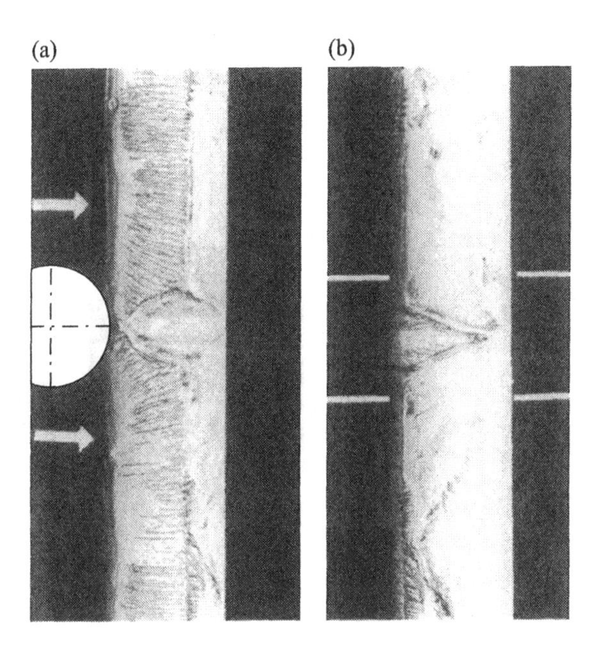

FIG. 26.88. Surface flow visualization on the downstream cylinder at $Re = 42\text{k}$, (a) side view, (b) rear view, Zdravkovich (1983J)

Zdravkovich (1983J) suggested that the flow topology consisting of two pairs of streamwise eddies sprang from the sides of cylinder U as tentatively sketched in Fig. 26.89(a). The early separation from cylinder D in the contact plane may be caused by the rapid 'widening' of the local cross section between the cylinders crossing, see Fig. 26.85. The direction of surface streamlines before and the secondary flow behind the separation indicate a swirling flow similar to that found in the horseshoe-swirl. The horseshoe-swirl crosses the separation lines at right angles as seen in Fig. 26.88(b). The velocity induced by the horseshoe-swirl keeps the flow attached to the surface over the rear part of cylinder D. The horseshoe-swirls gradually converge and form a characteristic wedge-like pattern. Finally, the converged horseshoe-swirls detach from both sides of cylinder D to form a cluster of four streamwise eddies, as depicted in Fig. 26.89(a).

Yamada *et al.* (1987J) carried out dye visualization in water at $Re = 700$. Figure 26.89(b) shows that eddy filaments shed along the span of both cylinders are cut off by the other cylinder. An intricate linking of the free ends of eddy filaments takes place analogously to those behind the free end cylinder.

The local drag coefficient C_d was evaluated from the measured C_p. Figure 26.90(a) shows that the lowest C_d appears at the contact plane, y/D, $z/D = 0$, and the highest C_d at the location of the blobs, $y/D = 1$. It would be expected that farther along the span, C_d eventually reaches $C_d = 1.02$ found for the 2-

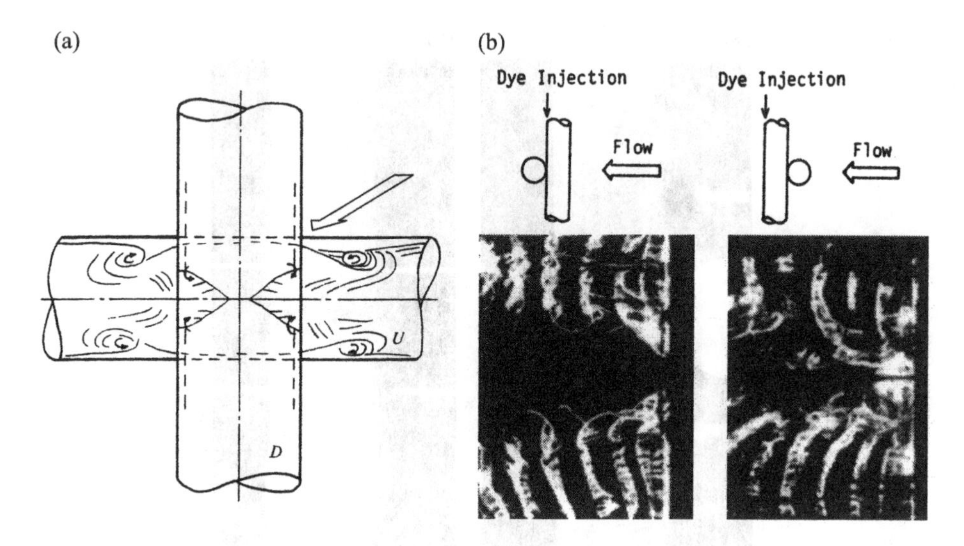

FIG. 26.89. Crossed cylinders, (a) flow topology, Zdravkovich (1983J), (b) eddy filament visualization, Yamada *et al.* (1987J)

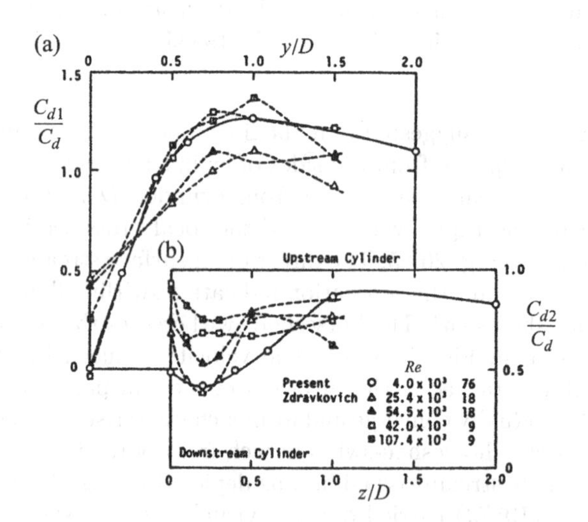

FIG. 26.90. Local normalized drag along (a) the upstream cylinder, (b) the downstream cylinder, Yamada *et al.* (1987J)

D cylinder. Gould *et al.* (1968P) also found $C_{d\max}$ near the free end while the rest of the span experienced lower C_d than that for the single cylinder. The C_d along cylinder D showed considerably less variation than that along cylinder

U. It is consistently less than that found for the single cylinder. Subsequent measurements by Yamada *et al.* (1987J) at $Re = 4$k are in good agreement with Zdravkovich (1983J).

26.6.4 *Two intersecting cylinders*

Early measurements of drag forces on intersecting struts and circular cylinders were reported by Biermann and Herrnstein (1933P). They established that intersecting cylinders forming a letter V with the angle γ, produce the same overall drag coefficient irrespective of the angle γ in the range $90° < \gamma < 180°$. A slight decrease in the overall C_D was found for the intersecting angles $15° < \gamma < 90°$ for Re = 60k, 106k, and 500k. Contrary to that, a significant increase in the overall C_D was found for streamlined struts in the range $15° < \gamma < 90°$. The maximum adverse interference took place for $\gamma = 30°$.

Zdravkovich (1985Ja) measured the pressure distribution around two intersecting cylinders at $\gamma = 90°$, $L/D = 8.85$, $D/B = 0.226$, $Ti = 0.4\%$, and 41k $< Re <$ 90k. The measured C_p distribution along the span was similar to C_{pU}. The surface flow visualization revealed the secondary flow, and four 'blobs', as sketched in Fig. 26.91(a). Their location coincided with the measured low $C_{p\min}$ region. Zdravkovich (1983J, 1985Ja) suggested that the blobs represented the 'feet' of swirls as shown in the tentative topology depicted in Fig. 26.91(a). This interpretation is confirmed by Osaka *et al.* (1982P). Figure 26.91(b) shows streamwise eddies measured by the hot-wire wake exploration. Although the pair of streamwise eddies in each quadrant is subjected to viscous diffusion and dissipation, it persists beyond $x/D = 90$.

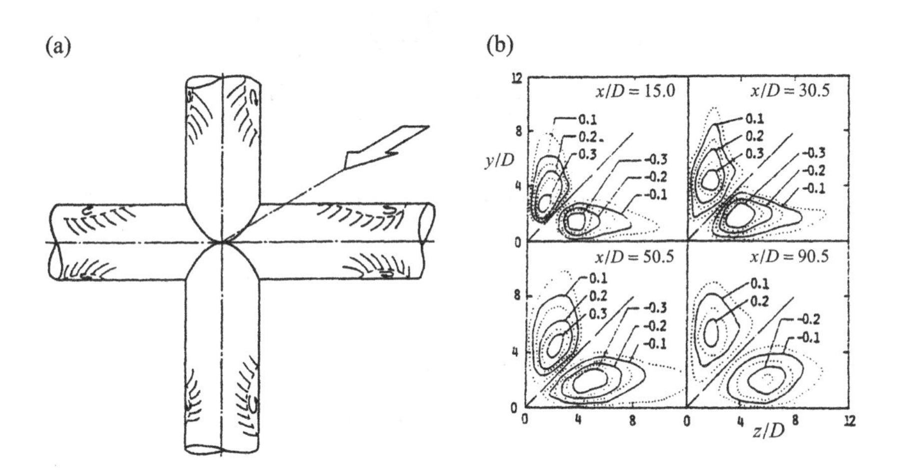

FIG. 26.91. (a) Flow topology, Zdravkovich (1985Ja), (b) persistence of streamwise eddy pair, Osaka *et al.* (1982P)

The rate of decay of the maximum velocity defect is shown in Fig. 26.92. The rate of decay of the centreline defect velocity, U_{d3}/V, was slower than that found in the two-dimensional wake, U_{d2}/V. This was due to the four superimposed pairs of streamwise eddies, as shown in Fig. 26.91(a). Even very far downstream, the three-dimensional wake did not behave as a single combination of two perpendicular two-dimensional wakes.

The most important parameter for engineering applications is the variation in the local C_d along the span. The local C_d was evaluated from the measured C_p, and is shown in Fig. 26.93. The peak coincided with the location of blobs in the surface flow visualization.

Osaka *et al.* (1982P) also measured the mean and fluctuating kinetic energy balance within the three-dimensional wake. Figure 26.94 shows the variation in diffusion, production, advection, and dissipation of the turbulent kinetic energy. As compared with a single cylinder, Fig. 5.58, p. 143, Vol. 1, it is evident that the curves for production and dissipation are similar, the curve for advection is reduced, and for diffusion is increased, respectively.

26.6.5 *Effect of gap between cylinders*

In some engineering applications two cylinders at right angles neither intersect nor are in contact, but instead they may be separated by spacing S. For example,

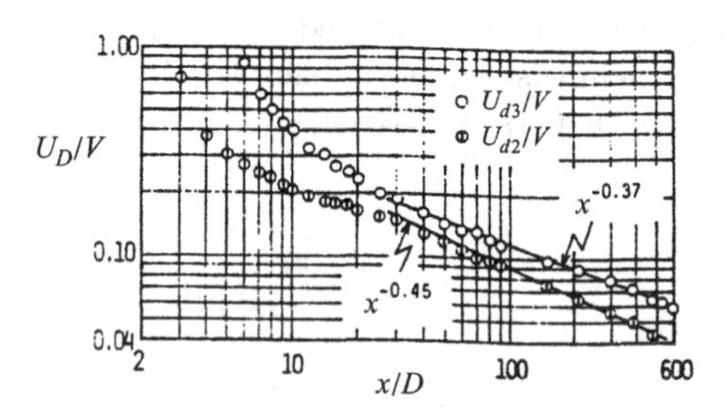

FIG. 26.92. Maximum velocity defect rate of decay along the wake, Osaka *et al.* (1983Ja)

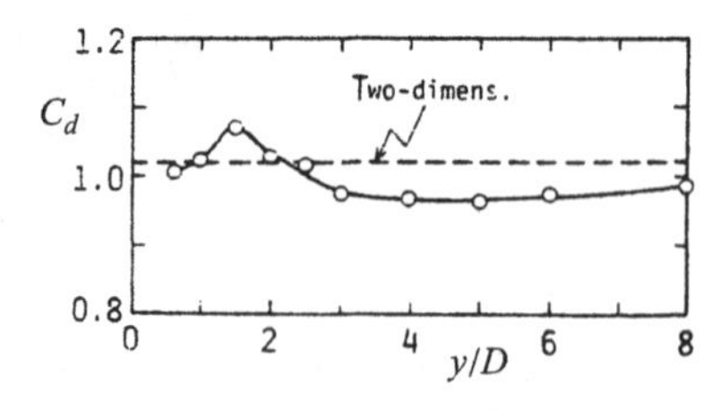

FIG. 26.93. Local drag variation along the span, Osaka *et al.* (1983Ja)

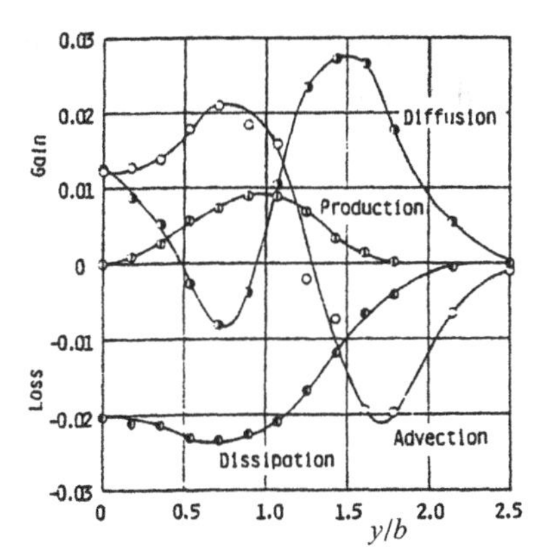

FIG. 26.94. Kinetic energy balance across the wake, Osaka *et al.* (1982P)

separated planar grids used to generate specific levels of turbulence.

Fox and Toy (1988J) investigated the effect of the gap in the range $1 < S/D \leqslant 10$, $L/D = 35.5/15.7$, $D/B = 0.022/0.028$, $Ti = 0.17\%$, and $Re = 20\text{k}$. Figure 26.95(a,b) shows the variation in C_{p01}, C_{p02}, C_{pb1}, and C_{pb2} in terms of S/D. The stagnation pressure coefficient for the upstream cylinder remains unaffected, $C_{p01} = 1$. The C_{p02} is reduced to 0.5 for $S/D = 10$. The variation in C_{pb1} and C_{pb2} differs considerably, as shown in Fig. 26.95(b). The C_{pb2} reaches a minimum -0.5 at $S/D = 3$, the latter value represents a critical ratio $(S/D)_c$. Fox and Toy (1988J) and Fox (1990J) carried out smoke visualization at $Re = 1\text{k}$ and 2k, and observed two types of flow regimes behind the upstream cylinder:

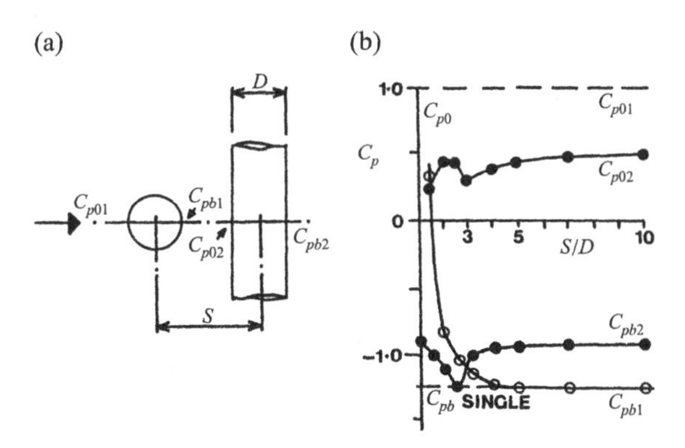

FIG. 26.95. Crossed separated cylinders, (a) definition sketch, (b) stagnation and base pressure, Fox and Toy (1988J)

(i) $1 < S/D < 3$, a pair of recirculating eddy is are trapped in the gap;

(ii) $S/D > 3$, periodic eddy shedding takes place in the gap.

27

CYLINDER CLUSTERS

27.1 Introduction

Cylinder clusters may be found in a wide range of engineering applications, such as a group of chimney stacks in civil engineering, multiple marine risers in offshore engineering, joint stranded conductors in electric transmission lines, etc. Chimney stacks, marine risers, and stranded conductors will be discussed under the common name *cylinders* in this book.

The main feature of all clusters mentioned so far is that they are erected, submerged, or suspended in an open space without confining walls or boundaries. Another category, which will be discussed in Chapter 28, is closely packed tubes in shell-tube heat exchangers where the flow is confined within the walls.

The cylinder clusters found in engineering applications may be grouped into at least three subgroups:

(i) the cylinders are arranged along a straight line and positioned in-line, staggered, or side by side relative to the free stream, Fig. 27.1(a,c);

(ii) the cylinders are arranged equidistantly to form a triangle for $n = 3$, square for $n = 4$, or polygons for $n > 4$, Fig. 27.1(b,d);

(iii) a 'satellite' arrangement, often used for multiple risers in offshore engineering, consists of a central tube D, surrounded circumferentially by n satellite tubes d, Fig. 27.1(e,f).

The irregular arrangements are usually avoided in applications but cylinders of different diameters are often employed.

27.2 Three cylinders

27.2.1 *In-line clusters*

It has been discussed in Chapter 26 that flow past tandem cylinders exhibits two basic flow regimes:

(i) $S/D < (S/D)_c$; the eddy shedding behind the upstream cylinder is suppressed;

(ii) $S/D > (S/D)_c$; both cylinders shed eddies.

An important aspect of the phenomenon is that $(S/D)_c$ strongly depends on the free stream turbulence. The third cylinder placed in-line with the other two is subjected to the additional turbulence generated behind the second cylinder. Hence, $(S/D)_c$ found for tandem cylinders is likely to be reduced from 3.8 to

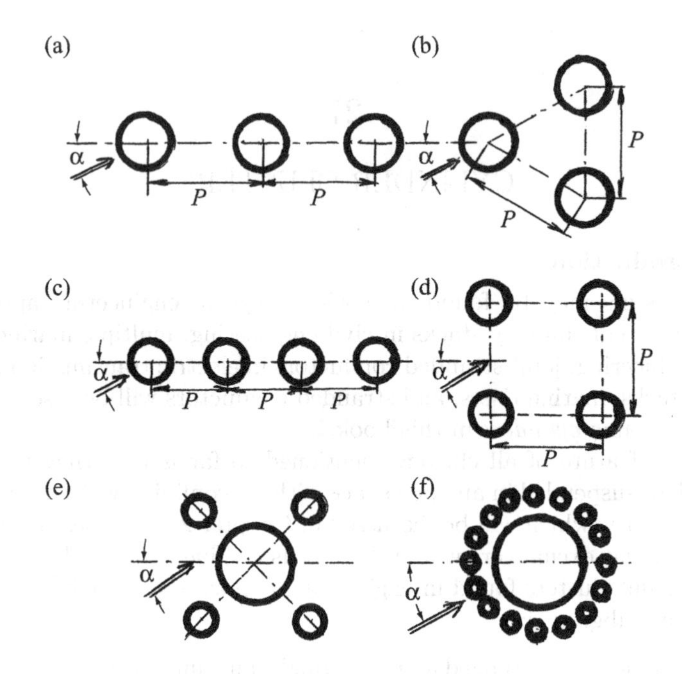

FIG. 27.1. Definition sketches of clusters, (a) three cylinders aligned, (b) triangle, (c) four cylinders aligned, (d) square, (e) satellite, (f) shroud

less than 2. This is illustrated in Fig. 27.2 for $S/D = 3.3$ and $Re = 2\text{k}$, Werle (1972J). Eddy shedding does not take place behind cylinder 1, but it does behind cylinder 2. The apparent paradox is that flow past identical cylinders having an equal spacing ratio develops different wakes behind cylinders 1 and 2.

Igarashi and Suzuki (1984J) carried out a detailed examination of flow past three in-line cylinders in the range $1 \leqslant S/D < 4$. The cylinders had $L/D = 4.41$, $D/B = 0.07$, $Ti = 0.6\%$, and $10.9\text{k} < Re < 39.2\text{k}$ for five S/D values. The measured C_p and C_p' on three cylinders at five spacing are compiled in

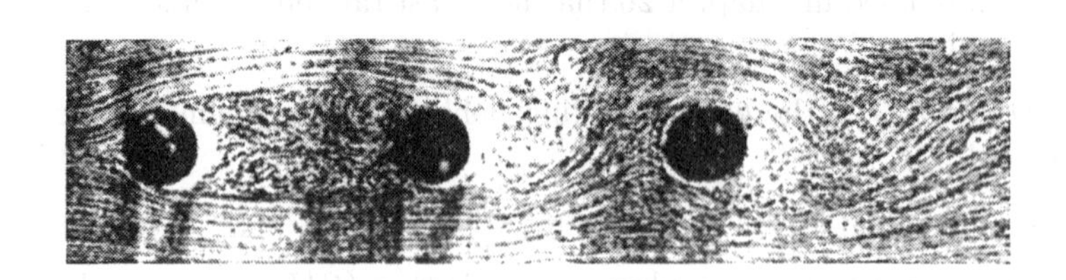

FIG. 27.2. Flow visualization around three aligned cylinders at $\alpha = 0°$, $Re = 2\text{k}$, $S/D = 3.3$, Werle (1972J)

Fig. 27.3(a,b). There are two types of mean C_p distribution with the stagnation at $\theta = 0°$, and with the reattachment peak at $60° < \theta < 80°$. Cylinder 1 is subjected to the first kind of C_p distribution for all five spacings. Cylinders 2 and 3, except for $S/D = 1.18$, are subjected to the reattachment regime, i.e. the C_p distribution of the second kind. However, for $S/D = 2.65$, cylinder 2 is still in the reattachment regime while cylinder 3 is facing the eddy street developed behind cylinder 2. Note that the flow past cylinder 2 in tandem (dashed line) is only slightly affected when cylinder 3 is added (full line) in Fig. 27.3(a), second column.

The fluctuating C_p' distribution is marked by the reattachment and separation peaks in Fig. 27.3(b). Cylinder 1 shows only the separation peak as being very small when eddy shedding is suppressed for $S/D < 3.82$. The reattachment peak is also absent on cylinders 2 and 3 for $S/D = 1.18$ in the no-reattachment regime. The reattachment and separation peaks are prominent on cylinder 2 for $1.18 < S/D < 3.82$. Note that for $S/D > 3.82$ the high C_p' at $\theta = 45°$ is due to the impingement of eddies shed behind cylinder 1.

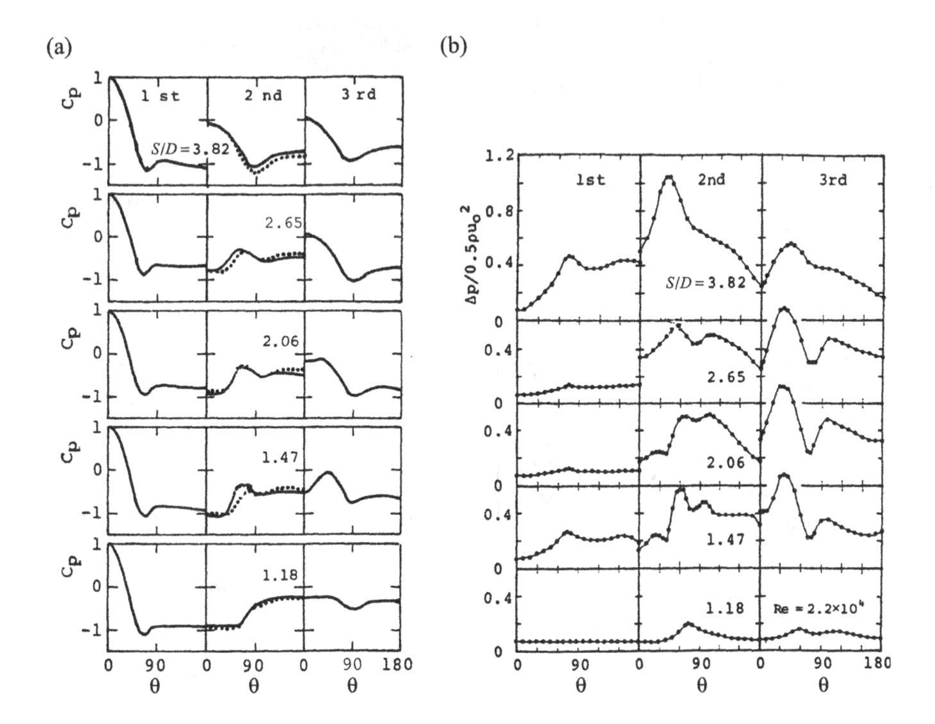

FIG. 27.3. Pressure distribution around three aligned cylinders $\alpha = 0°$, $Re = 22\text{k}$, (a) mean, (b) fluctuating, Igarashi and Suzuki (1984J)

Igarashi and Suzuki (1984J) complemented the C_p and C_p' measurements with smoke and surface visualization at $Re = 3$k. Figure 27.4(a,b) shows instant photos of the development of the reattachment regime behind cylinder 2, and Fig. 27.4(c,d) the eddy shedding behind cylinders 2 and 1, respectively.

The drag exerted on the three cylinders C_{d1}, C_{d2}, and C_{d3}, and mean value $\overline{C}_d$ were evaluated from C_{p1}, C_{p2}, and C_{p3} at $Re = 22$k, and shown in Fig. 27.5(a). The C_{d1} and C_{d2} are positive and negative up to $(S/D)_c$, respectively. Note there is another jump around $S/D = 1.3$ where the no-reattachment regime suddenly changes to the reattachment regime. Figure 27.5(b) shows the variation in St behind cylinder 3 at $Re = 22$k. A fall in St at $S/D = 1.3$–1.4 is caused by the sudden change from a narrow (wrapped) wake to a wide wake in the

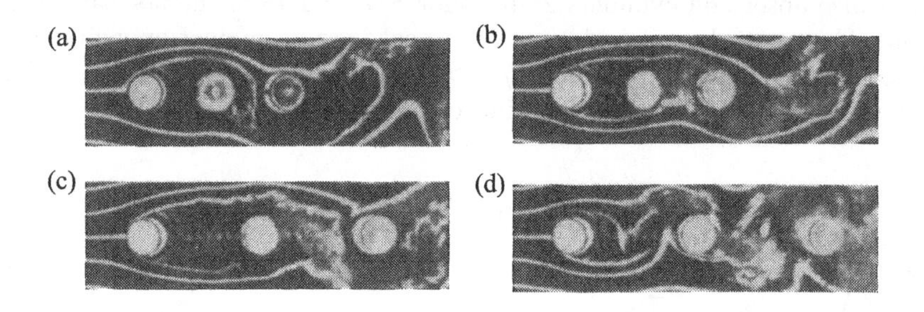

FIG. 27.4. Smoke visualization around three aligned cylinders at $Re = 13$k, (a) $S/D = 1.91$, (b) $S/D = 2.06$, (c) $S/D = 3.24$, (d) $S/D = 3.53$, Igarashi and Suzuki (1984J)

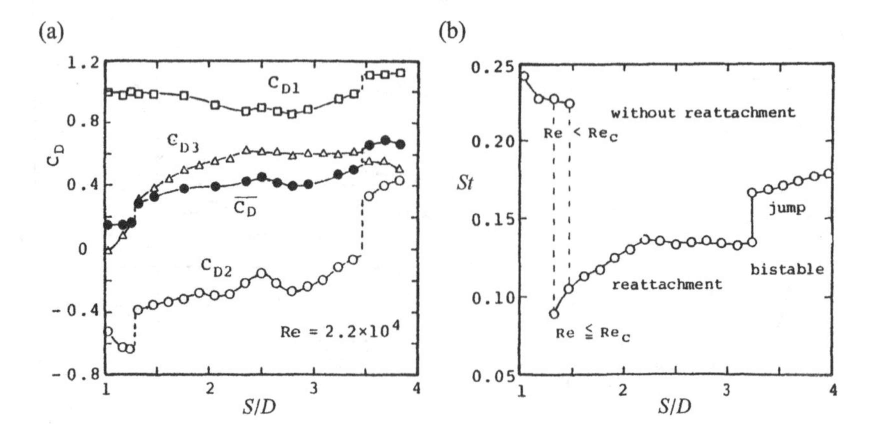

FIG. 27.5. Three aligned cylinders at $Re = 22$k, (a) drag coefficient, (b) Strouhal number, Igarashi and Suzuki (1984J)

reattachment regime. The jump at $(S/D)_c$ refers to the start of eddy shedding behind cylinder 1.

27.2.2 *Effect of tripping wires*

It was discussed in Section 22.6 that tripping wires attached to a smooth cylinder can promote transition in boundary layers to lower Re than for the smooth cylinder. A similar effect may be expected on free shear layers surrounding the two gaps between the three cylinders in the in-line arrangements.

Aiba *et al.* (1980Ja, 1980Jb) carried out tests on flow past, and heat transfer from, three in-line cylinders. The tripping wires $d/D = 0.005$ were attached at $\theta_w = \pm 65°$ on the first cylinder only. The three identical cylinders had $L/D = 8.65$, $D/B = 0.08$, $Ti = 0.1\%$, and $12\text{k} < Re < 52\text{k}$ in the range $1.3 \leqslant S/D \leqslant 4.5$. The tripping wires decrease $(S/D)_c$ from 4 to 2.8. The effect was analogous to the effect of free stream turbulence as discussed in Chapter 26.

The effect of the tripping wires on heat transfer from cylinder 2 expressed through $\overline{Nu}$ in terms of Re and S/D is shown in Fig. 27.6. There are three characteristic features;

(A) a linear increase in $\overline{Nu}$ up to about $Re = 2.5\text{k}$;

(B) a constant $\overline{Nu}$ for $S/D = 1.8$ up to $Re = 3.5\text{k}$;

(C) a 15%–20% jump in $\overline{Nu}$ for $Re > 4.6\text{k}$ above $\overline{Nu}$ for the plain cylinder.

The jump in $\overline{Nu}$ is caused by the upstream displacement of the reattachment line from around $\theta_r = 70°$ to $45°$. This has a considerable effect on the local drag coefficient C_{d2}.

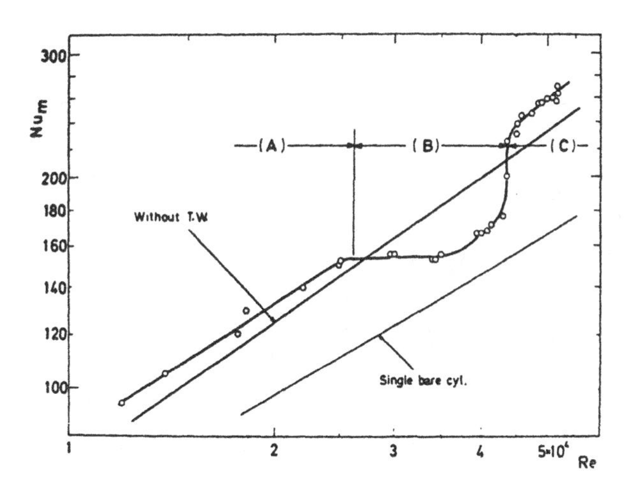

FIG. 27.6. Three aligned cylinders at $S/D = 1.8$ with and without tripping wire, mean Nusselt number versus Re, Aiba *et al.* (1980Ja, 1980Jb)

27.2.3 *Three cylinders of different diameters*

The Japanese Space Development Agency developed a rocket for launching a synchronous meteorological satellite. It consisted of a main rocket 48 m high having $D = 4$ m and two attached booster rockets, $d = 1.8$ m in diameter, with a gap of $G/D = 0.06$, as shown in Fig. 27.7. When on a launching pad the wind might come from any direction α.

Igarashi *et al.* (1988P) and Igarashi (1993J) carried out extensive tests on a 1:100 model of two-dimensional cylinders in the range 15k $< Re_D <$ 57k. The cylinders had $L/D = 3$, $D/B = 0.067$, $d/D = 0.45$, $G/D = 0.062$, and $Ti = 0.5\%$. The mean C_p distribution measured in the range $0° \leqslant \theta \leqslant 360°$ on all three cylinders in-line $\alpha = 0°$ and side-by-side $\alpha = 90°$ is shown in Fig. 27.8(a,b), respectively. Cylinder 1 exhibits two reattachment peaks at $\theta_r = \pm 40°$, while the flat C_p for cylinder 3 indicates that the latter is fully submerged in the

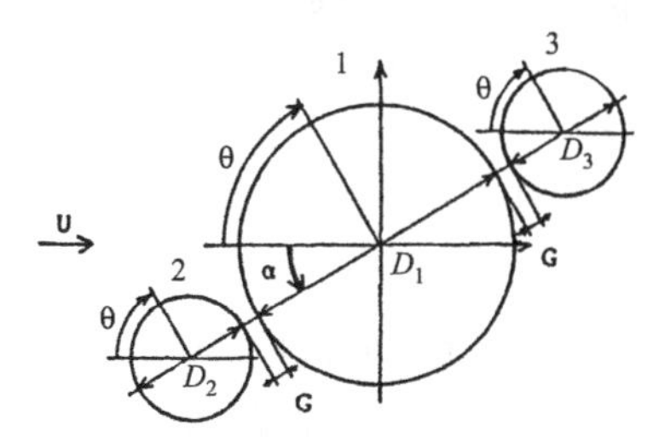

FIG. 27.7. Definition sketch of three different sized cylinders, Igarashi *et al.* (1988J)

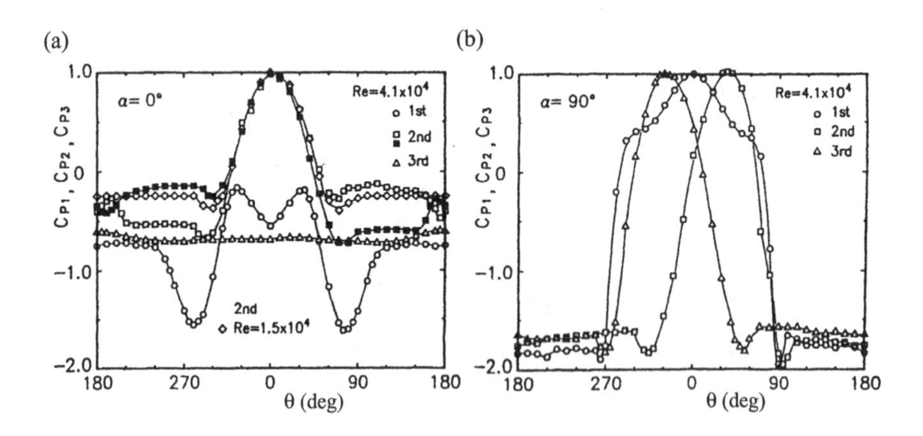

FIG. 27.8. Mean pressure distribution around three cylinders, (a) $\alpha = 0°$, (b) $\alpha = 90°$, Igarashi *et al.* (1988J)

big cylinder 1 wake. Upstream cylinder 2 showed two bistable modes of the C_p distribution.

The side-by-side arrangement is shown in Fig. 27.8(b). Both side cylinders 2 and 3 have a displaced stagnation point $\theta_0 = \pm 33°$ towards cylinder 1. There is a significant increase in the positive C_p range because C_{pmin} moves to $\pm 90°$, the minimum gap area. A gradual variation in the C_p distribution is found in the range $0° < \alpha < 90°$.

The local C_d and C_l on each cylinder were evaluated from the C_p distribution. Figure 27.9(a,b) shows the variation in C_d and C_l in terms of α. Cylinder 2 and cylinder 3 show high and low C_d up to $\alpha = 45°$, respectively, beyond which $C_{d2} = C_{d3}$. The values of C_{l2} and C_{l3} show different variation in terms of α in comparison with C_{l1}. An extremely low $C_{l2min} = -1.4$ is found for $\alpha = 22°$

Smoke visualization at $Re = 15$k revealed instantaneous streakline patterns in Fig. 27.10(a–d). For $\alpha = 5°$, cylinder 3 is fully shielded by cylinder 1. For $\alpha = 40°$, a narrow near-wake is seen behind cylinder 2. A strong gap flow is induced behind large cylinder 1 by the wide near-wake behind cylinder 3 for $\alpha = 60°$. Finally, for the side-by-side arrangement $\alpha = 90°$, cylinder 3 has a narrow near-wake, corresponding to a positive C_{l3} in Fig. 27.9(b).

Figure 27.11 shows variations in St in terms of α for gaps between cylinders open (open circles) and closed (closed circles). When the gap flow is blocked off, $G = 0$, the St is high for $\alpha < 20°$, and decreases for $\alpha > 30°$. The latter implies a wider wake and a higher C_D than for $G/D = 0.06$.

27.2.4 *Side-by-side and staggered clusters*

Ad hoc tests were carried out by Gerhardt and Kramer (1981J) on three finite cylinders, $H/D = 28.6$, in side-by-side arrangements having surface roughness $K/D = 0.6\%$ in order to simulate $Re = 15$M. Only three spacings were tested:

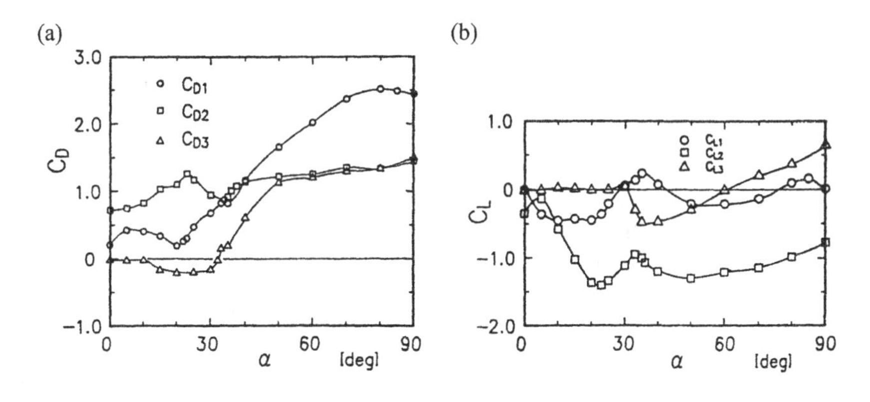

FIG. 27.9. Three unequal cylinders at $Re = 41$k, (a) local drag, (b) local lift, Igarashi (1993J)

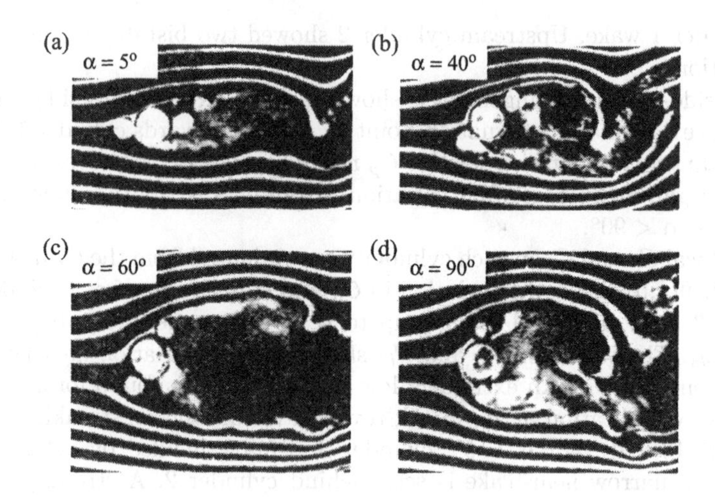

FIG. 27.10. Smoke visualization at Re = 15k, (a) $\alpha = 5°$, (b) $\alpha = 40°$, (c) $\alpha = 60°$, (d) $\alpha = 90°$, Igarashi (1993J)

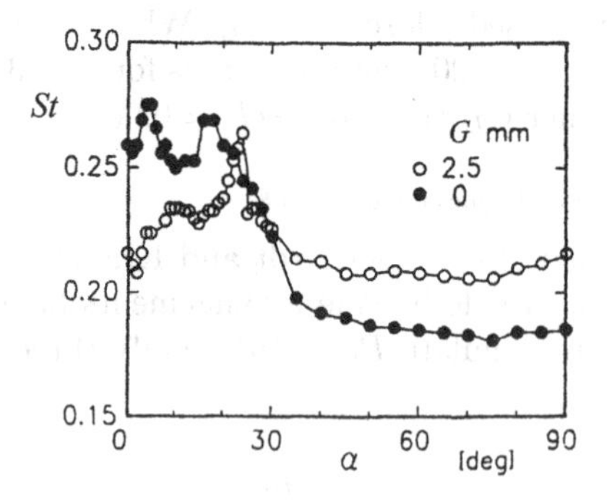

FIG. 27.11. Strouhal number behind three cylinders in terms of orientation, Igarashi (1993J)

$T/D = 1.5$, 2, and 3 and drag was evaluated from the C_p distribution separately on all three cylinders. The middle cylinder had the lowest drag for $T/D = 1.5$ and 2, and the highest for $T/D = 3$ as seen in Fig. 27.12. In addition to C_{d1}, C_{d2}, and C_{d3}, the $\overline{C}_d$ is also plotted (full bar).

Dalton and Szabo (1977J) measured only drag by using strain gauges on three aligned parallel cylinders having $L/D = 16$ and $D/B = 0.06$, in the range 27k < Re < 78k. Figure 27.13(a,b,c,d) shows the variation in C_D for three cylinders positioned at $\alpha = 0°$, $30°$, $60°$, and $90°$, respectively. A small effect of Re on C_D is found for all three cylinders in the in-line and staggered arrangements.

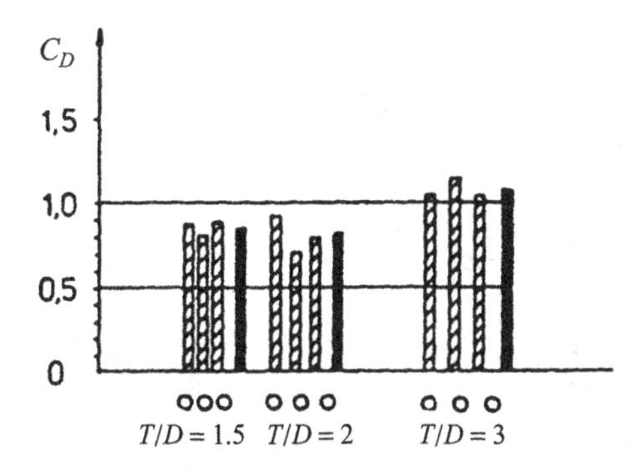

FIG. 27.12. Individual and overall drag for three side-by-side cylinders at $\alpha = 90°$, Gerhardt and Kramer (1981J)

Cylinder 2 displays a large variation in C_{D2} from -1 to $+0.5$ for $S/D = 1$–5. The staggered arrangements for $\alpha = 30°$ and $60°$ have a similar C_D variation, i.e. high C_{D1}, medium C_{D3}, and low C_{D2}. Finally, the side-by-side arrangements for $\alpha = 90°$ show a strong effect of Re and $C_{D1} > C_{D2}$.

27.2.5 *Triangle clusters at low Re*

The three cylinders forming a triangle matrix produce a combination of in-line, side-by-side, and staggered interference categories. The streamwise and transverse interferences develop up to a degree, which depends on the pitch ratio and orientation of the cluster. For closely spaced clusters, a single eddy street may form while for widely spaced clusters two or three interfering eddy streets form behind the cluster.

Zdravkovich (1968J) carried out smoke visualization tests on three cylinders in various triangle clusters having $L/D = 137$ and $D/B = 0.0036$ in the range $60 < Re < 300$, $5 \leqslant S/D \leqslant 21$, and $2 \leqslant T/D \leqslant 10$. Two kinds of interference flow patterns were observed for $\alpha = 0$:

(i) the formation of a single eddy street behind the cluster as a whole for $2 \leqslant T/D \leqslant 6$;

(ii) the formation of three eddy streets for $T/D > 6$.

The first kind of interference is shown in Fig. 27.14 for $S/D = 5$, followed by the $T/D = 2.5$. The eddy street is not formed behind the upstream cylinder, and two rows of eddies are seen behind the side-by-side cylinders. The unusual feature of the eddy street formed behind the cluster at $Re = 100$ is the large width between the rows, the spacing ratio $Ka = 0.7$–0.8. The rapid decay of the eddy street is followed by a trail instability, which leads to the roll-up of the remnants of two consecutive eddies and the formation of a secondary eddy street having $Ka = 0.35$. It appears that the initial spacing ratio is not stable,

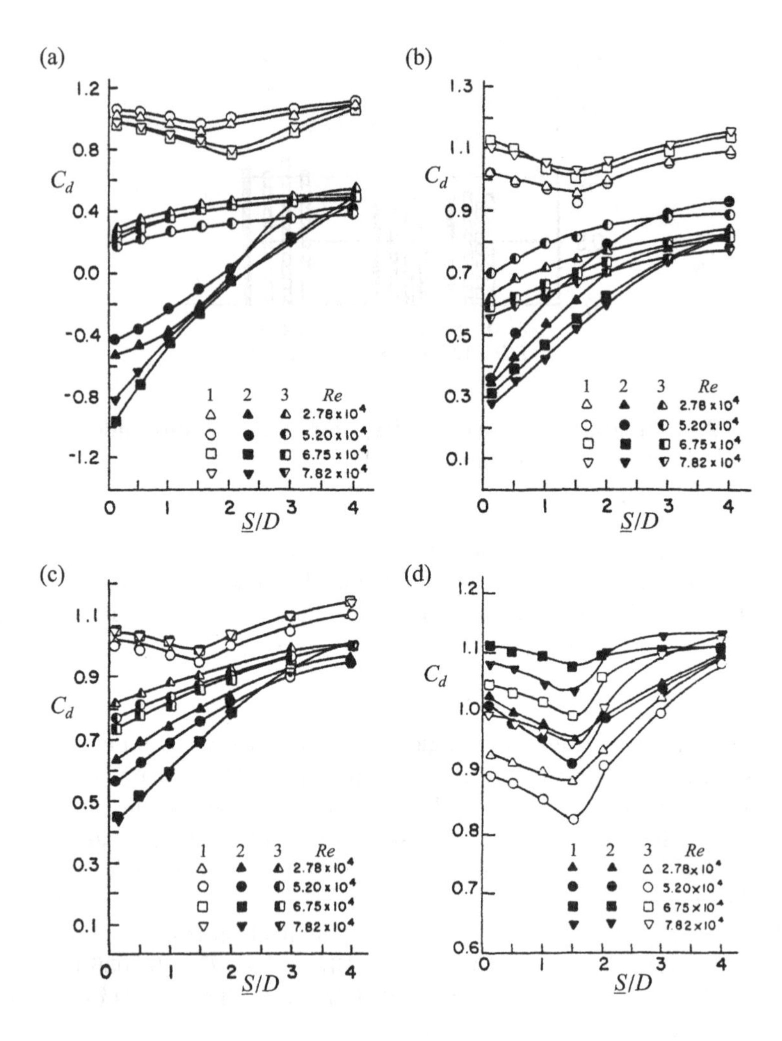

FIG. 27.13. Three aligned cylinders' drag, (a) $\alpha = 0°$, (b) $\alpha = 30°$, (c) $\alpha = 60°$, (d) $\alpha = 90°$, Dalton and Szabo (1977J)

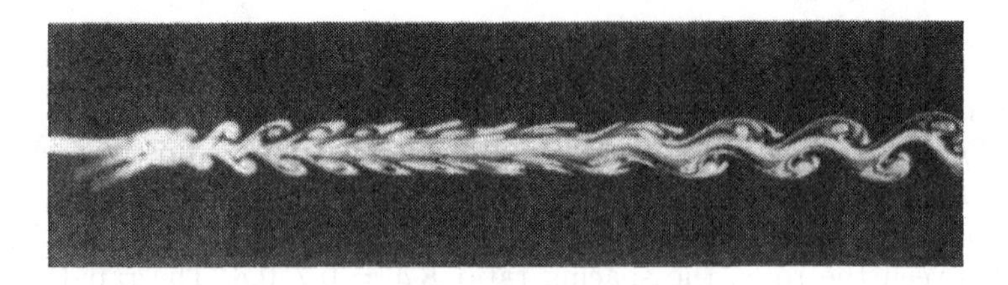

FIG. 27.14. Smoke visualization of triangle cluster at $Re = 100$, $S/D = 5$, $T/D = 2.5$, one cylinder upstream, Zdravkovich (1968J)

and after the rearrangement of the vorticity the secondary eddy street takes on a spacing ratio similar to that observed behind a single cylinder. The formation process seems to be analogous to the formation of the secondary eddy street far downstream of the single cylinder, as observed by Taneda (1959J).

Figure 27.15 shows the effect of an inverse arrangement where the cylinder pair $T/D = 2.5$ is followed by the downstream cylinder at $S/D = 5$ at the same $Re = 100$. The trail instability produced behind the downstream cylinder enhanced the formation of the secondary eddy street. Finally, Fig. 27.16 shows the effect of a small angle of incidence of the cluster, $\alpha = 5°$, relative to the free stream for $S/D = 5$ and $T/D = 2.5$ at $Re = 100$. The eddies are formed on one side of the cluster initially. However, the secondary eddy street had two different rows of eddies. The single-core eddies along the lower row are seen to be formed by a roll-up of the vorticity in the shear layer.

When $T/D > 6$, three Kármán–Bénard eddy streets are formed behind three cylinders. Figure 27.17(a) shows the arrangement $S/D = 19$ followed by $T/D = 10$. The side eddy streets show little variation while the middle eddy street is subjected to a considerable rearrangement. This is easy to follow in Fig. 27.17(b) where only the wake behind the upstream cylinder is made visible. The middle eddy street begins to interact immediately after passing between the downstream pair of cylinders in such a way that its two rows of eddies approach one another until all eddies lie in a single row on the wake axis. Further rearrangement is marked by the separation of these two rows again so that the eddies in the lower

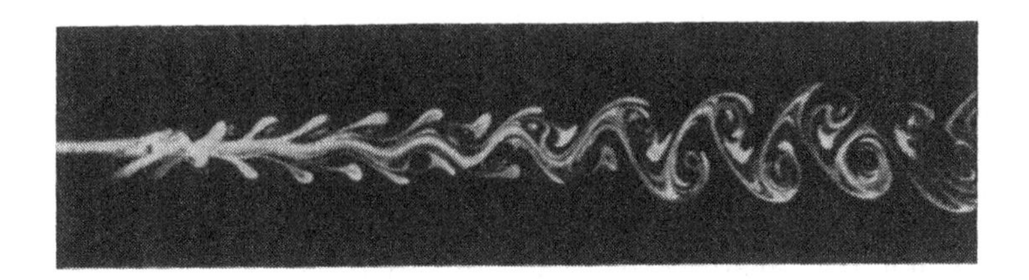

FIG. 27.15. Smoke visualization of triangle cluster at $Re = 100$, $T/D = 2.5$, $S/D = 5$, inverse arrangement, Zdravkovich (1968J)

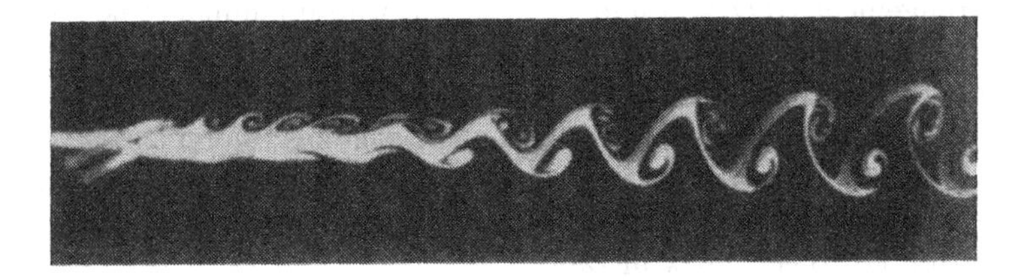

FIG. 27.16. Smoke visualization of triangle cluster at $Re = 100$, $S/D = 5$, $T/D = 2.5$, $\alpha = 5°$ (inverted T arrangement), Zdravkovich (1968J)

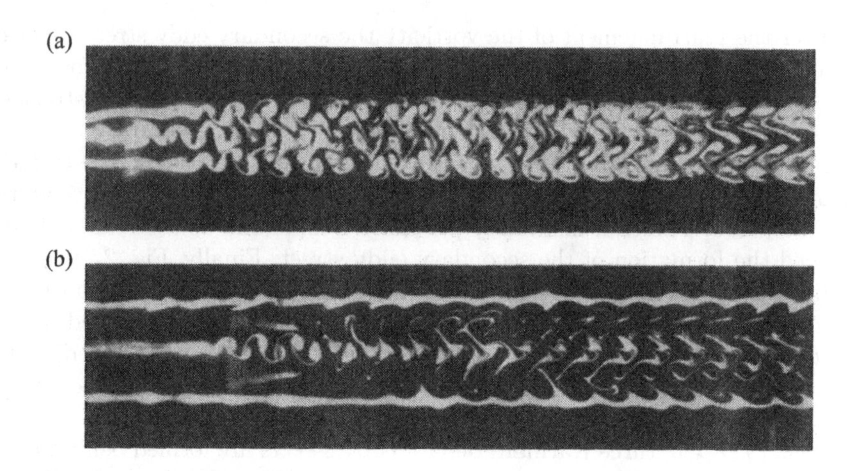

FIG. 27.17. Smoke visualization of triangle cluster at Re = 110, S/D = 19, T/D = 10, (a) three wakes, (b) one wake visible, Zdravkovich (1968J)

row of the middle eddy street move over to the upper side and vice versa. The spacing ratio after reaching zero becomes negative, i.e. the flow downstream of the middle upstream cylinder corresponds to that along a jet. The crossing of eddy rows in the middle eddy street takes place earlier and faster when the cluster is at a small angle α. The same cluster in the inverse arrangement with the cylinder pair upstream does not produce crossing of eddy rows in the middle eddy street.

The last part of the observation was directed at the wake transition to turbulence. Three different mechanisms were observed depending on the spacing ratio and orientation:

(i) the symmetric transverse oscillation of eddy rows;

(ii) oscillation of individual eddies;

(iii) the development of an asymmetric wake transition.

The arrangement S/D = 5 and T/D = 2.5, whose laminar wakes were described in detail at Re = 100, was used again at a higher Re = 180–220. The first mechanism of transition is illustrated in Fig. 27.18(a,b). The two rows of eddies oscillate symmetrically relative to the wake axis. The maximum and zero amplitude of oscillation are shown in Fig. 27.18(a,b), respectively. At a downstream node of oscillation the wake bursts into turbulence. Hence, the first mechanism may be attributed to eddy-row instability.

The second mechanism, the individual eddy oscillation, may be attributed to eddy filament instability. The mechanism is analogous to that observed behind the single cylinder in the transition-in-wake state, Vol. 1, Chapter 4.

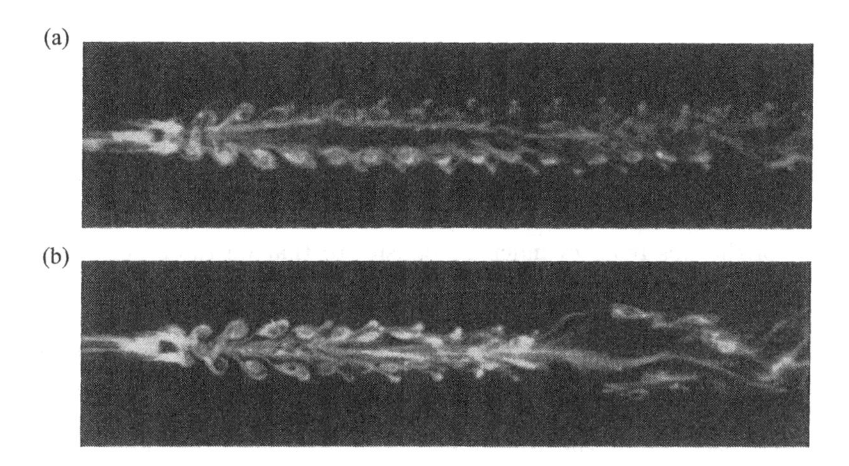

FIG. 27.18. Smoke visualization of triangle cluster at Re = 180, S/D = 5, $T/D = 2.5$, symmetric oscillation of eddy rows, (a) maximum amplitude, (b) minimum amplitude, Zdravkovich (1968J)

The third mechanism is unique for the triangle cluster when $\alpha > 0°$. The flow consists of that behind a single and tandem cylinders. Figure 27.19 shows the asymmetric transition along the wake. The eddies formed behind a single cylinder remained stable and laminar farther downstream than along the wake behind the tandem cylinders. The transition to turbulence behind the latter takes place earlier and faster than behind the single cylinder.

27.2.6 *Triangle clusters, forces*

Sayers (1987J) carried out systematic measurements of the C_p distribution around a monitored cylinder in triangle clusters $1.25 \leqslant P/D \leqslant 5$ at Re = 31.8k. The cylinders had $L/D = 11.7$ between end plates $D_e/D = 12.4$ in an open jet wind tunnel. The orientation was in the range $0° \leqslant \alpha \leqslant 180°$ in increments of 7.5°. The initial orientation $\alpha = 0°$ repeats itself for $\alpha = 120°$ and 240° when cylinders

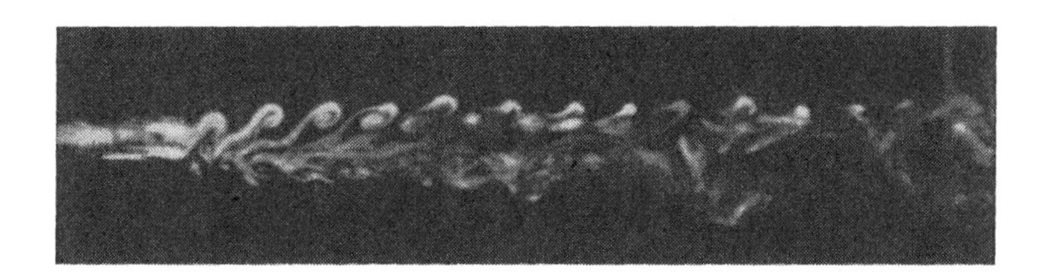

FIG. 27.19. Smoke visualization of triangle cluster at Re = 180, α = 28°, Zdravkovich (1968J)

3 and 2 are the leading ones, respectively.

Figure 27.20(a–d) shows the C_d and C_l curves for $S/D = 1.25$ and 2. The low and negative C_d values coincide with the tandem arrangements. The region of high C_l is always related to staggered arrangements.

Ahmed and Ostowari (1990J) carried out surface flow visualization on the end plate. Figure 27.21(a,b) shows the horseshoe patterns for $T/D = 2$ and $S/D = 1$ and 2, respectively. The sense of rotation of the horseshoe-swirl on one side of the upstream cylinder is opposite to that on the other side of the downstream cylinder. Two parallel streamwise swirls of opposite sign induce an upward velocity away from the surface. Hence, the trace of the horseshoe-swirls disappears behind the downstream cylinder in Fig. 27.21(a,b).

27.2.7 *Triangle clusters, Strouhal number*

It has been discussed in Vol. 1, Chapter 5, that St remained almost constant in the TrSL state due to the invariance in the wake width. There is a significant variation in wake shape and width induced by the three-cylinder interference in triangle clusters at different spacing P/D and orientation relative to the free stream velocity α. This is reflected in a considerable variation in St in terms of P/D and α.

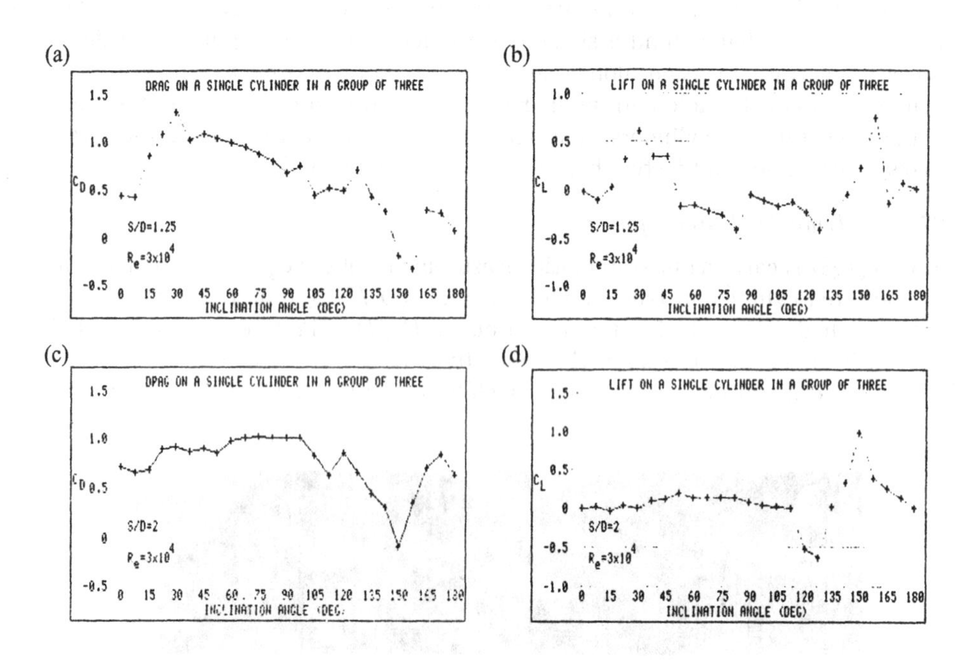

FIG. 27.20. Force on one cylinder in triangle cluster at $Re = 31.8$k, (a) $S/D = 1.25$, drag, (b) $S/D = 1.25$, lift, (c) $S/D = 2$, drag, (d) $S/D = 2$, lift, Sayers (1987J)

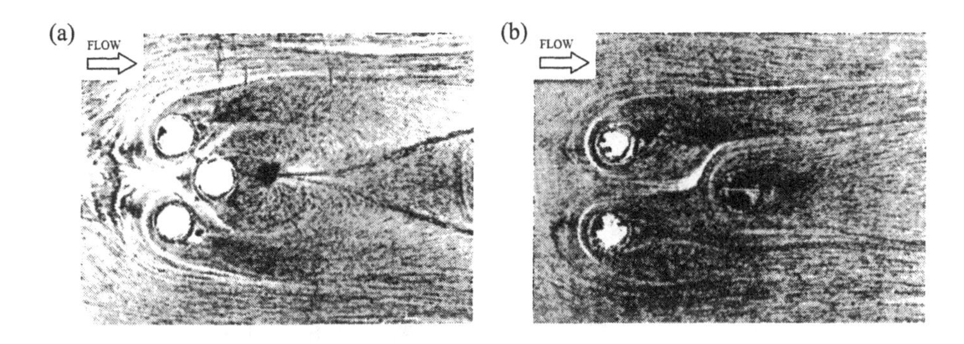

FIG. 27.21. Surface flow visualization on end plate at $Re = 18\text{k}$, (a) $T/D = 2$, $S/D = 1$, (b) $T/D = 2$, $S/D = 2$, Ahmed and Ostowari (1990J)

Lam and Cheung (1988J) carried out dye visualization in water. They estimated St at $Re = 2.1\text{k}$ and 3.5k in the TrSL2 regime for each cylinder in the range $1.27 \leqslant P/D \leqslant 5.43$. The cylinders had $L/D = 13.3$, $D/B = 0.075$, $Ti = 1\%$, with end plates $D_e/D = 6.6$. A typical flow pattern observed for $P/D = 2.29$ at $Re = 2.1\text{k}$ is shown in Fig. 27.22(a,b,c,d) for $\alpha = 0°$, $20°$, $30°$, and $60°$, respectively. All near-wakes are different in shape and width depending on orientation α. The gap flow between cylinders 1 and 3 is suppressed for $\alpha = 20°$. However, for $\alpha = 30°$, when cylinders 1 and 2 are actually in tandem, the biased gap flow develops as seen in Fig. 27.22(c).

The variation in St was measured in the range $0° < \alpha < 60°$ for $10°$ increments. Figure 27.23(a) shows St variation in terms of P/D for $\alpha = 0°$ at $Re = 2.1\text{k}$ (closed symbols) and 3.5k (open symbols). The wake of cylinder 1 becomes narrower as P/D decreases, and St trebles reaching 0.6. Cylinders 2 and 3 have negligible change in the wake width down to $P/D = 2.2$, hence a constant St. For $P/D < 2.2$, a biased bistable flow develops with narrow and wide wakes, and high and low St, respectively. This kind of St variation is typical for two side-by-side cylinders.

Figure 27.23(b) shows St variation in terms of P/D for $\alpha = 40°$ at $Re = 2.1\text{k}$ (closed symbols) and 3.5k (open symbols). Cylinders 1 and 2 are in tandem but eddy shedding from the upstream cylinder 1 continues unabated. For two cylinders in tandem, eddy shedding ceases at $(St)_c < 3.5$–3.8. Cylinder 3 deflects the upstream cylinder wake, and the biased wake leads to the rise in St, as seen in Fig. 27.22(c) for $\alpha = 30°$. The value of St is almost equal for cylinders 2 and 3 down to $P/D = 2.5$. Below that, for closer spacings, a wide wake develops behind cylinder 2 and a narrow one behind cylinder 3 leading to a fall and a continuous rise in St_2 and St_3, respectively.

Figure 27.23(c) shows St variation in terms of P/D for $\alpha = 60°$, side-by-side cylinders 1 and 3 are followed by cylinder 2 placed behind them. As expected, due to the presence of the downstream cylinder, the biased bistable flow is suppressed

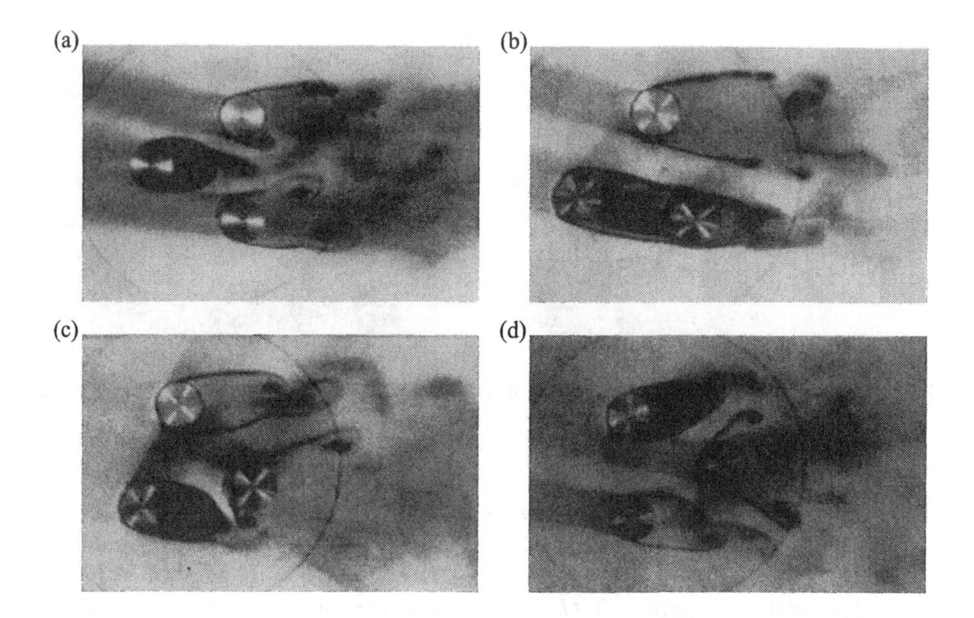

FIG. 27.22. Dye visualization behind three cylinders at $Re = 2.1\text{k}$, $P/D = 2.29$, (a) $\alpha = 0°$, (b) $\alpha = 20°$, (c) $\alpha = 30°$, (d) $\alpha = 60°$, Lam and Cheung (1988J)

and $St_1 = St_3$ throughout. The monotonous rise in St_1 and St_3 reflects the related narrowing of both wakes induced by the downstream cylinder 2. The latter develops a wide wake for $P/D < 4.5$, and this is reflected in a low and decreasing St_2.

Sayers (1990J) also measured St for triangle clusters at $Re = 30\text{k}$ in an open jet. $St_{\max} = 0.32$ for $P/D = 1.5$ is lower than found by Lam and Cheung (1988J) presumably due to zero blockage and higher Re.

27.2.8 *Irregular triangle clusters*

A family of irregularly staggered triangle clusters has been generated by keeping a side-by-side pair of cylinders at a fixed $T/D = 1.5$ and 4, and traversing the third cylinder across y/D at a fixed streamwise distance $S/D = \pm 1.5$. Price and Paidoussis (1984J) measured C_D and C_L on the traversing cylinder having $L/D = 24.4$, $D/B = 0.03$, and $Ti = 0.5\%$ at $Re = 51\text{k}$.

Figure 27.24(a,b) shows the variation in measured C_D and C_L when the third cylinder was upstream at $S/D = -1.5$ and downstream at $S/D = 1.5$ from the side-by-side pair at $T/D = 1.5$. As expected, the alignment of the downstream cylinder into a tandem arrangement produced $C_{D\min} = -0.7$ at $y/D = \pm 0.75$. When the traversing cylinder was the upstream one in the tandem arrangement a local $C_{D\max}$ was found. The variation in C_L, as shown in Fig. 27.24(b), is considerably different for the same stagger of the traversing cylinder. The peaks

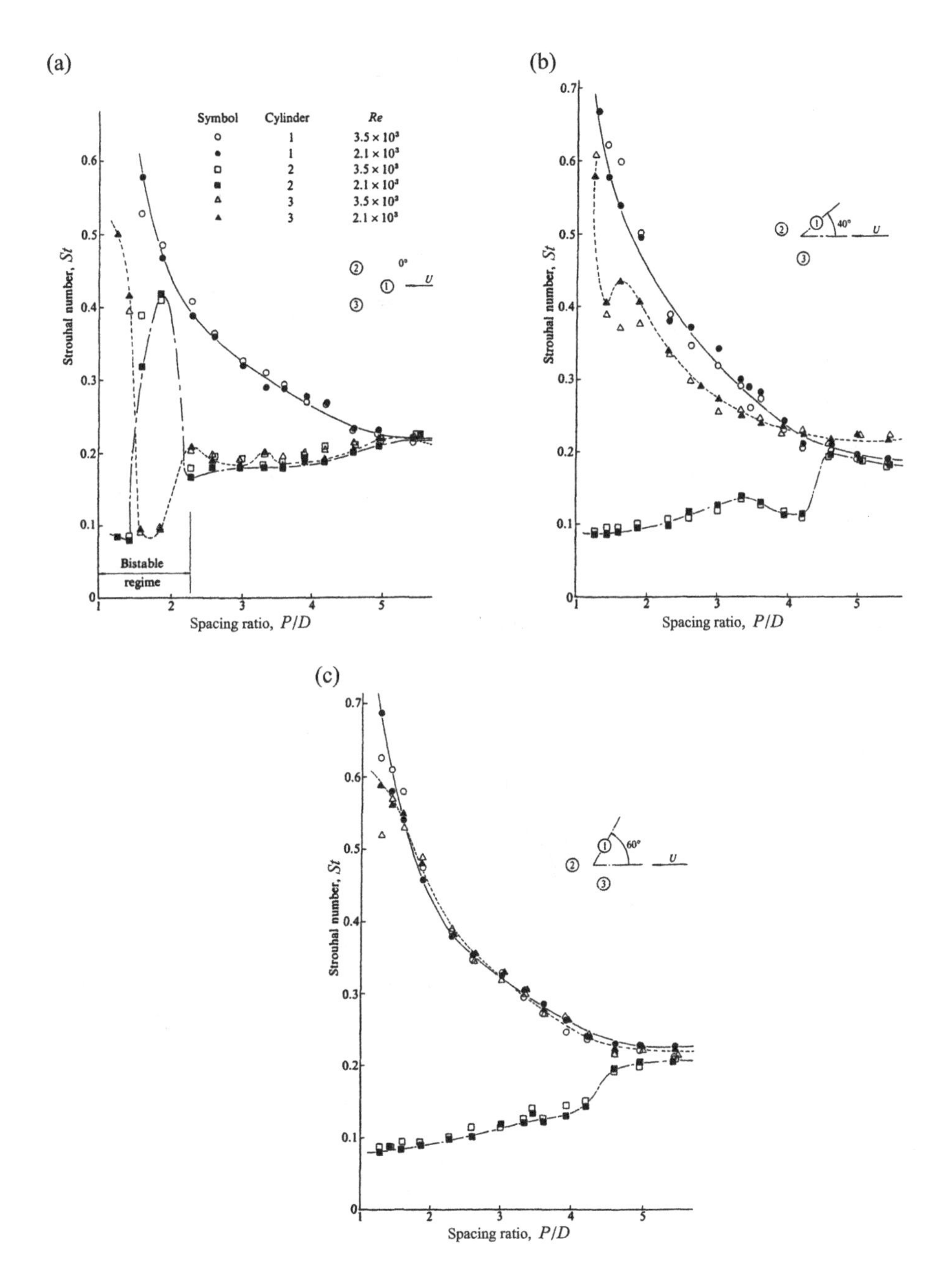

FIG. 27.23. Strouhal number behind three cylinders at $Re = 21\text{k}$ and 35k, (a) $\alpha = 0°$, (b) $\alpha = 40°$, (c) $\alpha = 60°$, Lam and Cheung (1988J)

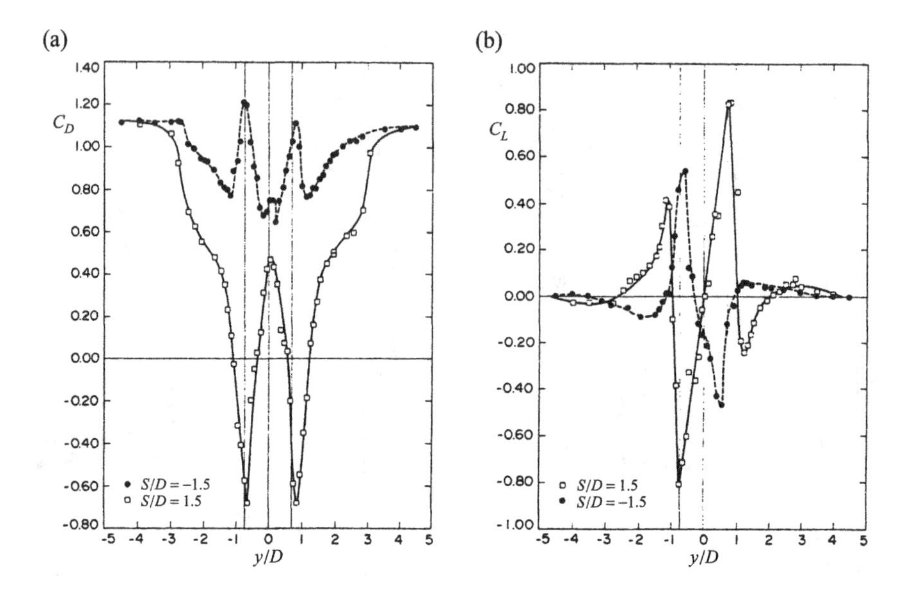

FIG. 27.24. Forces on triangle clusters, $S/D = \pm 1.5$, $T/D = \pm 1.5$, $Re = 51\text{k}$, (a) drag, (b) lift, Price and Paidoussis (1984J)

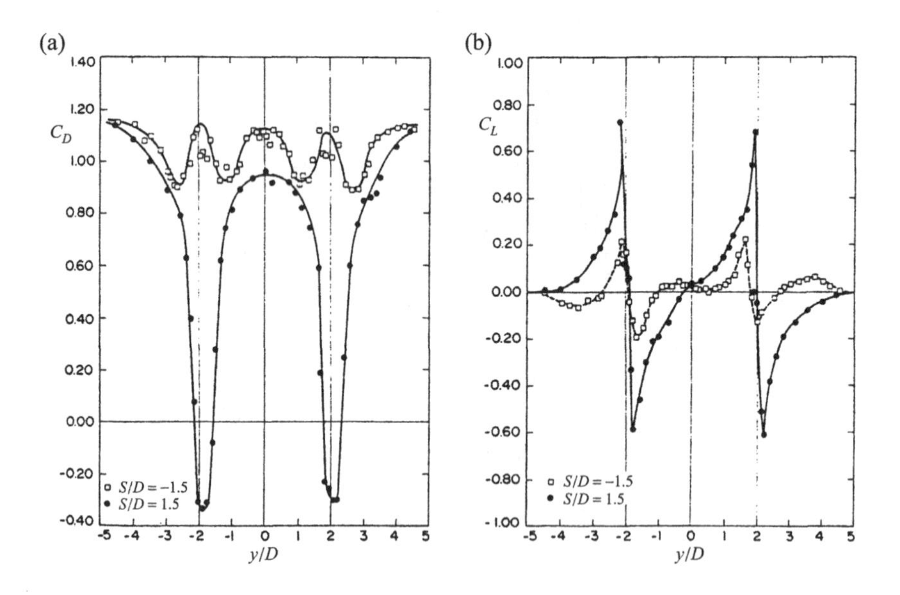

FIG. 27.25. Forces on triangle clusters, $S/D = \pm 1.5$, $T/D = 4$, $Re = 51\text{k}$, (a) drag, (b) lift, Price and Paidoussis (1988J)

in C_L strongly depend on the location of the traversing cylinder being either outside or inside the pair of cylinders. For example, the peaks in C_L are 0.4, −0.8, and −0.1, 0.5 for the downstream and upstream cylinders, respectively. The $C_{L\min}$ on the downstream cylinder and $C_{L\max}$ on the upstream cylinder occur in the tandem arrangement due to interference of the side cylinder.

Similar C_D and C_L variations were found for $T/D = 4$ and $S/D = \pm 1.5$, as shown in Fig. 27.25(a,b). It is notable that the peaks on both sides of the stagger are comparable. The increased distance of the side cylinder reduces the interference effect and $C_L = 0$ in the tandem arrangements.

27.3 Four cylinders

27.3.1 *In-line clusters*

A systematic experimental investigation of four cylinders of equal diameters arranged in-line was carried out for $1 \leqslant S/D \leqslant 5$ by Aiba *et al.* (1981J). Igarashi (1986Jb) carried out measurements of C_{pi} and C'_{pi}, and evaluated C_{di}, where $i = 1, 2, 3, 4$, in the range $1.18 \leqslant S/D < 2.65$ and $8.7\text{k} < Re < 35\text{k}$. The cylinders had $L/D = 4.1$, $D/B = 0.05$, and $Ti = 0.5\%$. Figure 27.26(a,b) is a compilation of C_{pi} and C'_{pi} distributions on four cylinders at $S/D = 1.18$, 1.32, 1.62, 1.91, and 2.50. Two distinctly dissimilar C_{p1} and C_{p2} distributions are contrasted by similar C_{p3} and C_{p4} ones in Fig. 27.26(a). The same can be said for C'_{p1}, C'_{p2} and C'_{p3}, C'_{p4}, as seen in Fig. 27.26(b).

A complementary smoke visualization at $Re = 27\text{k}$ is shown in Fig. 27.27(a–d). For $S/D = 1.32$ and 1.62, an intermittent alternate reattachment is seen behind the second and third cylinders, respectively. A periodic formation and shedding of large eddies behind the second and third cylinders are seen in Fig. 27.27(c,d) for $S/D = 1.91$ and 2.50. The observed flow patterns depend not only on S/D but also on Re.

The high and low St measured behind the fourth cylinder is shown in Fig. 27.28 in terms of Re for fixed values of S/D. In the no-reattachment regime, all cylinders are wrapped inside the free shear layers separated from the first cylinder as found for $S/D = 1.18$, 1.32, and 1.47 (open symbols). The latter persists up to $Re = 22\text{k}$ for $S/D = 1.18$ beyond which a low St appears related to a wide wake. A considerable hysteretic effect was observed in the change-over of the two flow regimes depending on whether the velocity increases or decreases. For $S/D \geqslant 1.62$, only the reattachment and eddy shedding regimes were found for the low St values.

Aiba *et al.* (1981J) extended the range of tested S/D up to 5. They evaluated C_{di} from the C_{pi} distributions for $i = 1, 2, 3, 4$. Figure 27.29(a) shows C_{di} in terms of S/D for $Re = 41\text{k}$. The drag exerted on the first and second cylinders represents two extremes, while C_{d3} and C_{d4} are close to each other and to the mean $\overline{C_d}$. The sudden start of the eddy shedding regime behind the first cylinder is marked by a jump in C_{d2} and C_{d1} at $S/D = 4$.

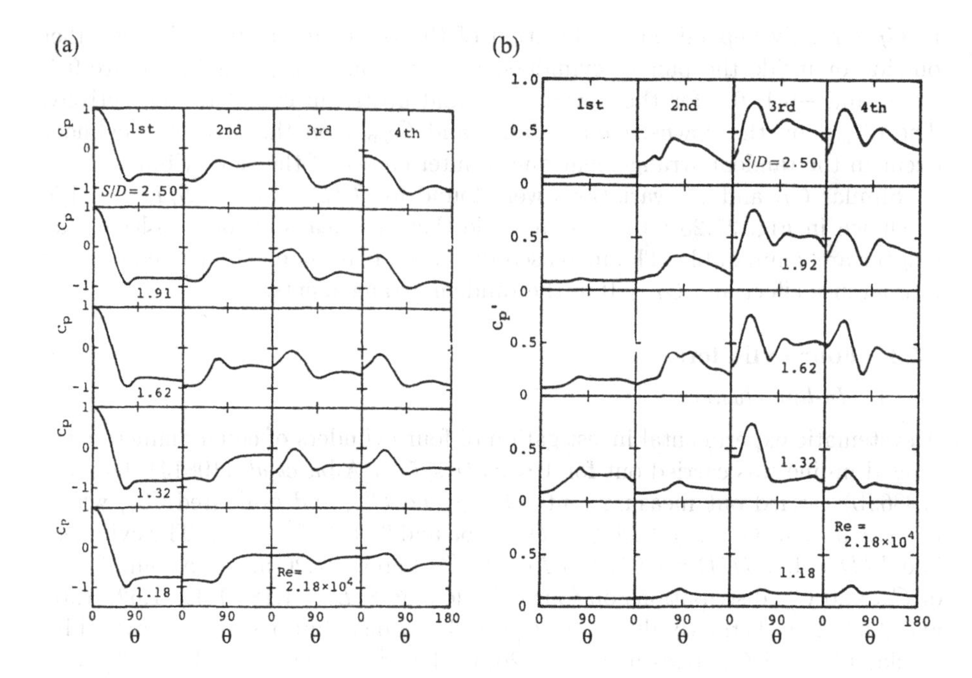

FIG. 27.26. Pressure distributions around four aligned cylinders at $Re = 21.8$k, (a) mean, (b) fluctuating, Igarashi (1986Jb)

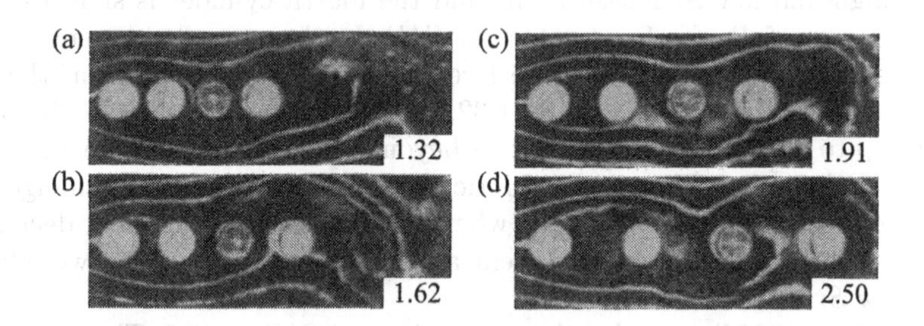

FIG. 27.27. Smoke visualization behind four aligned cylinders at $Re = 22$k, (a) $S/D = 1.32$, (b) $S/D = 1.62$, (c) $S/D = 1.91$, (d) $S/D = 2.50$, Igarashi (1986Jb)

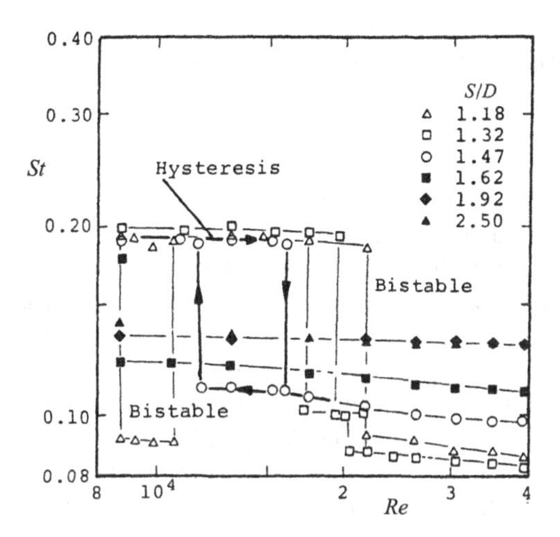

FIG. 27.28. Strouhal number behind four aligned cylinders in terms of Re, Igarashi (1986Jb)

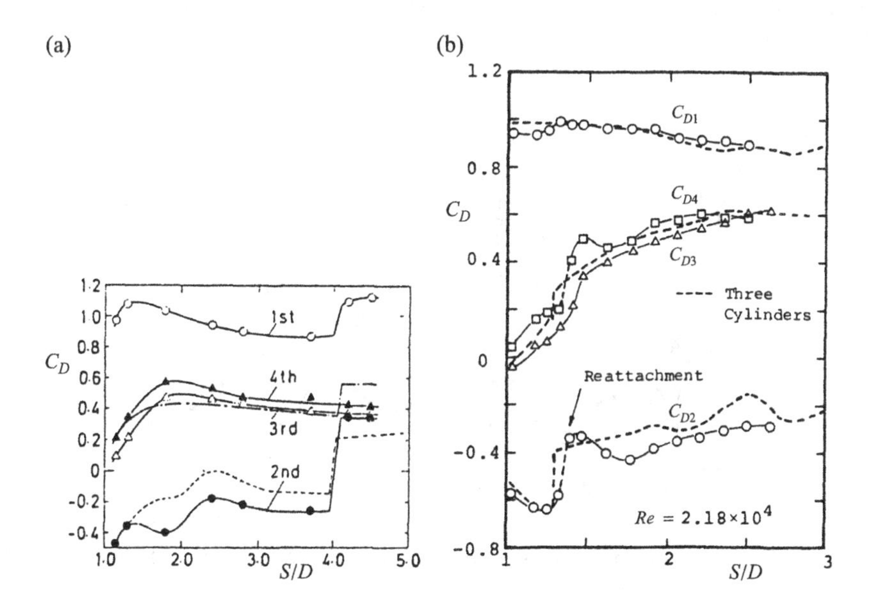

FIG. 27.29. Local drag of four cylinders, (a) $Re = 40$k, Aiba *et al.* (1981J), (b) $Re = 34$k, Igarashi (1986Jb)

A similar variation in C_{di} was found by Igarashi (1986Jb) at $Re = 21.8$k in $1 < S/D < 2.6$ as shown in Fig. 27.29(b). The change-over from the no-reattachment to the reattachment regime was more abrupt and affected C_{d4} variation as well.

27.3.2 *Heat transfer*

Aiba *et al.* (1981J) carried out tests on flow past and heat transfer from four cylinders arranged in-line. The cylinders had $L/D = 8.6$, $D/B = 0.08$, $Ti = 0.007\%$, $12\text{k} < Re < 50\text{k}$, and $1.15 \leqslant S/D \leqslant 4.5$. Figure 27.30(a,b) shows typical C_p and Nu distributions for $S/D = 1.8$ at $Re = 41$k. The upstream cylinder had a higher C_{pb} and lower Nu than the single cylinder (dash-dot line). The second cylinder with the reattachment and $Nu_{\max}$ peaks at $\theta_r = 70°$ showed a considerably higher $\overline{Nu}$ than for the first cylinder. Similar C_{p3} and C_{p4} distributions are seen behind cylinders 3 and 4 with the $Nu_{\max}$ at $\theta_r = 40°$. The intermittent eddy shedding in the gap between the third and fourth cylinders enhanced the heat transfer in the range $120° < \theta < 180°$. It may be concluded that both C_{p1}, C_{p2} and Nu_1, Nu_2 are anomalous in comparison with C_{p3}, C_{p4} and Nu_3, Nu_4.

27.3.3 *Side-by-side clusters*

Ad hoc tests were carried out by Gerhardt and Kramer (1981J) on four equal cylinders set side-by-side at $T/D = 1.5$, 2, and 3. Figure 27.31 shows the histogram of C_{di}, for $i = 1, 2, 3, 4$, and $\overline{C_d}$. The high C_d corresponds to a narrow wake and the low C_d to a wide wake.

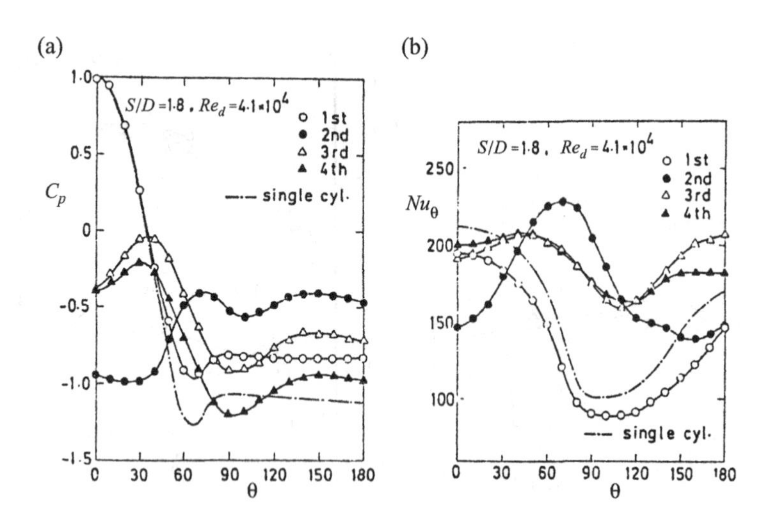

FIG. 27.30. Four cylinders at $Re = 41$k, $S/D = 1.8$, (a) mean pressure, (b) Nusselt number, Aiba *et al.* (1981J)

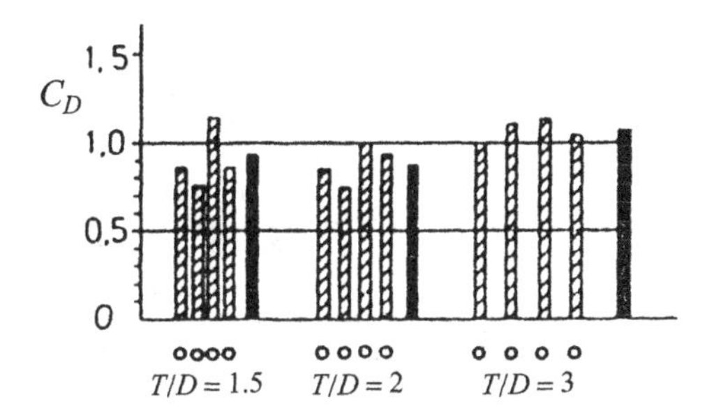

FIG. 27.31. Individual and overall drag on four side-by-side rough cylinders in terms of T/D, Gerhardt and Kramer (1981J)

27.3.4 *Square clusters, forces*

The main feature of square clusters is that four cylinders form a square matrix, as shown in Fig. 27.1(d). Two pairs of cylinders form two tandem arrangements at $\alpha = 0°$. The other extreme, $\alpha = 45°$, represents a diagonal tandem and symmetric stagger. All the other arrangements, $0° < \alpha < 45°$, are asymmetric staggered arrangements. The typical application of square clusters is found in offshore engineering (marine risers) and in electrical engineering (transmission lines), etc.

Pearcey *et al.* (1982P, 1985P) carried out extensive tests on 2×2, 3×3, and 4×4 square clusters of equidistantly spaced marine risers at fixed $P/D = 5$ in the range $40\text{k} < Re < 80\text{k}$. All cylinders had $L/D = 54$, $2D/B = 0.025$, fitted with tripping wires $d_w/D = 0.04$ at $\pm 40°$ and without end plates. The local drag was evaluated from the C_p distribution for the 2×2 cluster and given in Table 27.1. The highest local C_d was found on the downstream unshielded cylinder at $\alpha = 18°$.

Wardlaw *et al.* (1974P) measured the overall drag of stranded conductors for $P/D = 13$ at $Re = 120\text{k}$. Figure 27.32 shows the variation in mean C_D and C_L for the square conductor cluster in terms of wind direction α. It is interesting to point out that $C_{D\min}$ and $C_{D\max}$ are reduced when the free stream turbulence is increased to $Ti = 10\%$.

Table 27.1 *Drag coefficient for a 2×2 cluster*

Cluster	α		
2×2	0°	18°	45°
	0.61, 0.61	0.60, 0.63	0.62
C_{di}	0.44, 0.43	0.69, 0.71	0.64, 0.64
			0.44
Mean C_d	0.52	0.64	0.58

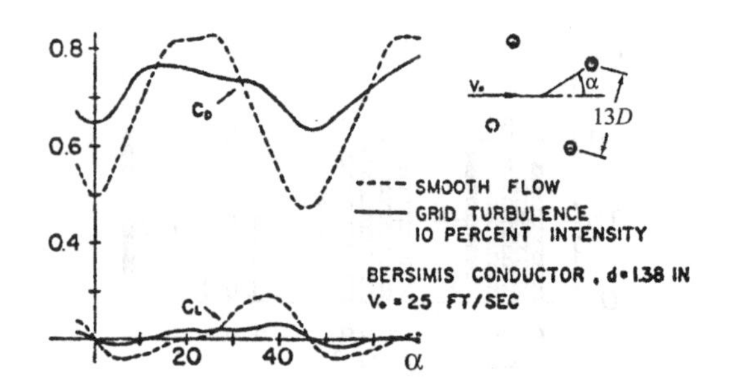

FIG. 27.32. Forces on square cluster $S/D = 15$ in terms of α for $Re = 187$k, (a) drag, (b) lift, Wardlaw *et al.* (1974P)

Lam and Fang (1995J) measured the C_p distribution on four cylinders in square cluster for spacings $1.26 \leqslant P/D \leqslant 5.8$ at $Re = 12.8$k. Cylinders had $L/D = 28.3$, $D/B = 0.036$, $Ti = 0.6\%$, without end plates. The local drag and lift coefficients C_l and C_d were evaluated from the measured C_p distribution. The orientation of clusters was in the range $0° \leqslant \alpha \leqslant 45°$ in increments of $15°$. Note that $\alpha = 0°$ corresponds to the in-line arrangement, i.e. two side-by-side tandems, while $\alpha = 45°$ is the symmetric stagger, i.e. the diagonal tandem sided by the two single cylinders. All the other arrangements within the range $0° < \alpha < 45°$ correspond to the asymmetric stagger.

An insight into the overall drag and lift is shown in Fig. 27.33. The lowest

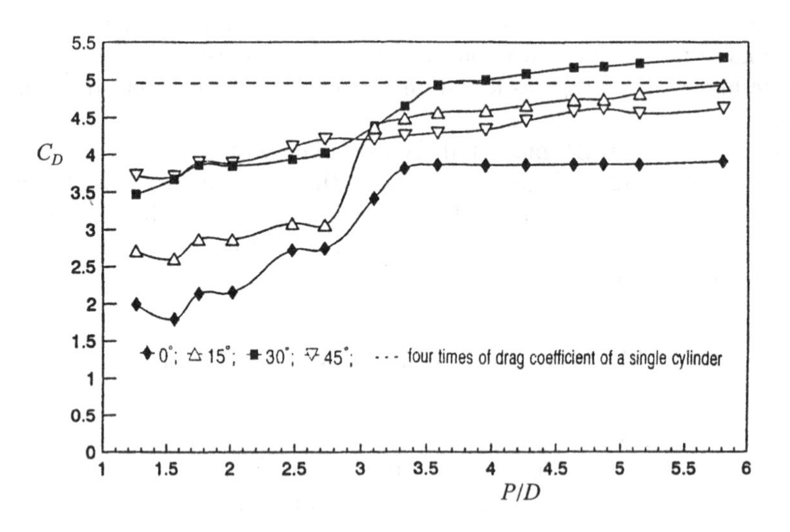

FIG. 27.33. Overall drag on square cylinders at $Re = 12.8$k in terms of P/D, Lam and Fang (1995J)

C_r is found for the in-line arrangement and the highest for $\alpha = 30°$. Note that a change in the flow regime occurs at $P/D = 2.8$ to 3 where eddy shedding behind the upstream cylinder commences for $\alpha = 0°$ and 15°. The overall lift force is negligible for $\alpha = 0°$ and 45° (Fig. 27.34). Sayers (1988P) also measured the C_p distribution in terms of α for $P/D = 1.25$, 1.5, 2, 2.5, 3, 4, and 5.

27.3.5 *Square clusters, Strouhal number*

Lam and Lo (1992J) carried out dye visualization and estimated frequency of eddy shedding by counting eddies. Figure 27.35(a–c) shows a selection of typical wake patterns for P/D from 2.13 to 3.08 and α from 0° to 40° at $Re = 2.1$k. The cylinders were $L/D = 21.5$, $2D/B = 0.094$, $D_e/D = 5.85$. Figure 27.35(a) shows no eddy formation behind the upstream cylinders and a 'mirrored' eddy shedding behind the downstream pair. There is an analogy to flow behind two side-by-side cylinders for $2 < T/D \leqslant 4$, as discussed in Chapter 26. A small asymmetric stagger at $\alpha = 20°$ is shown in Fig. 27.35(b). There is no eddy formation behind upstream cylinders although the initial wake deflection is evident. The narrow wakes behind the upstream cylinders are followed by the wide wakes behind the downstream cylinders. Another asymmetric stagger at $\alpha = 40°$ is shown in Fig. 27.35(c). The upstream cylinder produces the narrowest wake followed by the biased narrow wakes behind the side cylinders and a wide wake behind the downstream cylinder.

Lam and Lo (1992J) quantified the St by counting the eddies seen in dye visualization in water (low accuracy). Figure 27.36(a–c) shows the variation in

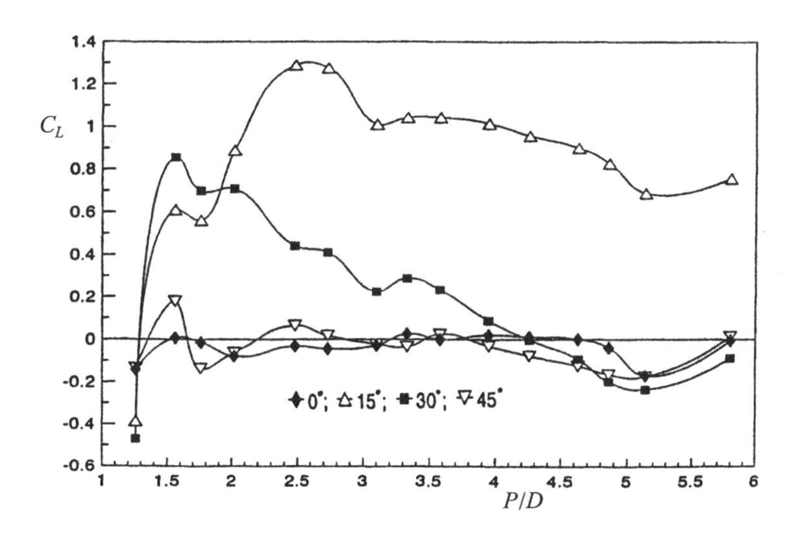

FIG. 27.34. Overall lift on square clusters at $Re = 12.8$k in terms of P/D, Lam and Fang (1995J)

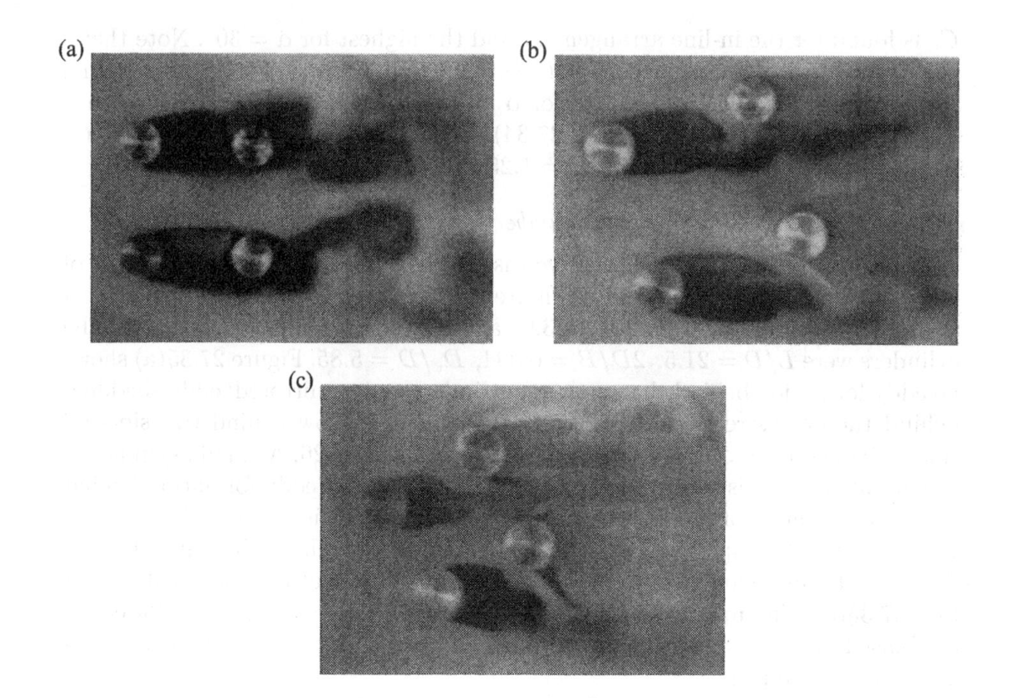

FIG. 27.35. Dye visualization behind square clusters at $Re = 2.1$k, (a) $P/D = 2.65$, $\alpha = 0°$, (b) $P/D = 3.04$, $\alpha = 20°$, (c) $P/D = 2.65$, $\alpha = 40°$, Lam and Lo (1992J)

St for three orientations α and $1.28 \leqslant P/D \leqslant 5.96$ at $Re = 2.1$k. The in-line orientation $\alpha = 0°$ exhibits St_1 and St_2 only for $P/D > 4$. For small spacings, $P/D < 1.9$, St_3 and St_4 branch due to the biased flow caused by wide and narrow wakes. The asymmetric stagger $\alpha = 20°$, in Fig. 27.36(b), exhibits three St curves for the two upstream and one side cylinder. The curves reflect three narrow wakes. Finally, the symmetric stagger $\alpha = 45°$ also shows three St curves: high for the narrowest wake behind the upstream cylinder, less high for two side cylinders, and low for the downstream one.

Another revealing plot is St in terms of the estimated wake width B_w. Figure 27.37 shows the scatter bound on the upper and lower sides by two dashed curves. An approximate mean St_m curve (not depicted) is

$$St_m = 0.264\,(B_w/D)^{-0.9} \tag{27.1}$$

where B_w is the estimated distance between the free shear layers at the end of the eddy formation region as seen in dye visualization. It should be noted that the in-line $\alpha = 0°$ orientation cannot be approximated by using eqn (27.1).

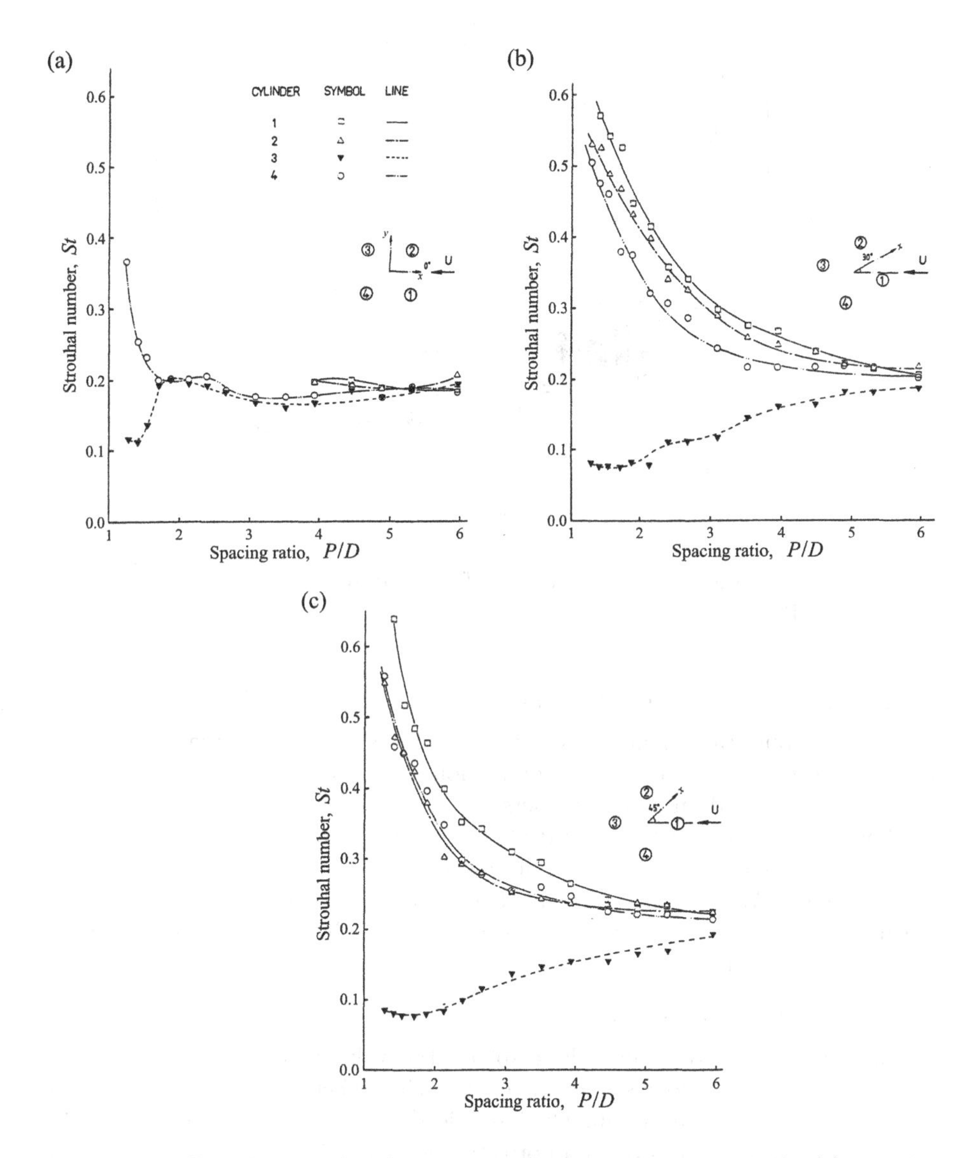

FIG. 27.36. Strouhal number behind square clusters at $Re = 2.1$k, (a) $\alpha = 0°$, (b) $\alpha = 30°$, (c) $\alpha = 45°$, Lam and Lo (1992J)

Lam and Lo (1992J) attempted to correlate the St data by replacing D with D_n, where the latter is the largest distance in the cluster. The attempt had a limited success. Zdravkovich (1997P) argued that the wake width is governed by the gap size and D should be replaced by G. Full discussions and results are given in Chapter 28.

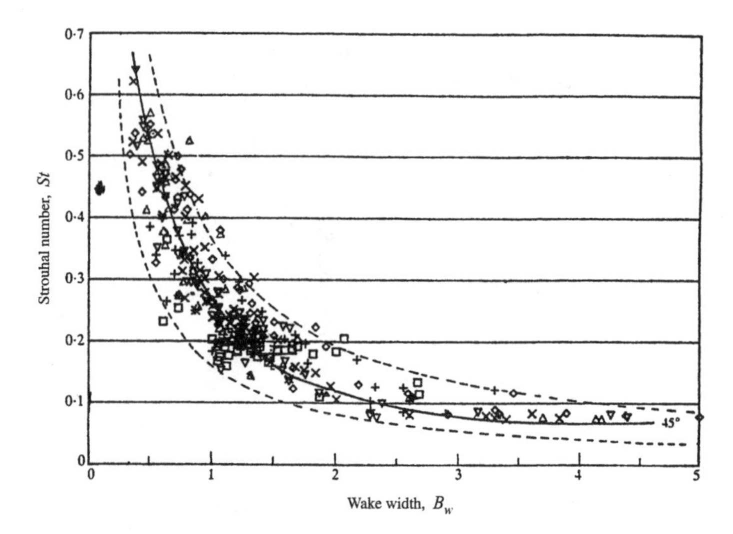

FIG. 27.37. Strouhal number in terms of estimated wake width for all α, Lam and Lo (1992J)

27.4 Cluster of $n > 4$ cylinders

Clusters[165] consisting of more than four cylinders have only occasionally been used in engineering. Some special applications are marine risers, overhead-stranded conductors, and antenna members.

The cylinders may be arranged along a straight line oriented at various α. The two extreme orientations are in-line $\alpha = 0°$, and side-by-side $\alpha = 90°$ arrangements. Another arrangement, common in offshore engineering for rigid marine risers, is 3×3, 4×4, etc. clusters. Finally, the least used and explored are polygonal clusters.

27.4.1 *Five cylinders, side-by-side clusters*

The base-bleed behind the cylinder displaces the wake structures until eventually they are carried away downstream. The stabilizing effect of the base-bleed will be discussed in Vol. 3, Chapter 36. The base bleed can be naturally induced behind porous bluff bodies. For a certain porosity, the near-wake may be displaced at a finite distance downstream of the cylinder.

Honji (1973Ja, 1973Jb, 1973Pa, 1973Pb) carried out hydrogen bubble visualization tests for five side-by-side cylinders at $\alpha = 90°$ in the range $0 \leqslant T/D \leqslant 2$, $10 < Re < 90$, $L/D = 225$, and $5D/B = 0.01$. Figure 27.38(a,b) shows the near-wake development in terms of the increasing T/D at $Re = 30$. When the five cylinders are in contact $T/D = 1$ a reversed flow is formed in the near-wake, the dark region in Fig. 27.38(a). For $T/D = 1.5$, a single row appears as a porous

[165]The term 'cluster' is restricted to any group of cylinders not confined between walls.

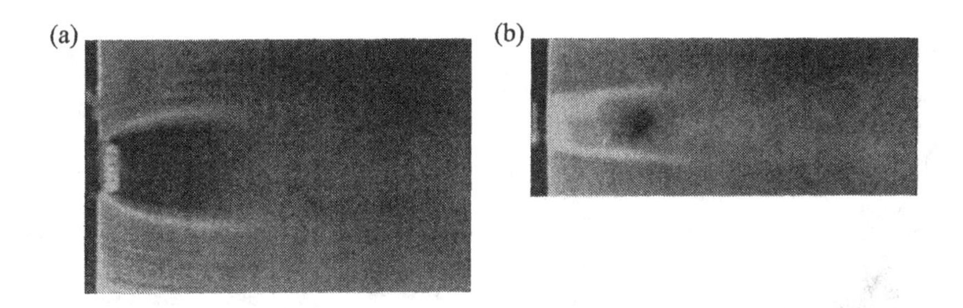

FIG. 27.38. Hydrogen bubble visualization behind five side-by-side cylinders at $Re = 30$, (a) $T/D = 1$, (b) $T/D = 1.5$, Honji (1973Jb)

plate, and the gap flows displace the reversed flow downstream the dark region in Fig. 27.38(b). For $T/D = 2$ and 3, the detached near-wake is displaced farther downstream, and carried away as seen in Fig. 27.39(a,b), respectively. Honji (1973Ja) observed the same sequence of events for $T/D = 1.5$ and Re increasing from 10 to 30. The onset of detachment of the reversed flow takes place at $Re \approx 16$.

The structure of the detached near-wake is shown in Fig. 27.40(a,b) by using a combined surface powder and electrolytic precipitation techniques. The gap-flows behind the row converge towards two sides, and form a stationary detached near-wake. The distance of the near-wake from the row depends on Re and T/D. Figure 27.40(a) shows the initial detached near-wake for $Re = 21.4$ and $T/D = 2$, while Fig. 27.40(b) displays the established detached near-wake.

Further tests with larger cylinders, $L/D = 90$, $5D/B = 0.03$, and $Re > 40$ revealed that each cylinder produced a separate near-wake. Figure 27.41(a,b) shows consecutive narrow and wide near wakes. Initially, for $Re = 42$ the wide wakes are formed behind cylinders 2 and 4. A very wide near-wake behind the middle cylinder 3 is seen in Fig. 27.41(b). This non-uniformity at low Re shows that the biased gap flow is not caused by eddy shedding, see Chapter 26.

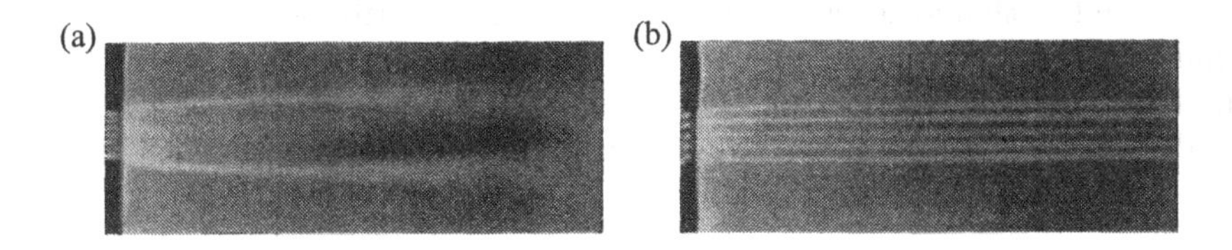

FIG. 27.39. Hydrogen bubble visualization around five side-by-side cylinders at $Re = 30$, (a) $T/D = 2$, (b) $T/D = 3$, Honji (1973Jb)

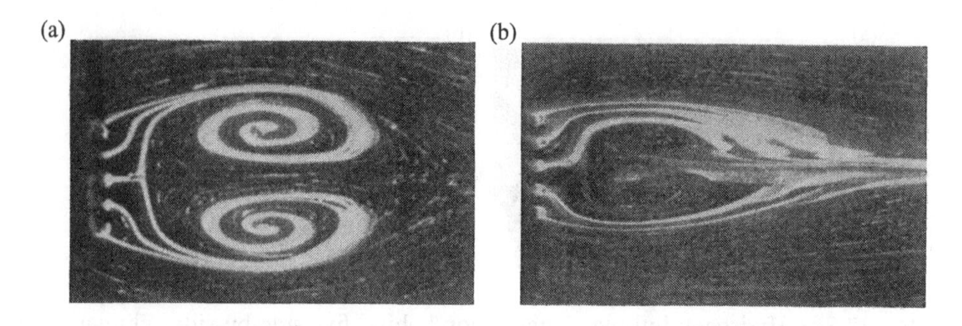

FIG. 27.40. Electrolytic precipitation visualization around five side-by-side cylinders, $Re = 21.4$, $T/D = 2$, (a) initial, (b) established, Honji (1973Pa)

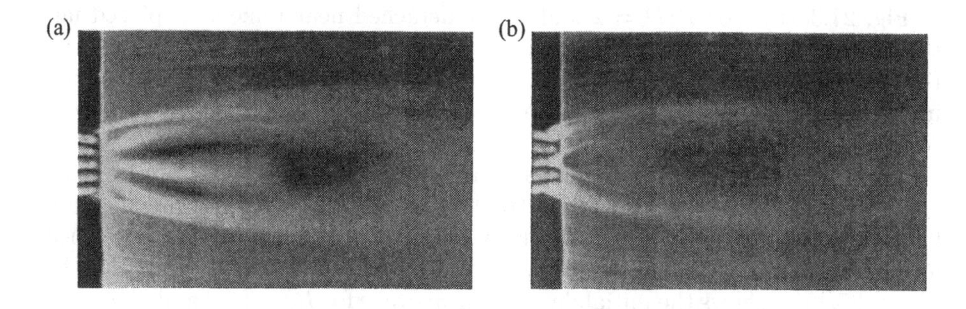

FIG. 27.41. Hydrogen bubble visualization around five side-by-side cylinders at $T/D = 1.5$, (a) $Re = 41.4$, (b) $Re = 89.2$, Honji (1973Jb)

27.4.2 *Seven cylinders, side-by-side clusters*

Further research on seven in-line cylinders $\alpha = 0°$, $S/D = 2$, and $Re = 34$k was reported by Pierce (1973P). Figure 27.42 shows that measured C_{pm} distributions[166] around the first and second cylinders are distinctly different from C_p around the fourth and seventh cylinders. The C_{pm} distribution around the first cylinder is similar to that behind a single cylinder while the C_{pm} for the second cylinder is dissimilar due to the reattached flow. By contrast, the C_{pm} around cylinders 4 and 7 are almost identical to each other. The reattachment takes place at $\theta_r = 37°$, and separation at $\theta_s = 99°$ and $105°$, respectively.

27.4.3 *Six and eight cylinders, polygonal clusters*

As the technology of electric power transmission over large distances has advanced, line voltages have exceeded 1 million volts. The need to suppress the Corona discharge at these high voltages required the use of clusters of stranded

[166] C_{pm} is based on the mean velocity $V_m = \frac{1}{2}(V + V_g)$.

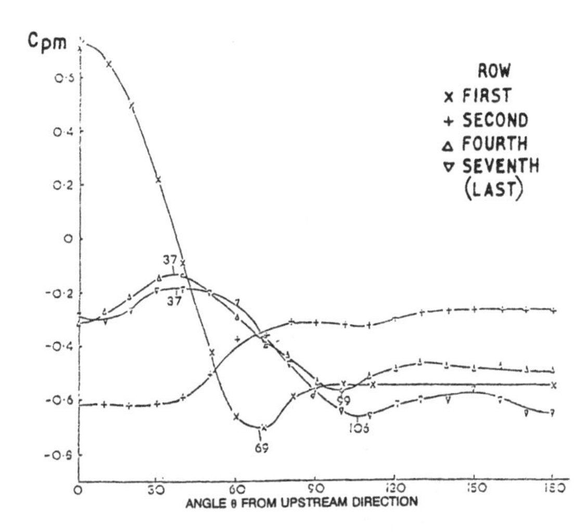

FIG. 27.42. Mean pressure distribution around seven aligned cylinders at $\alpha = 0°$, $Re = 34$k, $S/D = 2$, Pierce (1973P)

conductors, where cluster spacings were maintained by 'spacers' that divide each span into 'subspans'. It was established by field observation that wind often causes conductor oscillation, which in turn has prompted research.

Wardlaw *et al.* (1974P) measured the drag and lift forces exerted on clusters of six and eight cylinders being $10D$ apart for various wind orientations $0° \leqslant \alpha \leqslant 60°$ at $Re = 30$k and 60k. Figure 27.43(a,b) shows the overall drag and lift coefficients in terms of the orientations α. As the cluster was rotated in the wind tunnel the instrumented conductor was immersed, in turn, in the wakes of different upstream conductors with different spacing ratios. The variation of forces in each wake was found to be similar to those for twin cylinders.

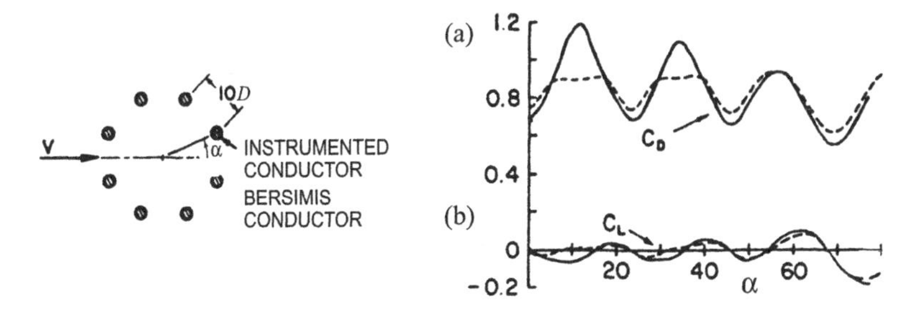

FIG. 27.43. Forces on an eight-cylinder cluster for $P/D = 10$, $Re = 187$k, (a) drag, (b) lift, Wardlaw *et al.* (1974P)

27.4.4 3×3, $4 \times 4, \ldots$, $n \times n$ *clusters*

2×2 clusters have already been discussed in Section 27.3.3. Pearcey *et al.* (1982P) also dealt with 3×3 and 4×4 clusters in uni-directional flows. Figure 27.44 shows the 3×3 cluster in the wind tunnel and Fig. 27.45 lists C_d values evaluated for each cylinder fitted with the tripping wires at $\theta_w = \pm 40°$ at $Re = 80$k for $P/D = 5$. For $\alpha = 0°$, $C_{d\max}$ is found in the first row and $C_{d\min}$ in the third row. For $\alpha = 18°$, $C_{d\max}$ is measured in the second row side cylinder and $C_{d\min}$ on the

FIG. 27.44. 3×3 cluster in the test section, Pearcey *et al.* (1982P)

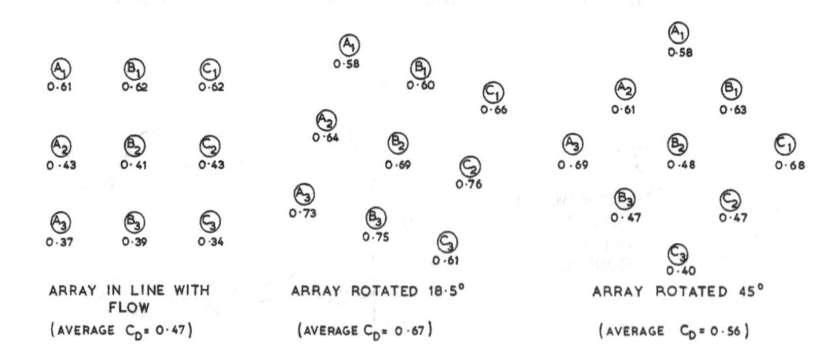

FIG. 27.45. 3×3 cluster individual drag coefficients for three orientations, Pearcey *et al.* (1982P)

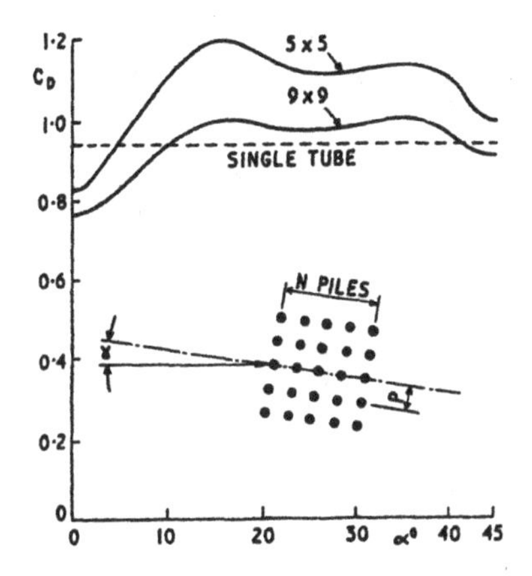

FIG. 27.46. Drag on 5×5 and 9×9 clusters at $P/D = 8$, $Re = 40\text{k}$, Ball and Hall (1980J)

most upstream cylinder. Finally, for $\alpha = 45°$, $C_{d\max}$ appears on the side cylinder in the third row and $C_{d\min}$ for the most downstream cylinder. The overall $C_{d\max}$ is found for $\alpha = 18°$.

Pearcey *et al.* (1982P) carried out tests on a 4×4 cluster where all cylinders were fitted with the tripping wires $d_w/D = 0.04$ at $\pm 40°$. For the 4×4 cluster, $\overline{C}_d = 0.43$, 0.62, and 0.59, for $\alpha = 0°$, 18°, and 45°, respectively.

Ball and Hall (1980J) tested 5×5 and 9×9 clusters at $Re = 10\text{k}$. Only the overall drag was measured, and the variation in $\overline{C}_D$ is shown in Fig. 27.46. The maximum $\overline{C}_D$ is again near $\alpha = 18°$.

The overall drag coefficient for any cluster cannot be taken as a representative of any individual cylinder in it. For example, the cylinder in the upstream row may experience up to 30% higher drag than $\overline{C}_D$, depending on the orientation of the cluster. There is also a lift component in all staggered arrangements, which has not been measured.

27.5 Satellite clusters

27.5.1 *Introduction*

Multiple marine risers have been developed to meet the ever-increasing demand for oil extraction. The design of marine risers was primarily guided by operational requirements and these determined the size and number of risers in the cluster. A typical riser cluster consisted of n tubes of diameter d arranged at a constant pitch P along a circle of radius R around a central tube of diameter D, as shown in Fig. 27.47. The actual grouping of tubes within the cluster attracted less consideration in the design stage. This resulted in a wide variation in R, n, and P for the marine risers employed in the range $300\text{k} < Re_D < 2\text{M}$.

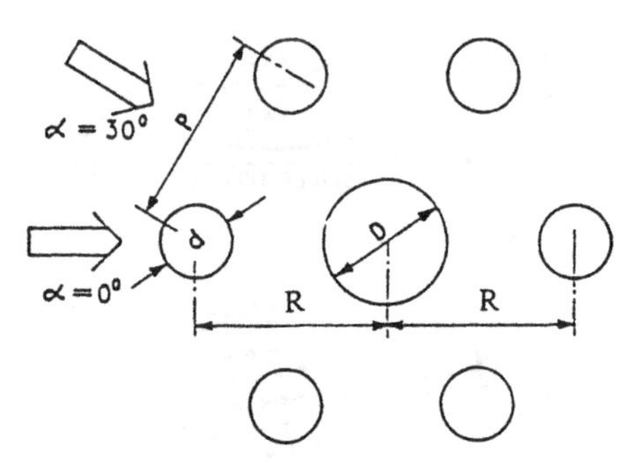

FIG. 27.47. Definition sketch of satellite cluster, Zdravkovich and Baldaro (1993P)

The hydrodynamic loading exerted by a current on satellite clusters depends strongly on the spacing of the riser tubes and the orientation of the cluster relative to the current direction. The flow interference within the satellite cluster may belong to various categories, namely some satellite tubes being in tandem, others in a staggered arrangement. The combined effect of all the interference categories results in the overall drag and lift forces, which have to be kept to a minimum.

27.5.2 *Interference parameters*

It has been discussed in Chapter 26 that flow interference past two parallel cylinders of equal diameters depends on the spacing ratio and orientation to the free stream. When two cylinders are unequal in diameter the ratio D/d becomes an additional interference parameter. The satellite cluster can be analysed as a combination of a number of two-cylinders interference categories such as:

(i) the satellite tube may be located upstream or downstream of the central tube for $\alpha = 0$, thus the interference parameters are R/d and R/D, respectively;

(ii) the mutual interference of satellite tubes being in a tandem or staggered arrangement depends on P/d and T/d, respectively;

(iii) the shielding of the central tube by the cluster of satellite tubes may be described as a shroud of porosity $nd/2\pi R$.

It may be argued that the overall drag coefficient, C_{Do}, depends on the following independent and dependent parameters:

$$C_{Do} = f\left(\alpha, D/d, R/D, R/d, P/d, T/d, \frac{nd}{2\pi R}, Re_D, Re_d, Ti, K/D\right) \quad (27.2)$$

where α is the cluster orientation, Re_D and Re_d are the Reynolds numbers based on D and d, respectively, Ti is the free stream turbulence, and K/D is the surface roughness. The drag coefficient of the cluster is defined as

$$C_{Do} = \frac{\text{overall drag force}}{\frac{1}{2}\rho V^2 L(D + \sum_{i=1}^{n} d_i)} \tag{27.3}$$

where L is the length of the riser exposed to the current. Note that the geometry of the satellite cluster repeats itself for each $\alpha = 180/n$. Hence, the C_{Do} variation within $0 \leqslant \alpha < 180/n$ also repeats itself. The in-line orientation is for $\alpha = 0$, the symmetric stagger is for $\alpha = 180/n$, and the asymmetric stagger is for all other orientations in between. The side force is zero only for $\alpha = 0$ and $180/n$.

27.5.3 *Force on satellite clusters*

Considerable research effort has been expended on quantifying the forces exerted on satellite clusters as a whole and on the individual tubes for a wide variety of $n, P/d, R/D, R/d$, and $nd/2\pi R$. Regrettably, only a limited number of papers have been published in the open literature, such as Ottesen-Hansen *et al.* (1979P), Demirbilek (1990J), etc. Zdravkovich (1991R) gave a comparative overview of the similarities and dissimilarities between marine risers and heat exchanger tube arrays. He emphasized that satellite clusters and tube arrays are intrinsically different due to the confinement within the walls of the latter.

An example of extensive tests carried out on a variety of satellite clusters was reported by Demirbilek and Halvorsen (1985J). Eleven tested satellite clusters are given in Table 27.2 in terms of the interference parameters discussed previously. The apparently random variation in C_{Do} may be explained and predicted by applying the interference analysis on tube pairs within the clusters.

When two satellite tubes are in tandem, the C_D will be considerably reduced on the downstream tube for $P/d < 3.8$ due to the absence of eddy shedding

Table 27.2 *Interference parameters for satellite risers*

No	D/d	R/d	R/D	P/D	$Nd/2\pi R$	C_{Do}
2	3.00	6.66	2.22	3.30	0.28	0.60
3	2.45	5.45	2.22	2.85	0.35	0.53
4	3.00	6.66	2.88	4.53	0.28	0.67
5	3.00	6.66	2.22	4.18	0.24	0.68
8	3.00	6.66	2.22	2.09	0.23	0.50
13	2.91	9.18	3.15	5.80	0.17	0.70 (max)
14a	4.18	5.45	1.30	2.15	0.35	0.51
15	4.55	6.66	1.46	3.49	0.28	0.58
17a	7.36	5.45	0.74	2.86	0.35	0.44 (min)
18	2.90	2.72	1.78	2.03	0.49	0.45
20	3.72	5.45	1.46	2.86	0.35	0.66

behind the upstream tube. This is the case for clusters 2, 3, 8, 14a, 15, 17a, 18, and 20, as evident in Table 27.2, column 5. The critical spacing ratio in terms of D/d is given in Fig. 27.48. Table 27.2, column 3, shows that clusters 14a, 17a, 18, and 20 have R/d less than critical.

Finally, the optimal porosity of the satellite shroud was found to be in the range $0.30 < nd/2\pi R < 0.36$ for suppressing eddy shedding behind the satellite clusters, Zdravkovich (1971P). This is the case for clusters 3, 14a, 17a, and 20 in Table 27.2, column 6. The minimum C_{Do} is expected to be for clusters 14a and 17a because all interference parameters are less than critical. The minimum measured C_{Do} is indeed found for cluster 17a, Table 27.2, column 7. Note also that the maximum C_{Do} is found for cluster 13 where all interference parameters are greater than critical.

Further evidence in support of the simple interference analysis is given by dye flow visualization of cluster 2 by Jacobsen (private communication, 1985). Figure 27.49(a) shows eddy shedding behind all upstream satellite tubes, and lack of it behind the downstream satellite tubes. The complementary Fig. 27.49(b) shows the measured drag on the individual satellite tubes, the central one, and the overall values. The high C_D on the upstream satellite tube is contrasted by low C_D on the downstream tubes.

An entirely different flow pattern is seen in Fig. 27.50. The water surface visualization shows flow past a satellite cluster consisting of 5+1 tubes. Most interference parameters are less than critical, and eddy shedding is absent within the cluster. The porosity $dn/2\pi R = 0.42$ is too high, and large-scale eddy shedding is seen behind the cluster. Hence, the 'internal' eddy shedding seen in Fig. 21.49(a) is replaced by 'external' eddy shedding. Both are detrimental as discussed by Zdravkovich (1991R).

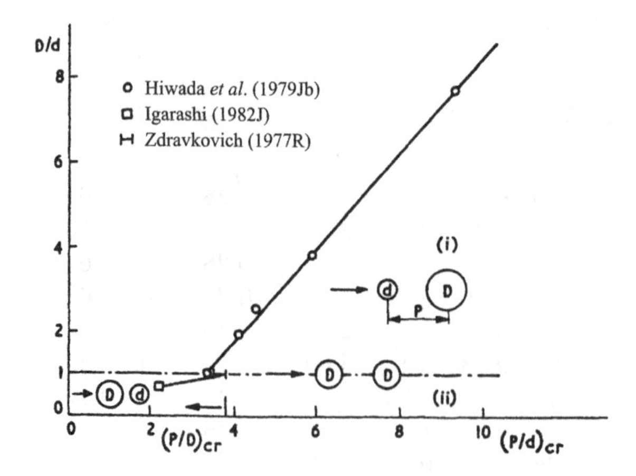

FIG. 27.48. Effect of tube diameter ratio on critical spacing, Zdravkovich (1991R)

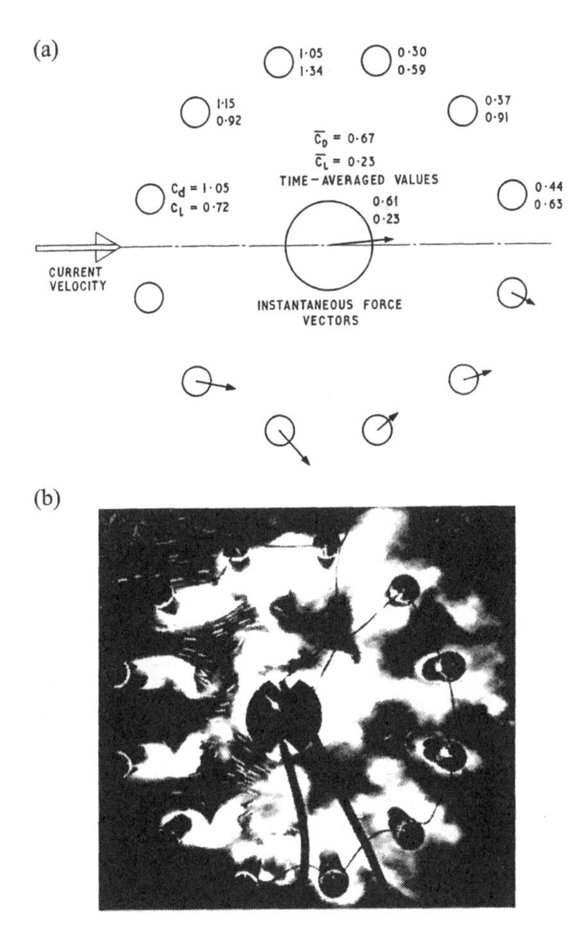

FIG. 27.49. A $12+1$ cluster at $\alpha = 1.5°$, (a) flow visualization, (b) drag and lift on satellite, central tube and cluster, by courtesy of Jacobsen, Danish Hydraulic Institute

27.5.4 *Effect of satellite tube spacing*

It might be expected that for a certain tube spacing both internal and external eddy shedding should be absent, resulting in a minimum C_{Do}. Johnson and Zdravkovich (1991P) and Zdravkovich and Baldaro (1993P) carried out a combined flow visualization and C_{Do} measurements for satellite clusters by systematically varying R in terms of α and Re. The clusters had $L/D = 22.7$ and 54.5, $D/B = 0.11$ and 0.23, $De/D = 5$, $D/d = 3$, $n = 6$, $Re = 3.1$k and 60k, respectively, in the range $1.7 < R/d < 4.5$. Table 27.3 shows the interference parameters for six satellite clusters from the widely spaced A to the tubes touching F.

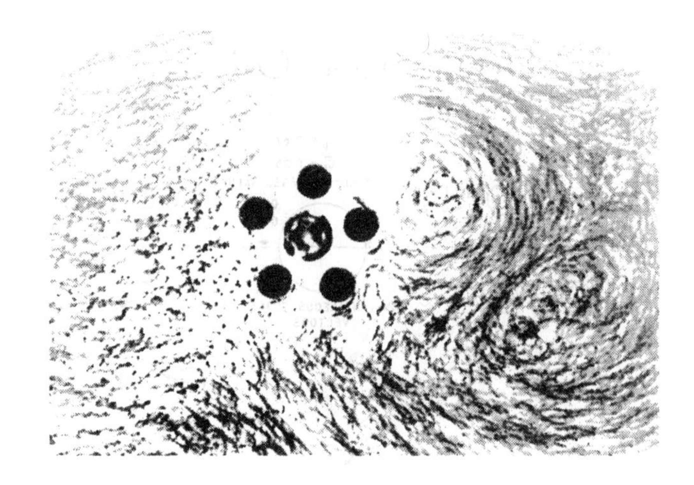

FIG. 27.50. Water surface visualization around a 5 + 1 satellite riser, Zdravkovich (1991R)

Table 27.3 *Interference parameters for $D/d = 3$ clusters*

Cluster	R/d	R/D	P/d	$Nd/2\pi R$	C_{Do}	Eddy shedding
A	4.5	$\underline{1.9}$	5.6	0.21	0.79	
B	3.9	1.6	$\underline{4.1}$	0.24	0.74	Internal
C	3.4	1.4	3.6	$\underline{0.28}$	$\underline{0.74}$	
D	2.9	1.2	3.0	0.33	0.72	None
E	2.3	1.0	2.4	0.41	0.71	
F	1.7	0.7	1.8	0.55	0.82	External
Critical	5.2	1.8	3.8	0.3	–	–

Figure 27.51(a–d) shows the water surface visualization around clusters A, C, D, and E at $\alpha = 0°$ for $Re = 3.4$k. Eddy shedding is absent behind the upstream and downstream satellite tubes, as well as behind the central tube in Fig. 27.51(a,b) for clusters A and C. Eddy shedding is absent behind all tubes in Fig. 27.51(c,d) for clusters D and E. Table 27.3 shows medium C_{Do} for clusters A, B, and C, and low C_{Do} for clusters D and E.

Figure 27.52(a–d) shows the water surface visualization around clusters A to E at $\alpha = 30°$, symmetric stagger, for $Re = 3.4$k. Eddy shedding is seen behind all satellite tubes in Fig. 27.52(a,b) for clusters A and C. Eddy shedding is seen only behind side satellite tubes in Fig. 27.52(c,d) for clusters D and E. Figure 27.53 shows high C_{Do} for clusters A and C and medium C_{Do} for clusters D and E.

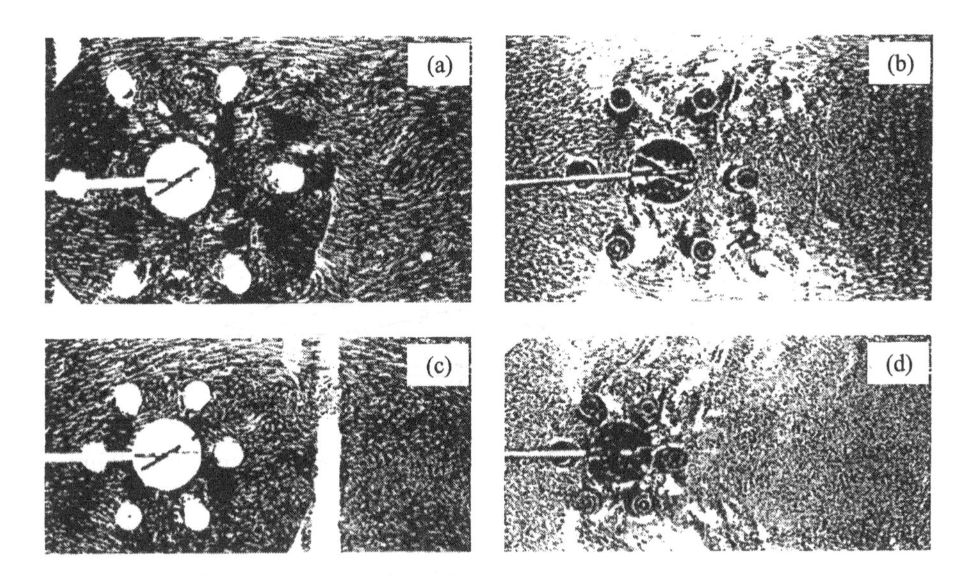

FIG. 27.51. Water surface visualization around a 6 + 1 cluster at $Re = 3.2$k, $\alpha = 0°$, (a) cluster A, (b) cluster C, (c) cluster D, (d) cluster E, Johnson and Zdravkovich (1991P)

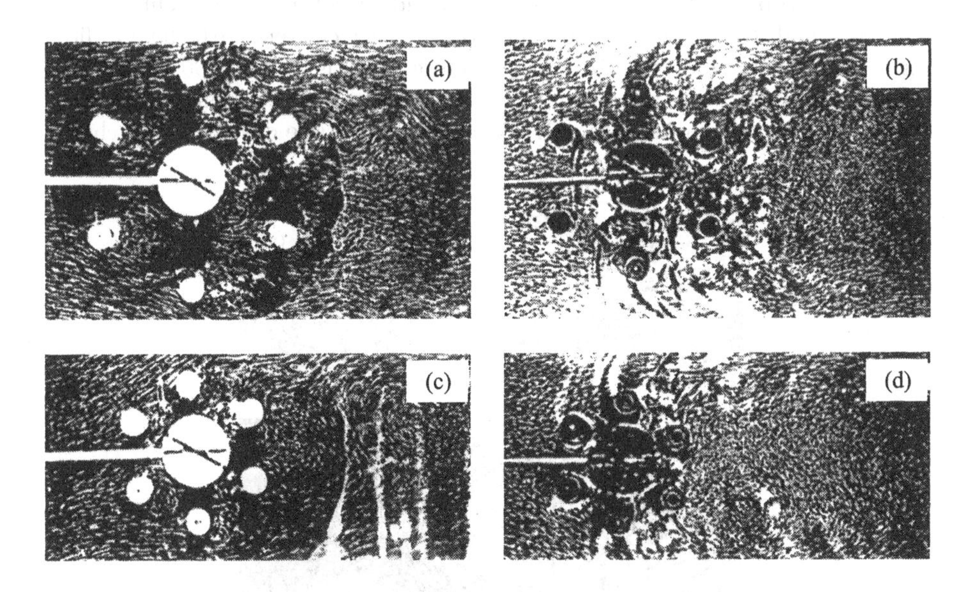

FIG. 27.52. Water surface visualization around a 6 + 1 cluster at $Re = 3.2$k, $\alpha = 30°$, (a) cluster A, (b) cluster C, (c) cluster D, (d) cluster E, Johnson and Zdravkovich (1991P)

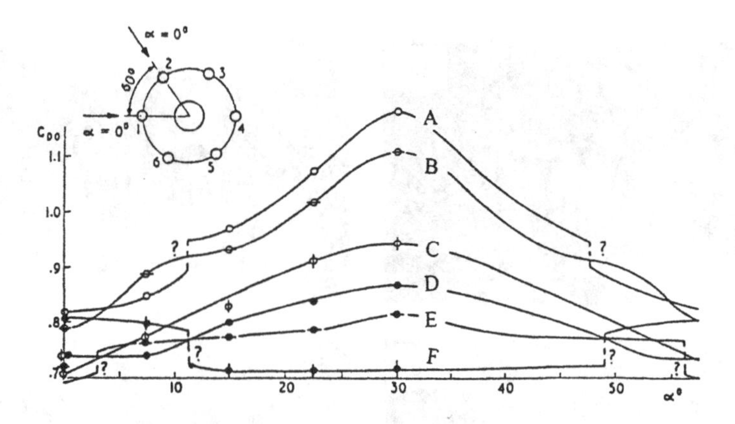

FIG. 27.53. Drag coefficient on 6 + 1 clusters A to F in terms of α, Johnson and Zdravkovich (1991P)

An anomalous variation in C_{Do} is found for cluster F when all satellite tubes are in contact with the central one. Figure 27.54(a,b) shows cluster F at $\alpha = 0°$ and 30°, respectively. Cluster F represents a bluff body formed by all satellite tubes being in contact with the central one. For $\alpha = 0°$, flow separates from the satellite tubes at $\theta = \pm 60°$ and forms a wide wake behind the cluster. The result is a high C_{Do} being greater than for all other clusters with the gap flow. For $\alpha = 30°$, flow separation from the satellite tubes is at $\theta = \pm 90°$, and forms a considerably less wide wake. The result is the least C_{Do}, as seen in Fig. 27.53. Note that question marks in Fig. 27.53 indicate a possible discontinuous change

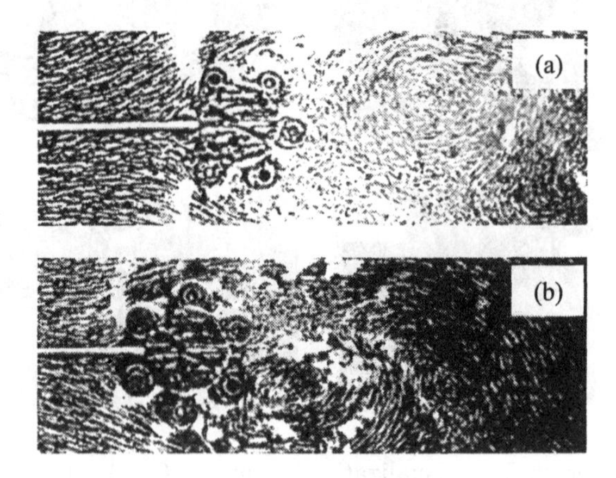

FIG. 27.54. Water surface visualization around 6 + 1 cylinders in contact, (a) $\alpha = 0°$, (b) $\alpha = 30°$, Johnson and Zdravkovich (1991P)

in flow regimes seen in Fig. 27.54(a,b).

The side force component, i.e. the lift force, is zero for $\alpha = 0$ and $30°$ due to symmetry. The C_L is measured for $\alpha = 15°$ and found to be 0.37, 0.36, 0.31, 0.27, 0.08, and -0.15 for clusters A, B, C, D, E, and F, respectively. The anomalous change in the C_L sign for cluster F may be due to the displacement of the stagnation point in the opposite direction.

28

MULTI-TUBE ARRAYS

28.1 Introduction

It has been discussed in Vol. 1 that a most prominent feature of flow past circular cylinders and tubes is that the transition from laminar to turbulent flow takes place through a succession of transition states over a large range of *Re*. Each transition state is sensitive to disturbances such as free stream turbulence, proximity to other tubes or walls, wake interference, surface roughness, and tube oscillation, to mention only those relevant to ***tube arrays***.[167] The complexity of interstitial flow in closely spaced tube arrays stems from the fact that the flow is in the transition states and most of the aforementioned disturbances are present.

Another complicating factor is that the disturbances may vary in space and time within the tube array. For example, the interstitial flow generates turbulence, row after row, until the rate of generation is eventually balanced by the rate of dissipation. The closely packed tubes are subjected to strong proximity and wake interference due to small transverse T/D, and streamwise S/D ratios, respectively. Some tubes are located near the side walls, and subjected to the wall proximity as well. All these features contribute to an amazing non-uniformity and variability of interstitial flows.

Engineering applications of multi-tube arrays are abundant in heat exchanger devices. These are widely used in chemical and mechanical engineering in general, and in fossil-fuel and nuclear power plants in particular. A heat exchanger's tube arrays may be arranged in four matrices: square, rotated square, equilateral triangle, and rotated triangle, as depicted in Fig. 28.1(a–d).

28.1.1 *Categorization of tube arrays*

Tube arrays may be grouped into two basic categories:

(i) The in-line category consists of square and rectangle tube arrays, where the interstitial flow is mostly straight through the arrays.

(ii) The staggered category consists of normal triangle, rotated square, and parallel triangle arrays.[168] Every even row is displaced in the transverse direction by $\frac{1}{2}T/D$, and the interstitial flow is forced along wavy paths through all staggered arrays.

The two categories represent intrinsically dissimilar interstitial flows. As an example, Fig. 28.2(a) shows Weaver and Abd-Rabo's (1985J) flow visualization

[167] The multi-row of tubes confined between walls will be called ***tube arrays*** for short.

[168] The rotated triangle arrays are referred to as ***parallel triangle*** arrays.

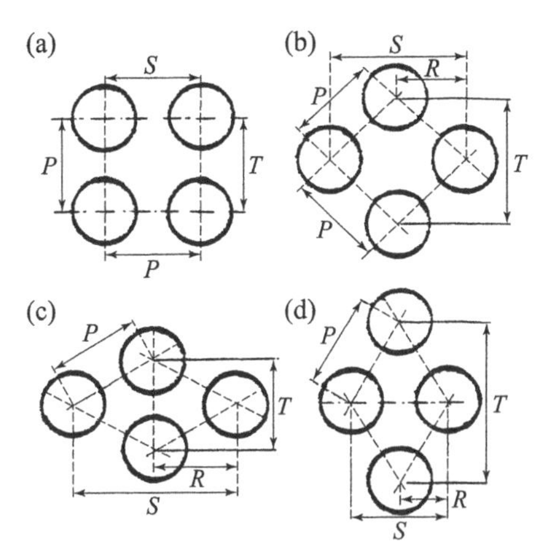

FIG. 28.1. Basic matrices of tube arrays, (a) square, (b) rotated square, (c) normal triangle, (d) parallel (rotated) triangle

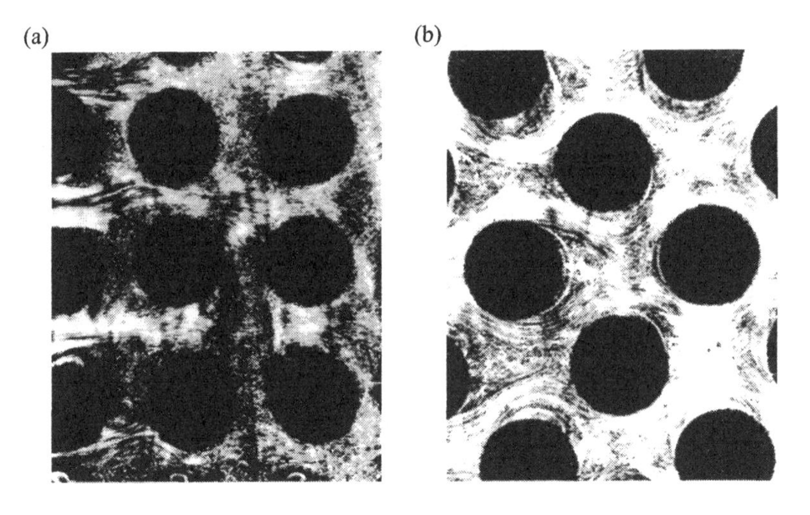

FIG. 28.2. Interstitial flow visualization, (a) square array $P/D = 1.5$, (b) rotated square $P/D = 1.5$, Weaver and Abd-Rabo (1985J)

of the interstitial flow of water in a square array $P/D = 1.5$ at $Re = 1.1$k. The interstitial flow consists of four jets bordered by tubes and streamwise gaps filled with slowly moving water. Figure 28.2(b) shows the interstitial flow in a rotated square array $P/D = 1.5$. The dominating feature is a wavy sinusoidal interstitial flow. The near-wakes are cusped in shape, narrowed in width, and connected to enlarged stagnation regions on the upstream side of the tubes.

For all square tube arrays, the matrix pitch ratio P/D is equal to the streamwise S/D and transverse pitch ratios T/D, as seen in Fig. 28.1(a). For all staggered tube arrays the streamwise pitch S/D refers to the distance between tubes in odd or even rows. The pitch between the successive rows R is equal to $S/2$. Also, the transverse pitch T is equal to P only for the normal triangle array. The general relationship is

$$\frac{T}{D} = K\frac{P}{D} \tag{28.1}$$

where $K = 1.0, 1.4$, and 1.7 for the normal triangle, rotated square, and parallel triangle tube arrays, respectively, as seen in Fig. 28.1(b–d).

The acceleration and deceleration of the interstitial flow from row to row in both categories depend on the size of the gap $T/D - 1$. The general equation for the ratio of the gap velocity V_g, and the free stream velocity V, is given in terms of the continuity equation as

$$\frac{V_g}{V} = \frac{T/D}{T/D - 1} \tag{28.2}$$

Equation (28.2) is valid *only* for uniform flow across each gap. The mean gap velocity is $1.5V$ for $T/D = 3$, and $6V$ for $T/D = 1.2$. The latter affects the values of the pressure and force coefficients by a factor of $36V^2$. This prompted some researchers to adopt the gap velocity as the reference one, whereas others retained the free stream velocity. Both types of data are compiled in this chapter, and the large difference in magnitude in the pressure and force coefficients is related to the adopted reference velocity.

28.2 Single row of tubes

Interstitial flow in closely spaced multi-tube arrays has provided a paradox.[169] In most heat exchangers all tubes are made identical in shape and size, and all are arranged at the same pitch in the transverse and streamwise directions. However, contrary to a preconceived expectation, the imposed regularity and uniformity do not produce uniform and regular interstitial flow.

28.2.1 *Gap flow jets*

The behaviour of parallel two-dimensional jets was studied by von Bohl (1940J). He produced jets in a closed test section by spanning 2, 3, and 4 sharp-edged flat plates having T/D ratios 3.24, 2.16, and 1.63, respectively. The velocity profiles behind the last grid ratio were very non-uniform, and remained so up to $x/D = 13.8$. Von Bohl concluded that parallel two-dimensional jets are stable for the first two grids, and become unstable for $T/D < 2$.

[169] A paradox is a statement that is seemingly absurd though, perhaps, actually well founded, *The Pocket Oxford Dictionary of Current English.*

A thorough investigation of the stability of two-dimensional flow through a grid of 'triangle' rods was carried out by Corrsin (1944P). Figure 28.3 shows the velocity profiles behind the grid, $T/D = 1.17$. The wide near-wakes are formed where the velocity peaks are displaced away from the gaps. This means that the jets are biased after leaving the gaps, despite the sharp corners and fixed separation points. Corrsin (1944P) concluded 'that the flow was grossly non-uniform and there was little evidence left downstream to show that the flow originated from a regular row of two-dimensional jets'.

The downstream coalescence of jets behind a single row of circular tubes was later examined by Moretti and Cheng (1987J). Figure 28.4 shows the velocity profile development downstream from the single row of nine tubes. The velocity peaks are of the same magnitude, but not spaced equidistantly. At $x/D = 1$, the peaks are not the same, and at $x/D = 1.5$ there are only five peaks. Eventually, at $x/D = 4.5$, only three peaks are seen.

Higuchi (1989J) and Takahashi and Higuchi (1988P) found the same phenomenon behind ribbon parachutes simulated by a concavely curved grid. The

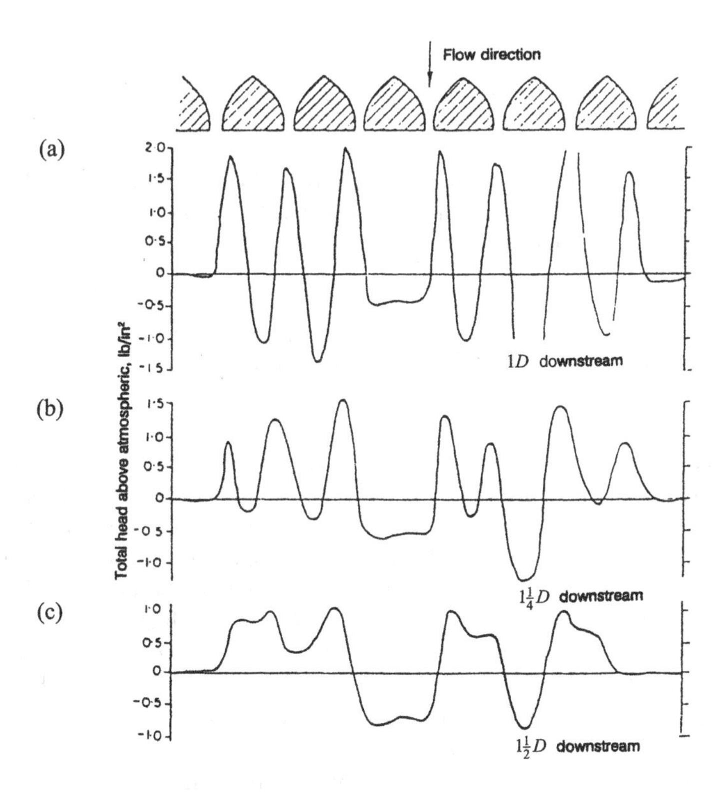

FIG. 28.3. Velocity profiles behind a single row of bluff bodies, $T/D = 1.17$, (a) $1D$, (b) $1.25D$, (c) $1.5D$, Corrsin (1944P)

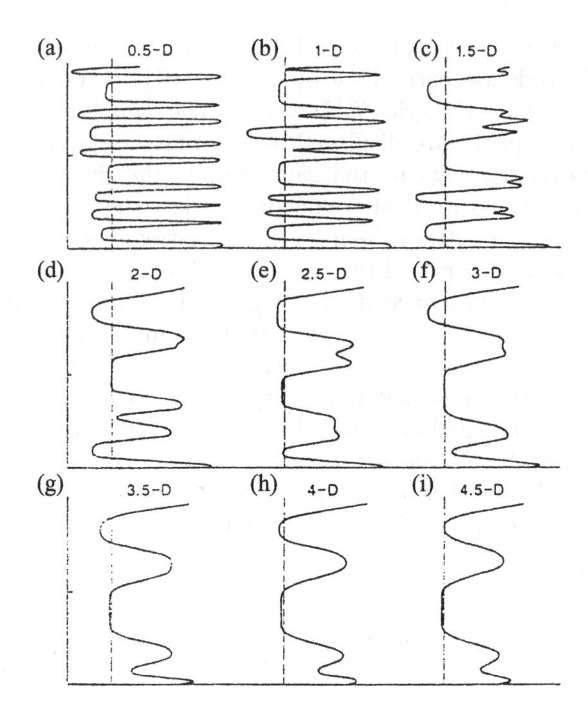

FIG. 28.4. Velocity profiles behind a single row of tubes, $T/D = 1.3$, $Re = 2.5$k, (a) $0.5D$, (b) $1D$, (c) $1.5D$, (d) $2D$, (e) $2.5D$, (f) $3D$, (g) $3.5D$, (h) $4D$, (i) $4.5D$, Moretti and Cheng (1987J)

two-dimensional jets behind any type of grid are intrinsically unstable when $T/D < 2$, and biased so as to merge with the adjacent jet(s).

28.2.2 *Structure of non-uniform flow*

Figure 28.5 shows Bradshaw's (1965J) smoke visualization of flow past a single row of tubes having a transverse pitch ratio $T/D = 1.5$ and $Re = 1.5$k. The jets emerging from the gaps between the tubes are, more often than not, biased. For example, narrow near-wakes are formed by the biased jets behind even tubes and wide near-wakes behind the odd tubes. There is a cell, however, formed by four biased and merged jets behind tubes 6, 7, and 8, which is different from a simple jet merging behind tubes 2 and 4. The biased jets induce a component of the aerodynamic force in the transverse direction. The existence of the lift force on a circular tube in the single row is a paradox, related to the non-uniform gap flows.

Cowdrey (1968Pa) was the first to carry out simultaneous base pressure measurements on all tubes in a single row at $T/D = 1.5$. Figure 28.6 shows an uneven C_{pb} distribution (full line), which intermittently changes over. Two other

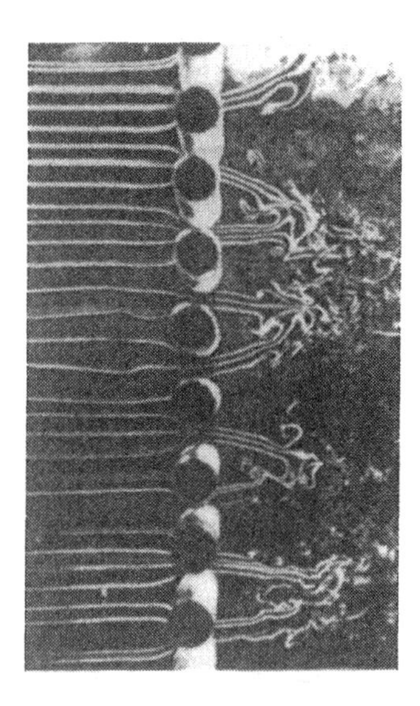

FIG. 28.5. Flow visualization around a single row of tubes, $T/D = 1.5$, $Re = 1.5\text{k}$, Bradshaw (1965J)

distributions are shown with dashed and dash-dot lines. The non-uniform base pressure is caused by biased jets forming wide and narrow near-wakes, as shown in Fig. 28.5. The minimum base pressure corresponds to narrow near-wakes.

Cowdrey (1968Pa) argued that the intermittent change-over of the base pressure is caused by a tendency of the flow to form quasi-stable cells. Such cells consist of four jets, as seen in Fig. 28.6 formed behind three tubes. In order to prove his argument, Cowdrey (1968Pa) displaced every fourth tube $1D$ upstream, and obtained a repeatable, stable, and uniform pressure distribution. He postulated that only the four-tube cell is stable, provided that the total number of tubes in the row is $4n + 1$, where n is any integer number.

Auger (1977P) and Auger and Coutanceau (1979J) systematically varied both the transverse pitch ratio, $1.14 < T/D < 3.8$, and the number of tubes in a row, $3 < n < 21$. Dust visualization showed that jets did not become biased when $T/D > 2$ for $20\text{k} < Re < 100\text{k}$. When $T/D < 2$, any number of tubes from 2 to 10 could be involved in forming a cell. Figure 28.7(a–d) shows four single rows having $(4n + 1)$ tubes. Only for 9 tubes, in Fig. 28.7(a), does the middle tube separate two four-tube cells. When the row consists of 13 tubes, two cells of 5 and 6 tubes may be formed, Fig. 28.7(b), or 5 and 3 tubes, Fig. 28.7(c). The two different flow patterns show a bistable jet flow. For 17 tubes, Fig. 28.7(d) shows a huge 9-tube cell. Cowdrey's (1968Pa) hypothesis of a prevalent four-tube cell is not confirmed for $n > 2$. Similar large cells were later found in water by Moretti and Cheng (1987J). It is evident that the cell pattern strongly depends on the number of tubes in the row.

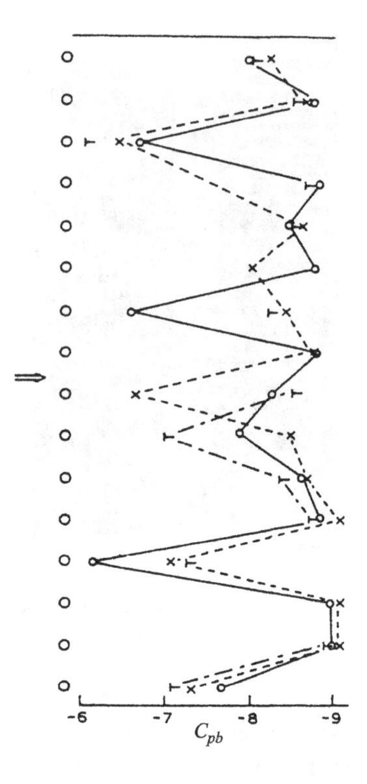

FIG. 28.6. Base pressure on a single row of tubes, T/D = 1.5, *Re* = 20k, Cowdrey (1968Pa)

Several fundamental features of the phenomenon have also been clarified; for example, whether the observed non-uniformity could persist in a laminar flow at low Reynolds numbers. Laminar flows are not encountered in heat exchangers, but it may be helpful to examine whether narrow and wide wakes are present there. Honji (1973Ja, 1973Jb) towed a single row of 5 tubes in a water tank. Figure 27.41(a,b) shows wide and narrow near-wakes arranged symmetrically relative to the middle tube. Figure 27.41(b) shows another possible arrangement of near-wakes where the wide wake is in the middle at *Re* = 70. It is evident that the laminar flow, *per se*, does not prevent non-uniform jet pairing.

Another question which is of practical importance is: could the non-uniformity be affected by disturbances, as, for example, by density fluctuations typical in two-phase flow. Hara (1987P) tested 5 tubes in water having T/D = 1.33 at *Re* = 72k. Figure 28.8(a) shows wide near-wakes behind the middle tube and narrow near-wakes symmetric to it. Hara (1987P) sketched the other four observed non-uniform patterns of the near-wakes: B, C, and D, Fig. 28.8(b). It is evident that neither air bubbles nor a limited number of tubes can suppress non-uniform jet pairing.

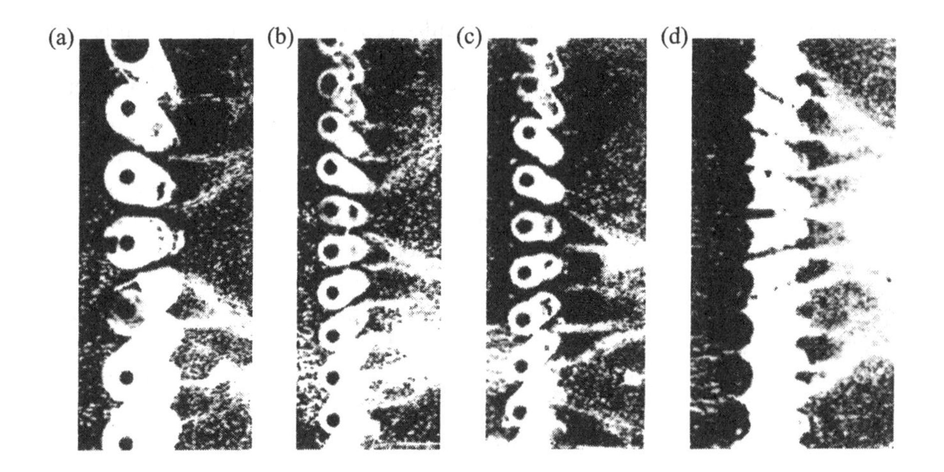

FIG. 28.7. Visualization of jet pairing behind a single row of tubes, (a) $T/D = 1.28$, $Re = 65\text{k}$, $n = 9$, (b) $T/D = 1.32$, $Re = 35\text{k}$, $n = 13$, (c) $T/D = 1.32$, $Re = 61\text{k}$, $n = 13$, (d) $T/D = 1.44$, $Re = 61\text{k}$, $n = 17$, Auger (1977P)

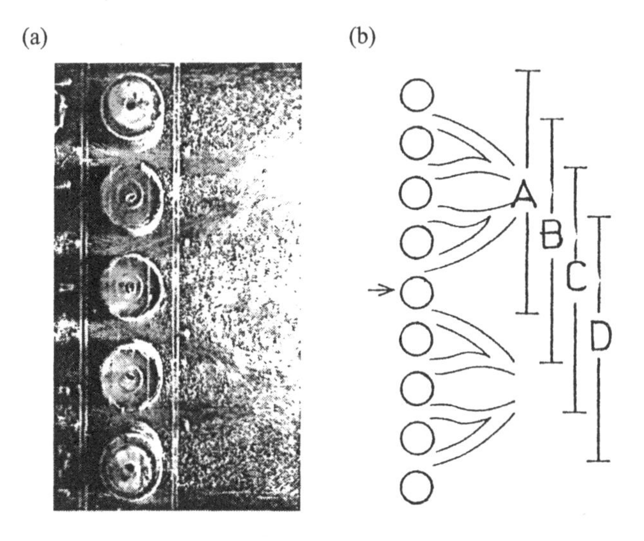

FIG. 28.8. Visualization of jet pairing behind five tubes, (a) one flow regime, (b) other flow regimes, Hara (1987P)

28.2.3 *Mean pressure distribution and forces*

Once it has been established that flow behind a single row of tubes is intrinsically non-uniform, the mean pressure and forces on identical tubes are expected to be different. Zdravkovich and Stonebanks (1990J) carried out simultaneous measurements of the mean base and side pressures on 11 tubes in a single row. The pitch-to-diameter ratio was varied in the range $1.2 < T/D < 2.1$. Figure 28.9(a) shows the base pressure distribution across the row. The non-uniformity is negligible for $T/D = 2.1$ and significant for $T/D = 1.2$. The non-uniformity is irregular, and Cowdrey's (1968Pa) 4-tube cells cannot be seen. Figure 28.9(b) shows the mean side pressure distribution C_{ps} measured simultaneously on 11 tubes at $\theta = 90°$ and $270°$. Only points between the adjacent tubes are connected because they represent the pressure gradient across each gap.

The mean pressure distributions measured simultaneously around four adjacent tubes in a single row by Zdravkovich and Stonebanks (1990J) are integrated, and the mean C_d and C_l evaluated, see Table 28.1. At $T/D = 1.2$, two different runs (1) and (2) reveal entirely different C_l and C_d values. The magnitude and variation in both C_d and C_l gradually decrease with rising T/D.

28.2.4 *Transition to turbulence in the TrBL state*

It is expected that flow past a single row of tubes should be similar to the flow past a single cylinder in the TrBL state. Achenbach (1971Jb) carried out tests in $40\text{k} < Re_g < 10\text{M}$, single row $T/D = 1.4$, 3 tubes per row, $L/D = 3.33$, $Ti = 0.7\%$, and $K/D = 0.11\%$ and 0.45%. He measured a pressure drop Δp through the tube array, which is proportional to the mean drag of tubes. Figure

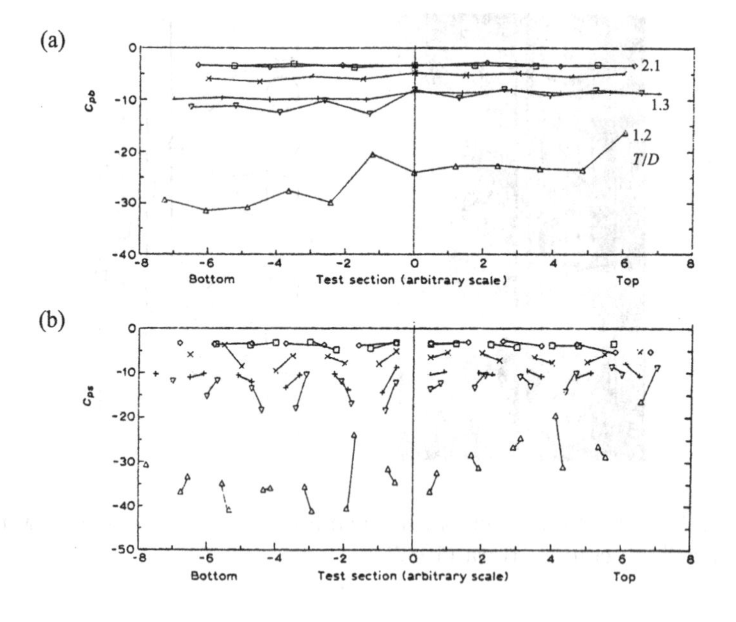

FIG. 28.9. Pressure on a single row of tubes, (a) base pressure, (b) side pressure, Zdravkovich and Stonebanks (1990J)

Table 28.1 *Drag and lift coefficients for a single row*

T/D	1.2(1)		1.2(2)		1.3		1.4		1.5	
Tube	C_d	C_l	C_d	C_l	C_d	C_l	C_d	C_l	C_d	C_l
$4T$	1.13	0.18	1.02	0.16	0.84	−0.01	0.96	−0.08	0.99	0.07
$3T$	1.16	0.13	0.86	−0.19	0.67	−0.05	0.59	0.01	0.75	0.00
$2T$	1.19	0.41	1.06	−0.31	0.75	0.07	0.75	0.03	0.81	0.01
C	0.87	0.00	0.99	−0.24	0.61	0.08	0.63	−0.01	0.37	0.08
Var.	±0.10	±0.21	±0.10	±0.24	±0.15	±0.09	±0.14	±0.08	±0.7	±0.08

28.10 shows remarkable similarities of the $C_{\Delta p}$ curves with a single cylinder C_D in the TrBL state. The effect of surface roughness is to reduce Re_c where $C_{\Delta p\min}$ occurs, and the post-critical regime is marked by a significant increase in $C_{\Delta p}$.

28.2.5 *Metastable states of flow*

Cowdrey (1968Pa) noted that a non-uniform flow behind a single row of tubes could change intermittently, and wrote that 'the commonly used word "*instability*" is thought to be inappropriate in this case'. The phenomenon of transient variation of state, after which another biased flow eventually settles down to a steady and stable pattern for some time, is termed *metastable* by Zdravkovich and Stonebanks (1990J).

Simultaneous measurements of the mean C_p distribution on three adjacent tubes subjected to three metastable states of flow are shown in Fig. 28.11. The three monitored tubes are part of an 11-tube row for $T/D = 1.2$. The tubes are located in the centre, C, and below the centre, $B1$ and $B2$, of the row. States 1 and 3 last longer than state 2, the latter state being more symmetric than the others. States 1 and 3 are asymmetrical and the values of $C_{p\min}$ differ considerably on either side of the tube. The calculated pressure drag C_d and lift C_l coefficients are given in Table 28.2. The cause of the metastable states

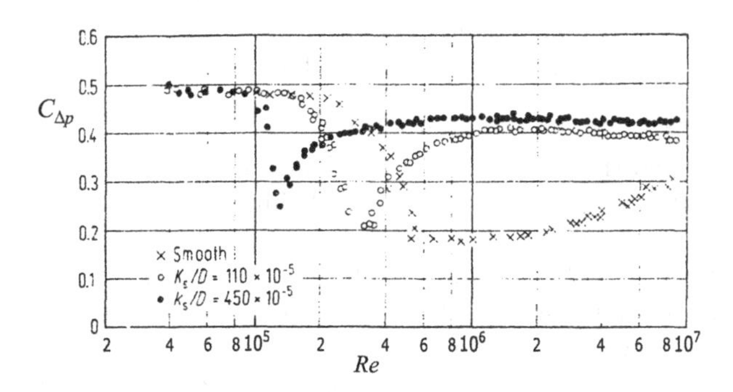

FIG. 28.10. Pressure drop coefficient across a single row, $T/D = 1.4$, Achenbach (1971Jb)

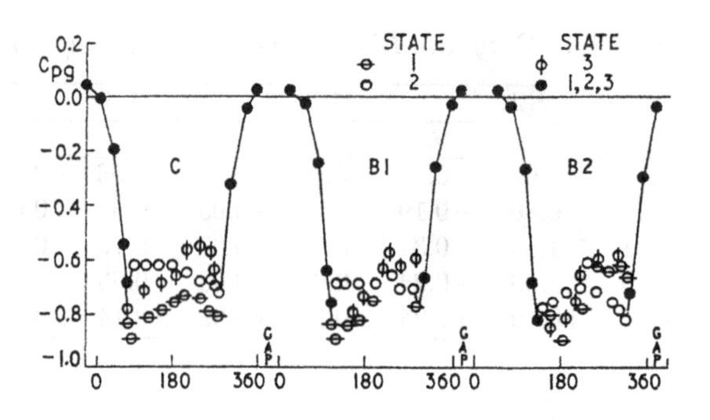

FIG. 28.11. Pressure distribution around three adjacent tubes, three metastable states, Zdravkovich (1992J)

Table 28.2 *Drag and lift coefficients for* $T/D = 1.2$

State	1		2		3	
Tube	C_d	C_l	C_d	C_l	C_d	C_l
$B2$	1.32	−0.18	1.02	0.16	1.32	0.05
$B1$	1.45	−0.43	1.14	0.04	1.35	−0.17
C	1.49	−0.06	1.11	0.01	1.36	−0.17

may be attributed to the non-uniform biased jets and their re-pairing in new combinations.

Zdravkovich (1992J) tried to trigger the change-over of states by introducing various kinds of disturbances. A rapid increase in the free stream velocity from rest produces the most frequent intermittent change-over of metastable states. Small disturbances, such as a person walking in front of the contraction of an open wind tunnel, have no noticeable effect on the change-over. However, the insertion of a pitot tube three diameters upstream of the array and perpendicular to the tube axis almost always triggers the change-over.

28.2.6 *Suppression of metastable states*

It has been inferred that the cause of the change-over of metastable states is an intermittent and irregular re-pairing of jets behind the tubes. The introduction of splitter plates behind the tubes may stabilize biased jets. Figure 28.12(a–c) shows the effect of a $10D$-long splitter plate placed $1D$ downstream of a single tube. It is evident that a single splitter plate placed consecutively behind tubes does not suppress two metastable states.

When all tubes are fitted with splitter plates, almost identical mean pressure distributions are measured, as seen in Fig. 28.13(a) around three adjacent tubes. The base pressure distribution measured simultaneously on all 11 tubes in Fig. 28.13(b) is uniform and stable. However, as soon as a single splitter plate is

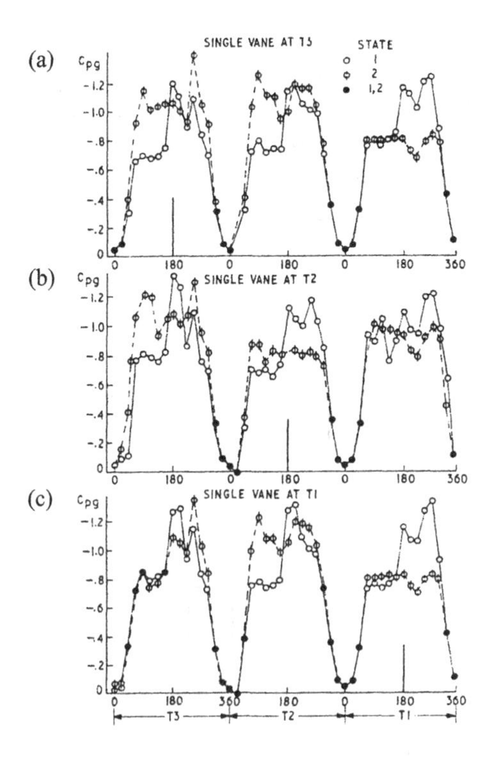

FIG. 28.12. Pressure distribution around three adjacent tubes, splitter plate behind, (a) tube T3, (b) tube T2, (c) tube T1, Zdravkovich (1992J)

withdrawn, Fig. 28.13(c), the non-uniform base pressure reappears locally. It is remarkable that once the metastable state is stabilized, the flow becomes uniform as well.

28.2.7 *Strouhal number*

It has been established experimentally that there are two modes of eddy shedding behind a single row of tubes:

(i) $T/D > 2$ to 2.2; coupled eddy streets across the row have the same St for all tubes;

(ii) $T/D < 2$ to 2.2; biased near-wakes produce *at least* two St values related to the wake width.

Early measurements by Dye (1973P) are reproduced in Fig. 28.14(a,b). St and St_g based on V and V_g, respectively, are plotted in terms of T/D. The single- and two-curve T/D ranges designate two modes of eddy shedding. Subsequent measurements by Chen (1967J), Borges (1969J), Ishigai and Nishikawa (1975J), and Zdravkovich and Stonebanks (1990J) are compiled in Fig. 28.15. Ishigai *et*

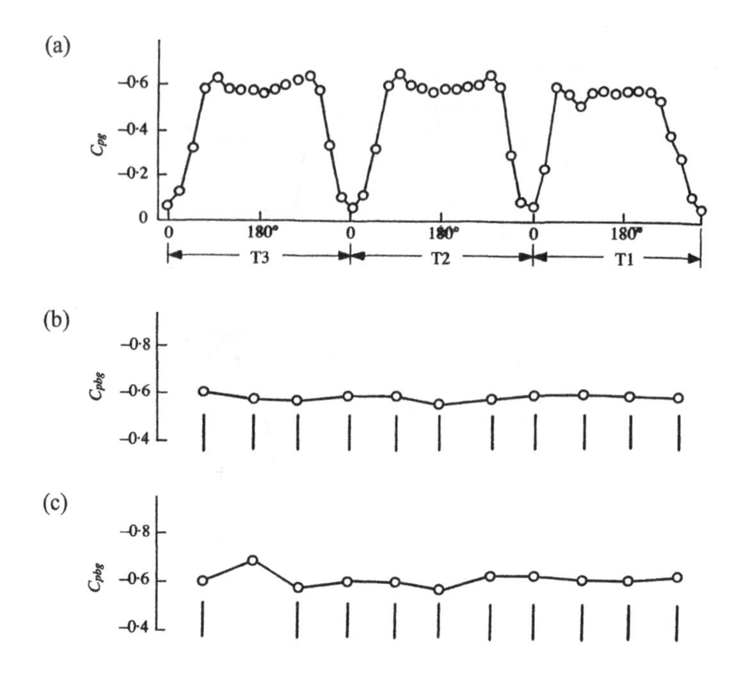

FIG. 28.13. Splitter plates behind all tubes, (a) pressure distribution, (b) base pressure on tubes, (c) one withdrawn, Zdravkovich (1992J)

al. (1975J) found four St curves for $T/D < 2.2$, and that was partly confirmed by Zdravkovich and Stonebanks' (1990J) data. The emergence of four St curves may be attributed to the intermittent change-over of the cell location along the single row. The most frequent narrow wake corresponds to a high St curve (full line) and the wide wake to the $St = 0.15$ curve. However, occasionally a very wide wake occurs, the $St = 0.05$ curve, and a normal wake may also appear $St = 0.22$ (dashed lines). Hence, the designer should be warned that a *range of* St is relevant for *any tube* in the single row due to the metastable states.

28.2.8 *Effect of tube proximity*

When a single tube of diameter D is placed between two parallel walls, the streamlines adjacent to the walls must be parallel to both walls. The constriction imposed by the walls increases as the breadth B between them decreases. The ratio D/B is the blockage ratio, and was discussed in Chapter 23 for a single cylinder. The tube proximity effect is similar to the blockage effect.

Figure 28.16(a) is an illustration of the distortion of the mean pressure distribution by the tube proximity. Turner (1978P) placed an odd number of tubes arranged in a single row in a test section, and measured the mean pressure distribution around the centre tube. For a single tube in the test section, the blockage is 0.07, and by inserting more and more tubes, the 'blockage' ratio increases as

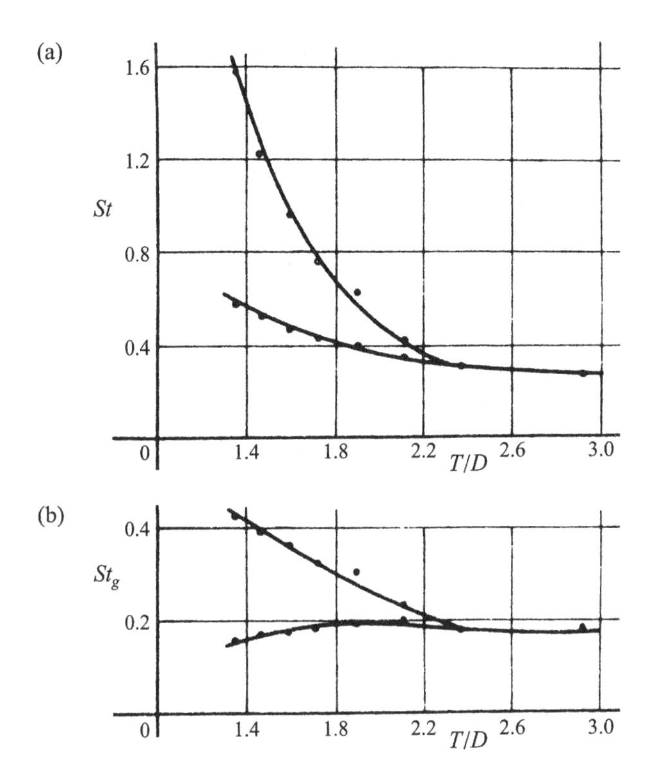

FIG. 28.14. Strouhal number in terms of transverse pitch ratio, (a) based on free stream velocity, (b) based on gap velocity, Dye (1973P)

shown in Table 28.3.

Figure 28.16(a) shows the effect of increasing blockage due to tube proximity on the C_p distribution:

(i) the favourable pressure gradient increases;

(ii) the minimum pressure is displaced to 90°;

(iii) the base pressure is significantly lowered.

An attempt to counteract the lowering of C_{pb} is to replace the free stream

Table 28.3 *Relation between pitch and blockage*

Tube	1	3	5	7	9	–
D/B	0.07	0.22	0.37	0.52	0.67	(0.84)
T/D	–	4.5	2.7	1.9	1.5	(1.2)

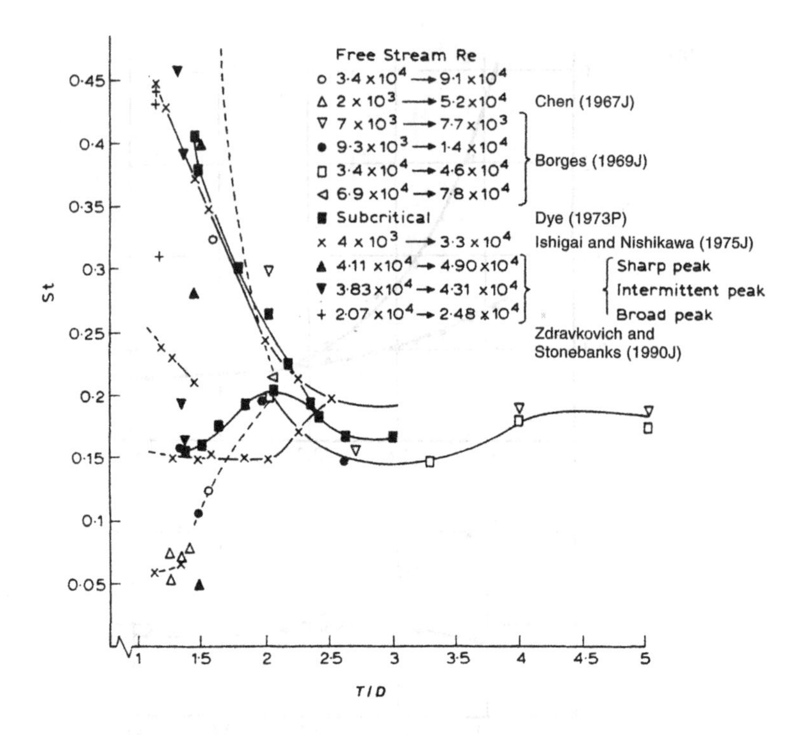

FIG. 28.15. Compilation of Strouhal number in terms of transverse ratio, Zdravkovich and Stonebanks (1990J)

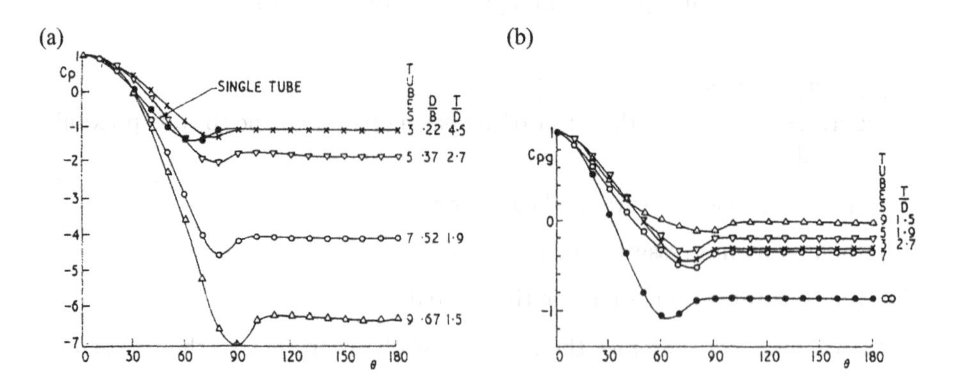

FIG. 28.16. Blockage due to proximity of tubes; pressure distribution based on (a) free stream velocity, (b) gap velocity, Turner (1978P)

velocity V by the gap velocity V_g, Fig. 28.16(b). The curves are closer together, but *do not* collapse onto a single curve. They are above the C_p curve for a single tube.

Another modification of the pressure coefficient by some researchers is

$$C^*_{pg} = C_{pg} - 1 \quad \text{or} \quad C^*_{pg} = 2(p - p_0)/\rho V_g^2 \tag{28.3}$$

where p_0 is the stagnation pressure. Note that for $p_0 = p$, $C^*_{pg} = 0$.

28.3 In-line tube arrays

The interstitial flow in in-line tube arrays is subjected to the combined effects of wake interference and proximity of neighbouring tubes. The former depends on the streamwise pitch ratio S/D and the latter on the transverse ratio T/D.

The effect of the tube proximity for a $P/D = 2$ square tube array at $Re =$ 67k on the interstitial flow can be appreciated from Fig. 28.17. The axis of symmetry along the flow direction (dash-dot line) represents an imaginary wall along which the local streamline is straight. Turner and Eastop (1979J) carried out measurements with an inclined hot wire so that both the local velocity and its direction are determined. The velocity vectors are shown by dashed lines, the separated regions by dotted lines, and the pressure distribution on tubes by open circles and a full line. Note the different flow patterns in the two gaps.

28.3.1 *Mean pressure distribution*

There are two subgroups of in-line tube arrays:

(i) square, $P/D = S/D = T/D$, equilateral;

(ii) rectangular, $S/D \gtrless T/D$, non-equilateral.

The mean C_p for small, medium, and large spacing ratios are described as follows:

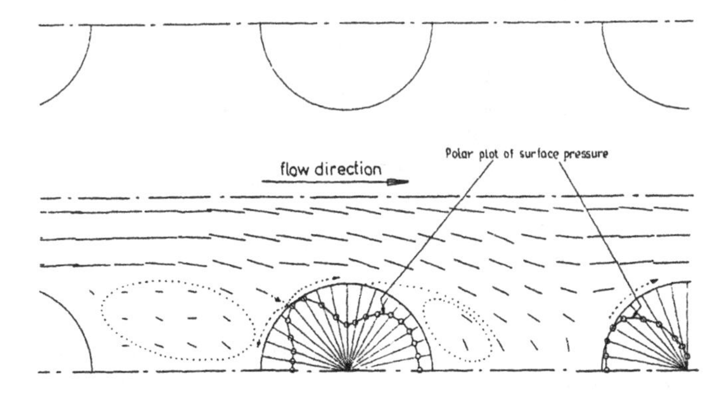

FIG. 28.17. Mean velocity vector in a $T/D = L/D = 2$ array three rows deep at $Re = 57$k, Turner and Eastop (1979J)

(i) $P/D = 1.2$, a closely packed square array, seven rows deep, having $\frac{1}{2} + 3 + \frac{1}{2}$, tubes per row. Figure 28.18(a) shows the mean C_p^* distributions measured by Aiba *et al.* (1982Ja) on both sides of the tube because of the asymmetric pressure distribution. The tube in the first row has $C_{p\min}^*$ at $\theta = \pm 90°$, and a non-uniform base pressure. The tube in the second row exhibits reattachment peaks considerably different in magnitude as caused by different impinging velocities on two sides of the tube. All other tubes show similar C_p^* distributions with uneven reattachment peaks.

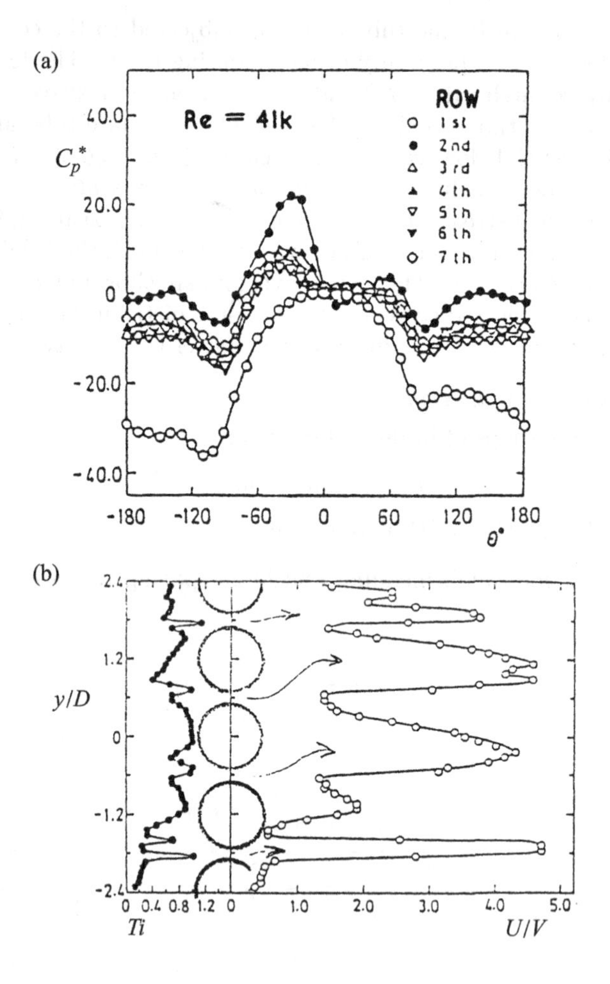

FIG. 28.18. Square array 1.2 × 1.2 at *Re* = 41k, (a) pressure distribution, (b) velocity and turbulence profiles between first and second rows, Aiba *et al.* (1982Ja)

The structure of the interstitial flow is revealed by the hot-wire traverse between the rows. Figure 28.18(b) shows the velocity and turbulence intensity profiles between the first and second rows ($x/D = 0.5$). The velocity peaks are displaced significantly from the gap location (arrows), and the velocity profile bears no resemblance to the regularly pitched tubes.

(ii) $P/D = 1.6$, a moderately packed array, seven rows deep, having $\frac{1}{2} + 3 + \frac{1}{2}$ tubes per row. Figure 28.19(a) shows the measured mean C_p^* around the

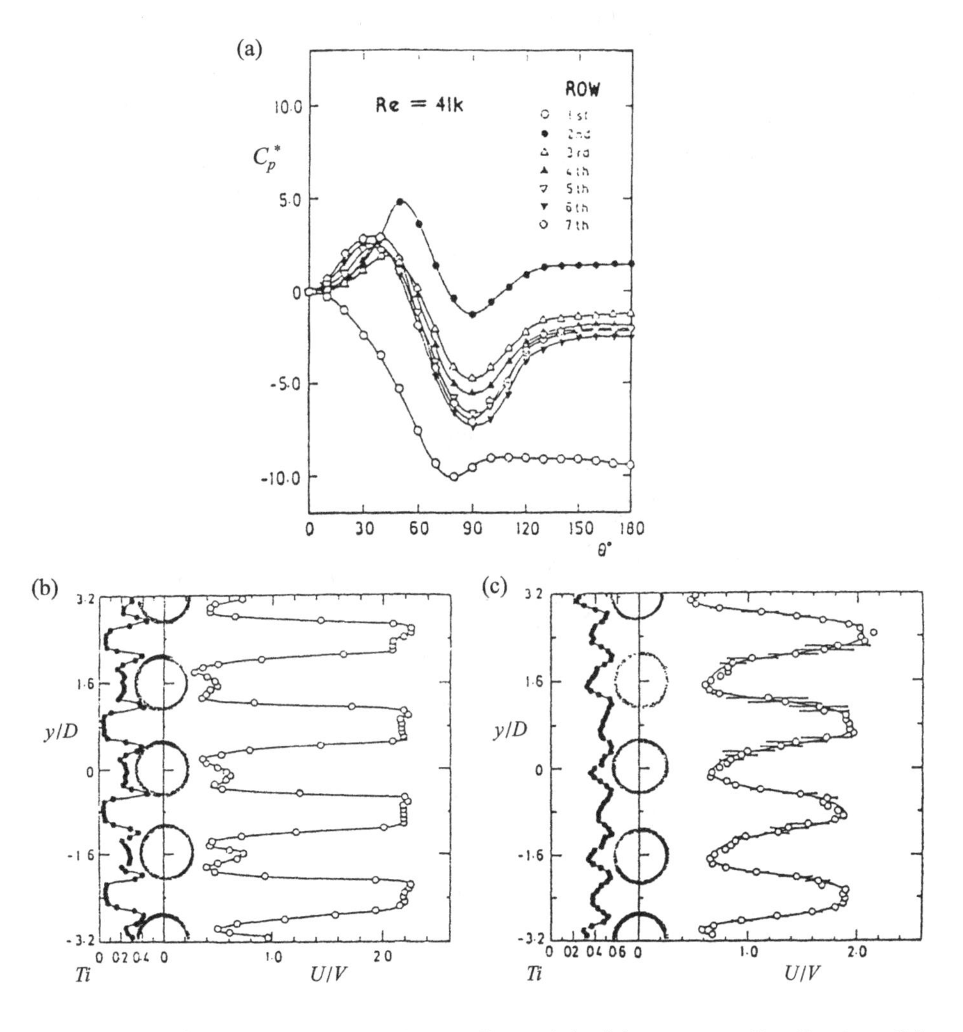

FIG. 28.19. Square array 1.6 × 1.6 at Re = 41k, (a) pressure distribution, (b) velocity and turbulence profiles between first and second rows, (c) between fourth and fifth rows, Aiba *et al.* (1982Ja)

monitored tube situated consecutively in all seven rows. The C_p^* distribution for the tube in the first row is similar to that for a single tube subjected to a high blockage. The tube in the second row shows a reattachment peak at $\theta = 50°$, and all other C_p^* distributions are close to each other.
The hot-wire traverse between the first and second rows, and the fourth and fifth rows, are shown in Fig. 28.19(b,c), respectively. The high velocity peaks match the location of the gaps. The uniformity of the velocity profiles is improved in comparison with the closely spaced in-line array in Fig. 28.18(b). The velocity and turbulence intensity curves mirror each other in such a way that the low turbulence corresponds to the high velocity and vice versa.

(iii) $P/D = 2$, a widely spaced square array, nine rows deep with 5 tubes per row. Batham (1973P) measured mean C_{pg} around a tube placed in odd-numbered rows. Figure 28.20(a) shows C_{pg} for a low free stream turbulence, $Ti < 0.5\%$. The curve for the tube in the first row has $C_{pg\min}$ at 80° and a flat base pressure distribution. All other curves have $C_{pg\min}$ at 90°, $\theta_s = 110°$, and flat C_{pbg} in the range $140° < \theta < 180°$. The reattachment peaks are wide.

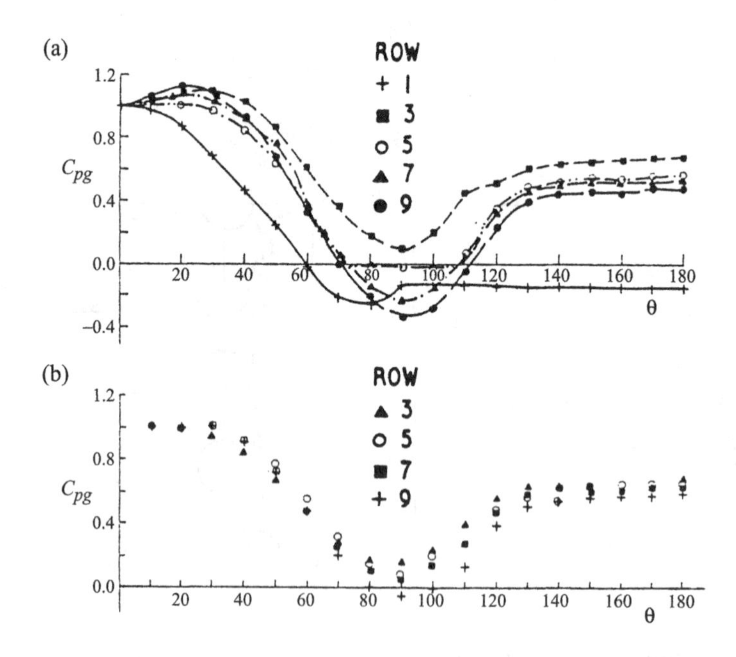

FIG. 28.20. Pressure distribution on tubes in a 2 × 2 array, (a) $Ti < 0.5\%$, (b) $Ti > 20\%$, Batham (1973P)

Figure 28.20(b) shows the effect of grid turbulence, $Ti = 20\%$. All odd-numbered C_{pg} curves collapse into a single curve except for the last row. This result confirms the argument that the different degree of transition to turbulence in consecutive rows is the cause of the dissimilarity of interstitial flows. The high turbulence in the free stream promotes an early transition to turbulence in the interstitial flow. The fully turbulent separated shear layers produce similar interstitial flows, row after row, and the result is an almost identical pressure distribution. The tube in the last row is different from the rest since there are no tubes behind it.

28.3.2 *Fluctuating forces*

Batham (1973P) also measured the fluctuating C'_p for $P/D = 2$ in the seventh row for $Ti = 0.5\%$ and 20%. For both Ti values, C'_p distributions are found to be similar. The fluctuating forces are $C'_l = 0.18$ and 0.11, and $C'_d = 0.11$ and 0.09 for the two Ti values, respectively. Batham (1973P) noted that the high Ti had little effect on a closely spaced square array $P/D = 1.25$.

Chen and Jendrzejczyk (1987J) carried out detailed measurements of fluctuating lift and drag forces on a tube in a square $P/D = 1.75$ array, seven rows deep, and 7 tubes per row. Figure 28.21(a,b) shows the variation in the fluctuating lift and drag coefficients in the range 12k < Re < 350k and for $Ti = 0.5\%$.

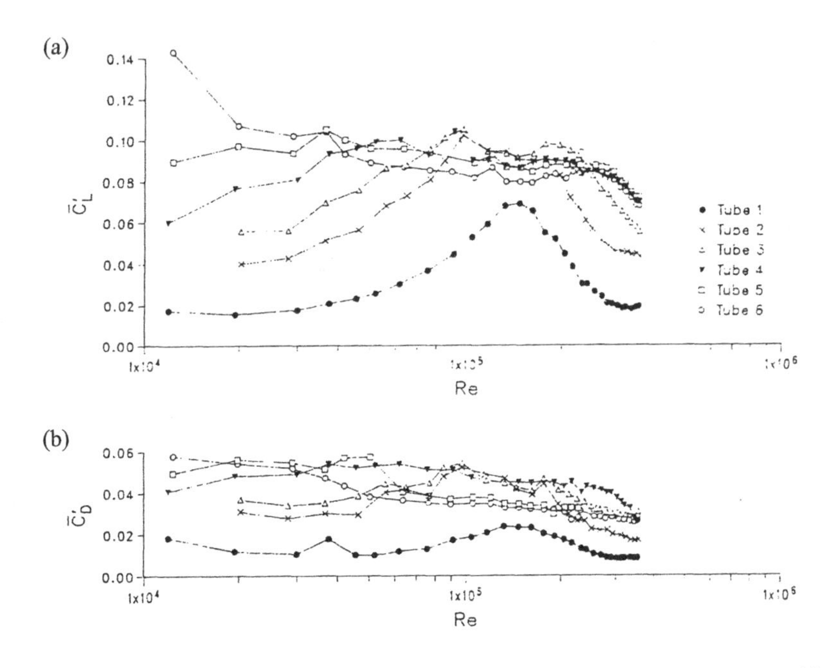

FIG. 28.21. Fluctuating forces in terms of Re for $P/D = 1.75$, $Ti = 0.5\%$, (a) lift, (b) drag, Chen and Jendrzejczyk (1987J)

The tube in the first row has the lowest C'_d and C'_l in comparison with the others.

Figure 28.22(a,b) shows the effect of a grid turbulence of $Ti = 10\%$ on the fluctuating lift and drag for the same array. All curves collapse into one, except for the first two rows, and the overall magnitude of C'_l is slightly lower than without turbulence. The C'_d coefficient is almost the same for all tubes from the third row to the last.

28.3.3 *Effect of surface roughness*

Achenbach (1971Jb) studied the effect of two relative surface roughness ratios $K/D = 0.11\%$ and 0.45% on the $T/D = 2$ and $S/D = 1.4$ in-line tube array across a wide range $40\text{k} < Re_g < 10\text{M}$ for 3 tubes per row and five rows deep. The measurements of skin friction and mean pressure distribution are shown in Fig. 28.23(a,b). The reattachment points for the tube in the fourth row moved from $\theta_r = 35°$ for smooth tubes to $50°$ for rough tubes. The separation points at $\theta_s = 120°$ moved upstream to $\theta_s = 105°$ and $100°$ for two roughnesses, respectively.

As a result of both, the size of the gap inflow is reduced for the rough tubes. Figure 28.24 shows the variation of the pressure drop coefficient $C_{\Delta p}$ in terms of Re_g. It is evident that in the post-critical range the smooth array produces greater $C_{\Delta p}$ than the two rough arrays. This paradox is clarified by Achenbach

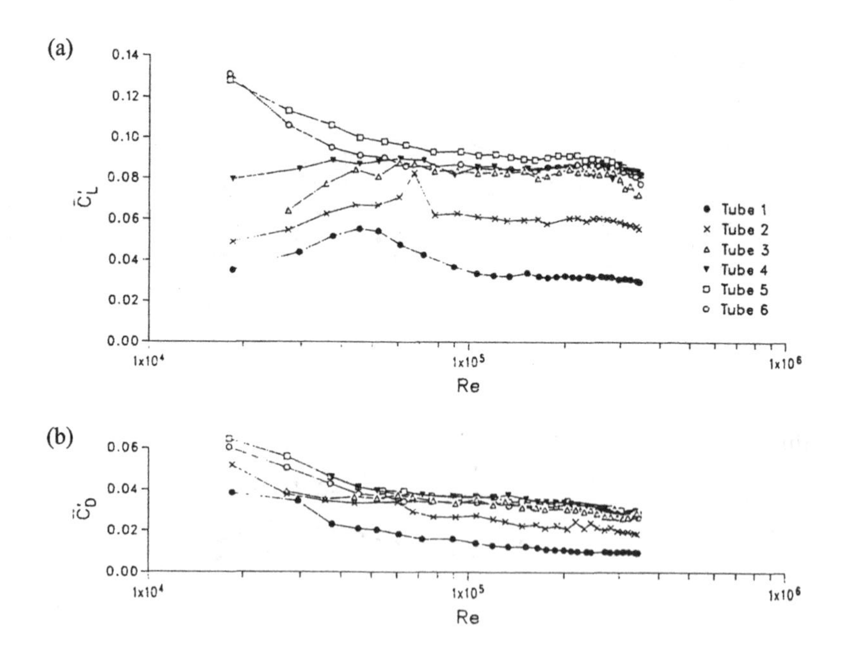

FIG. 28.22. Fluctuating forces in terms of *Re* for $P/D = 1.75$, $Ti = 10\%$, (a) lift, (b) drag, Chen and Jendrzejczyk (1987J)

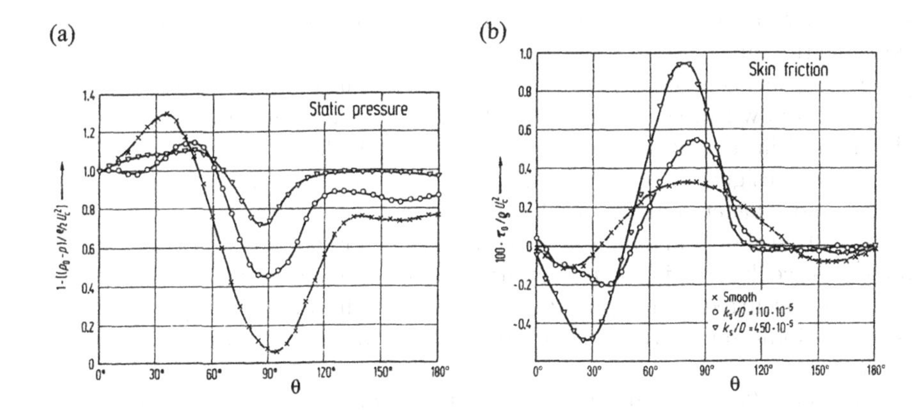

FIG. 28.23. Tube in the fourth row of the $T/D = 2$ and $S/D = 1.4$ array, (a) static pressure, (b) skin friction, Achenbach (1971Jb)

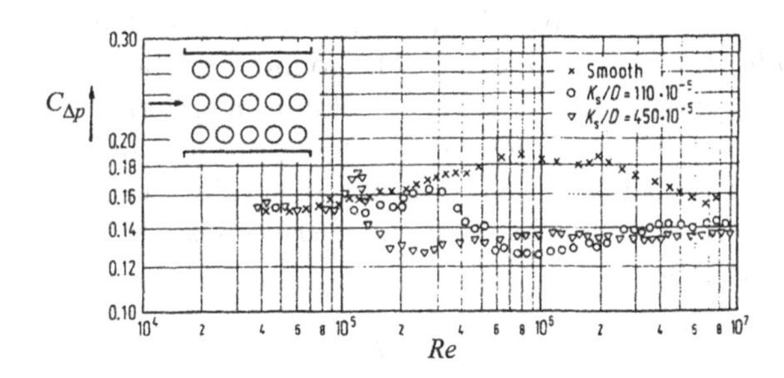

FIG. 28.24. Pressure drop coefficient in terms of *Re* for $T/D = 2$, $S/D = 1.4$, Achenbach (1971Jb)

(1971Jb): 'as the gap size and shape are increased it renders less distortion of the alley flow'.

28.3.4 *Acoustic resonance; historical background*

Excessive noise generation coupled with large pressure losses, shell-wall buckling, and tube failures have accompanied the development of larger and more compact heat exchangers. The associated high-gap velocities V_g increased the kinetic energy of the interstitial flow $\propto V_g^2$, the fluctuating forces exerted on tubes, and shell walls. It was hypothesized in the 1950s that the excessive noise appeared as a result of a resonance between the acoustic frequency inside the heat exchanger shell and the pressure fluctuation generated by the tube array. The latter was a priori assumed to be due to eddy shedding in tube arrays.

A further increase in size of heat exchangers in the 1960s and 1970s led to the relative decrease in the natural frequency of tubes, and to an increase in detrimental tube vibration, fretting, fatigue, and failure. It was argued that a resonance between the natural tube frequency and periodic fluctuations of the interstitial flow caused this type of vibration. Again, a priori the fluctuations in the interstitial flow were associated with eddy shedding. It seemed at the time that what designers needed was detailed charts of the Strouhal number for all in-line and staggered arrays used.

It will be shown in subsequent sections that for some tube arrays eddy shedding was weak, erratic, or non-existent, but severe noise and/or flow induced vibration occurred. The opposite was found for in-line arrays where some existing periodic fluctuations did excite neither the acoustic resonance nor tube vibration.

Weaver (1993R) summarized the state of affairs in the following way: 'Few topics have been as controversial as so-called vortex (eddy) shedding in tube arrays.'

28.3.5 *Speed of sound in tube arrays*

The speed of sound in compressible gases has been defined as the speed of propagation of *small pressure disturbances.* The sound wave propagates by the alternate compression and expansion of a gas column in spherical waves.

For an ideal gas, the thermodynamic equation of state is applicable. This means that the gas constant R and the ratio of specific heats $\gamma = c_p/c_v$ are constant. The additional assumption that the thermodynamic process is reversible, i.e. isentropic and adiabatic, yields

$$a = (\gamma RT)^{1/2} \tag{28.4}$$

where a is the local speed of sound and T is the absolute temperature in degrees Kelvin. The simple eqn (28.4) cannot be used for steam because the latter is not an ideal gas, i.e. γ and $R = f(T)$. The temperature varies along the heat exchanger and so does the local speed of sound.

Another limitation of eqn (28.4) is that it assumes a free space filled with the gas only. As the sound wave propagates through the tube array, it is locally scattered by the tubes. Parker (1978J) proposed a modified expression for the effective speed of sound a_{eff} propagating perpendicularly to the tube axes

$$a_{\text{eff}} = a(1+\sigma)^{-1/2} \tag{28.5}$$

where a is the speed of sound in the free space, and σ is the fraction of space occupied by solid tubes. Equation (28.5) is based on physical intuition and agrees with experiments. Burton (1980J) modified σ by multiplying it with the added mass coefficient m_a for water.

Blevins (1986J) derived a rigorous theoretical model starting with the exact differential equations of continuity and momentum for an inviscid fluid. The tubes were replaced by the forces they impose on the gas. The result was an equation of motion that represented sound propagation through a regular array of stationary solid tubes. The final equation was reduced to Parker's (1978J) empirical eqn (28.4).

As the solidity of the tube array σ increased, the effective speed of sound a_{eff} was reduced. Note that a_{eff} was independent of the tube pattern, i.e. the same for in-line and staggered tube arrays. It became weakly dependent on tube patterns when the added mass m_a was taken into account. Figure 28.25 shows a comparison of theory and experiment. The open symbols designate the triangle, square, and rotated square arrays. For example, the triangle pitch ratio $P/D = 1.2$ gives $\sigma = 0.63$, while $P/D = 4$ leads to $\sigma = 0.057$. The agreement between the theory and experiment is very good.

28.3.6 *Acoustic excitation and suppression*

The acoustic resonance may take place when the travel time of the sound wave between walls coincides with the time period of the pressure disturbance. Baird (1954J) wrote:

> The minute and randomly occurring instability forces arising from flow over parallel heat exchanger tubes need an acoustically resonant chamber to synchronize and amplify them. The vibration of the gas column then arises from the standing wave pressure variation acting directly upon the wall shells.

The acoustic resonance frequency across the shell is proportional to the effective speed of sound a_{eff}, and inversely proportional to the width B between

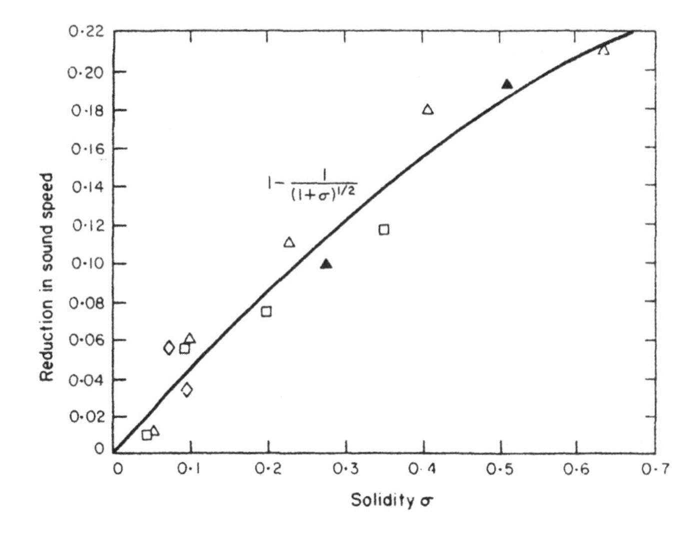

FIG. 28.25. Reduction in sound speed in terms of tube array solidity; comparison of theory and experiment, Blevins (1986J)

the side walls

$$f_a = \frac{na_{\text{eff}}}{2B} \tag{28.6}$$

where n is an integer number describing the frequency mode (higher harmonics). The lowest fundamental mode is $n = 1$ when half of the sound wavelength is equal to the width B as shown in Fig. 28.26(a,b).

The acoustic vibration of the gas column normal to the free stream and tubes axes has been referred to as a *standing pressure wave* by Baird (1954J), and elaborated further by Putnam (1959J, 1964R), Cohan and Dean (1965R), Barrington (1973R), etc. Figure 28.26(a–f) shows four possible modes of standing waves in a duct with parallel walls. For the fundamental mode $n = 1$, one-half wavelength, the maximum wave displacement of the gas column is in the middle where a zero node point takes place, Fig. 28.26(b). The transverse displacement of the gas column is prevented at both side walls where the maximum pressure occurs, Fig. 28.26(b). The first harmonic, $n = 2$, is shown in Fig. 28.26(c,d) having the maximum displacement at $\frac{1}{4}B$ and $\frac{3}{4}B$, and zero at $\frac{1}{2}B$. Similar $1\frac{1}{2}$ and 2 wave displacement modes for $n = 3$ and 4 are depicted in Fig. 28.26(e,f), respectively. In standing waves, the location of maximum particle displacement is 90° out of phase with the location of maximum pressure.

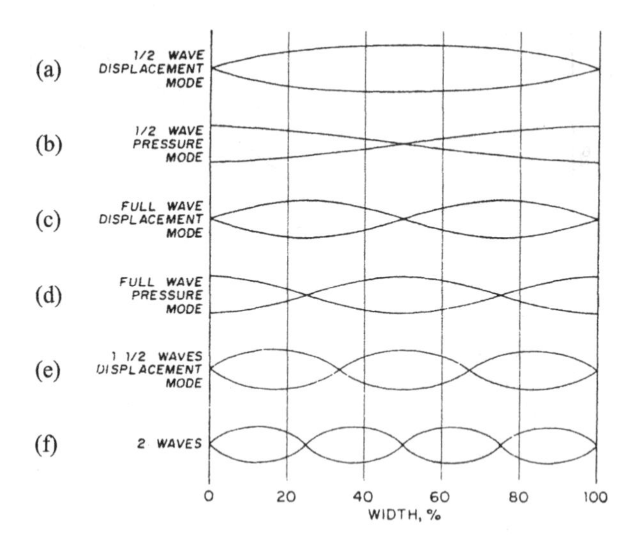

FIG. 28.26. Acoustic standing wave modes, (a,c,e,f) displacement, (b,d) pressure, Cohan and Dean (1965R)

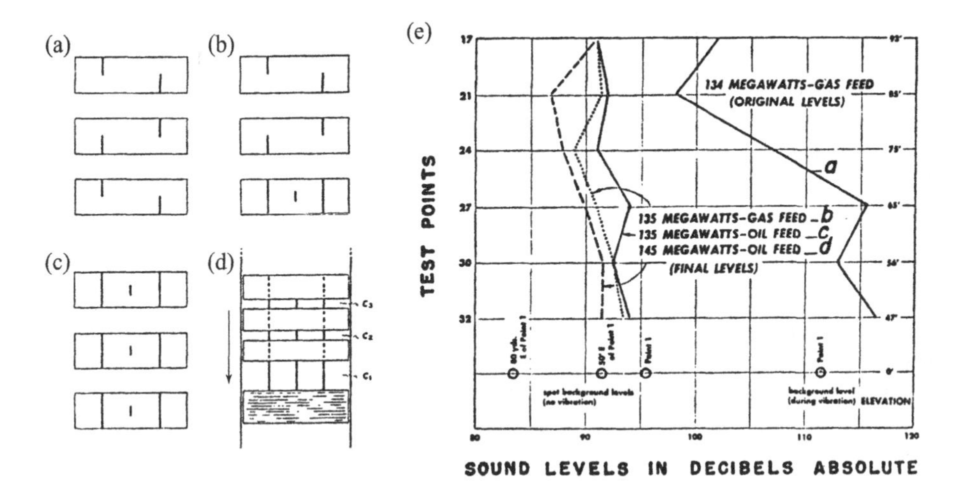

FIG. 28.27. Baffle installation, (a) original, (b) partly extended, (c) fully extended, (d) final, (e) test points elevation vs sound level in dB, Baird (1954J)

Baird (1954J) described a cure for the severe acoustic vibration in the Etiwanda boilers[170] designed for 145 MW. The measurement of pressure fluctuation and intensity of sound showed the presence of a standing wave $n = 2$ in Fig. 28.26(c,d). Baird (1954J) inserted short flat plates, called baffles, at the maximum gas column displacement at $\frac{1}{4}B$ and $\frac{3}{4}B$ as sketched in Fig. 28.27(a). This produced a noise reduction as seen in Fig. 28.27(e). The effect of varying baffle length is shown in Fig. 28.27(b–d) as seen in Fig. 28.27(e). Note that baffles at $\frac{1}{2}B$ are introduced to prevent the acoustic mode between the baffles.

Zdravkovich and Nuttall (1974J) showed that the partial or full withdrawal of tubes positioned at the particular location fully suppresses acoustic standing pressure wave.

28.3.7 *Owen's buffeting model*

An entirely new concept of flow instability was proposed by Owen (1965J), who abandoned the eddy shedding hypothesis as a suspected cause of interstitial flow induced vibration in heat exchangers. He succinctly wrote:

> Deep within a bank of tubes, the cumulative growth of random irregularities in the labyrinth-like high *Re* flow must lead to a state of almost complete incoherence on which it is difficult to imagine any superimposed regular pattern to be discernible.

[170]The Etiwanda Steam Power Plant is in southern California and was owned by the Southern California Edison Utility Company.

His main physical argument was that sufficiently far from the entry to the tube array, the rates of production and dissipation of turbulent energy must become equal. He considered conditions within a 'box' of gas shown in Fig. 28.28 between rows of tubes arranged at a transverse pitch T and a streamwise pitch S. The turbulent energy is produced in the front of the box, and is convected into it at an approximate rate

$$C_{\Delta p}\tfrac{1}{2}\rho V^3 T \qquad (28.7)$$

where $C_{\Delta p}$ is the coefficient of pressure drop along the tube array, $C_{\Delta p}\frac{1}{2}\rho V^2$ is the actual pressure drop, and VT is the volumetric flow rate per unit width of the array.

The kinetic energy of isotropic turbulence has three equal components in three dimensions $3q^2/2$, where q is the turbulent velocity component, and it resides in eddies of scale l. Its rate of dissipation is approximately $3q^3/2l$. Hence, the rate at which the turbulence in the box suffers decay is $3\rho q^3 TS/2l$. Following on from the assumption of the energy equilibrium

$$C_{\Delta p}\tfrac{1}{2}\rho V^3 = \tfrac{3}{2}\rho q^3 S/l \qquad (28.8)$$

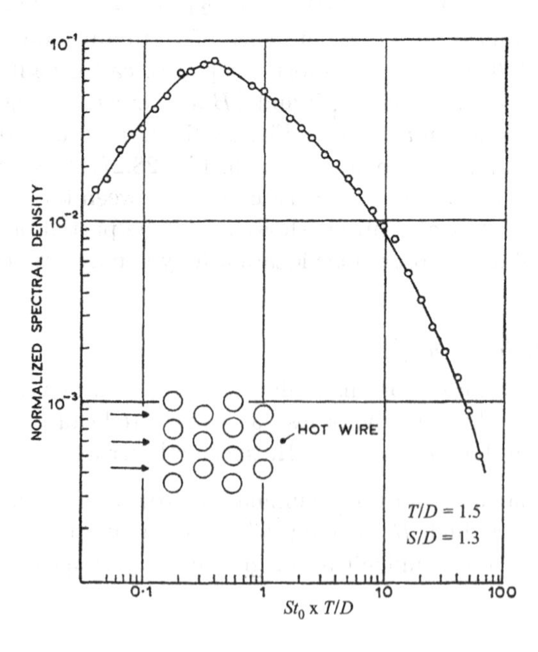

FIG. 28.28. Normalized buffeting power spectral density in terms of non-dimensional frequency in staggered tube array, Owen (1965J)

Adopting a suggestion of Wieghardt (1953J), $C_{\Delta p}$ can be related to the dimensions of the tube array by assuming, for each tube C_{Dg}, that it is invariant with respect to the transverse tube spacing T

$$C_{\Delta p} = C_{Dg}(1 - T/D)^{-1/2} \tag{28.9}$$

If the eddies of scale l are convected through the box in Fig. 28.28 with mean velocity V, then the frequency f_t with which they cross the plane containing the downstream row of tubes is f_t/l. Thus using eqns (28.5) and (28.6)

$$\frac{f_t S}{V_g}\frac{T}{D} = \frac{1}{3}C_{Dg}\left(\frac{V_g}{q}\right)^3\left(1 - \frac{D}{T}\right)^2 \tag{28.10}$$

where $f_t S/V_g$ is Owen's Strouhal number, St_0.

The basic restrictions of the model are stated by Owen (1965J):

> Both the mean flow and turbulence are assumed uniform in space between the rows. In fact, turbulence is not generated impulsively at a row of tubes nor does it thereafter exhibit a uniform energy density until the next row is reached, on the contrary, it arises partly as a result of the redistribution of mean velocity between successive rows, and neither q nor l is constant in that region.

It follows that eqn (28.10) can be rewritten as

$$St_0 = K\frac{D}{T}\left(1 - \frac{D}{T}\right)^2 \tag{28.11}$$

where St_0 is Owen's Strouhal number based on streamwise pitch S, the gap velocity V_g, and $K = 3.05$ is a semi-empirical constant being the same for in-line and staggered arrays.

A typical one-dimensional turbulent energy spectrum found in a staggered $T/D = 1.5$ and $S/D = 1.3$ tube array is shown in Fig. 28.28. Note that the normalized spectral density is plotted against $St_0, T/D$, and that the peak is at $St_0 = 0.24$.

Owen (1965J) compiled data by Grotz and Arnold (1956P) and Hill and Armstrong (1962J), who noted large acoustic excitation at the frequency f_a consistent with the transverse acoustic mode. He added the Imperial College measurements of the interstitial flow turbulence frequency spectrum. Figure 28.29 shows $St_0 T/D$ versus $(1 - D/T)^2$ having an unexpectedly small scatter in view of the generous limits of accuracy placed on f_t.

Finally, Owen (1965J) suggested a simple practical rule for the use of his theory:

> The dominant frequency of vibrations in a tube array, for which the ratio of the diameter to the transverse spacing ratio lies between 0.2 and 0.6 is equal to the interstitial gas velocity divided by twice the distance between the successive rows.

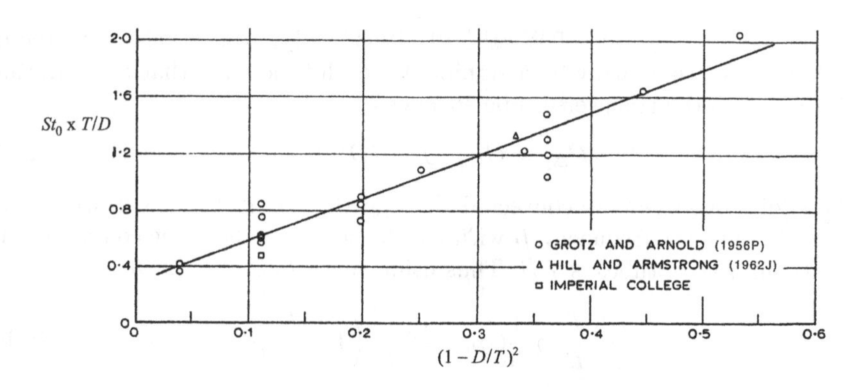

FIG. 28.29. Comparison of theory and experiment for St_0T/D vs $(1 - D/T)^2$, Owen (1965J)

28.3.8 *Interstitial flow, transition eddies*

It has been discussed in the previous section that Owen (1965J) disputed the possibility of eddy formation inside closely packed tube arrays. He argued that there was a lack of space and high level of turbulence in the interstitial flow. Subsequently, Chen (1968J), Borges (1969J), Funakawa and Umakoshi (1970J), Ishigai and Nishikawa (1975J), Weaver and Grover (1978J), Weaver and Abd-Rabo (1985J), etc. found a distinct periodicity peak in frequency spectra behind the front rows for closely packed tube arrays at relatively low *Re*. Flow visualization revealed weak and small eddies in the interstitial space. The term ***vorticity shedding*** was used to distinguish the weak small eddies from the Kármán–Bénard eddy shedding behind a single cylinder.

It was discussed in Vol. 1, Chapter 5, that laminar shear layers separated from a cylinder develop transition waves for $Re > 350$, followed by the formation of transition eddies for $Re > 1k$ in the TrSL1 and TrSL2 regimes, respectively. The latter, also known as Gerrard–Bloor eddies, appear as a chain of minute laminar eddies along the shear layer preceding transition to turbulence. Although symmetric at 180° out-of-phase on two sides of the cylinder, they roll-up into alternate turbulent eddies farther along the wakes in an asymmetric manner. The transition Gerrard–Bloor eddies are an order of magnitude less regular and weaker than the Kármán–Bénard eddies. The frequency peak in power spectra is small and wide-band in comparison with the alternate eddies. Being small and irregular, transition Gerrard–Bloor eddies are disturbance-sensitive over a wide range of frequencies. This means that either small *Ti* or external sound in the appropriate frequency range can considerably affect the transition eddies.

A considerable modification of transition eddies takes place in the interstitial flow for in-line tube arrays. Two distinct interference effects are:

(i) The transition eddies behind two side-by-side tubes develop along the neighbouring shear layers ($T/D - 1$) apart. They are coupled 180° out-of-phase, i.e. they develop symmetrically relative to the tube gap axis.

(ii) The presence of tubes in the subsequent row at S/D terminates the development of transition eddies, and prevents their roll-up. Instead, they are carried away between the tubes by the jet-like interstitial flow, causing the frequency peak in power spectra.

The extent of the interference effects strongly depends not only on T/D and S/D but also on Re and Ti, as well as on the number of rows and tubes per row. As T/D and S/D increase, the symmetric coupling of transition eddies decreases, and their development and strength are enhanced. The formation and development of transition eddies depend on Re, and can be triggered earlier by Ti.

28.3.9 *Instability of jet shear layers*

The main feature of all in-line tube arrays is a straight,[171] jet-like interstitial flow along 'alleys' between the tube columns, Fig. 28.1. For small spacings, $P/D < 1.5$, the width of each jet is smaller than the tube diameter. This allows easier interaction between the two shear layers across the jet than across the wake. It is expected that the laminar jet instability will couple two chains of transition eddies along its borders in an analogous manner to transition in shear layers behind a single cylinder, Vol. 1, Chapter 5, in the TrSL2 regime.

Ziada *et al.* (1989J), Ziada and Oengören (1992J, 1993J), and Oengören and Ziada (1992J) complemented frequency measurements in air by extensive interstitial flow visualization in water. Ziada *et al.* (1989J) restricted tests to a rectangular array, $S/D = 1.35$, $T/D = 1.6$, six rows deep with 13 columns of tubes, 1k $< Re <$ 12k, $Ti < 1\%$.

The onset of laminar jet instability was first observed for $Re >$ 1k behind downstream rows. Gradually, it moves upstream with increasing Re to reach the first row for $Re =$ 12k. Analogously to a single cylinder, the development of transition waves and eddies on two sides of the jet occurred out-of-phase so that the flow pattern appeared symmetric relative to the jet axis. This is shown in Fig. 28.30(a,b) at $Re =$ 3.7k for one transition eddy shedding cycle. Figure 28.30(a) shows the initial out-of-phase undulations of two shear layers. The partial roll-up is accomplished in Fig. 28.30(b) just before the third row of tubes is reached. The symmetric shedding of transition eddies behind the third row is seen in Fig. 28.30(c). The ejection of transition eddies into the small gap between the adjacent tubes generates a pulsating flow in the streamwise direction.

Sato (1960J) examined the instability of a single 2-D jet, and found two modes. The first mode was symmetric at $St_j = 0.23$ (based on the jet width). The second mode was antisymmetric at $St_j = 0.14$, the ratio of the two being 0.6. When the interstitial jet flow was confined between the tube columns, Ziada

[171]The straight interstitial flow may also occur for some medium spacings in rotated triangle tube arrays

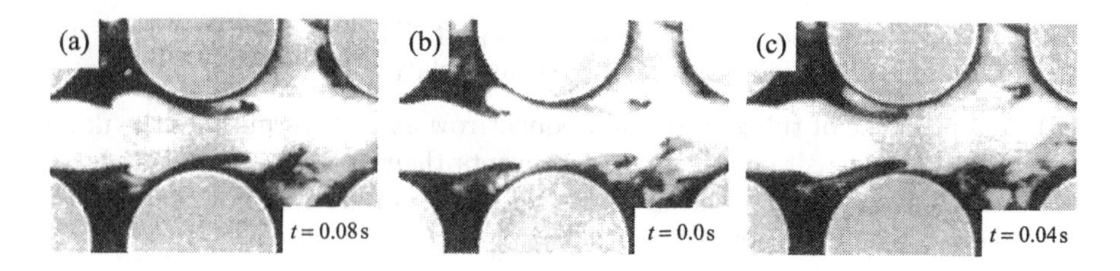

FIG. 28.30. Development of jet symmetric instabilities in shear layers at $Re = 3.7$k, (a) undulation, (b) roll-up, (c) eddy shedding, Ziada and Oengören (1992J)

et al. (1989J) found $St_j = 0.3$ and 0.21 for the two modes, respectively. They argued that the discrepancy with the free jet St_j was due to the impingement of the interstitial jet onto the side tubes.

Ziada and Oengören (1992J) carried out the jet-flow visualization for intermediate spacing, $S/D = 1.75$, $T/D = 2.25$, ten rows deep. The coupled symmetric eddies are seen in Fig. 28.31. The two neighbouring symmetric jets were also 180° out-of-phase. This produced the alternate formation of eddies behind each tube. The symmetric jets were observed only behind the two front rows. A rapid transition to turbulence occurred thereafter. The transition eddies behind the front rows were well organized and repeated themselves consistently up to $Re = 20$k. At higher Re, the flow became disorganized due to a sudden burst into turbulence in the shear layers, see Vol. 1, Chapter 5, the TrSL3 regime.

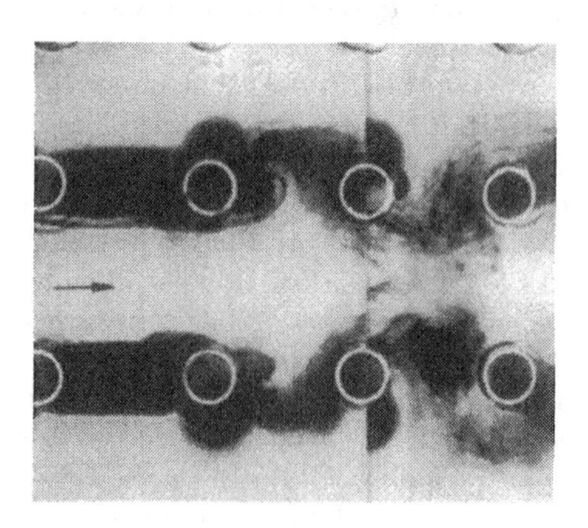

FIG. 28.31. Flow visualization of jet symmetric instability for $S/D = 3.25$, $T/D = 3.75$, $Re = 2$k, Ziada and Oengören (1993J)

The symmetric jet instability became weak at large spacing, $S/D = 3.25$, $T/D = 3.75$, seven rows deep. Ziada and Oengören (1993J) found that the symmetric jet instability was suppressed for $Ti > 1\%$, and replaced by the uncoupled eddies behind the tubes. The alternate eddy formation and shedding in the wake of each tube did not keep any phase relationship with other wakes.

28.3.10 *Acoustic synchronization mechanism*

Baird (1954J) inferred that the energy source for the acoustic standing wave in tube arrays is some form of flow instability. This correct supposition was altered by Putnam (1959J, 1964R) who hypothesized that the flow instability is in the form of the alternate eddy shedding behind tubes, similar to that behind a single cylinder. Regrettably, this hypothesis restricted acoustic excitation only to large spacings, which is incorrect. The most powerful excitation and synchronization to the acoustic frequency takes place at small and intermediate tube spacings in the absence of Kármán–Bénard eddies.

Hill and Armstrong (1962J) found that the acoustic standing wave never occurred behind a single row of tubes, despite eddy shedding being present. The acoustic standing wave, however, always generated intense sound when a second row of tubes was added. Their tests showed that the asymmetric shear layer instability between the two rows was a cause of the acoustic synchronization. The acoustic Strouhal number, St_a, based on the gap velocity and the acoustic frequency of the test section, decreased from 0.61 to 0.25 as the streamwise spacing between the rows increased from 0.95 to 2.30.

It was discussed in Vol. 1, Chapter 18, that external sound could strongly affect transition eddies behind a single cylinder. Peterka and Richardson (1969J) imposed external sound radiated perpendicular to the free stream and cylinder axis. Figure 28.32(a,b) shows frequency spectra without, 0 dB, and with the external sound of 135 dB and 140 dB. For 0 dB, the three peaks are due to the fan frequency, f_f, the eddy shedding, f_e, and the transition eddies, f_t, respectively. For 135 dB and 140 dB, the additional peaks are due to the external sound, f_s, its sub-harmonics, $1/2f_s$, and higher harmonics, $2f_s$. The weak transition eddy peak at 0 dB is replaced by the strong $1/2f_s$ peaks in Fig. 28.32(b). The sub-harmonic peak at $1/2f_s$ may be caused by the transition eddies pairing along the shear layer.

Figure 28.32(c) shows that at resonance $f_t = f_s$, the peak at $1/2f_s$ is considerably reduced, presumably due to the weakening of the eddy pairing. Finally for $f_t > f_s$, the $1/2f_s$ peak disappears and the $2f_s$ peak is enhanced, Fig. 28.32(d).

The unexpected feature in Fig. 28.32(b) is that the maximum enhancement of forced transition eddies is found at around $3/4f_s$, well before $f_t = f_s$. The synchronization starts to develop well before resonance. The term 'acoustic resonance' seems to be inappropriate for the external sound excitation of transition eddies.

An inverse phenomenon takes place in in-line tube arrays. The transition eddies become synchronized by the acoustic frequency mode and generate an

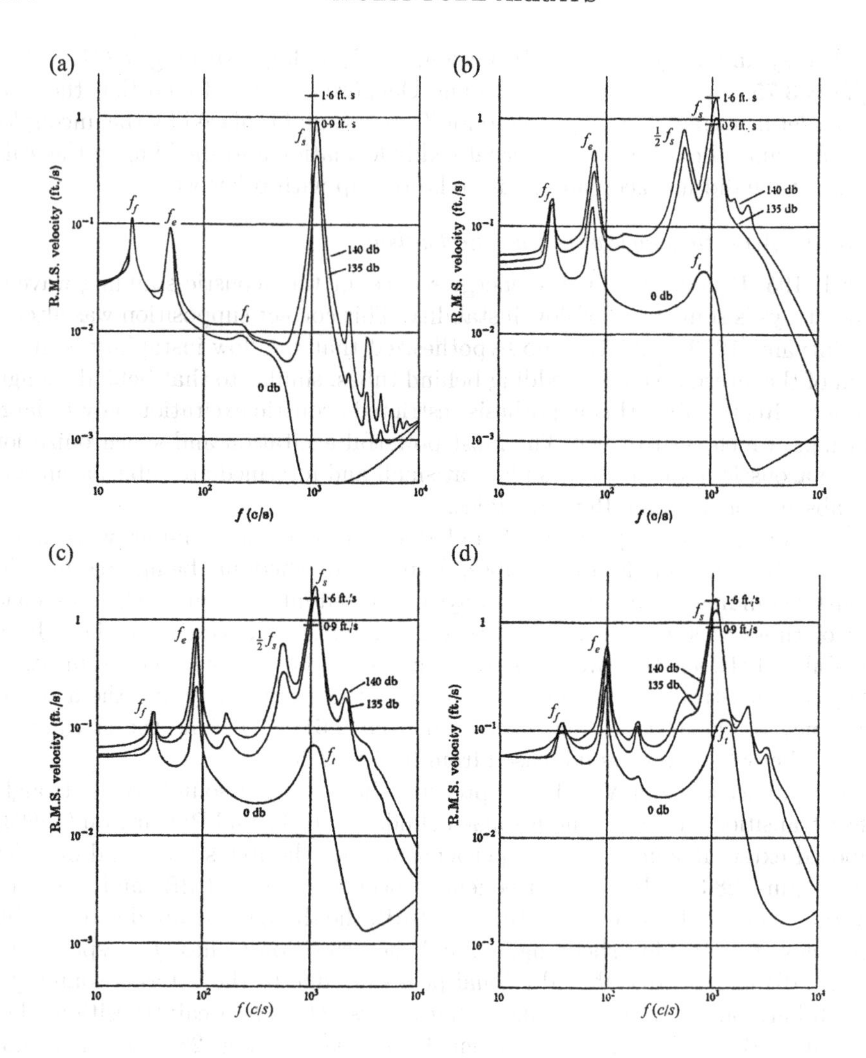

FIG. 28.32. Effect of external sound on transition eddies behind a single cylinder, (a) 5.5k, (b) 8.2k, (c) 9.5k, (d) 10.9k, Peterka and Richardson (1969J)

intense sound. Oengören and Ziada (1992J) simulated acoustic excitation in air by standing surface waves in water. The in-line tube array tested had $S/D = 1.25$, $T/D = 2.25$, six rows deep, five columns. Figure 28.33 shows a combined hot film spectra and dye visualization taken at different water levels. Figure 28.33(a) shows a full synchronization of the transition eddies to the water standing wave frequency at f_w and $1/2 f_w$. Figure 28.33(b,c) shows the gradual weakening of the synchronization due to the decrease in the transverse water displacement along the standing wave height away from the free surface.

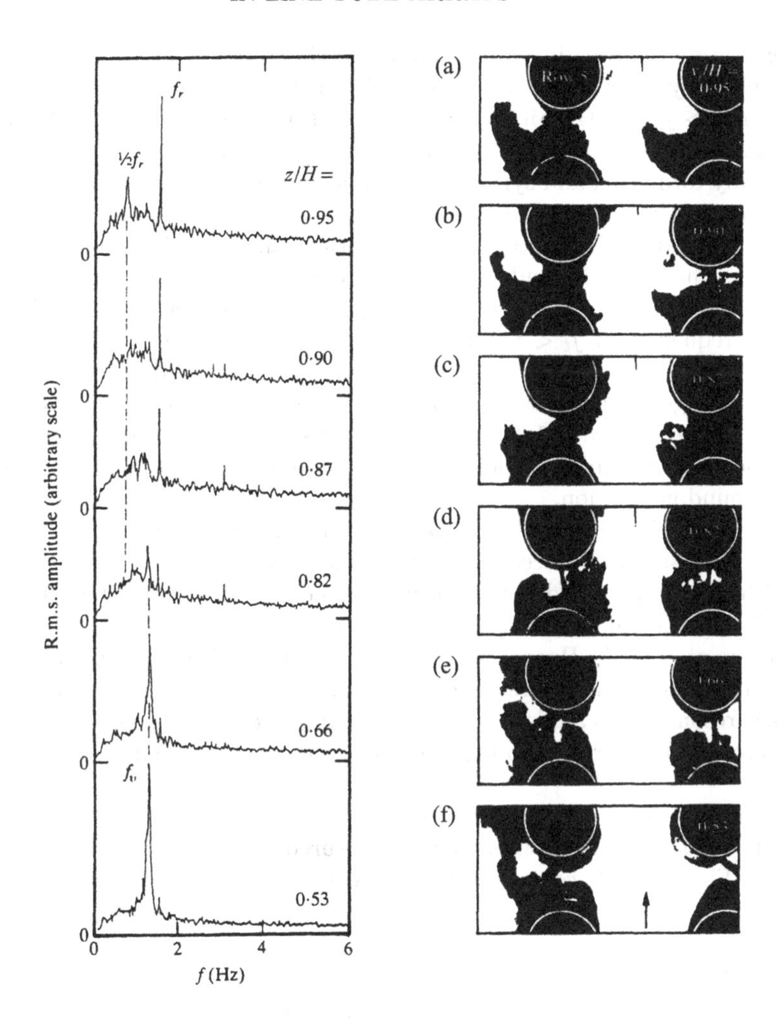

FIG. 28.33. Partial synchronization along the span for $S/D = 1.25, T/D = 2.25$, $Re = 5.3$k, (a) $z/H = 0.95$, (b) $z/H = 0.90$, (c) $z/H = 0.87$, (d) $z/H = 0.82$, (e) $z/H = 0.66$, (f) $z/H = 0.53$, Oengören and Ziada (1992J)

Note that the acoustic synchronization in air remains constant along the tube span, and varies only across the tube array. Figure 28.33(d–f) shows the gradual development of the symmetric jet instability.

Oengören and Ziada (1992J) described the same phenomenon in air tests:

> As the acoustic synchronization becomes stronger, several other spectral peaks appear around $1/2f_a$, which later in the synchronization range become a single peak with a centre frequency of *exactly* $1/2f_a$.

The excitation of acoustic modes cannot be related in any way to the transition eddy frequency prior to synchronization. Towards the end of the synchronization range for $f_t > f_a$, the transition eddies get weaker in strength and smaller in size. The latter can be attributed to the weakening of eddy-pairings. The topology of the eddy formation and synchronization for the jet instability to the acoustic excitation are sketched in Fig. 28.34(a,b).

The mechanism of acoustic excitation of the transition eddies may be inferred as follows: The small size and strength of irregular transition eddies are sensitive to disturbances over a wide range of frequencies. They can be excited by the acoustic frequency for $f_t < f_a$, and become regular, equal in size, and strength. However, the occasional pairing at $1/2f_a$ doubles the size and strength of eddies. This may trigger an increase in the transverse air displacement and enhance the standing acoustic wave at $1/2f_a$. Once the transition eddy pairing is established, a feedback mechanism is developed in which acoustic energy is provided for the intense sound generation.

It is remarkable that well-organized flow in the form of symmetric 180° out-of-phase jet instability precedes and succeeds the synchronization range characterized by the in-phase eddy formation and shedding from all tubes. The onset of synchronization at f_a is not related to f_t, and the termination of synchronization occurs at $f_t = f_a$. Hence, the onset of acoustic excitation and maintenance of the intense sound is not related to the resonance. The long-established term 'acoustic resonance' is a misnomer for small and intermediate spacings of in-line tube banks. It is restricted to huge spacings where the alternate eddy shedding occurs.

Blevins and Bressler (1993J) carried out tests in a wide range of $1.1 < S/D < 6$ and $1.1 < T/D < 6$. The maximum measured sound pressure level in dB is shown in Fig. 28.35. It centred in the range $1.1 < S/D < 4$ and $1.1 < T/D < 3$

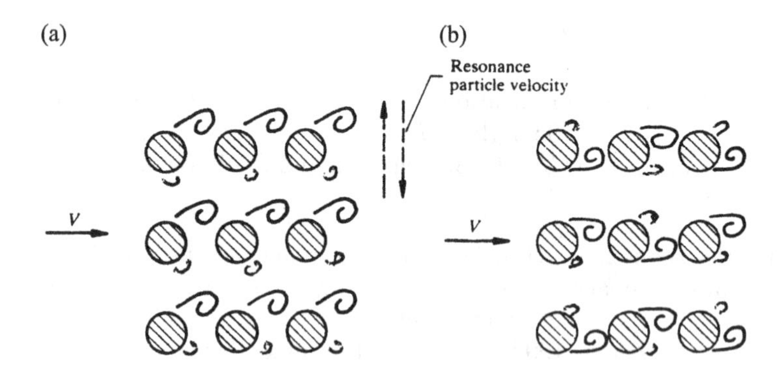

FIG. 28.34. Topology of eddy formation and synchronization, (a) asymmetric acoustic synchronization, (b) symmetric jet instability, Ziada and Oengören (1993J)

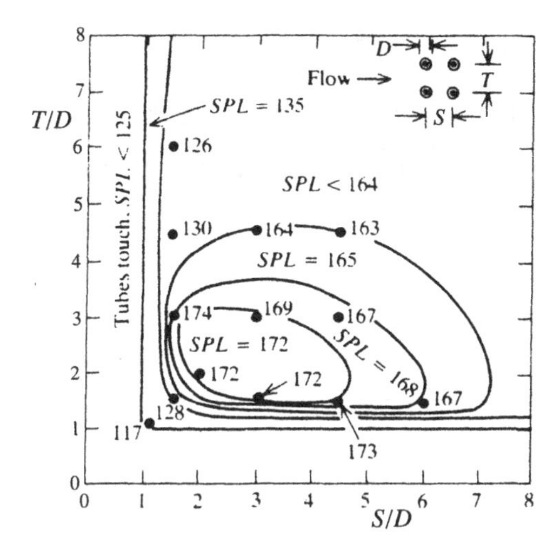

FIG. 28.35. Sound pressure level during acoustic synchronization in in-line arrays, Blevins and Bressler (1993J)

where the alternate eddy shedding is suppressed. The effect of the number of rows $1 < N < 15$ is to excite the acoustic sound at the lower reduced velocity.

28.3.11 *Interstitial flow and turbulence*

The interstitial flow in in-line tube arrays starts with a relatively low free stream turbulence. As the interstitial flow proceeds through the array along the straight alleys the turbulence builds up to a certain row depending on T/D, S/D, and *Re*. Beyond that row, the rate of turbulence generation is balanced by the turbulence dissipation.

Fitzpatrick and Donaldson (1980J) tested the $S/D = 1.73$, $T/D = 1.97$ in-line tube array 2, 3, 5, 7, and 11 rows deep in the range 5.6k < *Re* < 90k. Figure 28.36(a,b) shows the variation of the mid-gap mean velocity and intensity of turbulence, respectively. There is a steep rise in Ti in the first five rows associated with an erratic rise in the mean velocity. The $V_{g1}/V_g > 1$ designates that the mid-gap velocity is greater than the mean velocity calculated from the flow rate divided by the gap cross-sectional areas. Beyond the fifth row, the Ti_{max} and Ti_{min} mirror $(V_{g1}/V_g)_{\text{min}}$ and $(V_{g1}/V_g)_{\text{max}}$, respectively. The effect of *Re* is small and enhances the peaks and troughs when the TrSL3 regime is reached.

Fitzpatrick *et al.* (1988J) suggested that the periodicity arising in the free shear layer will dominate the alley flow in the entry region of the tube array. Owen's (1965J) type of turbulent buffeting occurs when turbulence generation and dissipation are balanced. Figure 28.37(a–i) shows frequency power spectra in a seven deep tube array $S/D = 1.73$, $T/D = 1.97$, and *Re* = 8.2k. The single frequency peak at $St_g = 0.23$ behind rows 3, 3.5, and 4 is being replaced by the

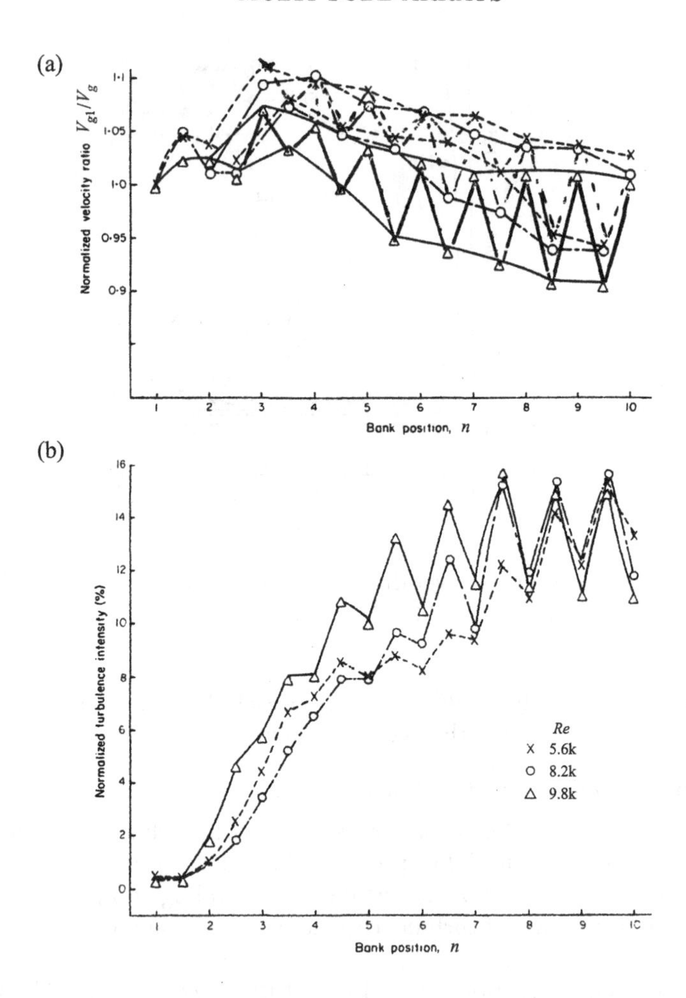

FIG. 28.36. Ten rows deep, $T/D = 1.97$, $S/D = 1.73$ array, (a) gap velocity ratio, (b) turbulence intensity, Fitzpatrick and Donaldson (1980J)

weak peaks behind rows 4.5, 5, and 5.5 and eventually developed wide peaks around $St = 0.19$ behind rows 5.5, 6, 6.5, and 7.

The effect of an initially high free stream turbulence was investigated by Price and Paidoussis (1989J) for a $P/D = 1.5$ square in-line tube array 10 rows deep. Figure 28.38(a,b) shows that the effect of $Ti = 20\%$ free stream turbulence is negligible on the mean mid-gap velocity. Figure 28.38(b) shows $Ti = 20\%$ is reduced to 2% inside the first row, and the Ti is slightly raised up to the fifth row. Beyond that a reversal is seen, i.e. the effect of the free stream turbulence is to reduce the generated turbulence in the interstitial flow.

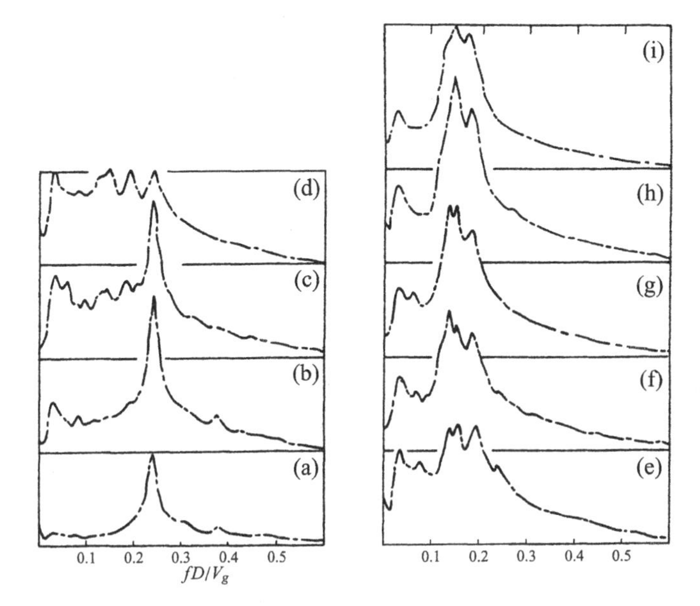

FIG. 28.37. Frequency power spectra at $Re = 8.2$k, (a) $n = 3$, (b) $n = 3.5$, (c) $n = 4$, (d) $n = 4.5$, (e) $n = 5$, (f) $n = 5.5$, (g) $n = 6$, (h) $n = 6.5$, (i) $n = 7$, Fitzpatrick and Donaldson (1980J)

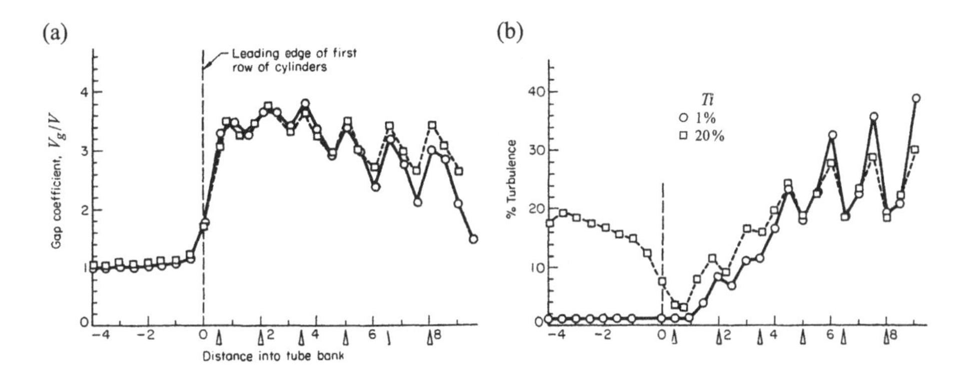

FIG. 28.38. Ten rows deep, $P/D = 1.5$ square array, $Ti = 1\%$ and 20%, (a) gap velocity ratio, (b) turbulence intensity, Price and Paidoussis (1989J)

28.3.12 *Classification of in-line tube arrays*

Ziada and Oengören (1992J) suggested the following classification based on the observed flow structures in in-line tube arrays:

(i) $P/D < 1.4$, small pitch ratios; the symmetric jet instability becomes weak, and it occurs at the upstream rows only. The antisymmetric jet instability is suppressed by the close proximity of symmetrically arranged tubes in subsequent rows.

(ii) $1.5 < P/D < 2.5$, intermediate pitch ratios; the dominant eddy pattern is generated by the symmetric jet instability, and it persists through the whole depth of the array. The antisymmetric jet instability is weak and intermittent.

(iii) $P/D > 2.5$, large pitch ratios; there are two different modes of eddy shedding:

 (a) at low free stream turbulence, the jet instability mode dominates and persists through the depth of the array;

 (b) at high free stream turbulence, the alternate eddy shedding is governed by the near-wake instability, and persists behind the upstream rows.

28.4 Staggered tube arrays

28.4.1 *Proximity effects*

An elementary model of a staggered tube array is a group of three tubes arranged to form a triangle. The streamwise and transverse interactions are present simultaneously to a degree that depends on the mutual spacing between the tubes, P/D.

28.4.2 *Mean and fluctuating pressure*

Systematic measurements of the mean and fluctuating pressure have been carried out by Zdravkovich and Namork (1979P) and Zdravkovich (1987R, 1993R). The pressure distributions are measured on a tube placed in a normal triangle array, $P/D = 1.375$, six rows deep. Figure 28.39 shows the mean and fluctuating (vertical bars) pressure distributions. The distributions differ considerably in the first two rows, and become similar in the other rows.

Figure 28.40 shows the replotted mean pressure distributions from Fig. 28.39 around the tube placed in the first three rows. The favourable pressure gradient is different for all three rows. The interstitial flow beyond separation is governed not only by the tubes in the second row, but also by the near-wakes formed behind the tubes in the first row, as sketched in Fig. 28.41. These near-wakes impose an additional reduction in the flow area. The related interstitial flow accelerates to produce $C^*_{p\text{min}}$ at 60° in the second row in Fig. 28.40. A short, adverse pressure gradient is followed by the second $C^*_{p\text{min}}$ at 90°, where the gap area is minimum.

The two abnormal pressure distributions measured in the first two rows are followed by the 'normal' C^*_p measured in the third row. The favourable pressure gradient does not form $C^*_{p\text{min}}$ at 60° in the third row but forms a kink instead.

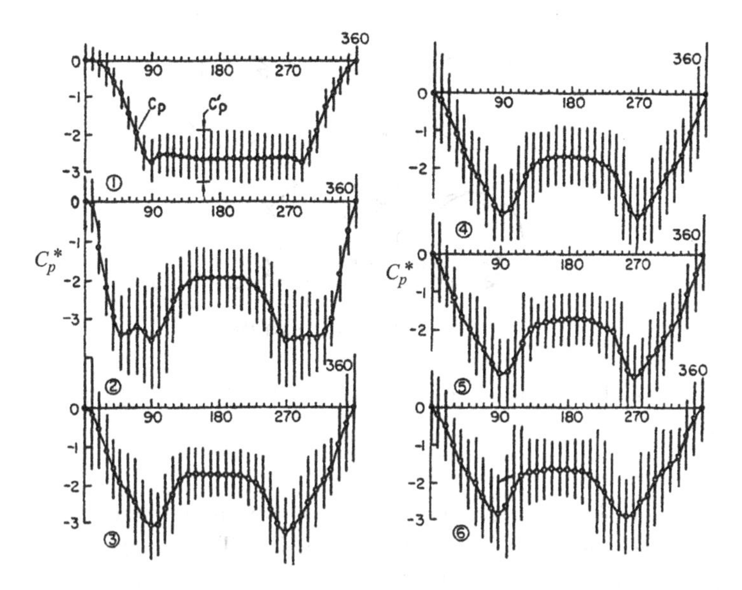

FIG. 28.39. Mean pressure (open circles) and fluctuating pressure (vertical bars) in a six-row deep normal triangle array, P/D = 1.375, Zdravkovich and Namork (1979P)

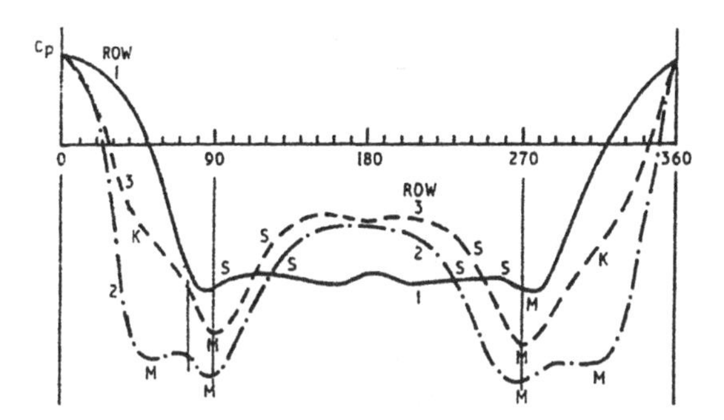

FIG. 28.40. Mean pressure distribution for a tube in rows 1, 2, and 3, Zdravkovich and Namork (1979P)

It is deduced that this feature is caused by a different shape of the near-wake behind the second row, as sketched in Fig. 28.41. Flow visualization by Abd-Rabo and Weaver (1986J) (rotated square array P/D = 1.41) in Fig. 28.2(b) shows not only near-wakes behind the first three rows, but also the retarded regions upstream of the tubes in the third and fourth rows, also sketched in Fig. 28.41.

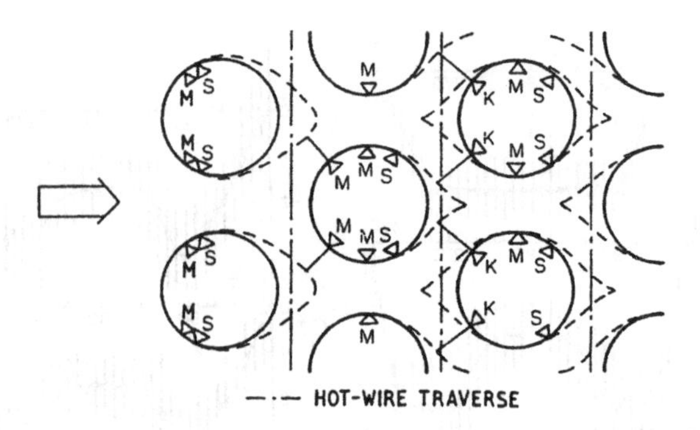

FIG. 28.41. Sketch of near-wakes, $M - C_{p\min}$, $S - \theta_s$, $K - C^*_{p\mathrm{kink}}$, Zdravkovich (1987R)

The effect of pitch size on the interstitial flow and related mean pressure distribution was studied by Aiba *et al.* (1982Jb). Two staggered tube arrays, seven rows deep, were tested: $P/D = 1.56$ and 2.08. Figure 28.42(a,b) shows the mean C_p^* around a tube located sequentially in all seven rows of both arrays. Figure 28.42(a) shows that the curves are similar to those measured in the normal triangle array, $P/D = 1.375$ in Fig. 28.39. Note that the curves for rows 4 and 5,

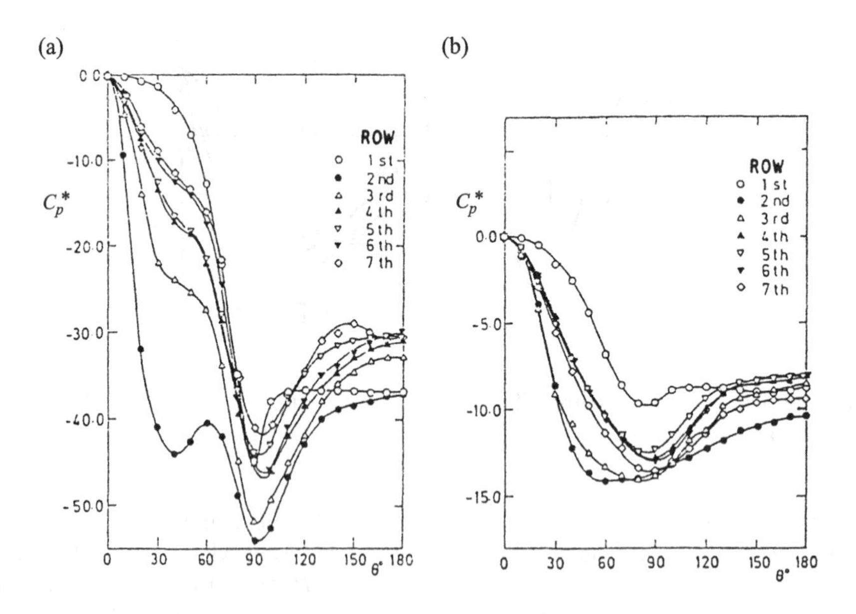

FIG. 28.42. Mean pressure distribution, seven rows deep rotated square array at $Re = 30$k, (a) $P/D = 1.56$, (b) $P/D = 2.08$, Aiba *et al.* (1982Jb)

and 6 and 7, are similar but form two curves. The second staggered array tested by Aiba *et al.* (1982Jb) has more widely spaced tubes, Fig. 28.42(b). This means that the effect of near-wakes and retarded regions on the interstitial flow will be small.

28.4.3 *Structure of interstitial flow*

The interstitial flow in tube arrays may be described by the mean velocity profile, turbulent intensity distribution, and frequency spectrum. It has been pointed out that the mean velocity and turbulence intensity Ti curves are antisymmetric and 'mirror' each other. A systematic hot-wire traverse across three pitch lengths and behind each of the rows from 1 to 11 is shown in Fig. 28.43. The velocity profile has two peaks in each gap behind the first row. These are accompanied by a single low-turbulence trough and high-turbulence crests. From the second to the last row, the two velocity peaks are always matched by turbulence troughs. This feature repeats itself, row after row, despite the non-uniformities that developed in the gaps in the same row. The relationship between the local velocity and tur-

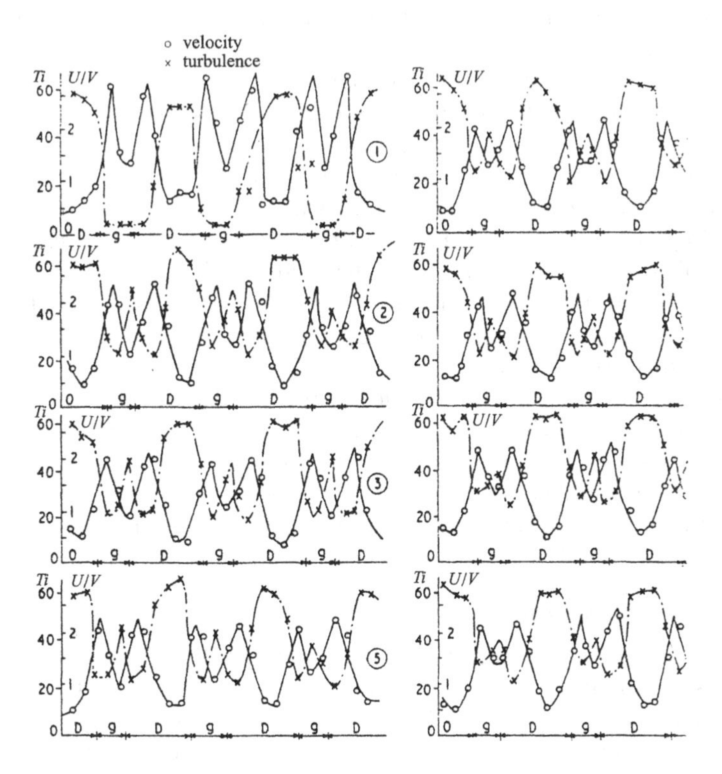

FIG. 28.43. Velocity and turbulence intensity behind rows 1 to 11, Zdravkovich (1993R)

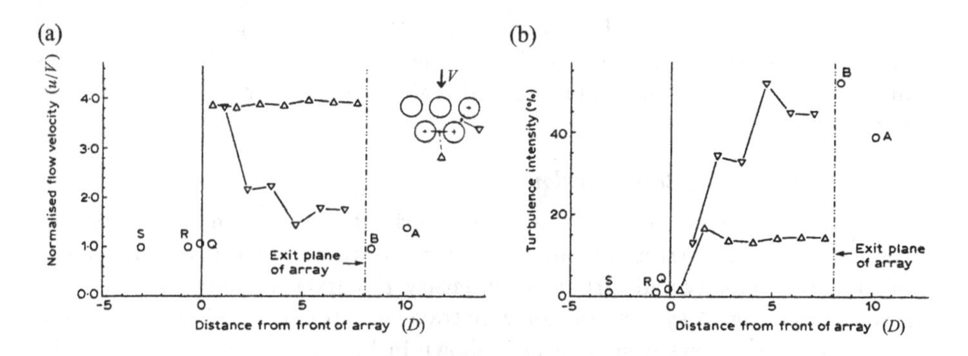

FIG. 28.44. Normal-triangle array, seven rows deep, $P/D = 1.375$, (a) gap velocity, (b) turbulence intensity, Price and Zahn (1991J)

bulence indicates that the periodic acceleration and deceleration of interstitial flow is closely related to the dissipation and generation of turbulence, respectively. The common feature is that the velocity peaks remain closer to the tubes than to the gap centreline.

Price and Zahn (1991J) tested a normal triangle array having the same pitch $P/D = 1.375$ as in Fig. 28.39. Figure 28.44(a,b) shows velocity and turbulence intensity measured in the middle of transverse and diagonal gaps at and behind all seven rows. Figure 28.44(a) shows that the velocity remains constant in the transverse gaps, whereas in the side gaps it varies considerably. It might be that the hot wire, which is placed near the near-wakes, is occasionally submerged in the near-wakes. Figure 28.44(b) shows the corresponding high-turbulence measurements. The turbulence intensity increases rapidly from the first to the second row. For the diagonal gaps, high $Ti = 50\%$ indicates that the hot wire might be inside the respective near-wakes.

The velocity and turbulence profiles in rotated square arrays $P/D = 1.56$ and 2.08 are shown in Fig. 28.45(a,b) and Fig. 28.46(a,b), respectively. The traverse between the fifth and sixth rows in Fig. 28.45(b) shows large velocity fluctuations (horizontal bars). The hot-wire traverse between the first and second rows, and the second and third rows for $P/D = 2.08$ are shown in Fig. 28.46(a,b), respectively.

28.4.4 *Effect of tube displacement*

The regular array of tubes arranged at a constant pitch becomes irregular when the tubes start to vibrate.[172] For large amplitudes of vibration, the pitch irregularity becomes considerable, particularly in closely spaced arrays. The effect of a displaced tube on the interstitial flow is strongly dependent on the ratio of amplitude of vibration and the size of the gap, A/G.

[172] Flow-induced vibration in flexible tube arrays will be discussed in Vol. 3, Chapter 39.

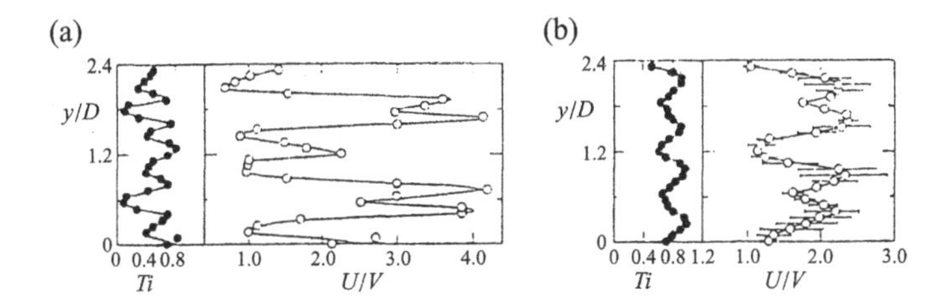

FIG. 28.45. Velocity and turbulence intensity profiles in rotated-square arrays, $P/D = 1.56$, (a) between rows 1 and 2, (b) between rows 5 and 6, Aiba *et al.* (1982Jb)

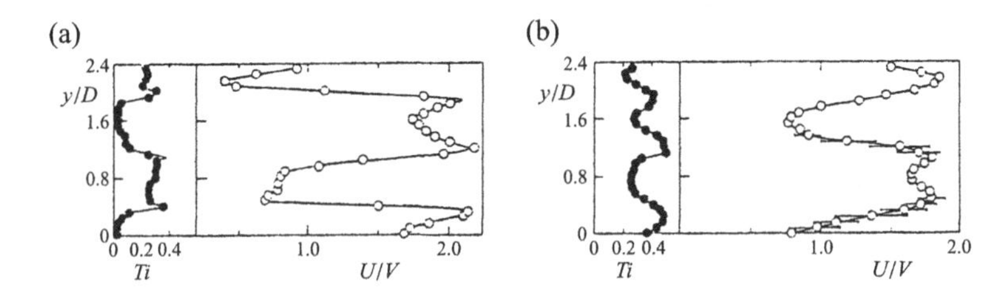

FIG. 28.46. Velocity and turbulence intensity profiles in rotated-square arrays, $P/D = 2.08$, (a) between rows 1 and 2, (b) between rows 2 and 3, Aiba *et al.* (1982Jb)

Zdravkovich *et al.* (1976P, 1987R) investigated the effect of the second row displacement in the streamwise and transverse directions. The initially regular staggered array had $T/D = 1.66$, $R/D = 2$, and 3 rows deep. Figure 28.47(a) shows the mean pressure distribution for the initial arrangement, the upstream displacement $R/D = 1.2$, and the downstream displacement $R/D = 2.5$. All three curves are similar, which means the interstitial flow is weakly affected by the streamwise displacement of the second row as a whole. Figure 28.47(b) shows the case when the second row is displaced in the transverse direction, $(T/D)_d = 1.1$. There is a significant change in mean pressure distribution, the stagnation point is at 340°, and the favourable pressure gradients are dissimilar on two sides of the tube. Figure 28.47(c) shows the effect of a further displacement of the second row in the transverse direction, which corresponds to $(T/D)_d = 0.55$.

All three curves show reattachment peaks at 40° to 60°, and stagnation peaks at 350°. The latter disappears in Fig. 28.47(d) when the array becomes in-line, $(T/D)_d = 0$. The reattachment peaks are formed on both sides of the tube, and the curves are again almost symmetric.

The mean drag and lift coefficients are evaluated for all 12 displacements

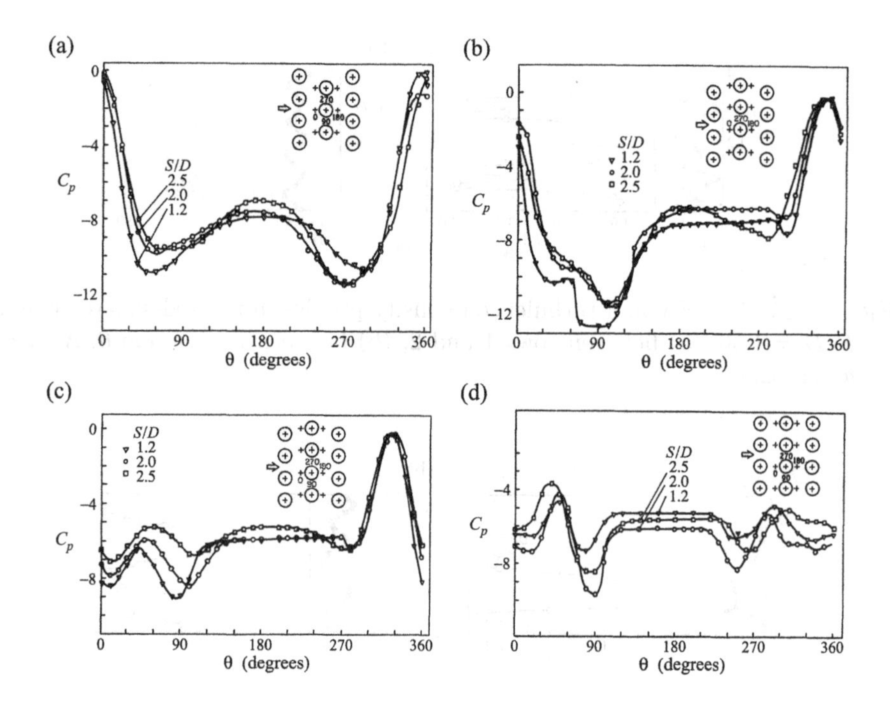

FIG. 28.47. Pressure distribution for displaced second row, $R/D = 1.2$, 2, 2.5, (a) $T/D = 1.66$, (b) $T/D = 1.10$, (c) $T/D = 0.55$, (d) $T/D = 0$, (in-line), Zdravkovich *et al.* (1976P)

of the second row. Table 28.4 shows that the tube displacement produces a significant lift coefficient caused by the asymmetric pressure distribution. The lift component of the force may be almost twice that of the drag component.

When a single tube in the array is displaced, the adjacent gaps become different in size. An irregular tube cluster of five stationary tubes was studied by Zdravkovich (1973P). It was found that at subcritical Reynolds numbers the flow rate through the large and small gaps adjusts itself to yield similar velocities in

Table 28.4 *Mean lift and drag coefficients for a two-row array*

S/D	1.2		2.0		2.5	
T/D	C_L	C_D	C_L	C_D	C_L	C_D
0	0.03	−0.12	0.00	−0.07	0.01	0.09
0.55	0.52	0.03	0.48	0.21	0.30	0.16
1.10	0.70	0.50	0.81	0.48	0.77	0.34
1.66	0.03	0.43	0.21	0.59	0.26	0.40

both gaps. At transition Reynolds numbers, $Re > 200k$, a large difference in the two gap velocities is found with the higher velocity in the small gap.

Paidoussis *et al.* (1991P) measured mean forces on the tube in the third row of a rotated triangle array $P/D = 1.375$, seven rows deep, consisting of 14 or 15 tubes per row. Figure 28.48 shows the measured drag and lift coefficients (based on the gap velocity) when the tube is statically displaced in the streamwise and transverse directions up to $\pm 0.23D$. The drag coefficient shows small variation except when the displaced tube approaches the downstream tube, $x/D = -0.23$ and $y/D = 0.23$. Figure 28.48 shows that the lift coefficient variation is strongly affected by the displacement in the transverse direction.

28.4.5 *Effect of surface roughness*

Achenbach (1971Jb) examined the effect of surface roughness $K/D = 0.11\%$ and 0.45% on flow in a staggered tube array $T/D = 2$, $R/D = 1.4$, five rows deep, across a wide range $40k < Re_g < 10M$. Figure 28.49 shows the pressure drop coefficient $C_{\Delta p}$ in terms of Re_g. For smooth tubes, $C_{\Delta p\min}$ occurs at $Re_g = 60k$. For two values of K/D, the $C_{\Delta p\min}$ moves to lower $Re_{gc} = 200k$ and 80k. Note the significant increase in $C_{\Delta p}$ induced by the surface roughness in the post-critical regime. Figure 28.49 is dissimilar to Fig. 28.24 for the in-line array, but similar to Fig. 28.10 for a single row of tubes.

28.4.6 *Strouhal number; historical*

Eddy shedding might or might not exist in staggered tube arrays depending on the matrix, pitch ratio P/D, number of rows n, Re, Ti, etc. Early studies assumed a priori that eddy shedding took place in all staggered tube arrays. The main effort was directed towards establishing St data for all staggered arrays.

An early measurement of St for two rows of staggered tube arrays was carried out by Borges (1969J). Two types of St curves are seen in Fig. 28.50:

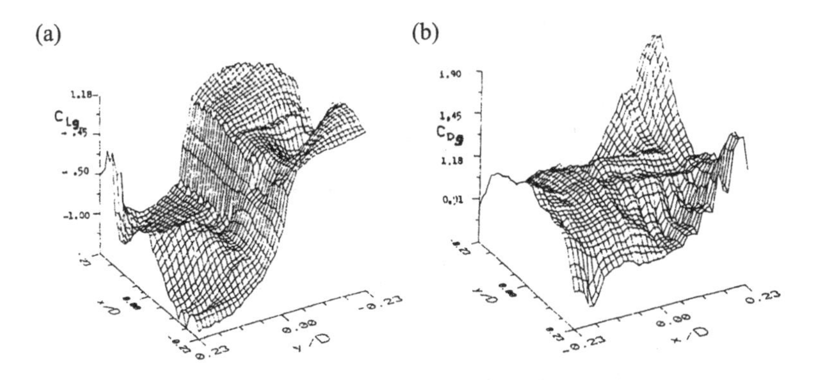

FIG. 28.48. Force coefficients for a displaced tube in a rotated-triangle array $P/D = 1.375$, (a) lift, (b) drag, Paidoussis *et al.* (1991P)

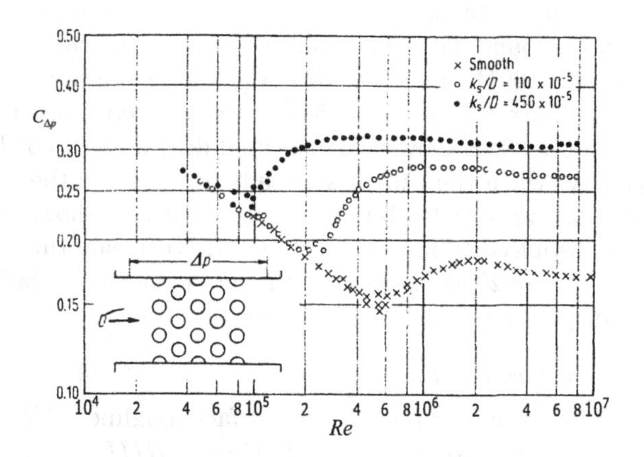

FIG. 28.49. Pressure drop in terms of Re_g for a normal-triangle array, $T/D = 2$, $R/D = 1.4$, Achenbach (1971Jb)

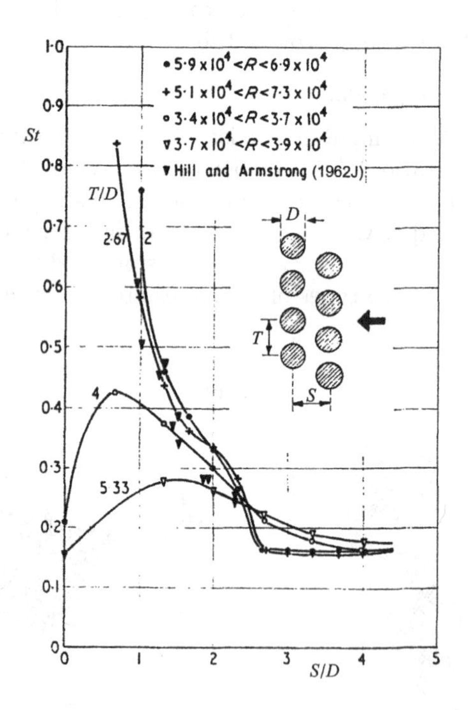

FIG. 28.50. Strouhal number in terms of S/D and T/D for a two-row deep array, Borges (1969J)

(i) $T/D > 3$, St reaches a peak at small R/D;

(ii) $T/D < 3$, St increases monotonously with decreasing R/D.

Chen (1968J) measured St in the range $1 \leqslant R/D \leqslant 1.97$ and $2 < T/D < 3$. He extrapolated his St curves below $T/D = 2$ by assuming that $St \to 0$ as $T/D \to 0$. Ishigai *et al.* (1973R) repeated measurements in the range $1.25 \leqslant R/D \leqslant 3$ and $1.25 \leqslant T/D \leqslant 3$. Figure 28.51 shows a compilation of data by Chen (1968J) and Ishigai *et al.* (1973R). The peaks assumed by Chen (1968J) were not confirmed.

28.4.7 *Parallel triangle tube arrays*

Ziada and Oengören (2000J) carried out extensive tests on parallel triangle arrays in the range $1.2 < P/D < 4.2$ and $940 < Re < 18$k, from 11 rows up to 23 rows deep, and from 5 to 15 columns. Flow visualization in water for an intermediate $P/D = 2.08$ is shown in Fig. 28.52(a,b). The antisymmetric and symmetric transition eddy shedding at $Re = 1.87$k is shown in Fig. 28.52(a,b).

The small pitch ratio $P/D = 1.44$ flow pattern appeared as a slightly wavy jet between the columns. The periodic vorticity was similar to that found for in-line tube arrays. For large pitch ratios the alternate eddy shedding occurred independently behind each tube.

Three different flow periodicities were found in the measured frequency spectra depending on Re and number of rows. Figure 28.53 shows the three St in terms of P/D. The highest St is the weakest and is caused by the transition

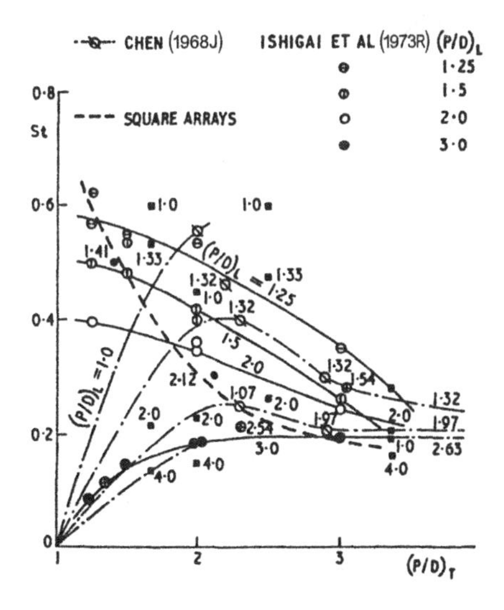

FIG. 28.51. Compilation of Strouhal number data for staggered arrays, Ishigai *et al.* (1973R)

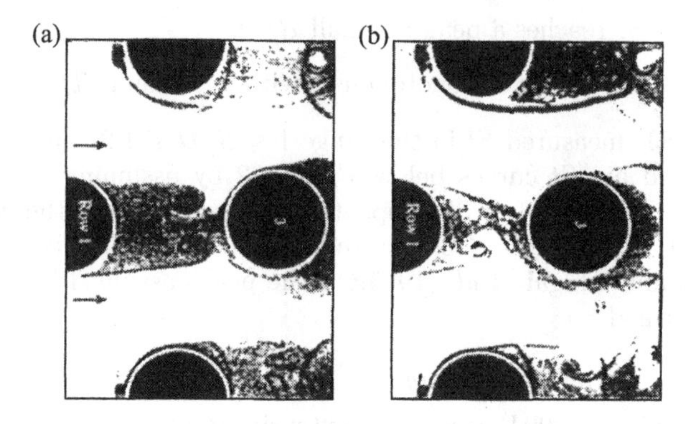

FIG. 28.52. Flow visualization in a parallel-triangle array, $P/D = 2.08$, $Re = 1.87\text{k}$, (a) symmetric, (b) antisymmetric, Ziada and Oengören (2000J)

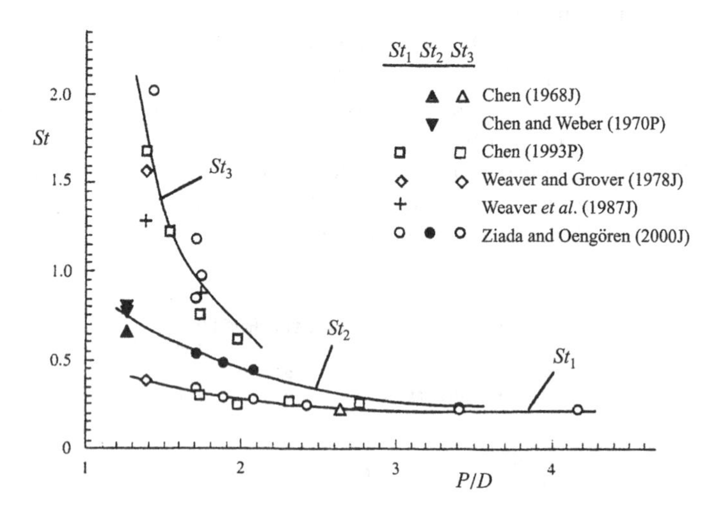

FIG. 28.53. Compilation of Strouhal number data for parallel-triangle arrays in terms of P/D, Ziada and Oengören (2000J)

eddy instabilities at low $Re < 5\text{k}$. The St_2 is associated with small-scale eddy shedding at the first row for $P/D < 2$. The lowest St_3 is the strongest at high Re. It is generated by the large-scale alternating eddy shedding at deeper rows and becomes dominant at all rows at high Re and P/D.

The acoustic excitation and subsequent synchronization strongly depend on P/D and Re. Ziada and Oengören (2000J) examined the acoustic response in a wide range $1.25 < P/D < 4.2$. Figure 28.54(a–c) shows the acoustic synchronization and response for small, intermediate, and large P/D. The main frequency

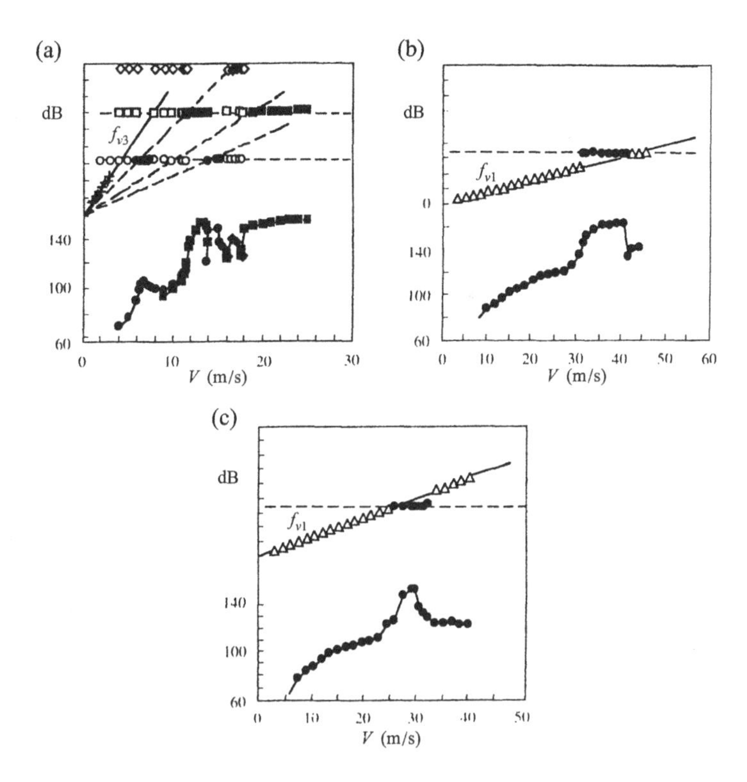

FIG. 28.54. Synchronization of acoustic and eddy frequency, (a) $P/D = 1.44$, (b) $P/D = 2.42$, (c) $P/D = 4.17$, Ziada and Oengören (2000J)

peaks observed are shown by full lines, and the dashed lines represent St_a based on acoustic frequency and the flow velocity at the onset of synchronization. Figure 28.54(a), $P/D = 1.44$, shows that the first acoustic mode of the standing wave is not excited by the acoustic frequency but at a substantially higher velocity (about 60% higher than the resonant velocity). As the velocity is increased further, the second and third acoustic modes are also excited. The excitation occurs according to an St_a relation (dashed line).

Figure 28.54(b) shows a typical acoustic response for the intermediate $P/D = 2.42$. The acoustic synchronization occurs at a flow velocity that is substantially lower than that corresponding to the frequency resonance. It terminates when $f_{v1} = f_{a1}$. Finally, Fig. 28.54(c), $P/D = 4.17$, shows the onset of acoustic response at the resonance $f_{v1} = f_{a1}$. Only for large spacings does the alternate eddy shedding trigger the acoustic resonance, i.e. $f_{v1} = f_{a1}$.

Ziada and Oengören (2000J) plotted St_a versus P/D in Fig. 28.55, which delineate the synchronization (high noise) and no-synchronization regions. Only the first acoustic mode (1) is observed for $P/D > 2.2$, and it corresponds to the

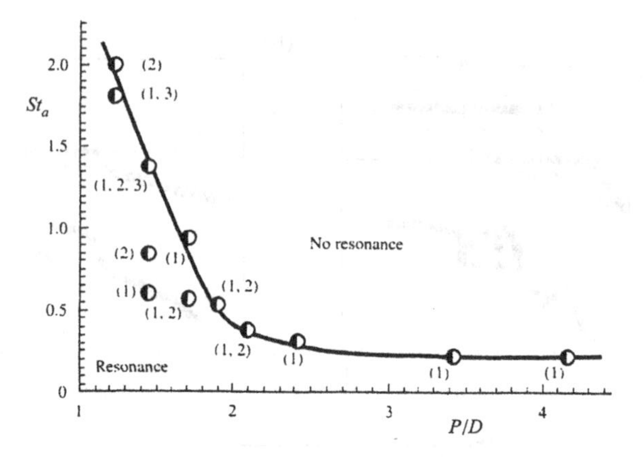

FIG. 28.55. Acoustic St_a in terms of P/D, Ziada and Oengören (2000J)

acoustic resonance. Both first and second modes (1,2) are observed for $P/D < 2.2$. Finally, the first three modes (1,2,3) are found for $P/D < 1.5$. The critical $St_a = f_a D/V_g$ values on the curve could be used in the design for predicting acoustic excitation in parallel triangle tube arrays.

28.4.8 *Rotated square arrays*

Weaver *et al.* (1993J) carried out dye visualization in water for a rotated square $P/D = 1.5$. Figure 28.56(a,b) shows two sequences of the formation and shedding of eddies behind the first and second rows, respectively. The free shear layers at $Re = 967$ ($Re_g = 1.3\text{k}$) are laminar up to the roll-up. Note that the wake width behind the first row tube is considerably narrower than that behind the

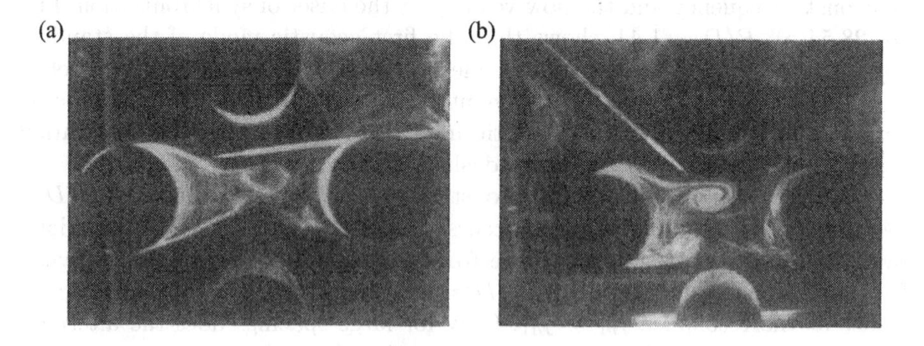

FIG. 28.56. Flow visualization of a rotated square array $P/D = 1.5$, $Re = 967$, (a) clockwise eddy shed behind first row, (b) behind second row, Weaver *et al.* (1993J)

second row. This resulted in two St values as found by Weaver *et al.* (1993J). They showed that if a universal St_g is based on the maximum gap velocity and estimated wake width from the smoke visualization, then a single $St_g = 0.22$ is obtained.

A compilation of two St values measured behind the first and second rows in the range $1.2 < P/D < 2.8$ was reported by Weaver *et al.* (1993J) and Weaver (1993R). Figure 28.57 shows St in terms of P/D for rotated square arrays. The experimental points form two distinct curves. Thus, what appeared to be a significant scatter in the data is largely attributed to the fact that the alternate eddy shedding occurs at two distinct frequencies for a given flow velocity.

28.4.9 *Normal triangle arrays*

Polak and Weaver (1995J) carried out a visualization of eddy shedding in normal triangle tube arrays. The tubes had $3.4 \leqslant L/D \leqslant 8$, $1.14 < P/D < 2.67$, and $760 < Re < 49\text{k}$. Figure 28.58(a,b) shows smoke visualization for $P/D = 2.67$. The alternate eddy shedding is evident in Fig. 28.58(a,b) at $Re = 3.3\text{k}$, behind the first and second rows, respectively.

Oengören and Ziada (1998J) also carried out experiments on normal triangle arrays in air and water, and found two modes of eddy shedding: alternate and symmetric. The tested arrays had $P/D = 1.61,\ 2.08$, and 3.41, $L/D = 6.5,\ 11.1$, and 9.1, number of rows, 10, 14, and 7, tubes per row $\frac{1}{2}+8+\frac{1}{2}$, $\frac{1}{2}+7+\frac{1}{2}$, and $\frac{1}{2}+5+\frac{1}{2}$, and $17.3\text{k} < Re < 52\text{k}$.

Dye visualization in water at $Re = 1.8\text{k}$ for $P/D = 2.08$ is shown in Fig. 28.59. The upper sequences show a symmetric mode of shedding behind the second row of tubes. The alternate eddy shedding behind the first row is coupled in-phase, so the eddies formed behind two adjacent sides of neighbouring tubes mirror

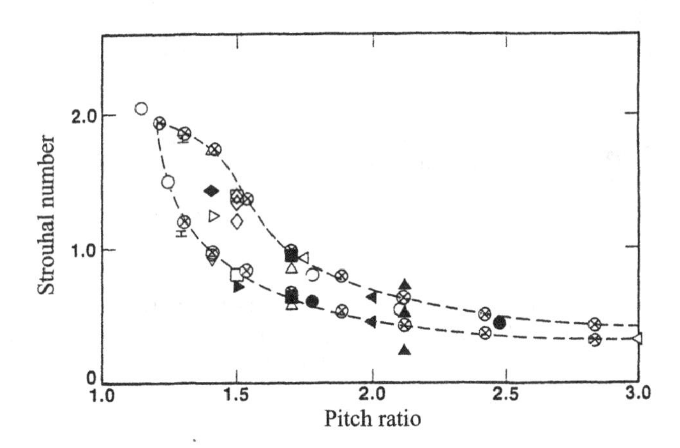

FIG. 28.57. Compilation of Strouhal number data for rotated-square arrays in terms of P/D, Weaver *et al.* (1993J)

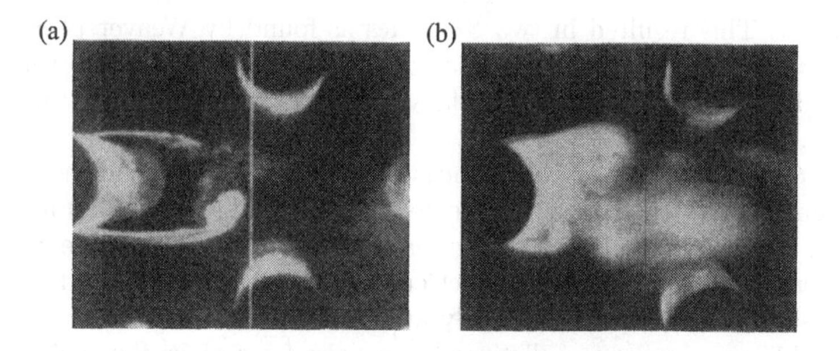

FIG. 28.58. Flow visualization for a normal-triangle array for $P/D = 2.67$, (a) row 1, $Re = 3.3$k, (b) row 2, $Re = 5.3$k, Polak and Weaver (1995J)

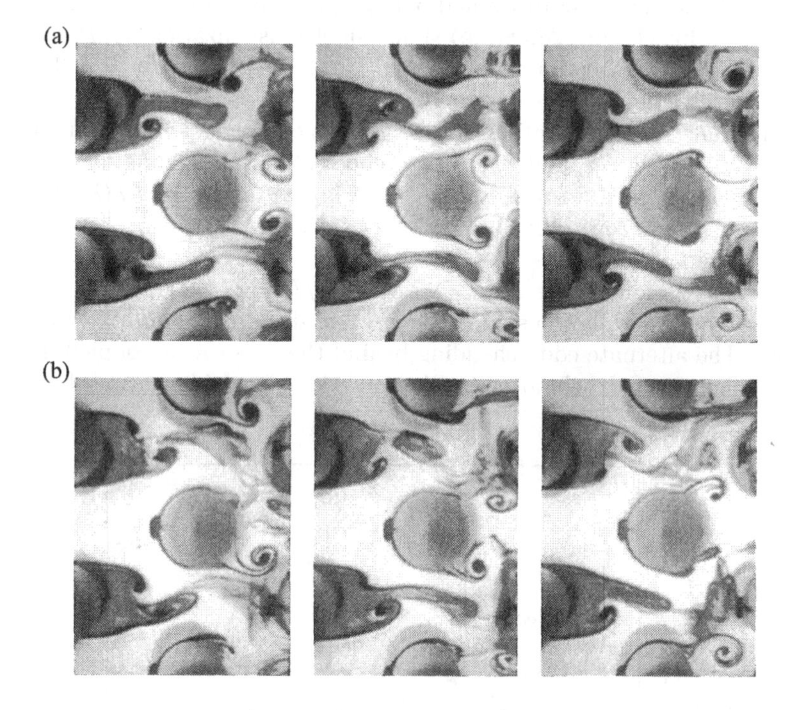

FIG. 28.59. Dye visualization of $P/D = 2.08$ normal-triangle array sequences from right to left at $Re = 1.8$k, (a) symmetric eddy shedding, (b) alternate eddy shedding, Oengören and Ziada (1998J)

each other relative to the gap axis of symmetry. These eddies trigger symmetric formation and shedding behind the tube in the second row. The eddies remain attached to the second row tubes while being formed, then join the passing eddies from the first row, and form symmetric eddy pairs. These impinge on the third row tubes and mix with the vorticity of opposite sign generated behind the third row tubes.

The symmetric mode of eddy shedding was intermittent and switches to an antisymmetric mode as shown in Fig. 28.59(b), lower sequences. Eddy pairing occurs alternately behind the second row tubes. Oengören and Ziada (1998J) noted that the symmetric mode was more persistent than the antisymmetric mode. Both modes occur at the same frequency, which corresponded to f_1.

Oengören and Ziada (1998J) compiled St_g in terms of P/D. Figure 28.60 shows that the data collapsed onto two curves. The upper curve corresponds to f_2, i.e. eddy shedding in the first row, and the lower curve to f_1 behind the second row.

The following empirical relationship was suggested:

$$St_{g1,2} = \frac{1}{A_{1,2}(P/D - 1)^{n_{1,2}}} \tag{28.12}$$

where $A_{1,2} = 3.62$ and 2.41, $n_{1,2} = 0.45$ and 0.41, for St_{g1} and St_{g2}, respectively. Equation (28.12) was recommended for designers to predict acoustic excitation.

28.4.10 *New universal St; a proposal*

It was shown in Chapter 27 that the St value for triangle, and rotated square clusters strongly depended on P/D. A decrease in P/D led to the narrowing of near-wakes, and the latter resulted in an increase in St. A similar trend was found for all staggered tube arrays. Zdravkovich (1997P) argued that the governing

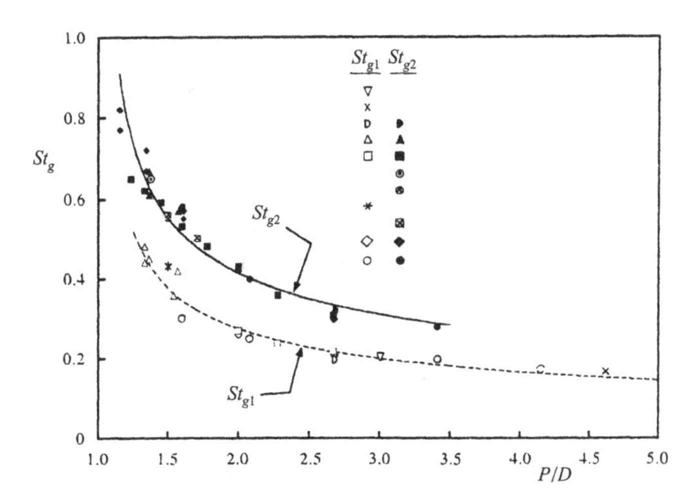

FIG. 28.60. Compilation of Strouhal number data in terms of P/D for a normal-triangle array, Oengören and Ziada (1998J)

parameter for St should be the gap G or pitch P, and not the tube diameter D. The diameter became the influencing parameter for staggered tube arrays.

Weaver and Fitzpatrick (1988R) and Weaver (1993R) showed that the St data for staggered tube arrays might be approximated by Owen's (1965J) semi-theoretical hyperbolic equation

$$St = \frac{1}{A(P/D - 1)} \tag{28.13}$$

where $A = 2, 1.73$, and 1.16 for the rotated square, normal, and parallel triangle tube arrays, respectively. More appropriate is St_g based on the gap velocity V_g described by eqn (28.2), so

$$St_g = \frac{T/D - 1}{A(P/D - 1)T/D} \tag{28.14}$$

Note that for the normal triangle tube array $T/D = P/D$ and $A = 1.73$, so that

$$St_g = \frac{0.58}{P/D} \tag{28.15}$$

or by introducing new $St_P = fP/V_g$ is a new universal St_g. Note that eqn (28.14) cannot be simplified for the rotated square and parallel triangle tube arrays because $P/D \neq T/D$. In the range $1.2 < P/D < 2.5$, there is $1.73 > St_P > 0.84$ and $4.63 > St_P > 1.91$ for rotated square and parallel triangle tube arrays, respectively.

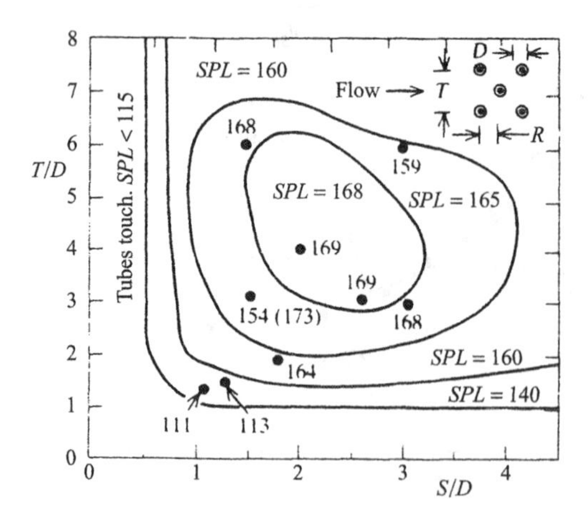

FIG. 28.61. Sound pressure level during acoustic synchronization for staggered arrays, Blevins and Bressler (1993J)

28.4.11 *Maximum sound level and its prediction*

The maximum sound pressure level in the acoustically excited staggered tube arrays was measured by Blevins and Bressler (1993J) and is shown in Fig. 28.61. The extent of high noise is in the range $1.1 < R/D < 4$ and $1.1 < T/D < 7$. This extent appears to be almost twice that of the in-line arrays shown in Fig. 28.35. This discrepancy is only apparent and may disappear if T is replaced by $T/2$, the distance between the tube columns.

A proliferation of empirical and semi-empirical criteria for predicting acoustic excitation has taken place in response to practical need in industry. They are referred to in chronological order: Grotz and Arnold (1956P), Chen (1968J), Parker (1978J), Fitzpatrick (1985J, 1986J), Blevins (1984R, 1986J), Blevins and Bressler (1987Ja, 1987Jb, 1993J), Ziada *et al.* (1989J), Ziada and Oengören (1992J), Weaver (1993R), etc. The criteria are based on: input energy, threshold acoustic pressure, maximum acoustic pressure, particle velocity, damping parameter, etc. They are all of limited accuracy.

28.5 Non-uniform flow in and behind arrays

28.5.1 *Historical introduction*

The puzzling paradox of non-uniform flow behind a single row of tubes has been discussed in Section 28.2. Figure 28.4 showed that biased jets merged in all possible ways, and formed irregular wide and narrow near-wakes. This phenomenon has been observed by many researchers, but has been ignored, presumably due to its paradoxical nature.

One of the first successful visualizations of interstitial flow was carried out by Lohrisch (1929P). Figure 28.62(a,b) shows the interstitial flow visualization made according to Thoma's proposal. The porous surfaces of the tubes are wetted with hydrochloric acid, and ammonia vapour is mixed with air. When the latter reaches the acid, a white vapour is formed. This method makes boundary layers and wakes visible. Figure 28.62(b) shows alternate narrow and wide near-wakes

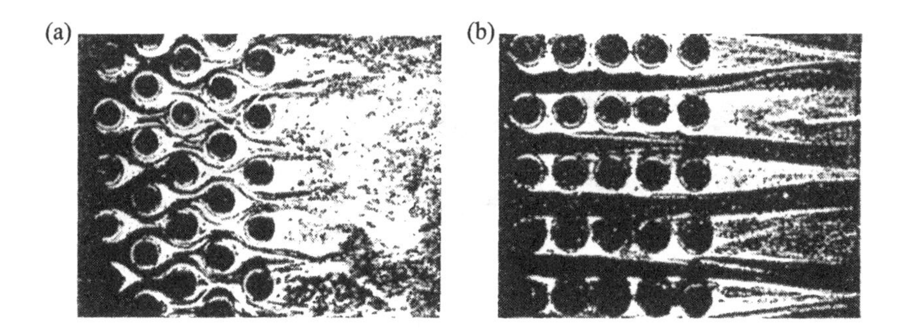

FIG. 28.62. Flow visualization in and behind, (a) normal-triangle array $P/D = 1.6$, (b) in-line $T/D = 2.1$, $S/D = 1.4$, Lohrisch (1929P)

behind the in-line array, and Fig. 28.62(a) shows one wide, two normal, and two narrow near-wakes behind the staggered array.

Another pre-war visualization in a water table was carried out by Wallis (1939J). A wide range of in-line and staggered arrays was tested, but, regrettably, the near-wakes behind the last row were not shown. An exception was Fig. 28.63(a) where the interstitial flow and part of the last row near-wakes are seen for $T/D = 1.5$ and $R/D = 2.6$, in a four-row deep in-line array. The wide near-wake behind the middle tube in the last row is surrounded by two narrow and biased near-wakes. A similar and more recent example of wide and narrow wakes is seen in Fig. 28.63(b) by Umeda and Yang (1999J) for $T/D = 2$, $R/D = 4.6$ at $Re = 15.7$k.

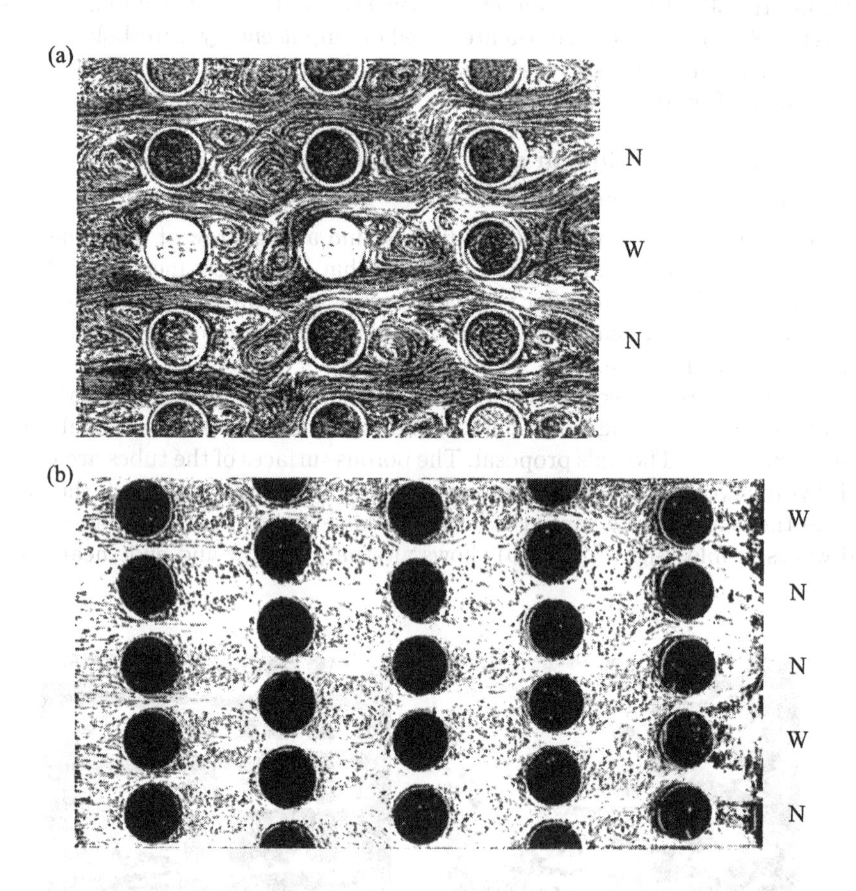

FIG. 28.63. Water visualization, (a) in-line array $T/D = 1.5$, $S/D = 2.75$, Wallis (1939J), (b) staggered array $T/D = 1.2$, $R/D = 2.25$, $Re = 15.7$k, Umeda and Yang (1999J)

28.5.2 *Non-uniform interstitial flow*

The fundamental question for practical applications is whether the non-uniformity observed behind arrays is also present within them. Tangemann (1979P) measured velocity and turbulence simultaneously in two adjacent transverse gaps by using two hot wires. The tube was situated in the second row of a square array, $P/D = 1.5$, seven rows deep, and with seven tubes per row. The two monitored gaps were around the middle tube in the second row.

Figure 28.64(a,b) shows the measured gap velocities, V_{g1} and V_{g2}, and the corresponding turbulence intensities, Ti_1 and Ti_2, in terms of the free stream velocity V. There are three features:

(i) the velocities on two sides of the tube are different;

(ii) the gap with high velocity has low turbulence and vice versa;

(iii) the change in gap velocities and Ti is always synchronous, and does not occur at the same velocity.

The last feature may be attributed to the irregular change-over of biased jets behind the last row.

Tangemann (1979P) examined the effect of increasing and decreasing the free stream velocities on the V_g jump. Figure 28.64(b) shows that the V_{g1} and V_{g2},

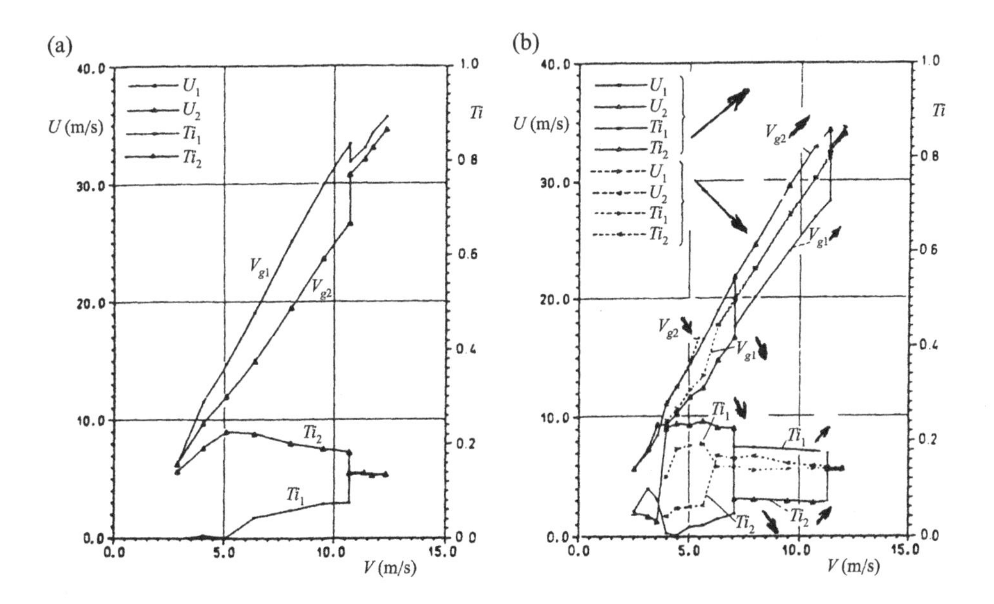

FIG. 28.64. Gap velocity and turbulence intensity in in-line $P/D = 1.5$, $5k < Re < 210k$, (a) increasing velocity, (b) increasing (full line) and decreasing (dashed line) velocity, Tangemann (1979P)

and Ti_1 and Ti_2 curves follow different paths when the velocity is increased and decreased. This feature confirms the metastable nature of interstitial flow in the tube array.

Tangemann (1979P) tried to equalize the interstitial flow in two gaps by attaching tripping wires at 102° to the tubes in the first row upstream of the two gaps. The differences in lift for the increasing and decreasing velocities were reduced in magnitude but not obliterated.

APPENDIX

A1 Glossary of terms

A2 Non-dimensional similarity parameters

A3 Epitome of flow regimes

A4 Abbreviations

A1 Glossary of terms

Aspect ratio	cylinder's length over diameter ratio, L/D
Bluff body*	associated with a large separation region
Blockage ratio	diameter over test section size ratio, D/B
Breakaway*	flow separation without adverse pressure gradient
Confluence point*	where four (2-D) or six (3-D) streamlines meet
Detachment line*	theoretical tangential separation
Discontinuity surface	theoretical step-jump in velocity
Disturbed region*	where the local velocity is different from the free stream velocity
Eddy*	real viscous rotational flow subject to viscous diffusion and dissipation (see vortex)
Eddy strength	circulation
Fingers*	reversed shear layer flow
Flow regime	distinct flow pattern of limited variation
Flow state	laminar, transitional, or turbulent
Free shear layer	separated boundary layer
Free stream	undisturbed upstream flow
Free end	finite cylinder protruding in flow
Free streamline	theoretical separation line
Governing parameter*	having strong effect on flow
Hollow vortex	circulation in the form of a ring
Influencing parameter*	weak effect on flow (see governing)
Instability	the cause of a change from steady to oscillatory flow
Nominally two-dimensional*	quasi 2-D for a finite cylinder between walls or end plates
Rankine vortex	having a solid-body rotation at the core
Reattachment	separated flow adheres to the surface
Resistance*	drag in creeping (low *Re*) flow
Separation	flow leaves the surface under the action of adverse pressure gradient
Separation bubble	flow between separation and reattachment
Spacing ratio	arrangement of eddies in wake (see Kármán number)
Swirl*	3-D individual eddy with axial velocity
Synchronization	equality of natural and eddy shedding frequency over a range of velocities
Theoretical model*	an approximate description of real flow restricted by simplifying assumptions
Topology	kinematic display of flow structures
Transition	change of flow states
Ultimate*	the last
Universal	common for all
Velocity defect	reduced velocity in wakes
Vortex*	theoretical inviscid eddy (see eddy)
Vorticity	elemental circulation
Vortex sheet	theoretical replacement of actual continuous vorticity sheet by discrete point vortices
Wake	flow region with velocity defect near-wake – close to bluff body far-wake – far downstream from bluff body

*Adopted by the author in this book.

A2 Non-dimensional similarity parameters

Name	Definition	Equation
Acosta*	$\frac{\text{Inertia flow}}{\text{Elastic flow}}$	$Ac = D^2/\tau\nu$
Frössling	$Nu/Re^{1/2}$	$Fr = Nu/Re^{1/2}$
Grashof	$\frac{\text{Buoyancy force}}{\text{Viscous force}}$	$\frac{D^3 g(T_\omega - T_0)}{\omega^2 T_0}$
Kármán*	$\frac{\text{Transverse spacing of rows}}{\text{Streamwise spacing of eddies}}$	$Ka = b/a$
Knudsen	$\frac{\text{Mean molecular free path}}{\text{Relevant body dimension}}$	$Kn = \frac{c}{D}$
Mach	$\frac{\text{Local velocity}}{\text{Speed of sound}}$	$Ma = V/a$
Nusselt	$\frac{\text{Heat transfer}}{\text{Thermal conductivity}}$	$Nu = \frac{\overline{\alpha} DA}{k}$
Prandtl	$\frac{\text{Molecular transport}}{\text{Heat transport}}$	$Pr = \frac{\rho\nu C_p}{k}$
Rayleigh	$GrPr$	$Ra = GrPr$
Reynolds	$\frac{\text{Inertia force}}{\text{Viscous force}}$	$Re = \frac{VD}{\nu}$
Riegels*	Aspect ratio $\times$ Re	$Ri = Reh/D$
Roshko*	$Re \times St$	$Ro = \frac{fD^2}{\nu}$
Sherwood	$\frac{\text{Mass transfer coefficient}}{\text{Diffusion coefficient}}$	$Sh = \frac{C_m}{C_{\text{dif}}}$
Strouhal	$\frac{\text{Eddy shedding frequency}}{\text{Time to pass distance } D}$	$St = \frac{Df}{V}$
Taylor*	Empirical	$Ta = Ti\left(\frac{D}{Ts}\right)$

* Not yet adopted.

A3 Epitome of disturbance-free flow regimes

State			Regime	*Re* ranges	L_w/L_f	C_D
L	Laminar	1	No-separation	0 to 4–5	None	↘
		2	Closed wake	4–5 to 30–48	↗	↘
		3	Periodic wake	30–48 to 180–200	↘	↗
TrW	Transition in wake	1	Far-wake	180–200 to 220–250	↘	↗
		2	Near-wake	220–250 to 350–400	↗	↘
TrSL	Transition in shear layers	1	Lower	350–400 to 1k–2k	↗	↘
		2	Intermediate	1k–2k to 20k–40k	↘	↗
		3	Upper	20k–40k to 100k–200k	Same	Same
TrBL	Transition in boundary layers	0	Precritical	100k–200k to 300k–340k	↗	↘
		1	Single bubble	300k–340k to 380k–400k	(?)	↓
		2	Two-bubble	380k–400k to 500k–1M	(?)	↓
		3	Supercritical	500k–1M to 3.5M–6M	None	↗
		4	Postcritical	3.5M–6M (?)	(?)	Same
T	Fully turbulent	1	Invariable	(?) to ∞	(?)	Same
		2	Ultimate		(?)	(?)

L_w = length of near-wake (only for the L2 regime),
L_f = length of eddy formation region (from L3 to T2 regimes),
Abbr. ↗ increase, ↘ decrease, ↓ rapid decrease, (?) not known.

A4 Abbreviations

For symbols see Nomenclature, p. 675.

2-D	Two-dimensional
3-D	Three-Dimensional
AGARD	Advisory Group for Aerospace Research and Development, Neuilly, France
AIAA	American Institute of Aeronautics and Astronautics
ARC	Aeronautical Research Council, London (publishes current papers, reports and memoranda)
ASCE	American Society of Civil Engineers
ASME	American Society of Mechanical Engineers
AVA	Aerodynamische Versuchanstalt, Göttingen, Germany
BBAA	Bluff Body Aerodynamics and Applications
BNES	British Nuclear Energy Society
calc.	Calculations
circ.	Circulation
comp.	Comparison
Comp.	Computation
conc.	Concentric
Conf.	Conference
corr.	Correlation
cyl.	Circular cylinder
dia.	Diameter
disturb.	Disturbance
eqn	Equation
exc.	Excitation
Exp.	Experimental
FV	Flow visualization
Hist.	Historical
IEEE	Institute of Electrical and Electronics Engineers
ISAS	Institute of Space and Aeronautical Sciences, University of Tokyo, Japan
IUTAM	International Union of Theoretical and Applied Mechanics
JSME	Japan Society of Mechanical Engineers
Ka	Kármán number
MIT	Massachusetts Institute of Technology
NACA	National Advisory Committee for Aeronautics
NASA	National Aeronautics and Space Administration
NMI	National Maritime Institute, Feltham, UK
NPL	National Physical Laboratory, Teddington, UK
NRL	Naval Research Laboratory, Washington, DC
ONERA	Office National d'Etudes et de Recherches Aerospatiales, Chattilon, France
osc.	Oscillation
OTC	Offshore Technology Conference, Houston, TX
PS	Power spectrum
Rev.	Review
rot.	Rotating
stag.	Stagnation
Th.	Theoretical
Ti	Intensity of turbulence

D. REFERENCES

Non multa sed multum.
(Not many but much)
Roman Proverb

Each reference is given in a conventional way: author(s), title, source, etc. However, this is often insufficient when further details are required. Hence, an extension is added in the form of a brief qualitative and quantitative description of the contents of each reference.

Another helpful innovation is the separation of books (B) and reviews (R) from papers. The latter are further split into two groups of papers: (J) those published in learned journals (easy access); (P) proceedings, reports, theses, etc. (harder access). Finally, the very last number at the end of each reference refers to the chapter where the reference is mentioned or, if in parentheses, it refers to the chapter related to and not mentioned.

Books (B)

Blake, W. K. (1986B). *Mechanics of Flow-Induced Sound and Vibration*, Vol. 2, Academic Press, New York. (Rev.). 28.

Drazin, P. G. and Reid, W. H. (1981B). *Hydrodynamic Stability*, Cambridge University Press. (Rev.). 24.

Dryden, H. L., Murnaghan, F. D., and Bateman, H. (1956B). *Hydrodynamics*, Dover Publ. Inc., New York. (Rev.). 8–13, 24.

Furman, T. (1981B). *Approximate Methods in Engineering*, Academic Press, London. (Quote).

Goldstein, S. ed. (1965B). *Modern Developments in Fluid Mechanics*, Dover Publ. Inc., New York. 21, 22, 23, 24.

Hoerner, S. F. (1956B). *Fluid Dynamic Drag*, Published by the author, Midland Park, NJ, USA. (Exp. Compilation). (24).

Kármán, Th. von. (1956B). *Collected Works of Theodore von Kármán*, 4 Vols, Butterworth, London. 22.

Landau, L. D. and Lifschitz, E. M. (1987B). *Fluid Mechanics*, 2nd Ed., Pergamon Press, London, Course of Theoretical Physics, Chapter III, Turbulence, 95–128. 21.

Milne-Thomson, L. M. (1967B). *Theoretical Hydrodynamics*, 5th Ed., Macmillan, London. (Th.). 23.

Muttray, N. (1932B). In: *Handbook of Experimental Physics*, in German, ed. Wien, W. and Harms, F., Vol. 4, Akademische Verlag, Leipzig. 10, 21.

Nakayama, Y. ed. in chief. (1984B). *Flow, a Photographic Album*, in Japanese, Maruzen, Tokyo. 26.

Naudascher, E. and Rockwell, D. (1994B). *Flow Induced Vibration, an Engineering Guide*, Hydraulics Structures Design Manual 7, A. A. Balkema, Rotterdam. 22, 26, 28.

Prandtl, L. (1952B). *Essentials of Fluid Mechanics* (translation W. M. Reaves), Blackie, London. (Hist. Rev.). 21, 22, 24.

Prandtl, L. (1961B). *Collected Works*, in German, ed. Tollmien, W., Schlichting, H., and Görtler, H., Springer, Berlin. 5, 6, 22, 25.

Prandtl, L. and Tietjens, O. G. (1957B). *Applied Hydro- and Aero-Mechanics* (translation J. Kestin), Dover, New York. (Hist. Rev.). 24.

Rankine, W. J. M. (1881B). *Miscellaneous Scientific Papers*, Griffin, London. 23.

Schlichting, H. (1979B). *Boundary Layer Theory* (translation J. Kestin), 7th Ed., McGraw Hill, New York. 22.

Scruton, C. (1981B). *An Introduction to Wind Effects on Structures*, Engg Design Series 40, Oxford University Press. 22.

Sumer, B. M. and Fredsøe, J. (1997B). *Hydrodynamics Around Cylindrical Structures*, World Scientific Publ. Co., Singapore. 27.

Thwaites, B. ed. (1960B). *Incompressible Aerodynamics*, Clarendon Press, Oxford. 24.

Van Dyke, M. (1982B). *An Album of Fluid Motion*, The Parabolic Press, Stanford, CA, USA. (Compilation of FV). 4, 6, 16, 20, 24, 26, 28.

Žukauskas, A. A., Ulinskas, R., and Katinas, V. I. (1988B). *Fluid Dynamics and Flow-Induced Vibrations of Tube Banks*, Hemisphere Publ., Washington, DC. 28.

Reviews (R)

Ackeret, J. (1925R). New investigation at the aerodynamic research institute Göttingen, in German. *Zeit. für Flugtechnik und Motorluft-Schiffahrt.*, **16**, 49–52. (Rev. rotating cylinder, Magnus effect). 24.

Anonymous, (1984R). Cylinder groups: mean forces on pairs of long circular cylinders. *Engg Sci. Data Unit*, ESDU 84015, 28 pp. (Rev. tandem, side-by-side, staggered, forces). (26).

Barrington, E. A. (1973R). Acoustic vibration in tubular exchangers. *Chem. Engg Progr.*, **69**, 62–8. (Rev. excitation, avoidance). (28).

Basu, R. I. (1985R). Aerodynamic forces on structures of circular cross-section. Part 1, Model-scale data obtained under two-dimensional conditions in low-turbulence stream. *J. Wind Engg Ind. Aero.*, **21**, 273–94. (Rev. surface roughness, C_D, St, C'_L). (5, 6, 7, 22).

Basu, R. I. (1986R). Aerodynamic forces on structures of circular cross-section. Part 2, The influence of turbulence and three-dimensional effects. *J. Wind Engg Ind. Aero.*, **24**, 33–59. (Rev. Ti, C_D, C_p, C_{pb}, u', C'_p, C'_L, H/D). (5, 6, 7, 13, 21, 22).

Bearman, P. W. (1998R). Developments in understanding of bluff body flows. *JSME Int. J.*, **41B**, 14–25. (Rev. 3-D flow behind nominally 2-D cyl., wavy plate, comp. PIV). 21.

Blevins, R. D. (1984R). Review of sound induced by vortex shedding from cylinders. *J. Sound Vib.*, **92(4)**, 455–70. (Rev. Th. Exp. single cyl. tube arrays). 28.

Blevins, R. D. (1993R). Turbulence-induced vibration. In: *Technologies for the '90s*, ed. Au-Yang, M. K., ASME, New York, 683–709. (Rev. random pressure–structure coupling, Th. Exp.). (28).

Cohan, L. J. and Dean, W. J. (1965R). Elimination of destructive, self-excited vibrations in large, gas and oil fired utility units. *ASME J. Engg Power*, **87**, 223–8. (Rev. acoustic exc.). 28.

Coutanceau, M. and Defaye, J. R. (1991R). Circular cylinder wake configurations: a flow visualisation survey. *Appl. Mech. Rev.*, **44**, 255–305. (Rev. FV). (3, 4, 5, 24, 28).

Eisenger, F. L., Sullivan, R. E., and Francis, J. T. (1994R). A review of acoustic vibration criteria compared to in-service experience with steam generator in line tube banks. *J. Pressure Vessel Technol.*, **116**, 17–23. (Rev. acoustic excitation). 28.

Ericsson, L. E. (1980R). Kármán vortex shedding and the effect of body motion. *AIAA J.*, **18**, 935–44. (Rev. Magnus effect, moving wall effect). (24).

Ericsson, L. E. and Reding, J. P. (1980R). Vortex-induced asymmetric loads in 2-D and 3-D flows. *AIAA 18th Aerospace Sci. Meeting*, Paper AIAA 80-00181. (Rev. asymmetric eddies, Re, Ma, spin, noise-induced asymmetric eddies, Th., side forces, alleviation of side loads, straight and helical strakes). 25.

Farell, C. (1981R). Flow around fixed circular cylinders: fluctuating loads. *ASCE J. Engg Mech. Div.*, **107**, 565–88. Disc. 1153–6. (Rev. surface roughness, turbulence, C'_L, C'_D). 11, 22.

Flettner, A. (1925R). Application of aerodynamic knowledge to wind-driven ships, in German. *Zeit. für Flugtechnik und Motorluft-Schiffahrt*, **16**, 52–65. Disc. 63–5. (Rev. rotating cylinder, C_L, C_D, Buchau ship). 24.
See also Flettner, A. (1925J). The Flettner rotor ship. *Engineering*, **19**, 117–120.

Glenny, D. E. (1966R). A review of flow around circular cylinders, stranded cylinders and struts inclined to the flow direction. *Australian Defence Sci. Service, Aero. Res. Lab.*, Mech. Engg Note 284. (Rev. 10k $< Re <$ 550k, C_D, C_{Dn}, C_L, C'_L, C'_D, C_p, L_c). (25).

Hall, M. (1966R). The structure of concentrated vortex cores. *Progr. Aero. Sci.*, **7**, Chapter 4, Pergamon Press. (Rev.).

Hasimoto. H. and Sano, O. (1980R). Stokeslets and eddies in creeping flow. *Annu. Rev. Fluid Mech.*, **12**, 335–61. (Th. Stokes flow, 2-D, confined eddies).

Hirschfeld, F. (1977R). Wind power dream or reality. *Mech. Engg*, **99**, 20–8. (Rev. windmills, rotors). 24.

Ishigai, S., Nishikawa, E., and Yagi, E. (1973R). Structure of gas flow in tube banks with tube axes normal to flow. *JSME Int. Symp. Marine Engg*, Tokyo, Nov. 1973, 1-5-23 to 1-5-33. (Exp. single tube, single row, two rows, in-line, stagger, $0 < T/D < 2$, $1 < R/D < 5$, 2.1k $< Re <$ 4k, FV, St, B/B_D, single column, $1.5 < S/D < 2.5$, 5k $< Re <$ 13k, St, flow patterns, classification, St, osc., δ, V_r, Sc). 28.

Iversen, J. D. (1979R). Autorotating flat-plate wings: the effect of the moment of inertia, geometry, and Reynolds number. *J. Fluid Mech.*, **92**, 327–48. (Rev. Th. Exp. C_L/C_D). (24).

King, R. (1977R). A review of vortex shedding research and its application. *Ocean Engg*, **4**, 141–72. (Rev.). 21, 26.

Mair, W. A. and Maull, D. J. (1971R). Aerodynamic behaviour of bodies in the wake of other bodies. *Phil. Trans. Roy. Soc., Lond.*, **269A**, 425–37. (Rev. D-cyl., H/D, tandem cyl., two rows, square prisms). (26).

Matsumoto, M. (1999R). Vortex shedding of bluff bodies: a review. *J. Fluids Struct.*, **13**, 791–811. (Rev. rect. cyl., bridge girders, yawed cables). (25).

Moretti, P. M. (1986R). Caught in a cross flow – The paradox of flow-induced vibrations. *Mech. Engg*, **108**, 56–61. (Popular rev., single row). (28).

Morgan, P. G. (1960R). The stability of flow through porous screen. *J. Roy. Aero. Soc.*, **64**, 359–62. (Rev.). 28.

Morkovin, M. V. (1964R). Flow around circular cylinders: a kaleidoscope of challenging fluid phenomena. *Proc. ASME Symposium on Fully Separated Flow*, Philadelphia, 102–18. (Rev. $0.1 < Re <$ 1M, L/D, D/B, Ti, Ts, BL, FV). 21.

Nielsen, J. N. (1979R). Missile aerodynamics – past, present, future. *AIAA Paper 79-1819.* (Rev.). 25.

Niemann, H. J. and Hölscher, N. (1990R). A review of recent experiments on the flow past circular cylinders. In: *Bluff Body Aerodynamics and its Applications*, ed. Ito, M., *et al.*, Elsevier, Amsterdam, 197–210. (Rev. smooth and rough cyl., C_D, St, C_p, K_s/D, Ti, free end). (22).

Oertel, H. (1990R). Wakes behind blunt bodies. *Annu. Rev. Fluid Mech.*, **22**, 539–64. (Rev.).

Ohya, Y., Okajima, A., and Hayashi, M. (1988R). Wake interference and vortex shedding. In: *Encyclopaedia of Fluid Mechanics*, **8**, ed. Cheremisinoff, N. P., Gulf Publ. Co., 323–89. (Rev. side-by-side, tandem, staggered, numerical analysis). 26.

Paidoussis, M. P. (1981R). Fluid-elastic vibration of cylinder arrays in axial and cross flow: state of the art. *J. Sound Vib.*, **76**, 329–60. (Rev. axial, in-line, staggered arrays). (28).

Prandtl, L. (1925R). Magnus effect and wind force ship, in German. *Die Naturwissenschaften*, **93**. See also English translation, *Nat. Adv. Committee Aero.*, NACA TM 367. (Rev. Magnus effect, rotor ship). (24).

Putnam, A. A. (1964R). Flow-induced noise and vibration in heat exchangers. *ASME WAM Paper 64 WA/HT21.* (Survey, in-line, stag., empirical correlation of data). (28).

Rockwell, D. (1990R). Active control of globally-unstable separated flows. *Inv. Lecture, ASME Symp. On Unsteady Flows*, Toronto, Canada. (Rev. active control, bluff-body wakes, suppression of eddy shedding, osc. cyl.). (4, 5, 24).

Rockwell D. (1998R). Vortex–body interactions. *Annu. Rev. Fluid Mech.*, **30**, 199–229. (Rev. parallel vortex shedding, leading edge instability, osc. body). (24).

Roshko, A. (1993R). Perspectives on bluff body aerodynamics. *J. Wind Engg Ind. Aero.*, **49**, 79–100. (Rev., C_{pb}, flow regimes, 3-D effects).

Sarpkaya, T. (1990R). On the effect of roughness on cylinders. *ASME J. Offshore Mech. Arctic Engg*, **112**, 334–40. (Rev. overview of six exps., osc. planar flow, C_d, C_m, C_L'). (22).

Sumer, B. M. and Fredsøe, J. (1995R). A review on vibrations of marine pipelines. *Int. J. Offshore Polar Engg*, **5**, 81–90. (Rev. scour, waves). 23.

Swanson, W. M. (1961R). The Magnus effect: a summary of investigations to date. *ASME J. Basic Engg*, **83**, 461–70. Disc. (Rev. Exp. 36k $< Re <$ 501k, $0 < V_r/V < 1$, C_L, C_D, δ, Bickley's calc.). 24.

Taneda, S. (1977R). Visual study of unsteady separated flows around bodies. *Progr. Aero. Sci.*, **17**, 287–348. (Rev. FV, flat plate, cyl. ellipse, rot. cyl.). 24.

Tobak, M. and Peak, D. J. (1982R). Topology of three-dimensional separated flow. *Annu. Rev. Fluid Mech.*, **14**, 51–85. (Rev.). 21.

Wardlaw, A. B. Jr. (1978R). High-angle of attack missile aerodynamics. Paper 5, AGARD LS-98. (Rev.). 25.

Weaver, D. S. (1993R). Vortex shedding and acoustic resonance in heat exchanger tube arrays. In: *Technologies for the '90s*, ed. Au-Yang, M. K., ASME, New York, 777–810. (Rev. *St*, eddy shedding, acoustic resonance, designer's guide). 28.

Weaver, D. S. and Fitzpatrick, J. A. (1988R). A review of cross-flow induced vibrations in heat exchanger tube arrays. *J. Fluids Struct.*, **2**, 73–93. (Rev. *St*). 28.

Willhofft, F. O. (1927R). Industrial applications of the Flettner rotor. *Mech. Engg*, **49**, 249–54. (Rev. Flettner's ship and windmill). 24.

Williamson, C. H. K. (1996R). Vortex dynamics in the cylinder wake. *Annu. Rev. Fluid Mech.*, **28**, 477–539. 21.

Zdravkovich, M. M. (1977R). Review of flow interference between two circular cylinders in various arrangements. *J. Fluids Engg*, **99**, 618–33. (Rev. tandem, side-by-side, staggered, $0 < S/D < 50$, C_p, C_D, u/V, St, FV, C_{pb}, C_{pg}, θ_r, topology). 26.

Zdravkovich, M. M. (1981R). Review and classification of various aerodynamic and hydrodynamic means for suppressing vortex shedding. *J. Wind Engg Ind. Aero.*, **7**, 145–88. (Rev. means for suppressing eddy shedding). (27).

Zdravkovich, M. M. (1987R). The effects of interference between circular cylinders in cross flow. *J. Fluids Struct.*, **1**, 239–61. (Rev. two cyl., side-by-side, tandem, staggered, three cyl., in-line, triangle, square cluster, tube arrays). (26, 27, 28).

Zdravkovich, M. M. (1990R). Conceptual overview of laminar and turbulent flows past smooth and rough circular cylinders. In: *Bluff Body Aerodynamics and its Application*, ed. Ito, M., *et al.*, Elsevier, Amsterdam, 53–62. (Classification of flow regimes, surface roughness, $1 < Re < 10^7$, C_D, C_L, C'_D, C'_L, St). 22.

Zdravkovich, M. M. (1991R). A comparative overview of marine risers and heat exchanger tube banks. *J. Offshore Mech. Arctic Engg*, **113**, 30–6. (Rev. flow regimes, St, C_{Do}, oscillation, damping, pressure drop, stability, C_D, C_L, FV). 27, 28.

Zdravkovich, M. M. (1993R). Interstitial flow field and fluid forces. In: *Technologies for the '90s*, ed. Au-Yang, M. K., ASME, New York, 595–658. (Rev. in-line, stag. arrays, C_p, C_D, C_L, FV). 28.

Zdravkovich, M. M. (1996R). Different modes of vortex shedding: an overview. *J. Fluids Struct.*, **10**, 427–37. (Rev. modes of eddy shedding behind stationary and osc. cyl.).

Žukauskas, A. A. (1972R). Heat transfer from tubes in cross flow. *Adv. Heat Transfer*, **8**, 109–60. (Rev. one cyl., tube banks, C_p, C_f, $\overline{Nu}$, Nu, Δp). 28.

Papers published in learned journals (J)

Abd-Rabo, A. and Weaver, D. S. (1986J). A flow visualization study of flow development in a staggered tube array. *J. Sound Vib.*, **106**, 241–56. (Exp. rotated square, $P/D = 1.41$, $L/D = 12$, $Ti = 0.2\%$, osc. A/D, excitation, frequency PS, FV). 28.

Abernathy, F. H. and Kronauer, R. E. (1962J). The formation of vortex streets. *J. Fluid Mech.*, **13**, 1–20. (Th. Comp. b/a, Ka). 12, 26.

Achenbach, E. (1969J). Investigations on the flow through a staggered tube bundle at *Re* up to 10^7. *Wärme- und Stoffübertragung*, **2**, 47–52. (Exp. staggered, $T/D = 2$, $R/D = 1.4$, $L/D = 3.3$, $Ti = 0.7\%$, 10k $< Re <$ 9M, C_p, C_f, V_g/V, C_{DP}, θ_s, K/D). (28).

Achenbach, E. (1971Ja). Influence of surface roughness on the flow through a staggered tube bank. *Wärme- und Stoffübertragung*, **4**, 120–6. (Exp. staggered, $T/D = 2$, $R/D = 1.4$, 40k $< Re <$ 10M, $L/D = 3.33$, $Ti = 0.7\%$, $K/D =$ 0.11%, 0.45%, C_p, C_f, C_{DP}, θ_s, single row $T/D = 2$, C_p, C_f). 28.

Achenbach, E. (1971Jb). On the cross-flow through in-line tube banks with regard to the effect of surface roughness. *Wärme- und Stoffübertragung*, **4**, 152–5. (Exp. in-line $T/D = 2$, $S/D = 1.4$, $L/D = 3.33$, 4k $< Re <$ 10M, C_{DP}, C_f, C_p). 28.

Achenbach, E. (1971Jc). Influence of surface roughness on the cross-flow around a circular cylinder. *J. Fluid Mech.*, **46**, 321–35. (Exp. 40k $< Re <$ 3M, 0.11% $< K_s/D <$ 0.9%, $L/D = 3.3$, $D/B = 0.16$, C_p, C_{pb}, C_d, θ_{tr}, θ_s). 22.

Achenbach, E. (1977J). The effect of surface roughness on the heat transfer from a circular cylinder to the cross-flow of air. *Int. J. Heat Mass Transfer*, **20**, 359–69. (Exp. 22k $< Re <$ 4M, 0 $< K_s/D <$ 0.9%, C_p, C_d, Fr, $\overline{Nu}$, θ_{tr}). 22.

Achenbach, E. and Heinecke, E. (1981J). On vortex shedding from smooth and rough cylinders in range of Reynolds numbers 6×10^3 to 5×10^6. *J. Fluid Mech.*, **109**, 239–51. (Exp. 6k $< Re <$ 5M, 0.075% $< K_s/D <$ 3%, $L/D = 3.4$, 6.75, $D/B = 0.16$, C_d, St). 22.

Ackeret, J. (1934J). Investigation of wind load on a gas container model, in German. *Schweiz. Vereins von Gas- und Wasser-fachmannern, No 8*, 1–6. (Exp. $Re = 1.2$M, $H/D = 1.13$, gas container with scaffolding). 21.

Ackeret, J. (1936J). Wind load on brick stack of circular section, in German. *Schweizerische Bauzeitung*, **108**, 25–26. (Exp. 100k $< Re <$ 1M, brick roughness, $L/D = 5$, end plates, $D/B = 0$, C_d). 22.

Agui, J. H. and Andreopoulos, J. (1992J). Experimental investigation of a three-dimensional boundary layer flow in the vicinity of an upright wall mounted cylinder. *J. Fluids Engg*, **114**, 566–76. (Exp. 100k $< Re <$ 220k, horseshoe-swirl, $H/D = 2$, PS, FV, C_{pw}, correlation). (21).

Ahlborn, F. (1929J). The Magnus effect in theory and practice, in German. *Zeit. Flugtechnik und Motorluftschif-fahrt*, **20**, 642–57. See also English translation, *Nat. Adv. Committee Aero.*, NACA TM 567. (Th. Exp. Disc. FV). 24.

Ahmed, A. and Ostowari, C. (1990J). Longitudinally and transversely spaced cylinders in cross flow. *J. Wind Engg Ind. Aero.*, **36**, 1095–104. (Exp. 3 cyl. 56k $< Re <$ 188k, $L/D = 10.3$, $2D/B = 0.28$, u, u', C_p, FV). (26, 27).

Aiba, S., Ota, T., and Tsuchida, H. (1979J). Heat transfer and flow around a circular cylinder with tripping wires, in English. *Wärme- und Stoffübertragung*, **12**, 221–31. (Exp. 12k $< Re <$ 52k, $L/D = 8.6$, $D/B = 0.08$, $d/D = 0.005$, C_p, C_D, St, Nu, $\overline{Nu}$, $u_{\max}$). 17, 22.

Aiba, S., Ota, T., and Tsuchida, H. (1980Ja). Heat transfer from tubes closely spaced in an in-line tube bank. *Int. J. Heat Mass Transfer*, **23**, 311–19. (Exp. 4 cyl. in-line, 12k $< Re <$ 60k, $1.3 < S/D < 4.5$, C_p, Nu, $\overline{Nu}$, C_d). 27, 28.

Aiba, S., Tsuchida, H., and Ota, T. (1980Jb). Heat transfer around a tube in a bank. *Bull. JSME*, **23**, 1163–70. (Exp. 3 cyl. tripping wire, $L/D = 8.65$, $D/B = 0.08$, $Ti = 0.1\%$, 12k $< Re <$ 52k, $1.3 \leqslant S/D \leqslant 4.5$, C_p, Nu, $\overline{Nu}$). 27.

Aiba, S., Tsuchida, H., and Ota, T. (1981J). Heat transfer around a tube in a bank. *Bull. JSME*, **24**, 380–7. (Exp. 4 cyl. 12k $< Re <$ 50k, $1.15 < S/D < 4.5$, $L/D = 8.6$, $D/B = 0.08$, $Ti = 0.07\%$, C_p, Nu, θ_r, $\overline{Nu}$, C_d, u, u', FV). 27.

Aiba, S., Tsuchida, H., and Ota, T. (1982Ja). Heat transfer around tubes in in-line tube banks. *Bull. JSME*, **25**, 919–26. (Exp. in-line 10k $< Re <$ 60k, $P/D = 1.2$, 1.6, $L/D = 9$, $Ti = 6\%$, C_p, Nu, $\overline{Nu}$, C_d, v/V, Ti). 28.

Aiba, S., Tsuchida, H., and Ota, T. (1982Jb). Heat transfer around tubes in staggered tube banks. *Bull. JSME*, **25**, 927–33. (Exp. stagger, $T/D = 1.2$, 1.6, $R/D = 1.2$, 1.6, $L/D = 9$, $Ti = 0.06\%$, 8.6k $< Re <$ 36k, C_p, $\overline{Nu}$, Nu, C_d). 28.

Albarede, P. and Provansal, M. (1995J). Quasi-periodic cylinder wakes and the Ginzburg–Landau model. *J. Fluid Mech.*, **291**, 191–222. (Th. Exp. calc. $Re < 180$, Landau's model, eddy filaments, large L/D). 21.

Albrecht, T., Barnes, F. H., Baxendale, A. J., and Grant, I. (1988J). Vortex shedding from two cylinders in tandem. *J. Wind Engg Ind. Aero.*, **28**, 201–8. (Exp. $1.05 < D_1/D_2 < 1.14$, $D_e/D = 4.7$, $L/D = 45.6$, $D/B = 0.02$, St, $5.2 < S/D_1 < 10.7$, 10.2k $< Re <$ 31k, $0° < \alpha < 75°$). 26.

Alemdaroglu, N., Rabillat, J. L., and Goethals, R. (1980J). An aeroacoustic coherence function method applied to circular cylinder flows. *J. Sound Vib.*, **69**, 427–39. (Exp. Th. 50k $< Re <$ 177k, 0.035% $< K/D <$ 1.2%, acoustic disturb. PS, St, L_c, Re). (22).

Angrilli, F., Bergamaschi, S., and Cossalter, V. (1982J). Investigation of wall-induced modifications to vortex shedding from a circular cylinder. *J. Fluids Engg*, **104**, 518–22. Disc. **105**, 241–2. (Exp. 2.9k $< Re <$ 7.6k, $0.4 < G/D < 5$, u/V, u', b/a). (23).

Archibald, F. S. (1975J). Self-excitation of an acoustic resonance by vortex shedding. *J. Sound Vib.*, **38**, 81–103. (Exp. Th.). (28).

Armitt, J. (1980J). Wind loading on cooling towers. *ASCE J. Struct. Div.*, **106**, 623–41. (Exp. model/full-scale, C_p, stress). (22).

Auger, J. L. and Coutanceau, M. (1979J). Cross flow of air through a tube array. *Entropie*, **86**, 13–23. (Exp. single row, FV). (18), 28.

Ayoub, A. and Karamcheti, K. (1982J). An experiment of the flow past a finite circular cylinder at high subcritical and supercritical Reynolds numbers. *J. Fluid Mech.*, **118**, 1–26. (Exp. Re = 85k, 180k, 770k, $H/D = 12$, PS, oscillograms). 21.

Baban, F. and So, R. M. C. (1991J). Recirculating flow behind and unsteady forces on finite-span circular cylinders in a cross flow. *J. Fluids Struct.*, **5**, 185–206. (Exp. Re = 46k, $1 < H/D < 2$, u/V, u'/V, PS, auto-cross correlation). (21).

Baban, F., So, R. M. C., and Otugen, M. U. (1988J). Unsteady forces on circular cylinders in a cross flow. *Exp. Fluids*, **7**, 293–302. (Exp. Re = 46k, $H/D = 2$, C'_L, C'_D, St). 21.

Bagnold, R. A. (1974J). Fluid forces on a body in shear flow; experimental use of 'stationary flow'. *Proc. Roy. Soc.*, **340A**, 147–71. (Exp. moving floor, $12 < Re < 200$, C_L, C_D, surface roughness). 14, (23).

Bahl, S. K. (1970J). Stability of viscous flow between two concentric rotating cylinders. *Defence Sci. J.*, **20**, 89–96. (Th. porous cyl.). 24.

Baines, W. D. and Peterson, E. G. (1951J). An investigation of flow through screens. *ASME Trans.*, **73**, 467–80. (Th. Exp. Ti). (22).

Baird, R. C. (1954J). Pulsation-induced vibration in utility steam generation units. *ASME J. Combustion*, **25**, 38–44. (Exp. industrial unit, acoustic resonance, avoidance, baffles as cure). 28.

Bairstow, L., Cave, B. M., and Lang, M. A. (1921J). The two-dimensional slow motion of viscous fluid. *Phil. Trans. Roy. Soc.*, **223**, 383–432. (Th. Stokes flow, Lamb, blockage). (23).

Baker, C. J. (1979J). The laminar horseshoe vortex. *J. Fluid Mech.*, **95**, 347–61. (Exp. horseshoe-swirl, $500 < Re < 7k$, $1 < D/\delta < 7$, FV, C_p, x_s, C_f, topology). 21.

Baker, C. J. (1980J). The turbulent horseshoe vortex. *J. Wind Engg Ind. Aero.*, **6**, 9–23. (Exp. horseshoe-swirl, $8k < Re < 80k$, $1.2 < D/\delta < 3.2$, FV, C_p, x_s). 21.

Baker, C. J. (1985J). The position of points of maximum and minimum shear stress upstream of cylinders mounted normal to flat plates. *J. Wind Engg Ind. Aero.*, **18**, 263–74. (Exp. Horseshoe vortex, dimensional analysis). 21.

Baker, C. J. (1991J). The oscillation of horseshoe vortex systems. *J. Fluids Engg*, **113**, 489–94. (Exp. horseshoe-swirl, $2k < Re < 14k$, $1 < D/\delta < 17$, St, Re_c). 21.

Balasubramanian, S. and Skop, R. A. (1996J). A non-linear oscillatory model for vortex shedding from cylinders and cones in uniform and shear flows. *J. Fluids Struct.*, **10**, 197–214. (Th. comp. taper, van der Pol osc.). (22).

Balasubramanian, S., Haan, F. L., Szewczyk, A. A., and Skop, R. A. (1998J). On the existence of a critical shear parameter for cellular vortex shedding from cylinders in non-uniform flow. *J. Fluids Struct.*, **12**, 3–15. (Exp. Th. Comp. taper, shear parameters, $Re = 60k$, $T/R = 0.04$, PS, cells, numerical simulation). (22).

Ball, D. J. and Hall, C. D. (1980J). Drag of yawed pile groups at low Reynolds numbers. *ASCE J. Waterw. Harbors Coastal Engg Div.*, **106**, 229–38. (Exp. 5×5, 9×9 clusters, $Re = 10k$, $S/D = T/D = 8$, C_{DO}). 27.

Barnes, F. H., Baxendale, A. J., and Grant, I. (1986J). A lock-in effect in the flow over two cylinders. *J. Roy. Aero. Soc.*, **90**, 128–38. (Exp. $1.75 < D_1/D_2 < 2.37$, $15k < Re < 50k$, $0° < \alpha < 25°$, St, f_1, f_2, PS, C_{pb}, C_p, C_p', C_d, C_l). 26.

Batham, J. P. (1973J). Pressure distribution on circular cylinders at critical Reynolds numbers. *J. Fluid Mech.*, **57**, 209–28. (Exp. $Re = 111k$, $230k$, $L/D = 6.65$, $D/b = 0.05$, $Ti = 0.5\%$, 12%, $K/D = 0.22\%$, C_p, C_p', St, L_z/D, R_{pp}). 22.

Batham, J. P. (1985J). Wind tunnel tests on scale models of a large power station chimney. *J. Wind Engg Ind. Aero.*, **18**, 75–90. (Exp. surface roughness simulation of high Re, C_p, C_p', PS). (22).

Baxendale, A. J., Grant, I., and Barnes, F. H. (1985J). The flow past two cylinders having different diameters. *J. Roy. Aero. Soc.*, **89**, 125–34. (Exp. $Re = 14.5k$, $D_1/D_2 = 0.5$, $0° < \alpha < 30°$, $1.3 < S/D < 3.88$, C_p, C_{pb}, St, C_D, C_L). (26).

Baylac, G. and Gregoire, J. P. (1975J). Acoustic phenomena in a steam generating unit. *J. Sound Vib.*, **42**, 31–48. (Th. Exp.). (28).

Bearman, P. W. (1967J). On the vortex street wakes. *J. Fluid Mech.*, **28**, 625–41. (Exp. Th. splitter plate, base bleed, b/a, St_B). 13, 21.

Bearman, P. W. and Harvey, J. K. (1976J). Golf ball aerodynamics. *Aero. Quart.*, **27**, 112–22. (Exp. golf ball, C_D, dimples). 22.

Bearman, P. W. and Harvey, J. K. (1993J). Control of circular cylinder flow by the use of dimples. *AIAA J.*, **31**, 1753–6. (Exp. spherical dimples, 20k $<$ Re $<$ 300k, C_D, St). 22.

Bearman, P. W. and Wadcock, A. J. (1973J). The interaction between a pair of circular cylinders normal to a stream. *J. Fluid Mech.*, **61**, 499–511. (Exp. side-by-side, Re = 25k, $D/B = 0.038$, $1 < T/D < 2$, C_p, C_{pb}, C_{pb2}, St, L_c, C_L, C_D, FV). 26.

Bearman, P. W. and Zdravkovich, M. M. (1978J). Flow around a circular cylinder near a plane boundary. *J. Fluid Mech.*, **89**, 33–47. (Exp. Re = 45k, $0 < G/D < 2$, $\delta/D = 0.8$, C_p, C_{pb}, FV). 23.

Beguier, C., Giralt, F., and Keffer, J. F. (1978J). Dynamic field around two identical cylinders with weak coupling, in French. *Compt. Rend. Acad. Sci.*, Paris, **286**, Ser. 243–6. (Exp. 1k $<$ Re $<$ 10k). (26).

Belik, L. (1973J). The secondary flow about circular cylinders mounted normal to a flat plate. *Aero. Quart.*, **29**, 47–54. (Exp. dimensional analysis, 36k $<$ Re $<$ 220k, FV, C_p, x_s, Re_o). 21.

Bénard, H. (1908J). Formation of centres of circulation behind a moving obstacle, in French. *Compt. Rend. Acad. Sci.*, **147**, 839–42. (Hist. Exp. FV). 1, 3, 5, 21.

Bénard, H. (1913J). On the zone of formation of alternate eddies behind an obstacle, in French. *Compt. Rend. Acad. Sci.*, **156**, 1003–5. (Hist. Exp. FV). 3, 21.

Bénard, H. (1926J). On the incorrectness, for real fluids, of Kármán's theory of stability of alternate vortices, in French. *Compt. Rend. Acad. Sci.*, **182**, 1523–5. (Exp. disagreement of b/a with Kármán's theory). 3, 23.

Betz, A. (1925J). The 'Magnus effect', the principle of the Flettner waltz, in German. *Zeit. Vereine Deutsche Ingenieur*, **69**, 9–14. See also English translation, *Nat. Adv. Committee Aero.*, NACA TM 310, 1925. (24).

Bickley, W. G. (1928J). The influence of vortices upon the resistance experienced by solids moving through a liquid. *Proc. Roy. Soc.*, **119A**, 146–51. (Th. point vortex, drag and lift, rotating cylinder). 24.

Bisshopp, F. E. (1963J). Asymmetric inviscid modes of instability in Couette flow. *Phys. Fluids*, **6**, 212–17. (Th. non-axisymmetric modes). (24).

Blackburn, H. M. (1994J). Effect of blockage on spanwise correlation in a circular cylinder wake. *Exp. Fluids*, **13**, 134–6. (Exp. Re = 4k, $0.05 < D/B < 0.40$, corr. length). 23.

Blevins, R. D. (1986J). Acoustic modes of heat exchanger tube bundles. *J. Sound Vib.*, **109**, 19–31. (Th. Exp., a_{eff}, in-line stagger, comparison exp.-th.). 28.

Blevins, R. D. and Bressler, M. M. (1987Ja). Acoustic resonance in heat exchanger tube bundles. Part I, Physical nature of the phenomena. *ASME J. Pressure Vessel Technol.*, **109**, 275–81. (Exp. acoustic resonance, baffles, Helmholtz resonator). (28).

Blevins, R. D. and Bressler, M. M. (1987Jb). Acoustic resonance in heat exchanger tube bundles. Part II, Prediction and suppression of resonance. *ASME J. Pressure Vessel Technol.*, **109**, 282–8. (Exp. in-line, $1.5 < P/D < 3.0$, stag. $1.2 < P/D < 4$, acoustic resonance, St, C_{bp}). 28.

Blevins, R. D. and Bressler, M. M. (1993J). Experiments on acoustic resonance in heat exchanger tube bundles. *J. Sound Vib.*, **164(3)**, 503–33. (Exp. single cyl., in-line, stagger arrays, V/fD, sound pressure level, prediction). 28.

Boasson, M. and Weihs, D. (1978J). Symmetric vortex shedding from a cylinder in confined flow. *Israel J. Technol.*, **16**, 56–63. (Th. streamlines, stability). (23).

Bohl, E. von. (1940J). On parallel air jets, in German. *Ingenieur Archiv.*, **11**, 295–305. (Exp. Hist. single row of flat plates, non-uniform flow). 28.

Bokaian, A. and Geoola, F. (1984J). Vortex shedding from two interfering circular cylinders. *ASCE J. Engg Mech.*, **110**, 623–8. (Exp. two cyl. $D_1/D_2 = 0.5$, 1, 2, St_1, St_2, 5.1k $< Re <$ 35k, $L/D = 18.7$, $D/B = 0.05$, $Ti = 6.5\%$, $1 < S/D < 5$, $0 < T/D < 3$). 26.

Bokaian, A. and Geoola, F. (1985J). Wake displacement as cause of lift force on cylinder pair. *ASCE J. Engg Mech.*, **111**, 92–9. (Exp. $D_1/D_2 = 0.25$, 4, $1.5 < S/D < 4$, $0 < T/D < 6$, C_{D1}, C_{D2}, C_{L2}, topology). 26.

Borges, A. R. J. (1969J). Vortex shedding frequencies of the flow through a two-row banks of tubes. *J. Mech. Engg Sci.*, **11**, 498–502. (Exp. two rows, St). 28.

Bradshaw, P. (1965J). The effect of wind tunnel screens on nominally two-dimensional boundary layers. *J. Fluid Mech.*, **22**, 679–87. (Exp. non-uniform flow, FV). 28.

Bragg, G. M., Kohli, H. S., and Sheshagiri, B. V. (1970J). Mean flow calculations behind arbitrarily spaced cylinders. *ASME J. Basic Engg*, **92**, 536–44. (Th. 1, 2, 3, 5 cyl. vel. profile prediction). (26).

Brede, M., Eckelmann, H., and Rockwell, D. (1996J). On secondary vortices in the cylinder wake. *Phys. Fluids*, **8**, 2117–24. (Exp. Comp. $160 < Re < 500$, PIV, FV, λ_2/D, $K/\pi DV$, $36 < L/D < 144$, topology). 21.

Brewster, D. B. and Nissan, A. H. (1958J). Hydrodynamics of flow between horizontal concentric cylinders. I, Flow due to rotation. *Chem. Engg*, **7**, 215–21. (Exp. Taylor–Dean problem). (24).

Brewster, D. B., Grossberg, P., and Nissan, A. H. (1959J). The stability of viscous flow between horizontal concentric cylinders. *Proc. Roy. Soc.*, **251A**, 76–91. (Exp. secondary Taylor's eddies). (24).

Brown, R. J. (1967J). Hydrodynamic forces on a submarine pipeline. *ASCE J. Pipeline Div.*, **92**, 9–19. Disc. PL3, 75–81. (Exp. $G/D = 0$, 60k < Re < 300k, C_p, C_d, C_l, spoiler). 23.

Bryson, A. E. (1959J). Symmetric vortex separation on circular cylinders and cones. *ASME J. Appl. Mech.*, **81**, 643–7. (Th.). (25).

Buresti, G. (1981J). The effect of surface roughness on the flow regime around circular cylinders. *J. Wind Engg Ind. Aero.*, **8**, 115–22. (Exp. 26k < Re < 280k, 0.1% < K_s/D < 1.2%, $L/D = 12$, 22, $D/B = 0$, C_d, St, flow regimes). 22.

Buresti, G. and Lanciotti, A. (1979J). Vortex shedding from smooth and roughened cylinders in cross-flow near a plane surface. *Aero. Quart.*, **30**, 305–21. (Exp. 85k < Re < 300k, $0 < G/D < 2.5$, $K/D = 0.1\%$, 0.35%, PS, St). (22, 23).

Burkhalter, J. E. and Koschmieder, E. L. (1973J). Steady supercritical Taylor vortex flow. *J. Fluid Mech.*, **58**, 547–60. (Exp. large Taylor's number, end effects, initial effects). (24).

Burton, T. E. (1980J). Sound speed in a heat exchanger tube bank. *J. Sound Vib.*, **71**, 157–60. (Th. added mass, σ). 28.

Busse, F. H. (1968J). Shear flow instability in rotating systems. *J. Fluid Mech.*, **33**, 572–89. (Th. non-linear instability). (24).

Carpenter, L. H. (1958J). On the motion of two cylinders in an ideal fluid. *J. Res. Nat. Bureau Standards*, **61**, 83–7. (Th. doublets, streamlines, plane boundary, tandem). 26.

Case, K. M. (1960J). Stability of inviscid plane Couette flow. *Phys. Fluids*, **3**, 143–8. (Th. Fourier–Laplace transform). (24).

Chandrasekhar, S. (1954J). The stability of viscous flow between rotating cylinders. *Mathematika*, **1**, 5–13. (Th. critical wave number). (24).

Chang, T. S. and Sartory, W. K. (1967J). Hydromagnetic stability of dissipative flow between rotating permeable cylinders. *J. Fluid Mech.*, **27**, 65–79. (Th. effect of radial flow). (24).

Cheeseman, I. C. (1968J). Circulation control and its application to stopped rotor aircraft. *J. Roy. Aero. Soc.*, **72**, 635–46. (Exp. C_p, C_L, C_D, C_T, wing). (24).

Cheeseman, I. C. and Seed, A. R. (1967J). The application of circulation control by blowing to helicopter rotors. *J. Roy. Aero. Soc.*, **71**, 451–63. Disc. 464–7. (Exp. blowing slot, C_L, Th.). (24).

Chen, P. Y. and Doepker, P. E. (1975J). Flow-induced forces on a cylinder for different wall confinements at high Reynolds numbers, $3.5 \times 10^5 < Re < 1.2 \times 10^6$. *J. Pressure Vessel Technol.*, **85**, 110–17. (Exp. 350k $< Re <$ 1.2M, $0.20 < D/B < 0.33$, $L/D = 3$, C_{Dg}, C'_{Lg}, C'_{Dg}, comparison). 23.

Chen, S. S. and Jendrzejczyk, J. A. (1987J). Fluid excitation forces acting on a square tube array. *J. Fluids Engg*, **109**, 415–23. (Exp. $S/D = T/D = 1.75$, 7 rows, $1\% < Ti < 11\%$, PS, C_D, C_L, C'_D, C'_L). 28.

Chen, Y. N. (1967J). Frequency of the Kármán vortex street in tube banks. *J. Roy. Aero. Soc.*, **71**, 211–14. (Exp. tube arrays, St). 28.

Chen, Y. N. (1968J). Flow induced vibration and noise in tube bank heat exchangers due to Kármán streets. *ASME J. Engg Ind.*, **90**, 134–46. Disc. Puchir, M. (1973J). *J. Engg Ind.*, **95**, 410–14. (Exp. St, acoustic resonance). 28.

Chiu, W. S. and Lienhardt, J. H. (1967J). On real fluid flow over yawed circular cylinders. *J. Basic Engg*, **89**, 851–7. Disc. (1969J), **91**, 132–3. (Exp. Th. boundary layer, St, C_D). (25).

Christensen, O. and Askergaard, V. (1978J). Wind forces on and excitation of a 130 m concrete chimney. *J. Ind. Aero.*, **3**, 61–77. (Exp. insitu., 6M $< Re <$ 108M, C_p, C'_p, C_D). 7, 21.

Clarkson, M. H., Malcolm, G. N., and Chapman, G. T. (1978J). A subsonic, high-angle-of-attack flow investigation at several Reynolds numbers. *AIAA J.*, **16**, 53–60. (Exp. 100k $< Re <$ 1.5M, $Ma = 0.25$, $45° < \alpha < 98°$, C_p, C_L, FV, C_{pn}). 25.

Cole, J. A. (1974J). Taylor vortices with short rotating cylinders. *J. Fluids Engg*, **96**, 69–70. (Exp.). (24).

Cole, J. A. (1976J). Taylor-vortex instability and annulus-length effects. *J. Fluid Mech.*, **75**, 1–15. (Th.). (24).

Coles, D. (1965J). Transition in circular Couette flow. *J. Fluid Mech.*, **21**, 385–425. (Exp. rot. conc. cyl., $0 < Re_o <$ 85k, $0 < Re_i <$ 15k, transition, spiral turbulence, FV). 24.

Couette, M. (1890J). Study of friction in liquids, in French. *Ann. Chim. Phys.*, **(6)21**, 433–510. (Exp. viscosity measurement between concentric cyl.). (24).

Coursimault, J. and Chabriere, J. (1971J). Vibration problems in nuclear power plant EDF3, in French. *La Houille Blanche*, No 5-1971, 429–40. (Exp. Th. sound field). (28).

Coutanceau, M. and Bouard, R. (1977J). Experimental determination of the main features of the viscous flow in the wake of a circular cylinder in uniform translation. Part 1, Steady flow. *J. Fluid Mech.*, **79**, 231–56. (Exp. $5 < Re < 40$, $D/B = 0.024, 0.09, 0.12$, $-(u/V)$, $-(u_{\max}/V)$, L_w/D, Re_{osc}, wake similarity). 23.

Dalton, C. (1971J). Allen and Vincenti blockage corrections in a wind tunnel. *AIAA J.*, **9**, 1864–5. (Th. calc. modified correction equation). 23.

Dalton, C. and Szabo, J. M. (1977J). Drag on group of cylinders. *J. Pressure Vessel Technol.*, **99**, 152–7. (Exp. 3 cyl., $1 < S/D < 5$, $2.7\text{k} < Re < 79\text{k}$, $\alpha = 0°$, $30°$, $60°$, C_{D1}, C_{D2}, C_{D3}). 27.

Dargahi, B. (1989J). The turbulent flow field around a circular cylinder. *Exp. Fluids*, **8**, 1–12. (Exp. horseshoe-swirl, $Re = 39\text{k}$, C_p, u, u', $0.57 < x/D < 8$, C_f, FV). 21.

Davey, A. (1962J). The growth of Taylor's vortices in flow between rotating cylinders. *J. Fluid Mech.*, **14**, 336–48. (Th. non-linear theory). 24.

Davey, A., Di Prima, R. C., and Stuart, J. T. (1968J). On the stability of Taylor's vortices. *J. Fluid Mech.*, **31**, 17–52. (Th. non-linear theory, bifurcation). 24.

Dayoub, A. H. (1982J). Aerodynamic forces on cylindrical structure in a wake of another cylindrical structure. *Appl. Sci. Res.*, **39(1)**, 3–20. (Exp. 2 cyls., $13\text{k} < Re < 72\text{k}$, C_{D2}, C_{L2}, C_p). 26.

Demirbilek, Z. (1990J). Experimental study of production risers in steady uniform flow. *ASCE J. Waterw. Harbors Coastal Engg Div.*, **116**, 575–91. (Exp. satellite risers, C_{DO}). (27).

Demirbilek, Z. and Halvorsen, T. (1985J). Hydrodynamic forces on multitube production risers exposed to current and waves. *J. Energy Resources Technol.*, **107**, 226–34. (Exp. satellite risers, C_D). 27.

Di Prima, R. C. (1959J). The stability of viscous flow between rotating concentric cylinders with a pressure gradient acting round the cylinder. *J. Fluid Mech.*, **6**, 462–8. (Th. pressure gradient). (24).

Di Prima, R. C. (1961J). Stability of non-rotationally symmetric disturbances for viscous flow between rotating cylinders. *Phys. Fluids*, **4**, 751–5. (Th. three-dimensional disturbances). (24).

Di Prima, R. C. and Eagles, P. M. (1977J). Amplification rates and torques for Taylor-vortex flows between rotating cylinders. *Phys. Fluids*, **20**, 171–5. (Th. Stuart and Thomson method). (24).

Diaz, F., Gavalda, J., Kawall, J. G., Keffer, J. F., and Giralt, F. (1983J). Vortex shedding from a spinning cylinder. *Phys. Fluids*, **26**, 3454–60. (Exp. rot. cyl., $Re = 9\text{k}$, $0 < V_r/V < 2.5$, $L/D = 50$, St, u', v'). 24.

Diaz, F., Gavalda, J., Kawall, J. G., Keffer, J. F., and Giralt, F. (1985J). Asymmetrical wake generated by a spinning cylinder. *AIAA J.*, **23**, 49–54. (Exp. rot. cyl., $Re = 9\text{k}$, $0 < V_r/V < 2.5$, $L/D = 30$, $D/B = 0.03$, u', v', C_d). 24.

Dimopoulos, H. G. and Hanratty, T. J. (1968J). Velocity gradients at the wall for flow around cylinder for Reynolds numbers between 60 and 360. *J. Fluid Mech.*, **33**, 303–19. (Exp. $60 < Re < 360$, C_f, St, BL). 3, 4, 25.

Donnelly, R. J. (1958J). Experiments on the stability of viscous flow between rotating cylinders. I. Torque measurements. *Proc. Roy. Soc.*, **246A**, 312–25. (Exp. wide range of ω_1/ω_2 and r_1/r_2). (24).

Donnelly, R. J. (1964J). Experiments on the stability of viscous flow between rotating cylinders. II. Enhancement of stability by modulation. *Proc. Roy. Soc.*, **281A**, 130–9. (Exp. rotary osc. of inner concentric cyl.). (24).

Donnelly, R. J. and Fultz, D. (1960J). Experiments on the stability of spiral flow between rotating cylinders. *Proc. Nat. Acad. Sci.*, Washington DC, **46**, 1150–4. (Exp. wide range of ω_1/ω_2, r_1/r_2). (24).

Donnelly, R. J. and Simon, N. J. (1960J). An empirical torque relation for supercritical flow between rotating cylinders. *J. Fluid Mech.*, **7**, 401–18. (Exp. torque in non-linear range). (24).

Drescher, H. (1956J). Measurement of time varying pressure on a cylinder in cross flow. *Zeit. Flugwiss.*, **4**, 17–21. (Exp. $Re = 112$k, $L/D = 3$, $D/B = 0.074$, C_p', C_D, C_L', C_D', St, FV). 5, 6, 25.

Dryden, H. L. and Hill, G. C. (1930J). Wind pressure on circular cylinders and chimneys. *Bureau of Standards J. Res.*, **5**, 653–93. (Exp. finite cyl. model, $H/D = 7.5$, $165\text{k} < Re < 330\text{k}$, C_p, wind: $H/D = 3$, C_p, C_D, $H/D = 11.36$, $Re = 17.5$M, C_p, C_d). 21.

Dunham, J. (1968J). A theory of circulation control by slot-blowing applied to a circular cylinder. *J. Fluid Mech.*, **33**, 495–514. (Th. blowing slot, vortex, source, C_j, δ, C_L). 24.

Dunham, J. (1970J). Experiments towards a circulation-controlled lifting rotor. *J. Roy. Aero. Soc.*, **74**, 91–103. (Exp. blowing slot, C_p, C_{pb}, C_L, Ma, C_D, C_j). (24).

Duty, R. L. and Reid, W. H. (1964J). On the stability of viscous flow between rotating cylinders. *J. Fluid Mech.*, **20**, 81–94. (Th. narrow-gap approximation). (24).

Eagles, P. M. (1974J). On the torque of wavy vortices. *J. Fluid Mech.*, **62**, 1–9. (Th. second wave instability). (24).

Eckerle, W. A. and Awad, J. K. (1991J). Effect of free stream velocity on the three-dimensional separated flow region in front of a cylinder. *J. Fluids Engg*, **113**, 37–44. (Exp. $960 < Re^{1/3}D/\delta < 2050$, C_p, u, v, FV, $K/\nu\delta$). 21.

Efthymiou, M. and Narayan, R. (1982J). Current-induced forces on submarine pipelines: a discrete vortex model. *Proc. Inst. Civil Eng.*, **73**, 109–23. (Th. Comp. wake development, C_L, C_D). (23).

Eiffel, G. (1912J). On the resistance of spheres in air motion, in French. *Compt. Rend. Acad. Sci.*, **195**, 1597. (Hist). 1, 6, 22.

Eisenger, F. L. (1980J). Prevention and cure of flow-induced vibration problems in tubular heat exchangers. *J. Pressure Vessel Technol.*, **102**, 138–45. (Exp. acoustic exc. Baffles, helical spacer). (28).

Eisenger, F. L., Francis, J. T., and Sullivan, R. E. (1996J). Prediction of acoustic vibration in steam generator and heat exchanger tube banks. *J. Pressure Vessel Technol.*, **118**, 221–36. (Emp. prediction, acoustic exc.). (28).

El-Taher, R. M. (1985J). Flow around two parallel circular cylinders in a linear shear flow. *J. Wind Engg Ind. Aero.*, **21**, 257–72. (Exp. two cyl., $Re = 34$k, $1 < S/D < 6$, $1.1 < T/D < 2$, C_p, C_{pb}, C_D). (15, 26).

Epik, E. Y. and Kozlova, L. G. (1974J). Effect of passage blockage and flow turbulence on the flow past circular cylinder. *Fluid Mech. – Sov. Res.*, **3**, 99–101. (Exp. $Re = 18$k, 75k, $1\% < Ti < 26\%$, C_p, u). 23.

Epshtein, L. A. (1977J). The hydroplaning of a rotating cylinder on water. *Fluid Mech. – Sov. Res.*, **6**, 68–77. (Th. Exp. C_L/C_D, $V_r/V = -1, 0, 1$). (24).

Ericsson, L. E. and Reding, J. P. (1979J). Criterion for vortex periodicity in cylinder wakes. *AIAA J.*, **17**, 1012–13. Disc. (1980J), **18**, 1408. (Disc. Straight separation line). (25).

Etzold, F. and Fiedler, H. E. (1976J). The near-wake structure of a cantilevered cylinder in a cross-flow. *Z. Flugwiss.*, **24**, 77–82. (Exp. $Re = 30$k, $1 \leqslant H/D \leqslant 10$, C_L, C_D, FV, L_w, topology). 21.

Farell, C. and Fedeniuk, S. K. (1988J). Effect of end plates on the flow around rough cylinders. *J. Wind Engg Ind. Aero.*, **28**, 219–30. (Exp. $42\text{k} < Re < 200\text{k}$, $L/D = 6$, $D/B = 0.12$, end plates, $K/D = 0.65\%$, 1%, C_p, C_{pb}, $C_{p\min}$, C_d, St, PS). 22.

Farell, C., Carrasquel, S., Guven, O., and Patel, V. C. (1977J). Effect of wind tunnel walls on the flow past circular cylinders and cooling tower models. *J. Fluids Engg*, **97**, 474–9. (Exp. cooling towers, D/B, C_p, C_{pb}, C_d, $C_{pb} - C_{p\min}$). 22.

Farell, C., Guven, O., and Maisch, F. (1976J). Mean wind loading on rough walled cooling towers. *ASCE J. Engg Mech. Div.*, **102**, 1059–81. (Exp. cooling towers, ribs, $150\text{k} < Re < 470\text{k}$, $D/B = 0.07$, $K/D = 0.4\%$, 0.16%, C_p, $C_{p\min}$, C_{pb}, $C_{pb} - C_{p\min}$). 22.

Farivar, D. (1981J). Turbulent uniform flow around cylinders of finite length. *AIAA J.*, **19**, 275–81. (Exp. $Re = 70$k, $Ti = 0.9\%$, $5 < H/D < 12.5$, C_p, C'_p, St). 21.

Fenstermacher, P. R., Swinney, H. L., and Gollub, J. P. (1979J). Dynamical instabilities and the transition to chaotic Taylor vortex flow. *J. Fluid Mech.*, **94**, 103–28. (Exp. final transition). (24).

Fiedler, H. E. and Wille, R. (1970J). Some observations in the near-wake of blunt bodies. *AIAA J.*, **8**, 1140–1. (Exp. Re = 40k, 240k, H/D = 4, 8, Ts, St). 21.

Filler, J. R., Marston, P. L., and Mih, W. C. (1991J). Response of the shear layers separating from a circular cylinder to small-amplitude rotational oscillation. *J. Fluid Mech.*, **231**, 481–99. (Exp. 250 < Re < 1k, rot. osc. cyl., FV, PS, f_{tr}/f, C_l). 24.

Fitzpatrick, J. A. (1982J). Acoustic resonances in in-line tube banks, Letter to Editor. *J. Sound Vib.*, **85**, 435–6. Paidoussis, M. P., Author's closure, detto, 437–41. (Acoustic resonance criteria, critical survey). (28).

Fitzpatrick, J. A. (1985J). The prediction of flow-induced noise in heat exchanger tube arrays. *J. Sound Vib.*, **99**, 425–35. (Exp. Th. prediction). (28).

Fitzpatrick, J. A. (1986J). A design guide proposal for avoidance of acoustic resonances in in-line heat exchangers. *J. Vib., Acoustics, Stress, Rel. Des.*, **108**, 296–300. (Design guide, acoustic excitation and avoidance) (28).

Fitzpatrick, J. A. and Donaldson, I. S. (1977J). A preliminary study of flow and acoustic phenomena in tube banks. *J. Fluids Engg*, **99**, 681–6. (Exp. in-line 1.3 < P/D < 1.97, 4k < Re < 10k, Re, St_a, St, sound intensity). (28).

Fitzpatrick, J. A. and Donaldson, I. S. (1980J). Row depth effects on turbulence spectra and acoustic vibrations in tube banks. *J. Sound Vib.*, **73**, 225–37. (Exp. in-line, P/D = 1.73, 1.97, L/D = 31.4, Ti = 0.6%, n = 1, 3, 5, 7, 10, 7.2k < Re < 15.9k, frequency PS, St, V_g/V, Ti_g). 28.

Fitzpatrick, J. A., Donaldson, I. S., and McKnight, W. (1988J). Strouhal numbers for flows in deep tube array models. *J. Fluids Struct.*, **2**, 145–60. (Exp. in-line, P/D = 1.73, 1.97, n = 20, L/D = 23.6, Ti = 0.6%, frequency PS, St, 4k < Re < 13k). 28.

Fox, T. A. (1990J). Flow visualisation at the center of a cross composed of tubes. *Int. J. Heat Fluid Flow*, **11**, 160–2. (Exp. Re = 10k, St, FV, 1 < S/D < 5). 26.

Fox, T. A. (1991J). Wake characteristics of two circular cylinders arranged perpendicular to each other. *J. Fluids Engg*, **113**, 45–50. (Exp. 2k < Re < 20k, 1 < S/D < 10, C_{pb}, topology, FV, vorticity, u'/v). 26.

Fox, T. A. and Toy, N. (1988J). The generation of turbulence from displaced cross-members in uniform flow. *Exp. Fluids*, **6**, 172–8. (Exp. Re = 20k, Ti = 0.17%, C_{ps1}, C_{ps2}, C_{pb1}, C_{pb2}, PS, 1 < S/D < 10, vorticity, u'/v). 26.

Fox, T. A. and West, G. S. (1990J). On the use of end plates with circular cylinders. *Exp. Fluids*, **9**, 237–9. (Exp. 3.3k < Re < 13.2k, 7 < B/D < 35, C_{pbz}, C_L'). 21.

Fox, T. A. and West, G. S. (1993Ja). Fluid-induced loading of cantilevered circular cylinder in a low-turbulence uniform flow. Part 1, Mean loading with aspect ratios in the range 4 to 30. *J. Fluids Struct.*, **7**, 1–14. (Exp. $Re = 44$k, $4 < H/D < 30$, C_p, C_d, St). 21.

Fox, T. A. and West, G. S. (1993Jb). Fluid-induced loading of cantilevered circular cylinder in a low-turbulence uniform flow. Part 2, Fluctuating loads on a cantilever cylinder of aspect ratio 30. *J. Fluids Struct.*, **7**, 15–28. (Exp. $Re = 44$k, $H/D = 30$, FV, C'_p, St). 21.

Fox, T. A. and West, G. S. (1993Jc). Fluid-induced loading of cantilevered circular cylinder in a low-turbulence uniform flow. Part 3, Fluctuating loads with aspect ratios 4 to 25. *J. Fluids Struct.*, **7**, 375–86. (Exp. $Re = 44$k, $4 < H/D < 25$, C'_p, C'_L, C'_d, St). 21.

Fredsøe, J. and Hansen, E. A. (1987J). Lift forces on pipelines in steady flow. *ASCE J. Waterw. Port, Coastal Ocean Engg*, **113**, 139–55. (Exp. Th. $Re =$ 25k, $0 \leqslant G/D < 1.3$, u/V, C_L, potential theory, comparison). 23.

Fredsøe, J., Hansen, E. A., Mao, Y., and Sumer, B. M. (1988J). Three-dimensional scour below pipelines. *J. Offshore Mech. Arctic Engg*, **110**, 373–9. (Exp. scour, physical model). 23.

Fujii, S. and Gomi, M. (1976J). A note on the two-dimensional cylinder wake. *J. Fluids Engg*, **98**, 318–20. (Exp. staggered, $Re = 20$k, u/v, $u'v'$). (26).

Fujita, H., Takahama, H., and Kawai, T. (1985J). Effects of tripping wires on heat transfer from a circular cylinder in cross flow. *Bull. JSME*, **28**, 80–7. (Exp. tripping wire, $Re = 50$k, $0.12\% < d/D < 4\%$, $15° < \theta_w < 90°$, $L/D = 8$, $D/B = 0.1$, C_p, C_{pb}, $C_{p\min}$, θ_s, δ, C_d, St). 22.

Funakawa, M. (1969J). The vibration of a cylinder caused by wake force in a flow. *Bull. JSME*, **12**, 1003–10. (Exp. single flexible cyl., V_r, Sc, FV, θ_s, A/D, Th.). (28).

Funakawa, M. and Umakoshi, R. (1970J). The acoustic resonance phenomena in the tube bank. *Bull. JSME*, **13**, 348–55. (Exp. Th. stag., $1.4 < T/D < 5$, $1.5 < S/D < 4$, St, FV, acoustic resonance). 28.

Fung, Y. C. (1960J). Fluctuating lift and drag acting on a cylinder in a flow at supercritical Reynolds numbers. *J. Aero. Sci.*, **27**, 801–14. (Exp. 300k $< Re <$ 1.4M, $L/D = 10$, $D/B = 1$, C'_L, C'_D, PS, forced osc.). 6, 23.

Furuya, Y. and Yoshino, F. (1975J). Aerodynamic force acting on a circular cylinder. *Bull. JSME*, **18**, 1002–10. (Exp. jet blowing, $Re = 32$k, $50° < \theta_j <$ $120°$, $L/D = 4$, $D/B = 0.14$, $0 < C_j < 0.46$, C_L, C_D, θ_s). 24.

Furuya, Y. and Yoshino, F. (1977J). Calculation of aerodynamic forces acting on a circular cylinder with tangential injection of air immersed in a uniform flow. *Bull. JSME*, **20**, 201–8. (Th. wall-jet flow, comp. with exp. C_{pb}, C_L). (24).

Gaster, M. (1969J). Vortex shedding from slender cone at low Reynolds numbers. *J. Fluid Mech.*, **38**, 565–76. (Exp. Th. cone, $0 < Re < 600$, $Tr = 0.023$, 0.055, modulation, *Ro*, van der Pol osc., FV). 22.

Gaster, M. (1971J). Vortex shedding from circular cylinders at low Reynolds numbers. *J. Fluid Mech.*, **46**, 745–56. (Exp. Th. cone, $Tr = 0.055, 0.027$, $75 < Re <$ 10k, *Ro*, van der Pol eqn, FV). 22.

Gerhardt, J. I. and Kramer, C. (1981J). Interference effects for groups of stacks. *J. Wind Engg Ind. Aero.*, **8**, 195–202. (Exp. 2, 3, 4, 5 cyl., $\bar{C}_{Dm}$, osc.). 27.

Gerich, D. and Eckelmann, H. (1982J). Influence of end plates and free ends on the shedding frequency of circular cylinders. *J. Fluid Mech.*, **122**, 109–21. (Exp. $50 < Re <$ 1.4k, $40 < L/D < 280$, PS, *St*, $10 < D_e/D < 150$). 21.

Gerrard, J. H. (1966Ja). The three-dimensional structure of the wake of a circular cylinder. *J. Fluid Mech.*, **25**, 143–64. (Exp. $Re = 85$, 235, 20k). 1, 3, 5, 21.

Gerrard, J. H. (1966Jb). The mechanism of the formation region of vortex shedding behind bluff bodies. *J. Fluid Mech.*, **25**, 401–13. (Exp. 2k $< Re <$ 50k, C_L', *St*, L_c/D, *Ti*). 5, 23.

Gerrard, J. H. (1978J). The wakes of cylindrical bluff bodies at low Reynolds numbers. *Phil. Trans. Roy. Soc.*, **288A**, 351–82. (Exp. $30 < Re <$ 2k, FV, *Ro*, *St*, f_{Tr}). 1, 3, 4, 5, 6, 23.

Glauert, M. B. (1957J). A boundary layer theory with applications to rotating cylinders. *J. Fluid Mech.*, **2**, 89–99. (Th. small perturbation). (24).

Gollub, J. P. and Swinney, H. L. (1975J). Onset of turbulence in a rotating fluid. *Phys. Rev. Lett.*, **35**, 927–30. (Exp. series of transitions). (24).

Golubovic, G. (1957J). Aerodynamic study of cooling tower of hyperboloid of revolution shape, in French. *Publ. Int. Assoc. Bridge and Struct. Engg*, **17**, 87–94. (Exp. cooling tower). 22.

Graham, J. M. R. (1969J). The effect of end-plates on the two-dimensionality of a vortex wake. *Aero. Quart.*, **20**, 237–47. (Exp. *D*-cylinder, end plates, parallel eddy shedding for $B/D \leqslant 4$, R_{uu}, PS, pressure gradient between end plates). 21.

Grass, A. J., Raven, P. W. J., Stuart, R. J., and Bray, J. A. (1984J). The influence of boundary layer velocity gradients and bed proximity on vortex shedding from free spanning pipelines. *ASME J. Energy Resources Technol.*, **106**, 70–8. (Exp. 2k $< Re <$ 4k, $0 < G/D < 2$, *St*, $2 < \delta_B/D < 6$). 23.

Griffiths, R. T. and Ma, C. Y. (1969J). Differential boundary layer separation effects in the flow over a rotating cylinder. *J. Roy. Aero. Soc.*, **73**, 524–6. (Exp. rotating cyl., 80k $< Re <$ 455k, $0 \leqslant V_r/V < 0.9$, C_p, C_L, C_D). (24).

Grove, A. S., Shair, F. H., Petersen, E. E., and Acrivos, A. (1964J). An experimental investigation of the steady separated flow past a circular cylinder. *J. Fluid Mech.*, **19**, 60–80. (Exp. Th. $25 < Re < 360$, C_{po} derivation, splitter plate, C_p, C_{po}, C_{Dp}, C_{pb}, θ_s, L_w/D). 23.

Grover, L. K. and Weaver, D. S. (1978J). Cross flow-induced vibrations in a tube bank-vortex shedding. *J. Sound Vib.*, **59(2)**, 263–76. (Exp. parallel triangle, $P/D = 1.375$, $L/D = 12$, $Ti = 0.2\%$, frequency PS, δ, osc., A/D, Ti_g, St). 28.

Gu, Z. F. (1996J). On interference between two circular cylinders at supercritical Reynolds number. *J. Wind Engg Ind. Aero.*, **62**, 175–90. (Exp. single cyl., two cyls, $220\text{k} < Re < 450\text{k}$, $1.1 < S/D < 3.5$, $1.1 < T/D < 3.5$, stag., $0° < \alpha < 90°$, C_p, C_d, C_l). 26.

Gu, Z. F. and Sun, T. F. (1999J). On interference between two circular cylinders in staggered arrangement at high sub-critical Reynolds numbers. *J. Wind Engg Ind. Aero.*, **80**, 287–309. (Exp. $220\text{k} < Re < 450\text{k}$, $1.1 < S/D < 3.5$, C_D, C_p, C_L). 26.

Gu, Z. F. and Sun, T. F. (2001J). Classification of flow pattern on three circular cylinders in equilateral triangular arrangements. *J. Wind Engg Ind. Aero.*, **89**, 553–68. (Exp. $1.7 < P/D < 5$, $Re = 55\text{k}$, FV, C_p, C_p', C_D, C_L). (27).

Gu, Z. F., Sun, T. F., He, D. X., and Zhang, L. L. (1993J). Two circular cylinders in high-turbulence flow at super-critical Reynolds number. *J. Wind Engg Ind. Aero.*, **49**, 379–88. (Exp. tandem, side-by-side, staggered, $1.05 < S/D < 5.0$, $1.05 < T/D < 5.0$, $Re = 650\text{k}$, $Ti = 10\%$, C_{p1}, C_{p2}, C_{p2}', C_{D1}, C_{D2}, C_{L1}, C_{L2}). 26.

Guven, O., Farell, C., and Patel, V. C. (1980J). Surface roughness effects on the mean flow past circular cylinders. *J. Fluid Mech.*, **98**, 673–701. (Exp. $70\text{k} < Re < 550\text{k}$, $L/D = 3.1$, $D/B = 0.18$, $0.25\% < K_s/D < 0.62\%$, C_p, C_d, C_{pb}, $C_{pb} - C_{p\min}$, δ). 22.

Guven, O., Farell, C., and Patel, V. C. (1983J). Boundary-layer development on a circular cylinder with ribs. *J. Fluids Engg*, **105**, 179–84. (Calc. comparison with Exp., C_p, δ/D, C_d, $C_{pb} - C_{p\min}$). (22).

Hamache, M. and Gharib, M. (1991J). An experimental study of the parallel and oblique vortex shedding from circular cylinders. *J. Fluid Mech.*, **232**, 567–90. (Exp. $65 < Re < 121$, control by two large diameter cylinders set normal, FV, C_p, C_{pb}). 21.

Hanson, A. R. (1966J). Vortex shedding from yawed cylinders. *AIAA J.*, **4**, 738–40. (Exp. $40 < Re < 150$, $0° < \Lambda < 70°$, St, Ro, Re_{osc}). 25.

Harris, D. L. and Reid, W. H. (1964J). On the stability of viscous flow between rotating cylinders. Part 2, Numerical analysis. *J. Fluid Mech.*, **20**, 95–101. (Th. calc.). (24).

Hayashi, M., Sakurai, A., and Ohya, Y. (1986J). Wake interference of a row of normal flat plates arranged side-by-side in a uniform flow. *J. Fluid Mech.*, **164**, 1–25. (Exp. 2, 3, 4, 5 flat plates, $1 < T/D < 3.5$, FV, jet bias, C_D, C_{pb}, St, comp.). 26.

Hayashi, T., Yoshino, F., Waka, R., Tanabe, S., and Kawamura, T. (1992J). Turbulent structure in a vortex wake shed from an inclined circular cylinder, in Japanese. *Trans. JSME*, **58**, 297–304. (Exp. $Re = 20\text{k}$, $0° < \Lambda < 20°$, PS, St, end plates, comp. streamlines). 25.

Higuchi, H. (1989J). Experimental investigation of the flow field behind grid models. *J. Aircraft*, **26**, 308–14. (Exp. ribbon parachutes, non-uniform flow), 28.

Higuchi, H., Kim, H. J., and Farell, C. (1989J). On flow separation and reattachment around a circular cylinder at critical Reynolds numbers. *J. Fluid Mech.*, **200**, 149–71. (Exp. $80\text{k} < Re < 200\text{k}$, spanwise cells, C_D, R_{pp}). 22.

Hill, R. S. and Armstrong, C. (1962J). Aerodynamic sound in tube banks. *Proc. Phys. Soc.*, **79**, 225–7. (Exp. 1 and 2 stag. rows, $T/D = 2.36$, $0.95 < S/D < 2.30$, St, acoustic resonance). 28.

Hirsch, G. and Ruscheweyh, H. (1975/76J). Full-scale measurements on steel chimney stacks. *J. Wind Engg Ind. Aero.*, **1**, 341–7. (Exp. full-scale chimney stack, $H = 145\,\text{m}$, $D = 6\,\text{m}$, $H/D = 24.2$). (22).

Hiwada, M. and Mabuchi, I. (1981J). Flow behaviour and heat transfer around the circular cylinder at high blockage ratios. *Heat Transfer Jap. Res.*, **10**, 17–39. (Exp. $10\text{k} < Re < 100\text{k}$, $D/B = 0.40, 0.80$, C_p, St, C_D, Sh, $Ti = 3.1\%$). 23.

Hiwada, M. and Taguchi, T. (1979J). Fluid flow and heat transfer around two cylinders of different diameters in cross flow. *Bull. JSME*, **22**, 715–23. (Exp. $0.13 < D_1/D_2 < 0.53$, $Re_2 = 50\text{k}$, $L/D = 7.9$, $(S/D)_c$, C_{p1}, C_{p2}, C_{D1}, C_{D2}, St, St_2, Sh, FV, Nu). 26.

Hiwada, M., Mabuchi, I., Kumada, M., and Iwakoshi, H. (1986J). Effect of turbulent boundary layer thickness on the flow characteristics around a circular cylinder near a plane surface, in Japanese. *Trans. Jap. Soc. Mech. Engg*, **52**, 2566–74. (Exp. $Re = 20\text{k}$, $0.17 < G/D < 3.33$, $0.25 < \delta/D < 2.82$, tripping wire, PS, St, C_d, C_L, θ_s). 23.

Hiwada, M., Mabuchi, I., and Yanagihara, H. (1982J). Fluid flow and heat transfer around two circular cylinders. *Bull. JSME*, **25**, 1737–45. (Exp. tandem, $20\text{k} < Re < 80\text{k}$, $1 < S/D < 6$, $1.5 \leq L/D < 8.8$, $0.06 < D/B < 0.13$, C_{p1}, C_{p2}, St, C_{d1}, C_{d2}, FV, Sh_2, $Nu_{\max}$, $\overline{Nu}$). 26.

Hiwada, M., Niwa, K., Kumada, M., and Mabuchi, I. (1979Ja). Effects of tunnel blockage on local mass transfer from a circular cylinder in cross-flow. *Heat Transfer Jap. Res.*, **8**, 37–51. (Exp. $10\text{k} < Re < 100\text{k}$, $0 < D/B < 0.43$, Nu, C_p, Sh, splitter plate). 23.

Hiwada, M., Taguchi, T., Mabuchi, I., and Kumada, M. (1979Jb). Fluid flow and heat transfer around two circular cylinders of different diameters in cross flow. *Bull. JSME*, **22**, 715–23. (Exp. $0.13 < D_1/D_2 < 0.53$, $3\text{k} < Re < 95\text{k}$, $(S/D)_C$, C_{p1}, C_{p2}, C_{d1}, C_{d2}, St, Sh, u', u/V, FV, SL). 26.

Homann, F. (1936J). Influence of high viscosity on flow around a cylinder, in German. *Forsch. Geb. Ing. Wes.*, **17**, 1–10. See also English translation, *Nat. Adv. Committee Aero.*, NACA TM 1334, 1952. (Exp. $4 < Re < 218$, C_p, FV). 1, 2, 4, 6, 23.

Honji, H. (1973Ja). Viscous flow past a group of circular cylinders. *J. Phys. Soc. Jap.*, **34**, 821–8. (Exp. 5 cyl., $10 < Re < 50$, $0.5 < T/D < 2$, $L/D = 150$, 375, $nD/B = 0.015$, 0.021, FV). 27.

Honji, H. (1973Jb). Formation of a reversed flow bubble in the time-mean wake of a row of circular cylinders. *J. Phys. Soc. Jap.*, **35**, 1533–6. (Exp. 5 to 15 cyl., $48 < Re < 150$, $0.5 < T/D < 2$, $L/D = 375$, $D/B = 0.003$, 0.007, FV). 27.

Hughes, T. H. and Reid, W. H. (1968J). The stability of spiral flow between rotating cylinders. *Phil. Trans. Roy. Soc.*, **263A**, 57–91. (Th.). (24).

Hunt, J. C. R., Abell, C., Peterka, J. A., and Woo, H. (1978J). Kinematical studies of the flows around free or surface mounted obstacles applying topology to flow visualisation. *J. Fluid Mech.*, **86**, 179–201. (Th. prism, FV). (22).

Igarashi, T. (1978J). Flow characteristics around a circular cylinder with a slit. *Bull. JSME*, **21**, 656–64. (Exp. slit, $13.8\text{k} < Re < 52\text{k}$, $S/D = 0.08$, 0.18, $10° < \beta < 90°$, $L/D = 4.4$, $D/B = 0.085$, FV, C_p, C_d, u/V, u'/V, St, PS). 22.

Igarashi, T. (1981J). Characteristics of the flow around two circular cylinders in tandem. *Bull. JSME*, **24**, 323–31. (Exp. tandem, $8.7\text{k} < Re < 52\text{k}$, $1.03 < S/D < 5$, $L/D = 4.4$, $D/B = 0.05$, C_p, C_{D1}, C_{D2}, St, FV, C'_{p1}, C'_{p2}, PS, classification). 26.

Igarashi, T. (1982J). Characteristics of a flow around two circular cylinders of different diameters arranged in tandem. *Bull. JSME*, **25**, 349–57. (Exp. $D_2/D_1 = 0.68$, $13\text{k} < Re < 58\text{k}$, $0.9 < S/D_1 < 4$, C_{p1}, C_{p2}, C_{d1}, C_{d2}, C'_{p1}, C'_{p2}, St, FV, topology). 26.

Igarashi, T. (1984J). Characteristics of the flow around two circular cylinders arranged in tandem. *Bull. JSME*, **27**, 2380–7. (Exp. tandem, $25\text{k} < Re < 65\text{k}$, $1 < S/D < 5$, $3 < L/D < 5$, PS, $D/B = 0.08$, St, C_{p1}, C_{p2}, C'_p, C_{d2}, FV). 26.

Igarashi, T. (1985J). Effect of vortex generators on the flow around a circular cylinder normal to an air stream. *Bull. JSME*, **28**, 274–82. (Exp. saw-blade, $8.7\text{k} < Re < 70\text{k}$, $L/D = 4.4$, $D/B = 0.085$, $0.6\% < K/D < 4.4\%$, $\theta_B = 40°$, 50°, 60°, 70°, St, C_p, C_d, C'_p, FV). 22.

Igarashi, T. (1986Ja). Effect of tripping wires on the flow around a circular cylinder normal to an air stream. *Bull. JSME*, **28**, 2917–24. (Exp. tripping wire, 1.3k $< Re <$ 90k, 0.08% $< d/D <$ 0.3%, $\theta_w = 50°$, 60°, $L/D = 4.4$, $D/B = 0.085$, St, C_d, C_p, FV, classification). 22.

Igarashi, T. (1986Jb). Characteristics of the flow around four circular cylinders arranged in-line. *Bull. JSME*, **29**, 751–7. (Exp. 4 cyl., $1.1 < S/D < 2.7$, 8.7k $< Re <$ 39.7k, $L/D = 4.1$, $D/B = 0.06$, $Ti = 0.5\%$, C_p, C_p', C_d, FV, St). 27.

Igarashi, T. (1993J). Aerodynamic forces acting on three circular cylinders having different diameters closely arranged in-line. *J. Wind Engg Ind. Aero.*, **49**, 369–78. (Exp. 3 cyl., $d/D = 0.45$, $L/D = 3$, $D/B = 0.067$, $G/D = 0.06$, $Ti = 0.5\%$, C_p, C_p', C_d, C_l, θ_o, C_{pb}, St, FV). 27.

Igarashi, T. and Iida, Y. (1988J). Fluid flow and heat transfer around a circular cylinder with vortex generators. *JSME Int. J.*, **31**, 701–8. (Exp. saw-blade trip, C_p, St, C_D, C_p', $\overline{Nu}$, Nu). 22.

Igarashi, T. and Suzuki, K. (1984J). Characteristics of the flow around three circular cylinders. *Bull. JSME*, **27**, 2397–404. (Exp. 3 in-line cyl., $1 < S/D < 4$, 10.9k $< Re <$ 39.2k, $L/D = 4.41$, $D/B = 0.07$, $Ti = 0.6\%$, C_p, C_p', C_d, C_l, St, FV). 27.

Igarashi, T., Nishida, K., and Mochizuki, S. (1988J). Aerodynamic forces acting on three circular cylinders having different diameters closely arranged in line, in Japanese. *Trans. JSME*, **54**, 2349–56. (Exp. 3 cyls, C_p, C_d, C_l, St, FV). 27.

Imaichi, K. and Ohmi, K. (1983J). Numerical processing of flow-visualisation pictures – measurement of two-dimensional vortex flow. *J. Fluid Mech.*, **129**, 283–311. (Exp. FV, stream function, vorticity, $Re = 100$, 200). (3).

Ingham, D. B. (1983J). Steady flow past a rotating cylinder. *Comp. Fluids*, **11**, 351–66. (Comp. Th. $5 < Re < 20$, $0 < V_r/V < 0.5$, streamlines, C_L, C_D, C_f). (24).

Ishigai, S. and Nishikawa, E. (1975J). On the structure of gas flow in single column, single row, and double row banks. *Bull. JSME*, **18**, 528–35. (Exp. $1.25 < T/D < 3$, $1.2 < S/D < 5$, 3k $< Re <$ 10k, C_p, St, γ, FV). 28.

Ishigai, S., Nishikawa, E., Nishimura, K., and Cho, K. (1972J). Experimental study on structure of gas flow in tube banks with tube axes normal to flow. *Bull. JSME*, **15**, 949–56. (Exp. tandem, side-by-side, staggered, 1.5k $< Re <$ 15k, $1.1 < S/D < 5$, $1.25 < T/D < 3$, C_{p1}, C_{p2}, St, FV). 26.

Iversen, J. D. (1973J). Correlation of Magnus force data for slender spinning cylinders. *J. Spacecraft*, **10**, 268–72. (Exp. rotating cyl., angle of incidence, Magnus force). 24.

Jacobson, I. D. (1974J). Contribution of a wall shear stress to the Magnus effect on nose shapes. *AIAA J.*, **12**, 1003–5. (Exp. 4.8M $< Re <$ 5.9M, $2 < Ma < 3$, $0.19 < V_r/V < 0.24$, boundary layer transition).(25).

James, D. F. and Truong, Q.-S. (1972J). Wind load on cylinder with spanwise protrusion. *ASCE J. Engg Mech. Div.*, **98**, 1573–89. (Exp. tripping wire, 15k $<$ $Re <$ 110k, $0.06\% < d/D < 6.3\%$, $15° < \theta_w < 90°$, $L/D = 14.2, 33$, $D/B =$ 0.07, 0.03, C_D, C_L). 22.

Jaminet, J. F. and Van Atta, C. W. (1969J). Experiments on vortex shedding from rotating circular cylinders. *AIAA J.*, **7**, 1817–19. (Exp. $50 < Re < 150$, $0 < V_r/V < 2$, St, Re_{osc}). 24.

Jensen, B. L., Sumer, B. M., Jensen, H. R., and Fredsøe, J. (1990J). Flow around and forces on a pipeline near a scoured bed in steady current. *ASME J. Offshore Mech. Arctic Engg*, **112**, 206–13. (Exp. scour, u/V, u'/V, C_D, C_L, C_p', FV). 23.

Johnson, T. R. and Joubert, P. N. (1969J). The influence of vortex generators on the drag and heat transfer from a circular cylinder normal to an air stream. *ASME J. Heat Transfer*, **91**, 91–9. (Exp. triangular eddy generator, 50k $<$ $Re <$ 400k, C_D, FV, Nu, $\overline{Nu}$). 22.

Joubert, P. N. and Hoffman, E. R. (1962J). Drag of circular cylinder with vortex generators. *J. Roy. Aero. Soc.*, **66**, 456–7. (Exp. 64k $< Re <$ 570k, $10° < \theta_e < 70°$, C_D). 22.

Kageyama, Y., Osaka, H., and Yamada, H. (1983J). The structure of turbulent wake behind a cruciform circular cylinder. Report 1. *Bull. JSME*, **26**, 356–63. (Exp. $Re =$ 4k, $L/D = 80$, $D/B = 0.12$, $Ti = 0.35\%$, u, u', v', $3 < x/D < 594$, b/D, C_p, C_d). 26.

Kamemoto, K. (1976J). Formation and interaction of two parallel vortex streets. *Bull. JSME*, **19**, 283–90. (Comp. side-by-side, vorticity clusters, FV). 26.

Karniadakis, G. E. M. and Triantafillou, G. S. (1992J). Three-dimensional dynamics and transition to turbulence in the wake of bluff objects. *J. Fluid Mech.*, **238**, 1–30. (Comp. N–S eqns, $Re = 225, 500$). 21.

Katinas, V. I., Markevicius, A. A., and Žukauskas, A. A. (1982Ja). Vortex shedding and velocity fluctuations in staggered tube bundles in cross flow of air. *Fluid Mech. – Sov. Res.*, **11**, 56–65. (Exp. single row $T/D = 1.34, 1.61$, and normal triangle, $1.15 < P/D < 2.68$, frequency PS, St, empirical approx.) (28).

Katinas, V. I., Perednis, E. E., and Žukauskas, A. A. (1982Jb). Hydraulic drag of tube bundles *vs* yaw angle of viscous fluid. *Fluid Mech. – Sov. Res.*, **11**, 66–74. (Exp. in-line $P/D = 1.34$, $L/D > 10.8$, 30k $< Re <$ 200k, $\Lambda = 25°, 40°, 60°, 75°$, C_d, $n = 10$). (28).

Kawaguti, M. (1964J). The flow of a perfect fluid across two moving bodies. *J. Phys. Soc. Jap.*, **19**, 1409–15. (Th. source, sink, doublet, C_p, spheres). (26).

Kawamura, T. and Hayashi, T. (1992J). Computation of flow around a yawed circular cylinder, in Japanese. *Trans. JSME*, **58**, 1071–8. (Comp. Th. Exp. $Re = 2$k, with and without end plates, C_p, C_{po}, C_{pb}, streamlines). 25.

Kawamura, T. and Hayashi, T. (1994J). Computation of flow around a yawed circular cylinder. *JSME Int. J.*, **37B**, 229–36. (Th. Comp. C_{pn}, C_{pno}, C_{pbn}, streamlines, surface flow). 25.

Kawamura, T., Hiwada, M., Hibino, T., Mabuchi, I., and Kumada, M. (1984Ja). Flow around finite circular cylinder on a flat plate. *Bull. JSME*, **27**, 2142–51. (Exp. $1 < H/D < 8$, $Re = 32$k, $\delta/D = 1$, C_p, FV, C_d, free end eddies, topology). 21.

Kawamura, T., Hiwada, M., Hibino, T., Mabuchi, I., and Kumada, M. (1984Jb). Heat transfer from a finite circular cylinder on the flat plate. *Bull. JSME*, **27**, 2430–9. (Exp. $H/D = 1, 8$, $Re = 32$k, FV, C_p, Sh, $\overline{Nu}$). 21.

Keefe, T. (1962J). Investigation of fluctuating forces acting on a stationary cylinder in subsonic stream and the associated sound. *J. Acoust. Soc. America*, **34**, 1711–14. (Exp. C_L, C_d, L/D, end plates). 14, 18, 21.

Kelly, H. R. and Van Aken, R. W. (1956J). The Magnus effect at high Reynolds numbers. *J. Aero. Sci.*, **23**, 1053–4. (Exp. rotating cyl., $101\text{k} < Re < 907\text{k}$, $0 < V_r/V < 1.2$, C_L). 24.

Keshavan, N. R. (1977J). Separation of turbulent boundary layer on a lifting cylinder. *AIAA J.*, **13**, 262–3. (Exp. Th. blowing slot, θ_s, empirical correlation). (24).

Kim, H. J. and Durbin, P. A. (1988J). Investigation of the flow between a pair of circular cylinders in the flopping regime. *J. Fluid Mech.*, **196**, 431–48. (Exp. side-by-side, partition plate, $T/D = 1.75$, $2\text{k} < Re < 7.5\text{k}$, C_{pb}, St, C'_{pb1}, C'_{pb2}, β, external sound, FV). 26.

Kim, M. S. and Geropp, D. (1998J). Experimental investigation of the ground effect on the flow around some two-dimensional bluff bodies with moving-belt technique. *J. Wind Engg Ind. Aero.*, **74–6**, 511–19. (Exp. $330\text{k} < Re < 990\text{k}$, $0.14 < G/D < 1.0$, C_p, C_{pb}, C_D, moving floor). 23.

Kimoto, S., Inoue, K., and Maki, V. (1986J). Study of flow characteristics of a circular cylinder in a rectangular conduit. *Bull. JSME*, **29**, 1462–70. (Exp. $7.8\text{k} < Re < 25.2\text{k}$, $D/B = 0.38$, C_p, C'_L, St, Nu, $\overline{Nu}$, $\overline{Nu_f}$, $\overline{Nu_s}$). (23).

Kimura, T. and Tsutahara, M. (1991J). Fluid dynamic effects of grooves on circular cylinder surface. *AIAA J.*, **29**, 7062–7. (Exp. spanwise groove, C_D, FV, comp). 22.

Kirchgässner, K. (1961J). Instability of flow between two counter-rotating cylinders against Taylor eddies with arbitrary gap, in German. *Zeit. Angew. Math. Phys.*, **12**, 14–30. (Th. calc.). (24).

Kirchgässner, K. and Sorger, P. (1969J). Branching analysis for the Taylor problem. *J. Mech. Appl. Math.*, **22**, 183–209. (Th. Liapunov–Schmidt method). (24).

Kiya, M. and Sasaki, K. (1983J). Structure of a turbulent separation bubble. *J. Fluid Mech.*, **137**, 83–113. (Exp. $Re = 26$k, u, u', PS, p', FV, R_p). 14, 21.

Kiya, M., Arie, M., Tamura, H., and Mori, H. (1980J). Vortex shedding from two circular cylinders in staggered arrangement. *J. Fluids Engg*, **102**, 166–73. (Exp. $Re = 16$k, $1 < S/D < 5.5$, $0° < \alpha < 180°$, St). 26.

Kiya, M., Sakamoto, H., and Arie, M. (1976J). Theory of inviscid shear flow over bluff bodies attached to normal plate and semi-circular projection. *Bull. JSME*, **19**, 513–30. (Th. circle and source, C_p, streamlines, semi-elliptical and arbitrary-shaped bodies). (23).

Ko, N. W. M. and Chan, A. S. K. (1990J). On the intermixing region behind circular cylinders with stepwise change of diameter. *Exp. Fluids*, **9**, 213–21. (Exp. step, $d/D = 0.5$, PS, u', v', w', L_f). 22.

Ko, N. W. M. and Wong, P. T. Y. (1992J). Flow past two circular cylinders of different diameters. *J. Wind Engg Ind. Aero.*, **41–4**, 563–4. (Exp. $Re = 50$k, $D_1/D_2 = 2$, $1 < S/D < 3$, $1.15 < T/D < 2$, C_{pb}, C_{pm}, C_d, C_l, f). 26.

Ko, N. W. M., Leung, W. L., and Au, H. (1982J). Flow behind two co-axial circular cylinders. *J. Fluids Engg*, **104**, 223–7. (Exp. step, $Re = 80$k, $d/D = 0.5$, $L/D = 51.5, 11.5$, PS, C_p). 22.

Ko, N. W. M., Leung, Y. C., and Chan, J. J. J. (1987J). Flow past V-grooved circular cylinders. *AIAA J.*, **25**, 806–11. (Exp. V-groove, $20\text{k} < Re < 160\text{k}$, $0.4\% < h/D < 1.2\%$, C_p, C_d, C_{pb}, $C_{pb} - C_{p\min}$, St, PS). 22.

König, M., Eisenlohr, H., and Eckelmann, H. (1990J). The fine structure in the Strouhal–Reynolds number relationship of the laminar wake of the circular cylinder. *Phys. Fluids*, **A2**, 1607–14. (Exp. $48 < Re < 160$, $56 < L/D < 140$, St, PS, z_c/D). 21.

König, M., Eisenlohr, H., and Eckelmann, H. (1992J). Visualisation of the spanwise cellular structure of the laminar wake of wall-bounded circular cylinders. *Phys. Fluids*, **A4**, 869–72. (Exp. $48 < Re < 160$, $L/D = 840$, St, FV). 21.

König, M., Noack, B. R., and Eckelmann, H. (1993J). Discrete shedding modes in the von Kármán vortex street. *Phys. Fluids*, **A5**, 1846–8. (Exp. $45 < Re < 160$, St, β, z/D, D_e/D). 21.

Koschmieder, E. L. (1979J). Turbulent Taylor vortex flow. *J. Fluid Mech.*, **93**, 515–27. (Exp. super-critical flow). (24).

Kostic, Z. G. and Oka, S. N. (1972J). Fluid flow and heat transfer with two cylinders in cross flow. *Int. J. Heat Mass Transfer*, **15**, 279–99. (Exp. tandem, 12k $< Re <$ 40k, $1.6 < S/D < 9$, $L/D = 5$, $D/B = 0.2$, C_{p1}, C_{p2}, Fr_1, Fr_2, C_{D1}, C_{D2}, topology, St). (26).

Kovasznay, L. (1949J). Hot wire investigation of the wake behind cylinders at low Reynolds numbers. *Proc. Roy. Soc.*, **198A**, 174–90. (Exp. $36 < Re <$ 10k, u', St). 1, 2, 3, 4, 5, 22.

Krahn, E. (1956J). Negative Magnus force. *J. Aero. Sci.*, **23**, 377–8. (Disc. rotating cylinder, equivalent Re, mechanism). 24.

Krueger, E. R. and Di Prima, R. C. (1962J). Stability of non-rotationally symmetric disturbances for inviscid flow between rotating cylinders. *Phys. Fluids*, **5**, 1362–7. (Th. non-axisymmetric disturbances). (24).

Krueger, E. R., Gross, A., and Di Prima, R. C. (1966J). On the relative importance of Taylor vortex and non-axisymmetric modes in flow between rotating cylinders. *J. Fluid Mech.*, **24**, 521–38. (Th. counter-rotating cyl.). (24).

Kruse, R. L. (1978J). Influence of spin rate on side force of an axisymmetric body. *AIAA J.*, **16**, 415-6. (Exp. $Re =$ 1M, $42.5° < \alpha < 60°$, C_L). 25.

Kumada, M., Hiwada, M., Ito, M., and Mabuchi, I. (1984J). Wake interference between three circular cylinders arranged side-by-side normal to a flow, in Japanese. *Trans. JSME*, **50**, 1699–707. (Exp. side-by-side, 3 cyl., 10k $< Re <$ 32k, $1.125 < T/D < 2.50$, C_p, u, St). (27).

Lafay, A. (1910J). On the inversion of Magnus phenomenon, in French. *Compt. Rend. Acad. Sci.*, **151**, 61. (Hist. Exp.). 24.

Lafay, A. (1912J). Experimental contribution on aerodynamics of a cylinder and phenomenon of Magnus, in French. *Rev. de Mécanique*, **30**, 417–42. (Exp. rotating cyl., $L/D = 4.5$, open jet, 88k $< Re <$ 198k, $0.2 < V_r/V < 1.3$, C_F, φ, C_L, C_M, FV). 24.

Lagally, M. (1929J). The frictionless flow outside two circles. *Zeit. Angew. Math. Mech.*, **9**, ZAMM, 299–305. (Th. ideal fluid, two circles of different diameter). (26).

Lam, K. and Cheung, W. C. (1988J). Phenomena of vortex shedding and flow interference of three cylinders in different equilateral arrangements. *J. Fluid Mech.*, **196**, 1–26. (Exp. 3 cyl., $Re =$ 2.1k, 3.5k, $2.3 \leqslant S/D \leqslant 4.6$, $L/D = 13.3$, $D/B = 0.075$, $Ti = 1\%$, $D_e/D = 6.6$, St, FV). 27.

Lam, K. and Fang, X. (1995J). The effect of interference of four equal-distanced cylinders in cross flow on pressure and force coefficients. *J. Fluids Struct.*, **9**, 195–214. (Exp. 4 cyl., $Re =$ 12.8k, $1.26 \leqslant S/D \leqslant 5.80$, $L/D = 28.8$, $D/B = 0.36$, $Ti = 0.6\%$, C_p, C_d, C_l, C_a, C_{pb} ,C_{dv}, C_{dr}, FV). 23.

Lam, K. and Lo, S. C. (1992J). A visualisation study on cross-flow around four cylinders in square arrangement. *J. Fluids Struct.*, **6**, 109–31. (Exp. 4 cyl., $Re = 2.1$k, $S/D = 1.3$ to 5.6, $L/D = 21.3$, $2D/B = 0.09$, St, FV). 27.

Lamont, P. J. (1982J). Pressure around an inclined ogive-cylinder with laminar, transitional, or turbulent separation. *AIAA J.*, **20**, 1492–9. (Exp. 200k $< Re <$ 4M, $20° < \alpha < 90°$, roll angle, C_s, $C_{S\max}$, C_p). 25.

Lamont, P. J. and Hunt, B. L. (1973J). Out-of-plane force on a circular cylinder at large angles of inclination to a uniform stream. *Aeronaut. J.*, **77**, 41–5. (Exp. $20° < \alpha < 70°$, impulsive, out-of-phase force, C_L). (25).

Lamont, P. J. and Hunt, B. L. (1976J). Pressure and force distributions on a sharp-nosed circular cylinder at large angles of inclination to a uniform subsonic stream. *J. Fluid Mech.*, **76**, 519–59. (Exp. missile, $L/D = 11$, $Re =$ 110k, $30° < \alpha < 75°$, C_{Dn}, C_{ds}, C_p). 25.

Lamont, P. J. and Hunt, B. L. (1977J). Prediction of aerodynamic out-of-plane forces on ogive-nosed circular cylinders. *J. Spacecraft and Rockets*, **14**, 38–44. (Prediction, Exp. out-of-plane force). (25).

Langston, L. S. and Boyle, M. T. (1982J). A new surface-streamline flow visualisation technique. *J. Fluid Mech.*, **125**, 53–7. (Exp. horseshoe vortex, $Re = 246$k). (21).

Lawson, T. (1982J). The use of roughness to produce high Reynolds number flows around circular cylinders at lower Reynolds numbers. *J. Wind Engg Ind. Aero.*, **10**, 381–7. (Th. calculations). (22).

Lee, S.-J. and Kim, H.-B. (1997J). The effect of surface protrusions on the near-wake of a circular cylinder. *J. Wind Engg Ind. Aero.*, **69–71**, 351–61. (Exp. helical wires, $d/D = 0.075$, $P/D = 5$, 10, 5k $< Re <$ 50k, FV, C_p, St, L_f/D, u/V). (22).

Lei, C., Cheng, L., and Kavanagh, K. (1999J). Re-examination of the effect of a plane boundary on force and vortex shedding of a circular cylinder. *J. Wind Engg Ind. Aero.*, **80**, 263–86. (Exp. $Re = 14$k, $0.14 < G/D < 2.89$, C_p, θ_s, C_{pb}, C_D, C_L, St). 23.

Lesage, F. and Gartshore, I. S. (1987J). A method of reducing drag and fluctuating side force. *J. Wind Engg Ind. Aero.*, **25**, 229–45. (Exp. control cyl., 10k $< Re <$ 70k, $2 < D/d < 5.8$, C_{p1}, C_{p2}, $1.5 < S/D < 7.5$, C_D, C_L'). 26.

Leung, Y. C. and Ko, N. W. M. (1991J). Near-wall characteristics of flow over grooved circular cylinder. *Exp. Fluids*, **10**, 322–32. (Exp. V-groove, 30k $<$ $Re <$ 150k, $L/D = 2$, 4.5, $D/B = 0.18$, C_d, St, C_f, τ, δ, δ^*/θ). 22.

Leung, Y. C., Ko, N. W. M., and Tang, K. M. (1992J). Flow past circular cylinder with different surface configurations. *J. Fluids Engg*, **114**, 170–7. (Exp. V-groove, 20k $< Re <$ 130k, $L/D = 4.3$, $D/B = 0.16$, $h/D = 1.4\%$, partial grooving, C_p, C_d, C_{pb}, θ_w, St, C_L, δ). 22.

Leung, Y. C., Wong, C. H., and Ko, N. W. M. (1997J). Characteristics of flows over an asymmetrically grooved circular cylinder in the transition of regimes. *J. Wind Engg Ind. Aero.*, **69–71**, 169–78. (Exp. 44k $< Re <$ 130k, grooves, C_p, C_D, C_L, St, f_{tr}/f, u/V). 22.

Lewis, C. G. and Gharib, M. (1992J). An exploration of the wake three-dimensionalities caused by a local discontinuity in cylinder diameter. *Phys. Fluids*, **A4**, 104–17. (Exp. step, $35 < Re < 200$, $0.51 < d/D < 0.88$, $110 < L/D < 194$, FV, f, vel. defect, v'). 22.

Linke, W. (1931J). New measurements on aerodynamics of cylinders, particularly their friction resistance, in German. *Physik. Zeit.*, **32**, 900–14. (Exp. 5k $< Re <$ 40k, tripping wire, C_{Dp}, C_{Df}, C_D, C_p). 1, 5. (22).

Ljungkrona, L., Norberg, C., and Sunden, B. (1991J). Free-stream turbulence and tube spacing effects on surface pressure fluctuations for two tubes in an in-line arrangement. *J. Fluids Struct.*, **5**, 701–27. (Exp. Re = 20k, $1.25 < S/D < 5$, $0.1\% < Ti < 3.2\%$, $L/D = 20$, $D/B = 0.04$, C_D, St, C_{p1}, C_{p2}, C'_{pb1}, C'_{pb2}, PS, FV). 26.

Lo, K. W. and Ko, N. W. M. (1995J). Effect of acoustic excitation on flow over a partially grooved circular cylinder. *Exp. Fluids*, **19**, 194–202. (Exp. groove, 44k $< Re <$ 122k, C_p, St, C_d, C_L, δ/D, u', f). (22).

Lo, K. W. and Ko, N. W. H. (1997J). Turbulent near-wake behind a partially grooved circular cylinder. *J. Fluids Engg*, **119**, 19–28. (Exp. Re = 126k, grooves, $u'v'$). (22).

Low, A. R. (1925J). The reaction of a stream of viscous fluid on a rotating circular cylinder. *J. Roy. Aero. Soc.*, **29**, 100–44. (Disc. Prandtl's paper, Flettner's rotor). (24).

Luo, S. C., Gan, T. L., and Chew, Y. T. (1996J). Uniform flow past one or two tandem finite length circular cylinders. *J. Wind Engg Ind. Aero.*, **59**, 69–93. (Exp. finite, Re = 33k, H/D = 4, 6, 8, $D/B = 0.05$, C_p, C_d, C_{d1}, C_{d2}). 26.

Mabey, D. G. (1965J). Aerodynamically-induced vibration in coolers. *J. Roy. Aero. Soc.*, **69**, 876–7. (Exp. normal triangle, 10 rows, $P/D = 2.6$, St, osc., baffles). (28).

Maekawa, T. (1964J). Study on wind pressure against ACSR double conductor. *Electr. Engg Jap.*, **84**, 21–8. (Exp. stranded cyls., 10k $< Re <$ 74k, $0° < \alpha < 30°$, $6 < S/D < 8$, C_D, C_L). 26.

Magnus, G. (1853J). On the deflection of projectile, in German. *Poggendorf's Annalen der Physik und Chemie*, **88**, 604–10. (Hist. Magnus effect). 24.

Mair, W. A., Jones, P. D., and Palmer, R. K. W. (1975J). Vortex shedding from finned tubes. *J. Sound Vib.*, **39**, 293–6. (Exp. 16k $< Re <$ 46k, fins, D_f/D = 1.2, 1.4, $0.02 < L/D < 0.06$, St). 22.

Mallock, A. (1896J). Experiments of fluid viscosity. *Phil. Trans. Roy. Soc.*, **187A**, 41–56. (Exp. viscosity measurements by using concentric cyl.). 24.

Martin, J. M. and Ingram, C. W. (1973J). Experimental correlation between flow and Magnus characteristics of a spinning ogive-nose cylinder. *AIAA J.*, **11**, 401–2. (Exp. $Re = 250\text{k}$, $0.25 < V_r/V < 0.5$, C_L, $2° < \alpha < 15°$, topology). (25).

Mathis, C., Provansal, M., and Boyer, L. (1984J). The Bénard–von Kármán instability; an experimental study near the threshold. *J. Phys. Lett.*, **45**, L-483–91. (Th. Exp. Landau's eqn, $40 < Re < 300$, Re_{osc}, $5 < L/D < 75$, v', knot, Re). 21.

Maull, D. J. (1966J). Wind forces on groups of structures. *Nature, London*, **211**, 1073–4. (Exp. 5 finite cylinders, C_p, α). 22.

Mazur, V. Y. (1970J). Motion of two circular cylinders in ideal fluid. *Izv. Akad. Nauk. SSR, Mekh. Zhidkosti Gaza*, No. 6, 80–4. (Th. inviscid, tandem circles, exact solution). (26).

Meister, B. (1962J). Taylor–Dean stability problem for arbitrary gap width, in German. *Zeit. Angew. Math. Phys.*, **13**, 83–91. (Th. pressure gradient). (24).

Meksyn. D. (1946J). Stability of viscous flow between rotating cylinders. *Proc. Roy. Soc.*, **187A**, 115–28, 480–91, 492–504. (Th. calc.). (24).

Melander, M. V. and Hussain, F. (1989J). Cross-linking of two anti-parallel vortex tubes. *Phys. Fluids*, **A1**, 633–5. (Comp. cross-linking of two eddies, $Re = 1\text{k}$). 21.

Melbourne, W. H., Cheung, J. C. K., and Goddard, C. R. (1983J). Response to wind action of 265 m Mount Isa stack. *ASCE J. Struct. Div.*, **109**, 2561–77. (Exp. insitu., PS, C_M, C_p, C_p'). 7, 21.

Miau, J. J., Wang, J. T., Chou, J. H., and Wei, C. Y. (1999J). Characteristics of low-frequency variations embedded in vortex-shedding process. *J. Fluids Struct.*, **13**, 339–60. (Exp. trapezoidal, circular cyl., $Re = 10\text{k}$, C_p'). (5).

Miller, B. L. (1976J). The hydrodynamic drag of roughened circular cylinders. *Trans. Roy. Inst. Naval Architects*, No 2, 55–70. (Exp. $200\text{k} < Re < 4\text{M}$, $1.5\% < K/D < 6.3\%$, C_D, Re_k). 22.

Miller, G. D. and Williamson, C. H. K. (1994J). Control of three-dimensional phase dynamics in a cylinder wake. *Exp. Fluids*, **18**, 26–35. (Exp. $45 < Re < 180$, end suction, St, FV, β). 21.

Miller, M. C. (1976J). Surface pressure measurements on a spinning wind tunnel model. *AIAA J.*, **14**, 1669–70. (Exp. rotating cylinder, $0.17 < V_r/V < 2.05$, $224\text{k} < Re < 449\text{k}$, C_p, C_l, C_d). 24.

Miller, M. C. (1979J). Wind-tunnel measurements of the surface pressure distribution on a spinning Magnus rotor. *J. Aircraft*, **16**, 815–22. (Exp. rotating cyl., vanes, $V_r/V = 0.16$, $Re = 264$k, $L/D = 1.6$, $D/B = 0.13$, $D_e/D = 1.93$, C_p, C_l, C_d). 24.

Miyazaki, T. and Hasimoto, H. (1980J). Separation of creeping flow past two circular cylinders. *J. Phys. Soc. Jap.*, **49**, 1611–18. (Th. Stokes flow, FV, θ_s, streamlines). (26).

Mizota, T., Zdravkovich, M. M., Graw, K. V., and Leder, A. (2000J). St Christopher and the vortex – a Kármán vortex in the wake of St Christopher's heels. *Nature*, **404**, 226. (Science in culture, FV).

Modi, V. J. and El-Sherbiny, S. E. (1977J). A free streamline model for bluff bodies in confined flow. *J. Fluids Engg*, **97**, 585–92. (Th. Parkinson–Jandali's model, comparison with exps). 23.

Moore, D. W. (1957J). The flow past a rapidly rotating circular cylinder in a uniform flow. *J. Fluid Mech.*, **2**, 541–50. (Th. small perturbations). (24).

Moore, J. and Forlini, T. J. (1984J). A horseshoe vortex in a duct. *ASME J. Engg Gas Turbines Power*, **106**, 668–76. (Exp. Th. Rankine half-body, FV, C_p, streamlines). (21).

Moretti, P. M. and Cheng, M. (1987J). Instability of flow through tube rows. *J. Fluids Engg*, **109**, 197–8. (Exp. single row, u/v, FV). 28.

Morton, J. B., Jacobson, I. D., and Saunders, S. (1976J). Experimental investigation of the boundary layer on a rotating cylinder. *AIAA J.*, **14**, 1458–63. (Exp. rot. cyl. boundary layer transition, stability curves, $0 < V_r/V < 0.6$). (24).

Müller, W. (1929J). Double source system in two-dimensional flow, related to flow past two circular cylinders. *Zeit. Angew. Math. Mech.*, **9**, 201–13. (Th. inviscid flow, two sources, tandem, side-by-side, flat boundary). (26).

Nakamura, I., Ueki, Y., and Yamashita, S. (1986J). The turbulent shear flow on a rotating cylinder in a quiescent fluid. *Bull. JSME*, **29**, 1704–9. (Exp. rot. cyl., $V = 0$, FV, eddy viscosity). (24).

Nakamura, Y. and Fukamachi, N. (1991J). Visualisation of the flow past a frisbee. *Fluid Dyn. Res.*, **7**, 31–5. (Exp. rotating frisbee, free flight, FV). 21.

Nakamura, Y. and Tomonari, Y. (1982J). The effects of surface roughness on the flow past circular cylinders at high Reynolds numbers. *J. Fluid Mech.*, **123**, 363–78. (Exp. 40k $< Re <$ 1.7M, $L/D = 3.3$, $D/B = 0.15$, $0.09\% < K/D < 1\%$, partial roughness, C_d, C_{pb}, St, C_p, $C_{pb} - C_{p\min}$). 22.

Ng, C. W. and Ko, N. W. M. (1995J). Flow interaction behind two circular cylinders of equal diameter – a numerical study. *J. Wind Engg Ind. Aero.*, **54–5**, 277–87. (Comp. side-by-side, $1.25 < T/D < 3.5$, vorticity, C_D, C_L). (26).

Niemann, H. J. (1980J). Wind effects on cooling tower shells. *ASCE J. Struct. Div.*, **106**, 643–61. (Exp. cooling tower, ribs, C_p, $C_{p\min}$, C_{pb}, C_{po}, $C_{pb} - C_{p\min}$). 22.

Niemann, H. J. and Pröpper, H. (1975/76J). Some properties of fluctuating wind pressures on a full-scale cooling tower. *J. Ind. Aero.*, **1**, 349–59. (Exp. full-scale cooling tower, PS, C_p, correction, C_p'). 22.

Nieuwstadt, F. and Keller, H. B. (1973J). Viscous flow past circular cylinders. *J. Comp. Fluids*, **1**, 59–71. (Comp. $1 < Re < 40$, C_p, C_f, C_D). 8, 23.

Nishioka, M. and Sato, H. (1974J). Measurements of velocity distributions in the wake of a circular cylinder at low Reynolds number. *J. Fluid Mech.*, **65**, 97–112. (Exp. $50 < Re < 160$, Ro, L/D, C_D). 21.

Nisi, H. (1925J). Experimental studies on eddies in air. *Jap. J. Phys.*, **4**, 1–11. (Exp. $Re = 15$, $L/D = 8.5$, FV). 21.

Noack, B. R. and Eckelmann, H. (1992J). On chaos in wakes. *Physica D*, **56**, 151–64. (Th. Hilbert space, SL, St). 21.

Noack, B. R. and Eckelmann, H. (1994Ja). A global stability analysis of the steady and periodic cylinder wake. *J. Fluid Mech.*, **270**, 297–330. (Th. Galerkin method, stability analysis, Landau's eqn, SL). 21.

Noack, B. R. and Eckelmann, H. (1994Jb). A low-dimensional Galerkin method for the three-dimensional flow around a circular cylinder. *Phys. Fluids*, **6**, 124–43. (Th. Galerkin method, L_c/D, $u(x, y)$, St). 21.

Noack, B. R., König, M., and Eckelmann, H. (1993J). Three-dimensional stability analysis of the periodic flow around a circular cylinder. *Phys. Fluids*, **A5**, 1279–81. (Th. Comp. Exp. $Re = 200$, FV). 21.

Noack, B. R., Ohle, F., and Eckelmann, H. (1992J). Construction and analysis of different equations from experimental time series of oscillatory systems. *Physica D*, **56**, 389–405. (Th. Exp. diff. eqn, stability analysis). 21.

Norberg, C. (1994J). An experimental investigation of the flow around a circular cylinder: influence of aspect ratio. *J. Fluid Mech.*, **258**, 287–316. (Exp. $50 < Re < 40$k, $2 < L/D < 5000$, end plate, St, C_{ps}, β, FV, PS). 21.

Oberkampf, W. L. and Bartel, T. J. (1980J). Symmetric body vortex wake characteristics in supersonic flow. *AIAA J.*, **18**, 1289–97. (Exp. 480k $< Re <$ 1.75M, $Ma = 2, 3$, $10° < \Lambda < 25°$, $K/\pi DV$, eddy centre location). (25).

Oengören, A. and Ziada, S. (1992J). Vorticity shedding and acoustic resonance in an in-line tube bundle. Part II, Acoustic resonance. *J. Fluids Struct.*, **6**, 293–309. (Exp. in-line, $S/D = 1.75$, $T/D = 2.25$, 10 rows, $L/D = 10$, acoustic resonance, p', u', St, frequency PS, FV, spanwise variation). 28.

Oengören, A. and Ziada, S. (1998J). An in-depth study of vortex shedding acoustic resonance and turbulent forces in normal-triangle tube arrays. *J. Fluids Struct.*, **12**, 717–58. (Exp. normal-triangle stag., $1.61 < P/D < 3.41$, $2.7k < Re < 78k$, p', PS, St_g, simulation of resonance, FV, lift and drag PS). 28.

Ohle, F. and Eckelmann, H. (1992J). Modelling of a von Kármán vortex street at low Reynolds numbers. *Phys. Fluids*, **A4**, 1707–14. (Exp. $Re = 143$, FV, St, PS, modelling, comp.). 21.

Ohle, F., Lehman, P., Roesch, E., Eckelmann, H., and Hubler, A. (1990J). Description of transient states of von Kármán vortex streets by low-dimensional differential equations. *Phys. Fluids*, **A2**, 479–81. (Th. Comp.). 21.

Okajima, A. (1979J). Flow around two tandem circular cylinders at very high Reynolds numbers. *Bull. JSME*, **22**, 504–11. (Exp. tandem, $40k < Re < 620k$, $1.07 < S/D < 6.3$, $L/D = 6.6$, $D/B = 0.075$, smooth, $K/D = 0.9\%$, C_{D1}, C_{D2}, St, FV, θ_s). 26.

Okajima, A. and Sugitani, K. (1980J). Mean aerodynamic characteristics of two side-by-side cylinders at high Reynolds numbers, in Japanese. *Appl. Mech. Inst. Rep.*, Kyushu University, **53**, 37–64. (Exp. side-by-side, $120k < Re < 450k$, $1.12 < Re < 3$, C_p, FV, C_D, C_{D2}, C_{L1}, C_{L2}, separation bubble, u/V, PS, St). 26.

Okajima, A. and Sugitani, K. (1984J). Flow around a circular cylinder immersed in a wake of an identical cylinder, in Japanese. *Trans. JSME*, **50**, 2531–8. (Exp. tandem, $16k < Re < 440k$, $6 < S/D < 60$, $L/D = 12.5$, $D/B = 0.04$, C_{p2}, C_{D2}, $K/D = 0.3\%$, St, St_2, C_{ps2}, C_{pb2}). 26.

Okajima, A., Sugitani, K., and Mizota, T. (1986J). Flow around a pair of circular cylinders arranged side-by-side at high Reynolds numbers, in Japanese. *Trans. JSME*, **52**, 2844–50. (Exp. side-by-side, $35k < Re < 450k$, $1 < T/D < 3$, C_D, C_L, FV, St, C_p, u/V). 26.

Okamoto, S. and Sunabashiri, Y. (1992J). Vortex shedding from a circular cylinder of finite length placed on a ground plane. *J. Fluids Engg*, **114**, 512–21. (Exp. $25k < Re < 47k$, $0.5 < H/D < 7$, C_p, C_D, FV, L_w, horseshoe-swirl, St, PS, u/v, U/V). 21.

Okamoto, T. and Takeuchi, M. (1975J). Effect of side walls of wind tunnel on flow around two-dimensional circular cylinder and its wake. *Bull. JSME*, **18**, 1011–17. (Exp. $Re < 32k$, 58k, $0.05 < D/B < 0.34$, C_p, FV, θ_s, K/V, b/a, C_D, $u/u_{\max}$, b/D). 23.

Okamoto, T. and Yagita, M. (1973J). The experimental investigation of the flow past a circular cylinder of finite length placed normal to the plane surface in a uniform stream. *Bull. JSME*, **16**, 805–14. (Exp. $1.7k < Re < 13k$, $1 < H/D < 12.5$, C_p, C_{po}, C_{pb}, C_d, C_D, St, FV, u). 21.

Osaka, H., Nakamura, I., Yamada, H., Kuwata, Y., and Kageyama, Y. (1983Ja). The structure of a turbulent wake behind a cruciform circular cylinder. *Bull. JSME*, **26**, 356–63. (Exp. $Re = 4$k, u/V, secondary flow, C_p, C_d). 26.

Osaka, H., Yamada, H., Nakamura, I., Kuwata, Y., and Kageyama, Y. (1983Jb). The structure of turbulent wake behind a cruciform circular cylinder. *Bull. JSME*, **26**, 521–8. (Exp. $Re = 4$k, u/V, u'/V, v/V, v'/V, $u'v'$, w'/V). (26).

Owen, P. R. (1965J). Buffeting excitation of boiler tube vibration. *Inst. Mech. Eng., J. Mech. Engg Sci.*, **7**, 431–9. (Th. Exp. turbulence equilibrium, PS, buffeting, St, prediction). 28.

Owen, P. R. and Zienkiewicz, K. H. (1957J). The production of uniform shear flow in a wind tunnel. *J. Fluid Mech.*, **2**, 521–31. (Exp. Th. C_p, grid). 15, 22.

Paidoussis, M. P., Price, S. J., Nakamura, T., Mark, B., and Mureithi, W. N. (1989J). Flow-induced vibrations and instabilities in a rotated square cylinder array in cross-flow. *J. Fluids Struct.*, **3**, 229–54. (Exp. rotated square, $P/D = 1.5$, $L/D = 36$, $Ti = 0.5\%$, osc. frequency PS, V_r, Sc). (28).

Palmer, M. D. and Keffer. J. F. (1972J). Experimental investigation of an asymmetrical turbulent wake. *J. Fluid Mech.*, **53**, 593–610. (Exp. side-by-side, $1.25 < D_1/D_2 < 1.67$, $0.3 < G/D_1 < 0.4$, $Re = 5$k, u/V, u'/V, $u'v'$, turbulent energy reversal). 26.

Parker, R. (1978J). Acoustic resonances in passages containing banks of heat exchanger tubes. *J. Sound Vib.*, **57(2)**, 245–60. (Exp. Th. normal triangle, $P/D = 2$, 1.2, $L/D = 5$, 8, a_{eff}, comp.). 28.

Parkin, J. H. (1927J). Research on channel wall interference. *Roy. Aero. Soc. J.*, **31**, 110–41. (Hist. Exp.). (24).

Parkinson, G. V. and Jandali, T. (1970J). A wake source model for bluff body potential flow. *J. Fluid Mech.*, **40**, 557–94. (Th. wake sources, C_p). 10, 23.

Pechstein, W. (1942J). Effect of natural wind on large circular cylinder. *Ingenieurforschung, VDI Zeit.*, **86**, 220–2. (Exp. $H/D = 4$, $1.2\text{M} < Re < 4\text{M}$, C_p, C_p', C_d). 21.

Peller, H. (1986J). Thermofluid dynamic experiments with a heated and rotating circular cylinder in cross-flow. Part 2.1, Boundary layer profiles and location of separation points. *Exp. Fluids*, **4**, 223–31. (Exp. rot. cyl., $Re = 48$k, $V_r/V = 0.5$, 2, $L/D = 1.25$, $D/B = 0.3$, δ, θ_s). 24.

Peller, H., Lippig, V., Straub, D., and Waibel, R. (1984J). Thermofluid dynamic experiments with a heated and rotating circular cylinder in cross flow. *Exp. Fluids*, **2**, 113–20. (Exp. rot. cyl., $8.3\text{k} < Re < 71\text{k}$, $0 < V_r/V < 2.5$, $\overline{Nu}$). 24.

Perry, A. E. and Steiner, T. R. (1987J). Large-scale vortex structures in turbulent wakes behind bluff bodies. Part 1, Vortex formation processes. *J. Fluid Mech.*, **174**, 233–70. (Th. Exp. topology, fence, streamlines, flat plate, inclined). (3, 5, 21).

Peterka, J. A. and Richardson, P. D. (1969J). Effect of sound on separated flows. *J. Fluid Mech.*, **37**, 265–87. (Exp. sound, St, St_{Tr}, St_s). 28.

Piccirillo, P. S. and Van Atta, C. W. (1993J). An experimental study of vortex shedding behind linearly tapered cylinders at low Reynolds numbers. *J. Fluid Mech.*, **246**, 163–99. (Exp. taper, $32 < Re < 179$, $0.1 < Tr < 0.2$, shedding cell, St, Re_{osc}, modulation, FV, wavelet). 22.

Pierce, F. J. and Tree, I. K. (1990J). The mean flow structure in the symmetry plane of a turbulent horseshoe vortex. *J. Fluids Engg*, **112**, 16–22. (Exp. $Re =$ 183k, horseshoe-swirl, streamlines, FV, C_p). (21).

Polak, D. R. and Weaver, D. S. (1995J). Vortex shedding in normal triangular tube arrays. *J. Fluids Struct.*, **9**, 1–17. (Exp. normal triangle, $1.14 \leqslant P/D \leqslant 2.67$, $760 < Re <$ 49k, $L/D = 8.0$, 3.4, $Ti = 1\%$, frequency PS, St, FV). 28.

Prasad, A. and Williamson, C. H. K. (1997J). A method of reduction of bluff body drag. *J. Wind Engg Ind. Aero.*, **69–71**, 155–67. (Exp. control plate, $0.03 < P/D < 1.00$, $Re =$ 50k, $0 < G/D < 6$, C_D, FV, C_p, C_p'). 26.

Price, S. J. (1976J). The origin and nature of the lift force on the leeward of two bluff bodies. *Aero. Quart.*, **26**, 154–68. (Exp. smooth, stranded cyls, 10k $< Re <$ 80k, C_D, $6 < S/D < 18$, C_L, C_p, FV, $0.5 < D_1/D_2 < 2$). 26.

Price, S. J. and Paidoussis, M. P. (1984J). The aerodynamic forces acting on groups of two and three circular cylinders when subject to cross-flow. *J. Wind Engg Ind. Aero.*, **17**, 329–47. (Exp. 2, 3 cyl., $Re =$ 51k, $T/D = 0.75$, 2, $S/D = \pm 1.5$, 5, $L/D = 24.4$, $D/B = 0.03$, $Ti = 0.5\%$, C_D, C_L). 27.

Price, S. J. and Paidoussis, M. P. (1989J). The flow-induced response of a single flexible cylinder in an in-line array of rigid cylinders. *J. Fluids Struct.*, **3**, 61–82. (Exp. in-line, $P/D = 1.5$, $L/D = 24$, 20.5, V_j/V, Ti_g, frequency PS, St, osc., A/D, Sc). (28).

Price, S. J. and Zahn, M. L. (1991J). Fluid-elastic behaviour of a normal triangular array subject to cross-flow. *J. Fluids Struct.*, **5**, 259–78. (Exp. normal triangle, $P/D = 1.375$, osc. acoustic resonance, u_g/V, u_g'/V). 28.

Price, S. J., Mark, B., and Paidoussis, M. P. (1986J). An experimental stability analysis of a single flexible cylinder positioned in an array of rigid cylinders and subject to cross-flow. *ASME J. Pressure Vessel Technol.*, **108**, 62–72. (Exp. two and three rows deep normal triangle, $P/D = 2.1$ and square $P/D = 1.5D$, $L/D = 36$, $Ti = 0.5\%$, osc., frequency PS, excitation, A/D). (28).

Price, S. J., Sumner, D., Smith, J. G., Leung, K., and Paidoussis, M. P. (2002J). Flow visualization around a circular cylinder near to a plane wall. *J. Fluids Struct.*, **16**, 175–91. (Exp. 1.2k $< Re <$ 4.9k, $0 < G/D < 2$, $\delta/D = 0.45$, FV, PS, St). (23).

Probert, S. D., Hasoon, M. A., and Lee, T. S. (1973J). Boundary layer separation for winds flowing around large fixed-roof oil storage tanks with small aspect ratios. *Atmos. Environ.*, **7**, 651–4. (Natural wind, 4M $< Re <$ 27M, $H/D = 1.4$, θ_s). 21.

Provansal, M., Mathis, C., and Boyer, L. (1987J). Bénard–von Kármán instability: transient and forced regimes. *J. Fluid Mech.*, **182**, 1–22. (Exp. Th. Landau's eqn, amplification, forced oscillation, Ro). 21.

Putnam, A. A. (1959J). Flow-induced noise in heat exchangers. *ASME J. Engg Power*, **81**, 417–22. (Review, correlation of noise data, Disc.). 28.

Qadar, A (1981J). The vortex scour mechanism at bridge piers. *Proc. Inst. Civil Eng.*, **71**, 739–57. (Exp. horseshoe-swirl, scour, sediment size, criterion). (21).

Quadflieg, H. (1977J). Vortex induced loads on a pair of cylinders in incompressible flow at high Reynolds numbers, in German. *Forsch. Ing. Wes.*, **43**, 9–18. (Exp. side-by-side, Th. potential flow, 10k $< Re <$ 200k, $1.1 < T/D < 2.0$, FV, C_p, C_p', θ_o, θ_s, St, C_{pb}, C_D, C_L, osc., A/D). 26.

Ramamurthy, A. S. and Lee, P. M. (1973J). Wall effects on flow past bluff bodies. *J. Sound Vib.*, **31**, 433–51. (Exp. 69k $< Re <$ 204k, $D/B < 0.31$, C_p, C_D, FV). 23.

Ramamurthy, A. S. and Ng, C. P. (1973J). Effect of blockage on steady force coefficients. *ASCE J. Engg Mech. Div.*, **99**, EM4, 755–72, Disc. Shaw, T. L., EM6, 1050–1. (Exp. 13k $< Re <$ 230k, $0.07 < D/B < 0.70$, C_D, St). (23).

Ramberg, S. E. (1983J). The effects of yaw and finite length upon vortex wakes of stationary and vibrating circular cylinders. *J. Fluid Mech.*, **128**, 81–107. (Exp. 160 $< Re <$ 1.1k, $-10° < \Lambda < 60°$, $20 < L/D < 100$, FV, β, St_n, L_f, b, C_{pb}, a/D). 25.

Raney, D. C. and Chang, T. S. (1971J). Oscillating modes of instability for flow between rotating cylinders with a transverse pressure gradient. *Zeit. Angew. Math. Phys.*, **22**, 680–90. (Exp. Taylor–Dean problem). (24).

Rayleigh, Lord (J. W. Strutt) (1916J). On the dynamics of revolving fluids. *Proc. Roy. Soc.*, **93A**, 148–54. (Th. instability of rotating inviscid flows). (24).

Reid, W. H. (1960J). Inviscid modes of instability in Couette flow. *J. Math. Anal. Appl.*, **1**, 411–22. (Th. inviscid instability). (24).

Richardson, E. G. (1930J). On the flow of air adjacent to the surface of the rotating cylinder in a stream. *Brit. Aero. Res. Council*, Rep. and Memo. 1368. (Exp. rot. cyl., 30k $< Re <$ 203k, $0 < V_r/V < 7$, $L/D = 24$, $D/B = 0.04$, u, u'). (24).

Richter, A. and Naudascher, E. (1976J). Fluctuating forces on a rigid circular cylinder in confined flow. *J. Fluid Mech.*, **78**, 561–76. (Exp. 20k $< Re <$ 300k, $0.12 < D/B < 0.50$, St, C_D, C_D', C_L'). 23.

Rosenhead, L. (1931J). The formation of vortices from a surface of discontinuity. *Proc. Roy. Soc.*, **134A**, 170–92. (Th. roll-up, discrete vortices). 12, 26.

Rosenhead, L. and Schwabe, M. (1930J). An experimental investigation of the flow behind circular cylinders in channels of different breadths. *Proc. Roy. Soc.*, **129A**, 115–35. (Exp. $20 < Re < 1k$, $0.05 < D/B < 0.33$, $L/D = 13, 6.5$, FV, V_c/V, b/D, a/D, b/a). 23.

Roshko, A. (1955J). On the wake and drag of bluff bodies. *J. Aero. Sci.*, **22**, 124–32. (Exp. Th. C_D, St). 3, 4, 5, 23.

Roshko, A. (1961J). Experiments on the flow past a circular cylinder at very high Reynolds numbers. *J. Fluid Mech.*, **10**, 345–56. (Exp. $1.8M < Re < 8.3M$, C_D, C_{pb}, St, PS). 1, 6, 7, 22.

Ruscheweyh, H. (1975/76J). Wind loading on hyperbolic natural draught cooling towers. *J. Wind Engg Ind. Aero.*, **1**, 335–40. (Exp. full-scale cooling tower, $Re = 60M$, Ti, C'_p, correlation). 22.

Ruscheweyh, H. (1976J). Wind loadings on the television tower, Hamburg, Germany. *J. Ind. Aero.*, **1**, 315–33. (Exp. *in situ.*, $6M < Re < 18M$, C_p, C_{pb}, St, C'_L, C'_D, C_D). 7, 21.

Ruscheweyh, H. and Fischer, K. (1979J). Aerodynamic effects of large natural draught cooling towers on the atmospheric dispersion from a stack. *J. Ind. Aero.*, **4**, 399–413. (Exp. effluent concentration, FV). 22.

Sakamoto, H. and Arie, M. (1983J). Vortex shedding from a rectangular prism and a circular cylinder placed vertically in a turbulent boundary layer. *J. Fluid Mech.*, **126**, 142–65. (Exp. $4.1k < Re < 7.5k$, $1.5 < H/D < 8$, $0.4 < H/D < 1.4$, St, FV). 21.

Sakamoto, H. and Haniu, H. (1994J). Optimum suppression of fluid forces acting on a circular cylinder. *J. Fluids Engg*, **116**, 221–7. (Exp. control cyl., $Re = 65k$, $0.34 < G/d < 3.2$, $D/d = 16.3$, $0° < \theta_L < 180°$, C_D, C_p, C_L, bistable flow, FV). 26.

Sakamoto, H. and Oiwake, S. (1984J). Fluctuating forces on a rectangular prism and a circular cylinder placed vertically in a turbulent boundary layer. *J. Fluids Engg*, **106**, 160–6. (Exp. $1 < H/D < 5$, $28k < Re < 169k$, C_p, C'_p, C'_L, C'_D, PS, St). 21.

Sanada, S. and Matsumoto, S. (1992J). Full-scale measurements of wind force acting on and response of a 200 m concrete chimney. *J. Wind Engg Ind. Aero.*, **43**, 2165–76. (Exp. *in situ.*, $10M < Re < 40M$, C_D, $H/D = 13.5$, St, C_p, C'_L, C'_D, A/D). 7, 21.

Sarode, R. S., Gai, S. L., and Ramesh, S. K. (1980J). Flow around circular- and square-section models of finite height in a turbulent shear flow. *J. Wind Engg Ind. Aero.*, **8**, 223–30. (Exp. $Re = 22k$, $1.14 < H/D < 10$, 2 cyl., C_p, C_D, Ti). 21, 26.

Sarpkaya, T. (1966J). Separated flow about lifting bodies and impulsive flow about cylinders. *AIAA J.*, **4**, 414–20. (Exp. 15k $< Re <$ 120k, C_D, FV). (25).

Sato, H. (1960J). The instability and transition of a two-dimensional jet. *J. Fluid Mech.*, **7**, 53–80. (Exp. jet, transition, St). 28.

Sayers, A. T. (1987J). Flow interference between three equi-spaced cylinders when subject to cross flow. *J. Wind Engg Ind. Aero.*, **26**, 1–14. (Exp. 3 cyl., $Re =$ 34k, $1.25 < S/D < 5$, $L/D = 11.3$, open jet, $D_e/D = 12.4$, C_p, C_d, C_l, $\bar{C}_d$, $\bar{C}_l$). 27.

Sayers, A. T. (1988J). Flow interference between four equi-spaced cylinders when subject to cross flow. *J. Wind Engg Ind. Aero.*, **31**, 9–28. (Exp. 4 cyl., $1.1 < P/D < 5$, $Re =$ 30k, $0° < \alpha < 180°$, C_p, C_d, C_l, C_{pb}). 27.

Sayers, A. T. (1990J). Vortex shedding from groups of three and four equi-spaced cylinders situated in a cross flow. *J. Wind Engg Ind. Aero.*, **34**, 213–21. (Exp. 3, 4 cyl., $Re =$ 30k, $1.5 < S/D < 5$, $L/D = 22.6$, open jet, St). 27.

Sayers, A. T. and Saban, A. (1994J). Flow over two cylinders of different diameters: lock-in effect. *J. Wind Engg Ind. Aero.*, **51**, 43–54. (Exp. $1.2 < D_1/D_2 < 2.4$, $5.8 < S/D < 8.2$, $13° < \alpha < 28°$, St_1/St_2, C_p, C_{pb}). (26).

Schewe, G. (1983J). On the force fluctuations acting on a circular cylinder in cross flow from a subcritical up to transcritical Reynolds numbers. *J. Fluid Mech.*, **133**, 265–85. (Exp. 20k $< Re <$ 7.1M, $L/D = 10$, $D/B = 0.1$, $Ti =$ 0.4%, C_D, C'_L, St, PS). 6, 21.

Schlichting, H. (1930J). On the plane wake problem, in German. *Ing. Archiv.*, **1**, 533–71. (Th. far-wake). 3, 5, 23.

Schmidt, L. V. (1965J). Measurements of fluctuating air loads on a circular cylinder. *J. Aircraft*, **2**, 49–55. (Exp. 200k $< Re <$ 500k, $H/D = 8.1$, C'_L, C'_D, $L_c\,(C'_L, x_s)$, $L_c\,(C'_D, x_s)$, PS, A/D). 21.

Schultz-Grünow, F. (1959J). On stability of Couette flow. *Zeit. Angew. Math. Phys.*, **7**, 101–10. (Th. Exp.). (24).

Schultz-Grünow, F. and Hein, H. (1956J). Contribution to Couette flow. *Zeit. Flugwissenschaften*, **4**, 20–30. (Exp. FV, laminar and turbulent Taylor eddies). (24).

Sears, W. R. (1948J). The boundary layer of yawed cylinders. *J. Aero. Sci.*, **15**, 49–52. (Th. BL eqn, profiles). 25.

Shair, F. H., Grove, A. S., Petersen, E. E., and Acrivos, A. (1963J). The effect of confining walls on the stability of the steady wake behind a circular cylinder. *J. Fluid Mech.*, **17**, 546–50. (Exp. $50 < Re < 140$, $0.05 < D/B < 0.20$, Re_{osc}). 23.

Shaw, T. L. (1971J). Effect of side walls on flow past bluff bodies. *Proc. ASCE J. Hydr. Div.*, **97**, 65–79. (Exp. flat plate, bluff bodies, C_p, C_d, C_L). 23.

Shirakashi, M., Hasegawa, A., and Wakiya, S. (1986J). Effect of secondary flow on Kármán vortex shedding from a yawed cylinder. *Bull. JSME*, **29**, 1124–8. (Exp. $800 < Re < 55$k, $0° < \Lambda < 45°$, PS, u/V, FV). (25).

Shirakashi, M., Ueno, S., Ishida, Y., and Wakiya, S. (1984J). Vortex excited oscillation of a circular cylinder. *Bull. JSME*, **27**, 1120–6. (Exp. $31 < Re <$ 53k, $0° < \Lambda < 60°$, St, PS, St_n). (25).

Sin, V. K. and So, R. M. C. (1987J). Local force measurements on finite-span cylinders in a cross flow. *J. Fluids Engg*, **109**, 139–43. (Exp. $Re = 48$k, $0.5 < H/D < 2$, C_p, C_{pb}, C_D, C'_L, C'_D, St). (21).

Slaouti, A. and Gerrard, J. H. (1981J). An experimental investigation of the end effects on the wake of a circular cylinder towed through water at low Reynolds numbers. *J. Fluid Mech.*, **112**, 297–314. (Exp. $100 < Re < 142$, $24 < L/D < 35$, free surface, contamination, free end, FV, yawed cylinder). 21.

Slaouti, A. and Stansby, P. K. (1992J). Flow around two circular cylinders by the random-vortex method. *J. Fluids Struct.*, **6**, 641–70. (Comp. side-by-side, tandem, streamlines, $1.5 < T/D < 3$, C_D, C_L, St, $1.2 < S/D < 10$, vorticity). 26.

Smith, A. M. D. and Murphy, J. S. (1955J). A dust method for locating the separation point. *J. Aero. Sci.*, **22**, 273–4. (Exp. FV). 22.

Smith, A. R., Moon, W. T., and Kao, T. W. (1972J). Experiments on flow about yawed circular cylinder. *J. Basic Engg*, **94**, 771–6. (Exp. 2k $< Re <$ 10k, $0° < \Lambda < 60°$, C_{pn}, C_D, u/v, L_f, PS). 25.

Snyder, H. A. (1960J). Stability of rotating Couette flow. *Phys. Fluids*, **11**, 1599–605. (Exp. supercritical Taylor eddies). (24).

Snyder, H. A. (1969J). Wave-number selection at finite amplitude in rotating Couette flow. *J. Fluid Mech.*, **35**, 273–98. (Exp. supercritical Taylor eddies). (24).

Snyder, H. A. and Lambert, R. B. (1966J). Harmonic generation in Taylor vortices between rotating cylinders. *J. Fluid Mech.*, **26**, 545–62. (Exp. supercritical Taylor eddies). (24).

So, R. M. C. (1977J). Lift on a rotating porous cylinder. *J. Fluids Engg*, **99**, 753–7. (Th. rot. cyl., circulation and sinks). (24).

Spivack, H. M. (1946J). Vortex frequency and flow pattern in the wake of two parallel cylinders at varied spacing normal to an air stream. *J. Aero. Sci.*, **13**, 289–301. (Exp. side-by-side, 5k $< Re <$ 93k, $1 < T/D < 6$, St, u/V). 26.

Stager, R. and Eckelmann, H. (1991J). The effect of end plates on the shedding frequency of circular cylinders in the irregular range. *Phys. Fluids*, **A3**, 2116–21. (Exp. $100 < Re < 1.6$k, $9 < D_e/D < 34$, St, PS). 21.

Strykovski, P. J. and Sreenivasan, K. R. (1990J). On the formation and suppression of vortex 'shedding' at low Reynolds number. *J. Fluid Mech.*, **218**, 71–107. (Exp. Comp. control cyl., $3 < D/d < 20$, $14 < L/D < 60$, $46.2 < Re < 120$, Ro, u/V). 26.

Sumner, D., Price, S. J., and Paidoussis, M. P. (2000J). Flow-pattern identification for two staggered circular cylinders in cross-flow. *J. Fluid Mech.*, **411**, 253–303. (Exp. $1 \leqslant P/D \leqslant 5$, $850 < Re < 1.9\text{k}$, FV, K/VD, St). (26).

Sumner, D., Wong, S. S. T., Price, S. J., and Paidoussis, M. P. (1999J). Fluid behaviour of side-by-side cylinders in steady cross-flow. *J. Fluids Struct.*, **13**, 308–38. (Exp. Comp. 2 and 3 side-by-side cyl., $500 < Re < 3\text{k}$, $1 < T/D < 6$, $L/D = 27$, $D/B = 0.07$, St, PS, FV, PIV, vorticity, δ, K/VD). (26).

Sun, T. F. and Gu, Z. F. (1995J). Interference between wind loading on group of structures. *J. Wind Engg Ind. Aero.*, **54–5**, 213–25. (Exp. tandem, staggered, side-by-side, $220\text{k} < Re < 650\text{k}$, $1 < S/D < 5$, $0° < \alpha < 90°$, C_{p1}, C_{p2}, C_{D1}, C_{D2}, C_{L1}, C_{L2}). 26.

Sun, T. F. and Zhou, L.-M. (1983J). Wind pressure distribution around a ribless hyperbolic cooling tower. *J. Wind Engg Ind. Aero.*, **14**, 181–92. (Exp. model and full-scale cooling tower, $0.9\text{M} < Re < 1.08\text{M}$, $54\text{M} < Re < 67\text{M}$, C_p, surface roughness, C_{pi}). 22.

Sun, T. F., Gu, Z. F., He, D. X., and Zhang, L. L. (1992Ja). Fluctuating pressure on two circular cylinders at high Reynolds numbers. *J. Wind Engg Ind. Aero.*, **41–4**, 577–88. (Exp. $325\text{k} < Re < 650\text{k}$, $2.2 < S/D < 4$, $\alpha = 0°$, $12.5°$, $90°$, C'_{p1}, C'_{p2}). 26.

Sun, T. F., Gu, Z. F., Zhou, L.-M., Li, P. H., and Cai, G. L. (1992Jb). Full-scale measurement and wind-tunnel testing of wind loading on two neighbouring cooling towers. *J. Wind Engg Ind. Aero.*, **41–4**, 2213–24. (Exp. cooling towers, full-scale $Re = 75\text{M}$, 1:450 model, $Re = 142\text{k}$, $1 < S/D < 3.5$, C_{p1}, C_{p2}, C'_p, $0° < \alpha < 90°$). 26.

Sung, H. J., Chun, C. K., and Hyun, J. M. (1995J). Experimental study of uniform shear flow past a rotating cylinder. *J. Fluids Engg*, **117**, 62–7. (Exp. rot. cyl., $600 \leq Re < 1.2\text{k}$, $-2 < V_r/V < 2$, u/V, PS, St, shear parameter). (24).

Szechenyi, E. (1975J). Supercritical Reynolds number simulation for two dimensional flow over circular cylinders. *J. Fluid Mech.*, **70**, 529–42. (Exp. $21\text{k} < Re < 6.5\text{M}$, $0.015\% < K/D < 0.2\%$, $C_{L\max}$, C_d, St, Re_k, L_c). 22.

Szepessy, S. (1993J). On the control of circular cylinder flow by end plates. *Environ. J. Mech., B/Fluids*, **12**, 217–44. (Exp. $4\text{k} < Re < 48\text{k}$, $0.25 < B/D < 27.6$, C_{pb}, C_p, FV). 21.

Szepessy, S. and Bearman, P. W. (1992J). Aspect ratio and end plate effects on vortex shedding from a circular cylinder. *J. Fluid Mech.*, **234**, 191–217. (Exp. $86\text{k} < Re < 140\text{k}$, $0.25 < B/D < 12$, C'_L, C_p, L_f, PS, L_c). 21.

Tanaka, H. and Nagano, S. (1973J). Study of flow around a rotating circular cylinder. *Bull. JSME*, **16**, 234–43. (Exp. rot. cyl., 48k $< Re <$ 311k, $0 < V_r/V < 1.1$, $L/D = 2.6$, $D/B = 0.18$, C_L, C_D, St, θ_s, b, u', u). 24.

Tanaka, S. and Murata, S. (1999J). An investigation of the wake structure and aerodynamic characteristics of a finite circular cylinder. *JSME, Int. J.*, **42B**, 178–87. (Exp. $Re =$ 37k, $1.25 < L/D < 10$, u/V, streamwise vorticity, streamlines). 21.

Taneda, S. (1956J). Experimental studies of the lift on two equal circular cylinders placed side-by-side in a uniform stream at low Reynolds numbers. *J. Phys. Soc. Jap.*, **12**, 419–22. (Exp. side-by-side, $0.01 < Re < 1.6$, $4.6 < T/D < 50$, C_L, Th.). 26.

Taneda, S. (1959J). Downstream development of wakes behind cylinders. *J. Phys. Soc. Japan*, **14**, 843–8. (Exp. $61 < Re < 149$, $58 < x/D < 140$, FV). 3, 27.

Taneda, S. (1964J). Experimental investigation of the wall-effect on a cylindrical obstacle moving in a viscous fluid at low Reynolds number. *J. Phys. Soc. Jap.*, **19**, 1024–30. (Exp. $4 \times 10^{-5} < Re < 5 \times 10^{-4}$, C_D, C_R, C_L, FV, $0.1 < G/D < 0.6$). 23.

Taneda, S. (1965J). Experimental investigation of vortex streets. *J. Phys. Soc. Jap.*, **20**, 1714–21. (Exp. $20 < Re < 186$, near wall, $D/B = 0.33$, 0.50, 0.66, taper, osc. whirl). 23.

Taneda, S. (1974J). Necklace vortices. *J. Phys. Soc. Jap.*, **36**, 298–303. (Exp. horseshoe-swirl at water surface, $0.7 < Fr < 2.8$, FV, ship models). (21).

Taneda, S. (1978J). Visual observations of the flow past a circular cylinder performing rotating oscillations. *J. Phys. Soc. Jap.*, **45**, 1038–43. (Exp. rot. osc. $30° < \theta_{\text{osc}} < 90°$, $36 < Re < 300$, $6.8 < L/D < 30$, $0 < f_{\text{osc}}$, $D/NV < 2.8$). 24.

Taneda, S. (1979J). Visualisation of separating Stokes flows. *J. Phys. Soc. Jap.*, **46**, 1935–40. (Exp. tandem, side-by-side, boundary, $Re = 0.01$, $1 < S/D < 2.5$, $T/D = 1.2$, $0 < G/D < 0.6$, FV, comp. Th.). 26.

Taneda, S. (1980J). Visualisation of steady flows induced by a circular cylinder performing a rotary oscillation about an eccentric axis. *J. Phys. Soc. Jap.*, **49**, 2038–41. (Exp. rot. osc. $15° < \theta_{\text{osc}} < 30°$, $0.02 < (\nu/\pi N D^2)^{1/2} < 2$, $e/D = 0.25$, FV, flow classification). 24.

Taneda, S., Honji, H., and Tatsuno, M. (1974J). The behaviour of tracer particles in flow visualisation by electrolysis of water. *J. Phys. Soc. Jap.*, **37**, 784–8. (Exp. $650 < Re <$ 12k, horseshoe-swirl, FV, tracer-generated wire). 21.

Tanida, Y., Okajima, A., and Watanabe, Y. (1973J). Stability of a circular cylinder oscillating in uniform flow or in a wake. *J. Fluid Mech.*, **61**, 769–84. (Exp. 2 cyl., C_D, C_L', C_D', St, forced osc.). 3, 26.

Taniguchi, S. and Miyakoshi, K. (1990J). Fluctuating fluid forces acting on a circular cylinder and interference with a plane wall. *Exp. Fluids*, **9**, 197–204. (Exp. $Re = 94$k, $0 < G/D < 3$, $0.34 < \delta/D < 1.05$, C'_D, C'_L, $(G/D)_c$, FV, C'_p, R_{12}). 23.

Taniguchi, S., Sakamoto, H., and Arie, M. (1981Ja). Flow around circular cylinders of finite height placed vertically in turbulent boundary layer. *Bull. JSME*, **24**, 37–44. (Exp. $460 < Re < 33$k, $0.75 < H/D < 5$, $0.5 < H/D < 2.5$, C_p, C_{po}, C_{pb}, C_D, FV). 21.

Taniguchi, S., Sakamoto, H., and Arie, M. (1981Jb). Flow around a circular cylinder vertically mounted in a turbulent boundary layer. *Bull. JSME*, **24**, 1130–6. (Exp. $1.93\text{k} < Re < 6.55$k, $H/D = 1$, $\delta/H = 1$, C_p, C_{pb}, C_{po}, q, C_d). 21.

Taniguchi, S., Sakamoto, H., and Arie, M. (1982J). Interference between two circular cylinders of finite height vertically immersed in a turbulent boundary layer. *J. Fluids Engg*, **104**, 529–36. Disc. **105**, 368–9. (Exp. $1.2 < S/D < 4.0$, $H/D = 3$, $Re = 15.5$k, $Ti = 0.3\%$, $\delta/H = 0.87$, C_{p1}, C_{p2}, θ_m, FV, C_D, C_L, C_L/C_D). 26.

Tatsuno, M. (1989J). Steady flows around two cylinders at low Reynolds numbers. *Fluid Dyn. Res.*, **5**, 49–60. (Exp. tandem, side-by-side, staggered, $0.02 < Re < 0.16$, $1 \leqslant D_2/D_1 < 5.7$, $L/D = 20$, $D/B = 0.025$, $1.7 < S/D < 3$, $1.5 < T/D < 2$, $1.1 < S/D < 2$, FV, θ_s). 26.

Tatsuno, M., Takayama, T., Amamoto, H., and Ishii, K. (1990J). On the stable posture of a triangular or a square cylinder about its central axis in a uniform flow. *Fluid Dyn. Res.*, **6**, 201–7. (Exp. $23\text{k} < Re < 35$k, C_p, C_D, C_L, C_M, FV). 27.

Taylor, G. I. (1921J). Experiments with rotating fluids. *Proc. Camb. Phil. Soc.*, **20**, 326–9. (Exp. FV). (24).

Taylor, G. I. (1923J). Stability of a viscous flow contained between two rotating cylinders. *Phil. Trans. Roy. Soc.*, **223A**, 289–343. (Th. Exp. viscous flow, small gap, FV). 24.

Taylor, G. I. (1936J). Fluid friction between rotating cylinders. *Proc. Roy. Soc.*, **157A**, 546–64. (Exp. conc. cyl.). (24).

Teissié-Solier, M., Castagneto, L., and Sabathe, M. (1937J). On the beating that accompanies the formation of Bénard–Kármán alternate eddies, in French. *Compt. Rend. Acad. Sci.*, **205**, 23–5. (Exp. $40 < Re < 200$, St). 2, 3, 21.

Thomas, D. G. and Kraus, K. A. (1964J). Interaction of vortex streets. *J. Appl. Phys.*, **35**, 3458–9. (Exp. tandem $200 < Re < 500$, $3.6 < S/D < 16$, FV). 26.

Thomson, K. B. and Morrison, D. F. (1971J). The spacing, position, and strength of vortices in the wake of slender cylindrical bodies at large incidence. *J. Fluid Mech.*, **50**, 751–83. (Exp. Th. 43k $< Re <$ 230k, $0.2 < Ma < 2.8$, St, p_o, b/a, FV, K/VD, C_D). 25.

Thwaites, B. (1948J). The production of lift independently of incidence. *Roy. Aero. Soc. J.*, **52**, 117–24. (Exp. suction, Thwaites flap, aerofoil, C_L). 24.

Ting, D. S. K., Wang, D. J., Price, S. J., and Paidoussis, M. P. (1998J). An experimental study on the fluid elastic forces for two staggered circular cylinders in cross-flow. *J. Fluids Struct.*, **12**, 259–94. (Exp. two staggered cyls, 40k $< Re <$ 200k, $15 < V_r < 300$, $1.3 < S/D < 5$, $0.2 < T/D < 0.4$, C_D, C_L, ϕ). (26).

Tobak, M., Schiff, L. B., and Peterson, V. L. (1969J). Aerodynamics of bodies of revolution in coning motion. *AIAA J.*, **7**, 95–9. (Exp. 208k $< Re <$ 729k, $Ma = 1.4$, $11° < \alpha < 45°$, $n_c = 600\,\text{rpm}$, FV, C_m). (25).

Tokumaru, P. T. and Dimotakis, P. E. (1991J). Rotary oscillation control of a cylinder wake. *J. Fluid Mech.*, **224**, 77–90. (Exp. rot. osc. cyl., $Re =$ 15k, $L/D = 4.5$, $D/B = 0.22$, $0.17 < N_{\text{osc}} < 3.3$, C_D, FV, phase). 24.

Tournier, C. and Py, B. (1978J). The behaviour of naturally oscillating three-dimensional flow around a cylinder. *J. Fluid Mech.*, **85**, 161–86. (Exp. $Re =$ 13.9k, split-film, Th., $0° < \Lambda < 30°$, surface velocity, R_{zz}, surface streamlines). 25.

Triantafillou, G. S. (1992J). Three-dimensional flow patterns in two-dimensional wakes. *J. Fluids Engg*, **114**, 356–61. (Th. time asymptotic instability, average flow with slow modulation along span). (21).

Tsutsui, T., Igarashi, T., and Kamemoto, K. (1997J). Interactive flow around two circular cylinders of different diameters at close proximity; experiment and numerical analysis. *J. Wind Engg Ind. Aero.*, **69–71**, 279–81. (Exp. Comp. $Re =$ 10k, 41k, $D_1/D_2 = 2.22$, $90° < \theta_c < 180°$, C_D, C_L, $\mathrm{d}K/\mathrm{d}tV^2$, θ_s, St, FV, vorticity). (26).

Turner, J. R. and Eastop, T. D. (1979J). A hot-wire anemometry method for the flow patterns in an array of heat exchanger tubes. *Trans. Inst. Chem. Eng.*, **57**, 139–42. (Exp. in-line $P/D = 2$, 45k $< Re <$ 111k). 28.

Tyler, E. (1928J). Vortices behind aerofoil sections and rotating cylinders. *Phil. Mag. Roy. Soc.*, 7th Ser., **5**, 449–63. (Exp. rot. cyl., $Re =$ 4k, $V_r/V = 0.5$, 1.0, u'). (24).

Uematsu, Y. and Yamada, M. (1995J). Effects of aspect ratio and surface roughness on the time-averaged aerodynamic forces on cantilevered circular cylinders at high Reynolds numbers. *J. Wind Engg Ind. Aero.*, **54/55**, 301–12. (Exp. 38k $< Re <$ 140k, $Ti = 0.7\%$, 6.4%, $1 < H/D < 4$, $0.28\% < K/D < 1\%$, C_D, C_{pb}, $C_{pb} - C_{p\text{min}}$, θ_s). (22).

Umeda, S. and Yang. W.-J. (1999J). Interaction of von Kármán vortices and intersecting main streams in staggered tube bundles. *Exp. Fluids*, **26**, 389–96. (Exp. staggered $1.2 < T/D < 2.5$, $1.5 < R/D < 4.1$, $L/D = 0.54$ to 1.16, $200 < Re < 6k$, FV, frequency PS, St). 28.

Van Atta, C. W. (1968J). Experiments on vortex shedding from yawed circular cylinders. *AIAA J.*, **6**, 931–3. (Exp. $50 < Re < 150$, $0° < \Lambda < 78°$, St, St_n). 25.

Vasantha, R. and Nath, G. (1985J). Unsteady non-similar laminar incompressible boundary layer flow over a yawed infinite circular cylinder. *ASME J. Appl. Mech.*, **52**, 496–8. (Th. bound layer eqns calc.). (25).

Vickery, B. J. and Clark, A. W. (1972J). Lift across wind response of tapered stack. *ASCE J. Struct. Div.*, **98**, 1–20. (Exp. taper, $H/D = 20$, $20k < Re < 70k$, St, C'_p, C_{pb}, C'_L). 22.

Votaw, C. W. and Griffin, O. M. (1971J). Vortex shedding from smooth cylinders and stranded cables. *ASME J. Basic Engg*, **93**, 457–60. (Exp. $Re = 240$, 500, St, stranded cable). 22.

Waka, R., Yoshino, F., and Hayashi, T. (1985J). Properties of flow near a side wall of a circular cylinder. *Bull. JSME*, **28**, 1069–76. (Exp. slot shape, C_{pb}, C_D, C_p). (24).

Waka, R., Yoshino, F., Hayashi, T., and Iwasa, T. (1983J). The aerodynamic characteristics at the mid-span of a circular cylinder with tangential blowing. *Bull. JSME*, **26**, 755–62. (Exp. Th., circulation, C_D, C_j, C_{pb}). (24).

Waldeck, J. L. (1992J). The measured and predicted response of a 300 m concrete chimney. *J. Wind Engg Ind. Aero.*, **41–4**, 229–40. (Exp. Th.). 7, 21.

Walker, W. M. and Reising, G. F. S. (1988J). Flow-induced vibrations in cross-flow heat exchangers. *Chemical and Process Engg*, **49**, 95–103. (Exp. in-line and staggered, $1.4 < T/D < 3.5$, $0.8 < P/D < 1.5$, finned, acoustic resonance, St, sound intensity, tube removal). 28.

Wallis, P. R. (1939J). Photographic study of fluid flow between banks of tubes. *Engineering*, **147**, 423–6. (Exp. tube arrays, FV). 28.

Weaver, D. S. and Abd-Rabo, A. (1985J). A flow visualisation study of a square array of tubes in water cross-flow. *J. Fluids Engg*, **107**, 354–63. (Exp. in-line, $P/D = 1.5$, $L/D = 12$, $Ti = 0.5\%$, $110 < Re < 6.5k$, FV, A/D, FS). 28.

Weaver, D. S. and Grover, L. K. (1978J). Cross-flow induced vibrations in a tube bank–turbulent buffeting and fluid elastic instability. *J. Sound Vib.*, **59(2)**, 277–94. (Exp. parallel triangle, $P/D = 1.375$, $L/D = 12$, $Ti = 0.2\%$, frequency PS, St, Ti_g, osc., A/D, Sc, V_r). 28.

Weaver, D. S. and Yeung, H. L. (1984J). The effect of tube mass on the flow-induced response of various tube arrays in water. *J. Sound Vib.*, **93(3)**, 409–25. (Exp. square, normal triangle, parallel triangle, rotated square, $P/D = 1.5$, $L/D = 11.8$, osc., A/D, tube mass, Sc). (28).

Weaver, D. S., Fitzpatrick, J. A., and El-Kashlan, M. (1987J). Strouhal numbers for heat exchanger tube arrays in cross-flow. *ASME J. Pressure Vessel Technol.*, **109**, 219–23. (Compilation of St data). 28.

Weaver, D. S., Lian, H. Y., and Huang, X. Y. (1993J). Vortex shedding in rotated square arrays. *J. Fluids Struct.*, **7**, 107–21. (Exp. rotated square, $1.21 \leqslant P/D \leqslant 2.83$, $925 < Re < 1.6$k, frequency PS, St, FV, V_g/V, St_G). 28.

Weaver, W. (1961J). Wind-induced vibration in antenna members. *Proc. ASCE J. Engg Mech. Div. EM1*, **87**, 141–65. (Exp. Th. helical wires, $Re = 117$k, St, A/D, C'_L, pitch, number of windings). (22).

Weihs, D. (1980J). Approximate calculations of vortex trajectories of slender bodies at incidence. *AIAA J.*, **18**, 1402–3. (Th. Föppl vortices). (25).

Weizel, R. (1973J). Potential flows on N circles, in German. *Zeit. Angew. Math. Mech.*, **53**, 463–74. (Th. potential flow, streamlines). (26).

Wendt, F. (1933J). Turbulent flow between rotating concentric cylinders, in German, *Ingen.-Archiv.*, **4**, 577–95. (Exp. torque). (24).

Werle, H. (1972J). Flow past tube banks, in French. *Revue Francais de Mecanique*, **41**, 7–19. (Exp. FV). 28.

Werle, H. (1979J). Vortices in slender bodies at high angle of attack, in French. *L'Aéronautique et L'Astronautique*, **6**, 3–22. (Exp. FV, $2\text{k} < Re < 4\text{k}$, $0° < \alpha < 38°$, attached and detached eddies). 25.

West, G. S. and Apelt, C. J. (1982J). The effects of tunnel blockage and aspect ratio on the mean flow past a circular cylinder with Reynolds numbers between 10^4 and 10^5. *J. Fluid Mech.*, **114**, 361–77. (Exp. $3\text{k} < Re < 100\text{k}$, $0.06 < D/B < 0.16$, $4 < L/D < 10$, C_{pb}, C_D, St). 21.

West, G. S. and Apelt, C. J. (1993J). Measurement of fluctuating pressure and forces on a circular cylinder in the Reynolds number range 10^4 to 2.5×10^5. *J. Fluids Struct.*, **7**, 227–44. (Exp. $10\text{k} < Re < 250\text{k}$, $0.04 < D/B < 0.08$, $0.2\% < Ti < 7.5\%$, C'_p, C'_L, C'_d). 21, 23.

West, G. S. and Apelt, C. J. (1997J). Fluctuating lift and drag forces on finite lengths of a circular cylinder in the sub-critical Reynolds number range. *J. Fluids Struct.*, **11**, 135–58. (Exp. $10\text{k} < Re < 210\text{k}$, $D/B = 0.042, 0.06, 0.09$, $0.2\% < Ti < 7.5\%$, corr. R_{pp}, C'_L, C'_D). 21, 23.

White, C. M. (1946J). The drag of cylinders in fluids at low speeds. *Proc. Roy. Soc.*, **186A**, 472–9. (Exp. $10^{-5} < Re < 100$, $0.0015 < D/B < 0.15$, C_R). 23.

Wieghardt, K. E. G. (1953J). On the resistance of screens. *Aero. Quart.*, **4**, 186–91. (Exp. screens). 28.

Wieselsberger, C. (1922J). Further data on the law of liquid and air drag, in German. *Phys. Zeit.*, **23**, 219–24. (Exp. $200 < Re < 600$k, $L/D = 5$). 21.

Williamson, C. H. K. (1985J). Evolution of a single wake behind a pair of bluff bodies. *J. Fluid Mech.*, **159**, 1–18. (Exp. side-by-side, $50 < Re < 150$, $D/B = 0.003$, $L/D = 140$, $1 < T/D < 5$, FV, St). 26.

Wood, W. W. (1964J). Stability of viscous flow between rotating cylinders. *Zeit. Angew. Math. Phys.*, **15**, 313–14. (Th. proof). (24).

Wu, J., Sheridan, J., Soria, J., and Welsh, M. C. (1994J). An experimental investigation of streamwise vortices in the wake of a bluff body. *J. Fluids Struct.*, **8**, 621–35. (Exp. $50 < Re < 1.8$k, streamwise eddies, FV, λ/D). 21.

Wu, J., Sheridan, J., Welsh, M. C., and Hourigan, K. (1996J). Three-dimensional structure in a cylinder wake. *J. Fluid Mech.*, **312**, 201–22. (Exp. streamwise eddies, $200 < Re < 550$, comp. z/D). 21.

Yagita, M., Kajima, Y., and Matsuzaki, K. (1984J). On vortex shedding from circular cylinder with a step. *Bull. JSME*, **27**, 426–31. (Exp. step $0 < d/D < 1$, $800 < Re < 10$k, $10 < L/D < 27$, St, FV). 22.

Yamada, H., Osaka, H., Kageyama, Y., and Takeda, O. (1987J). The flow around crossed two circular cylinders normal to uniform stream, in Japanese. *Trans. JSME*, **53**, 333–40. (Exp. $2\text{k} < Re < 8$k, $L/D = 40$, $D/B = 0.024$, C_p, FV, θ_o, C_{pb}, C_d, C_D, Ro). 26.

Yang, E. E., Rahai, H. R., and Nakayama, A. (1994J). Mean pressure distribution and drag coefficient of wire-wrapped cylinders. *J. Fluids Engg*, **116**, 376–8. (Exp. $33\text{k} < Re < 53$k, $0.25 < P/D < 2$, C_{pb}, $C_{p\text{min}}$, $C_{pb} - C_{p\text{min}}$, C_D). (22).

Yang, X. and Zebib, A. (1989J). Absolute and convective instability of a cylinder wake. *Phys. Fluids*, **A1**, 689–96. (Th. Comp. linear stability theory, $Re < 45$). 21.

Yih, C.-S. (1972J). Spectral theory of Taylor vortices. *Arch. Rat. Mech. Anal.*, **46**, 218–40. (Th.). (24).

Yokoi, Y. and Kamemoto, K. (1993J). Initial stage of three-dimensional vortex structure existing in a two-dimensional boundary layer separation flow. *JSME Int. J.*, **36B**, 201–7. (Exp. $230 < Re < 1.3$k, FV, spanwise cell, z_c/D). 21.

Yoshino, F. and Furuya, Y. (1974J). The wall-jet on circular cylinder immersed in uniform flow. *Bull. JSME*, **17**, 1030–8. (Th. Exp. jet blowing, δ). (24).

Yoshino, F. and Hayashi, T. (1984J). The numerical solution of flow around a rotating circular cylinder in uniform shear flow. *Bull. JSME*, **27**, 1850–7. (Comp. Th. $20 < Re < 80$, C_L, C_D, streamlines, ζ). (24).

Yoshino, F., Waka, R., Iwasa, T., and Hayashi, T. (1981J). The effect of side wall on aerodynamic characteristics of a circular cylinder with tangential blowing. *Bull. JSME*, **24**, 926–33. (Th. Exp. Comp. C_L, C_D, C_{pb}, θ_u). (24).

Zdravkovich, M. M. (1967J). Note on transition to turbulence in vortex street wakes. *J. Roy. Aero. Soc.*, **71**, 886–90. (Exp. transition $100 < Re < 200$, wake instability). (26).

Zdravkovich, M. M. (1968J). Smoke observation of the wake of three cylinders at low Reynolds number. *J. Fluid Mech.*, **32**, 339–57. (Exp. 3 cyl., $60 < Re < 300$, $L/D = 137$, $D/B = 0.004$, $5 \leqslant S/D \leqslant 21$, $2 \leqslant T/D \leqslant 10$, FV). 27.

Zdravkovich, M. M. (1972J). Smoke observations of wakes of tandem cylinders at low Reynolds numbers. *J. Roy. Aero. Soc.*, **76**, 108–14. (Exp. tandem, $40 < Re < 250$, $1 < S/D < 12$, $25° < \alpha < 45°$, FV, C_D). (26).

Zdravkovich, M. M. (1973J). Smoke visualisation of three-dimensional flow patterns in a nominally two-dimensional wake. *J. Mécanique*, **12**, 173–7. (Exp. $50 < Re < 200$, two, three, cyl., FV). (26, 27).

Zdravkovich, M. M. (1980J). Aerodynamics of two parallel circular cylinders of finite height at simulated high Reynolds numbers. *J. Wind Engg Ind. Aero.*, **6**, 59–71. (Exp. $H/D = 5$, $S/D = 1.325$, $\alpha = 15°$, $45°$, $90°$, $K/D = 0.017\%$, $Re = 200$k, $Re_K = 400$, C_{p1}, C_{p2}, C_d, C_l, C_r). 26.

Zdravkovich, M. M. (1983J). Interference between two circular cylinders forming a cross. *J. Fluid Mech.*, **128**, 231–46. (Exp. $21\text{k} < Re < 107\text{k}$, $L/D = 9.18$, $D/B = 0.056$, 0.113, C_{p1}, C_{p2}, C_{D1}, C_{D2}, C_{pb1}, C_{pb2}, C_{ps1}, C_{ps2}). 26.

Zdravkovich, M. M. (1985Ja). Flow around two intersecting cylinders. *J. Fluids Engg*, **107**, 505–11. (Exp. $41\text{k} < Re < 90\text{k}$, $D/B = 0.22$, $L/D = 9$, C_{p1}, C_{p2}, C_{pb}, FV, C_D, topology). 26.

Zdravkovich, M. M. (1985Jb). Forces on a circular cylinder near a plane wall. *Appl. Ocean Res.*, **7**, 197–201. (Exp. $48\text{k} < Re < 300\text{k}$, $0.12 < \delta/D < 0.97$, $0 < G/D < 2$, C_D, C_L, tripping wires). 23.

Zdravkovich, M. M. (1993J). On suppressing metastable interstitial flow behind a tube array. *J. Fluids Struct.*, **7**, 245–52. (Exp. single row, C_{pb}, C_{ps}, C_{pg}, guide vane). 28.

Zdravkovich, M. M. and Nuttall, J. A. (1974J). On the elimination of aerodynamic noise in a staggered tube bank. *J. Sound Vib.*, **34**, 173–7. (Exp. stag. acoustic resonance, tube removal suppressing noise). (28).

Zdravkovich, M. M. and Pridden, D. L. (1977J). Interference between two circular cylinders; series of unexpected discontinuities. *J. Ind. Aero.*, **2**, 255–70. (Exp. $67\text{k} < Re < 220\text{k}$, $L/D = 8.9$, 15.7, $D/B = 0.06$, 0.10, $1 < S/D < 6$, u/V, C_{pb}, C_{pg}, C_{ps}, $C_{pg} - C_{pb}$, θ_r, C_D, C_L, C_p). 26.

Zdravkovich, M. M. and Stonebanks, K. L. (1990J). Intrinsically non-uniform and metastable flow in and behind tube arrays. *J. Fluids Struct.*, **4**, 305–19. (Exp. single and two rows, $37\text{k} < Re < 74\text{k}$, $1.2 < T/D < 2.1$, C_p, C_{pb}, C_{ps}, C_D, C_L, FV, PS, St). 28.

Zdravkovich, M. M., Brand, V. P., Mathew, G., and Weston., A. (1989J). Flow past short circular cylinders with two free ends. *J. Fluid Mech.*, **203**, 557–75. (Exp. 60k $< Re <$ 260k, $1 < L/D < 10$, C_D, topology, FV, C_p, C_y, C_r, St). 21.

Zdravkovich, M. M., Flaherty, A. J., Pahle, M. G., and Skelhorne, I. A. (1998J). Some aerodynamic aspects of coin-like cylinders. *J. Fluid Mech.*, **360**, 73–84. (Exp. 200k $< Re <$ 600k, $0.02 < L/D < 0.9$, C_p, C_D, C_{DS}, FV, topology). 21.

Zhang, H. and Melbourne, W. H. (1992J). Interference between two circular cylinders in tandem in turbulent flow. *J. Wind Engg Ind. Aero.*, **41–4**, 589–600. (Exp. tandem, $Re =$ 110k, $0.4\% < Ti < 11.5\%$, $L/D = 8$, $D/B = 0.05$, C_{pi}, C'_{pi}, C'_{po}, C_{pb}, C'_{pb}). 26.

Zhang, H.-Q., Fey, U., Noack, B. R., König, M., and Eckelmann, H. (1995J). On the transition of the cylinder wake. *Phys. Fluids*, **7**, 779–94. (Comp. Exp. $50 < Re < 300$, $1.33 < L/D < 50$, St, C_D, C'_L, 3-D cells, z/D, fingers, FV). 21.

Ziada, S. and Oengören, A. (1992J). Vorticity shedding and acoustic resonance in an in-line tube bundle. Part I, Vorticity shedding. *J. Fluids Struct.*, **6**, 271–92. (Exp. in-line, $S/D = 1.75$, $T/D = 2.25$, $L/D = 10$, 6.8k $< Re <$ 20k, u', St, coherence, phase, FV, symmetric jet instability). 28.

Ziada, S. and Oengören, A. (1993J). Vortex shedding in an in-line tube bundle with large tube spacings. *J. Fluids Struct.*, **7**, 661–87. (Exp. $T/D = 3.75$, $S/D = 3.25$, $800 < Re <$ 9.6k, $L/D = 10$, $Ti = 1\%$, frequency PS, St_g, p', V_g/V, Ti_g, FV). 28.

Ziada, S. and Oengören, A. (2000J). Flow periodicity and acoustic resonance in parallel triangle tube bundles. *J. Fluids Struct.*, **14**, 197–217. (Exp. parallel-triangle stag., $1.2 < P/D < 4.2$, 6k $< Re <$ 90k, p', PS, u', FV, St_g, acoustic response, St_a). 28.

Ziada, S., Oengören, A., and Buhlmann, E. T. (1989J). On acoustical resonance in tube arrays. Part I, Experiments. Part II, Damping criteria. *J. Fluids Struct.*, **3**, 293–324. (Exp. in-line $T/D = 1.6$, $S/D = 1.35$, $L/D = 6.7$, normal triangle $P/D = 1.6$, $L/D = 10$, frequency PS, St_g, p', V_g/V, Ti_g, FV, damping criterion). 28.

Žukauskas, A. A., Daujotas, P. M., and Ilgarubis, V. S. (1982J). Effect of flow turbulence and channel blockage on the flow pattern over a circular cylinder in cross-flow of water at critical *Re*. *Fluid Mech. - Sov. Res.*, **11**, 38–47. (Exp. 35k $< Re <$ 1M, $1\% < Ti < 7\%$, $2 < L/D < 5$, C_p, C_{pb}, C_D). (23).

Žukauskas, A. A., Katinas, V. I., Perednis, E. E., and Sobolev, V. A. (1980J). Viscous flow over inclined in-line tube bundles and vibrations induced in the latter. *Fluid Mech. - Sov. Res.*, **9**, 1–12. (Exp. in-line $P/D = 1.34$ square, $\Lambda = 25°$, $40°$, $60°$, $75°$, 2k $< Re <$ 100k, osc., A/D). (28).

Papers published in proceedings, reports, theses, etc. (P)

Allen, H. J. and Perkins, E. W. (1951P). A study of effect of viscosity on flow over slender inclined bodies of revolution. *Nat. Adv. Committee Aero.*, NACA Rep. 1048. (Exp.). 25.

Allen, H. J. and Vincenti, W. G. (1948P). Wall interference in a two-dimensional-flow wind tunnel, with consideration of the effect of compressibility. *Nat. Adv. Committee Aero.*, NACA Tech. Rep. 782. (Th. blockage correction, aerofoil, circle, source, images, compressibility, comparison with exp.). 23.

Aode, H., Tatsuno, M., and Taneda, S. (1985P). Visual studies of wake structure behind two cylinders in tandem arrangement. In: *Rep. Res. Inst. Appl. Mech.*, Kyushu University, **32**, 1–20. (Exp. $100 < Re < 1\text{k}$, $1.5 < S/D < 10$, $A/D = 0.1, 0.3$, FV, St, synchr.). 26.

Arie, M., Kiya, M., Suzuki, Y., and Yoshimura, H. (1978P). Characteristics of flow around a yawed circular cylinder subjected to an interference of the plain wall, in Japanese. *Bull. Fac. Engg*, Hokkaida University, **87**, 11–21. (Exp.). (23).

Armitt, J. (1968P). The effect of surface roughness and free stream turbulence on the flow around a model of cooling tower at critical Reynolds numbers. In: *Symp. Wind Effects on Buildings and Structures*, ed. Johns, D. J., Loughborough University, Paper 6. (Exp. $85\text{k} < Re < 390\text{k}$, $Ti = 3\frac{1}{2}\%$, $5\frac{1}{2}\%$, $10\frac{1}{2}\%$, $K/D = 0.09\%, 0.19\%, 0.38\%$). (22).

Armitt, J., Counihan, J., Millborrow, D. J., and Richards, D. J. W. (1967P). Wind tunnel measurements of the surface pressures on models of the Ferrybridge C cooling towers. *Central Electricity Res. Lab.*, Rep. RD/L/R/1430. (Exp. cooling towers, C_p, C_d, C_p'). 22.

Auger, J. L. (1977P). Study on vibration and acoustic phenomena in heat exchangers in cross flow, in French. *Ph.D. Thesis*, University of Poitiers, France. (Exp. single row, FV). 28.

Bailac, G., Bai, D., and Gregoire, J. P. (1973P). Study of flow and acoustic phenomena in a tube bank. In: *Proc. BNES Symp. Vibration Problems in Industry*, Keswick, UK. (Exp. Th. in-line arrays). (28).

Bardowicks, H. (1982P). A new six-component balance and applications on wind tunnel models of slender structures. In: *Colloquium on Industrial Aero.*, ed. Gerhardt and Kramer. (Exp. 3 cyl., $H/D = 10, 2D/B = 0.067, 2 \leqslant P/D \leqslant 5$). (27).

Batham, J. P. (1973P). Pressure distributions on in-line tube arrays in cross flow. In: *Proc. BNES Symp. Vibration Problems in Industry*, ed. Wakefield, J., Keswick, UK. (Exp. in-line, $P/D = 2$, 1.25, $28\text{k} < Re < 100\text{k}$, $L/D = 9$, 18, $Ti = 3.5\%$, 10%, $n = 9$, C_p, C_p', L_c, frequency PS). 28.

Betz, A. (1961P). History of boundary layer research in Germany. In: *Boundary Layer and Flow Control, Vol. 1*, ed. Lachmann, G. V., 5–11, Pergamon Press, Oxford. (Hist. Prandtl's early work). 24.

Biermann, D. and Herrnstein, W. H. (1933P). The interference between struts in various combinations. *Nat. Adv. Committee Aero.*, NACA TR 468. (Exp. tandem, side-by-side, 60k $< Re <$ 150k, $1 < S/D < 9$, $1 < T/D < 6$, C_{D1}, C_{D2}). 26.

Bokaian, A. and Geoola, F. (1985P). Hydrodynamic forces on a pair of cylinders. In: *Offshore Tech. Conf.*, OTC 5007. (Exp. 2.6k $< Re <$ 5.9k, $Ti = 6.5\%$, $1.09 < S/D < 5$, $1 < T/D < 7$, C_{D1}, C_{D2}, C_{L1}, C_{L2}, topology). 26.

Borthwick, A. G. L., Chaplin, J. R., Burrows, R., and Drossopoulos, G. M. (1991P). Measurements in the near-wake of a K-node immersed in a steady unidirectional flow. In: *Proc. 1st Int. Offshore Polar Engg Conf.*, Edinburgh, UK, Vol. 3, 301–7. (Exp. K-side-by-side at 45°, $D_2/D_1 = 2$, u/V, PS, u'/V). (26).

Bryce, W. B., Wharmby, J. S., and Fitzpatrick, J. A. (1978P). Duct acoustic resonances induced by flow over coiled and rectangular heat exchanger test banks of plain and finned tubes. In: *Proc. BNES Vibration in Nuclear Power Plants*, Keswick, UK, Paper 3:5. (Th. Exp. in-line, acoustic resonance). (28).

Buresti, G. (1982P). On the evaluation of universal wake numbers for roughened circular cylinders in cross-flow. In: *ASME, Winter Annu. Meeting*, Washington, DC, Paper 81-WA/FE-23. (Universal St, Bearman's, Griffin's, and Chen's St). (22).

Burnsall, J. W. and Loftin, L. K. (1951P). Experimental investigation of the pressure distribution about a yawed circular cylinder in the critical Reynolds number range. *Nat. Adv. Committee Aero.*, NACA TN 2463. (Exp. 60k $< Re <$ 500k, $0° < \Lambda < 60°$, C_p, C_{pn}, Re_n, C_{Dn}). 25.

Busemann, A. (1932P). Measurement on rotating cylinders, in German. *Ergenisse der Aerodynamik Versuchanstalt*, Göttingen, IV Lieferung, 101. (Exp.). 24.

Chen, Y. (1993P). Strouhal periodicity in parallel triangular tube arrays. *M.Sc. Thesis*, McMaster University, Hamilton, Canada. (Exp.). (28).

Chen, Y. N. (1980P). Turbulence as excitation source in staggered tube bundle heat exchangers. In: *Proc. ASME Pressure Vessel and Piping Conf.*, ed. Au-Yang, M. K., PVP, **41**, 45–63. (Exp. Rev.). (28).

Chen, Y. N. and Weber, M. (1970P). Flow-induced vibration in tube bundle heat exchangers with cross and parallel flow. In: *Proc. ASME Symp. Flow-Induced Vibration in Heat Exchangers*, 55–77. (Exp. Rev.). (28).

Cheng, M., Tsuei, H. E., and Chow, K. L. (1994P). Experimental study of flow interference phenomena of cylinder/cylinder and cylinder/plane arrangements. In: *Proc. ASME Symp. Flow-Induced Vibration*, PVP, **272**, 173–84. (Exp. $Re = 500$, $0.125 < T/D < 3.75$, St, γ, b_w/D, $0 < G/D < 5$, St, FV). (23).

Cooper, K. R. (1973P). Wind tunnel and theoretical investigations into the aerodynamic stability of smooth and stranded twin-bundled power conductors. *Nat. Res. Council*, Canada, Tech. Rep. LTR-LA-115. (Exp. Th. 2 cyl. osc., C_D, C_L, FV, instability region). 26.

Cooper, K. R. (1974P). Wind tunnel measurements of the steady aerodynamic forces on a smooth circular cylinder immersed in the wake of an identical cylinder. *Nat. Res. Council*, Canada, Nat. Aero. Est., LTR-LA-119. (Exp. tandem, stag., $10\text{k} < Re < 125\text{k}$, $1.3 < S/D < 50.5$, $1 < T/D < 6$, C_L, C_D, C_p, u/V). 26.

Cooper, K. R. and Wardlaw, R. L. (1971P). Aeroelastic instabilities in wakes. In: *Wind Effects on Buildings and Structures*, Tokyo, IV, 1–9. (Exp. osc. C_D, C_L, A/D). 26.

Corrsin, S. (1944P). Investigation of the behaviour of parallel two-dimensional air jets. *Nat. Adv. Committee Aero.*, NACA W-90 Adv. Conf. Rep. (Exp. single row, triangular tubes, jet merging, u/V). 28.

Counihan, J. (1963P). Lift and drag measurements on stranded cables. *Imperial College Aero. Dept.*, London, Rep. 117. (Exp. stranded conductor, $N = 18$, 24, 42, strands, $52\text{k} < Re < 56\text{k}$, C_D, C_L, yaw). 22, 25, 26.

Coutanceau, M. and Menard, C. (1983P). Visualisation of the flow development around a circular cylinder impulsively subjected to a combined motion of rotation and translation. In: *Flow Visualisation III*, ed. Yang, W. J., 54–8. (Exp. $Re = 200$, $0.3 < V_r/V < 2$, eddy path, FV). 24.

Cowdrey, C. F. (1968Pa). Some observations on the flow of air through a single row of parallel closely spaced cylinders. *(Brit.) Nat. Phys. Lab.*, NPL Aero. Note 1064. (Exp. single row, C_{pb}). 28.

Cowdrey, C. F. (1968Pb). Part 1. Application of Maskell's theory of wind-tunnel blockage to some large solid models, Part 2. Design of velocity profile grids. In: *Proc. Symp. Wind Effects on Buildings and Structures*, ed. Johns, D. J., Loughborough University, Paper 29. (Exp. Th. extension). (23).

Cowdrey, C. F. and O'Neill, P. G. G. (1956P). Report of tests on a model cooling tower for CEA: pressure measurements at high Reynolds numbers. *Nat. Phys. Lab.*, Aero. Rep. 316a. (Exp. cooling tower, $Re = 9.1\text{M}$, 14.1M, $K/D = 0.016\%$, C_p, C_{po}, C_{pb}). 22.

Davenport, A. G. and Isyumov, N. (1966P). The dynamic and static action of wind on hyperbolic cooling towers. *Eng. Sci. Res. Rep., Univ. Western Ontario*, London, Canada, BLWT-1-66, BLWT-2-66. (Exp. cooling tower, model with surface roughness, C_p). 22.

Diaz, F., Gavalda, J., Kawall, J. G., and Giralt, F. (1982P). Interpretation of the complex turbulent flow generated by a rotating cylinder. In: *Structure of Complex Turbulent Shear Flow*, IUTAM Symp., Marseille, ed. Dumas, R., Springer, Berlin, 175–83. (Exp. rot. cyl. $V_r/V = 1$, 2.5, far wake, C_D, u/V, w/V, v'/V, $u'v'/V^2$, cross, autocorrelation). 24.

Dye, R. C. F. (1973P). Vortex-induced vibration of a heat exchanger tube row in cross-flow. In: *Proc. Int. Symp. Vibration Problems in Industry*, ed. Wakefield, J., Keswick, UK. (Exp. single row, St). 28.

Ebner, H. (1968P). Investigation of influences on force due to wind loading on Balcke cooling tower Mengele power plant, in German. *Tech. Hochsch. Aachen, Lehrstuhl für Leichtbau*, Bericht No. 21. (Exp. cooling tower, ribs, $Re = 640$k, C_p, C_{pb}). 22.

Ebner, H. and Ruscheweyh, H. (1974P). Wind loading on hyperbolic cooling towers, in German. *Forschungsbericht des Landes Nordrhein-Westfallen*, Rep. (Exp. correlation probability). 22.

Fage, A. (1929P). On the two-dimensional flow past a body of symmetrical cross-section mounted in a channel of finite breadth. *Brit. Aero. Res. Council*, Rep. and Memo. 1223. (Th. Rankine's oval, circle, Exp. $0.05 < D/B < 0.20$, C_D). 23.

Fage, A. and Warsap, G. H. (1929P). The effects of turbulence and surface roughness on the drag of a circular cylinder. *Brit. Aero. Res. Council*, Rep. and Memo. 1283. (Exp. 25k < Re < 240k, $1\% < Ti < 3.2\%$, $0.28\% < K/D < 2\%$, $L/D = 8$, 20, 2, $D/B = 0.125$, 0.05, $2.1\% < Ti < 6.4\%$, $0.5 < TS/D < 1$). 14, 22.

Farell, C. and Arroyave, J. (1989P). On uniform flow around rough circular cylinders at critical Reynolds numbers. In: *Proc. 6th US Wind Engg Conf.*, Houston, TX. (Exp. 59k < Re < 147k, $K/D = 0.45\%$, C_p, PS). 22.

Fiechter, M. (1966P). On vortex system on slender bodies of revolution and their influence on aerodynamic forces, in German. *Deutsch-Französisches Forschungs Institut*, Bericht 10/66, St Louis, France. (Exp. FV). 25.

Fitzhugh, J. S. (1973P). Flow-induced vibration in heat exchangers. In: *Proc. Int. Symp. Vibration Problems in Industry*, Keswick, UK, Paper 427. (Survey, in-line, stag. tube arrays, compilation of St). 28.

Fitzpatrick, J. A., Donaldson, I. S., and McKnight, W. (1978P). Some observations of the pressure distribution in a tube bank for conditions of self-generated acoustic resonance. In: *Proc. Symp. Vibration in Nuclear Plant*, Keswick, UK, Paper 316. (Exp. in-line, $P/D = 1.73$, 26 rows, C_p^*, 3.6k < Re < 9.2k). (28).

Fujisawa, N., Ikemoto, K., and Nakabayashi, T. (1997P). Feedback control of vortex shedding from a circular cylinder. In: *5th Triennial Int. Symp. FLUCOME*, Hayama, Japan, 693–8. (Exp. rot. cyl. osc., C_R, C_p, FV). (24).

Funakawa, M. (1973P). Vibration of tube banks by wake force. In: *Int. Symp. Vibration Problems in Industry*, ed. Wakefield, J., Keswick, UK, Paper 428. (Exp. staggered, $1.66 < T/D < 2.5$, $1 < R/D < 4$, osc., A/D, St). (28).

Gerich, D. (1986P). On the changes in the Kármán vortex street with end plates on a circular cylinder, in German. *Mitteilungen Max Planck Institut für Strömmungsforschung*, Göttingen, Rep. 81. (Exp. $50 < Re < 3.47$k, u/V, z/D, u'/V, St, FV). 21.

Gerich, D. (1987P). Visualisation of transition from laminar to turbulent wake by using a continuously operating smoke wire, in German. *Mitteilungen Max Planck Institut für Strömmungsforschung*, Göttingen, Rep. 104. (Exp. $150 < Re < 370$, FV). 21.

Gerich, D. (1989P). The development of three-dimensionality in the near- and the far-wake due to changes of the boundary conditions. In: *10th Australasian Fluid Mechanics Conf.*, University of Melbourne, Paper 13D-3, 39–42. (Exp. $Re = 136$, u/V, u'/V, FV). (21).

Glauert, H. (1933P). The interference of a wind tunnel on a symmetrical body. *Brit. Aero. Res. Council*, Rep. and Memo. 1544. (Th. semi-empirical, comparison exp.). 23.

Göktun, S. (1975P). The drag and lift characteristics of a cylinder placed near a plane surface. *M.Sc. Thesis*, Naval Postgraduate School, Monterey, CA. (Exp. $0 < G/D < 3$, 90k $< Re <$ 400k, C_p, C_L, C_D, St). 23.

Gould, R. W., Raymer, W. G., and Ponsford, P. J. (1968P). Wind tunnel tests on chimneys of circular section at high Reynolds numbers. In: *Symp. Wind Effects on Buildings and Structures*, Loughborough University, Paper 10. (Exp. 140k $< Re <$ 5.4M, $H/D = 6$, 12, FV, C_p, C_d, L_c). 21.

Gowen, F. E. and Perkins, E. W. (1953P). Drag of circular cylinders for a wide range of Reynolds and Mach numbers. *Nat. Adv. Committee Aero.*, NACA TN 2960. (Exp. 50k $< Re <$ 1M, $0.3 < Ma < 2.9$, FV, C_p, C_D). 16, 25.

Grotz, B. J. and Arnold, F. R. (1956P). Flow-induced vibration in heat exchangers. *Dept. Mech. Engg, Stanford University*, Tech. Rep. 31. (Exp. St, correlation). (28).

Hara, F. (1987P). Unsteady fluid dynamic forces acting on a single row of cylinders in a cross flow. In: *ASME Symp. Flow-Induced Vibrations – 1987*, PVP, **122**, 51–8. (Exp. one row, $T/D = 1$, 33, $Re =$ 72k, five tubes per row, osc., A/D, ϕ, FV). (28).

Hayn, F. (1967P). Pressure distribution on model of power plant Scholven, in German. *Deutsche Versuchanstalt für Luft und Raumfahrt*, E. V., Inst. Angew. Gasdynamik, Port-Wahn, Bericht AM511. (Exp. cooling tower ribs, C_p). 22.

Heinecke, E. (1973P). Stationary and unstationary flow phenomena in and behind staggered and in-line banks. In: *Int. Symp. Vibration Problems in Industry*, ed. Wakefield, J., Paper 412. (Exp. staggered, $1.35 < T/D < 2.61$, $1 < R/D < 3$, 10k < Re < 100k, FV, $100 < Re < 750$, frequency PS, St, θ_s). (28).

Heinecke, E. and Mohr, K. H. (1982P). Investigations on fluid borne forces in heat exchangers with tubes in cross flow. In: *3rd Keswick Symp. Vibration in Nuclear Plant*, Paper 36. (Exp. in-line, $1.15 < P/D < 2.9$, 10k < Re < 100k, $L/D = 22.5$, $Ti = 0.7\%$, C_p, C_p', C_D, C_D', C_L, C_L', osc., A/D). (28).

Hetz, A. A., Dhaubhadel, M. N., and Telionis, D. P. (1990P). The hydrodynamic response of five in-line cylinders. *Virginia Poly. Inst. State University*, Blacksburg, VA, Rep. (Exp. 5 cyl., 10k < Re < 50k, $1.1 \leqslant S/D \leqslant 1.8$, $L/D = 5.9$, $D/B = 0.12$, $Ti = 0.5\%$, FV). 27.

Honji, H. (1973Pa). Wake characteristics of a finite row of circular cylinders normal to the stream at low Reynolds numbers, in Japanese. In: *Bull. Res. Inst. Appl. Mech.*, Kyushu University, **39**, 1–36. (Exp. steady and impulsive start, 5, 7, 9, 19 cyl., $0.5 < T/D < 2$, $10 < Re < 48$, $L/D = 150$, 375, $D/B = 0.025$, 0.021, FV, St, $\alpha = 45°$, x_s/D, x_b/D). 27.

Honji, H. (1973Pb). Average flow fields around a group of circular cylinders, in Japanese. In: *Bull. Res. Inst. Appl. Mech.*, **39**, 17–20 plates. (Exp. 5 cyl., $10 < Re < 48$, $0.5 \leqslant T/D \leqslant 2$, $L/D = 375$, 100, $D/B = 0.015$, FV). 27.

Hori, E. (1959P). Experiments on flow around a pair of parallel circular cylinders. In: *Proc. 9th Japan Nat. Congress Appl. Mech.*, 231–4. (Exp. 2k < Re < 12k, $S/D = 0.2$, 1, 2, 6, $0° < \alpha < 180°$, C_{pb2}, St, C_p, u/V). 26.

Hsiao, F. B., Pan, J. Y., and Chiang, C. H. (1992P). The study of vortex shedding frequencies behind tapered circular cylinders. In: *ASME Symp. Flow-Induced Vibration and Noise*, **6**, FED 138. (Exp. taper, $0.025 < TR < 0.125$, 10k < Re < 14k, end plates, St, PS). 22.

Hurley, D. G. and Thwaites, B. (1951P). An experimental investigation of the boundary layer on a porous circular cylinder. *Brit. Aero. Res. Council*, Rep. and Memo. 2829. (Exp. Th. Thwaites flap, BL, V/V_o, FV, δ_s, θ). 24.

Igarashi, T. (1985P). Fluid flow around a bluff body used for a Kármán vortex flow meter. In: *Fluid Control and Measurement*, ed. Harada, M., Pergamon Press, 1017–22. (Exp. bluff bodies, FV, u', p', PS, St, C_p'). 22.

Igarashi, T. (1986P). Correlation between heat transfer and fluctuating pressure in the separated region of a bluff body. In: *Proc. 8th Heat Transfer Conf.*, San Francisco, CA, 1023–8. (Exp. 11k $< Re <$ 60k, 3.75 $< L/D <$ 6, 0.06 $< D/B <$ 0.10, Nu_b, Fr, C_{pb}, control cyl., 0.12 $< d/D <$ 0.3, splitter plate). (26).

Igarashi, T. (1992P). Visualisation of flow control around a circular cylinder by a new method. In: *Flow Visualisation VI*, ed. Tanida, Y. and Miyashiro, H., Springer, Berlin, 312–16. (Exp. saw-blade, 13k $< Re <$ 52k, $L/D = 4.4$, $D/B = 0.088$, $Ti = 0.5\%$, $0.9\% < h/D < 3.8\%$, C_p, $\overline{Nu}$, FV, St, C_p', C_d, C_l, Fr). 27.

Igarashi, T. and Iida, Y. (1987P). Fluid flow and heat transfer around a circular cylinder with vortex generations. In: *ASME-JSME Thermal Engg Joint Conf.*, Honolulu, 143–9. (Exp. saw-blade, 13k $< Re <$ 52k, $L/D = 4.4$, $D/B = 0.088$, $Ti = 0.5\%$, $0.9\% < h/D < 3.8\%$, C_p, St, C_p', C_d, Fr, $\overline{Nu}$). 22.

Igarashi, T. and Tsutsui, T. (1994P). Fluid forces acting on a circular cylinder controlled by a small rod. In: *Proc. 3rd JSME-KSME, Fluids Engg Conf.*, Sendai, Japan, 571–6. (Exp. control cyl., $D/d = 20$, $2 < G/d < 3$, $90° < \theta_c < 180°$, St, C_p, C_{pb}, FV, C_d, C_l, C_p'). (26).

Igarashi, T. and Yamasaki, H. (1992P). Fluid flow and transfer around two circular cylinders arranged in tandem. In: *Proc. 2nd JSME-KSME Thermal Engg Conf.*, 7–12. (Exp. tandem, $3 < L/D < 5$, $0.07 < D/B < 0.12$, C_{p1}, C_{p2}, C_{d2}, St, C_{p1}', C_{p2}', Fr_1, Fr_2, $Nu_{\max}$). (26).

Igarashi, T., Nishida, K., and Mochizuki, S. (1988P). Aerodynamic forces acting on three circular cylinders having different diameters closely arranged in line, in Japanese. In: *Trans. JSME*, 54-505 B, 2349–56. (Exp. 3 cyl., $d/D = 0.45$, $L/D = 3$, $D/B = 0.067$, $G/D = 0.06$, $Ti = 0.5\%$, C_{pi}, $i = 1, 2, 3$, C_d, C_l, θ_o, St, FV). 27.

Igarashi, T., Tsutsui, T., and Tajima, N. (1997P). A non-steady flow around two circular cylinders closely arranged in tandem. In: *5th Triennial Int. Symp. FLUCOME*, Hayama, Japan, 675–80. (Exp. tandem, 20.5k $< Re <$ 128k, $1.2 < S/D < 1.3$, St, C_{p1}, C_{p2}, C_{p1}', C_{p2}', FV, unsteady $\tilde{C}_D$, $\tilde{C}_l$). (26).

Jensen, B. L. (1987P). Large-scale vortices in the wake of a cylinder placed near a wall. In: *2nd Int. Conf. Laser Anemometry*, Univ. Strathclyde, Glasgow. (Exp. $Re =$ 7k, $L/D = 16$, $0 \leqslant G/D < 0.37$, u/v, u'/v, FV, St, PS). (23).

Jensen, B. L. and Sumer, B. M. (1986P). Boundary layer over a cylinder placed near a wall. *Inst. Hydrodyn. and Hydraulic Engg*, Tech. Univ., Denmark, Progress Rep., 64, 31–9. (Exp. $0.1 < G/D < 2$, $Re =$ 45k, C_f, δ_D, θ_s, FV). (23).

Johnson, S. P. and Zdravkovich, M. M. (1991P). Optimal arrangement of 6 + 1 satellite riser in a current. In: *Proc. 1st Int. Offshore Polar Engg Conf.*, Edinburgh, **2**, 164–9. (Exp. 6 + 1 riser, 50k $< Re <$ 85k, $0° < \alpha < 30°$, $D/d = 2.4$, $1.7 < P/D < 4.5$, C_{Do}, C_{Lo}, FV). 27.

Kiya, M., Mochizuki, O., Ido, Y., Suzuki, T., and Arai, T. (1992P). Flip-flopping flow around two bluff bodies in tandem arrangement. In: *Bluff-Body Wakes and Instabilities*, IUTAM Symp., ed. Eckelmann, H., *et al.*, 15–18. (Exp. 20k $< Re <$ 40k, u', R_{ff}). 26.

Knisely, C. W. and Kawagoe, M. (1990P). Force-displacement measurements on closely spaced tandem cylinders. In: *Bluff Body Aerodynamics and its Applications*, ed. Ito, M., *et al.*, 81–90. (Exp. two cyl. staggered, C_L, C'_L, delay). (26).

Kraemer, K. (1965P). Pressure distribution measurements related to three-dimensional end perturbations, in German. *Mitteilungen Max Planck Institut für Strömmungsforschung*, Göttingen, Rep. 32. (Exp. 6k $< Re <$ 600k, $0.5 < L/D < 220$, C_p, C_{pb}, end plates, $1.5 < D_e/D < 8$). 21.

Krause, R. L., Keener, E. R., Chapman, G. T., and Claser, G. (1979P). Investigation of the asymmetric aerodynamic characteristics of cylindrical bodies of revolution with variations in nose geometry and rotational orientation at $\alpha = 58°$ and Mach numbers to 2. *NASA T 78533*. (Exp. rot. missile or part thereof, C_n). 25.

Lamont, P. J. (1980P). Pressure measurements on an ogive-cylinder at high angles of attack with laminar, transition, or turbulent separation. In: *AIAA Atmospheric Fluid Mechanics Conf.*, AIAA-80-1556-CP. (Exp. 200k $< Re <$ 4M, $20° < \Lambda < 90°$, C_{DN}, C_{DS}, γ, C_p, classification). 25.

Landweber, L. (1942P). Flow about a pair of adjacent, parallel cylinders normal to a stream. *Navy Dept., The David W. Taylor Model Basin*, Rep. 485. (Th. FV, $1 < T/D < 3$, stability of two rows, V_{in}, V_{out}, comp. th.-exp.). 26.

Lewis, C. G. and Gharib, M. (1993P). The effect of axial motion on the wake of a cylinder in steady uniform flow. In: *Bluff-Body Wakes, Dynamics, and Instabilities*, IUTAM Symp., Göttingen, 1992, ed. Eckelmann, H., *et al.*, 345–8. (Exp. axially moving cyl., $Re = 100$, FV, $0° < \psi < 84°$, β, St). (25).

Ljungkrona, L., Norberg, C., and Sunden, B. (1991P). Flow around two tubes in an in-line arrangement: flow visualisation and pressure measurements. In: *Experimental Heat Transfer, Fluid Mechanics and Thermodynamics*, Elsevier, 333–40. (Exp. tandem 3.3k $< Re <$ 12k, $1.25 < S/D < 4$, $0.1\% < Ti < 3.2\%$, FV, St, C_{D1}, C_{D2}, C_{p1}, C_{p2}, C'_{p1}, C'_{p2}). (26, 28).

Lock, C. N. H. (1929P). The interference of a wind tunnel on a symmetrical body. *Brit. Aero. Res. Council*, Rep. and Memo. 1275. (Th. inviscid flow, source images, Rankine oval, ellipse, circle, three-dimensional). 25.

Lohrisch, W. (1929P). Determination of heat transfer coefficients by a diffusion method, in German. In: *VDI Mitteilungen Forschungsarbeit*, 46–68. (Exp. tube arrays, FV). 28.

Lowe, H. J. and Richards, D. J. W. (1967P). Aerodynamic implications of the Ferrybridge investigations. In: *Proc. Symp. Ferrybridge Cooling Towers and After*, Inst. Civil Eng., Session 3, 79–84. (Exp. cooling towers, C_p, stress). 22.

Luo, S. C. (1991P). Bistable flow associated with two tandem unequal diameter cylinders. In: *Proc. 1st Int. Offshore Polar Engg Conf.*, Edinburgh, UK, **3**, 326–9. (Exp. $D_1/D_2 = 0.33$, $35k < Re < 88k$, C_p, C_p', intermittency, PS). (26).

Maskell, E. C. (1963P). A theory of the blockage effects on bluff bodies and stalled wings in a closed wind tunnel. *Roy. Aircraft Establ.*, Farnborough, RAE Rep. 2685, also Rep. and Memo. 3400. (Th. momentum balance). 23.

Matsui, T. (1977P). The flow near the surface of a cylinder rotating in a uniform flow. In: *Euromech 90 Colloquium*, Nancy, France, Session V1.b.2. (Exp. $214 < Re < 238$, $1.7 < V_r/V < 2.4$, transition). 24.

Mitry, R. T. (1977P). Wall confinement effects for circular cylinders at low Reynolds numbers. *Ph.D. Thesis*, Dept. Mech. Engg, University of British Columbia. (Exp. $5 < Re < 20k$, $0.02 < D/B < 0.50$, C_p, C_{pm}, C_{pb}, St, C_D, FV, θ_s). 23.

Mizuno, S. (1970P). Effects of three-dimensional roughness elements on the flow around circular cylinder. In: *J. Sci., Ser. B, Hiroshima University*, **25**, 215–58. (Exp. tripping spheres, $N = 1, 2, 4, 5, 11$, $60k < Re < 290k$, $L/D = 3$, $D/B = 0.23$, $Ti < 0.1\%$, $0.36\% < d/D < 0.9\%$). 22.

Modi, V. J. and El-Sherbiny, S. E. (1971P). Effect of wall confinement on aerodynamics of stationary circular cylinders. In: *Proc. 3rd Inst. Conf. Wind Effects on Buildings and Structures*, Tokyo, 365–75. (Exp. $0.032 < D/B < 0.38$, $30k < Re < 100k$, C_p, C_{pb}, C_D, St, a/D, b/D, b/a, C_{DC}). 23.

Modi, V. J. and El-Sherbiny, S. E. (1973P). On the wall confinement effects in the industrial aerodynamics studies. In: *BNES Vibration in Nuclear Plant*, ed. Wakefield, J., Keswick, UK, Paper 116. (Exp. $10k < Re < 120k$, $0.03 < D/B < 0.35$, C_D, St, C_{pmax}, b/a, Glauert's and Maskell's corrections). 23.

Modi, V. J. and El-Sherbiny, S. E. (1975P). Wall confinement effects on bluff bodies in turbulent flows. In: *Proc. 4th Int. Conf. Wind Effects on Buildings and Structures*, ed. Eaton, Heathrow, 121–32. (Exp. $20k < Re < 160k$, $0.07\% < Ti < 12\%$, C_p, C_{pb}, C_D). 23.

Morsy, M. G. (1975P). Skin friction and form pressure loss in tube bank condensers. In: *Proc. Inst. Mech. Eng.*, **189**, 49–75. (Exp. normal triangle, $P/D = 1.5$, C_p, C_f, C_d, suction, C_{Dp}, C_{ps}, C_{fs}). 28.

Mujumdar, A. S. and Douglas, W. T. M. (1970P). The unsteady wake behind a group of three parallel cylinders. In: *ASME*, Paper 70-Pet-8. (Exp. 5k $< Re <$ 10k, $1 < S/D < 4$, $0° < \alpha < 15°$, $L/D = 14.7$, $D/B = 0.07$, $Ti = 0.12\%$, St). (27).

Nakagawa, K., Fujino, T., Arita, Y., and Shima, T. (1963P). An experimental study of aerodynamic devices for reducing wind-induced oscillatory tendencies of stacks. In: *Int. Conf. Wind Effects on Buildings*, ed. Scruton, C., London, 774–95. (Exp. 15k $< Re <$ 1.5M, $L/D = 7$, $D/B = 0.14$, osc., C_D, St, A/D, helical wires). 22.

Naumann, A. and Quadflieg, H. (1968P). Aerodynamic aspects of wind effects on cylindrical buildings. In: *Symp. Wind Effects on Buildings and Structures*, ed. Johns, D. J., Loughborough Univ. (Exp. $0.1 < Ma < 0.45$, shock waves, separation wire, taper, FV). 22.

Naumann, A. and Quadflieg, H. (1974P). Vortex generation on cylindrical buildings and its simulation in wind tunnels. In: *Flow-Induced Structural Vibration*, ed. Naudascher, E., Springer, Berlin, 730–7. (Exp. 20k $< Re <$ 420k, staggered wire, step change in dia., FV, C_D, St, C_p, C_p', θ_s). 6, 22.

Naumann, A., Morsbach, M., and Kramer, C. (1966P). The conditions of separation and vortex formation past cylinders. In: *AGARD, CP4 Separation Flow*, 539–74. (Exp. $Re =$ 680k, $0.1 < Ma < 0.6$, staggered separation wire, FV). 1, 16, 22.

Niemann, H. J. (1971P). Stationary wind load on hyperbolic cooling towers. In: *Wind Effects on Buildings*, Tokyo, 335–44. (Exp. cooling tower, full-scale, C_p, θ_{cpm}, θ_{cpb}, C_d). (22).

Nikuradze, J. (1933P). Flow in rough tubes, in German. *Forsch. Arb. Ing. Wes.*, No. 361. (Exp. sand roughness, C_{Dp}). 22.

Norberg, C. (1992P). An experimental study of the flow around cylinders joined with a step in diameter. In: *11th Australasian Fluid Mech. Conf.*, Hobart, 507–10. (Exp. step, 3k $< Re <$ 13k, $0.5 \leqslant d/D \leqslant 1$, FV, St, PS, $p_{\max}$). 22.

Norton, D. J., Heideman, J. C., and Mallard, W. W. (1981P). Wind tunnel tests of inclined circular cylinders. In: *Offshore Tech. Conf.*, OTC 4122. (Exp. 200k $< Re <$ 2M, $0° < \Lambda < 40°$, $0.037\% < K/D < 3.5\%$, C_{Dn}, C_{Da}, C_D, St). 25.

Novak, J. (1974P). Strouhal number of square prism, angle iron and two circular cylinders arranged in tandem. In: *Acta Tech. Czech. Acad. Sci.*, 361–73. (Exp. 4.5k $< Re <$ 10.5k, $L/D = 8$, $D/B = 0.04$, $1 < S/D < 9$, St_1, St_2). (26).

Novak, J. (1975P). Strouhal number of two cylinders of different diameter arranged in tandem. In: *Acta Tech. Czech. Acad. Sci.*, 366–74. (Exp. 4k $< Re <$ 12.6k, $D_1/D_2 = 0.5, 2.0$, $0.5 < S/D < 10$, St). (26).

Ohmi, K. and Imaichi, K. (1992P). Vortex wake visualisation of two circular cylinders in parallel. In: *Flow Visualisation V*, ed. Tanida, Y., 323–6. (Exp. PIV, $40 < Re < 500$, $2 < S/D < 20$, $1 < T/D < 5$, $L/D = 15$, 30, $D/B =$ 0.03, 0.06, FV, streamlines, vorticity). (26).

Okajima, A. (1977P). The aerodynamic characteristics of stationary tandem cylinders at high Reynolds numbers, in Japanese. In: *Bull. Res. Inst. Appl. Mech.*, Kyushu University, **46**, 111–27. (Exp. tandem, $40\text{k} < Re < 640\text{k}$, $1.07 < S/D < 6$, FV, C_{D1}, C_{D2}, St, θ_s, smooth, $K/D = 0.9\%$, upstream, downstream, both). 26.

Okajima, A. and Nakamura, Y. (1973P). Flow around rough-surfaced cylinder at high Reynolds numbers, in Japanese. In: *Bull. Res. Inst. Appl. Mech.*, Kyushu University, **40**, 387–400. (Exp. $400\text{k} < Re < 2\text{M}$, $L/D = 3.3$, $D/B =$ 0.15, $K/D = 0.11\%$, 0.45%, C_D, PS, St, C_{pb}). 22.

Okajima, A. and Sugitani, K. (1980P). Aerodynamic characteristics of stationary tandem cylinders at high Reynolds numbers, in Japanese. In: *Bull. Res. Inst. Appl. Mech.*, Kyushu University, **53**, 111–16. (Exp. tandem, $Re = 118\text{k}$, 236k, 436k, $3 < S/D < 21$, C_{p2}, C_{ps2}, C_{pb2}, C_{D2}). (26).

Okajima, A., Takata, H., and Asanuma, T. (1975P). Viscous flow around a rotationally oscillating circular cylinder. In: *Inst. Space and Aero. Sci.*, University of Tokyo, ISAS Rep. 532 (Vol. 40, No. 12), 311–38. (Th. Exp. $Re = 40$, 80, $0.02 < N_{\text{osc}} < 0.3$, $0.2 < V_r/V < 1.0$, N–S eqn, C_p, C_D, C'_L, φ, $40 < Re <$ 1.6k). 24.

Okui, K., Iwabuchi, M., Shimada, K., and Harashima, K. (1997P). Characteristics of vortex-shedding frequency in finned tube banks with in-line arrangement. In: *Proc. Exp. Heat Transfer, Fluid Mech., Thermodyn.*, ed. Giot, M., *et al.*, 1451–8. (Exp. $25\text{k} < Re < 600\text{k}$, in-line, $2.1 < P/D < 3.2$, helical fins, St, pressure PS). (28).

Osaka, H., Yamada, H., and Nakamura, I. (1982P). Three-dimensional structure of the turbulent wake behind an intersecting cruciform circular cylinder. In: *IUTAM Symp. Three-Dimensional Boundary Layers*, ed. Fernholz, H. H., Springer, Berlin. (Exp. $Re = 4\text{k}$, $15 < x/D < 475$, u/V, secondary flow, streamwise eddies, kinetic energy). 26.

Osaka, H., Yamada, H., and Nakamura, I. (1985P). Statistical characteristics of the turbulent wake behind an intersecting cruciform circular cylinder. In: *4th Int. Symp. Turbulent Shear Flows*, ed. Bradbury, L. J. S., *et al.*, Karlsruhe 1983, Springer, Berlin. (Exp. $Re = 4\text{k}$, u/V, corr. PS). (26).

Ottesen-Hansen, N. E. and Justesen, P. (1987P). Shielding in pipe arrays in waves and current. In: *Proc. 6th Int. OMAE Symp.*, Houston, Texas, **1**, 127–36. (Exp. satellite risers, C_{Do}, C_{Lo}, C_D, C_L). (27).

Ottesen-Hansen, N. E., Jacobsen, V., and Lundgren, M. (1979P). Hydrodynamic forces on composite risers and individual cylinders. In: *Proc. Offshore Techn. Conf.*, Houston, Texas, Paper OTC 3451. (Exp. satellite risers, current, osc. flow, combined, C_{Do}, C_{Lo}, C_D, C_L, FV). 27.

Owen, P. R. (1967P). Some aerodynamic problems of grouped cooling towers. In: *Proc. Symp. Ferrybridge Cooling Towers and After*, Inst. Civil Eng., Session 3, 73–8. (Th. Exp. inviscid streamlines, C_p, FV, Disc.). 22.

Paidoussis, M. P., Price, S. J., and Mureithi, W. N. (1991P). Non-linear dynamics of a single flexible cylinder within a rotated triangle array. In: *Proc. Inst. Mech. Eng.*, IMechE 1991-6, 121–8. (Exp. Th. C_D, C_L, bifurcation). 28.

Pankhurst, R. C. and Thwaites, M. A. (1950P). Experiments on the flow past a porous circular cylinder fitted with a Thwaites' flap. *(Brit.) Aero. Res. Council*, Rep. and Memo. 2787. (Exp. suction, C_p, u/V, C_D, C_L, C_Q, FV). 24.

Pannell, J. R., Griffiths, E. A., and Coales, J. D. (1915P). Experiments on the interference between pairs of aeroplane wires of circular and lenticular cross-section. *(Brit.) Aero. Res. Council*, Rep. and Memo. 206. (Exp. tandem and staggered, $Re = 10$k, $1 < S/D < 6$, $0° < \alpha < 20°$, C_D). 26.

Pearcey, H. H., Cash, R. F., Salter, I. J., and Boribond, A. (1982P). Interference effects on the drag loading for groups of cylinders in uni-directional flow. *National Maritime Institute*, Feltham, UK, Rep. NMI R130. (Exp. 4, 9, 16 cyl., $S/D = T/D = 5$, 40k $< Re <$ 80k, $L/D = 54$, $2D/B = 0.025$, tripping wire, $d_w/D = 0.04$, at $\pm 40°$, C_p, C_D). 22, 27.

Pearcey, H. H., Singh, S., Cash, R. F., and Matten, R. B. (1985P). Fluid loading on roughened cylindrical members of circular cross-section. *National Maritime Institute*, Feltham, UK, NMI Rep. 191. (Exp. 500k $< Re <$ 8M, $0.01\% < K/D < 3\%$, C_p, C_d, St, C_{pb}, $V_{\max}^2 - V_s^2/V_{\max}^2$, vorticity dispersion ratio). 22.

Pierce, H. R. (1973P). Noise and vibration in heat exchangers. *Ph.D. Thesis*, Oxford University. (Exp. one row, one column, tube arrays, C_p, St). 28.

Pineau, G., Texier, A., Coutanceau, M., and Loc, T. P. (1992P). Experimental and numerical visualisation of the 3-D flow round a short circular cylinder fitted with end plates. In: *Flow Visualisation V*, ed. Tanida, Y., 343–7. (Exp. $Re = 1$k, $L/D = 5$, FV, impulsive start, freefall). (21).

Polidori, G., Pineau, G., Abed-Meraim, K., and Coutanceau, M. (1993P). Shedding process of the initial vortices from impulsively started cylinders at $Re = 1000$; end and body geometry. In: *Bluff-Body Wakes, Dynamics, and Instabilities*, IUTAM Euromech Symp., Göttingen, 1992, ed. Eckelmann, H., 285–8. (Exp. freefall, $Re = 1$k, $L/D = 5$, FV, vorticity propagation). (21).

Pris, M. R. (1959P). Aerodynamic study I: Hyperboloid cooling tower, in French. In: *Annales de l'Institute Techn. Batiment et Travaux Public*, No. 139, 147–67. (Exp. cooling tower, C_D). 22.

Pris, M. R. (1961P). Aerodynamic study IV: Reservoirs and chimneys – Drag of polygonal and rough circular cylinders. In: *Annales de l'Institute Techn. Batiment et Travaux Public*, No. 163–4, 736–56. (Exp. C_D, surface roughness, C_p). 22.

Ramberg, S. E. (1978P). The influence of yaw angle upon the vortex wakes of stationary and vibrating cylinders. *Naval Res. Lab.*, Washington, DC, NRL Mem. Rep. 3822. (Exp. $150 < Re < 1.1$k, $10° < \Lambda < 60°$, end plates, St, St_n, FV, β, L_f, b, C_{pb}, C_{pbn}, PS, a/D, V-shape cyl.). 25.

Rao, D. M. (1978P). Side-force alleviation on slender, pointed forebodies at high angle of attack. In: *AIAA*, Paper No. 78-1339. (Exp. missile, $5° < \alpha < 50°$, tripping wires, straight, helical, C_s). 25.

Reid, G. E. (1924P). Tests on rotating cylinders. *Nat. Adv. Committee Aero*, NACA TN, 209. (Exp. Hist.). (24).

Relf, E. F. and Powell, C. H. (1917P). Tests on smooth and stranded wires inclined to the wind direction, and a comparison of results on stranded wires in air and water. *(Brit.) Aero. Res. Council*, Rep. and Memo. 307. (Exp. 4.4k $< Re <$ 26k, $0° < \Lambda < 90°$, C_D). (25).

Richardson, E. G. (1930P). On the flow of air adjacent to the surface of a rotating cylinder in a stream. In: *(Brit.) Aero. Res. Council*, Rep. and Memo. 1368, 160–70. (Exp. $0.5 < V_r/V < 3.33$, velocity field, K_L, K_D). (24).

Rizzo, R. (1925P). The Flettner rotorship in the light of the Kutta–Joukovsky theory and of experimental results. *Nat. Adv. Committee Aero.*, NACA TN 228. (Th. Exp.). (24).

Roshko, A. (1993P). Free shear layers, base pressure, and bluff body drag. In: *Proc. Symp. Developments in Fluid Mech. and Aerospace Engg*, Bangalore, India. (Rev. C_{pb}, C_D). 21.

Roshko, A., Steinolfson, A., and Chattoorgoon, V. (1975P). Flow forces on a cylinder near a wall or near another cylinder. In: *Proc. 2nd Nat. Wind Conf. on Wind Engg Res.*, Fort Collins, CO, USA. (Exp. side-by-side, boundary, C_D, C_L, FV). 26.

Sallet, D. W. (1980P). Suppression of flow-induced motions of a submerged moored cylinder. In: *Practical Experiences with Flow-Induced Vibration*, ed. Naudascher, E. and Rockwell, D., Springer, Berlin, 587–94. (Exp. splitter plate, helical rope, 40k $< Re <$ 670k, $L/D = 5$, pendulum osc.). (22).

Salter, C. and Raymer, W. G. (1962P). Pressure measurements at high Reynolds number on a model cooling tower shielded by second tower. *Nat. Phys. Lab.*, NPL Aero. Rep. 1027. (Exp. cooling towers, $Re = 1.3$M, 1.6M, $S/D_m = 3.6$, $H/D_m = 2.2$, C_p). (22).

Sawyer, R. A. (1973P). Wake and gust loading on cooling towers. In: *Int. Symp. Vibration Problems in Industry*, Keswick, UK, Paper 117. (Exp. tripping strips $\pm 50°$, $Re = 120$k, C_p, C'_p, PS). 22.

Sayers, A. T. (1988P). Flow interference between groups of three and four equispaced cylinders. *Ph.D. Thesis*, University of Cape Town, South Africa. (Exp. Th. 3, 4 cyl., C_p, C_d, C_l, potential flow). 27.

Scanlan, R. H. and Sollenberger, N. J. (1976P). Pressure differences across the shell of a hyperbolic natural draught cooling tower. In: *4th Int. Conf. Wind Effects on Buildings and Structures*, ed. Eaton, Cambridge University Press, 143–60. (Exp. full-scale cooling tower, C_p, St, discussion Niemann). (22).

Schmidt, L. V. (1966P). Fluctuating force measurements upon a circular cylinder at Reynolds numbers up to 5×10^6. In: *NASA TMX 57, 779*, Paper 19. (Exp. 300k $< Re <$ 5M, $H/D = 8.1$, C_D, C_{pb}, C'_L, C'_D, L_{C1}). (21).

Schwind, R. (1962P). The three-dimensional boundary layer near a strut. *MIT Gas Turbine Lab.*, Rep. 67. (Exp. class.). 21.

Scruton, C. (1967P). The problems of estimating wind loading on structures with special reference to cooling towers. In: *Proc. Symp. Ferrybridge Cooling Towers*, Inst. Civil Eng., Session 3, 85–9. (Exp. wind shear, C_p, C_d). 22.

Shaw, T. L. (1971P). Wake dynamics of two-dimensional structures in confined flows. In: *Proc. 14th Congress Int. Conf. Assoc. Hydraulic Research*, Paris, France, **2**, 41–8. (Exp. $0.05 < D/B < 0.30$, St, also flat plate, sluice gate, square cyl.). 23.

Shojaee Fard, M. H., Bagheri, H., and Rahai, H. R. (1997P). On the drag and shedding frequency of wire-wrapped cylinders. In: *Experimental Heat Transfer and Thermodynamics*, ed. Giot, M., *et al.*, Edizioni ETC, 1563–70. (Exp. 16.5k $< Re <$ 84k, wire pitch $P/D = 0.25, 0.5, 1.0$, $d/D = 0.0035$, C_p, C_{pb}, C_D, St, u/V, $b_{1/2}$). (22).

Spalding, D. B. (1964P). Unified theory of friction, heat transfer and mass transfer in the turbulent boundary layer and wall jet. *Aero. Res. Council*, Current Papers 829. (Comp.). 24.

Stevenson, G. C. and Tang, T. T. (1974P). Flow-induced noise and vortex shedding in tube-bank systems. In: *Proc. 5th Australasian Conf. Hydro. Fluid Mech.*, University of Canterbury, New Zealand, 617–702. (Exp. normal triangle, $1.5 < P/D < 4.0$, 11k $< Re <$ 57k, $L/D = 20$, frequency, PS, St). (28).

Suzuki, N., Sato, H., Iuchi, M., and Yamamoto, S. (1971P). Aerodynamic forces acting on circular cylinders arranged in longitudinal row. In: *Proc. 3rd Int. Symp. Wind Effects on Buildings and Structures*, Tokyo, 377–87. (Exp. 30k $<$ $Re <$ 63k, $0° < \alpha < 10°$, 1, 2, 3, 6 cyls, $1.05 < S/D < 10$, C_{D1}, C_{D2}, C_{pb}, C_{p2}, $0.5 < D_1/D_2 < 2$, flat boundary, FV). 26.

Szechenyi, E. (1974P). Simulation of high Reynolds numbers on a cylinder in wind tunnel tests, in French. *ONERA La Recherche Aerospatial*, **3**, 155–64. (Exp. 96k $< Re <$ 4.2M, 0.015% $< K/D <$ 0.2%, C_L, C_d, St, Re_k, FV). 22.

Takahashi, F. and Higuchi, H. (1988P). Flow past two-dimensional ribbon parachute models. In: *AIAA*, Paper 88-2524. (Exp. Comp. single row, non-uniform flow). 28.

Taneda, S. (1952P). An experimental study on the structure of the vortex street behind a circular cylinder of finite height. In: *Rep. Res. Inst. Appl. Mech.*, Kyushu University, **1**, 131–43. (Exp. $39 < Re < 75$, $40 < H/D < 220$, free end, FV, topology). 21.

Taneda, S. (1980P). Definition of separation. In: *Rep. Res. Inst. Appl. Mech.*, Kyushu University, **28**, 73–81. (Exp. $26 < Re < 100$, streamlines and streaklines, circular, elliptic 2 cyls, rot. cyl., $V_r/V = 0.63, 2.1$). 24.

Tangemann, H. (1979P). Fluid-elastic oscillations of one tube in cross flow tube bundle, in German. *Ph.D. Thesis*, University of Hannover, Germany. (Exp. C_p, u, u', St, osc.). 28.

Tatsuno, M. and Ishii, K. (1983P). Flow visualisation and force measurements on two cylinders at low Reynolds numbers. In: *Flow Visualisation*, ed. Yang, W. J., Hemisphere Publ., Washington, DC, 392–6. (Exp. 2 cyl., $0.02 < Re < 1$, $1.2 < S/D < 2$, C_{D1}, C_{D2}, FV). 26.

Thom, A. (1926Pa). Experiment on the air forces on rotating cylinders. *(Brit.) Aero. Res. Council*, Rep. and Memo. 1018. (Exp. $0 < V_r/V < 6.6$, K_L, K_D, end effect, sand roughness). (24).

Thom, A. (1926Pb). The pressure round a cylinder rotating in an air current. In: *(Brit.) Aero. Res. Council*, Rep. and Memo. 1082, 66–73. (Exp. $0 < V_r/V < 4$, 7.7k $< Re <$ 38k, C_p, K_L, K_D). 24.

Thom, A. (1931P). Experiments on the flow past a rotating cylinder. In: *(Brit.) Aero. Res. Council*, Rep. and Memo. 1410, 212–23. (Exp. $Re =$ 18k, $V_r/V = 2$, streamlines, velocity fields, C_p). (24).

Thom, A. (1934P). On the effect of discs on the air forces on a rotating cylinder. In: *(Brit.) Aero. Res. Council*, Rep. and Memo. 1623, 376–85. (Exp. end plates, K_L, K_D, $0 < V_r/V < 7$, fins). 24.

Thom, A. (1943P). Blockage corrections in a closed high-speed tunnel. *(Brit.) Aero. Res. Council*, Rep. and Memo. 2033. (Th. compressibility). (23).

Thom, A. and Sengupta, S. R. (1932P). Air torque on a cylinder rotating in an air stream. *(Brit.) Aero. Res. Council*, Rep. and Memo. 1520. (Exp. rotating cyl. torque). (24).

Toebes, G. H. (1971P). The frequency of oscillatory forces acting on bluff cylinders in constricted passages. In: *Proc. 14th Congress Int. Hydraulic Res.*, Paris, France, **2**, B-7-58. (Exp. $0.05 < D/B < 0.45$, C'_L, St). 23.

Toebes, G. H. and Ramamurthy, A. S. (1970P). Lift and Strouhal frequency for bluff shapes in constricted passages. In: *Proc. Flow-Induced Vibrations in Reactor System Components*, Argonne Nat. Lab, ANL 7685, 225–47. (Exp. $Re = 10$k, $0.05 < D/B < 0.45$, C_D, C'_L, St). 23.

Tunstall, M. J. (1974P). Some measurements of the wind loading on Fawley Generating Station chimney. In: *Proc. Symp. Full-Scale Fluid Dynamics Measurements*, Leicester University, 26–41. (Exp. full-scale). 21, 22.

Turner, J. R. (1978P). Heat transfer and pressure distribution in tube banks. *Ph.D. Thesis*, Wolverhampton Polytechnic, UK. (Exp. C_p, Nu). 28.

Uematsu, Y., Yamada, M., and Ishii, K. (1983P). Some effects of free-stream turbulence on the flow past a cantilevered circular cylinder. In: *Bluff Body Aerodynamics and its Applications*, ed. Ito, Matsumoto, and Shirashi, 1988, Elsevier, 43–52. (Exp. 58k $< Re <$ 150k, $0.9\% < Ti < 10.6\%$, $1 < H/D < 4$, free end, C_p, C_D, PS, St, Ro, θ_s, C_D, St). 21.

Van Atta, C. W. and Piccirillo, P. S. (1990P). Topological defects in vortex streets behind tapered circular cylinders at low Reynolds numbers. In: *New Trends in Non-Linear Dynamics and Pattern-Forming Phenomena*, ed. Coullet, P., *et al.*, Plenum Press, New York, 243–50. (Exp. taper, $0.03 < TR < 0.08$, $Re = 170$, Ro, FV). 22.

Vanden Berghe, T. M., Dhaubhadel, M. N., Gundappa, M., Diller, T. E., and Telionis, D. P. (1988P). Pulsating flow and heat transfer over three in-line cylinders. In: *ASME Int. Symp. Flow-Induced Vibrations and Noise*, ed. Paidoussis, M., *et al.*, **1**, 113–26. (Exp. $Re = 49$k, $1.1 < S/D < 4.7$, FV, C_p, C'_p, C_f, C'_f, PS, $Nu/Re^{1/2}$). (27).

Wallis, Sir B. N. (1964P). The 'dam busting' weapon. *Aero. Res. Dev. Dept.*, Weybridge, UK, Science Museum Library, London, RNW D9/2. (Background notes issued on the occasion of the release of previously classified information on the design, development, and use of the weapon). 24.

Wardlaw, R. L. and Cooper, K. R. (1973P). A wind tunnel investigation of the steady aerodynamic forces on smooth and stranded twin bundled power conductors for the aluminium company in America. *Nat. Res. Council*, Canada, Nat. Aero. Est., LTR-LA-117. (Exp. 2 cyl., stag., 20k $< Re <$ 150k, $1.2 < S/D < 35$, smooth, stranded, conductor, C_D, C_L, FV). 22, 26.

Wardlaw, R. L., Cooper, K. R., Ko, R. G., and Watts, J. A. (1974P). Wind tunnel and analytical investigations into the aeroelastic behaviour of bundled conductors. In: *IEEE Summer Meeting Energy Resources Conf.*, Anaheim, CA, T74, 368–77, 1–20. (Exp. Th. 4, 6, 8 cyls, $Re = 120$k, C_D, C_L, Ti). 27.

Werle, H. (1963P). Ground effect simulation in a water tunnel, in French. *La Recherche Aerospatial*, **95**, 7–15. (Exp. FV, aerofoil). (23).

Werle, H. (1979P). Vortices on slender bodies at high angles of attack. In: *L'Aeronautique et l'Astronautiqu, No. 79*, 3–22. (Exp. various missiles, symmetric and asymmetric eddies, FV). 25.

West, G. S. and Apelt, C. J. (1990P). Measurements of fluctuating effects on a circular cylinder in a uniform flow. *University of Queensland, Dept. Civil Engg*, Res. Rep. No. CE110 (see also West and Apelt, 1997J). (Exp. $10 < B/D < 50$, $10\text{k} < Re < 250\text{k}$, $0.2\% < Ti < 7.5\%$, R_{pp}, C'_L, C'_D). 21.

Zdravkovich, M. M. (1971P). Circular cylinder enclosed in various shrouds. In: *ASME Vibration Conf.*, Paper 71-VIBR-28. (Exp. axial rod shroud, $Re = 100\text{k}$, D_s/D, C_p, FV). 27.

Zdravkovich, M. M. (1973P). Flow-induced vibrations in irregular tube bundles. In: *Proc. Int. Symp. Vibration Problems in Industry*, ed. Wakefield, J., Keswick, UK, Paper 413. (Exp. irregular cluster, C_p, A/D). 28.

Zdravkovich, M. M. (1982P). Intermittent flow separation from flat plate induced by a nearby circular cylinder. In: *Flow Visualisation II*, ed. Merzkirch, W., Hemisphere Publ. Corp., Washington, DC, 265–70. (Exp. towing cyl., $Re = 2.5\text{k}$, $\delta/D = 0$, FV). 23.

Zdravkovich, M. M. (1985P). Observation of vortex shedding behind a towed circular cylinder near a wall. In: *Flow Visualisation III*, ed. Yang, W. J., Hemisphere Publ. Corp., Washington, DC, 423–7. (Exp. towed cyl., $Re = 3.5\text{k}$, $L/D = 12$, $\delta/D = 0$, FV). 23.

Zdravkovich, M. M. (1986P). A circular cylinder partly submerged in three turbulent boundary layers. In: *5th Int. Symp. Offshore Mech. and Arctic Engg*, Tokyo, Japan. (Exp. $Re = 50\text{k}$, 200k, $0.17 < \delta/D < 0.68$, class., C_p, C'_p). 23.

Zdravkovich, M. M. (1995P). Interaction of bistable/metastable flows and stabilizing devices. In: *Proc. 6th Int. Conf. Flow-Induced Vibration*, ed. Bearman, P. W., 431–9, Balkema, Rotterdam. (Exp. biased flow, splitter/partition plates, C_p, C_{pb}). 26.

Zdravkovich, M. M. (1997P). Vortex shedding in tube arrays. In: *JSME Centennial Congress, Int. Conf. Fluids Engg*, Vol. 1, 293–6. (Th. St, St_g, St_p, in-line, staggered). 22, 28.

Zdravkovich, M. M. and Baldaro, J. L. (1993P). Suppression of vortex shedding from satellite risers in a current. In: *Symp. Wave Kinematics and Environmental Forces*, **29**, 239–47, Soc. Underwater Technol., London. (Exp. $6 + 1$, $8 + 1$ risers, $120\text{k} < Re < 150\text{k}$, $D/d = 2.4$, 3, C_{Do}, $0° < \alpha < 22.5°$, FV). 27.

Zdravkovich, M. M. and Namork, J. E. (1979P). Structure of interstitial flow between closely spaced tubes in staggered array. In: *ASME Symp. Flow-Induced Vibration*, ed. Chen, S. S., *et al.*, 41–6. (Exp. normal triangle, $P/D = 1.375$, six rows deep, $L/D = 9$, $Ti = 1\%$, $Re = 100\text{k}$, C_p, C'_p, C_d, V_g/V, Ti). 28.

Zdravkovich, M. M. and Namork, J. E. (1980P). Excitation, amplification, and suppression of flow-induced vibrations in heat exchangers. In: *Practical Experiences with Flow-Induced Vibration*, ed. Naudascher, E. and Rockwell, D., Springer, Berlin, 109–17. (Exp. stag., $P/D = 1.375$, C_p, C_p', FV, tripping wires, forces). 28.

Zdravkovich, M. M. and Stanhope, D. J. (1972P). Flow pattern in the gap between two cylinders in tandem. *Internal Rep. 15/72*, University of Salford. (Exp. tandem, $Re = 100k$, 210k, $1 < S/D < 7$, C_p, C_{pb}, C_{pg}, $(u/V)_{\text{gap}}$). (26).

Zdravkovich, M. M., Singh, S., Nuttall, J. A., and Causon, D. M. (1976P). Flow-induced vibration in staggered tube banks. In: *Inst. Mech. Eng. 6th Thermo. and Fluid Mech. Convention*, Durham, UK, 237–43. (Exp. tube array, C_p, displaced rows, A/D). 28.

Zhang, H. (1993P). Flow interference between two circular cylinders in turbulent flow. *Ph.D. Thesis*, Monash University, Australia. (Exp. 2 cyl., $Re = 110k$, $0.4\% < Ti < 11.5\%$, $H/D = 8$, $1 < S/D < 10$, $1 < T/D < 4$, C_p, C_p', C_d, C_d', PS). 26.

Zhang, H. and Melbourne, W. H. (1989P). Interference between two two-dimensional circular cylinders in turbulent flow. In: *Proc. 10th Australasian Fluid Mech. Conf.*, University of Melbourne, 13.13–13.16. (Exp. tandem, staggered, side-by-side, C_{Di}, C_L, PS). (26).

Zhang, H. and Melbourne, W. H. (1992P). Flow interference between two three-dimensional circular cylinders in turbulent flow. In: *Proc. 11th Australasian Fluid Mech. Conf.*, University of Tasmania, Hobart, 783–6. (Exp. $H/D = 8$, $Re = 110k$, $0.4\% < Ti < 11.5\%$, C_{Di}, $1 < S/D < 10$, $T/D = 1$, C_{Di}', PS). 26.

Ziada, S., Bolleter, U., and Chen, Y. N. (1984P). Vortex shedding and acoustic resonance in a staggered-yawed array of tubes. In: *ASME Symp. Flow-Induced Vibration*, ed. Paidoussis, M. P., **2**, 227–42. (Exp. stag., $T/D = 2$, $R/D = 1.85$, p', PS, $\Lambda = 15°$, $30°$, C_L', acoustic resonance). (28).

Žukauskas, A. A. and Katinas, V. I. (1980P). Flow-induced vibration in heat-exchanger tube banks. In: *Practical Experiences of Flow-Induced Vibration*, ed. Naudascher, E. and Rockwell, D., Springer, Berlin, 188–96. (Exp. square, $1.15 < P/D < 1.61$, normal triangle, $1.15 < P/D < 2.00$, osc., A/D, frequency, PS, Sc, V_r). 28.

AUTHOR INDEX

Key to page numbering

Roman number = the first (second) author(s) mentioned in text
Italic number = the author(s) cited in figure captions
Bold number = the full reference is given

Abd-Rabo, 1118, *1119*, 1146, 1157, **1188**, **1226**
Abed-Meraim, **1242**
Abell, **1204**
Abernathy, 1032, **1188**
Achenbach, 749, *750*, 752–3, *754–5*, 758–9, *761–2*, 767, 795, 1126, *1127*, 1138, *1139*, 1163, *1164*, **1188**
Ackeret, 738, *738–9*, 760, **1183**, **1188**
Acrivos, **1202**, **1220**
Agui, **1189**
Ahlborn, 917, *919*, **1189**
Ahmed, 1090, *1091*, **1189**
Aiba, 1081, *1081*, 1095, *1097–8*, 1134, *1134–5*, 1158, *1158–9*, *1161*, **1189**
Albarede, **1189**
Albrecht, **1189**
Alemdaroglu, **1190**
Allen, 859–60, 862–4, *865*, 868, 917, *919*, 984, **1231**
Amamoto, **1224**
Andreopoulos, **1189**
Angrilli, **1190**
Aode, 1009, **1231**
Apelt, 706, *707*, **1227**, **1247**
Arai, **1238**
Archibald, **1190**
Arie, 683, 728–9, *728*, **1208**, **1219**, **1224**, **1231**
Arita, **1240**
Armitt, *811*, 815, *817*, 817, **1190**, **1231**
Armstrong, 1145, 1149, **1203**
Arnold, 1145, 1173, **1235**
Arroyave, 763, **1234**
Asanuma, **1241**
Askergaard, 736, **1195**
Au, **1208**
Auger, 1123, *1125*, **1190**, **1231**
Awad, 686, **1197**
Ayoub, 733–4, **1190**

Baban, **1190**
Bagheri, **1244**
Bagnold, **1190**
Bahl, **1190**
Bai, **1231**
Bailac, **1231**
Baines, 751, **1190**
Baird, 1141–3, *1143*, 1149, **1190**
Bairstow, **1190**
Baker, 681–3, *681–2*, *685*, 686, *687*, **1191**
Balasubramanian, **1191**
Baldaro, *1110*, 1113, **1247**
Ball, *1109*, 1109, **1191**
Bardowicks, **1231**
Barnes, 1058, *1058*, **1189**, **1191**
Barrington, 1145, **1183**
Bartel, **1214**
Basu, **1184**
Bateman, **1182**
Batham, 756, 1136–7, *1136*, **1191**, **1231**
Baxendale, *1057–8*, 1057, **1189**
Baylac, **1191**
Bearman, 703, 705–6, *705–6*, 717, 789–90, 874–5, *874*, *879–80*, 880, 882, *883*, 1023–5, 1027, **1184**, **1192**, **1222**
Beguier, **1192**
Belik, 684, *685*, 686, **1192**
Bénard, 708, **1192**
Bergamaschi, **1190**
Betz, 893, 899, 917, *918*, **1192**, **1232**
Bickley, 899–901, *900–1*, **1192**
Biermann, 1002, *1003*, 1009, 1023–4, *1024*, 1073, **1232**
Bisshopp, **1192**
Blackburn, 847, *847*, **1192**
Blake, **1182**
Blevins, 1140, *1141*, 1152, *1153*, *1172*, 1173, **1184**, **1193**
Boasson, **1193**
Bohl, see Von Bohl
Bokaian, 1012, 1039, **1193**, **1232**
Bolleter, **1248**
Borges, 1129, 1146, 1163, *1164*, **1193**

Boribond, **1242**
Borthwick, **1232**
Bouard, 829–30, *830–2*, 832–3, **1195**
Boyer, **1212**, **1218**
Boyle, **1210**
Bradshaw, 1122, *1123*, **1193**
Bragg, **1193**
Brand, **1230**
Bray, **1201**
Brede, 700, *700–1*, **1193**
Bressler, 1152, *1153*, 1172–3, **1193**
Brewster, **1193–4**
Brown, 875, *876*, **1194**
Bryce, **1232**
Bryson, **1194**
Buhlmann, **1230**
Buresti, 755, *756*, 758, 767, **1194**, **1232**
Burkhalter, **1194**
Burnsall, 976, *977*, **1232**
Burrows, **1232**
Burton, 1140, **1194**
Busemann, 899, **1232**
Busse, **1194**

Cai, **1222**
Carpenter, **1194**
Carrasquel, **1198**
Case, **1194**
Cash, **1242**
Castagneto, **1224**
Causon, **1248**
Cave, **1190**
Chabriere, **1195**
Chan, A. S. K., **1208**
Chan, J. J. J., **1208**
Chandrasekhar, **1194**
Chang, **1194**, **1218**
Chaplin, **1232**
Chapman, **1195**, **1238**
Chattoorgoon, **1243**
Cheeseman, 944, *949*, **1194**
Chen, P. Y., 852, *852*, 854, *854*, **1195**
Chen, S. S., 1137, *1137–8*, **1195**
Chen, Y., **1232**
Chen, Y. N., 1129, 1146, 1165, 1173, **1195**, **1232**, **1248**
Cheng, L., **1210**
Cheng, M., 1121, *1122*, 1123, **1213**, **1233**
Cheung, J. C. K., 707, **1212**
Cheung, W. C., 1091–2, *1092–3*, **1209**
Chew, **1211**
Chiang, **1236**
Chiu, **1195**
Cho, **1205**
Chou, **1212**
Chow, **1233**
Christensen, 736, **1195**
Chun, **1222**
Clark, **1226**
Clarke, 809, *810*
Clarkson, **1195**
Claser, **1238**
Coales, **1242**
Cohan, 1142, *1142*, **1184**
Cole, **1195**
Coles, 938, 940–3, *941–4*, **1195**
Cooper, 781, 1009, 1039, 1044, 1049, *1049*, **1233**, **1246**
Corrsin, 1121, *1121*, **1233**
Cossalter, **1190**
Couette, 937–9, *941*, *944*, **1195**
Counihan, 781, *782*, *979–80*, 980, 1009, 1049, **1231**, **1233**
Coursimault, **1195**
Coutanceau, 829–30, *830–3*, 832–3, 896, *897*, 1123, **1184**, **1190**, **1195**, **1233**, **1242**
Cowdrey, 812–13, 815, 817, 1122–3, *1124*, 1126–7, **1233**

D'Alembert, 898, 946
Dalton, 863, *864*, 1084, *1086*, **1196**
Dargahi, 686, *686–7*, **1196**
Daujotas, **1230**
Davenport, 817, 821, **1234**
Davey, **1196**
Dayoub, 1039, 1044, *1045*, **1196**
Dean, 1142, *1142*, **1184**
Defaye, **1184**
Demirbilek, 1111, **1196**
Dhaubhadel, **1236**, **1246**
Di Prima, 942, **1196**, **1209**
Diaz, 917, 920, *921–2*, **1196**, **1234**
Diller, **1246**
Dimopoulos, 969, **1196**
Dimotakis, 936–7, *936–7*, **1225**
Doepker, 852, *852*, 854, *855*, **1195**
Donaldson, 1153, *1154–5*, **1199**, **1234**
Donnelly, **1197**
Douglas, **1240**
Drazin, 938, **1182**
Drescher, 989, **1197**
Drossopoulos, **1232**
Dryden, 734–6, *736–7*, 938, **1182**, **1197**
Dunham, *949*, 950, *951–3*, 952, **1197**
Durbin, 1027, *1028–9*, **1207**
Duty, **1197**
Dye, 1129, *1131*, **1234**

Eagles, **1196–7**
Eastop, 1133, *1133*, **1225**
Ebner, 817, **1234**

Eckelmann, 695–8, *695–6*, **1193**, **1201**, **1208**, **1214–15**, **1221**, **1230**
Eckerle, 686, **1197**
Efthymiou, **1197**
Eiffel, 769, **1197**
Eisenger, **1184**, **1198**
Eisenlohr, **1208**
El-Kashlan, **1227**
El-Sherbiny, 838, 841, 844, 848, *849*, 850, 860, *861*, 866–7, *867–9*, **1213**, **1239**
El-Taher, **1198**
Epik, **1198**
Epshtein, **1198**
Ericsson, 981, 993, **1184**, **1198**
Etzold, 717–18, *718–19*, 741, **1198**

Fage, 748, 751–2, *752*, 758, *760*, 765, *766*, 768, *770–1*, 770, 781, 795, 838, 855–7, *858*, 860, 863, **1234**
Fang, 1100, *1100*, **1209**
Farell, 763, *764*, 817–19, *818*, *820*, 863, *865*, 866, **1184**, **1198**, **1202–3**, **1234**
Farivar, 724, *724–5*, 727, *727*, 743, **1198**
Fedeniuk, 763, *764*, **1198**
Fenstermacher, **1198**
Fey, **1230**
Fiechter, 981, *983*, **1234**
Fiedler, 717, *718–19*, 741, **1198–9**
Filler, **1199**
Fischer, **1219**
Fitzhugh, **1234**
Fitzpatrick, 1153, *1154–5*, 1172–3, **1187**, **1199**, **1227**, **1232**, **1234**
Flaherty, **1230**
Flettner, 923, *923–5*, 925, **1185**
Font, 875, *876*
Forlini, **1213**
Föttinger, 925
Fox, 725, *726–7*, 727, 1075, *1075*, **1199–200**
Francis, **1184**, **1198**
Fredsøe, 889, *889–90*, **1183**, **1186**, **1200**, **1206**
Fujii, **1200**
Fujino, **1240**
Fujisawa, **1235**
Fujita, 770, *771*, *774–5*, **1200**
Fukamachi, 745, **1213**
Fultz, **1197**
Funakawa, 1146, **1200**, **1235**
Fung, 852, *855*, **1200**
Furman, 673, **1182**
Furuya, 953, **1200**, **1228**

Gai, **1219**
Gan, **1211**
Gartshore, 1060, *1061*, **1210**
Gaster, 803–4, *804–5*, 807, *807*, **1201**
Gavalda, **1196**, **1234**
Geoola, 1012, 1039, **1193**, **1232**
Gerhardt, 1083, *1084–5*, 1098–9, **1201**
Gerich, 695–6, *695–9*, 698–9, 701, **1201**, **1235**
Geropp, **1207**
Gerrard, 695, *707*, 708–9, *709*, 716, *717*, 831, **1201**, **1221**
Gharib, 799–800, *799–801*, **1202**, **1211**, **1238**
Giralt, **1192**, **1196**, **1234**
Glauert, H., 855, 860, *861*, **1235**
Glauert, M. B., **1201**
Glenny, **1185**
Goddard, **1212**
Goethals, **1190**
Göktun, 875, 882, **1235**
Goldstein, 896, **1182**
Gollub, **1198**, **1201**
Golubovic, 817, **1201**
Gomi, **1200**
Gould, *722*, 730, *732–3*, 733, 1070, 1072, **1235**
Gowen, **1235**
Graham, 694, **1201**
Grant, **1189**, **1191**
Grass, **1201**
Graw, **1213**
Gregoire, **1191**, **1231**
Griffin, 779, *781*, **1226**
Griffiths, E. A., **1242**
Griffiths, R. T., **1201**
Gross, **1209**
Grossberg, **1194**
Grotz, 1145, 1173, **1235**
Grove, 829, *836*, **1202**, **1220**
Grover, 1146, **1202**, **1226**
Gu, 1013, *1014*, 1039, 1043, 1047, *1048*, 1051, *1051*, **1202**, **1222**
Gundappa, **1246**
Guven, 752, *753*, **1198**, **1202**

Haan, **1191**
Hall, C. D., 1109, *1109*, **1191**
Hall, M., **1185**
Halvorsen, 1111, **1196**
Hamache, **1202**
Haniu, 1064, *1064–6*, **1219**
Hanratty, 969, **1196**
Hansen, **1200**
Hanson, 956, *956–7*, 958–9, **1202**
Hara, 1124, *1125*, **1235**

Harashima, **1241**
Harris, **1202**
Harvey, 789–90, **1192**
Hasegawa, **1221**
Hasimoto, 998, **1185**, **1213**
Hasoon, **1218**
Hayashi, M., **1186**, **1203**
Hayashi, T., 966–7, *967–8*, 969, *970*, 973, *974–6*, 976, 979, **1203**, **1207**, **1226**, **1228**
Hayn, 817, **1236**
He, **1202**, **1222**
Heideman, **1240**
Hein, 941, **1220**
Heinecke, 758–9, *761*, **1188**, **1236**
Herrnstein, 1002, *1003*, 1009, 1023–4, *1024*, 1073, **1232**
Hetz, **1236**
Hibino, **1207**
Higuchi, 1122, **1203**, **1245**
Hill, G. C., 734–6, *736–7*, **1197**
Hill, R. S., 1145, **1203**
Hirsch, **1203**
Hirschfeld, *926–7*, **1185**
Hiwada, 838–9, *840*, 841, *842*, 843, 845–6, *845–6*, 875, 880, *880*, 884–6, *884–5*, 1007, 1052–4, *1052–4*, **1203–4**, **1207**, **1209**
Hoerner, **1182**
Hoffman, 785, *786–7*, **1206**
Hölscher, 768, *768*, **1186**
Homann, 829, **1204**
Honji, 1104–5, *1105–6*, 1124, **1204**, **1223**, **1236**
Hori, 1003, *1004*, 1009, 1024, 1036, *1037*, **1236**
Hourigan, **1228**
Hsiao, 808, *808–9*, **1236**
Huang, **1227**
Hubler, **1215**
Hughes, **1204**
Hunt, B. L., *986–8*, 986–7, **1210**
Hunt, J. C. R., **1204**
Hurley, 945, 948, **1236**
Hussain, **1212**
Hyun, **1222**

Ido, **1238**
Igarashi, 773, *775–6*, 788, *788–9*, 790, *791–3*, 792, 1004, *1005–6*, 1006, 1008, *1008*, 1017–18, *1017*, 1026, 1054, *1055–7*, 1056, 1065, *1066–7*, 1078, *1079–80*, 1082, *1082–4*, 1094, *1096*, 1097, **1204–5**, **1225**, **1236–7**
Iida, **1205**, **1237**
Ikemoto, **1235**
Ilgarubis, **1230**
Imaichi, **1205**, **1241**
Ingham, **1205**
Ingram, **1212**
Inoue, **1207**
Ishida, **1221**
Ishigai, 1002, *1003*, 1018, *1019*, 1023, 1036, *1036*, 1044, *1045*, 1129, 1146, 1165, *1165*, **1185**, **1205**
Ishii, 998–9, *999–1000*, 1038, *1038*, **1224**, **1245–6**
Isyumov, 817, 821, **1234**
Ito, **1209**
Iuchi, **1244**
Iversen, **1185**, **1205**
Iwabuchi, **1241**
Iwakoshi, **1203**
Iwasa, **1226**, **1228**

Jacobsen, 1112, *1113*, **1242**
Jacobson, **1206**, **1213**
James, 769, *772–3*, **1206**
Jaminet, 903–4, *903–4*, 918, **1206**
Jandali, 867, **1216**
Jendrzejczyk, 1137, *1137–8*, **1195**
Jensen, B. L., 889, *891–2*, **1206**, **1237**
Jensen, H. R., **1206**
Johnson, S. P., 1113, *1115–16*, **1238**
Johnson, T. R., 785, *787*, **1206**
Jones, **1211**
Joubert, **1206**
Justesen, **1241**

Kacker, *707*
Kageyama, **1206**, **1216**, **1228**
Kajima, **1228**
Kamemoto, 700, 1032, *1033*, **1206**, **1225**, **1228**
Kao, **1221**
Karamcheti, 733–4, **1190**
Kármán, 748, 834–5, **1182**
Karniadakis, 712, *713*, **1206**
Katinas, **1183**, **1206**, **1230**, **1248**
Kavanagh, **1210**
Kawagoe, **1238**
Kawaguti, **1207**
Kawai, **1200**
Kawall, **1196**, **1234**
Kawamura, *718*, 718, *723*, 969, **1203**, **1207**
Keefe, 705, *706*, 707, **1207**
Keener, **1238**
Keffer, 1059, *1060*, **1192**, **1196**, **1216**
Keller, **1214**

Kelly, 894, 911, **1207**
Keshavan, **1207**
Kim, H.-B., **1210**
Kim, H. J., 1027, *1028–9*, **1203**, **1207**
Kim, M. S., **1207**
Kimoto, **1207**
Kimura, 798, *798*, **1207**
King, **1185**
Kirchgässner, **1208**
Kiya, 744, 1028, 1045, *1046*, **1208**, **1231**, **1238**
Knisely, **1238**
Ko, N. W. M., 795, *795–6*, 801, **1208**, **1210–11**, **1213**
Ko, R. G., **1246**
Kobayashi, 1006, *1007* (see Nakayama)
Kohli, **1193**
König, 689, *690*, **1208**, **1214**, **1230**
Koschmieder, **1194**, **1208**
Kostic, **1209**
Kovasznay, 800, **1209**
Kozlova, **1198**
Kraemer, *691–4*, 691, 693–4, 703–4, *703–5*, 706–7, *708*, **1238**
Krahn, 894, 910–11, **1209**
Kramer, 1083, *1084–5*, 1098–9, **1201**, **1240**
Kraus, 1000, **1224**
Krause, 993, **1238**
Kronauer, 1032, **1188**
Krueger, **1209**
Kruse, **1209**
Kumada, **1203–4**, **1207**, **1209**
Kuwata, **1216**

Lachmann, 949, **1232**
Lafay, 894, 905, 909–10, 917, **1209**
Lagally, **1209**
Lam, 1091–2, *1092–3*, 1100–2, *1100–4*, **1209–10**
Lambert, **1221**
Lamont, 986–7, *986–90*, **1210**, **1238**
Lanciotti, **1194**
Landau, 710, *711*, 712, **1182**
Landweber, 1029–31, *1030*, **1238**
Lang, **1190**
Langston, **1210**
Lawson, **1210**
Leder, **1213**
Lee, P. M., 839, *841*, 850, 869, *871*, **1218**
Lee, S.-J., **1210**
Lee, T. S., **1218**
Lehman, **1215**
Lei, 885, *886*, **1210**
Lesage, 1060, *1061*, **1210**
Leung, K., **1217**
Leung, W. L., **1208**
Leung, Y. C., 795–7, *796–7*, **1208**, **1210–11**
Lewis, 799–800, *799–801*, **1211**, **1238**
Li, **1222**
Lian, **1227**
Lienhardt, **1195**
Lifschitz, 710, *711*, 712, **1182**
Linke, **1211**
Lippig, **1216**
Ljungkrona, 1012, *1013*, **1211**, **1238**
Lo, K. W., **1211**
Lo, S. C., 1101, *1102–4*, 1103, **1210**
Loc, **1242**
Lock, 855, 857, 859–60, 907, **1238**
Loftin, 976, *977*, **1232**
Lohrisch, 1123, *1123*, **1239**
Low, **1211**
Lowe, *816*, **1239**
Lundgren, **1242**
Luo, 1016, *1017*, **1211**, **1239**

Ma, **1201**
Mabey, **1211**
Mabuchi, 838–9, *840*, 841, *842*, 843, *844–6*, 845, **1203–4**, **1207**, **1209**
Madaras, 926 (see Hirschfeld)
Maekawa, 1044, **1211**
Magnus, 893–4, 909–11, **1211**
Mair, 793, *794*, 1044, **1185**, **1211**
Maisch, **1198**
Maki, **1207**
Malcolm, **1195**
Mallard, **1240**
Mallock, 938, **1212**
Mao, **1200**
Mark, **1216–17**
Markevicius, **1206**
Marston, **1199**
Martin, **1212**
Maskell, 863–4, *865*, 867, **1239**
Mathew, **1230**
Mathis, **1212**, **1218**
Matsui, 904, *905*, **1239**
Matsumoto, M., **1185**
Matsumoto, S., 738, **1219**
Matsuzaki, **1228**
Matten, **1242**
Maull, 824, *825*, 1044, **1185**, **1212**
Mazur, **1212**
McGregor, *707*
McKnight, **1199**, **1234**
Meister, **1212**
Meksyn, **1212**
Melander, **1212**

Melbourne, *707*, 736, 1012, *1013*, 1039, 1051, **1212**, **1230**, **1248**
Menard, 896, *897*, **1233**
Miau, **1212**
Mih, **1199**
Millborrow, **1231**
Miller, B. L., **1212**
Miller, G. D., **1212**
Miller, M. C., *751*, *764–5*, 765, *909–10*, 909, **1212–13**
Milne-Thomson, 877, **1182**
Mitry, 832, 835, *837*, **1239**
Miyakoshi, 885, *887–8*, **1224**
Miyazaki, 998, **1213**
Mizota, (frontispiece), **1213**, **1215**
Mizuno, 781, 783, *783–5*, **1239**
Mochizuki, O., **1238**
Mochizuki, S., **1205**, **1237**
Modi, 838, 841, *844*, 848, *849–50*, 850, 860, *861*, 866–7, *867–9*, **1213**, **1239**
Mohr, **1236**
Moon, **1221**
Moore, D. W., **1213**
Moore, J., **1213**
Moretti, 1121, *1122*, 1123, **1185**, **1213**
Morgan, **1185**
Mori, **1208**
Morkovin, 679, 985, **1185**
Morrison, 983, *984–5*, 985, *990–1*, 991–2, **1225**
Morsbach, **1240**
Morsy, **1239**
Morton, **1213**
Mujumdar, **1240**
Müller, **1213**
Murata, **1223**
Mureithi, **1216**, **1242**
Murnaghan, **1182**
Murphy, **1221**
Muttray, 741, **1183**

Nagano, 914, *916–17*, 916–17, **1223**
Nakabayashi, **1235**
Nakagawa, 778–9, *779*, **1240**
Nakamura, I., **1213**, **1216**, **1241**
Nakamura, T., **1216**
Nakamura, Y., 745, 766, *766*, **1213**, **1241**
Nakayama, A., **1228**
Nakayama, Y., **1183**
Namork, 1156, *1157*, **1247–8**
Narayan, **1197**
Nath, **1226**
Naudascher, 841, 843, *843*, 850–3, *851*, *853–4*, **1183**, **1218**
Naumann, 773–4, *777–8*, 777, 801, *803*, 808–9, **1240**
Ng, C. P., 842, **1218**
Ng, C. W., **1213**
Nielsen, 981, **1185**
Niemann, 818, *818*, 821, *822–3*, **1186**, **1214**, **1240**
Nieuwstadt, **1214**
Nikuradze, 753, **1240**
Nishida, **1205**, **1237**
Nishikawa, 1129, 1146, **1185**, **1205**
Nishimura, **1205**
Nishioka, 688, *689*, **1214**
Nisi, 687–8, *688*, **1214**
Nissan, **1193–4**
Niwa, **1203**
Noack, 712, **1208**, **1214**, **1230**
Norberg, 700, *701–2*, 702–3, *704*, 706, *707*, 801, **1211**, **1214**, **1238**, **1240**
Norton, 978–9, *978*, **1240**
Novak, 1056, **1240**
Nuttall, 1143, **1229**, **1248**

O'Neill, *812–13*, 815, 817, **1233**
Oberkampf, **1214**
Oengören, 1147–51, *1148–9*, *1151–2*, 1155, *1166–8*, 1167, 1169, *1170–1*, 1171, 1173, **1214–15**, **1230**
Oertel, **1186**
Ohle, **1214–15**
Ohmi, **1205**, **1241**
Ohya, 1089, **1186**, **1203**
Oiwake, **1219**
Oka, **1209**
Okajima, 766, *766*, 932–3, *933–5*, 1008, *1009*, 1011–12, *1011*, 1015, *1015*, 1023, *1023*, 1031, *1032–4*, 1033–4, 1041, *1042*, **1186**, **1215**, **1223**, **1241**
Okamoto, S., 718, *719*, *729*, **1215**
Okamoto, T., 720, *721–3*, 726–8, 741, 838, *839*, 847, *848*, **1215**
Okui, **1241**
Osaka, 1073–4, *1073–5*, **1206**, **1216**, **1228**, **1241**
Ostowari, 1090–1, **1189**
Ota, **1189**
Ottesen-Hansen, 1111, **1241–2**
Otugen, **1190**
Owen, 819, 822–3, *825*, 1143, *1144*, 1145–6, *1146*, 1153, 1172, **1216**, **1242**

Pahle, **1230**

Paidoussis, 1039, 1041, 1092, *1094*, 1154, *1155*, 1163, *1163*, **1186**, **1216–17**, **1222**, **1225**, **1242**
Palmer, M. D., 1059, *1060*, **1216**
Palmer, R. K. W., **1211**
Pan, **1236**
Pankhurst, 945–7, *946–8*, **1242**
Pannell, 1001, *1002*, 1009, 1036, **1242**
Parker, 1140–1, 1173, **1216**
Parkin, **1216**
Parkinson, 867, **1216**
Patel, **1198**, **1202**
Peak, **1186**
Pearcey, 757, *759*, 759, 1099, 1108, *1108*, **1242**
Pechstein, 736, **1216**
Peller, 914, *915*, **1216**
Perednis, **1206**, **1230**
Perkins, 984, **1231**, **1235**
Perry, **1216**
Peterka, 1149, *1150*, **1204**, **1217**
Petersen, **1202**, **1220**
Peterson, E. G., 757, **1190**
Peterson, V. L., **1225**
Piccirillo, 804, *804*, 806, *806*, **1217**, **1246**
Pierce, F. J., **1217**
Pierce, H. R., 1106, *1107*, **1242**
Pineau, **1242**
Polak, 1169, *1170*, **1217**
Polidori, **1242**
Ponsford, **1235**
Powell, 780, 954, **1243**
Prandtl, 893–4, *895*, 896, *897*, 898–900, *899*, 904, 916–17, 944, *944–5*, 955, 984, **1183**, **1186**
Prasad, 1060, **1217**
Price, 1039, 1041, 1044, *1092*, *1094*, 1154, *1155*, *1160*, 1160, **1216–17**, **1222**, **1225**, **1242**
Pridden, 1024, *1025*, 1039, **1229**
Pris, 818, **1242–3**
Probert, 740, *740*, **1218**
Pröpper, 821, **1214**
Provansal, 711, *711*, **1189**, **1212**, **1218**
Putnam, 1142, 1149, **1186**, **1218**
Py, 969, *971–2*, **1225**

Qadar, **1218**
Quadflieg, 777, *778*, 801, *803*, 808, *808*, 1025, *1026*, **1218**, **1240**

Rabillat, **1190**
Rahai, **1228**, **1244**
Ramamurthy, 839, *841–3*, 841, 850, 869, 871, **1218**, **1246**
Ramberg, *955*, 959–60, *960–4*, 962–4, **1218**, **1243**
Ramesh, **1219**
Raney, **1218**
Rankine, 856, 907, **1183**
Rao, 992, *993*, **1243**
Raven, **1201**
Rayleigh, 938–9, **1218**
Raymer, 813, **1235**, **1243**
Reding, 981, 993, **1184**, **1198**
Reid, G. E., 899, 917, *918*, **1243**
Reid, W. H., 938, **1182**, **1197**, **1202**, **1204**, **1218**
Reising, **1226**
Relf, 780, 954, **1243**
Richards, *816*, **1231**, **1239**
Richardson, E. G., **1218**, **1243**
Richardson, P. D., 1149, *1150*, **1217**
Richter, 841, *843*, 850–3, *851*, *853–4*, **1218**
Rizzo, **1243**
Rockwell, **1183**, **1186**, **1193**
Roesch, **1215**
Rosenhead, 831, *832*, 833–5, *834–5*, *837*, 1031, **1219**
Roshko, 745, 777, 875, 956, 981, **1186**, **1219**, **1243**
Ruscheweyh, 736, 821, **1203**, **1219**, **1234**

Saban, 1058–9, *1059*, **1220**
Sabathe, **1224**
Sakamoto, 683, 728, *728*, 1064, *1064–6*, **1208**, **1219**, **1224**
Sakurai, **1203**
Sallet, **1243**
Salter, C., 813, **1243**
Salter, I. J., **1242**
Sanada, 736, **1219**
Sano, **1185**
Sarode, **1219**
Sarpkaya, 986, **1186**, **1220**
Sartory, **1194**
Sasaki, 744, **1208**
Sato, 688, *689*, 1147, **1214**, **1220**, **1244**
Saunders, **1213**
Sawyer, 819, *821*, **1244**
Sayers, 1058–9, *1059*, 1089, *1090*, 1092, 1101, **1220**, **1244**
Scanlan, **1244**
Schewe, *706–7*, **1220**
Schiff, **1225**
Schlichting, 847, 1044, **1183**, **1220**
Schmidt, 734, *735*, 832, 855, **1220**, **1244**
Schultz-Grünow, 941, **1220**
Schwabe, **1219**
Schwind, 682, **1244**

Scruton, *814*, 824, **1183**, **1244**
Sears, 954–5, **1220**
Seed, **1194**
Sengupta, **1245**
Shair, 833, *833*, **1202**, **1220**
Shaw, 882, **1220**, **1244**
Sheridan, **1228**
Sheshagiri, **1193**
Shima, **1240**
Shimada, **1241**
Shirakashi, 965–6, *966*, 973–4, **1221**
Shojaee Fard, **1244**
Simon, **1197**
Sin, **1221**
Singh, **1242**, **1248**
Skelhorne, **1230**
Skop, **1191**
Slaouti, 706, *709*, 709, 716, *717*, **1221**
Smith, A. M. D., 782, **1221**
Smith, A. R., 967, *967*, 973, *975*, 976, **1221**
Smith, J. G., **1217**
Snyder, **1221**
So, **1190**, **1221**
Sobolev, **1230**
Sollenberger, **1244**
Sorger, **1208**
Soria, **1228**
Spalding, 951, **1244**
Spivack, 1022–3, *1022*, **1221**
Sreenivasan, 1061–2, *1062*, **1222**
Stager, 702, **1221**
Stanhope, **1248**
Stansby, **1221**
Steiner, **1216**
Steinolfson, **1243**
Stevenson, **1244**
Stonebanks, **1229**
Straub, **1216**
Strutt, **1218**
Strykovski, 1061–2, *1062–3*, **1222**
Stuart, J. T., **1196**
Stuart, R. J., **1201**
Sugitani, 1008, *1009*, 1031, *1032*, 1041, *1042*, **1215**, **1241**
Sullivan, **1184**, **1198**
Sumer, **1183**, **1186**, **1200**, **1206**, **1237**
Sumner, **1217**, **1222**
Sun, 824, *826*, 1034, *1035*, 1039, 1043, 1047, *1048*, 1051, *1051*, **1202**, **1222**
Sunabashiri, 718, *719*, *729*, **1215**
Sunden, *707*, **1211**, **1238**
Sung, **1222**
Sutton, 680
Suzuki, K., 1078, *1079–80*, **1205**
Suzuki, N., 1009, 1038–9, *1040*, **1244**
Suzuki, T., **1238**
Suzuki, Y., **1231**
Swanson, 894, 899, 901, *902*, 911, *912–14*, 913–14, 918, **1186**
Swinney, **1198**, **1201**
Szabo, 1084, *1086*, **1196**
Szechenyi, 752, 757, *758*, 767–8, *768*, **1222**, **1245**
Szepessy, 705–6, *705–6*, **1222**
Szewczyk, **1191**

Taguchi, **1203–4**
Tajima, **1237**
Takahama, **1200**
Takahashi, 1122, **1245**
Takata, **1241**
Takayama, **1224**
Takeda, **1228**
Takeuchi, 838, *839*, 847, *848*, **1215**
Tamura, **1208**
Tanabe, **1203**
Tanaka, H., 914, *916–17*, 916–17, **1223**
Tanaka, S., **1223**
Taneda, 714, *715–17*, 716, 799, 829, 831–2, *836*, 869, *870*, 882, 904, *905*, 929–30, *929–31*, **1186**, **1223**, **1231**, **1245**
Tang, K. M., **1210**
Tang, T. T., **1244**
Tangemann, 1175–6, *1175*, **1245**
Tanida, **1223**
Taniguchi, 683, *684*, *730–1*, 885, *887–8*, 1016, *1016*, 1050, *1050*, **1224**
Tatsuno, 998, *999–1000*, 1019, *1019*, 1038, *1038*, **1223–4**, **1231**, **1245**
Taylor, 938–41, *940*, 943, **1224**
Teissié-Solier, 695, **1224**
Telionis, **1236**, **1246**
Texier, **1242**
Thom, 905–8, *906–8*, 913, 918–20, **1245**
Thomas, 1000, **1224**
Thomson, 983, *984–5*, 985, *990–1*, 991–2, **1225**
Thwaites, B., 680, 945–6, *948*, 948, 951, **1183**, **1225**, **1236**
Thwaites, M. A., 945–6, *946–8*, **1242**
Tietjens, 894, *897*, *899*, **1183**
Ting, **1225**
Tobak, **1186**, **1225**
Toebes, 841, *843*, 850, **1245–6**
Tokumaru, 936–7, *936–7*, **1225**
Tomonari, **1213**
Tournier, 969, *971–2*, **1225**
Toy, 1075, *1075*, **1199**
Tree, **1217**

Triantafillou, 712, *715*, **1206**, **1225**
Truong, 769, *772–3*, **1206**
Tsuchida, **1189**
Tsuei, **1233**
Tsutahara, 798, *798*, **1207**
Tsutsui, 1065, *1066–7*, **1225**, **1237**
Tunstall, 736, **1246**
Turner, 1130, *1132–3*, 1133, **1225**, **1246**
Tyler, **1225**

Ueki, **1213**
Uematsu, **1225**, **1246**
Ueno, **1221**
Ulinskas, **1183**
Umakoshi, 1146, **1200**
Umeda, 1174, *1174*, **1226**

Van Aken, 894, **1207**
Van Atta, 804, *804*, 806, *806*, 903–4, *903–4*, 918, 957–9, *958*, 962, **1206**, **1217**, **1226**, **1246**
Van Dyke, **1183**
Vanden Berghe, **1246**
Vasantha, **1226**
Vickery, 809, *810*, **1226**
Vincenti, 859–60, 862–4, *865*, *868*, **1231**
Von Bohl, 1120, **1193**
Votaw, 779, *781*, **1226**

Wadcock, 1023–5, 1027, **1192**
Waibel, **1216**
Waka, **1203**, **1226**, **1228**
Wakiya, **1221**
Waldeck, 736, **1226**
Walker, **1226**
Wallis, B. N., 927–9, *928*, **1246**
Wallis, P. R., 1174, *1174*, **1226**
Wang, D. J., **1225**
Wang, J. T., **1212**
Wardlaw, A. B. Jr, 981, **1187**
Wardlaw, R. L., 781, 1009, 1038, 1049, *1049*, 1099, *1100*, 1107, *1107*, **1233**, **1246**
Warsap, 748, 751–2, *752*, 758, *760*, 765, *766*, 769–70, *770–1*, 781, 795, **1234**
Watanabe, **1223**
Watts, **1246**
Weaver, D. S., 1118, *1119*, 1140, 1146, 1149, 1168–9, *1169–70*, 1172–3, **1187–8**, **1202**, **1217**, **1226–7**
Weaver, W., 778, *780*, **1227**
Weber, **1232**
Wei, **1212**
Weihs, **1193**, **1227**
Weizel, **1227**
Welsh, **1228**
Wendt, **1227**
Werle, 981, *982–3*, 1078, *1078*, **1227**, **1246–7**
West, 706, *707*, 725–8, *727*, **1199–200**, **1227**, **1247**
Weston, **1230**
Wharmby, **1232**
White, 828, *828–9*, **1227**
Wieghardt, **1227**
Wieselsberger, 740–1, *741–2*, 851, **1228**
Wille, **1199**
Willhofft, 925, *925*, **1187**
Williamson, 997, 1020–1, *1020–2*, 1030, 1032, 1060, **1187**, **1212**, **1217**, **1228**
Wong, C. H., **1211**
Wong, P. T. Y., **1208**
Wong, S. S. T., **1222**
Woo, **1204**
Wood, **1228**
Wu, 700, **1228**

Yagi, **1185**
Yagita, 720, *721–2*, 723, 726–7, 741, *747*, 801, *802*, **1215**, **1228**
Yamada, H., 1071, *1072*, 1073, **1206**, **1216**, **1228**, **1241**
Yamada, M., **1225**, **1246**
Yamamoto, **1244**
Yamasaki, 1017–18, *1017*, **1237**
Yamashita, **1213**
Yanagihara, **1203**
Yang, E. E., **1228**
Yang, W.-J., 1174, *1174*, **1226**
Yang, X., **1228**
Yeung, **1227**
Yih, **1228**
Yokoi, 700, **1228**
Yoshimura, **1231**
Yoshino, 953, **1200**, **1203**, **1226**, **1228**

Zahn, 1160, **1217**
Zdravkovich, 741, *742–7*, 778, *873–4*, 874–5, 877, *879–83*, 880, 882, 885–6, 1000, *1001*, 1009, *1010*, 1024, *1025*, 1027, 1036–9, *1039*, *1041*, 1042, *1043*, 1044, 1050, *1051*, 1068, *1068–71*, 1070–1, 1073, *1073*, 1085, *1086–9*, 1103, 1111–14, *1112*, *1115–16*, 1126–30, *1126*, *1128–9*, *1132*, 1143, 1156, *1157–9*, 1161–2, *1162*, 1171, **1187**, **1192**, **1213**, **1229–30**, **1238**, **1247–8**
Zebib, **1228**

Zhang, H., 1012, *1013*, 1035, *1036*, 1039, 1048, 1051, **1230**, **1248**
Zhang, H.-Q., 713, *714*, **1230**
Zhang, L. L., **1202**, **1222**
Zhou, 824, *826*, **1222**
Ziada, 1147–51, *1148*, *1151–2*, 1155, *1166–8*, 1167, 1169, *1170–1*, 1171, 1173, **1214–15**, **1230**, **1248**
Zienkiewicz, 819, **1216**
Žukauskas, **1183**, **1188**, **1206**, **1230**, **1248**

SUBJECT INDEX

acoustics, 1027, 1139–43, 1145, 1149–53, 1166–8, 1171–3
added mass, 1141
aeronautical engineering, 994, 1001
aligned, 1078–81, 1084, 1086, 1096–7, 1107
angle of incidence, 981, 984
antenna, 1104
aspect ratio, 679–80, 687, 691, 693–4, 700, 702, 706–7, 713, 720, 728, 738, 740–1, 744–5, 752, 758, 811, 813, 815, 841, 843, 848, 917, 955, 957, 969, 1050, 1059
AWACS, 680

baffle, 1143
base bleed, 703, 1104
beating, 1000
Bernoulli, 866
biased, 997, 1013, 1020–1, 1023, 1025–9, 1034–6, 1042, 1054, 1059, 1091, 1102, 1105, 1121–3, 1127–9, 1173, 1175
binary eddy street, 997, 1006, 1021, 1056
bistable flow, 995, 997–8, 1006, 1008, 1017–18, 1023–9, 1036, 1042–3, 1059, 1061, 1083, 1123
blockage ratio, 751–2, 758, 813, 827–31, 833–4, 835–43, 845–50, 853, 855–6, 860, 862, 865–7, 869, 914, 930, 937, 1092, 1130–2, 1136
blowing slot, 790, 792–3, 893, 944, 949, 992–3
boundary layer, 679–81, 683, 686–7, 694–5, 720, 722, 730, 748–9, 751, 754, 757, 767, 780, 782, 788, 790, 792, 795–8, 812, 820, 846, 848, 850, 855, 872, 874–5, 878, 880, 884–6, 893–4, 902, 911, 913–15, 930, 944–5, 948–50, 953, 955, 970–1, 1011–12, 1015–16, 1047, 1060, 1064, 1081
bridge pier, 680
buffeting, 1142, 1144, 1153

cable, stranded, 779, 980
chimney stack, 679, 714, 736, 994–5, 1016, 1077
circulation, 890, 894–900, 906, 946, 1031–2
civil engineering, 994, 1052, 1077
coalescence, 1121
Coanda effect, 1026–7
Coles spiral, 938
concentric rotating cylinder, 937–8
conductor, stranded, 779–80, 1049, 1077, 1099, 1104, 1107
confluence point, 688, 829, 831, 838, 929, 1063
control cylinder, see cylinder
convection velocity, 1029
cooling tower
  model, 809–13, 815–16, 818–26, 1051–2
  full scale, 1051–2
correction, 827, 855, 860–8, 954
correlation length, 706, 733–4, 756–7, 846–7, 955, 958, 963, 973, 975
creeping flow, 828, 998–9, 1018–20, 1038
crossed cylinders, 1067–71, 1075
cylinder
  finite, 679–80, 716, 718–19, 720–1, 724–6, 728, 730, 734, 788–9, 811, 995, 998, 1015, 1050, 1070
  free end, 679, 714–16, 718, 720–2, 726–8, 730, 733–4, 741, 744, 799, 806, 809, 815, 960–2, 1016, 1050, 1070–2
  near boundary, 827, 870–2, 996, 1018–19, 1024–5, 1118
  near cylinder, 1025, 1118, 1130–3, 1156

dam buster, 927–8
detachment, 744–5, 1105
diffusion, 700, 804, 922, 936, 1012, 1038, 1073–4
dissipation, 1073–4, 1118, 1144, 1153, 1160
disturbance, 768–70, 772–3, 776, 778–82, 785–6, 789–90, 794–5, 797, 809, 815, 822, 824, 1033, 1054, 1118, 1124, 1128, 1141, 1146, 1152
drag
  fluctuating, 705, 707, 726, 734, 852–5, 885–7, 934–5, 1035, 1064, 1071–2, 1074, 1137–9
  interference, 1002, 1004, 1006–7, 1013, 1016, 1018, 1024, 1036–7

local, 721–3, 731–2, 735–6, 755, 759, 810, 822, 824, 849–50, 875, 877, 884, 908–10, 917, 926–8, 987–8, 1033–4, 1041, 1047, 1051, 1053, 1055, 1065, 1080–1, 1083–4, 1090–4, 1097–100, 1126–7, 1162
mean, 730–1, 736, 740–3, 751–3, 757–8, 761, 763–6, 835, 838–41, 850–2, 856–9, 861–4, 866–7, 869–71, 874, 880–1, 890–1, 898, 900–2, 913, 917–20, 934–5, 937, 944–5, 950, 971, 978–80, 986, 998, 1001–3, 1009–12, 1015–17, 1023, 1025, 1039–41, 1043, 1045, 1047, 1049–51, 1055, 1060, 1064, 1066, 1073, 1080, 1084–6, 1090, 1092, 1095, 1099, 1107, 1109–14, 1116, 1161–5

eddy
asymmetric, 981–3, 986, 992
cell, 701, 724, 763, 799–800, 803–4, 974
filament, 688–9, 709, 724–5, 741, 778, 780, 791–2, 798, 959, 961–2, 991, 1071–2, 1088
formation region, 703, 716, 724–5, 741, 778, 780, 791–2, 798, 847, 853, 894, 936, 963–4, 970, 1002, 1102
shedding, 688–9, 695, 699, 703, 706, 717, 726–9, 733–4, 755, 774, 776, 778, 793, 802–3, 807–8, 810, 841–2, 844, 872–3, 878, 880, 882, 884, 886, 892, 896, 903–4, 916, 923, 930, 932, 934–7, 956–7, 959, 962, 965, 970, 973–5, 984, 990, 994, 997, 1000, 1003, 1005, 1007–9, 1012, 1015–16, 1021–2, 1047, 1050, 1053–6, 1059–61, 1076–81, 1091, 1095, 1098, 1101, 1105, 1111–14, 1129, 1139–40, 1143, 1149, 1152–3, 1156, 1163, 1165–7, 1169–71
streamwise, 700–1, 713, 745, 786, 788, 792, 1070–1, 1073, 1075, 1090
street, 688, 696, 698, 713, 728, 799, 801–2, 834–5, 845, 936, 997, 1000–1, 1003, 1006–7, 1013, 1018, 1021, 1023, 1038, 1058, 1061–2, 1079, 1085, 1087–8, 1129
strength, 985, 991
symmetric, 981–4, 986, 992
end
effect, 679, 692, 694, 855, 908, 919, 939, 955, 959
plate, 679, 694–6, 698–700, 702–4, 707, 763–4, 781, 798, 801, 803, 806, 847, 899–900, 903, 909, 912, 918, 923, 928, 936, 959–60, 963, 966–9, 973–4, 976, 980, 1016, 1061, 1091, 1099–100
excitation, 1141, 1145, 1149, 1152, 1166–8, 1171, 1173

fin, 919
fingers, 699–700, 713
flow regime
2-D, 679, 688, 694, 713, 716, 718, 720, 725, 769, 779, 800, 823, 825, 856–7, 889, 917–18, 955, 1056, 1074, 1082, 1121–2
3-D, 679–80, 687, 712, 720, 733, 774, 778, 781, 799, 904, 917, 959, 966, 973, 981, 1050, 1067–8, 1074
L1, 823, 830, 998, 1018, 1038
L2, 687, 828–9, 831, 930, 956, 998
L3, 688, 694, 702, 799, 803, 833, 902, 956, 973, 998, 1019, 1038
nominal 2-D, 679, 693–5, 718, 720, 724–5, 733, 740, 799–800, 815, 855, 959, 1000
TrBL0, 703, 706–7, 733, 752–3, 757, 759–60, 769, 783, 851–2, 863, 910–11, 913, 980, 1015
TrBL1, 706, 740, 752, 757, 850–1, 911, 1011, 1015–16
TrBL2, 706–7, 722, 734, 757, 759–60, 764, 850, 852, 976, 989, 1011, 1015–16
TrBL3, 706–7, 728, 733–4, 752–3, 755, 757–8, 760, 783, 842, 850, 854, 911, 989, 1011, 1015
TrBL4, 706, 730, 734, 752–4, 757, 759, 767, 773, 850, 989, 1015, 1127, 1138, 1163
TrSL1, 692, 703, 842, 919, 973, 1007, 1018, 1044
TrSL3, 692, 705, 730, 740, 752, 760, 769, 780, 827, 853, 863, 884, 1007, 1044
TrW1, 700–1, 998, 1000
TrW2, 700–1, 728–9
fluids engineering, 1067
free stream turbulence, 694, 731, 748, 815, 848–9, 874, 995, 998, 1009, 1012–13, 1035, 1048, 1067, 1077, 1081, 1111, 1118, 1136, 1153–4, 1156

free water surface, 680
frequency power spectrum, 695–6, 701, 734, 794, 820–1, 844, 852, 883, 886, 904, 917, 965, 973–5, 1147, 1149, 1155, 1159, 1165
friction, c_f, 686, 722, 748, 932, 969–70, 973–5
frisbee, 680
fuel storage tank, 679, 738–40

gap flow, 998, 1018, 1021, 1025–8, 1034–6, 1038, 1042–3, 1047, 1049, 1054, 1083, 1091, 1105, 1116, 1120, 1154–5, 1169
grid, 1067, 1120–1, 1137–8
gust, 819–21

harmonic mode, 1021, 1142
heat exchanger, 1077, 1111, 1118, 1124, 1139–40, 1143
heat transfer, 758, 760, 762, 785, 787, 793, 810, 1016–18, 1052, 1054, 1081, 1098, 1141
helical strake, 778, 992
Helmholtz's law, 708, 714–15
high-rise building, 680
horseshoe swirl, see swirl
hydronautical engineering, 994
hysteresis, 784, 942–3, 956, 1095

ideal gas, 1140
impulsive flow, 984–6
in-line, 1077, 1081–2, 1085, 1095, 1098, 1102, 1104, 1106, 1111, 1118, 1133, 1136, 1138, 1140–1, 1149–50, 1154–5, 1161, 1163, 1165, 1173–4
independence principle, 954–6, 958–9, 962–3, 967–8, 970, 975–6
instability, 683, 688, 699–700, 710, 712–13, 831, 833, 930, 936, 938–9, 942, 945, 956, 992, 1061, 1063, 1085, 1087–8, 1141, 1143, 1147, 1149, 1152, 1156, 1166
interference, 813, 820, 872, 882, 969, 973, 975, 994–6, 998–1000, 1041–4, 1049–53, 1055, 1065, 1067, 1085, 1090, 1095, 1109–14, 1116, 1118
intersecting cylinders, 1073–4
interstitial flow, 1118–20, 1133, 1135, 1137, 1139–40, 1143, 1145–8, 1153–4, 1156, 1158–61, 1173–6

jet instability, 1147–9, 1151, 1156
jet pairing, 1124–5, 1128
jetty, 994

kinetic energy, 1074–5, 1139, 1144
knot, 689, 696–7, 709, 799, 804, 807

lift
  fluctuating, 705–7, 724–6, 734–5, 757, 767, 809–10, 852, 854–5, 885–7, 890–1, 932–4, 979–80, 1035, 1055, 1137–9
  mean, 778, 780, 815, 869, 872, 874–5, 877, 880–1, 885–7, 890–1, 899–902, 906, 908–13, 916–20, 946, 950, 952, 998, 1019–20, 1023–4, 1033–4, 1037, 1039–43, 1047, 1049–51, 1055–6, 1083, 1090, 1092, 1094–5, 1100, 1107, 1109–10, 1126–8, 1161
lighthouse, 679

Magnus effect, 893–4, 909–11, 916, 926, 929
mass transfer, 843–6, 854–6
mechanical engineering, 1118
metastable state, 1127–9, 1176
missile, 954, 981–2
mooring, 779

Navier–Stokes equation, 712, 856, 932, 938–9, 954
noise, 1139–40, 1173
nuclear power station, 1118

oceanology, 779
offshore platform, structure, 749, 763, 954, 994, 1052, 1067, 1077
orientation, 994, 1085, 1090–1, 1099–100, 1102, 1104, 1107–11
oscillation, 1118, 1160
Oseen, 1019

parameter
  governing, 681, 688, 736, 741, 811, 828, 995, 998, 1048, 1052, 1172
  influencing, 694, 752, 767, 798–9, 811, 872, 874, 880, 902, 957, 983, 987
partition plate, 965, 974, 1027–9
phase lag, 973–4, 997, 1018
pipeline, 871–2, 994
pitch ratio, 1085, 1091–2, 1099, 1101, 1108, 1120, 1122–3, 1126, 1131, 1141, 1156, 1160, 1163, 1165
porous, 1104, 1110, 1112, 1123
power plant, Madaras, 926
pressure

base, 692–4, 700–5, 716, 722, 730–1, 740–1, 809, 813–14, 816–17, 835, 838, 849, 864–5, 874, 882, 945, 955, 963–4, 967–8, 969–70, 975, 997, 1004, 1024, 1027–8, 1057–8, 1064, 1068, 1075, 1122–4, 1126, 1128–31, 1134
drop, 1126–7, 1138–9, 1144, 1163–4
fluctuating, 724–5, 727, 733–4, 762, 792–3, 801, 815, 820, 841, 844, 886–8, 932, 1005–6, 1014, 1017, 1024, 1034–5, 1047–8, 1055–6, 1078–80, 1095–6, 1137, 1139, 1143, 1156
gradient, adverse, 680, 683, 745, 757, 815, 818, 820, 873–5, 1003, 1044, 1126, 1137, 1156, 1161
instant, 691, 730–1, 813–14, 816
mean, 683–5, 708, 720–1, 724, 736–40, 744–6, 835–6, 838, 840, 842–9, 865, 869, 871, 874–6, 879, 882–3, 905–7, 909–10, 945–8, 951, 966–7, 975–6, 980–1, 986–90, 1003–5, 1013–14, 1016, 1024–5, 1038–40, 1042–3, 1048, 1050–2, 1055, 1064–5, 1068–9, 1073, 1078–80, 1082–3, 1089, 1095–6, 1098, 1100, 1106–7, 1123, 1125–7, 1129–37, 1139, 1156–8, 1161–2
recovery, 745, 906, 945, 1068
side, 1126
stagnation, 691, 730–1, 813–14, 816, 835, 864, 879, 894, 898, 906, 913–14, 945, 948, 955, 967–71, 1003, 1025–6, 1039, 1044, 1068, 1075, 1133, 1161
standing wave, 1142–3, 1150, 1152, 1167
wave, 1142, 1149
proximity to
cylinder, 1025, 1037, 1041, 1118, 1130–3, 1156
wall, 827, 870–2, 996, 1018–19, 1024–5, 1118

quasistable flow, 1023, 1123

Rankine's oval, 856, 858, 908
Rayleigh, 938–9
reattachment, 744–5, 771, 773, 775–6, 788, 971–3, 996, 1002–6, 1008, 1012–13, 1016–18, 1043, 1047, 1054–5, 1065, 1067–8, 1079–82, 1095, 1098, 1106, 1134, 1136, 1138, 1161
rectangle, 1118, 1133
resistance, 783–4, 790, 798, 825, 828–9, 869–70, 999–1000
resonance, 1139–41, 1149, 1152, 1167–8
rib, 812, 816–18, 1051
ribbon parachute, 1121
riser, 1077, 1099, 1104, 1109, 1111, 1114
rocket booster, 994, 1082
Roshko's number, 956–7
rotary oscillation, 929–34, 936–7
rotating cylinder, 893–6, 902–6, 909, 911–20, 922, 926, 928–9, 934, 944, 992–3
rotor, 923–7
rotorship, Flettner, 923–4

satellite tube, 1077–8, 1082, 1109–14, 1116
scatter, 1140, 1145, 1169
scour, 871, 888–9
screen, 1067
secondary flow, 1068–71, 1073, 1085
separation
angle, 680, 719, 740, 745, 754–5, 757, 760, 765, 769–73, 775, 777, 782, 829–30, 835, 837–8, 848, 864, 875, 880, 894, 896, 914, 916, 929, 944–5, 950, 955–6, 963, 971, 973, 981, 984, 989, 998–9, 1004–6, 1011, 1013, 1019–20, 1026–7, 1034, 1044, 1047, 1065, 1068–71, 1079, 1095, 1106, 1116, 1133, 1137–8, 1156
bubble, 700, 744–5, 783, 788, 852, 873, 878, 973, 1012, 1015, 1017, 1029, 1032
line, 700, 721–2, 725, 733, 744–5, 758, 823
serrated blade, 788
shear stress, 686, 798, 804
shock wave, 986, 990–1
shroud, 1078, 1110, 1112
side force, 986–9, 992–3
side-by-side, 695–6, 997, 1002, 1018–22, 1025, 1027, 1030–6, 1038, 1042, 1047–8, 1050, 1059, 1077, 1082–3, 1085, 1091–2, 1098, 1101, 1104–6
skin friction, 748, 753–6, 1138–9
slanted angle, 801, 959, 962
slit, 790–3
solidity, 1191
space engineering, 994
spacing ratio, 810, 834, 936, 986, 992, 995, 997–8, 1003, 1018, 1036,

1040, 1053, 1085, 1087, 1107, 1110, 1112, 1133, 1145
speed of sound, 1140–1, 1143, 1149–50, 1152–3, 1172–3
splitter plate, 945, 1008, 1128–30
square, 999–1002, 1118–20, 1133–7, 1141, 1154–5, 1157, 1161, 1168–9, 1171–2, 1175
stability, 714, 1001, 1021, 1027
staggered, 994–5, 997, 1001, 1035–9, 1047–8, 1050–1, 1057–8, 1077, 1083, 1085, 1090, 1092, 1099, 1101–2, 1110–11, 1118, 1120, 1140–1, 1156, 1158–9, 1161, 1163, 1171–4
step change in diameter, 798–9, 801–3, 1021
strake, see helical strake
stranded
  cable, see cable
  conductor, see conductor
Strouhal number, 688–91, 699–702, 713, 716, 726–7, 729, 734, 743, 756, 759, 761, 763, 766–8, 771, 773, 775–6, 779, 788–90, 793–4, 797, 801, 809, 841–3, 852–3, 884–5, 889–91, 903, 916–17, 932–3, 937, 956, 958–60, 962–3, 965, 1007–9, 1011, 1015, 1022–3, 1045–6, 1056–7, 1066–7, 1080, 1084, 1090–3, 1097, 1101–4, 1129–32, 1140, 1145, 1148–9, 1163–6, 1169, 1171
strut, 680, 1001
suction, 790, 792–3, 893, 944–7, 950
suppression, 680, 703, 706, 728, 774, 776, 778, 904, 916, 930, 992, 1050–1, 1061–2, 1091, 1106, 1112, 1128, 1141, 1143, 1149
surface roughness, 738, 748–50, 754, 760, 764–5, 777, 795, 812, 815, 820, 919, 978, 980, 992, 995, 998, 1015, 1083, 1111, 1118, 1127, 1138, 1163
  abrasive, 749–53
  brick, 760–1, 763
  dimples, 757, 784, 789
  eddy generator, 751, 786–8
  equivalent, 753, 758, 760
  groove, 794–8
  marine, 749–51, 783–5
  pyramidal, 749–53
  tripping sphere, 751, 781, 1081, 1099, 1108–9
  tripping wire, 751, 767, 769–74, 777–8, 884, 992, 1176
  turbulence, 748–50, 752, 757, 765, 767, 769, 785
  wire gauze, 703, 1067
  wire helical, 993
swirl, 679–87, 694, 720–4, 730, 733, 855, 973, 981, 989, 1015–16, 1070–1, 1073, 1090
synchronization, 793, 903, 932–4, 936–7, 957, 959, 997, 1017, 1029, 1056–60, 1141, 1149–53, 1166–7, 1172

tandem, 994–6, 998, 1000–4, 1006–14, 1016–17, 1035, 1037–9, 1047, 1052–4, 1056, 1077, 1089–91, 1095, 1099, 1110–11
taper, 738, 799, 802–4, 806, 808–10, 812–13, 815
Taylor's eddy, 938–40
thermodynamic equation, 1140
Thwaite's flap, 945–6, 948
topology, 680, 688–70, 716–17, 741, 747, 773, 775, 783, 1043, 1070–3, 1152
toroidal (ring) eddy, 904–5, 918, 939
transition (Gerrard–Bloor) eddy, 749, 752, 754–5, 757, 767–8, 812, 1146–50, 1152, 1165
transition
  in boundary layer, TrBL, 700, 827, 850, 976, 992, 1011, 1032, 1047, 1126–7
  in shear layer, TrSL, 691, 701–2, 801–2, 808, 827, 830, 905, 963, 1007, 1038
  in wake, TrW, 699–700, 702, 827, 902, 973, 1000, 1038, 1077, 1085–9
  state, 1118
transmission line, 779, 954, 994, 1049, 1077, 1099, 1106
triangle, 1090, 1092, 1118–20, 1141, 1156–8, 1160, 1163–7, 1169–74
tripping wire, see surface roughness
turbomachinery blade, 680
turbulence
  energy, 1144–5
  intensity, 686–7, 697–8, 748–9, 803, 820, 922–3, 1153–5, 1159–61, 1175
  relative, 748–9, 753, 758, 817, 822
  scale, 748–9
  texture, 748–9, 753, 758, 817, 822
turbulent state, T, 755, 765, 825, 998
TV tower, 679
two-phase flow, 1124

underwater acoustics, 779

van der Pol theory, 807
velocity
  defect, 718, 800, 920–2, 1009, 1044, 1059–60, 1074
  fluctuation, 696–8, 718, 727, 739, 800, 805, 807, 813, 922, 979, 1059–60, 1135–6, 1160
  gap, 1120, 1131–3, 1139, 1149, 1160, 1172, 1175
  modulated, 695, 804, 806–7
  profile, 719, 730, 921, 967, 972, 1059–60, 1120, 1122, 1135–6, 1159, 1161
vortex
  discrete, 899, 920, 950, 957, 1029–33
  sheet, 1031–2
  street, 1029–30
  strength, 896, 985, 1029
vorticity, 700, 723, 756–9, 824, 896, 899, 936–7, 1027, 1062, 1087, 1146, 1165, 1171

wake
  displacement, 998, 1042, 1044, 1091
  near-, 683, 688, 696, 700, 706–7, 713, 716, 721, 730, 741, 767, 783, 790, 797, 799, 820, 829–31, 834, 841, 849, 866–7, 872, 879, 894–6, 916, 964, 966–7, 997, 1018, 1020–1, 1029, 1061–3, 1068, 1083, 1104–5, 1119–20, 1122–4, 1129, 1156–60, 1171, 1173–4
  width, 832–3, 842–8, 930, 964, 966–7, 1044–5, 1059, 1091–4, 1129, 1168–9
wall proximity, see proximity to wall
wind, 735–6, 810, 995, 1082, 1107
wind engineering, 1052
windmill, Flettner, 915
wing, 1001
wing–fuselage junction, 680

yaw, 781, 954, 956, 958–61, 963, 965–6, 968–71, 973–4, 976, 978, 980–1, 984–5, 991, 993

Printed in the USA/Agawam, MA
May 3, 2024

865425.015